PLANT SURFACE MICROBIOLOGY

Springer

Berlin
Heidelberg
New York
Hong Kong
London
Milan
Paris
Tokyo

Ajit Varma
Lynette Abbott
Dietrich Werner
Rüdiger Hampp (Eds.)

Plant Surface Microbiology

With 138 Figures, 2 in Color

Springer

Professor Dr. Ajit Varma
Jawaharlal Nehru University
School Life Science
New Delhi 110067
India
email: ajitvarma73@hotmail.com

Professor Dr. Lynette Abbott
School of Earth and Geographical Sciences
The University of Western Australia
Nedlands WA 6009
Australia
email: labbott@cyllene.uwa.edu.au

Professor Dr. Dietrich Werner
FG Zellbiologie und Angewandte Botanik
Philipps Universität Marburg
35032 Marburg
Germany
email: djg.werner@gmx.de

Professor Dr. Rüdiger Hampp
Physiological Ecology of Plants
University of Tübingen
72116 Tübingen
Germany
email: ruediger.hampp@uni-tuebingen.de

ISBN 3-540-00923-X Springer-Verlag Berlin Heidelberg New York

Library of Congress Cataloging-in-Publication Data.

Plant surface microbiology / Ajit Varma … [et al.], (eds.).
p. cm.
Includes bibliographical references and index.
ISBN 3-540-00923-X (hbk. : alk. paper)
1. Plant surfaces--Microbiology. I. Varma, A. (Ajit), 1939-

QP351.P5833 2004
579'.178--dc22 2003064760

Springer-Verlag is a part of Springer Science+Business Media

springeronline.com

Printed in Germany

Production and typesetting: Friedmut Kröner, 69115 Heidelberg, Germany
Cover design: *design & production* GmbH, 69126 Heidelberg, Germany

31/3150 YK – 5 4 3 2 1 0 – Printed on acid free paper

Preface

The complexity of plant surface microbiology is based on combinations. A large number of microbial species and genera interact with several hundred thousand species of higher plants. At the same time, they interact with each other. Therefore, this book describes only some very important model interactions which have been studied intensively over the last years. The methods developed for some important groups of microorganisms can be used for a large number of other less studied interactions and combinations. The pace of discovery has been particularly fast at two poles of biological complexity, the molecular events leading to changes in growth and differentiation, as well as the factors regulating the structure and diversity of natural populations and communities.

The area of plant surfaces is enormous. A single maize plant has a leaf surface of up to 8000 cm^2, a single beech tree has a leaf surface of around 4.5 million cm^2. The leaf area index (LAI) varies from 0.45 in tundra areas up to 14 in areas with a dense vegetation. Calculated for all plant surfaces above ground, the surface area is more than 200 million km^2. This area is still surpassed by the below ground surface areas of plants, especially those with an extensive root hair system. For a single rye plant, a root hair surface of around 400 m^2 has been calculated. Even if this is an exceptional case, it can be assumed that in many plants the root and root hair surface is ten times larger than the surfaces of the above ground plant parts. This means that more than 2000 million km^2 of plant surface is present underground. Taking both figures together, it exceeds the land surface area of the planet Earth of 149 million km^2 by more than a factor of 10.

This volume summarizes and updates both the state of knowledge and theories and their possible biotechnological applications. It will thus be of interest to a diverse audience of researchers and instructors, especially biologists, biochemists, agronomists, foresters, horticulturists, mycologists, soil scientists, ecologists, plant physiologists, plant molecular biologists, geneticists, and microbiologists.

In the planning of the book, invitations for contributions were extended to leading international scientists working in the field of plant surface microbiology. The basic concepts in plant surface microbiology are discussed at length in 30 chapters including a few specialized and innovative methodologies and novel techniques. The editors would like to express deep appreciation to each contributor for his/her work, patience and attention to detail during the entire production process. It is hoped that their reviews, interpretations, and basic concepts will stimulate further research. We are confident that the joint efforts of the authors and editors will contribute to a better understanding of the advances in the study of the challenging area of surface microbiology and will further stimulate progress in this field.

It has been a pleasure to edit this book, primarily due to the stimulating cooperation of the contributors. We would like to express sincere thanks to all the staff members of Springer-Verlag, Heidelberg, especially, Drs. Dieter Czeschlik and Jutta Lindenborn for their help and active cooperation during the preparation of the book.

New Delhi, India — *Ajit Varma*
Nedlands, Australia — *Lynette Abbott*
Marburg, Germany — *Dietrich Werner*
Tübingen, Germany — *Rüdiger Hampp*

July 2003

Contents

Section B

Section C

Contributors

ABBOTT, LYNETTE K.
School of Earth and Geographical Sciences
Faculty of Natural and Agricultural Sciences
The University of Western Australia
Crawley, WA 6009
Australia
(e-mail: labbott@cyllene.uwa.edu.au)

ANDRADE, GALDINO
State University of Londrina, CCB
Dept of Microbiology
Microbial Ecology Laboratory
PO Box 6001
86051-990 Londrina, PR
Brazil
(e-mail: andradeg@uel.br)

AZCÓN, ROSARIO
Departamento de Microbiología del Suelo y Sistemas Simbióticos
Estación Experimental del Zaidín
CSIC
Prof. Albareda 1
18008 Granada
Spain

AZCÓN-AGUILAR, CONCEPCIÓN
Departamento de Microbiología del Suelo y Sistemas Simbióticos
Estación Experimental del Zaidín
CSIC
Prof. Albareda 1
18008 Granada
Spain

BACON, CHARLES W.
Department of Agriculture
Agriculture Research Service
Athens, Georgia
USA

BAREA, JOSÉ-MIGUEL
Departamento de Microbiología del Suelo y Sistemas Simbióticos
Estación Experimental del Zaidín
CSIC
Prof. Albareda 1
18008 Granada
Spain
(e-mail: josemiguel.barea@eez.csic.es)

BAUER, ROBERT
Universität Tübingen
Lehrstuhl Spezielle Botanik und Mykologie
Auf der Morgenstelle 1
72076 Tübingen
Germany
(e mail: robert.bauer@uni-tuebingen.de)

BAZZICALUPO, MARCO
Dipartimento di Biologia Animale e Genetica 'Leo Pardi'
Via Romana 17
50125 Firenze
Italy
(email: marcobazzi@dbag.unifi.it)

Belanger, Faith
Department of Plant Biology and Pathology
Cook College-Rutgers University
New Brunswick, New Jersey
USA

Biondi, Emanuele G.
Dipartimento di Biologia Animale e Genetica 'Leo Pardi'
Via Romana 17
50125 Firenze
Italy

Bloemberg, Guido V.
Leiden University
Institute of Biology
Wassenaarseweg 64
2333 AL Leiden
The Netherlands

Botton, Bernard
University Henri Poincaré Nancy 1
Faculty of Sciences and Techniques
UMR INRA-UHP no. 1136
B.P. 236
54506 Vandoeuvre-Les-Nancy Cedex
France
(e-mail: Bernard.Botton@scbiol.uhp-nancy.fr)

Brun, Annick
University Henri Poincaré Nancy 1
Faculty of Sciences and Techniques
UMR INRA-UHP no. 1136
B.P. 236
54506 Vandoeuvre-Les-Nancy Cedex
France

Buscot, François
Institute of Ecology
Department of Environmental Sciences
University of Jena
Dornburger Strasse 159
07743 Jena, Germany
Present address: Institute of Botany
Department of Terrestrial Ecology
University of Leipzig
Johannisallee 21
04103 Leipzig
Germany

Chalot, Michel
University Henri Poincaré Nancy 1
Faculty of Sciences and Techniques
UMR INRA-UHP no. 1136
B.P. 236
54506 Vandoeuvre-Les-Nancy Cedex
France

Chin-A-Woeng, Thomas F.C.
Leiden University
Institute of Biology
Wassenaarseweg 64
2333 AL Leiden
The Netherlands
(email: chin@rulbim.leidenuniv.nl)

Conrad, Ralf
Max-Planck-Institut für Terrestrische Mikrobiologie
Marburg, Germany
(e-mail: conrad@staff.uni-marburg.de)

Dazzo, Frank B.
Center for Microbial Ecology
Department of Microbiology and Molecular Genetics
Michigan State University
East Lansing, MI 48824,
USA
(e-mail: dazzo@msu.edu)

Declerck, Stephané
Unité de Microbiologie
Mycothèque de l'Université catholique de Louvain
Université catholique de Louvai
3 Place Croix du Sud
1348 Louvain-la-Neuve
Belgium

Gantner, Stephan
GSF–National Research Center for Environment and Health
Institute of Soil Ecology
Ingolstädter Landstrasse 1
85764 Neuherberg/München
Germany

GHIMIRE, SITA R.
Centre for Research in Fungal Diversity
Department of Ecology and Biodiversity
The University of Hong Kong
Pokfulam Road, Hong Kong
Hong Kong SAR
PR China

GLICK, BERNARD R.
Department of Biology
University of Waterloo, Waterloo
Ontario, Canada N2L 3G1
(e-mail: glick@sciborg.uwaterloo.ca)

HAMPP, RÜDIGER
Institute of Botany
Department of Physiological Ecology of Plants
University of Tübingen
Auf der Morgenstelle 1
72076 Tübingen
Germany
(e-mail: ruediger.hampp@uni-tuebingen.de)

HARTMANN, ANTON
GSF–National Research Center for Environment and Health
Institute of Soil Ecology
Ingolstädter Landstrasse 1
85764 Neuherberg/München
Germany
(e-mail: anton.hartmann@gsf.de)

HEHL, SOLVEIG
Application Specialist
Advanced Imaging Microscopy
Carl Zeiss Jena GmbH
Carl-Zeiss-Promenade 10
07745 Jena
Germany

HERRMANN, SYLVIE
Institute of Ecology
Department of Environmental Sciences
University of Jena
Dornburger Strasse 159
07743 Jena
Germany

HYDE, KEVIN D.
Centre for Research in Fungal Diversity
Department of Ecology and Biodiversity
The University of Hong Kong
Pokfulam Road, Hong Kong
Hong Kong SAR
PR China
(e-mail: kdhyde@hkucc.hku.hk)

JAVELLE, ARNAUD
University Henri Poincaré Nancy 1
Faculty of Sciences and Techniques
UMR INRA-UHP no. 1136
B.P. 236
54506 Vandoeuvre-Les-Nancy Cedex
France

KALDORF, MICHAEL
Institute of Ecology
Department of Environmental Sciences
University of Jena
Dornburger Strasse 159
07743 Jena, Germany
Present address: Institute of Botany
Department of Terrestrial Ecology
University of Leipzig
Johannisallee 21
04103 Leipzig
Germany
(e-mail: kaldorf@rz.uni-leipzig.de)

KNOLL, DANIEL
Institut für Allgemeine Botanik
Angewandte Molekularbiologie der Pflanzen
Universität Hamburg
Ohnhorststrasse 18
22609 Hamburg
Germany

KOST, GERHARD
FB Biologie
Spezielle Botanik und Mykologie
Philipps-Universität Marburg
35032 Marburg
Germany

KOTHAMASI, DAVID MANOHAR
Department of Microbiology
University of Delhi South Campus
Benito Juarez Road
New Delhi 110 021, India

KOTTKE, INGRID
Fakultät für Biologie
Botanisches Institut
Spezielle Botanik
Mykologie und Botanischer Garten
Universität Tübingen
Auf der Morgenstelle 1
72076 Tübingen
Germany
(e-mail: ingrid.kottke
@uni-tuebingen.de)

KRIMM, URSULA
Institut für Zelluläre und Molekulare
Botanik (IZMB)
Abteilung Ökophysiologie
Universität Bonn
Kirschallee 1
53115 Bonn
Germany

KUHAD, RAMESH CHANDER
Department of Microbiology
University of Delhi South Campus
Benito Juarez Road
New Delhi 110 021
India
(e-mail: kuhad@hotmail.com)

KUMARI, RINA
School of Life Sciences
Jawaharlal Nehru University
New Delhi 110067
India

LAGOPODI, ANASTASIA L.
Leiden University
Institute of Biology
Wassenaarseweg 64
2333 AL Leiden
The Netherlands

LANFRANCO, LUISA
Dipartimento di Biologia Vegetale dell'Università
Viale Mattioli 25
10125 Torino
Italy

LEWIS, ELIZABETH
Department of Plant Biology
and Pathology
Cook College-Rutgers University
New Brunswick, New Jersey
USA

LUGTENBERG, BEN J.J.
Leiden University
Institute of Biology
Wassenaarseweg 64
2333 AL Leiden
The Netherlands

LUIS, PATRICIA
Institute of Ecology
Department of Environmental Sciences
University of Jena
Dornburger Strasse 159
07743 Jena
Germany

MAIER, ANDREAS
Institute of Botany
Department of Physiological
Ecology of Plants
University of Tübingen
Auf der Morgenstelle 1
72076 Tübingen
Germany

MALLA, RAJNI
School of Life Sciences
Jawaharlal Nehru University
New Delhi 110067
India

MENGONI, ALESSIO
Dipartimento di Biologia Animale e
Genetica 'Leo Pardi'
Via Romana 17
50125 Firenze
Italy

METZ, SIGRUN
GSF–National Research Center for Environment and Health
Institute of Soil Ecology
Ingolstädter Landstrasse 1
85764 Neuherberg/München
Germany

MEYER, WILLIAM
Department of Plant Biology
and Pathology
Cook College-Rutgers University
New Brunswick, New Jersey
USA

MITTAG, MARIA
7Institute of General Botany
Friedrich-Schiller-University Jena
Am Planetarium 1
07743 Jena
Germany

MOGGE, BERNHARD
GSF–National Research Center for Environment and Health
Institute of Soil Ecology
Ingolstädter Landstrasse 1
85764 Neuherberg/München
Germany

MOY, MELINDA
Department of Plant Biology
and Pathology
Cook College-Rutgers University
New Brunswick, New Jersey
USA

MULDERS, INE H.M.
Leiden University
Institute of Biology
Wassenaarseweg 64
2333 AL Leiden
The Netherlands

NEHLS, UWE
Physiologische Ökologie der Pflanzen
Universität Tübingen
Auf der Morgenstelle 1
72076 Tübingen
Germany
(e-mail: uwe.nehls@uni-tuebingen.de)

NIINI, SARA
Department of Biosciences
Plant Physiology
P.O. Box 56
00014 Helsinki University
Finland

OBERWINKLER, FRANZ
Universität Tübingen
Lehrstuhl Spezielle Botanik
und Mykologie
Auf der Morgenstelle 1
72076 Tübingen
Germany
(e mail: franz.oberwinkler
@uni-tuebingen.de)

OELMÜLLER, RALF
Institute of General Botany
Department of Environmental Sciences
University of Jena
Dornburger Strasse 159
07743 Jena
Germany

PENROSE, DONNA M.
Department of Biology
University of Waterloo, Waterloo
Ontario, Canada N2L 3G1

PESKAN, TANJA
Institute of General Botany
Department of Environmental Sciences
University of Jena
Dornburger Strasse 159
07743 Jena
Germany

PHAM, GIANG HUONG
School of Life Sciences
Jawaharlal Nehru University
New Delhi 110067
India

PODILA, GOPI K.
Department of Biological Sciences
University of Alabama
Huntsville, AL-35899
USA
(e-mail: podilag@email.uah.edu)

PRASAD, RAM
School of Life Sciences
Jawaharlal Nehru University
New Delhi 110067
India

PUKALL, RÜDIGER
DSMZ–German Collection of Microbes and Cell Cultures GmbH
Mascheroder Weg 1b
38124 Braunschweig
Germany

RAUDASKOSKI, MARJATTA
Department of Biosciences
Plant Physiology
P.O. Box 56
00014 Helsinki University
Finland
(e-mail: marjatta.raudaskoski@helsinki.fi)

REXER, KARL-HEINZ
FB Biologie
Spezielle Botanik und Mykologie
Philipps-Universität Marburg
35032 Marburg
Germany

ROTHBALLER, MICHAEL
GSF–National Research Center for Environment and Health
Institute of Soil Ecology
Ingolstädter Landstrasse 1
85764 Neuherberg/München
Germany

SACHDEV, MINU
School of Life Sciences
Jawaharlal Nehru University
New Delhi 110067
India

SAXENA, ANIL KUMAR
Division of Microbiology
Indian Agricultural Research Institute
New Delhi 110012
India

SCHLOTER, MICHAEL
GSF–National Research Center for Environment and Health
Institute of Soil Ecology
Ingolstädter Landstrasse 1
85764 Neuherberg/München
Germany

SCHREIBER, LUKAS
Institut für Zelluläre und Molekulare Botanik (IZMB)
Abteilung Ökophysiologie
Universität Bonn
Kirschallee 1
53115 Bonn
Germany
(e mail: lukas.schreiber@uni-bonn.de)

SINGH, AJAY
Department of Biology
University of Waterloo, Waterloo
Ontario N2T 2J3
Canada

SINGH, ANJANA
School of Life Sciences
Jawaharlal Nehru University
New Delhi 110067
India

SOLAIMAN, M. ZAKARIA
Soil Science and Plant Nutrition
School of Earth and Geographical Sciences
Faculty of Natural and Agricultural Sciences
The University of Western Australia
Crawley, WA 6009
Australia

STABENTHEINER, EDITH
Institute for Plant Physiology
Karl-Franzens University Graz
University Street 51
8010 Graz
Austria

SULLIVAN, RAYMOND
Department of Plant Biology and Pathology
Cook College-Rutgers University
New Brunswick, New Jersey
USA

TARKKA, MIKA
Universität Tübingen
Botanisches Institut
Auf der Morgenstelle 1
72076 Tübingen
Germany

Tripathi, K. K.
Department of Biotechnology
Ministry of Science and Technology
C.G.O. Complex, Lodi Road
New Delhi110 003
India

Varma, Ajit
School of Life Sciences
Jawaharlal Nehru University
New Delhi 110067
India
(email: ajitvarma73@hotmail.com)

Werner, Dietrich
Fachbereich Biologie
Fachgebiet Zellbiologie und
Angewandte Botanik
Philipps-University Marburg
Germany
(e-mail: werner@mailer.uni-marburg.de)

White Jr., James F.
Department of Plant Biology
and Pathology
Cook College-Rutgers University
New Brunswick, New Jersey
USA
(e-mail: jwhite@AESOP.Rutgers.edu)

Zhang, Chi
Institute of Botany
Department of Physiological Ecology
of Plants
University of Tübingen
Auf der Morgenstelle 1
72076 Tübingen
Germany

1 The State of the Art

AJIT VARMA, LYNETTE K. ABBOTT, DIETRICH WERNER
and RÜDIGER HAMPP

As we enter the second century of research on associative and symbiotic microorganisms, it is heartening to see that attention is increasingly focused on the functions of these organisms in the natural and semi-natural systems in which it evolved. This volume, while encapsulating the spirit of the new adventure, also provides two further opportunities. It enables us to assess the strength of the platform from which we launch into this challenging area and to identify which experimental approaches might provide the most realistic evaluation of the roles played by surface microorganisms in natural communities. The long and difficult climb towards understanding the impacts of the microflora upon the species composition and dynamics, above and below ground, of plant communities is just beginning. This volume demonstrates both the strength and the weakness of the position from which we launch into the future. The strength may be that we have much precise information about microbial function under simplified conditions. The weakness, on the other hand, is that we have, as yet, little reliable information about the extent to which these functions are expressed under relevant, essentially multi-factorial circumstances of the kind that prevail in nature. The plant carries its major microbial community on its entire exposed surfaces, from apical tip to root cap. These plant surfaces represent an oozing, flaking layer of integument which discharges a wide range of substances that support a vast number of spatially discrete and specialized microbial communities, including parasites and symbionts, which can have a major impact on plant growth and development. In today's scenario the plant surface is considered as a dynamic adaptable envelope, flexible in both its own right and the first barrier between the moist, concentrated, balanced plant cell and a hostile ever-changing external environment. It is well known that the microbial diversity on the plant surface and in the soil habitats is much greater compared to the insight using cultivation techniques. Manipulation of the plant surface microflora to improve its health is a desirable and much needed goal in plant microbiology. However, efforts to exploit this type of biological control have frequently been impeded because of major

Plant Surface Microbiology
A. Varma, L. Abbott, D. Werner, R. Hampp (Eds.)

technical difficulties that must be overcome in order to fully understand the microbial ecology of this ecosystem, especially the lack of ability to extract in situ data that are both informative and quantifiable at spatial scales relevant to the ecological niches of the microorganisms involved. The entire volume is divided into five broad sections.

The combining aspect of the chapters in sections A and B are microbial communities in their interactions with higher plants. The communities are mainly dominated by a few species, however, a large number of other species may be equally important, although they are present only in the range of 1 % of the total population or less. Experimental studies concentrate, of course, on the major components of the communities. These representatives are also used for biotechnology purposes such as seed inoculation by *Pseudomonas* and *Bacillus* control strains (Chap. 2). The interactions of methanogens and methanotrophs independent of the plant photosynthesis and the plant root ecology is a major contribution to the global CH_4 cycle. These communities are especially present in anoxic sites in wetlands such as flooded rice fields. The different carbon sources affect the CH_4 to CO_2 ratio, an important aspect for the impact of different root components on the microbial communities in the rhizosphere, as described in Chapter 3. Abiotic factors also influence the colonization of *Pseudomonas fluorescens* on seeds and include, besides growth substrates, also temperature, soil humidity and pH (Chap. 3). The dynamics of microorganism populations in the rhizosphere is a topic where a large number of research groups worldwide are involved. This is related to the huge amount of organic carbon exudated from plant roots into the rhizosphere, in the order of 10 % or more of the total carbon assimilation by photosynthesis in higher plants. All major nutrient cycles such as the carbon cycle, the nitrogen cycle, the sulfur cycle, the phosphorus cycle and the cycle for micronutrients are much more active in this rhizosphere soil compared to the bulk soil. The enormous diversity in this microhabitat is increased by the fact that many different plant families and species exudate different sets of components into the soil. In addition, the composition of lignins and hemicellulose in the cell walls can be quite different, leading to a different composition of the rhizosphere communities (Chap. 4). More information on the major groups of microorganisms in soils in general are covered in Chapter 5, describing especially the impact of microorganisms on plant development by mycorrhiza species, actinorhiza species, plant growth-promoting rhizobacteria (PGPR), phosphate-solubilizing microorganisms and the important group of lignocellulolytic microorganisms.

Biotic signals from the microsymbionts inducing symbiosis and nodule development in legumes are even more specific in determining the interaction of the plants with their specific associated bacteria such as *Bradyrhizobium japonicum*, *Mesorhizobium loti*, *Sinorhizobium meliloti*, *Rhizobium tropici* or *Rhizobium etli*. Flavonoids and nod factors (lipochitooligosaccharides) are the major components of the chemical language, in which the microsymbionts and the host plants communicate to each other. The signalling concept studied in this type of symbiosis is equally complicated as the mammalian notch homologues and the integrin-adhesion-receptor signalling in other multicellular organisms (Chap. 6). A large stimulus for ongoing and future research in the area of plant surface microbiology will be available from the use of already completed genome projects and on-going genome projects for prokaryotic and eukaryotic organisms. At present, about 145 genome projects are finished and more than 580 projects are on-going (http://wit.integratedgenomics.com/GOLD/gold.html). A list of completed genomes present in the public data bases, available in June 2003, is presented in Table 1. It is interesting to note that plant symbiotic and parasitic bacteria such as *Bradyrhizobium japonicum*, *Mesorhizobium loti*, *Sinorhizobium meliloti* and *Pseudomonas synringae* have the largest procaryotic genomes. On the other side, there are some animal pathogenic organisms like *Rickettsia*

Table 1. Complete genomes present in the public DataBases, June 2003 (http://wit.integratedgenomics.com/GOLD/gold.html)

Organism	Size (kb)	ORF number
Archaeal		
Methanosarcina mazei	4.096	3,371 orfs MAP
Methanobacterium thermoautotrophicum	1.751	1,918 orfs MAP
Bacterial		
Bradyrhizobium japonicum	9.105	8.317 orfs MAP
Mesorhizobium loti	7.596	6.752 orfs MAP
Sinorhizobium meliloti	6.690	6.205 orfs MAP
Nostoc sp. PCC 7120	6.413	5.366 orfs MAP
Pseudomonas synringae	6.397	5.615 orfs MAP
Pseudomonas aeruginosa	6.264	5.570 orfs MAP
Escherichia coli 0157:H7, Sakai	5.594	5.448 orfs MAP
Xanthomonas campestris pv. Campestris	5.076	4.182 orfs MAP
Agrobacterium tumefaciens	4.915	5.402 orfs MAP
Bacillus subtilis	4.214	4.099 orfs MAP
Escherichia coli 0157:H7, EDI.933	4.100	5.283 orfs MAP
Nitrosomonas europeae	2.812	2.573 orfs MAP
Borrelia burgdorferi B 31	1.230	1.256 orfs MAP
Rickettsia prowazekii	1.111	834 orfs MAP
Chlamydia trachomatis	1.042	896 orfs MAP
Eukaryal		
Orysa sativa L. ssp. *indica*	420.000	50.000 orfs
Oryza sativa ssp. *japonica*	420.000	50.000 orfs
Arabidopsis thaliana	115.428	25.498 orfs
Neurospora crassa	43.000	10.082 orfs
Schizosaccharomyces pombe	14.000	4.824 orfs
Saccharomyces cerevisiae	12.069	6.294 orfs

prowazekii with only 1.1 Mb, *Chlamydia trachomatis* with 1.04 Mb and *Borellia burgdorferi* with 1.23 Mb.

Bacillus thuringiensis and Bt transgenic plants are examples for biotechnology concentrated on a small number of well-studied soil microorganisms. The bio-insecticide protein is present only at a certain stage of sporulation in these organisms. Under natural conditions the spores have only a very limited survival time with less than 20 % present after 24 h (Chap. 7). The toxin from *Bacillus thuringiensis* released from transgenic plants in the soil is much more stable with 25 % still present after 120 days. The toxin is protected from degradation by linkage and adsorption to clay minerals. Many other important signal molecules produced by plants and microorganisms in the soil may also have very different half-life times by specific adsorption to soil minerals. The impact of increasing concentrations of these toxins in soils due to this bio-control technique has not been sufficiently studied. Increases and decreases of specific subpopulations of soil microorganisms have been reported (Chap. 7). The other side of interactions, promotion instead of inhibition, is a topic of Chapter 8, which studies the mechanisms of plant growth-promoting rhizobacteria by phytohormones such as auxin and ethylene. An intermediate of ethylene synthesis is 1-aminocyclopropane-1-carboxylic acid (ACC). Microorganisms with an ACC deaminase gene increase stress tolerance of several plant species (Chap. 8). Compared to the rhizosphere, the communities in the phyllosphere have been studied less. The main reason is that the plant exudation from the rhizodermis is much larger than from the epidermis, due to the cuticles limiting carbon supply to the leaf surfaces. In contrast to bacteria, fungi have the ability to penetrate the cuticles and get access to carbon supplies (Chap. 9). Future work may concentrate especially on conditions where oligotrophic situations persist and genotypes adapted to these conditions may be present and not been recognized so far. The presence of animals in the interface of plants and microorganisms is another important aspect of communities, with the example of the Clavicipitaceae. It is very interesting to note that species of this family predominantly infect insects or the ancestors of grass-infecting species (Chap. 10). By sophisticated mechanisms, the fungi modify the plant tissues for nutrient acquisition. The shift from pathogenic interaction to mutualistic interaction in some species is a general aspect related to symbiosis and phytopathology. A completely new field of research has been developed, using the interaction of genetically modified plants (GMP) with microbial communities or specific microorganisms (Chap. 11). In the list of GMP species, important crop plants such as potatoes, maize, cotton, tobacco and alfalfa are used. The aspect of horizontal gene transfer (HGT) from GMP plants to associated bacterial species and fungal species is a topic for several biotechnology research projects.

Section C deals with interactions between plants, fungi, and bacteria. The plant root constitutes an environment which forms the basis for multiple relationships with microorganisms. Fine roots of most plants are associated with symbiotic fungi, which facilitate uptake of nutrients and water. An example of such a symbiotic interaction (termed mycorrhiza), which occurs mainly with roots of trees in temperate and alpine regions is ectomycorrhiza. The formation of the resulting symbiotic structure is commonly associated with changes in root morphology. Properties of the root surface are obviously an important parameter which determines the establishment of the physical contact with soil fungi. Chapter 12 gives an overview about the current knowledge on this topic with regard to the interaction of soil bacteria and ectomycorrhiza-forming fungi. This includes recent data on the effects of a co-cultivation of a range of soil bacteria (Actinomycetes) with an important and widely distributed ectomycorrhiza-forming fungus, *Amanita muscaria*, as part of a model system. A specific topic is the interference of a bacterial strain, which highly promotes fungal growth with the protein complement of the latter. Chapter 13 deals with respective root properties such as type of root (long/short root) and surface chemistry. Here, hydrophobic cuticle layers obviously play an important role in hyphal attachment. In addition, compatible fungi are able to penetrate and digest this layer. How far this process is involved in altering the morphology of fungal hyphae when inside the root cortex (Hartig net formation) is discussed. As the data presented in this chapter originate mainly from ultrastructural investigations, possible pitfalls of such studies are also addressed.

An integral part of root–fungus associations are soil bacteria. These can support the development of the root/fungus interaction by improving fungal root colonization, the availability of nutrients, or by producing exudates (e.g., antibiotics) which can prevent attacks of pathogenic microorganisms. While ectomycorrhizas only constitute a small fraction of all root/fungus interactions known, another form of this symbiosis, namely endomycorrhiza, dominates by far, and facilitates nutrient uptake of many crop plants. Fungi forming this type of mycorrhiza can usually not be cultured in the absence of a plant root. Chapter 14 focuses on structural studies of the interaction of these fungi with their host plants. Electron microscopy reveals interaction-specific structures such as fungal deposits and interactive vesicles, which can be used for diagnostic purposes. *Piriformospora indica* is possibly an exception because this fungus can be cultivated separately and forms structures comparable to those of endomycorrhizas. Chapter 15 deals with the diverse interactions of this fungus with roots from a variety of plants (from bryophytes to a wide range of angiosperms) and various groups of soil microorganisms, including bacteria of the rhizosphere (compare also Chap. 12) and other soil fungi such as *Aspergillus* or *Gaeumannomyces* (root pathogen).

Interactions between smut fungi and their plant hosts are another topic of Section C. The term "smut fungus" characterizes fungi sharing similar organization and life strategies. As these fungi can considerably reduce crop yields, they are of economic importance. Most of them are members of the Ustilaginomycetes, which comprise a large number of species. Fungi can also parasite on other fungi. Basidiomycetes, e.g., include saprobes, mycorrhiza-forming fungi, plant parasites, but also fungi which are parasites of other fungi. Hosts are both Basidiomycetes and Ascomycetes. Ultrastructural investigations of this kind of organismic interaction (Chap. 16) revealed two main types, the formation of colacosomes and the fusion between pathogen and host fungus cells. Colacosomes are unique organelles, which appear at the interface between parasite and host while fusion is based on specialized interactive cells (haustoria), which establish a direct cytoplasmic connection.

Many microorganisms coexist with plants in ways that do not lead to plant disease, symbiosis, or other specific interactions. Some fungi or bacteria can be latent pathogens. Some have little or no influence on the plant, but may form toxic compounds that are damaging to grazing animals. Microorganisms that form more or less benign associations with plants are generally termed 'endophytes' and are genetically diverse. A large number of fungal endophytes can be difficult to identify because they include a high proportion with sterile mycelia (Chap. 17). Overall, the roles of many of these organisms are poorly understood.

The mechanisms for entry of endophytic organisms into plants can be investigated using methodologies such as those applied to elucidate the cytoskeletal rearrangements of plant cells and fungal hyphae at the plant–microbe interface during colonization of roots by mycorrhizal fungi (Chap. 18). Invading organisms have been shown to influence the expression of plant genes for some filamentous structures within the cell cytoskeleton. Indirect immunofluorescence microscopy has been used to investigate the cytoskeleton of some mycorrhizal associations demonstrating the separation and invagination of the plasma membrane from the plant cell wall in response to growth of fungi inside the cell wall.

Colonization of plants by related and unrelated groups of microorganisms may occur simultaneously. For example, saprophytes, pathogens and mycorrhizal fungi may be associated with the same root systems and colonize roots to different degrees. Several species of arbuscular mycorrhizal fungi can simultaneously colonize the same sections of root, although they are generally separated in different cells or parts of the root cortex. Prior colonization by one organism can influence sequential colonization by other organisms. This occurs to varying degrees for different groups of plant endophytes, symbionts and pathogens. The relative extent to which roots become colonized by several species of arbuscular mycorrhizal fungi present in the same soil depends on the relative abundance of propagules of the fungi in the soil, the developmental stage of the hyphae associated with fungal propagules, the susceptibility of the roots to invasion and the physiological responses of the root to different species of fungi (Chap. 19). Investigations of the molecular communication between these fungi and their host plants during root colonization and nutrient acquisition are now beginning to be understood in terms of gene expression in plants and fungi. This provides a basis for predicting physiological responses of plants to colonization by communities of arbuscular mycorrhizal fungi comprising species with different capacities to take up phosphorus from soil, transport it along hyphae and transfer it to the plant.

When microbial communities are established in association with roots, they may be affected by changes in rooting patterns and exudates (Chap. 20). Introduction of plant growth promoting rhizobacteria (PGPRs) into the soil/plant/microbial environment can influence organisms already present (e.g., pathogenic and mycorrhizal fungi) in addition to the roots themselves. Techniques for microbial community fingerprinting are being adapted for assessment of PGPRs, in addition to in situ methods such as confocal laser scanning microscopy, to understand root – microbial associations from the perspective of communities of organisms that perform different, and sometimes contrasting, functions.

Nutrients introduced into the rhizosphere from plants and decaying organic matter can influence physiological responses of microorganisms and their interactions with plants. Gene regulation in some ectomycorrhizal fungi has been shown to be altered in nutrient-limiting environments and this could have consequences for nutrient uptake and transfer to plants. For example, regulation of gene expression associated with some sugars has been shown to depend on the concentration of specific carbohydrates in the medium with threshold responses identified (Chap. 21). Expression of ammonium transporter genes can be stimulated for some fungi grown under nitrogen-limiting conditions and this could have important consequences for plant establishment in nitrogen-limiting natural ecosystems. Different patterns of gene regulation have been identified for the ectomycorrhizal fungus *Amanita muscaria* in relation to carbon and nitrogen nutrition. Some genes are regulated by both nitrogen and carbon nutrition, while others by either nitrogen or carbon (Chap. 21). Recent advances in the adaptation of molecular techniques to studies of plant and fungal biochemistry have contributed to understanding nitrogen metabolism in plants and microorganisms (Chap. 22). For some time, studies of nitrogen assimilation by ectomycorrhizal fungi have investigated nitrate and nitrite uptake kinetics, ammonium transport and amino acid transport. Techniques such as immunogold and ^{14}C labelling can now be combined with gene cloning to clarify physiological processes involved in nitrogen assimilation in ectomycorrhizal fungi to highlight their differences from saprophytic and pathogenic fungi.

Section E deals with the sophisticated and novel techniques to formulate critical experiments and their design in order to retrieve excellent and reliable results. Background information for the selection of beneficial properties of *Pseudomonas* and *Bacillus* strains from the rhizospheric antagonistic to phytopathogenetic community requires elaboration, evaluation and bioassay (Chap. 23). After the selection of strains, these can be marked with a reporter gene and used to study cellular and molecular interactions between one or more beneficial microbes. These strains can also serve as a tool to study the interaction with soil-borne phytopathogens in the rhizosphere of their host plants. Autofluorescent proteins can be used for the noninvasive study of rhizosphere interactions using epifluorescence and confocal laser scanning microscopy (CSLM). Autofluorescent proteins have become an outstanding and convenient tool for studying rhizosphere and other in situ environmental interactions and have allowed microbiologists to visualize the spatial distribution of various microorganisms. The advent of fluorescent proteins offers a broad range of applications to track bacteria and study gene expression in the rhizosphere. The whole procedure of isolation, screening of antifungal activity, determining disease suppression in bioassays, preparation and transformation of protoplasts, allows fast isolation of potential biocontrol strains. The gnotobiotic test system has proven to be a valuable test system to study interactions between biocontrol bacteria, phytopathogen, and host plant. Combined with the use of autofluorescent proteins, it provides us with an extraordinary opportunity to study the intricate cellular and molecular interactions that the key players use to mediate their actions in the rhizosphere. In depth characterization of bacterial communities residing in environmental habitats has been greatly stimulated by the application of molecular phylogenetic tools such as 16S ribosomal RNA-directed oligonucleotide probes derived from extensive 16S rDNA sequence analysis. These phylogenetic probes are successfully applied in diverse microbial habitats using the fluorescent in situ hybridization (FISH) technique. In addition, the application of the immunofluorescence techniques to detect specific subpopulations or enzymes and of fluorescence marker-tagged bacteria or reporter constructs enables a highly resolving population and functional analysis. Phylogenetic in situ studies of the population structure can thus be supplemented with functional or phenotypic in situ investigation approaches. Two experimental approaches to investigate root-associated bacterial communities are presented in Chapter 24. On one hand, population and functional studies can be conducted directly in the rhizoplane (in situ) by combining specific fluorescence probing with confocal laser scanning microscopy yielding detailed information about the localization and small scale distribution of bacterial cells and their activities on the root surface. On the other hand, the separated rhizosphere compartments and the bacteria extracted from these different compartments allow a variety of subsequent ex situ studies. The separation into the three compartments, bulk soil, ectorhizosphere and rhizoplane/endorhizosphere, has to be performed with great care and actually needs an optimization for each plant and soil type under study. The degree by which adhering soil particles (ectorhizosphere) are included in the rhizosphere studies considerably influences the outcome of the study, since these soil particles are carrying a microbial community resembling, to a varying extent, the soil situation as compared to the root surface or rhizoplane situation. Certainly, in situ and ex situ studies (with separated rhizosphere compartments) both complement each other to give a more comprehensive picture. Although the microscopic in situ approach has the great advantage of providing detailed spatial information about root surface colonization, quantitative and qualitative data about the structural and functional diversity of root colonization can be obtained by a variety of complimentary ex situ approaches.

The plant cuticle forms the solid surface environment for epiphyllic microorganisms. Detailed analysis of a variety of microbe – cuticle interactions combining physicochemical, ecophysiological and microbial aspects are presented in Chapter 25. Isolated cuticles are excellent model surfaces to study the mechanisms of such interactions. Using the in vitro system, even minor changes in cuticular wax composition or permeability can be examined in relation to microbial growth. Working with entire leaves such changes would probably be masked by the physiological influence of the leaf. Therefore, this new approach might be very helpful to reveal possible mechanisms of interactions that occur, in reality, only in the scale of microhabitats. The impact of cuticular features will help us to understand the observed heterogeneous colonization of the leaf habitat and the formation of micro-colonies. Vice-versa the capacity of microbial cells to change cuticular properties might be of crucial importance for a successful colonization of the leaf surfaces and could contribute substantially to microbial fitness of individual epiphyllic species. Changes in cuticular properties in relation to microbial growth can be assessed in vitro under controlled conditions. *Pseudomonas putida* GR12-2, a well-known plant growth promoting strain, contains the enzyme 1-aminocyclopropane-1-carboxylic acid (ACC) deaminase. This enzyme hydrolyses ACC, the immediate precursor of ethylene in plant tissues. Ethylene is required for seed germination and the rate of ethylene production increases during germination and seedling growth. One model has been suggested where ACC deaminase containing growth-promoting bacteria can lower ethylene levels and thus stimulate plant growth. A rapid and novel procedure for the isolation of ACC deaminase-containing bacteria has been described in Chapter 26. In order to be able to test the model, a method for measuring ACC in plant tissues is described. Since all of the available methods for ACC quantification had problems and limitations associated with their use, Waters AccQ.Tag Method, designed to measure amino acids, was successfully applied for ACC analysis. This procedure is simple and relatively sensitive.

The protocol for understanding *Rhizobium*-legume root nodule symbiosis has been taken up by various microscopy techniques including bright-field, phase contrast, Nomarski interference contrast, polarized light, real time and time-lapse video, dark-field, conventional and laser scanning confocal epifluorescence, scanning electron, transmission electron, and field-emission scanning/transmission electron microscopies combined with visual counting techniques and manual interactive applications of image analysis. A new generation of innovative, customized image analysis software-CMEIAS (Center for Microbial Ecology Image Analysis System), designed specific digital images of microbial populations and communities and extracted all the informative, quantitative data of in vitro microbial ecology from them at spatial scales relevant to the microbes themselves. New computer-assisted imaging technology has been successfully applied to the fascinating field of plant surface microbiology (Chap. 27). CMEIAS software can "count what really counts" to enhance the quantitative analysis of microbial communities and populations in situ without cultivation.

Knowledge of genetic diversity in the bacterial population has increased considerably over the last 15 years, due to the application of molecular techniques to microbial ecological studies. Among the molecular methods, the PCR-based techniques provide a powerful and high throughput approach for the study of genetic diversity in bacterial populations. Some of the most commons are the PCR-RFLP of specific sequences (16S rDNA, intergenic transcribed spacer, ITS), the repetitive extragenic palindromic-PCR and the BOX-PCR based on the presence of repetitive elements within the bacterial genome, the DNA amplification fingerprintings, RAPDs (random amplified polymorphic DNA, and AFLPs (amplified fragment length polymorphism). ITS, RAPD and AFLP have been shown to be particularly relevant for the study of genetic diversity within populations of bacteria belonging to the same or closely related species (Chap. 28). AFLP shows some advantages over the other methods due to high stringency PCR conditions which give reproducibility and easy application to plant, animal and bacterial genomic DNA. AFLP has a high informational content per single reaction, in fact, up to 100 different bands can be displayed in a single lane and the scoring can be done with an automatic sequencer.

While there is a considerable amount of knowledge based on the ecology and physiology of mycorrhizal fungi and their uses, the knowledge about cellular and molecular aspects leading to the growth and development of the mycorrhizal fungus, as well as the establishment of a functioning symbiosis is still limited. An appropriate approach to the study of these special fungi is to understand the molecular process leading to the host recognition, development and functioning of mycorrhiza through the analysis of expressed sequences. With the advent of many highly sophisticated techniques that have been successfully applied to the functional analysis of genes from many organisms, it is now possible to apply similar strategies to study the various aspects of the mycorrhizal symbiosis (Chap. 29). The protocol describes expressed sequence tags (EST) and macroarray techniques. These approaches provide efficient tools for mycorrhizal symbiosis research. They have the resolution and ability to obtain a more comprehensive view of various stages of mycorrhiza development or treatment effects due to nutritional changes or differences due to host responses. Data can be exchanged and compared between different laboratories and eventually will provide a platform to understand the key players (genes) that are markers for ectomycorrhizal and AM fungal symbiosis. A large number of media compositions are available in the literature for the cultivation of various groups of fungi, but almost no literature is available for axenic cultivation of symbiotic fungi. Chapter 30 deals with the possible methods and the tested media composition to cultivate *Piriformospora indica*. These media can be utilized to understand the morphological and functional properties, or to test possible biotechnological applications.

Finally, for many groups of microorganisms, growth in axenic conditions is not yet possible. New methodologies for producing axenic cultures of the symbiotic fungus *Piriformospora indica* provide avenues for advancing the study of growth of other symbiotic organisms separately from their hosts. This is an important avenue of further studies, because it will allow us to understand a wider range of interactions between plants and can more closely reflect the enormous diversity of plant/microbe associations that exist in every environment.

2 Root Colonisation Following Seed Inoculation

THOMAS F.C. CHIN-A-WOENG and BEN J.J. LUGTENBERG

1 Introduction

This chapter provides protocols for the use of a gnotobiotic sand system to study root colonisation after seed inoculation. The complete experimental setup for a gnotobiotic system to grow plants for 7–14 days in the presence of inoculated bacteria or fungi is described. Subsequently, rhizosphere interactions and the in situ behaviour of inoculated organisms is visualised using autofluorescent proteins or other reporter systems. The behaviour of a good root-colonising *Pseudomonas* strain in this gnotobiotic system is described in terms of distribution, localisation, and root colonisation strategies as observed by microscopy.

2 Bacterial Root Colonisation

Microbial attachment to and proliferation on roots is generally referred to as root colonisation. Root colonisation is an important factor in plant pathogenesis of soil-borne microorganisms as well as in beneficial interactions used for microbiological control, biofertilisation, phytostimulation, and phytoremediation.

Various methods for studying rhizosphere colonisation under axenic as well as under field soil conditions have been described and the experimental approaches taken often depend on the problems studied. In this chapter, we describe a method for studying bacterial colonisation of the plant root system after introduction by seed inoculation. This simple system can be extended to study the influence of individual biotic and abiotic factors such as those present in potting soil.

Root colonisation is influenced by many variables. These factors can be biotic, such as genetic traits of the host plant and the colonising organism. For example, the possession of certain colonisation genes such as *sss* and/or *colS*/*colR* is necessary for efficient competitive root colonisation. In addition,

Plant Surface Microbiology
A. Varma, L. Abbott, D. Werner, R. Hampp (Eds.)

abiotic factors, such as growth substrate, soil humidity, soil and rhizosphere pH, and temperature heavily influence root colonisation. The study of the molecular mechanism of root colonisation of a host plant by one or more bacterial strains is complicated due to many biotic and abiotic field-soil variables which can be difficult to control. The use of a gnotobiotic system limits the biological variation and results in more reliable and reproducible experimental data. However, since the purpose of colonisation studies is to learn about the processes which occur under realistic conditions, we always test interesting gnotobiotic results in field or potting soil. With only one exception, the gnotobiotic results also appear to be the case in soil.

Various visualisation systems, including light and electron microscopy and confocal laser scanning microscopy (CLSM) combined with reporter systems such as those using genes for autofluorescent proteins, β-glucuronidase, and β-galactosidase allow us to determine numbers of bacteria on the root and follow the fate of inoculant bacteria in the spermosphere after seed inoculation and along the root system after growth.

In this chapter, we will also focus on the genetic and metabolic burdens in the rhizosphere as a consequence of genetic modification of the organisms required to enable the marking, tracking, recovery, and selection of bacteria in and from the rhizosphere. The gnotobiotic system provides a reproducible method to study root colonisation in terms of strategies and competition. Afterwards, the data should be verified under more natural conditions as emphasised before. Various growth substrates including sand, potting soil, field soil, and stonewool have been successfully used in the root colonisation system presented in this chapter. The system has been extended by introducing soil-borne pathogens, which allows the study of interactions between pathogen, microbes, and host plants at the cellular level which may be important for applications such as biocontrol.

3 Analysis of Tomato Root Tip Colonisation After Seed Inoculation Using a Gnotobiotic Assay

3.1 Description of the Gnotobiotic System

To assay colonisation, a gnotobiotic sand system comprised of two glass tubes is used. A silicone ring of 15 mm, cut from a silicone tube (25x35 mm, Rubber BV, Hilversum, The Netherlands), is placed around the top tube (outer diameter 25 mm, inner diameter 21 mm, length 200 mm) at 5 cm from the end (Fig. 1). The same end is closed with gauze using a rubber band. This end is placed in a bottom tube (outer diameter 40 mm, inner diameter 35 mm, height 95 mm) that contains 3 ml of water to prevent the tube content from desiccation. Subsequently, high quality quartz sand (quartz sand 0.1–0.3 mm; Wessem BV, Wessem, The Netherlands) is moisturised with plant nutrient

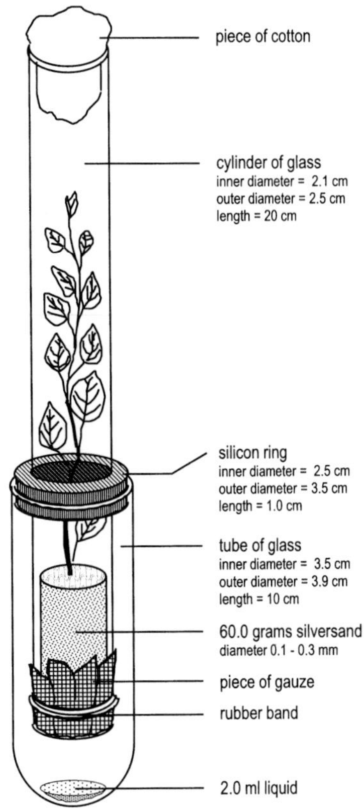

Fig. 1. Colonisation tube system (for explanation, see text)

solution (PNS: 1.25 mM $Ca(NO_3)_2$, 1.25 mM KNO_3, 0.50 mM $MgSO_4$, 0.25 mM KH_2PO_4 and trace elements (0.75 mg/l KI, 3.00 mg/l H_3BO_3, 10.0 mg/l $MnSO_4{\cdot}H_2O$, 2.0 mg/l $ZnSO_4{\cdot}5H_2O$, 0.25 mg/l $Na_2MoO_4{\cdot}2H_2O$, 0.025 mg/l $CuSO_4{\cdot}5H_2O$, 0.025 mg/l $CoCl_2{\cdot}6H_2O$, pH adjusted to 5.8; 10 % v/w). After thorough mixing, the top tubes are loosely filled with about 60 g of moisturised sand and closed with a cotton plug. The entire system is sterilised at 120 °C for 20 min.

3.2 Seed Disinfection

Many ways have been described to disinfect the surface of seeds of various crop plants without causing notable decreased seed germination efficiency. Common household bleach (sodium hypochlorite) or ethanol is often used for seed surface treatments. Most bacteria and fungi on the seed coat are killed after treatment with these disinfectants. Higher concentrations of up to

50 % (v/v) sodium hypochlorite can be prepared from commercial stocks. The effectiveness of a certain procedure is dependent upon the species and source of the seeds. To ensure sterility, checks should be performed by placing the disinfected seeds on rich agar medium. Care should be taken to remove traces of the disinfectant since this may influence germination efficiency as well as the survival of the bacteria after coating or inoculation of the seed. Sterilised tomato (*Lycopersicon esculentum*) seeds are obtained by rinsing tomato seeds with household bleach (adjusted to approximately 5 % sodium hypochlorite) and stirring in a sterile flask for 3 min. Not all seeds sink to the bottom of the flask despite stirring. After 3 min, sterilised demineralised water is added and most, if not all, seeds will then sink to the bottom of the flask. Seeds that remain floating are discarded. The hypochlorite is removed by washing the seeds five times extensively with 20 ml sterile water, followed by 2-h washing in sterile water during which the water is replaced at least three times. Contamination checks, carried out by placing the disinfected seeds on King's medium B agar (KB), show whether the seeds are free of contaminating microorganisms. For colonisation assays, this method is a reliable disinfection method. For disinfection of grass and wheat seeds, NaOCl/0.1 % SDS solutions can be used.

If seedlings are used instead of seeds, the surface disinfected seeds are placed on PNS solidified with 1.8 % Bacto Agar and placed in the dark to allow germination. Prior to transfer to a suitable temperature for germination (e.g. 28 °C for tomato), the seeds are incubated overnight at 4 °C, which often improves the germination efficiency and enhances synchronous germination of the seeds. For seeds such as tomato, wheat, or radish, it subsequently takes 1–2 days before 3–5-mm root tips appear. Seeds are inspected for proper germination and seedlings with the same length of root tips are selected.

3.3 Growth and Preparation of Bacteria

Liquid cultures of bacterial strains are grown overnight on a rotary shaker. For colonisation experiments with a mixture of strains (e.g. wild type versus mutant) a suspension of washed bacteria is prepared in a 1:1 ratio. A volume of 1.0 ml of an overnight culture is sedimented by centrifugation and the supernatant is discarded. The cells are washed with 1 ml phosphate buffered saline (PBS: 20 mM sodium phosphate, 150 mM NaCl, pH 7.4) and resuspended in PBS. The concentration of bacteria in this suspension is determined by measuring the optical density ($OD_{600\ nm}$). The strains are diluted to a concentration of $1 \cdot 10^8$ CFU/ml. If a mixture of strains is to be used for inoculation, the cells are mixed prior to inoculation of the seeds or seedlings, e.g. in a 1:1 ratio. The suspension is vortexed vigorously to yield a homogenous suspension of two strains.

3.4 Seed Inoculation

Seeds are placed in the bacterial suspension with sterile forceps and shaken gently for a few seconds. After approximately 10 min, the inoculated seeds are aseptically planted in the sand column of the gnotobiotic system, 5 mm below the sand surface. At a concentration of the inoculation mixture of 10^8 CFU/ml, the number of bacteria attaching to tomato seeds or seedlings is close to saturation (approximately 10^4 CFU/seed) and lowering the inoculation concentration to 10^4 CFU/ml does not appear to have an effect on the numbers and distribution of bacteria on the root system after 7 days of growth. Care should be taken not to damage the roots of the seedling since this will induce formation of lateral roots. The seedlings are grown in a climate-controlled chamber (19 °C, 16/8 h day/night cycles, 70 % relative humidity) for 7 days, or until the root tips penetrate the gauze.

The gnotobiotic system can be used to study the root colonisation behaviour of bacteria or be used to test strains for their competitive colonisation abilities. To screen for mutants that are impaired in competitive root colonisation, two mutants can be employed. Depending on the selectable properties of the strains (one strain must be marked with an antibiotic resistance or a reporter) the suspension can be plated on an appropriate selective medium to check the ratio of the strains. The use of Tn*5lacZ* marked strains allows the discrimination between wild type and mutant on 5-bromo-4-chloro-3-indolyl-β-galactopyranoside (X-gal) plates after reisolation of the bacteria from part of the root system. Since chances are small that two randomly picked mutants are both colonisation mutants, one Tn*5* (white) mutant can be tested against a Tn*5lacZ* mutant (blue), which allows faster screening for colonisation mutants, after which each mutant has to be tested against the wild-type strains.

3.5 Analysis of the Tomato Root Tip

To reisolate bacteria from the rhizosphere, the complete sand column is carefully removed from the tube. Most of the still adhering rhizosphere sand is removed and a length of 1–2 cm root tip is cut off with caution to prevent cross-contamination from upper root parts. If the complete root system is to be analysed, the root can be divided into segments. The root segments are shaken in 1 ml sterile PBS in the presence of the adhering rhizosphere sand or sterile glass beads to release tightly associated bacteria from the root surface on an Eppendorf shaker for 20 min. The bacterial suspension thus obtained is diluted with PBS and plated using a spiral plater on solid medium supplemented with X-gal when *lacZ* is used as a marker. The use of an automatic plating system and counter usually allows fast and accurate bacterial counts covering five orders of magnitude using a single dilution step.

With *P. fluorescens* strains WCS365 and WCS365::Tn*5lacZ*, a 10^4 dilution of the resuspended bacteria is plated with a spiral plater on KB medium containing X-gal (40 μg/ml). After growth, the numbers of white and blue colonies are determined. Since the bacteria are lognormally distributed in the rhizosphere, the data are $\log_{10}$(CFU+1)/cm transformed prior to statistical analysis with ANOVA followed by the non-parametric Wilcoxon-Mann-Whitney U-test to test significance between sample data. Details of the statistical approaches when handling these experimental data have been reviewed.

Alternatively, root sections can be prepared for visualisation by light, electron, or confocal laser scanning microscopy to obtain details of the distribution pattern of the bacteria on the root surface.

3.6 Confocal Laser Scanning Microscopy

Autofluorescent proteins have been successfully expressed in bacterial cells and are widely used to monitor the localisation of bacterial cells or gene expression in cells. Autofluorescent proteins can be detected in living cells without staining or invasive detection methods and require no cofactors. Furthermore, the generation and discovery of various forms of autofluorescent proteins, such as BFP, CFP, YFP, DsRed, with differing luminescent and spectral properties have spurred additional interest in the use of these proteins as reporters. Autofluorescent protein-labelled strains have been used to study microbial communities in various environmental applications such as the study of dynamics and distribution of bacteria in soil, water systems, rhizospheres, activated sludges, biodegradation/bioremediation, biofilms, and root nodulation. The protein can also be used to study gene expression and gene transfer in bacterial populations.

The analysis of autofluorescent proteins using CLSM is a very powerful technique to visualise microorganisms in complex environments such as in biofilms and the rhizosphere.

Computer-assisted CLSM provides high resolution imaging under non-invasive conditions. With software for three-dimensional image analysis, a spatial arrangement of the distribution of labelled bacteria can be determined.

4 Genetic Tools for Studying Root Colonisation

4.1 Marking and Selecting Bacteria

While antibiotic resistance can be very well applied as a marker to select bacteria in vitro, field conditions often require other or additional selection methods. There are numerous ways to track bacteria in the rhizosphere, associated habitats, and phyllosphere. Commonly used marker genes include the *gusA*, *lacZ*, *phoA*, *xylE*, *luxAB*, *luc*, and *celB* genes (Table 1).

The use of reporter genes such as β-galactosidase or β-glucuronidase as reporter genes has greatly facilitated the localisation of bacteria on the root surface. For β-galactosidase staining, roots or root sections can be directly fixed in 1.25% (v/v) glutaraldehyde in Z buffer (10 mM KCl, 1 mM $MgSO_4$, 50 mM KH_2PO_4, 50 mM K_2HPO_4, pH 7.0) for 30 min. Subsequently, the roots are washed twice in Z buffer for 30 min and stained overnight at 28 °C in a solution of X-Gal (0.8 mg/ml). The roots can be mounted for light microscopic analysis after thorough rinsing in Z buffer. The use of cross-linking fixation immobilises the bacteria on the root surface and the enzyme in the tissue. Although plants are known to possess endogenous β-galactosidase activity, this method gives no background of β-galactosidase activity from a number of plant root systems including tomato and *Arabidopsis*, since endogenous plant β-galactosidases are inactivated at high temperatures. By making cross-sections of roots after staining, the method can also be used to study bacterial-root associations in which bacteria penetrate deeper into the root tissue. In a similar way bacteria carrying a β-glucuronidase gene can be detected on the root system after staining with 5-bromo-4-chloro-3-indolyl-β-D-glucuronide. The major advantage of the use of β-glucuronidase is that plants do not possess endogenous β-glucuronidase activity.

The optimal reporter system should provide an easy and non-invasive way to follow the fate of individual cells in the rhizosphere. In addition, it should provide the possibility to quantify the activity of specific promoters in the rhizosphere. Many of the reporters have several drawbacks and restrictions, which limit their application. Some make use of specific substrates, have high background signals, or require sophisticated and expensive equipment for detection (Table 1). Compared to these reporters, autofluorescent proteins possess several advantages and have been shown to be good tools for the detection of cells (see Chap. 23, Visualisation of rhizosphere interactions of *Pseudomonas* and *Bacillus* biocontrol strains), and are promising tools for the measurement of gene activities in the rhizosphere. Nowadays, an argon laser (488-nm wavelength) is often used to excite red-shifted *gfp*-variants. An epifluorescence microscope equipped with a standard fluorescein isothiocyanate filter is effective for the detection of *gfp* red-shifted mutants which have excitation and emission maxima at 488 and 510 nm, respectively. A DAPI (4′6-diamidino-2-phenylindole) filter set with excitation at 330–380 nm and barrier filters at 435 nm can be used to detect wild-type Gfp. Autofluorescently labelled colonies on agar plates can be detected under a hand-held UV-lamp or a low-resolution binocular microscope equipped with a UV lamp. Other methods such as flow cytometry can be used to quantify *gfp*-labelled bacteria. Individual cells can be detected, quantified, and sorted with high speed and accuracy. On media without added iron, fluorescent pseudomonads tend to emit background fluorescence, which can obscure the GFP fluorescence. For

Table 1. Reporter genes commonly used for the detection of bacteria in environmental applications

Gene	Gene product or function	Advantages and disadvantages for use in the rhizosphere	References
gusA	β-Glucuronidase	No background in rhizobia and plants. Requires substrate.	Sharma and Signer (1990); Streit et al. (1992)
lacZ	β-Galactosidase	High background in most plants and bacteria.	Drahos et al. (1986); Katupitiya et al. (1992); Krishnan and Pueppke (1992)
phoA	Alkaline phosphatase	High background in most plants and bacteria.	Reuber et al. (1991)
xylE	Catechol 2,3-dioxygenase	Soluble end product.	Winstanley et al. (1991)
luxA, *luc*	Luciferase	Amplification and or photographic exposure for detection.	O'Kane et al. (1988); de Weger et al. (1991);
		Low resolution.	Silcock et al. (1992); de Weger et al. (1997)
celB	β-Glucosidase	Detection after denaturation of endogenous enzymes.	Voorhorst et al. (1995)
tfdA	2,4-Dichlorophenoxyacetate monooxygenase	Low resolution.	King et al. (1991)
gfp, *bfp*, *yfp*, *cfp*, *rfp*	Autofluorescent protein	High resolution. Real-time application. Requires oxygen for proper folding.	Chalfie et al. (1994)
Antibiotic resistance	Antibiotic resistance	Requires plate counting.	Hagedorn (1994)
Heavy metal resistance	Heavy metal resistance	Requires plate counting.	de Lorenzo (1994)

selection using GFP-expressing bacteria, this can be easily overcome by the addition of 0.45 mM $FeSO_4 \cdot H_2O$. Since most GFP gene sequences are known, *gfp*-tagged cells can also be detected by molecular methods such as gene probing, DNA hybridisation, or PCR.

4.2 Rhizosphere-Stable Plasmids

To understand the biological significance of genes and mutations, they need to be studied or expressed in the context in which they are assumed to function. Also, the complementation of rhizosphere-expressed mutations and expression of reporter genes need to be performed in situ. One consideration when studying processes in complex living systems, such as under soil or rhizosphere conditions, is that antibiotic selection often cannot be applied. In addition, bacteria in the rhizosphere are assumed to be covered by a mucigel layer or form biofilms which are known to have increased resistance to antibiotics. Therefore, field and rhizosphere studies often require the use of rhizosphere-stable plasmids, e.g. for complementation of mutations or for tracking bacteria. While naturally occurring plasmids are often stably maintained within a bacterial population in the absence of selection pressure, many cloning vectors disappear without the appropriate selection. Plasmids with genes for complementation or reporter studies should therefore be stably maintained in strains without antibiotic pressure or be integrated into the chromosome.

The *Pseudomonas* replicon pVS1 is stably maintained in many genera including *Pseudomonas*, *Agrobacterium*, *Rhizobium*, *Burkholderia*, *Aeromonas*, and *Comamonas*. Cloning vectors harbouring a 3.8-kb region of pVS1 with functions for replication (*rep*) and stability (*sta*) also appear to be stably maintained. pVS1 derivatives pVSP41, pWTT2081, pME6010, pME6030, pME6040, and derivatives have been shown to be completely stable in various rhizosphere bacteria in the rhizospheres of various crop plants. Although the incompatibility group of pVS1 has not been determined, the replicon appears to be compatible with IncP-1, IncP-4, IncP-8, IncP-10, and IncP-11 plasmids in *P. aeruginosa*.

4.3 Genetic and Metabolic Burdens

Another consideration when introducing foreign or additional DNA on plasmids into bacterial strains is a plasmid or metabolic burden. The presence of a plasmid may confer a metabolic burden on the cells because of the presence of additional DNA and/or the expression of the reporter gene. Although the effects are often not visible under laboratory conditions, the presence of a plasmid may very well cause a genetic or metabolic burden in the rhizosphere and e.g. negatively affect the colonisation ability of a strain, as was shown for the presence of the rhizosphere-stable plasmid pWTT2081 in *P. fluorescens* WCS365 in the tomato rhizosphere. In competitive colonisation studies it is, therefore, of crucial importance to restore the balance by introducing the same empty vector in other strains when they are compared. Similarly, some biocontrol strains marked with autofluorescent proteins show decreased control of disease compared to the wild type such as in the control of seed-borne net blotch by *Pseudomonas chlororaphis* MA 342. *E. coli* cells harbouring DsRed also appear to be smaller than untransformed bacteria.

5 Behaviour of Root-Colonising Pseudomonas Bacteria in a Gnotobiotic System

5.1 Colonisation Strategies of Bacteria

Using light, electron, or confocal laser scanning microscopy, bacteria can be directly visualised on the root surface and as such allow determination of distribution and colonisation patterns. Although light microscopy offers an easy way of visualising bacteria on the root, the resolution is often just below that necessary for detailed studies. More recently, CLSM has provided much more detailed information on the distribution and interactions in the rhizosphere. The number of bacteria present on the root system can also be simply followed by dilution plating of cell suspensions of bacteria that have been reisolated from root sections. On many plant root systems bacteria appear to be distributed lognormally rather than in a uniform way. In a typical bioassay with tomato seedlings grown for 7 days in a gnotobiotic sand system bacteria also appear to be distributed lognormally. High bacterial numbers are found at the root base (10^7–10^8 CFU/cm) which rapidly decrease to 10^3–10^4 CFU/cm at the root tip. Under the same growth conditions, bacterial numbers on one of the many roots of wheat are one order of magnitude higher, whereas in competition with indigenous rhizobacteria the numbers are usually one order of magnitude lower.

The pattern of microbial occupation of root sites by bacteria varies considerably with plant species and conditions under which plants are grown, but the percentage of root surface covered is usually estimated less than 10%. Often, the distribution within a small area of the plant root surface appears to consist of heavily populated areas, whereas other parts are practically devoid of bacteria. *Pseudomonas* cells on the tomato root are mainly present as elongated stretches on indented areas, such as junctions between epidermal cells and the deeper parts of the root epidermis, and root hairs.

Transmission (TEM) and scanning electron microscopy (SEM) of the root–soil interface can reveal more details regarding the spatial relationships of microorganisms, soil, and roots than light microscopy. After removal of the plant roots from the sand, these can be directly fixed and prepared according to standard protocols for TEM or SEM analyses. A prominent feature observed with these techniques is the mucilage or biofilm which surrounds the root and in which microorganisms develop. This biofilm is believed to provide a contact between soil and roots for diffusion of nutrients and may give some protection from other microorganisms. Although the film is also produced by the plant under axenic conditions, it appears to be thicker in non-sterile roots, where bacterial capsular material such as exopolysaccharides (EPS) may contribute significantly to this layer. The biofilm can also be visualised using confocal laser scanning microscopy combined with fluorescently marked bacteria. The encapsulation of bacteria in a mucigel may have considerable consequences for the action of certain diffusible compounds such as autoinducer molecules involved in quorum sensing. This phenomenon also complicates proper visualisation of marked bacteria that have penetrated deeper into the surface layers of the root. CLSM usually can cope with these difficulties since the system can focus on multiple planes of the specimen.

An in-depth study of the stages of root colonisation by CLSM has shown that *P. fluorescens* WCS365 microcolonies on the root surface are usually formed from one single cell, since mature microcolonies that have been visualised on the root surface usually consist of one type of bacterium. The log-normal distribution of bacteria on the root tip indicates that most bacteria remain close to the inoculation site after seed inoculation. It is believed that occasionally, single cells detach from older parts of the root and travel along the growing root tip to establish new colonies. In later stages, mixed microcolonies can be observed with CLSM, indicating that other bacteria can join at some stage of microcolony formation.

5.2 Competitive Colonisation Studies

For a long-lasting effect, biocontrol bacteria must compete with the native microflora and establish themselves for several months at a high level in the rhizosphere. Successful colonisation of the plant root is often considered to be important for the success of various applications for beneficial purposes and for suppression of plant diseases. When studying colonisation traits in our laboratory, we therefore determine competitive root colonisation of two or more strains on the root.

It was assumed that various bacterial traits contribute to the ability of a bacterial strain to colonise the rhizosphere and that loss of such a trait reduces the ability to establish itself effectively in the rhizosphere and, hence, also reduces its beneficial effects. Using initially competitive root tip colonisation in the gnotobiotic system as the assay, various competitive colonisation genes and traits were identified. One of the identified traits involved in colonisation is chemotaxis towards root exudate. *cheA*$^-$ chemotaxis mutants of various *P. fluorescens* strains appear to be strongly reduced in competitive root colonisation (de Weert et al. 2002). Chemotaxis was also suggested to be the first step in establishment of bacterial seed and root colonisation.

Flagella-less *Pseudomonas* strains, when tested in competition with the wild type after application on seeds, are severely impaired in colonisation of the root tip of potato and tomato. A non-motile mutant of the *Fusarium oxysporum* f. sp. *radicis-lycopersici* (*F.o.r.l.*) antagonist *P. chlororaphis* PCL1391, was 1000-fold impaired in competitive tomato root tip colonisation.

Agglutination and attachment of *Pseudomonas* cells to plant roots are likely to play a role in colonisation. Compounds that can mediate attachment or agglutination are adhesins, fimbriae, pili, cell surface proteins, and polysaccharides. The degree of attachment to tomato roots is correlated with the number of type 4 fimbriae on bacterial cells of *P. fluorescens* WCS365. The outer membrane protein OprF of *P. fluorescens* OE28.3 is involved in attachment to plant roots. A root-surface glycoprotein agglutinin was shown to mediate agglutination of *P. putida* isolate Corvallis, but had no effect on colonisation.

Various *Pseudomonas* mutant derivatives lacking the O-antigen side chain of lipopolysaccharide (LPS) are impaired in colonisation. The colonisation defect in strains with defective LPS can be explained by assuming that for the optimal functioning of nutrient uptake systems, an intact outer membrane is required.

Genes for the biosynthesis of amino acids and vitamin B1 and for utilisation of root exudate components such as organic acids are also important for colonisation of *P. fluorescens* WCS365 on tomato roots (Simons et al. 1997; Wijfjes et al. in preparation) and *P. chlororaphis* PCL1391. Putrescine is an important root exudate component of which the uptake level must be carefully regulated. *P. fluorescens* mutants with an increased putrescine level have a decreased growth rate resulting in a colonisation defect.

Other traits that are likely to influence colonisation include generation time, osmotolerance, resistance to predators, host plant cultivar, and soil type.

Genes of which the role in colonisation were not obvious were identified after screening of a random Tn*5* mutant library of *P. fluorescens* WCS365 in competition with the parental strain. They include the *nuoD* gene which is part of a 14-gene operon encoding NADH dehydrogenase NDH-1 (Camacho et al. 2002). The biocontrol strain *P. fluorescens* WCS365 possesses two NADH dehydrogenases, and apparently, the absence of NDH-1 cannot be adequately compensated for by the other NADH dehydrogenase under rhizosphere conditions, resulting in lower fitness on the root.

A two-component regulatory system consisting of the *colS* and *colR* genes, which have homology to sensor kinases and response regulators, respectively, was also shown to be involved in efficient root colonisation of strain *P. fluorescens* WCS365. It was concluded that an environmental stimulus is important for colonisation, but neither the nature of the stimulus, nor the target genes are known.

The *sss* gene, encoding a protein of the lambda integrase gene family of site-specific recombinases, to which XerC and XerD also belong, is necessary for adequate root colonisation of *P. fluorescens* WCS365 and *P. chlororaphis* PCL1391. It was postulated that a certain bacterial subpopulation, which expresses an as yet unknown cell surface component regulated by a site-specific recombinase, is important for competitive colonisation of *P. fluorescens* WCS365.

For some strains the production of secondary metabolites contributes to the ecological competence of strains as was indeed shown for the phenazine-producing strains *P. fluorescens* 2–79 and *P. aureofaciens* 30–84 using phenazine biosynthetic mutants. Phenazine-minus strains had a reduced survival and a diminished ability to compete with the resident microflora. However, production of the antifungal factor 2,4-diacetylphloroglucinol in *P. fluorescens* strain F113 did not influence its persistence in the soil.

5.3 Monocots Versus Dicots

Differences in colonisation of bacterial strains may be attributed to different root exudate compositions of the host plant. Sugars, organic acids, and amino acids are considered to be the major readily metabolisable exudate compounds. The role of root exudate composition in colonisation behaviour was studied for a number of plants including tomato and grass. The amount of organic acid in tomato root exudate appears to be five times higher than that of exudate sugars. Using mutants of *P. fluorescens* WCS365, it was shown that organic acids are the nutritional basis for tomato root colonisation by this strain (Wijfjes et al. 2002, in prep.), whereas sugars appear to be less essential for colonisation. For monocots such as wheat and grass, a ten times higher number of *Pseudomonas* bacteria was found on the root compared to dicots such as tomato, radish, or potato. Since dicots and monocots have different organic acid and sugar compositions, increased root colonisation efficiency by certain strains might be related to a better growth on root exudates of monocots.

6 Influence of Abiotic and Biotic Factors

6.1 Abiotic Factors

Commercial inoculants are mostly attached to the seed or are applied in the furrow where the bacteria can reach the seedling. However, for laboratory studies, bacterisation of seedlings instead of seeds will increase reproducibility since the experiments start with a homogenous set of seedlings and this eliminates problems associated with irregular seed germination. The use of a sterile system not only ensures more reproducible bacterial numbers on the root system, but also results in higher numbers on the root due to the absence of competition by indigenous soil bacteria.

Various environmental conditions influence root colonisation efficiency in the gnotobiotic sand system. The effect of a number of biotic and abiotic factors on colonisation was determined in a tomato-*P. fluorescens* WCS365 system. These factors include growth substrate, temperature, soil humidity, pH, and the presence of (competing) indigenous bacteria. Usually, ten times lower bacterial numbers are found on the tomato root system when experiments are performed in non-sterile potting soil instead of sterile quartz sand, which might be explained by the presence of indigenous competing organisms. The choice of material to sustain growth of seedlings is mainly determined by the system of interest. The use of chemically clean sand ensures a reliable experimental approach, but cannot be applied for studies requiring field conditions. Sand can be replaced by potting or field soil and the soil can be practically freed from indigenous organisms by gamma irradiation. Rockwool drained in plant nutrient solution also supports plant growth and bacterial colonisation. For more compact soil systems, such as clay-containing soils, the soil can be amended with sand to facilitate the recovery of roots from the system. The gnotobiotic system has been tested for tomato, radish, potato, cucumber, grass, and wheat, and may well be suitable for growth of other plant species. Although our seedlings in the gnotobiotic sand system are normally grown for 7 days, they can be grown for up to 14 days without watering.

To determine the influence of a number of abiotic factors on colonisation in the gnotobiotic system, *P. fluorescens* WCS365 was marked with a β-glucuronidase reporter and singly inoculated on tomato seedlings. The overall bacterial distribution of the marked bacteria was determined using dilution plating and visualised using root prints (unpublished data). Increasing fluid content from 10 up to 20 % (v/w) in sand results in an overall increase of bacterial numbers on the tomato root tip. The increased colonisation may be due to increased motility or passive transport of bacteria down the root. Utilisation of 5 % (v/w) nutrient solution severely limits plant growth and consequently, bacterial numbers are lower. Temperatures at which plants are grown need to be selected depending on the plant species. Although we grow tomato seedlings at an intermediate temperature of 19 °C, growth is significantly enhanced at higher temperatures (e.g. 28 °C). This is also reflected in the number of bacteria sustained by the plant root system, possibly due to the effect of increased root exudation on bacterial growth.

6.2 Biotic Factors

In potting soil, numbers of inoculated bacteria on the root system are usually ten-fold lower. Decreased root colonisation is not only caused by competition with soil-borne bacteria since numbers of inoculated bacteria on roots in non-sterilised and gamma-irradiated soil are comparable.

Sometimes plant roots grown in potting soil are difficult to remove from the glass colonisation tube. In such cases, a mixture of potting soil/sand (1:3 w/w) can be used as a compromise between the wish to use potting soil and that to experimentally study colonisation.

When a fungal pathogen is included in the system, it is possible to determine biocontrol abilities of strains under controlled conditions. In our lab, bioassays with tomato and the fungal pathogens *Fusarium oxysporum* f. sp. *radicis-lycopersici* (*F.o.r.l.*), *Rhizoctonia solani*, and *Pythium ultimum* systems have been successfully employed to determine antifungal abilities of pseudomonads and bacilli (Lagopodi et al. unpublished data) and to perform microscopic analyses of rhizosphere interactions (see Chap. 23, Visualisation of Rhizosphere Interactions of *Pseudomonas* and *Bacillus* Biocontrol Strains). The pathogen can be introduced together with the biocontrol agent onto the seed or mixed with the sand as a spore or mycelium suspension, depending on the question under study. For the tomato-*F.o.r.l.* system, spores are collected from a 3-day-old culture of *F.o.r.l.* grown in liquid Czapek-Dox medium. Mycelium obtained from a PDA agar culture was used for inoculation of the culture. Spores are collected after passage through a miracloth filter, washed with water, and resuspended in PNS. Numbers of spores can be determined using a haemocytometer. Finally, the spores are mixed through the sand to a final concentration of 50 CFU/g sand.

P. ultimum is grown for 3–4 weeks in clarified V8-medium (20 % V8 vegetable juice [Campbell Foods, Inc.], 25 mM $CaCO_3$, 30 µg/ml cholesterol). Prior to use, V8 is clarified by sedimentation at 6000 rpm for 30 min. Alternatively, the fungus can be cultured in hemp (*Cannabis* sp.) seed extract for 1–2 weeks. Oospores that are abundantly produced during incubation are collected and freed from the mycelium. The fungal mycelium is washed three times in sterile water and blended in 0.1 M sucrose for 1–2 min. The culture is incubated for 2 h at 130 rpm at 28 °C. The suspension is sedimented by centrifugation at 4000 rpm for 10 min., resuspended in 1 M sucrose, and incubated at –20 °C for 12 h to kill the mycelium fragments. After washing with water, the suspension is layered over 1 M sucrose and centrifuged at 2351 rpm for 1 min. Consecutive washing steps remove most of the mycelium fragments. Oospores are added to the sand to a final concentration of 3–24 oospores/g sand.

Plants are judged according to a fixed disease index based upon disease symptoms (Table 2). The presence of the fungus on diseased plants can be confirmed by dipping suspected diseased parts in 0.05 % household bleach

Table 2. *Pythium ultimum* and *Fusarium oxysporum* f. sp. *radicis-lycopersici* disease indices

Disease symptoms	Disease index
No visible symptoms	0
Small brown spots on the main root and/or the crown	1
Brown spots on the central root and extensive discoloration of crown	2
Damping-off or wilting	3
Dead plant	4

for 30 s and rinsing in sterile water to surface-disinfect the sample, followed by incubation on PDA agar medium.

7 Conclusions

The sand gnotobiotic system has proven to be a good tool to study rhizosphere interactions. Environmental and biotic conditions can be more carefully controlled in this system than in natural soils. Under controlled conditions, it also allows the enrichment of strains for particular traits. In our lab mutant screening has resulted in numerous mutants involved in root colonisation, which subsequently have been genetically characterised. Combined with the use of autofluorescent proteins and CLSM, the gnotobiotic system is a powerful tool to study the interactions between biocontrol bacteria, the pathogen, and the host plant. Unstable fluorescent proteins provide the tools for study of gene expression in the rhizosphere. Rhizosphere-associated phenomena such as bacterial cell-to-cell signalling events and signalling between pathogens and rhizosphere bacteria can be investigated in a clean and reproducible way.

References and Selected Reading

Bahme JB, Schroth MN (1987) Spatial-temporal colonization patterns of a rhizobacterium on underground organs of potato. Phytopathology 77:1093–1100

Bloemberg GV, O'Toole GA, Lugtenberg BJJ, Kolter R (1997) Green fluorescent protein as a marker for *Pseudomonas* spp. Appl Environ Microbiol 63:4543–4551

Bloemberg GV, Wijfjes AH, Lamers GE, Stuurman N, Lugtenberg BJ (2000) Simultaneous imaging of *Pseudomonas fluorescens* WCS365 populations expressing three different autofluorescent proteins in the rhizosphere: new perspectives for studying microbial communities. Mol Plant-Microbe Interact 13:1170–1176

Bowen GD, Rovira AD (1976) Microbial colonization of plant roots. Annu Rev Phytopathol 14:121–144

Buell CR, Anderson AJ (1993) Expression of the *aggA* locus of *Pseudomonas putida* in vitro and in planta as detected by the reporter gene, *xylE*. Mol Plant-Microbe Interact 6:331–340

Bull CT, Weller DM, Thomashow LS (1991) Relationship between root colonization and suppression of *Gaeumannomyces graminis* var. *tritici* by *Pseudomonas fluorescens* strain 2-79. Phytopathology 81:954–959

Camacho MM (2001) Molecular characterization of type 4 pili, NDHI and PyrR in rhizosphere colonization of *Pseudomonas fluorescens* WCS365. PhD Thesis, Universiteit Leiden, Leiden

Camacho Carvajal MM, Wijfjes AHM, Mulders IHM, Lugtenberg BJJ, Bloemberg GV (2002) Characterization of NADH dehydrogenases of *Pseudomonas fluorescens* WCS365 and their role in competitive root colonisation. Mol Plant-Microbe Interact 15:662–671

Campbell R, Rovira AD (1973) The study of the rhizosphere by scanning electron microscopy. Soil Biol Biochem 5:747–752

Caroll H, Moënne-Loccoz Y, Dowling D, O'Gara F (1995) Mutational disruption of the biosynthesis genes coding for the antifungal metabolite 2,4-diacetylphloroglucinol does not influence the ecological fitness of *Pseudomonas fluorescens* F113 in the rhizosphere of sugar beets. Appl Environ Microbiol 61:3002–3007

Chalfie M, Tu Y, Euskirchen G, Ward WW, Prasher DC (1994) Green fluorescent protein as a marker for gene expression. Science 263:802–805

Chin-A-Woeng TFC, de Priester W, van der Bij AJ, Lugtenberg BJJ (1997) Description of the colonization of a gnotobiotic tomato rhizosphere by *Pseudomonas fluorescens* biocontrol strain WCS365, using scanning electron microscopy. Mol Plant-Microbe Interact 10:79–86

Chin-A-Woeng TFC, Bloemberg GV, van der Bij AJ, van der Drift KMGM, Schripsema J, Kroon B, Scheffer RJ, Keel C, Bakker PAHM, Tichy HV, de Bruijn FJ, Thomas-Oates JE, Lugtenberg BJJ (1998) Biocontrol by phenazine-1-carboxamide-producing *Pseudomonas chlororaphis* PCL1391 of tomato root rot caused by *Fusarium oxysporum* f. sp. *radicis-lycopersici*. Mol Plant-Microbe Interact 11:1069–1077

Chin-A-Woeng TFC, Bloemberg GV, Mulders IHM, Dekkers LC, Lugtenberg BJJ (2000) Root colonization by phenazine-1-carboxamide-producing bacterium *Pseudomonas chlororaphis* PCL1391 is essential for biocontrol of tomato foot and root rot. Mol Plant-Microbe Interact 13:1340–1345

Christensen BB, Sternberg C, Molin S (1996) Bacterial plasmid conjugation on semi-solid surfaces monitored with the green fluorescent protein (GFP) from *Aequorea victoria* as a marker. Gene 173:59–65

Clarholm M (1984) Heterothrophic, free-living protozoa: neglected microorganisms with an important task in regulating bacterial populations. In: Klug MJ, Reddy CA (eds) Current perspectives in microbial ecology. American Society of Microbiology, Washington, DC, pp 321–326

Cormack BP, Valdivia RH, Falkow S (1996) FACS-optimized mutants of the green fluorescent protein (GFP). Gene 173:33–38

Davies KG, Whitbread R (1989) A comparison of methods for measuring the colonisation of a root system by fluorescent pseudomonads. Plant Soil 116:239–246

de Lorenzo V (1994) Designing microbial systems for gene expression in the field. Trends Biotechnol 12:365–371

de Weert S, Vermeiren H, Mulders IHM, Kuiper I, Hendrickx N, Bloemberg GV, Vanderleyden J, DeMot R, Lugtenberg BJJ (2002) Flagella-driven chemotaxis towards exudate components is an important trait for tomato root colonization by *Pseudomonas fluorescens*. Mol Plant-Microbe Interact 15:1173–1180 de Weger LA, Bakker PAHM, Schippers B, van Loosdrecht MCM, Lugtenberg BJJ (1989) *Pseudomonas* spp. with mutational changes in the O-antigenic side chain of their lipopolysaccharide are affected in their ability to colonize potato roots. In: Lugtenberg BJJ (ed) Signal molecules in plants and plant-microbe interactions. NATO ASI Series H, Springer, Berlin Heidelberg New York, pp 197–202

Dekkers LC (1997) Isolation and characterization of novel rhizosphere colonization mutants of *Pseudomonas fluorescens* WCS365. PhD Thesis, Leiden University, Leiden, The Netherlands

Dekkers LC, van der Bij AJ, Mulders IHM, Phoelich CC, Wentwood RAR, Glandorf DCM, Wijffelman CA, Lugtenberg BJJ (1998a) Role of the O-antigen of lipopolysaccharide, and possible roles of growth rate and of NADH:Ubiquinone oxidoreductase (*nuo*) in competitive tomato root-tip colonization by *Pseudomonas fluorescens* WCS365. Mol Plant-Microbe Interact 11:763–771

Dekkers LC, Phoelich CC, van der Fits L, Lugtenberg BJJ (1998b) A site-specific recombinase is required for competitive root colonization by *Pseudomonas fluorescens* WCS365. Proc Natl Acad Sci USA 95:7051–7056

Dekkers LC, Bloemendaal CP, de Weger LA, Wijffelman CA, Spaink HP, Lugtenberg BJJ (1998 c) A two-component system plays an important role in the root-colonizing ability of *Pseudomonas fluorescens* strain WCS365. Mol Plant-Microbe Interact 11:45–56

Dekkers LC, Mulders IH, Phoelich CC, Chin-A-Woeng TFC, Wijfjes AH, Lugtenberg BJ (2000) The *sss* colonization gene of the tomato-*Fusarium oxysporum* f. sp. *radicis-lycopersici* biocontrol strain *Pseudomonas fluorescens* WCS365 can improve root colonization of other wild-type *Pseudomonas* spp. bacteria. Mol Plant-Microbe Interact 13:1177–1183

DeMot R, Veulemans B, Vanderleyden J (1991) Root-adhesive protein of *Pseudomonas fluorescens* OE28-3. In: Keel C, Knoller B, Défago G (eds) Plant growth-promoting rhizobacteria. Progress and prospects. International organization for biological and integrated control of noxious animals and plants. Proceedings of the 2nd International Workshop on PGPR. WPRS Bulletin XIV/8, 308–312

de Weger LA, Dunbar P, Mahafee WF, Lugtenberg BJJ, Sayler G (1991) Use of bioluminescence markers to detect *Pseudomonas* spp. in the rhizosphere. Appl Environ Microbiol 57:3641–3644

de Weger LA, Kuiper I, van der Bij AJ, Lugtenberg BJJ (1997) Use of a *lux*-based procedure to rapidly visualize root colonisation by *Pseudomonas fluorescens* in the wheat rhizisphere. Anton Leeuw Int J G 72:365–372

Drahos DJ, Hemming BC, McPherson S (1986) Tracking recombinant organisms in the environment: β-galactosidase as a selectable non-antibiotic marker for fluorescent pseudomonads. Bio/Technology 4:439–444

Errampalli D, Okamura H, Lee H, Trevors JT, van Elsas JD (1998) Green fluorescent protein as a marker to monitor survival of phenanthrene-mineralizing *Pseudomonas* sp. UG14Gr in creosote-contaminated soil. FEMS Microbiol Ecol 26:181–191

Errampalli D, Leung K, Cassidy MB, Kostrzynska M, Blears M, Lee H, Trevors JT (1999) Applications of the green fluorescent protein as a molecular marker in environmental microorganisms. J Microbiol Meth 35:187–199

Foster RC (1986) The ultrastructure of the rhizoplane and rhizosphere. Annu Rev Phytopathol 24:211–234

Glandorf DCM, Sluis I, Anderson AJ, Bakker PAHM, Schippers B (1994) Agglutination, adherence, and root colonization by fluorescent pseudomonads. Appl Environ Microbiol 60:1726–1733

Greaves MP, Darbyshire JF (1972) The ultrastructure of the mucilaginous layer on plant roots. Soil Biol Biochem 4:443–449

Habte M, Alexander M (1977) Further evidence for the regulation of bacterial populations in soil by protozoa. Arch Microbiol 113:181–183

Hagedorn C (1994) Spontaneous and intrinsic antibiotic resistance markers. In: Weaver RS, Angle S, Bottomley P (eds) Methods of soil analysis, Part 2, Microbiological and chemical properties. Soil Science of America, Inc, Madison, WI, pp 575–591

Heeb S, Itoh Y, Nishijyo T, Schnider U, Keel C, Wade J, Walsh U, O'Gara F, Haas D (2000) Small, stable shuttle vectors based on the minimal pVS1 replicon for use in gram-negative, plant-associated bacteria. Mol Plant-Microbe Interact 13:232–237

Hoffland E, Findenegg GR, Nelemans JA (1989) Solubilization of rock phosphate by rape. Plant Soil 113:161–165

Howie WJ, Cook RJ, Weller DM (1987) Effect of soil matric potential and cell motility on wheat root colonization by fluorescent pseudomonads suppressive to take-all. Phytopathology 77:286–292

Itoh Y, Haas D (1985) Cloning vectors derived from the *Pseudomonas* plasmid pVSP1. Gene 36:27–36

Itoh Y, Watson JM, Haas D, Leisinger T (1984) Genetic and molecular characterization of the *Pseudomonas* plasmid pVS1. Plasmid 11:206–220

Jakobs S, Subramaniam V, Schonle A, Jovin TM, Hell SW (2000) EFGP and DsRed expressing cultures of *Escherichia coli* imaged by confocal, two-photon and fluorescence lifetime microscopy. FEBS Lett 479:131–135

Jenny H, Grossenbacher K (1963) Root-soil boundary zones as seen in the electron microscope. Soil Sci Soc Am Proc 27:273–277

Katupitiya S, New PB, Elmerich C, Kennedy IR (1992) Improved N_2-fixation in 2,4-D-treated wheat roots associated with *A. lipoferum*: studies of colonisation using reporter genes. Soil Biol Biochem 27:447–452

King EO, Ward MK, Raney DE (1954) Two simple media for the demonstration of pyocyanin and fluorescein. J Lab Clin Med 44:301–307

King RJ, Short KA, Seidler RJ (1991) Assay for detection and enumeration of genetically engineered microorganisms which is based on the activity of a deregulated 2,3-dichlorophenoxyacetate monooxygenase. Appl Environ Microbiol 57:1790–1792

Kloepper JW, Beauchamp CJ (1992) A review of issues related to measuring colonization of plant roots by bacteria. Can J Microbiol 38:1219–1232

Knudsen IMB, Hockenhull J, Jensen DF, Gerhardson B, Hokeberg M, Tahvonen R, Teperi E, Sundheim L, Henriksen B (1997) Selection of biological control agents for controlling soil and seed-borne diseases in the field. Eur J Plant Pathol 103:775–784

Krishnan HB, Pueppke SG (1992) A *nolC-lacZ* gene fusion in *Rhizobium fredii* facilitates direct assessment of competition for nodulation of soybean. Can J Micriobiol 38:515–519

Kuiper I, Bloemberg GV, Lugtenberg BJ (2001a) Selection of a plant-bacterium pair as a novel tool for rhizostimulation of polycyclic aromatic hydrocarbon-degrading bacteria. Mol Plant-Microbe Interact 14:1197–1205

Kuiper I, Bloemberg GV, Noreen S, Thomas-Oates JE, Lugtenberg BJJ (2001b) Increased uptake of putrescine in the rhizosphere inhibits competitive root colonization by *Pseudomonas fluorescens* strain WCS365. Mol Plant-Microbe Interact 14:1096–1104

Lagopodi AL, Ram AFJ, Lamers GEM, Punt PJ, van den Hondel CAMJJ, Lugtenberg BJJ, Bloemberg GV (2002) Novel aspects of tomato root colonization and infection by *Fusarium oxysporum* f. sp. *radicis-lycopersici* revealed by confocal laser scanning microscopic analysis using the green fluorescent protein as a marker. Mol Plant-Microbe Interact 15:172–179

Lewis K (2001) Riddle of biofilm resistance. Antimicrob Agents Chemother 45:999–1007

Loper JE, Suslow TV, Schroth MN (1984) Lognormal distribution of bacterial populations in the rhizosphere. Phytopathology 74:1454–1460

Loper JE, Haack C, Schroth MN (1985) Population dynamics of soil pseudomonads in rhizosphere of potato (*Solanum tuberosum* L.). Appl Environ Microbiol 49:416–422

Lugtenberg BJJ, de Weger LA, Schippers B (1994) Bacterization to protect seed and rhizosphere against disease. BCPC Monograph 57:293–302

Lugtenberg BJJ, Dekkers LC, Bansraj M, Bloemberg GV, Camacho M, Chin-A-Woeng TFC, van den Hondel C, Kravchenko L, Kuiper I, Lagopodi AL, Mulders I, Phoelich C, Ram A, Tikhonovich I, Tuinman S, Wijffelman C, Wijfjes A (1999a) *Pseudomonas* genes and traits involved in tomato root colonization. In: de Wit PJGM, Bisseling T, Stiekema WJ (eds) 1999 IC-MPMI Congress Proceedings: Biology of plant-microbe interactions, volume 2. International Society for Molecular Plant-Microbe Interactions, St. Paul, MN, pp 324–330

Lugtenberg BJJ, Kravchenko LV, Simons M (1999b) Tomato seed and root exudate sugars: composition, utilization by *Pseudomonas* biocontrol strains and role in rhizosphere colonization. Environ Microbiol 1:439–446

Lugtenberg BJJ, Dekkers LC, Bloemberg GV (2001) Molecular determinants of rhizosphere colonization by *Pseudomonas*. Annu Rev Phytopathol 39:461–490

Mah TF, O'Toole GA (2001) Mechanisms of biofilm resistance to antimicrobial agents. Trends Microbiol 9:34–39

Mazzola M, Cook RJ, Thomashow LS, Weller DM, Pierson LS (1992) Contribution of phenazine antibiotic biosynthesis to the ecological competence of fluorescent pseudomonads in soil habitats. Appl Environ Microbiol 58:2616–2624

McLoughlin AJ (1994) Plasmid stability and ecological competence in recombinant cultures. Biotechnol Adv 12:279–324

Miller WG, Lindow SE (1997) An improved GFP cloning cassette designed for prokaryotic transcriptional fusions. Gene 191:149–153

Möller S, Steinberg C, Andersen JB, Christensen BB, Ramos JL, Givskov M, Molin S (1998) In situ gene expression in mixed-culture biofilms evidence of metabolic interactions between community members. Appl Environ Microbiol 64:721–732

Nordström K (1989) Mechanisms that contribute to the stable segregation of plasmids. Annu Rev Genet 23:37–69

O'Kane DJ, Lingle WL, Wampler JE, Legocki RP, Szalay AA (1988) Visualization of bioluminescence as a marker of gene expression in rhizobium-infected soybean root nodules. Plant Mol Biol 10:387–399

Okker RJ, Spaink H, Hille J, van Brussel TA, Lugtenberg B, Schilperoort RA (1984) Plant-inducible virulence promoter of the *Agrobacterium tumefaciens* Ti plasmid. Nature 312:564–566

Palmer RJJ, Sternberg C (1999) Modern microscopy in biofilm research: confocal microscopy and other approaches. Curr Opin Biotechnol 10:263–268

Partridge JE, Smith FD, Harpending PR, Rasmussen JL, Sanford JC (1991) Preparation of mycelium-free suspensions of oospores of *Phytophtera megasperma* var. *sojae*. Appl Environ Microbiol:480–485

Prosser JI (1994) Molecular marker systems for detection of genetically engineered micro-organisms in the environment. Microbiology 140:5–17

Rengel Z, Ross G, Hirsch P (1998) Plant genotype and micronutrient status influence colonization of wheat roots by soil bacteria. J Plant Nutr 21:99–113

Reuber TL, Long SL, Walker GC (1991) Regulation of *Rhizobium meliloti exo* genes in free-living cells and in planta examined using Tn*phoA* fusions. J Bacteriol 173:426–434

Rhodes DJ, Powell KA (1994) Biological seed treatments – the development process. BCPC Monograph 57:303–310

Rovira AD, Sands DC (1974) Quantitative assessment of the rhizoplane microflora by direct microscopy. Soil Biol Biochem 6:211–216

Rovira AD, Campbell R (1975) A scanning electron microscope study of interactions between micro-organisms and *Gaeumannomyces graminis* (syn. *Ophiobolus graminis*) on wheat roots. Microbial Ecol 3:177–185

Ryder MH, Pankhurst CE, Rovira AD, Corell RL, Ophel Keller KM (1994) Detection of introduced bacteria in the rhizosphere using marker genes and DNA probes. In: O'Gara F, Dowling DN, Boesten B (eds) Molecular ecology of rhizosphere microorganisms. VCH, Weinheim, pp 29–47

Scher FM, Kloepper JW, Singleton CA (1985) Chemotaxis of fluorescent *Pseudomonas* spp. to soybean seed exudates in vitro and in soil. Can J Microbiol 31:570–574

Sharma SB, Signer ER (1990) Temporal and spatial regulation of the symbiotic genes of *Rhizobium meliloti* in planta revealed by transposon Tn*5-gusA*. Genes Dev. 4:344–356

Silcock DJ, Waterhouse RN, Glover LA, Prosser JI, Killham K (1992) Detection of a single genetically engineered modified bacterial cell in soil by using charge coupled device-enhanced microscopy. Appl Environ Microbiol 58:2444–2448

Simons M, van der Bij AJ, Brand J, de Weger LA, Wijffelman CA, Lugtenberg BJJ (1996) Gnotobiotic system for studying rhizosphere colonization by plant growth-promoting *Pseudomonas* bacteria. Mol Plant-Microbe Interact 9:600–607

Simons M, Permentier HP, de Weger LA, Wijffelman CA, Lugtenberg BJJ (1997) Amino acid synthesis is necessary for tomato root colonization by *Pseudomonas fluorescens* strain WCS365. Mol Plant-Microbe Interact 10:102–106

Singleton LL, Mihail JD, Rush CM (1992) Methods for research on soil-borne phytopathogenic fungi. American Phytopathological Society Press, St. Paul, Minnesota

Sokal RR, Rohlf FJ (1981) Biometry: the principles and practice of statistics in biological research. Freeman, New York, pp 432–436

Stanisich VA, Bennett PM, Richmond MH (1977) Characterization of a translocation unit encoding resistance to mercuric ions that occurs on a nonconjugative plasmid in *Pseudomonas aeruginosa*. J Bacteriol 129:1227–1233

Streit W, Kosch K, Werner D (1992) Nodulation competitiveness of *Rhizobium leguminosarum* bv. *phaseoli* and *Rhizobium tropici* strains measured by glucuronidase (*gus*) gene fusion. Biol Fertil Soils 14:140–144

Teeri TH, Lehvaslaiho H, Franck M, Uotila J, Heino P, Palva ET, Van Montagu M, Herrera-Estrella L (1989) Gene fusions to *lacZ* reveal new expression patterns of chimeric genes in transgenic plants. EMBO J 8:343–350

Timms-Wilson TM, Bailey MJ (2001) Reliable use of green fluorescent protein in fluorescent pseudomonads. J Microbiol Meth 46:77–80

Tombolini R, van der Gaag DJ, Gerhardson B, Jansson JK (1999) Colonization pattern of the biocontrol strain *Pseudomonas chlororaphis* MA 342 on barley seeds visualized by using green fluorescent protein. Appl Environ Microbiol 65:3674–3680

van der Bij AJ, de Weger LA, Tucker WT, Lugtenberg BJJ (1996) Plasmid stability in *Pseudomonas fluorescens* in the rhizosphere. Appl Environ Microbiol 62:1076–1080

Voorhorst WGB, Eggen RIL, Luesink EJ, de Vos WM (1995) Characterization of the *celB* gene coding for β-glucosidase from the hyperthermophilic archaeon *Pyrococcus furiosus* and its expression and mutation analysis in *Escherichia coli*. J Bacteriol 177:7105–7111

Ward DM (1989) Molecular probes for analysis of microbial communities. In: Characklis WG, Wilderer PA (eds) Structure and function of biofilms. Wiley, New York, pp 145–155

Warren TM, Williams V, Flechcher M (1992) Influence of solid surface, adhesive ability, and inoculum size on bacterial colonization in microcosm studies. Appl Environ Microbiol 58:2954–2959

Weller DM (1988) Biological control of soilborne plant pathogens in the rhizosphere with bacteria. Annu Rev Phytopathol 26:379–407

Winstanley G, Morgan JAW, Pickup RW, Saunders JR (1991) Use of a *xylE* marker gene to monitor survival of recombinant *Pseudomonas putida* populations in lake water by culture on nonselective media. Appl Environ Microbiol 57:1905–1913

3 Methanogenic Microbial Communities Associated with Aquatic Plants

Ralf Conrad

1 Introduction

Methanogenic microbial communities are typically active at anoxic sites that are depleted in electron acceptors other than CO_2 and H^+. At these sites CH_4 is one of the major products of degradation of organic matter. The degradation products of cellulose, for example, which has an oxidation state of zero, would be CH_4 and CO_2 in a ratio of 1:1. Organic matter with a higher or lower oxidation state would yield respectively less or more CH_4 (Yao and Conrad 2000). Consequently, anoxic methanogenic habitats can be significant sources in the global CH_4 cycle. The global CH_4 cycle is important with respect to atmospheric chemistry and climate, since CH_4 is an important greenhouse gas and has tripled in abundance over the last two centuries (Cicerone and Oremland 1988; Ehhalt 1999). The most important individual source for atmospheric CH_4 is wetlands (including flooded rice fields), which account for about 175 Tg CH_4 per year or 33 % of the total atmospheric CH_4 budget (Conrad 1997; Aulakh et al. 2001). The general microbiology and that of methanogenic microbial communities in flooded soils has recently been reviewed in detail (Kimura 2000; Liesack et al. 2000; Conrad and Frenzel 2002). In the following I will concentrate on methanogenic microbial communities associated with aquatic plants.

2 Role of Plants in Emission of CH_4 to the Atmosphere

Aquatic plants are an integral part of wetland ecosystems that emit CH_4 into the atmosphere. Aquatic plants interact in three different ways with the microbial CH_4 cycling, i.e., by serving as gas conduits, by supplying O_2 to the rhizosphere and by supplying organic substrates to the soil (Fig. 1).

Aquatic plants live in anoxic soil habitats and thus have to make sure that their roots are supplied with O_2. The supply of O_2 is accomplished by vascular gas transport and aerenchyma systems. These systems and their mode of

Plant Surface Microbiology
A. Varma, L. Abbott, D. Werner, R. Hampp (Eds.)

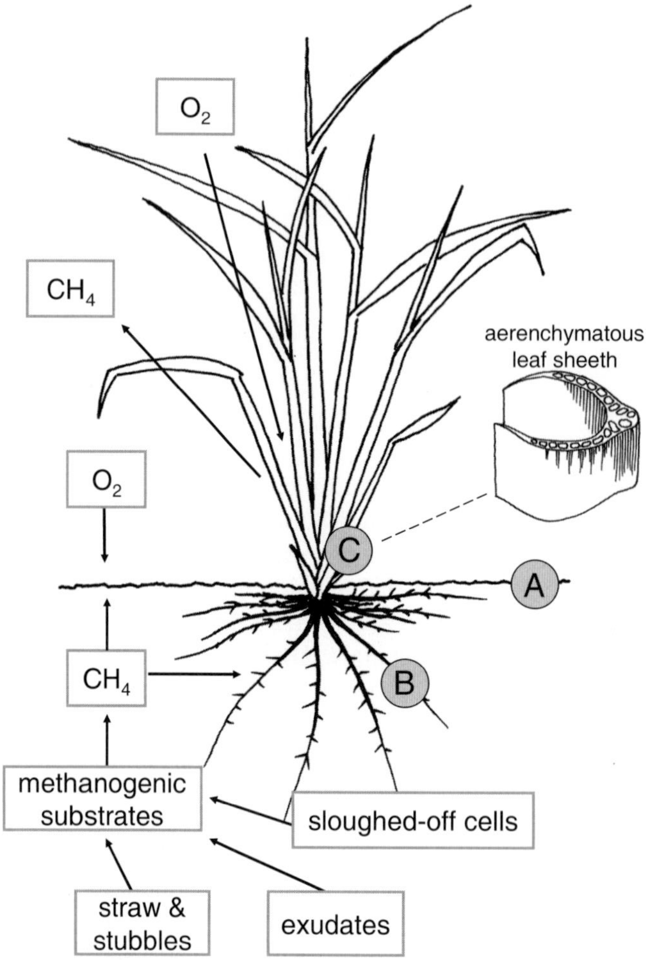

Fig. 1. Role of aquatic plants for cycling of CH_4 by serving as gas conduits (*C*): cross section through an aerenchymatous leaf sheath, by supplying O_2 to the rhizosphere (*B*), and by supplying organic substrates to the soil; *A*, *B* and *C* are sites where O_2 is available (taken from Frenzel 2000)

operation can be different in the different plant species (Armstrong 1979; Grosse et al. 1996; Jackson and Armstrong 1999). However, they all allow for transport of O_2 to the roots and vice-versa allow for the transport of CH_4 from the anoxic soil into the atmosphere. In rice fields, up to about 90 % of total CH_4 emission can be accomplished by ventilation through the rice plants (Holzapfel-Pschorn et al. 1986; Aulakh et al. 2001). The exact contribution of rice plants to the transport of CH_4 from the soil into the atmosphere depends on the size of the rice plants and their capacity for gas transport (Aulakh et al.

2001). Other aquatic plants have similar features (Chanton and Dacey 1991; Grosse et al. 1996). Due to ventilation through aquatic plants, only a few bubbles accumulate in the soil and the ratio of CH_4 to N_2 in soil gas is relatively low (Chanton and Dacey 1991). Transport through the aquatic plants results in the fractionation of the stable isotope composition of CH_4 (delaying transport of heavy carbon and hydrogen), the extent being dependent on the mode of gas transport, e.g., by molecular diffusion or thermo-osmosis (Chanton and Dacey 1991; Chanton and Whiting 1996).

By supplying O_2 to the rhizosphere, aquatic plants create a habitat there that is partially oxic. The presence of O_2 and increase in the redox potential have been demonstrated in the rhizosphere of aquatic plants (Frenzel 2000). However, O_2 availability may be spatially and temporarily restricted. Leakage of O_2 from the roots only occurs at specific sites, e.g., at the tips and where lateral roots emerge (Armstrong 1967). As the root grows, the soil sites which are affected by O_2 leakage change also (Flessa and Fischer 1992; Flessa 1994). A further factor, which affects the availability of O_2, is microbial respiration of organic substrates, which also varies in time and space. Availability of useful substrates can dramatically limit the availability of O_2 in the rhizosphere (Van Bodegom et al. 2001a, b). The creation of oxic microhabitats may have dramatic effects on methanogenic microbial communities that also occur in the rhizosphere. First, O_2 is probably toxic to most of the anaerobic microorganisms and to methanogenic ones in particular (see below). Second, availability of O_2 allows the microbial and/or chemical oxidation of reduced inorganic compounds such as ammonia, sulfide and ferrous iron. These oxidation activities in turn result in the availability of inorganic oxidants and an increase of the redox potential in the rhizosphere beyond the zone where molecular O_2 is available. The availability of nitrate, sulfate and ferric iron, in turn, allows the operation of microbial nitrate reduction, sulfate reduction and iron reduction interfering with the activity of methanogenesis (Conrad 1993; Conrad and Frenzel 2002). Most of all, however, availability of O_2 allows the partial oxidation of CH_4 produced by the methanogenic microbial community. In fact, a significant percentage of the CH_4 produced in the anoxic soil and/or the rhizosphere is oxidized by methanotrophic bacteria (Frenzel 2000). The methanotrophic bacteria live by oxidation of CH_4 with O_2 to CO_2 and thus depend on the availability of both CH_4 and O_2. The methanotrophic activity in the rhizosphere of aquatic plants scavenges a significant part of the produced CH_4 which otherwise would be emitted into the atmosphere. The percentage of oxidized CH_4 varies with circumstances, but is typically in the order of 30 % of the CH_4 produced (Frenzel 2000). The CH_4 that escapes oxidation is generally enriched in isotopically heavy carbon (Chanton et al. 1997; Tyler et al. 1997; Krüger et al. 2002). The methanotrophs live directly on the root surface, partially even penetrating into the root (Gilbert et al. 1998), but may also be active a short distance away from the root surface if O_2 is available (Van Bodegom et al. 2001a).

Finally, aquatic plants stimulate the methanogenic microbial communities in the rhizosphere and the bulk soil by providing additional organic substrates that can be methanogenically degraded. Theoretically, we may expect two different classes of organic substrates that originate from the plants, i.e., soluble exudates that are released from the roots briefly after being generated through photosynthesis, and structural organic matter provided by plant debris.

3 Role of Photosynthates and Plant Debris for CH_4 Production

Field observations suggested that root exudate-driven CH_4 production might play a major role in CH_4 emission from flooded rice fields (Holzapfel-Pschorn et al. 1986). Another preliminary indication of photosynthesis affecting CH_4 production came from field observations that CH_4 emission from various wetlands correlates with primary productivity (Whiting and Chanton 1993; Joabsson and Christensen 2001) and that CO_2 enrichment of the atmosphere results in increased CH_4 emission (Dacey et al. 1994; Hutchin et al. 1995; Megonigal and Schlesinger 1997; Ziska et al. 1998). However, CO_2 enrichment and increased temperature caused in Florida rice fields a decreased CH_4 emission, probably because of enhanced delivery of O_2 into the rhizosphere (Schrope et al. 1999). This study is in contrast to that by Ziska et al. (1998) on rice fields in the Philippines, and shows that field observations have to be interpreted with care due to the highly complex interactions in the ecosystem.

However, there are more direct indications that plant photosynthesis affects the methanogenic microbial community in the rhizosphere. For example, CH_4 production is correlated to the extent of root exudation in rice (Aulakh et al. 2001). Pulse labeling studies with rice and other aquatic plants have shown that different percentages (<1 to <6 %) of photo-assimilated CO_2 are subsequently converted to CH_4 (Dannenberg and Conrad 1999; Megonigal et al. 1999; King and Reeburgh 2002). Briefly after the labeled CO_2 has been photo-assimilated, the radioactivity is detected in soil organic substrates and is emitted as CH_4 (Fig. 2). The specific radioactivity per mol of substrate, which decreased in the order lactate>propionate>acetate>CH_4, indicates the degradation pathway of the excreted organic substrate to CH_4.

Other pulse-labeling studies have shown that photosynthate-derived CH_4 contributes more than 50 % to the total CH_4 emission from flooded rice fields (Minoda et al. 1996; Watanabe et al. 1999). These studies also confirmed the speculations from earlier field work (Holzapfel-Pschorn et al. 1986; Schütz et al. 1989) that seasonal peaks in CH_4 emission were due to decomposition of rice straw, followed by stimulation through root exudation and finally through decay of roots (Fig. 3).

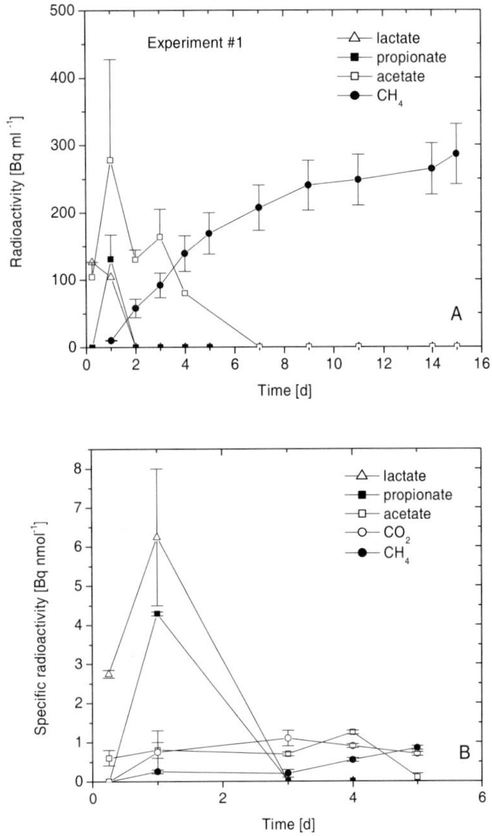

Fig. 2. Transfer of carbon via rice plants to the soil and into CH_4. *Above* After pulse labeling of the rice plants with $^{14}CO_2$, radioactivity transiently accumulates in soil organic compounds and is ultimately converted to $^{14}CH_4$; *below* specific radioactivities indicate that radioactive compounds are converted in the sequence lactate>propionate>acetate>CH_4 (taken from Dannenberg and Conrad 1999)

Besides the more direct effect of photosynthesis through root exudation, CH_4 production in wetlands is furthermore stimulated by plant debris. This may be decaying roots or dead aboveground plant material. In rice fields, for example, rice straw from the previous season is often plowed under to improve soil quality. Also, composted plant material is used as soil fertilizer. There are numerous studies which show that addition of such organic matter dramatically increases CH_4 emission rates (Denier van der Gon 1999). Decomposition of isotopically labeled rice straw contributes significantly to production and emission of CH_4 during the early season (Chidthaisong and Watanabe 1997; Watanabe et al. 1998, 1999)

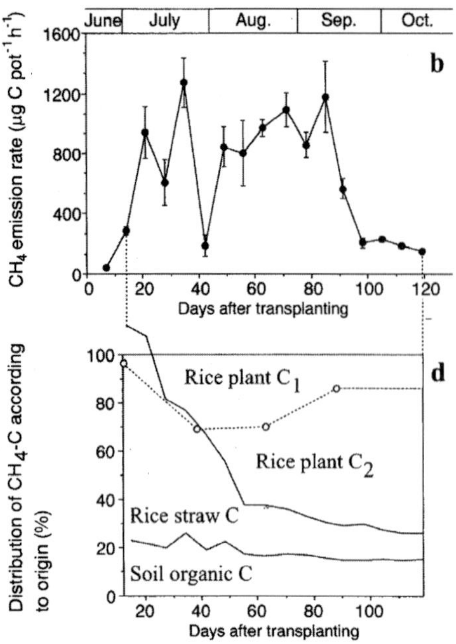

Fig. 3. Emission of CH_4 from straw-fertilized, planted rice field soil and partitioning of the primary carbon from which CH_4 was formed, determined by using ^{13}C-labeled CO_2, rice straw and soil organic matter; rice plant C1 is carbon released within 2 weeks after assimilation of $^{13}CO_2$; rice plant C2 is other plant-derived carbon, presumably from sloughed-off root cells or decaying roots (taken from Watanabe et al. 1999)

4 Methanogenic Microbial Communities on Plant Debris

Straw and decomposing roots are important plant debris in flooded rice fields. Rice straw mainly consists of cellulose, hemicellulose and lignin and is encrusted with silica (Tsutsuki and Ponnamperuma 1987; Watanabe et al. 1993). After a rapid mineralization of 80–90 % of the straw during the first year, a more resistant fraction of organic matter remains. The latter is degraded slowly with a half-life of about 2 years (Neue and Scharpenseel 1987). Rice straw is colonized by microorganisms and the structure of the leave blades and sheaths gradually disintegrates. The degradation process becomes visually apparent after about 3 weeks (Kimura and Tun 1999; Tun and Kimura 2000) with a dry weight loss of about 50 % during the first 30 days (Glissmann and Conrad 2002). Degradation of rice straw proceeds via hydrolysis, fermentation of the released sugars, syntrophic conversion of primary fermentation products to acetate, CO_2 and H_2, and conversion of acetate and H_2/CO_2, respectively, to CH_4 (Glissmann and Conrad 2002). The same degradation pathway is generally found in methanogenic environments such as lake sediments or anaerobic digestors (Zinder 1993). The methanogenic degradation pathway of rice straw is similar to that of the organic matter present in flooded soil to which no rice straw was added, but the rate of CH_4 production is lower in the unamended soil (Glissmann and

Conrad 2000). Under steady state conditions, the conversion of rice straw to CH_4 is limited by the hydrolysis of the straw polysaccharides, which become increasingly recalcitrant to decomposition (Glissmann and Conrad 2002). It is likely that rice straw is in this way gradually converted to soil organic matter and humus.

The bacteria that colonize and degrade rice straw mainly consist of clostridia (Weber et al. 2001a) which belong to the same taxonomic clusters as found in unamended soil (Chin et al. 1999; Hengstmann et al. 1999; Lüdemann et al. 2000). On the other hand, the community of methanogenic archaea on rice straw is less diverse and abundant than in the bulk soil (Weber et al. 2001b). The genus *Methanosaeta*, in particular, was lacking in degrading straw. This genus is common in bulk soil (Grosskopf et al. 1998) and especially becomes abundant at limiting acetate concentrations (Fey and Conrad 2000). Consistent with the low abundance of methanogens on rice straw is the observation that straw pieces retrieved from the soil mainly exhibit fermentative production of H_2 and fatty acids, while the subsequent conversion of the fatty acids to CH_4 takes place in the bulk soil to where the fatty acids are released (Glissmann et al. 2001). Hence, methanogenic degradation of rice straw is compartmentalized in a way that methanogenesis occurs in the soil at some distance to the microbial community that colonizes the straw (Fig. 4).

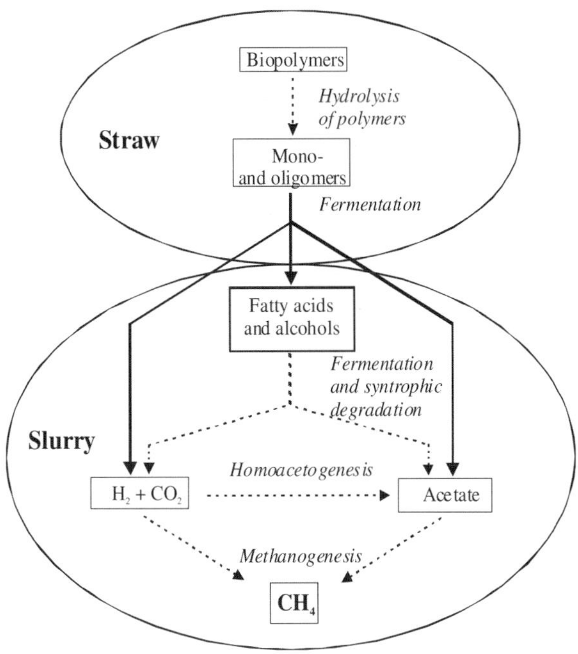

Fig. 4. Conceptual model of methanogenic degradation of rice straw, and the localization of the major processes either on the straw or in the soil slurry (taken from Glissmann et al. 2001)

The general colonization patterns of rice roots with microorganisms and their potential involvement in degradation of the dead roots have been reviewed by Kimura (2000). It is likely that dead roots are decomposed in a similar way to rice straw, but detailed studies are lacking.

5 Methanogenic Microbial Communities on Roots

Plant roots are apparently colonized by methanogenic microorganisms. This evidence came from incubation of excised roots of aquatic plants under anoxic conditions resulting in substantial CH_4 production (Kimura et al. 1991; King 1994; Frenzel and Bosse 1996). Subsequently, it was shown by Grosskopf et al. (1998) that rice roots are indeed inhabited by a diverse community of methanogenic archaea, which can be retrieved by DNA extraction, and amplification of the archaeal SSU rRNA genes. Archaeal diether lipids were also detected on rice roots (Reichardt et al. 1997).

Production of CH_4 on rice roots is dominated by H_2/CO_2-utilizing methanogens (Lehmann-Richter et al 1999; Conrad et al. 2000). The most prominent group of methanogens on rice roots is that of the uncultivated rice cluster I (Grosskopf et al. 1998). Since this cluster is also dominant in soils in which CH_4 is exclusively produced from H_2/CO_2 (Fey et al. 2000) and in methanogenic enrichment cultures on H_2/CO_2 (Lueders et al. 2001), it is likely that it is responsible for the observed H_2/CO_2 dependent methanogenesis on rice roots. However, methanogens belonging to Methanobacteriaceae and Methanomicrobiaceae, i.e., groups that are able to utilize H_2/CO_2, have also been detected (Grosskopf et al. 1998). Populations of acetoclastic *Methanosarcina*, on the other hand, only developed at a later stage of anoxic incubation of excised rice roots, when sufficient acetate had accumulated and only in the absence of phosphate. Phosphate concentrations higher than 10 mM were found to prohibit the activity of acetoclastic methanogenesis (Conrad et al. 2000). Collectively, these observations suggest that the methanogenic flora in situ produces CH_4 mainly from H_2/CO_2 rather than from acetate. This is a major difference to the behavior in the soil, where acetate is the dominant methanogenic substrate.

Consequently, the stable isotopic signature of the produced CH_4 was found to be different for the methanogenic microbial communities in the soil and on the roots (Conrad et al. 2000, 2002). This fact may have implications for estimates dealing with the budget of atmospheric CH_4 and the global CH_4 cycle, for which the stable isotopic signature of CH_4 is an important constraint (Stevens 1993). Unfortunately, we presently do not know how much the methanogenic microbial community on rice roots, or on the roots of aquatic plants in general, contribute to the CH_4 source strength of wetland ecosystems compared to the methanogenic microbial communities in the anoxic soil.

Another implication of the observations concerns the structure of the methanogenic microbial community on the roots, which seem to be very simple, consisting only of H_2-producing fermenting and H_2-consuming methanogenic microorganisms. However, experiments with excised rice roots have demonstrated a more complex community of fermenting bacteria including vigorous fermentative production of acetate, propionate and butyrate (Conrad and Klose 1999, 2000). Interestingly, a significant percentage (up to 60 %) of these fatty acids was produced by reduction of CO_2. The stable isotope signature of the produced acetate was consistent with the production by CO_2 reduction (Conrad et al. 2002). Acetate production from CO_2 indicates that homoactogenic bacteria were active, a likely conclusion, since homoacetogenic *Sporomusa* are members of the rice root microflora (Rosencrantz et al. 1999). Homoactogens have also been found on the roots of sea grass (Küsel et al. 1999, 2001). Approximately 30 % of the root epidermal cells of sea grass were colonized with microorganisms that hybridized with an archaeal probe suggesting the presence of methanogens (Küsel et al. 1999). Presently, little is known about the fate of the produced fatty acids. Propionate and butyrate can potentially be further converted to acetate, CO_2 and H_2 by syntrophic bacteria, which are present in the anoxic rice soil, followed by H_2/CO_2-dependent methanogenesis (Krylova et al. 1997). Syntrophic oxidation of acetate, however, is unlikely since [2-^{14}C]acetate was hardly turned over in root preparations (Lehmann-Richter et al. 1999). The most likely fate of the acetate produced by the root microflora is its escape into the bulk soil where it is methanogenically decomposed (Fig. 5). Alternatively, acetate may be a substrate for anaerobic bacteria using nitrate, ferric iron or sulfate as electron

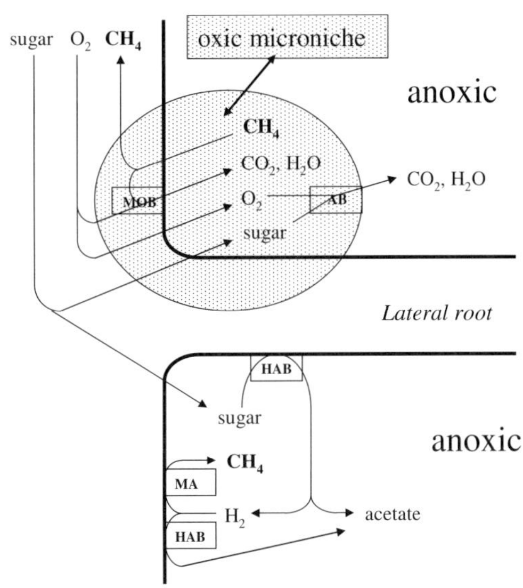

Fig. 5. Conceptual model of the localization of methanogenic archaea (*MA*), homoacetogenic bacteria (*HAB*), methane-oxidizing bacteria (*MOB*) and aerobic bacteria (*AB*) in vicinity of rice roots and to each other, and the flow of organic carbon. The insertion of lateral roots (and root tips) are the most likely sites where O_2 and organic substrates (e.g., sugars) are released into the soil

acceptor. However, little is known about the activity of these functional groups on the roots of aquatic plants (Bodelier et al. 1997; Nijburg and Laanbroek 1997; King and Garey 1999; Küsel et al. 1999; Wind et al. 1999; Arth and Frenzel 2000).

6 Interaction of Methanogens and Methanotrophs

Although it has become evident that methanogenesis is stimulated by plant photosynthesis (see above), it has been a rather unexpected result that methanogenic activity is obviously localized directly on the root surface. This result was surprising, since textbook knowledge suggests that methanogenic archaea need an absolutely O_2-free environment, which the root surface does not provide (see above). Quite the contrary, roots have been shown to be the site of methanotrophic activity (Frenzel 2000). This discrepancy between roots being colonized by both aerobic and anaerobic microorganisms has not been completely reconciled.

One possible explanation is a spatially heterogeneous colonization of the roots. The aerobic methanotrophs would colonize only those parts where O_2 is leaking from the roots and the methanogens only those that stay anoxic. However, methanogens would probably only colonize the roots if provision of the substrate (H_2) is better there than in the bulk soil. Production of H_2 requires microbial fermentation activity and this in turn requires the provision of a degradable substrate. Hence, colonization of roots by methanogens is most likely at the sites with high leakage rates of organic substrates. The localization of sites with exudation of organic substrates along the root is not quite clear, but the root tips were found to be most actively excreting sucrose in an annual grass (Jaeger et al. 1999). However, root tips are also the most active sites of O_2 leakage (Armstrong 1967). Thus, we have to expect that the optimal sites for colonization by methanotrophs and methanogens are the same.

Another possible explanation is that the methanogens are largely protected from O_2, because they are living in the vicinity of O_2-consuming methanotrophs. A similar close association of methanotrophs and methanogens has been hypothesized for pelagic microbial assemblages, thus explaining the formation of CH_4 in oxic ocean surface water (Sieburth 1991). Although Sieburth's hypothesis has so far not been confirmed in pelagic microbial flocs (Ploug et al. 1997), experiments in microbial chemostat cultures have shown that anaerobic methanogens can co-exist with aerobic microorganisms under aerated conditions (Gerritse and Gottschal 1993).

Moreover, at least some of the species of methanogens seem to be more resistant to exposure to O_2 than generally expected. For example, methanogens in rice field soil have been found to survive desiccation of the soil and prolonged exposure to air (Fetzer et al. 1993). *Methanosarcina bark-*

eri was found to be able to initiate CH_4 production despite a positive redox potential of the medium (Fetzer and Conrad 1993). Methanogenic activity has been detected in just that part of the termite gut that is oxygenated (Brune and Friedrich 2000), and *Methanobrevibacter* isolates were able to grow slowly despite the presence of low O_2 concentrations (Leadbetter and Breznak 1996). Recently, some methanogenic species were found to contain catalase and superoxide dismutase to protect against oxidative stress (Shima et al. 1999, 2001; Brioukhanov et al. 2000). Unfortunately, we do not know the O_2 resistance of the methanogenic species that inhabit rice roots, in particular the uncultivated rice cluster I methanogens (Lueders et al. 2001).

As methanogens and methanotrophs live in the same environments in close vicinity, it might be possible that they communicate with each other not only by transfer of CH_4, but also more directly through gene exchange. R.K. Thauer (Germany) and his group have recently put forward this idea. It is indeed intriguing that methanotrophic bacteria contain genes and coenzymes, which had been postulated to be specific for methanogens. Thus, methanotrophs seem to contain tetrahydromethanopterin in addition to tetrahydrofolate, and utilize tetrahydromethanopterin-dependent enzymes for catabolic C1 transfer reactions similarly to the methanogens (Chisdoserdova et al. 1998; Vorholt et al. 1999). It is likely that methanotrophs have acquired the necessary genes from methanogens. The other way round, methanogens seem to have acquired genes that are of bacterial origin. Thus, *Methanosarcina* and *Methanobrevibacter* species contain a monofunctional catalase. Such an enzyme is unexpected for *Archaea*, which generally contain a bifunctional catalase (Shima et al. 2001). Roots of aquatic plants would be a possible habitat where such gene transfers between methanogenic *Archaea* and methanotrophic *Bacteria* might occur.

Acknowledgements. I thank Peter Frenzel and Rolf Thauer for discussion.

References and Selected Reading

Armstrong W (1967) The use of polarography in the assay of oxygen diffusing from roots in anaerobic media. Physiol Plantarum 20:540–553

Armstrong W (1979) Aeration in higher plants. Adv Bot Res 7:226–332

Arth I, Frenzel P (2000) Nitrification and denitrification in the rhizosphere of rice: the detection of processes by a new multi-channel electrode. Biol Fertil Soils 31:427–435

Aulakh MS, Wassmann R, Rennenberg H (2001) Methane emissions from rice fields – Quantification, mechanisms, role of management, and mitigation options. Adv Agron 70:193–260

Bodelier PLE, Wijlhuizen AG, Blom CWPM, Laanbroek HJ (1997) Effects of photoperiod on growth of and denitrification by *Pseudomonas chlororaphis* in the root zone of *Glyceria maxima*, studied in a gnotobiotic microcosm. Plant Soil 190:91–103

Brioukhanov A, Netrusov A, Sordel M, Thauer RK, Shima S (2000) Protection of *Methanosarcina barkeri* against oxidative stress: identification and characterization of an iron superoxide dismutase. Arch Microbiol 174:213–216

Brune A, Friedrich M (2000) Microecology of the termite gut: structure and function on a microscale [Review]. Curr Opin Microbiol 3:263–269

Chanton JP, Dacey JW (1991) Effects of vegetation on methane flux, reservoirs, and carbon isotopic composition. In: Rogers JE, Whitman WB (eds) Trace gas emissions by plants. Academic Press, San Diego, pp 65–92

Chanton JP, Whiting GJ (1996) Methane stable isotopic distributions as indicators of gas transport mechanisms in emergent aquatic plants. Aquat Bot 54:227–236

Chanton JP, Whiting GJ, Blair NE, Lindau CW, Bollich PK (1997) Methane emission from rice: Stable isotopes, diurnal variations, and CO_2 exchange. Global Biogeochem Cycles 11:15–27

Chidthaisong A, Watanabe I (1997) Methane formation and emission from flooded rice soil incorporated with ^{13}C-labeled rice straw. Soil Biol Biochem 29:1173–1181

Chin KJ, Hahn D, Hengstmann U, Liesack W, Janssen PH (1999) Characterization and identification of numerically abundant culturable bacteria from the anoxic bulk soil of rice paddy microcosms. Appl Environ Microbiol 65:5042–5049

Chistoserdova L, Vorholt JA, Thauer RK, Lidstrom ME (1998) C-1 transfer enzymes and coenzymes linking methylotrophic bacteria and methanogenic archaea. Science 281:99–102

Cicerone RJ, Oremland RS (1988) Biogeochemical aspects of atmospheric methane. Global Biogeochem Cycles 2:299–327

Conrad R (1993) Mechanisms controlling methane emission from wetland rice fields. In: Oremland RS (ed) The biogeochemistry of global change: radiative trace gases. Chapman and Hall, New York, pp 317–335

Conrad R (1997) Production and consumption of methane in the terrestrial biosphere. In: Helas G, Slanina J, Steinbrecher R (eds) Biogenic volatile organic carbon compounds in the atmosphere. SBP Academic Publ, Amsterdam, pp 27–44

Conrad R, Klose M (1999) Anaerobic conversion of carbon dioxide to methane, acetate and propionate on washed rice roots. FEMS Microbiol Ecol 30:147–155

Conrad R, Klose M (2000) Selective inhibition of reactions involved in methanogenesis and fatty acid production on rice roots. FEMS Microbiol Ecol 34:27–34

Conrad R, Frenzel P (2002) Flooded soils. In: Bitton G (ed) The encyclopedia of environmental microbiology. Wiley, New York, pp 1316.1333

Conrad R, Klose M, Claus P (2000) Phosphate inhibits acetotrophic methanogenesis on rice roots. Appl Environ Microbiol 66:828–831

Conrad R, Klose M, Claus P (2002) Pathway of CH_4 formation in anoxic rice field soil and rice roots determined by 13C-stable isotope fractionation. Chemosphere 47:797–806

Dacey JWH, Drake BG, Klug MJ (1994) Stimulation of methane emission by carbon dioxide enrichment of marsh vegetation. Nature 370:47–49

Dannenberg S, Conrad R (1999) Effect of rice plants on methane production and rhizospheric metabolism in paddy soil. Biogeochemistry 45:53–71

Denier van der Gon H (1999) Changes in CH_4 emission from rice fields from 1960 to 1990s – 2. The declining use of organic inputs in rice farming. Global Biogeochem Cycles 13:1053–1062

Ehhalt DH (1999) Gas phase chemistry in the troposphere. In: Zellner R (ed) Global aspects of atmospheric chemistry. Springer, Berlin Heidelberg New York, pp 21–109

Fetzer S, Conrad R (1993) Effect of redox potential on methanogenesis by *Methanosarcina barkeri*. Arch Microbiol 160:108–113

Fetzer S, Bak F, Conrad R (1993) Sensitivity of methanogenic bacteria from paddy soil to oxygen and desiccation. FEMS Microbiol Ecol 12:107–115

Fey A, Conrad R (2000) Effect of temperature on carbon and electron flow and on the archaeal community in methanogenic rice field soil. Appl Environ Microbiol 66:4790–4797

Flessa H (1994) Plant-induced changes in the redox potential of the rhizospheres of the submerged vascular macrophytes *Myriophyllum verticillatum* L and *Ranunculus circinatus* L. Aquat Bot 47:119–129

Flessa H, Fischer WR (1992) Plant-induced changes in the redox potentials of rice rhizospheres. Plant Soil 143:55–60

Frenzel P (2000) Plant-associated methane oxidation in rice fields and wetlands [Review]. Adv Microb Ecol 16:85–114

Frenzel P, Bosse U (1996) Methyl fluoride, an inhibitor of methane oxidation and methane production. FEMS Microbiol Ecol 21:25–36

Gerritse J, Gottschal JC (1993) Two-membered mixed cultures of methanogenic and aerobic bacteria in O_2-limited chemostats. J Gen Microbiol 139:1853–1860

Gilbert B, Assmus B, Hartmann A, Frenzel P (1998) In situ localization of two methanotrophic strains in the rhizosphere of rice plants. FEMS Microbiol Ecol 25:117–128

Glissmann K, Conrad R (2000) Fermentation pattern of methanogenic degradation of rice straw in anoxic paddy soil. FEMS Microbiol Ecol 31:117–126

Glissmann K, Conrad R (2002) Saccharolytic activity and its role as a limiting step in methane formation during the anaerobic degradation of rice straw in rice paddy soil. Biology and Fertility of Soils 35:62–67

Glissmann K, Weber S, Conrad R (2001) Localization of processes involved in methanogenic degradation of rice straw in anoxic paddy soil. Environ Microbiol 3:502–511

Grosse W, Armstrong J, Armstrong W (1996) A history of pressurised gas-flow studies in plants. Aquat Bot 54:87–100

Großkopf R, Stubner S, Liesack W (1998) Novel euryarchaeotal lineages detected on rice roots and in the anoxic bulk soil of flooded rice microcosms. Appl Environ Microbiol 64:4983–4989

Hengstmann U, Chin KJ, Janssen PH, Liesack W (1999) Comparative phylogenetic assignment of environmental sequences of genes encoding 16S rRNA and numerically abundant culturable bacteria from an anoxic rice paddy soil. Appl Environ Microbiol 65:5050–5058

Holzapfel-Pschorn A, Conrad R, Seiler W (1986) Effects of vegetation on the emission of methane from submerged paddy soil. Plant and Soil 92:223–233

Hutchin PR, Press MC, Lee JA, Ashenden TW (1995) Elevated concentrations of CO_2 may double methane emissions from mires. Global Change Biol 1:125–128

Jackson MB, Armstrong W (1999) Formation of aerenchyma and the processes of plant ventilation in relation to soil flooding and submergence [review]. Plant Biol 1:274–287

Jaeger CH III, Lindow SE, Miller S, Clark E, Firestone MK (1999) Mapping of sugar and amino acid availability in soil around roots with bacterial sensors of sucrose and tryptophan. Appl Environ Microbiol 65:2685–2690

Joabsson A, Christensen TR (2001) Methane emissions from wetlands and their relationship with vascular plants: an Arctic example. Global Change Biol 7:919–932

Kimura M (2000) Anaerobic microbiology in waterlogged rice fields. In: Bollag JM, Stotzky G (eds) Soil biochemistry, vol 10. Marcel Dekker, New York, pp 35–138

Kimura M, Tun CC (1999) Microscopic observation of the decomposition process of leaf sheath of rice straw and colonizing microorganisms during the cultivation period of paddy rice. Soil Sci Plant Nutr 45:427–437

Kimura M, Murakami H, Wada H (1991) CO_2, H_2, and CH_4 production in rice rhizosphere. Soil Sci Plant Nutr 37:55–60

King GM (1994) Associations of methanotrophs with the roots and rhizomes of aquatic vegetation. Appl Environ Microbiol 60:3220–3227

King GM, Garey MA (1999) Ferric iron reduction by bacteria associated with the roots of freshwater and marine macrophytes. Appl Environ Microbiol 65:4393–4398

King JY, Reeburgh WS (2002) A pulse-labeling experiment to determine the contribution of recent plant photosynthates to net methane emission in arctic wet sedge tundra. Soil Biol Biochem 34:173–180

Krüger M, Eller G, Conrad R, Frenzel P (2002) Seasonal variation in pathways of CH_4 production and in CH_4 oxidation in rice fields determined by stable isotopes and specific inhibitors. Global Change Biol 8:265–280

Krylova NI, Janssen PH, Conrad R (1997) Turnover of propionate in methanogenic paddy soil. FEMS Microbiol Ecol 23:107–117

Küsel K, Pinkart HC, Drake HL, Devereux R (1999) Acetogenic and sulfate-reducing bacteria inhabiting the rhizoplane and deep cortex cells of the sea grass *Halodule wrightii.* Appl Environ Microbiol 65:5117–5123

Küsel K, Karnholz A, Trinkwalter T, Devereux R, Acker G, Drake HL (2001) Physiological ecology of *Clostridium glycolicum* RD-1, an aerotolerant acetogen isolated from sea grass roots. Appl Environ Microbiol 67:4734–4741

Leadbetter JR, Breznak JA (1996) Physiological ecology of *Methanobrevibacter cuticularis* sp. nov and *Methanobrevibacter curvatus* sp. nov, isolated from the hindgut of the termite *Reticulitermes flavipes.* Appl Environ Microbiol 62:3620–3631

Lehmann-Richter S, Großkopf R, Liesack W, Frenzel P, Conrad R (1999) Methanogenic archaea and CO_2-dependent methanogenesis on washed rice roots. Environ Microbiol 1:159–166

Liesack W, Schnell S, Revsbech NP (2000) Microbiology of flooded rice paddies [Review]. FEMS Microbiol Rev 24:625–645

Lüdemann H, Arth I, Liesack W (2000) Spatial changes in the bacterial community structure along a vertical oxygen gradient in flooded paddy soil cores. Appl Environ Microbiol 66:754–762

Lueders T, Chin KJ, Conrad R, Friedrich M (2001) Molecular analyses of methyl-coenzyme M reductase alpha-subunit (*mcrA*) genes in rice field soil and enrichment cultures reveal the methanogenic phenotype of a novel archaeal lineage. Environ Microbiol 3:194–204

Megonigal JP, Schlesinger WH (1997) Enhanced CH_4 emissions from a wetland soil exposed to elevated CO_2. Biogeochemistry 37:77–88

Megonigal JP, Whalen SC, Tissue DT, Bovard BD, Albert DB, Allen AS (1999) A plant-soil-atmosphere microcosm for tracing radiocarbon from photosynthesis through methanogenesis. Soil Sci Soc Am J 63:665–671

Minoda T, Kimura M, Wada E (1996) Photosynthates as dominant source of CH_4 and CO_2 in soil water and CH_4 emitted to the atmosphere from paddy fields. J Geophys Res 101:21091–21097

Neue HU, Scharpenseel HW (1987) Decomposition pattern of ^{14}C-labeled rice straw in aerobic and submerged rice soils of the Philippines. Sci Total Environ 62:431–434

Nijburg JW, Laanbroek HJ (1997) The fate of ^{15}N-nitrate in healthy and declining *Phragmites australis* stands. Microb Ecol 34:254–262

Ploug H, Kühl M, Buchholz-Cleven B, Joergensen BB (1997) Anoxic aggregates – an ephemeral phenomenon in the pelagic environment? Aquat Microb Ecol 13:285–294

Reichardt W, Mascarina G, Padre B, Doll J (1997) Microbial communities of continuously cropped, irrigated rice fields. Appl Environ Microbiol 63:233–238

Rosencrantz D, Rainey FA, Janssen PH (1999) Culturable populations of *Sporomusa* spp. and *Desulfovibrio* spp. in the anoxic bulk soil of flooded rice microcosms. Appl Environ Microbiol 65:3526–3533

Schrope MK, Chanton JP, Allen LH, Baker JT (1999) Effect of CO_2 enrichment and elevated temperature on methane emissions from rice, *Oryza sativa.* Global Change Biology 5:587–599

Schütz H, Holzapfel-Pschorn A, Conrad R, Rennenberg H, Seiler W (1989) A 3-year continuous record on the influence of daytime, season, and fertilizer treatment on methane emission rates from an Italian rice paddy. J Geophys Res 94:16405–16416

Shima S, Netrusov A, Sordel M, Wicke M, Hartmann GC, Thauer RK (1999) Purification, characterization, and primary structure of a monofunctional catalase from *Methanosarcina barkeri*. Arch Microbiol 171:317–323

Shima S, Sordel-Klippert M, Brioukhanov A, Netrusov A, Linder D, Thauer RK (2001) Characterization of a heme-dependent catalase from *Methanobrevibacter arboriphilus*. Appl Environ Microbiol 67:3041–3045

Sieburth JM (1991) Methane and hydrogen sulfide in the pycnocline: a result of tight coupling of photosynthetic and "benthic" processes in stratified waters. In: Rogers JE, Whitman WB (eds) Microbial production and consumption of greenhouse gases: methane, nitrogen oxides, and halomethanes. American Society for Microbiology, Washington, DC, pp 147–174

Stevens CM (1993) Isotopic abundances in the atmosphere and sources. In: Khalil MAK (ed) Atmospheric methane: sources, sinks, and role in global change, Springer, Berlin Heidelberg New York, pp 62–88

Tsutsuki K, Ponnamperuma FN (1987) Behavior of anaerobic decomposition products in submerged soils . Effects of organic material amendment, soil properties, and temperature. Soil Sci Plant Nutr 33:13–33

Tun CC, Kimura M (2000) Microscopic observation of the decomposition process of leaf blade of rice straw and colonizing microorganisms in a Japanese paddy field soil during the cultivation period of paddy rice. Soil Sci Plant Nutr 46:127–137

Tyler SC, Bilek RS, Sass RL, Fisher FM (1997) Methane oxidation and pathways of production in a Texas paddy field deduced from measurements of flux, $\delta^{13}C$, and δD of CH_4. Global Biogeochem Cycles 11:323–348

Van Bodegom P, Goudriaan J, Leffelaar P (2001a) A mechanistic model on methane oxidation in a rice rhizosphere. Biogeochem 55:145–177

Van Bodegom P, Stams F, Mollema L, Boeke S, Leffelaar P (2001b) Methane oxidation and the competition for oxygen in the rice rhizosphere. Appl Environ Microbiol 67:3586–3597

Vorholt JA, Chistoserdova L, Stolyar SM, Thauer RK, Lidstrom ME (1999) Distribution of tetrahydromethanopterin-dependent enzymes in methylotrophic bacteria and phylogeny of methenyl tetrahydromethanopterin cyclohydrolases. J Bacteriol 181:5750–5757

Watanabe A, Katoh K, Kimura M (1993) Effect of rice straw application on CH_4 emission from paddy fields. 2. contribution of organic constituents in rice straw. Soil Sci Plant Nutr 39:707–712

Watanabe A, Yoshida M, Kimura M (1998) Contribution of rice straw carbon to CH_4 emission from rice paddies using ^{13}C-enriched rice straw. J Geophys Res 103:8237–8242

Watanabe A, Takeda T, Kimura M (1999) Evaluation of origins of CH_4 carbon emitted from rice paddies. J Geophys Res 104:23623–23629

Weber S, Stubner S, Conrad R (2001a) Bacterial populations colonizing and degrading rice straw in anoxic paddy soil. Appl Environ Microbiol 67:1318–1327

Weber S, Lueders T, Friedrich MW, Conrad R (2001b) Methanogenic populations involved in the degradation of rice straw in anoxic paddy soil. FEMS Microbiol Ecol 38:11–20

Whiting GJ, Chanton JP (1993) Primary production control of methane emission from wetlands. Nature 364:794–795

Wind T, Stubner S, Conrad R (1999) Sulphate-reducing bacteria in rice field soil and on rice roots. Syst Appl Microbiol 22:269–279

Yao H, Conrad R (2000) Electron balance during steady-state production of CH_4 and CO_2 in anoxic rice soil. Eur J Soil Sci 51:369–378

Zinder SH (1993) Physiological ecology of methanogens. In: Ferry JG (ed) Methanogenesis: ecology, physiology, biochemistry and genetics. Chapman and Hall, New York, pp 128–206

Ziska LH, Moya TB, Wassmann R, Namuco OS, Lantin RS, Aduna JB, Abao E, Bronson KF, Neue HU, Olszyk D (1998) Long-term growth at elevated carbon dioxide stimulates methane emission in tropical paddy rice. Global Change Biology 4:657–665

4 Role of Functional Groups of Microorganisms on the Rhizosphere Microcosm Dynamics

GALDINO ANDRADE

1 Introduction

This chapter discusses the role of functional microorganism groups that live in the rhizosphere and contribute to nutrient cycling. Soil ecology has much to contribute to our knowledge of important processes at the ecosystem level, such as how plant growth is affected by the rhizosphere biota, organic matter dynamics and nutrient cycling, and soil structure dynamics (Brussaard 1998).

Many groups work directly on plant nutrition, such as rhizobia and mycorrhiza fungi which are symbiotic. These groups have been studied extensively in the last few decades, but little has been investigated about the relationship between other functional groups, notwithstanding that many other interactions exist in the rhizosphere that are ecologically important to maintain life on Earth and consequently in the soil, since this is a part of the whole.

Many steps of nutrient cycling are made exclusively by microorganism populations, and some of them may participate in one or more biogeochemical cycles. The understanding of these interactions between different populations according to specific phenotypes could give a better perspective about the processes that are occurring. A percentage of the microbial community can be grown in culture medium under laboratory conditions, if cultured microorganisms are considered as a sample of microbial community in soil microcosms. Grouping the microbial communities by phenotypes is more realistic than determining the species that are involved in these process. Although only a small amount of high quality data can be obtained, it is possible to monitor the effects of hazardous chemical products, environmental disturbance, and disturbances in nutrient cycling and soil fertility controlled by these organisms, and also ecosystem health.

Functionality aspects are more important than biodiversity in natural or sustainable agriculture systems. Some questions could be raised concerning biodiversity. The first question that should be asked is: what is more important to the Earth? The number of species that compose the functional group

Plant Surface Microbiology
A. Varma, L. Abbott, D. Werner, R. Hampp (Eds.)

or the transformation power of one group? On the other hand, other questions could be asked such as: what is the importance of one species inside the biological dynamic system? What is the capacity of one species to influence nutrient cycling? What does a species represent within the biological dynamic? What importance can one species have in nutrient cycling? These questions could lead us to conclude that we need to review our vision of the soil microcosm, extend our understanding of the biological processes and interactions that occur in the soil – plant system, assuming that these processes are a whole, and each functional group is only a small fraction of the whole. Only in this way can we improve the determination of the environmental impact of any disturbance effect on the soil microbial community, not only on one specific group of microorganisms. The several functional groups which take part in different stages of the carbon, phosphorus, nitrogen and sulphur biogeochemical cycles should be assessed, looking for a correlation between them and a response in plant growth.

Microflora biodiversity is important for other objectives, such as searching for specific microbial phenotypes to use in food or the pharmaceutical industry. Its importance in the environment should also be investigated, since molecular biology does not permit assessment of the microbial interaction mechanisms in the soil microcosm.

2 General Aspects of Functional Groups of Soil Microorganisms

In a soil microbial ecosystem individual cells grow and form populations (Fig. 1). Metabolically related populations constitute groupings called functional groups, and sets of functional groups conducting complementary physiological processes interact to form microbial communities. Microbial communities then interact with communities of macroorganisms to define the whole biosphere.

We can define the functional groups of microbial populations that take part in the same transformation of nutrients in the soil, where the same population of microorganisms may participate in different steps in different

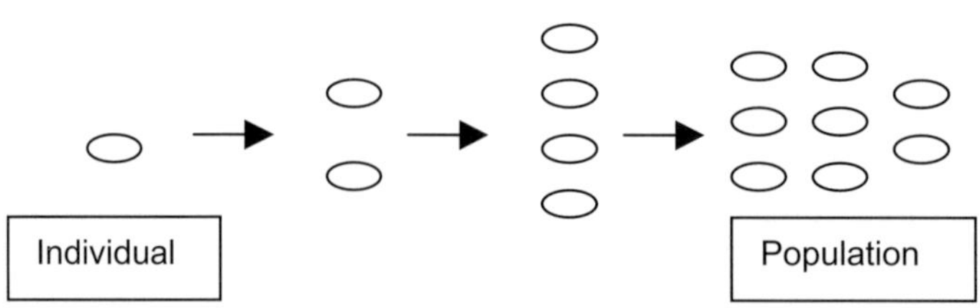

Fig. 1. The individual cells grow and form populations

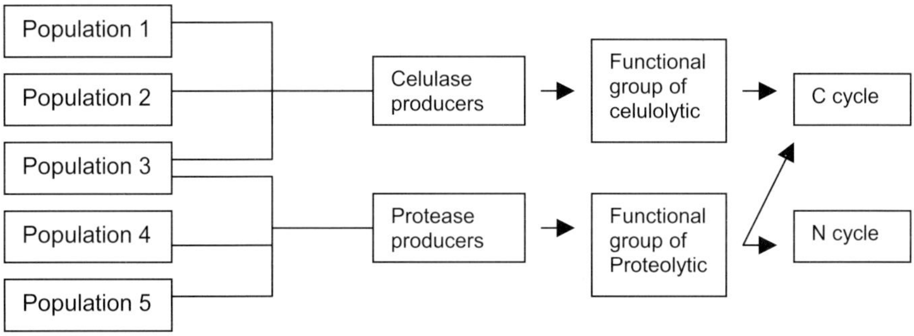

Fig. 2. Many populations of microorganisms may participate in one or more biogeochemical cycling

cycles (Fig. 2). An example is the cellulolytic functional group; if the soil suspension is inoculated in Petri dishes with selective culture media for cellulolytic microorganisms, where cellulose is the only carbon source, and the culture is incubated at 28 °C for 3 days, some colonies will form halos around the colonies after staining with Congo red. If we count the different organisms by decreasing order of numbers, we can observe colonies forming units of fungi, actinomycetes and then bacteria. Many species will be observed within the fungi group, as will also occur with actinomycetes and bacteria populations.

The number of colony forming units (CFU) and the ratio between colony size and degradation halo diameter should be considered in an evaluation study, while assessing the cellulolytic activity. These parameters determine the community size and/or the activity of the individuals that compose it. The biodiversity of the fungi, actinomycetes and bacteria that form this functional group are secondary parameters when assessing the functionality of the biogeochemical cycle under study.

3 Carbon Cycle Functional Groups

The largest carbon reservoir is present in the sediments and rocks of the Earth, but the turnover time is so long that flow from this compartment is relatively insignificant on a human scale. From the viewpoint of living organisms, a large amount of organic carbon is found in land plants. This represents the carbon of forests and grasslands and constitutes the major site of photosynthetic CO_2 fixation. However, more carbon is present in dead organic material, called humus, than in living organisms (Madigan et al. 2000)

Plant residues are the largest fraction of all organic carbon entering the soil. Plants contain 15–60 % cellulose, 10–30 % hemicellulose, 2–30 % lignin, and 2–15% protein. Soluble substances, such as sugars, amino sugars, organic acids, and amino acids, can constitute 10% of the dry weight (Paul and Clark 1989). Soil microbes use residue components as substrates for energy and also as carbon sources in the synthesis of new cells. The presence or absence of substrates can increase or decrease the populations.

Microorganism populations capable of cellulose, starch and both animal and plant protein hydrolisation can be assessed in the carbon cycle. These polymers are broken into smaller units of sugars and amino acids, respectively (Fig. 3).

The functional group of cellulotic microorganisms is formed by fungi, actinomycetes and bacteria. These microorganisms can produce exoenzymes called cellulases. The term cellulase describes a diversity of enzyme complexes that act in two distinct stages. First, there is a loss of the crystalline

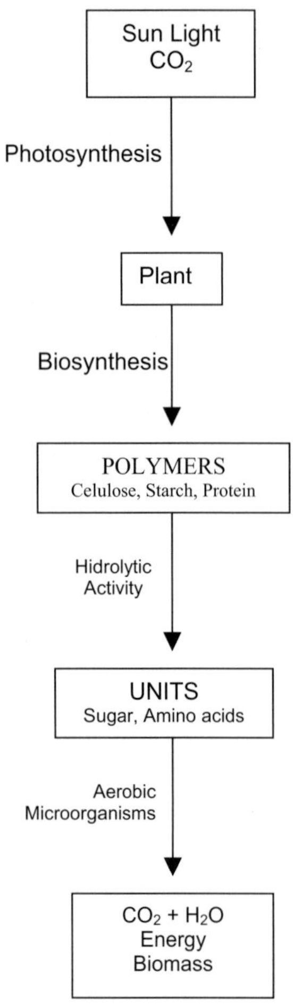

Fig. 3. The activity of some functional groups of microorganisms in the carbon cycle

structure, and then the depolymerisation itself occurs. The resultant disaccharide, cellobiose, is hydrolysed by the enzyme cellobiase to glucose (Paul and Clark 1989).

The amylolytic group hydrolyses starch, which is a common reserve of polysaccharide that serves as an energy storage product in plants. Starch is called amylose when it is a linear polymer of glucose linked in the α1-4 position. The α1-4 linkage facilitates a more rapid breakdown rate than the β1-4 linkage found in cellulose. Glucose can also be found linked in α1-6 positions to produce a polymer known as amylopectin.

Extracellular enzymes known as amylases are produced by numerous fungi, actinomycetes and some bacteria. α-amylases hydrolyse both amylose and amylopectin to units consisting of several glucose molecules. β-amylase reduces amylose to maltose (two glucose units), subsequent hydrolysis of maltose by an α1-4 glucosidase (maltase) yields glucose, and amylopectin is broken down to a mix of maltose and dextrins.

The proteolytic functional group can act in both the carbon and nitrogen cycles, described later. Many microorganisms, such as fungi, actinomycetes and bacteria may produce extra-cell enzymes called proteinases and peptidases. The proteinases degrade proteins releasing peptides which in turn are attacked by the peptidases releasing amino acids which are transported inside the cells (Fig. 3).

The amino acids may be used as a source of either carbon or nitrogen. In the carbon cycle, the amino acids are catabolised into various compounds, as intermediate metabolites of the glucolytic path or tricarboxylic acid cycle. In this conversion, the amino acid undergoes a de-amination process where the amine group is removed and converted into ammonia (NH_3^+) which may be excreted by the cells. The carboxylic group can enter in the tricarboxylic acid cycle or undergo a process of de-carboxylisation (removal of COOH) and de-hydrogenisation, releasing carbon dioxide and nitrogen compounds, such as amines and di-amines.

4 Functional Groups of Microrganisms of the Nitrogen Cycle

Plants, animals, and most microorganisms require combined forms of nitrogen for incorporation into cellular biomass, but the ability to fix atmospheric nitrogen is restricted to a limited number of bacteria and symbiotic associations. Whereas many habitats depend on plants for a supply of organic carbon that can be used as a source of energy, all organisms depend on the bacterial fixation of atmospheric nitrogen (Atlas and Bartha 1993).

Several functional groups in the nitrogen cycle can be used as bioindicators of disturbances in the soil. Among these, the groups to be considered are the symbiotic or free-living nitrogen fixers for legumes and non-legumes plants, respectively, and others which participate in the mineralisation process of the organic nitrogen in the soil such as free-living ammonifiers and protozoans, which also have an important function of mobilisation and mineralisation of nitrogen compounds (Fig. 4). The choice of these groups within the nitrogen cycle was based on their ability to produce ammonia as an end product. Both the nitrogen fixers and the ammonifiers such as the protozoa release ammonia into the rhizosphere. However, the pathway production is different: (1) the first group uses atmospheric nitrogen that by biological fixation produces ammonia, (2) the second group takes part in the mineralisation process of nitrogen organic compounds and, (3) the third group, such as microorganism predators, obtain proteins from their prey and excrete ammonia, among other substances.

Atmospheric nitrogen fixation is a fundamental process for the maintenance of the biosphere, as all organisms require proteins. Nitrogenase is an enzyme complex which is responsible for nitrogen fixation and requires great quantities of energy for its activity. Non-symbiotic biological fixation of nitrogen is carried out by some free-living bacteria genera which are associ-

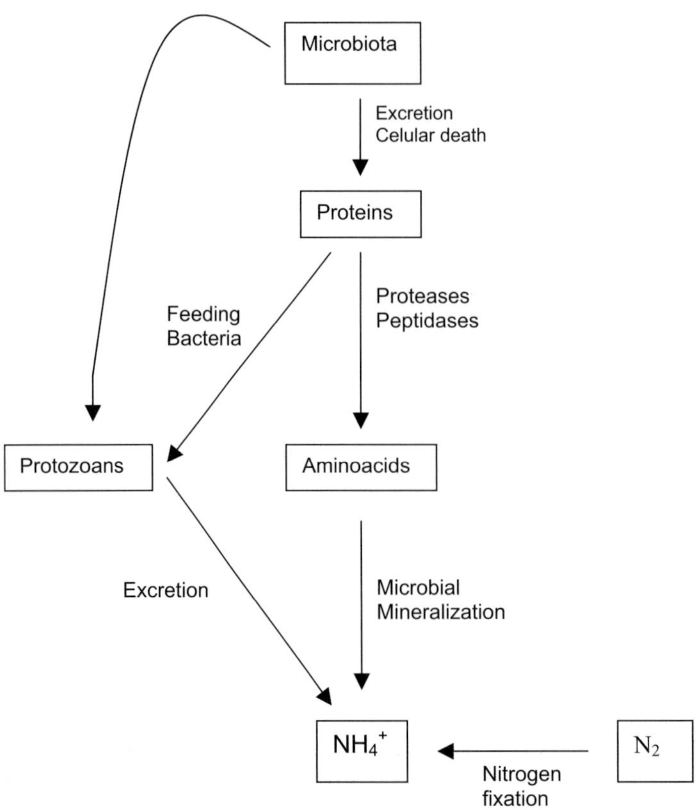

Fig. 4. The activity of some functional groups of microorganisms in the nitrogen cycle

ated with the plant rhizosphere. The symbiotic association of microorganisms and legumes is the most effective in terms of the quantity of nitrogen fixed. The quantity of nitrogen fixed per year by these microorganism groups is much greater than free-living fixing.

The mineralisation of nitrogen compounds in the soil (ammoniation and nitrification) is an essentially microbiological process. The two phases are equally important because the plants are capable of absorbing the nitrogen in the two forms (NH_3^+ and NO_3^-). When there is no addition of nitrogen fertilisers, as in the case of natural areas, nitrification depends on the ammoniation rate for the supply of NH_3^+ substrate.

Ammoniation which occurs in the de-amination process of nitrogen organic compounds is carried out by a large variety of heterotrophic microorganisms that can use amino acids as a source of nitrogen and carbon.

The protozoans are composed of the three groups flagellates, amoebae and ciliates and are important in maintaining plant-available nitrogen and the mineralisation process. The role of protozoa in the soils is still unclear, but evidence for their central position is now accumulating. Protozoans can consume 150–900 g of bacteria $m^{-2}year^{-1}$, which is equal to a production of 15–85 times standing crop (Stout and Heal 1967). This means that preying on bacteria is an important mechanism in nutrient uptake, resulting in greater mineralisation and higher nitrogen release by plants (Juma 1993; Fig. 4).

The correlation among these functional groups is obvious and very important in maintaining the nitrogen cycle and soil fertility. Any factor which alters the populations of these groups will have an immediate response in plant growth.

5 Functional Groups of Microrganisms of the Sulphur Cycle

Plants, algae, and many heterotrophic microorganisms assimilate sulphur in the form of sulphate. For incorporation into amino acids biosynthesis as cysteine, methionine and coenzymes in the form of sulphydril (S-H) groups, sulphate needs to be reduced to the sulphide level by assimilatory sulphate reduction.

The stages assessed in the sulphur cycle involve the organic sulphur mineralisers and the sulphate reducers. These two functional groups participate at different stages of the sulphur cycle and have hydrogen sulphide (H_2S) formation as an end product. Hydrogen sulphide, which is volatile, may decrease the sulphur concentration if it does not complex with other compounds in the soil (Fig. 5).

Mineralisation of organic sulphur in soil is greatly mediated by microbial activity. Carbon-linked sulphur is mineralised either though oxidative (aerobic) decomposition or a desulphirisation (anaerobic) process. The mineralisation process may be direct (cell-mediated), involving enzymes such as sul-

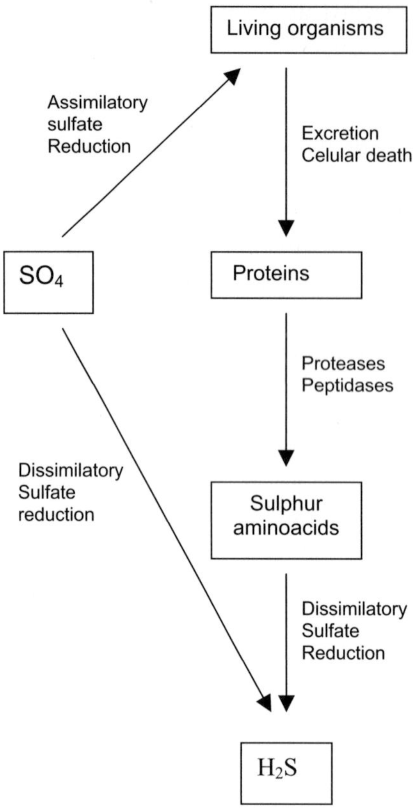

Fig. 5. The activity of some functional groups of microorganisms in the sulphur cycle

phatases where elements such as nitrogen and sulphur-linked carbon mineralised by microorganisms oxidize the organic carbon compounds to obtain energy. The heterotrophic soil microorganisms decompose organic sulphur to form sulphide. In the case of indirect mineralisation, those elements that exist as sulphate esters are hydrolysed by endo or exoenzymes. This process occurs by positive feedback or negative control (Sylvia et al. 1998).

The activity of these microorganisms may be aerobic or anaerobic. Anaerobic microorganisms exist in fairly low numbers in the rhizosphere of plants which live in non-flooded soils. Bearing in mind that sulphate is fundamental for plant metabolism and that the turnover of organic to inorganic sulphate implies availability of the nutrient for plant growth, the study of these populations may complement the analysis of functional microorganism groups as indicators of environmental impact or of biotic fertility indexes in sustainable agricultural systems or natural areas.

6 Functional Groups of Microrganisms of the Phosphorus Cycle

The main functional groups of the phosphorus cycle are the mycorrhizal fungi and the inorganic phosphate solubiliser microorganisms. The interaction between these two microbial groups is fundamental for the nutrition of the majority of native plants and is also of agronomic interest.

The phosphate solubiliser functional group can include fungi, actinomycetes and bacteria that are capable of solubilizing inorganic phosphate by production and excretion of organic and inorganic acids, of a phosphatase group of enzymes and of carbon dioxide (CO_2) in the rhizosphere soil solution. Carbon dioxide can cause the solubilisation of calcium, magnesium and

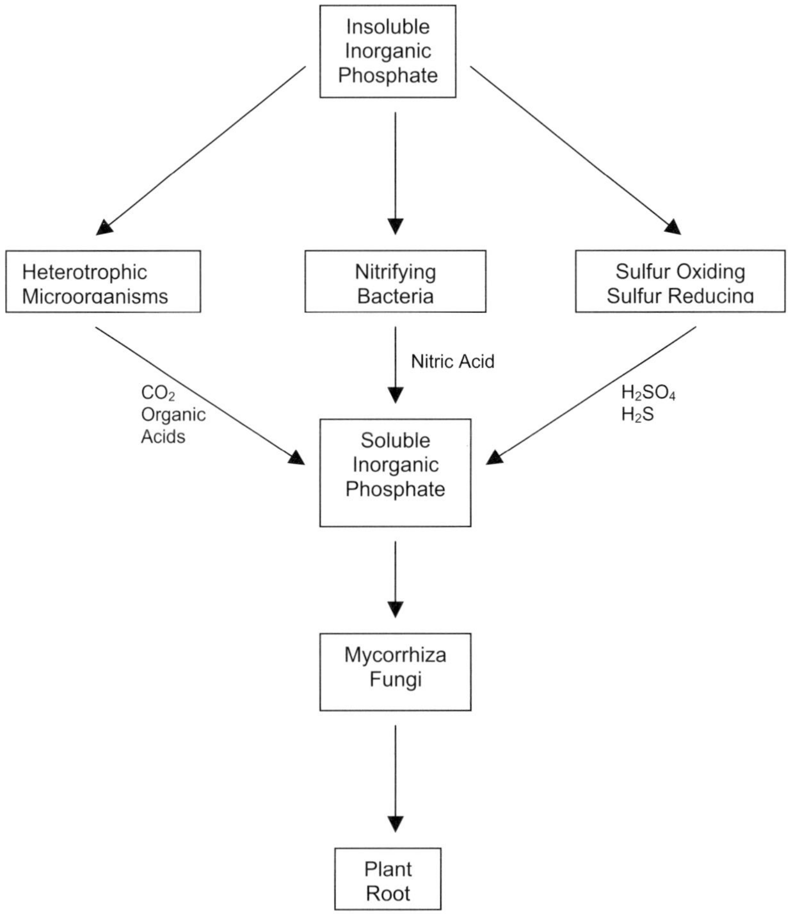

Fig. 6. The activity of some functional groups of microorganisms in the phosphorus cycle

phosphate compounds. The nitrifying, sulphur oxidants and sulphur- reducing bacteria can also solubilise insoluble phosphate salts and produce H_2S under anaerobic conditions. Many microorganisms and plants can produce organic acids by acting as solubilizing agents and quelants and releasing orthophosphate in the soil solution (Sylvia et al. 1998). Soluble phosphate in the soil solution can be absorbed and transported to the plant by arbuscular mycorrhizal (AM) fungus mycelia. The interaction between the phosphate solubilisers and the mycorrhizae can stimulate mycorrhizal colonisation and/or plant growth by increasing the phosphorus levels (Fig. 6).

The arbuscular mycorrhizal fungi are symbiotic fungi of plant roots. This symbiosis is present in almost all plants in the most different ecosystems (Hayman 1982). The symbiotic relationship between plant roots and mycorrhizal fungi improves plant mineral nutrient acquisition from the soil, especially immobile elements such as P, Zn and Cu, but also more mobile ions such as S, Ca, K, Fe, Mg, Mn, Cl, Br and N (Tinker 1984). In soils where such elements may be deficient or less available, mycorrhizal fungi increase efficiency of mineral uptake, resulting in increased plant growth (Linderman 1988).

The mycorrhizal complex (AM fungi and root) changes the nutritional and physicochemical conditions of the rhizosphere, and has a large negative or positive impact on the functional microorganism groups. This effect depends on the cycle to which the functional group belongs. However, in spite of the importance of the mycorrhizae, these groups should not be assessed in isolation.

7 Dynamics of the Rhizosphere Functional Groups of Microrganisms

The interaction of specific biological systems, in a ecosystem or microcosm, depends on the interplay of three general factors – environment, biological community structure (biodiversity), and biological activity (function). The role of diversity, particularly of microorganisms, and the relationship between microbial diversity and function is largely unknown (Griffiths et al. 1997). As can be seen, each functional group can interact with different biogeochemical soil cycles and the environmental impact caused by an agent can be determined by the changes observed in the populations, as a determined environmental condition can affect the microbial activity without affecting the community biodiversity (Griffiths et al. 1997). The dynamic behaviour of perturbed communities is a branch of general ecology closely related to the study of natural and artificial disturbances in microbial habitats. Another important factor is the relationship between resistance and resilience, whose combined effects determine the ecosystem stability. Resistance is the inherent capacity of the system to hold disturbance, whereas resilience is the capacity to recover after disturbance.

8 Relationship Among r and k Strategist Functional Groups

The determination of the r and k strategists (Andrews 1984) is also related to soil disturbance, resilience and health. The r strategist microorganism has a high reproductive rate with few competitive adaptations. On the other hand, the k strategist microorganism reproduces more slowly than the r strategist, and is usually a more stable and permanent member of the community.

Fungi and actinomycetes are normally k strategists and are involved in the carbon cycle degrading cellulose and structural proteins among other macromolecules. The stability of these compounds and the slow k strategist growth rate render them not very sensitive to swift environmental changes. On the other hand, the r strategists such as bacteria are more sensitive to quick environmental changes. Heterotrophic bacteria populations are affected by the lack of carbohydrates which occurs due to changes in the rhizosphere carbon flow in the photosynthesis function variations between day and night. Only heterotrophic bacteria populations that have metabolic diversity and can manage to use other compounds, such as amino acids for obtaining carbon and energy, will keep their numbers in the rhizosphere. The other populations decrease their CFU number. Plants begin photosynthesis at daybreak with a consequent increase in carbon concentration in the exudates, and the heterotrophic bacteria community returns to its previous composition.

9 Arbuscular Mycorrhizal Fungi Dynamics in the Rhizosphere

The MA can also be considered k strategists and influence several biogeochemical soil cycles: (1) the carbon cycle due to alterations in the flow of carbon compounds from the exudates, (2) the phosphorus cycle due to stimulus to phosphate-solubilising bacteria activity and absorption of soluble phosphorus by plants (Toro 1998), (3) the nitrogen cycle due to stimulus to symbiotic (Toro 1998) and non-symbiotic fixation (Vosátka and Gryndler 1999) and to the rhizosphere ammoniation process (Amora-Lazcano et al. 1998). The sulphur cycle is also influenced by alterations in the autotrophic sulphur oxidising and sulphate reducing bacteria populations (Amora-Alzcano and Azón 1997).

The term mycorrhizosphere (Oswald and Ferchau 1968) refers to the zone of influence of the mycorrhiza (fungus-root) in the soil. The mycorrhizosphere has two components. One is the rhizosphere, a thin layer of soil that surrounds the root and is under the joint, direct influence of the root, root hairs, and AM hyphae adjacent to the root. The other, the hyphosphere, is not directly influenced by the root. The hyphosphere is a zone of AM hypha-soil interactions (Marschner 1995), and may be more or less densely permeated by the AM soil mycelium.

In our laboratory, hypha colonisation of some MA fungus species by bacteria in spores germinated in 1% agar-water medium on a Petri dish was observed. These bacteria had as their single nutrient source the products excreted by the MA mycelia in the medium (Fig. 7). The bacteria formed a dense cell layer around the hypha in an experiment with *Glomus etunicatum*. From this layer, as the exudate excretion increased the medium nutrient to optimum levels, the bacteria developed and colonised the remaining mycelia (Fig. 8).

In an axenic conditions experiment with maize plants and colonised *Glomus clarum* mycelia, the bacteria which colonised the *G. clarum* mycelia without plants continued to prefer products excreted by the mycelia, and no colonies were observed in the plant roots (Fig. 9). These results seem to indicate that the fungus mycelia produce some growth factor essential for the bacteria growth, which is not found in the maize root exudates. However, the mechanisms involved in this interaction are not yet known.

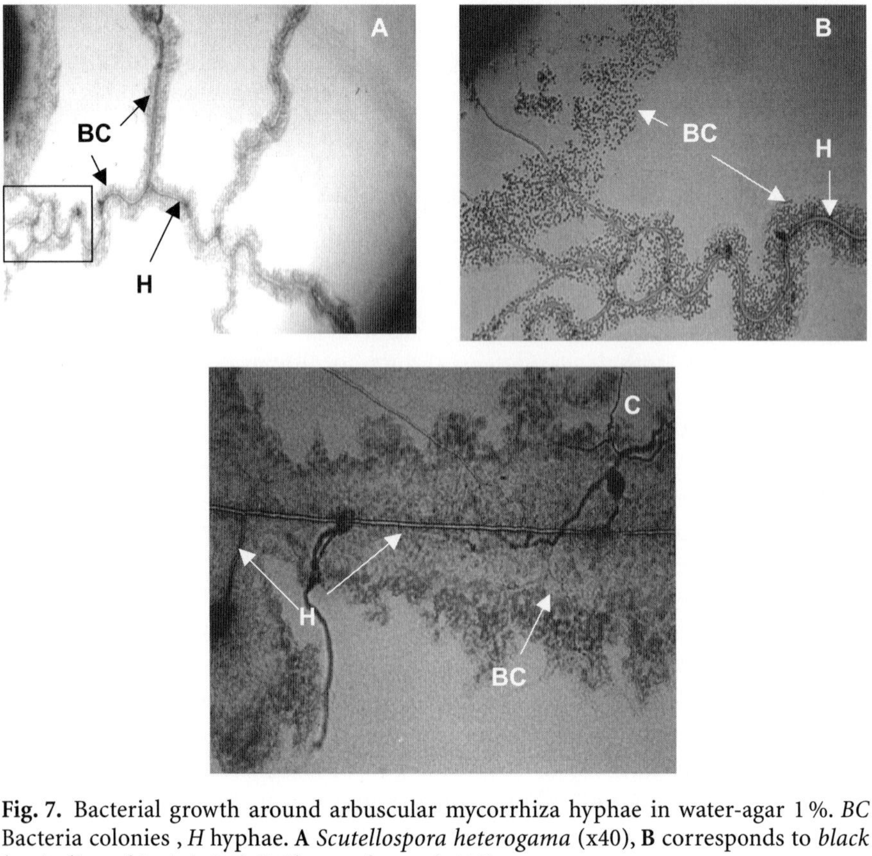

Fig. 7. Bacterial growth around arbuscular mycorrhiza hyphae in water-agar 1%. *BC* Bacteria colonies , *H* hyphae. **A** *Scutellospora heterogama* (x40), **B** corresponds to *black box* indicated in **A** (x100) **C** *Glomus clarum* (x100)

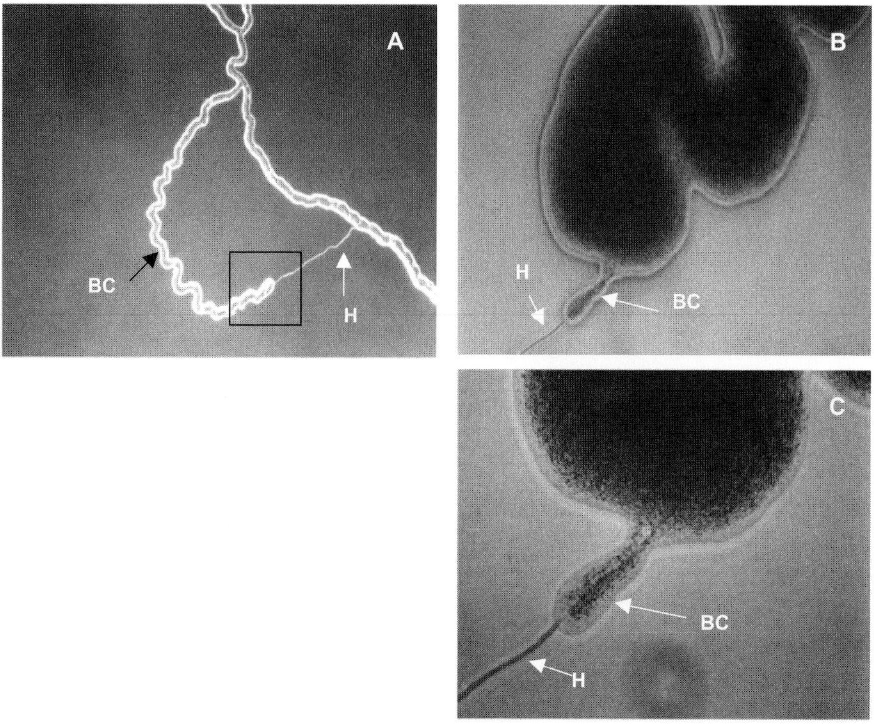

Fig. 8. Bacterial colonising AM hyphae of *Glomus etunicatum* in 1 % water-agar. *BC* Bacteria colonies, *H* hyphae. **A** General aspects of mycelia colonised by bacteria (x20), **B** corresponds to *black box* indicated in the **A**, where bacteria is growing around hyphae (x100), **C** bacteria growing around hyphae (x400)

In the soil, Andrade et al. (1997) observed sorghum plants inoculated with several exotic or native *Glomus* species either exotic or native to the test soil. The soils adhering to the root were considered rhizosphere or not adhering to the root were considered hyphosphere. Bacterial numbers were greater in rhizo- than in hyphosphere soil. Isolates of *Bacillus* and *Arthrobacter* were most frequent in hyphosphere and *Pseudomonas* in rhizosphere soils. More bacterial species were found in hyphosphere than in rhizosphere soil, and bacterial communities varied within and among AM treatments. The development of the AM mycelium in soil had little influence on the composition of the microflora in the hyphosphere, while AM root colonisation was positively related with bacterial numbers in the hyphosphere and with the presence of *Pseudomonas* in the rhizosphere.

In another experiment, Andrade et al. (1998) inoculated *Alcaligenes eutrophus* and *Arthrobacter globiformis* in sorghum plants. The first is an isolate of the *Glomus mosseae* hyphosphere and the second an isolate of the *G. mosseae* and *G. intraradices* mycorrhizosphere. Ten days after inoculation,

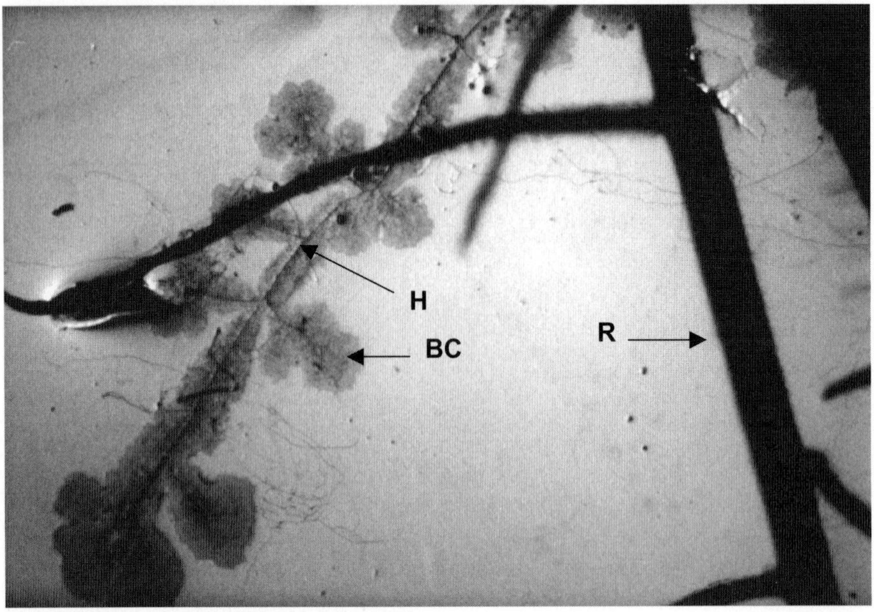

Fig. 9. Bacteria colonising mycelia of *Glomus clarum* in the hyphosphere of maize plants grown under axenic conditions in 1 % water-agar. Bacteria did not colonise maize roots, colonies were observed only around mycelia (x40). *BC* Bacteria colonies, *H* hyphae, *R* root

the *A. globiformis* population present in bulk soil, in the rhizosphere and hyphosphere were similar, but that present in the mycorrhizosphere was larger. *A. eutrphus* was dependent on the presence of *G. mosseae* in the soil, indicating that even in soil some bacteria may depend on MA-excreted metabolic products.

These results show that the MA-plant system is very complex and the influence of these microorganisms is fundamental for the regulation of the biogeochemical cycles in the rhizosphere system. On the other hand, the microorganisms of other cycles also influenced the mycorrhizal activity and root infection with direct consequences on the plant growth and soil fertility.

In degraded areas of tropical regions, the soil is compacted displaying minimum aeration and draining capacity, aluminium and manganese toxicity and low fertility indices especially for nitrogen, phosphorus and organic matter. In these areas, the re-vegetation process is directly related to the interaction between the plant roots and the functional microorganism groups. The pioneer plants are the first to colonise these low fertility areas, and they are very dependent on AM for phosphorus. The pioneer plants in this process are r strategists which improve the physicochemical characteristics of the soil and fertility levels with time, allowing other groups of more demanding plants (k strategists) to establish in the area and to form a forest in equilibrium.

The pioneer plants can survive under adverse conditions due to the presence in the rhizosphere of microorganisms which supply nutrients for their metabolism, and in turn, their exudates maintain these rhizosphere microorganisms. The k strategist mycorrhizae are sufficiently stable to maintain the required nutrient levels for this plant group. In this sense, groups of r strategist microorganisms succeed each other, maintaining the dynamic of the system and the reconstitution of other biogeochemical cycles until the system equilibrium is reached with the establishment of late secondary and climax plant groups.

10 Dynamics Among the Functional Microrganism Groups of the Carbon, Nitrogen, Phosphorus and Sulphur Cycles

There are several stages in each biogeochemical cycle, and many microorganisms can take part in one or more cycles depending on the diversity of their metabolic path (Fig. 10). Microbiota metabolic versatility makes a single bacteria species able to use various carbohydrates, such as glucose, fructose and saccharose, as a carbon and energy source, and in their absence they can use amino acids or other compounds.

The biosphere is composed of all living organisms which depend on matter transformation for their maintenance. The functional microorganism groups are inserted in this system which transforms matter and maintains the levels of nutrients available on Earth. Due to their functional importance, they can be used as biological indicators to determine any natural or artificial impact which may occur in the soil. It is obvious that the complexity of the biological interactions occurring on the soil–plant interface must be simplified to allow quick and accurate assessment of these microorganism populations. Thus, only those stages of the biogeochemical cycles which directly influence plant growth should be chosen. However, different stages can be selected according to the experimental objective.

Autotrophic organisms have the important function of matter de-mineralisation and transform it into organic molecules. In this group are plants that de-mineralise carbon, i.e. transform carbon dioxide (CO_2) into glucose, which is then polymerised mainly into starch, cellulose, hemicellulose and lignin. Plants are also responsible for transforming NO_3^-, NH_3^+, and SO_4^{2-} into amino acids, PO_4^{2-} into nucleic acids while ATP, NADP, and SO_4^{2-} can be transformed into glutathione.

In a simplified way, plants can be considered as nutrients from the soil solution plus solar energy accumulated in chemical form. Plants generally release organic molecules into the soil in two ways: (1) by depositing dead plant material to form the litter; and, (2) by exuding excretion and lysates into

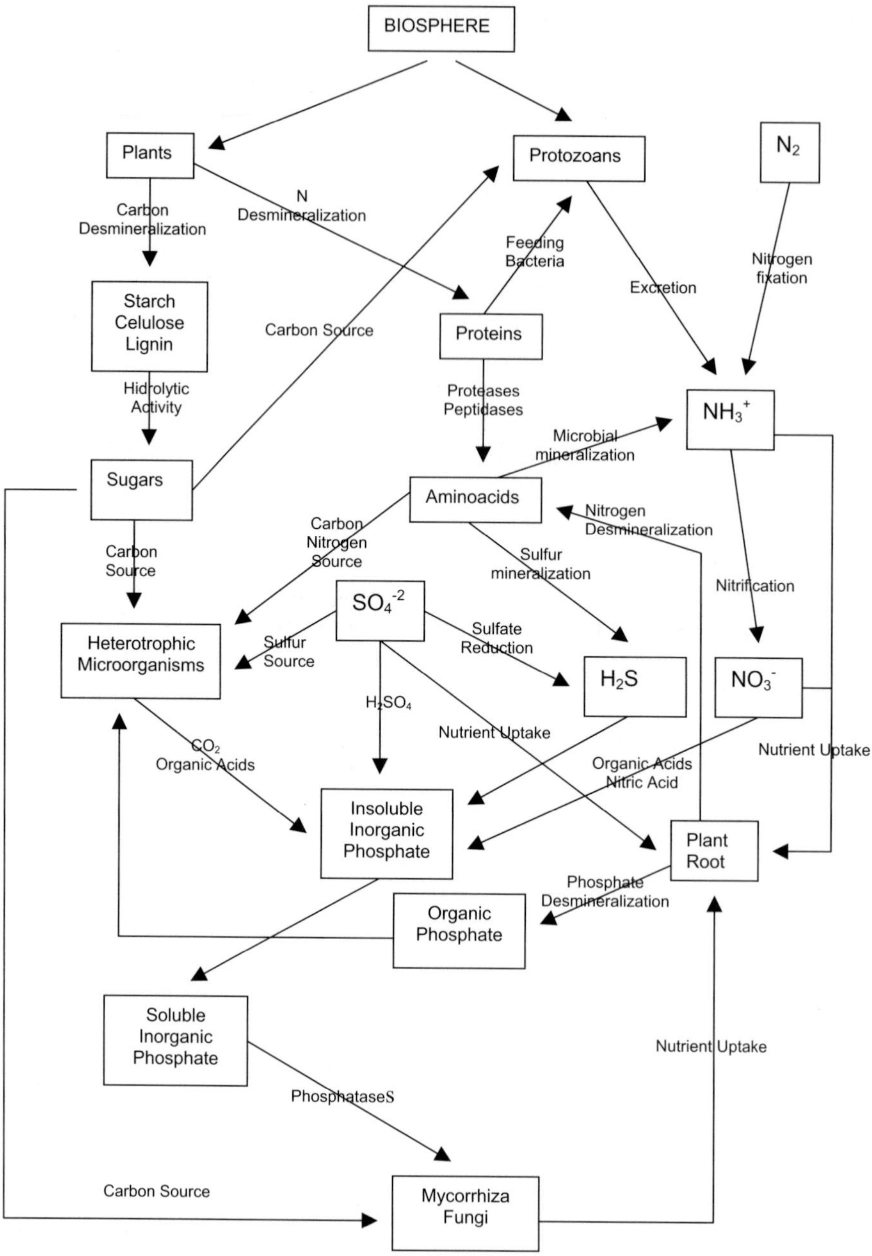

Fig. 10. The interaction among functional groups of microorganisms in the carbon, nitrogen, phosphorus and sulphur cycles

the rhizosphere, a phenomenon known as rhizodeposition. These compounds, which are continuously released into the soil, constitute the main nutrient sources, maintain the microbiota, the fertility and participate in the maintenance of the soil structure.

Microorganisms are classified into several categories according to the carbon and energy source used, but only some groups will be considered in this chapter. The heterotrophic microorganisms can use glucose or amino acids as carbon sources. Glucose can be obtained from some macromolecules, such as cellulose and starch, which undergo lytic action by enzymes produced by the cellulose and starch-reducing microorganisms (Fig. 10).

Proteins are degraded to amino acids by proteolytic organisms, which can use these compounds as carbon or nitrogen sources. On the other hand, sulphur amino acids such as cystine and cysteine can also be used to obtain sulphur which is used in the biosynthesis of other compounds necessary for cell metabolism. The amino acids can also be used by the cell without lysis of the molecule, as many microorganism species are not able to biosynthesise all the amino acids required by the cell.

Protozoa, such as amoebas, ciliates and flagellates, are organisms which have the function of immobilising and mineralising the nitrogen in the rhizosphere system. Bacteria are their main nutrient source, and they obtain nitrogen and other nutrients for their metabolism from them. Some of these nitrogen compounds are released into the soil as inorganic NH_3^+ and can be absorbed by the root or by other microorganism groups such as nitrifiers, sulphate reducers or oxidisers or phosphate solubilisers. Biological nitrogen fixation is very important in the introduction of NH_3^+ molecules into the rhizosphere (free-living N fixers) or in the plant (symbiotic N fixers). These fixed molecules can be transformed in NO_3^- or used in the biosynthesis of amino acids that will form the cell proteins when polymerised. Sulphur amino acids may be synthesised from SO_4^{2-} obtained by the oxidation of S by sulphur cycle bacteria. NO_3^- and NH_3^+ can be used in amino acid biosynthesis and also as final receptors of electrons for some groups of facultative anaerobic bacteria.

Phosphate exists in the soil mainly in the soluble inorganic form. Several solubilisation mechanisms have been described and many microorganisms produce compounds which can solubilise phosphates. The nitrogen cycle functional group, the nitrifiers, produces NO_3^- that can form nitric acid. The sulphur cycle functional group can produce SO_4^{2-} that can form H_2SO_4 or reduce it to H_2S, which will also solubilise insoluble inorganic phosphate. In the degradation of sulphur amino acids, proteolytic microorganisms release H_2S or CO_2, which can form carbonic acid. Both molecules can also solubilise inorganic phosphate. The carbon cycle microorganisms form CO_2 and organic acids as end products of their catabolism, and both compounds are responsible for pH reduction and inorganic phosphate solubilisation.

Soluble inorganic phosphate is absorbed mainly by the mycorrhizal fungi that transport these molecules to the plant, which in turn transform them into organic phosphate. This phosphate is deposited in the soil by rhizodeposition or absorbed in the organic form by heterotrophic microorganisms which take part in the nitrogen, carbon or sulphur cycles (Fig. 10).

References and Selected Reading

Amora-Lazcano E, Azcón R (1997) Response of sulfur cycling microorganisms to arbuscular mycorrhizal fungi in the rhizosphere of maize. Appl Soil Ecol 6:217–222

Amora-Lazcano E, Vázquez MM, Azcón R (1998) Response of nitrogen-transforming microorganisms to arbuscular mycorrhiza fungi. Biol Fertil Soils. 27:65–70

Andrade G, Linderman RG, Bethlenfalvay GJ (1998) Bacterial associations with the mycorrhizosphere and hyphosphere of the arbuscular mycorrhizal fungus *Glomus mosseae*. Plant Soil 202:79–87

Andrade G, Mihara KL, Linderman RG, Bethlenfalvay GJ (1997) Bacteria from rhizosphere and hyphosphere soils of different arbuscular mycorrhizal fungi. Plant Soil 192:71–79

Andrews JH (1984) Relevance of r and k theory to the ecology of plant pathogens. In: Klug MJ, Reddy CA (eds) Current perspectives in microbial ecology. American Society for Microbiology, Washington, pp 1–7

Atlas RM, Bartha R (eds) (1993) Microbial ecology: fundamentals and applications, 3rd edn. The Benjamin/Cummings Publishing Company, California, 563 pp

Brussaard L (1998) Soil fauna, guilds, functional groups and ecosystem processes. Appl Soil Ecol 9:123–135

Griffiths BS, Ritz K, Wheatley RE (1997) In: Insan H, Ranger A (eds) Microbial communities: functional versus structural approaches. Springer, Berlin Heidelberg New York, pp 1–10

Hayman DS (1982) Influence of soils and fertility on activity and survival of vesicular-arbuscular mycorrhizal fungi. Phytopathology 72:1119–1125

Juma NG (1993) Interactions between soil structure/texture, soil biota/soil organic matter and crop production. Geoderma 57:3–30

Linderman RG (1988) Mycorrhizal interactions with the rhizosphere microflora. The mycorrhizosphere effect. Phytopathology 78:366–371

Madigan TM, Martinko JM, Parker J (eds) (2000) Microbial ecology. In: Brock biology of microorganisms, 9th edn, Prentice Hall, New Jersey, pp 642–719

Marschner H (ed) (1995) The soil–root interface (rhizosphere) in relation to mineral nutrition. Mineral nutrition of higher plants, 2nd edn. Academic Press, London, pp 537–595

Oswald ET, Ferchau HA (1968) Bacterial associations of coniferous mycorrhizae. Plant Soil 28:187–192

Paul EA, Clark FE (eds) (1989) Carbon cycling and soil organic mater. In: Soil microbiology and biochemistry. Academic Press, San Diego, pp 93–116

Stout JD, Heal OW (1967) Protozoa. In: Burgues A, Raw F (eds) Soil biology. Academic Press, New York, pp 149–195

Sylvia DM, Fuhrman JJ, Hartel PG, Zuberer DA (eds) (1998) Principles and applications of soil microbiology. Prentice Hall, Englewood Cliffs, pp 346–367

Tinker PB (1984) The role of microorganisms in mediating and facilitating the uptake of plant nutrients from soil. Plant Soil 76:77–91

Toro M, Azcón R, Barea JM (1998) The use of isotopic dilution techniques to evaluate the interactive effects of *Rhizobium* genotype, mycorrhizal fungi, phosphate-solubilizing rhizobacteria and rock phosphate on nitrogen and phosphorus acquisition by *Medicago sativa*. New Phytol 138:265–273

Vosátka M, Gryndler M (1999) Treatment with culture fractions from *Pseudomonas putida* modifies the development of *Glomus fistolosum* mycorrhiza and response of potato and maize plants to inoculation. Appl Soil Ecol 11:245–251

5 Diversity and Functions of Soil Microflora in Development of Plants

RAMESH CHANDER KUHAD, DAVID MANOHAR KOTHAMASI,
K. K. TRIPATHI and AJAY SINGH

1 Introduction

Soil is a dynamic and complex system consisting of living organisms interacting with inorganic mineral particles and organic matter. A wide range of functions is performed by soil that directly or indirectly sustains the world's human population. Soil plays a vital role in food production, as a reservoir for water and filter for pollutants. Soils store almost twice as much carbon as the atmosphere does and are important links in the natural cycle that determines atmospheric carbon dioxide level (O'Donnell and Görres 1999). Soils sustain an immense diversity of microbes, which exceeds that of eukaryotic organisms (Torsvik and Øvreås 2002). Microorganisms exist in every conceivable place on earth, even in extreme environments. One gram of soil may harbor up to 10 billion microorganisms of possibly thousands of different species.

It is widely accepted that the extent of microbial diversity has not been adequately explored. Some bacteriologists believe that about 100,000 to 1 billion bacterial species actually exist in the earth environment and only about 4000 species have been described (Staley 1997). Mycologists estimate that there are more than 1.5 million species of fungi of which only 72,000 species have been isolated or described (Hawksworth 1997). Microorganisms can exist either in an active or in a dormant yet persistent form. The ratio of viable counts to direct counts reflects the ratio between the numbers of the active (dividing) cells and the quiescent cells, and most bacteria in soil are in the latter form (Hattori et al. 1997). The tropics are considered to be richer in microbial diversity than boreal or temperate environments (Hunter-Cevera 1998). Some microbiologists believe that there is an equal amount of microbial diversity in the deserts. Actinomycetes with motile spores appear to be widely distributed in littoral zones and arid environments.

Analysis of microbial functional diversity is important when considering the ability of the ecosystem to respond to changing environmental conditions, links between ecosystem processes and functional diversity and the

Plant Surface Microbiology
A. Varma, L. Abbott, D. Werner, R. Hampp (Eds.)

need to conserve the microbial gene pool (Prosser 2002). Fortunately, with the development of advanced molecular in situ methods and improved cultivation procedures, more accurate estimates of the microbial functional diversity on earth can be predicted and their role in the soil ecosystem can be thoroughly evaluated. In this chapter, the interaction and functional diversity of microorganisms in the soil environment related to plant growth and development is discussed.

2 Functional Diversity of Soil Microflora

The microbial functional diversity encompasses a range of activities and has been assumed to influence ecosystem stability, productivity and resilience towards stress and disturbances. Typically, microorganisms decrease with depth in the soil profile, as do the plant roots and soil organic matter. Differences in microbial community structures reflect the ability of microorganisms to respond to specific environmental controls and substrates (Paul and Clark 1998). For example, the arbuscular mycorrhizal fungus, *Glomus*, occurs worldwide on a variety of agricultural plants. Examination of the crop rotations shows that strains of this fungus change with the type and nutrition of the host crop. The fluorescent pseudomonads are attracted to plant roots and show genetic and physiological divergence between soil and plant surfaces. While *Penicillium* is abundant in temperate and cold climates, *Aspergillus* predominates in warm areas. Cyanobacteria are commonly found in neutral to alkaline soils, but rarely under acidic conditions. Depending on the preferred metabolites present in the soil, nitrogen-fixing, sulfur- and hydrogen-oxidizing and nitrifying bacteria are often found in addition to the denitrifiers, sulfate-reducers and methanogens. Various microbial processes in soil, which directly or indirectly influence plant development, are shown in Table 1.

Microbiologists are continually learning that microbial function in the ecosystem is as diverse as the microbes themselves. In studying functional relationships between agricultural plants and microbes, Shen (1997) reported that *Pseudomonas* and *Bacillus* spp. enable plants to remain healthy and help improve growth yields. Microbially digested organic waste enhances plant growth and improves soil structure and nutrients (Shen 1997). Denitrifying bacteria can utilize nitrous oxides (NO_x) as the terminal electron acceptor. Many denitrifiers produce NO_x reductase and can metabolize NO_x in aerobic and anaerobic conditions (Stepanov and Korpelal 1997).

Soil comprises a variety of microhabitats with different physicochemical gradients and discontinuous environmental conditions. Microbes adapt to the microhabitat and live together in consortia with more or less clear boundaries, interacting with each other and with other parts of the soil biota (Yin et al. 2000; Tiedje et al. 2001). Competitive interactions are also thought to be a

Table 1. Major processes of soil microflora influencing plant growth

Microbial process	Examples of microbes
Organic matter decomposition	*Trichoderma, Fusarium, Bacillus, Streptomyces, Clostridium*
Symbiotic nitrogen fixation	*Rhizobium, Bradyrhizobium, Frankia, Anabaena*
Nonsymbiotic nitrogen fixation	*Azotobacter, Beijerinckia, Aerobacter, Chlorobium, Nostoc*
Nitrogen mineralization	*Bacillus, Pseudomonas, Serratia*
Nitrification	*Nitrobacter, Nitrosomonas*
Denitrification	*Achromobacter, Pseudomonas*
Phosphate solubilization	*Azotobacter, Enterobacter, Bacillus, Aspergillus, Penicillium, Rhizoctonia, Trichoderma*
Sulfur transformation	*Desulfovibrio, Thiobacillus*
Iron transformation	*Ferribacterium, Leptothrix*
Phytohormone production	*Azotobacter, Azospirillum, Pseudomonas, Rhizobium, Bacillus, Flavobacterium, Actinomyces, Nocardia, Fusarium, Gibberella, Aletrnaria, Penicillium*
Siderophore production	*Neurospora, Trichoderma, Agaricus, Fusarium, Penicillium*, ericoid mycorrhizal fungi, *Nocardia, Pseudomonas, Bacillus, Aeromaonas, Erwinia*
Biotic control	*Pseudomonas, Bacillus, Strepetomyces*

key factor controlling microbial community structure and diversity. The impact of soil structure and spatial isolation on microbial diversity and community structure has been clearly demonstrated (Staley 1997; Pankhurst et al. 2002). More than 80 % of the bacteria were found located in micropores of stable soil microaggregates (2–20 µm) in soils subjected to different fertilization treatments (Ranjard and Richaume 2001). Such microhabitats offer most favorable conditions for microbial growth with respect to water and substrate availability, gas diffusion and protection against predation. Soil structure and water regime influence competitive interactions by causing spatial isolation within communities. A high diversity in soil with high spatial isolation may also have been caused by a higher heterogeneity of carbon resources in the soil. Particle size and other factors like pH and type and amount of available organic compound may highly impact microbial diversity and community structure (De Fede et al. 2001). Soil microbes are also subjected to considerable seasonal fluctuations in environmental conditions such as temperature, water content, and nutrient availability (Smit et al. 2001).

Catabolic diversity has been used to investigate the effect of stress and the disturbance on the soil biodiversity. The catabolic response profile (CRP), a measure of short-term substrate-induced respiration, has been used to calculate the diversity and catabolic functions expressed in situ (Degens et al. 2001). When soils from long-term managed environments were subjected to stress and disturbances, microbial communities with low catabolic evenness (crop fields) were less resistant to stress and disturbance than were communities with high catabolic evenness (pasture). After a major disturbance (landslides, volcanic eruptions, etc.), marked changes in catabolic functional diversity has been reported in developing soil ecosystems (Schipper et al. 2001).

Most members of the soil biota are organotrophs. The major source of carbon input for soil organisms are the plant roots and organic residues contributed during and following plant growth. The proportion of nitrogen, carbon and other organic matter changes with both plant types and landscape, which in turn, alter microbial mass, activity and diversity (Paul and Clark 1998).

Microorganisms play an essential role in functioning and sustainability of all natural ecosystems including biogeochemical cycling of nutrients and biodegradation. Most soils are exposed to fluctuating environmental conditions and the high diversity of organic substrate is likely to have a positive effect on the function. Interactions between different trophic levels were elucidated in a simple ecosystem model in which primary producers (plants) and decomposers (microorganisms) were linked through cycling of a limiting nutrient factor for the primary producers (Loreau 2001). The model predicts that microbial diversity has a positive effect on nutrient cycling efficiency, and contributes to increased ecosystem processes. However, in interacting microbial consortia, a small linear change in diversity may result in nonlinear changes in the process, therefore relationship between microbial diversity and soil processes may not necessarily be linear.

Biochemical quality of the substrate and the physical availability of those components to the degradative microorganisms are key determinants of the rate of decomposition processes in soils, and reflects a number of interacting components (Bending et al. 2002). In the case of crop residues, nitrogen content and structural polymers such as lignin interact to control microbial nitrogen mineralization-immobilization processes during decomposition. The types of nutritional substrates available are different in soils with varying soil organic matter quality, and directly affect the microbial communities active in the soil. Native soil organic matter content may also significantly affect the enzyme diversity, which is found greater in high organic matter soil. Organic acids, such as malate, citrate and oxalate, have also been proposed to be involved in many rhizospheric processes, including nutrient acquisition, metal detoxification, alleviation of anaerobic stress in roots, mineral weathering and pathogen attraction (Jones 1998).

The ecological relevance of the community structure for the function of systems is the main reason to study the microbial diversity. There is no single technique available today that can reveal the entire diversity of a microbial community. Several approaches are available for assessment of microbial diversity (Bridge and Spooner 2001; Dahllöf 2002; Prosser 2002). Time-consuming cultivation-based assessment of microbial diversity has been widely used (Torsvik et al. 1996). With advanced methods, identification can be accelerated by automated methods, e.g., Biolog; phospholipid fatty acid (PLFA) profiling, fatty acid methyl ester profiling (FAME), DNA-hybridization and re-association. However, potential limitations of this approach are widely accepted. Separation of biomass from particulate material varies between species, and, with growth form (spore, cells, and mycelia), introduces bias. It is almost impossible to design growth media and cultivation conditions that are suitable for all members of the microbial community. The approach of identification using traditional methods, based on phenotypic characteristics, is also limited for analysis of diversity in complex environments, such as soil when quantification of the diversity is required.

The importance and need to study the vast biodiversity in different environments has stimulated the development of molecular methods for culture-independent study of microbial communities. These methods have employed a combination of analysis of genes and microscopy. Analysis of 16S rRNA genes is now widely used for the analysis of bacterial populations and analysis of 18S rRNA genes and internal transcribed spacer (ITS) regions are increasingly used for fungal population analysis (Hunter-Cevera 1998; Bridge and Spooner 2001; Torsvik and Øvreås 2002). Ribosomal RNA genes are ideal for this purpose because they possess regions with sequences conserved between all bacteria or fungi, facilitating alignment of sequences when making comparisons, while other regions exhibit different degrees of variation, enabling distinction between different groups. These differences provide the basis for a phylogenic taxonomy and enable quantification of evolutionary differences between different groups. Polymerase chain reaction (PCR)-based fingerprinting techniques provide a rapid analysis of changes in whole community structure with high resolution. These fingerprinting techniques, such as denaturind gradient gel electrophoresis (DGGE), amplified rDNA restriction analysis (ARDRA), terminal restriction fragment length polymorphism (T-RFLP) and ribosomal intergenic spacer analysis (RISA), provide information on the species composition, and can be used to compare common species present in samples. Sequence information can also be used to design and construct fluorescent-labelled oligonucleotide probes specific for particular microbial groups using fluorescence in situ hybridization (FISH technique). For a comprehensive description and discussion on potential and limitations of various molecular approaches, excellent reviews by Bridge and Spooner (2001), Kozdroj and van Elsas (2001), Dahllöf (2002), Prosser (2002) and Torsvik and Øvreås (2002) may be consulted.

3 Role of Soil Microflora in Plant Development

3.1 Mycorrhiza

Fungi, which form a symbiotic association with plant roots, are referred to as mycorrhizal fungi and the association itself is called as mycorrhiza. There are five broad groups of mycorrhiza: the ectomycorrhizae, the arbuscular mycorrhizae, the ericaceous mycorrhizae, the ectendomycorrhizae, and the orchidaceous mycorrhizae (Bagyaraj and Varma 1995; Hodge 2000). The most common mycorrhizal association found in cultivated crop plants throughout the world is the arbuscular mycorrhizal (AM) fungi. Ectomycorrhiza (EM), formed by fungi belonging to basidiomycetes and ascomycetes, are commonly associated with temperate trees, whereas ericoid mycorrhiza are found in the plants from the family Ericaceae and plant communities at high latitude and altitude (Perotto et al. 2002; Koide and Dickie 2002). Orchid mycorrhizae are associated with orchids. The AM and ectendomycorrhizal fungi are more prevalent in the tropics and arid/semiarid regions. AM, the most prevalent plant-fungus association, comprise about 150 species, belonging to the order Glomales of Zygomycotina (Simon 1996; Myrold 2000).

Most angiosperm, gymnosperm, fern and bryophyte families form mycorrhizae. It is believed that plants growing in aquatic, water logged and saline habitats usually do not form mycorrhizae. However, AM colonization in the mangrove plants of the Great Nicobar Islands in India has been reported in the past. Among the monocots, Cyperaceae and Juncaceae often do not form mycorrhizal associations. In the dicots, Brassicaceae, Chenopodiaceae, Proteaceae, Restionaceae, Zygophylaceae, Lecythidaceae, Sapotaceae and all families of Centrospermae do not form mycorrhizae. Families rich in glucosinolates predominantly lack mycorrhizae because of the inhibitory action on fungal growth (Vierheilig et al. 2000).

Mycorrhizae form the connecting link between the biotic and geochemical portions of the ecosystem (Miller and Jastrow 1994). Mycorrhizae aid the plant in better growth by assisting it in absorbing useful nutrients from the soil, in the competition between plants and in increasing the diversity of a given area (Koide and Dickie 2002; Perotto et al. 2002). Owing to their role in nutrient cycling, mycorrhizae keep more nutrients in the biomass and, thereby increase the productivity of the ecosystem. Mycorrhizal links between seedlings and mature trees may help the seedlings in establishing themselves by providing them with the required nutrients.

AM form hyphal links between plants of different species which could be involved in the transfer of nutrients between plants. At the plant community level, AM hyphae form a network – the wood-wide web that facilitates carbon exchange between the host and the symbiont, uptake of nutrients and their movement between plants (Watkins et al. 1996; Fitter et al. 1998; Helgason et al. 1998; Sen 2000). AM are present in most soils and are generally not considered to be host-specific. However, population sizes and species composition is highly variable and influenced by plant characteristics. A number of environmental factors such as temperature, soil pH, soil moisture, P and N levels, heavy metal concentration (Boddington and Dodd 1999), the presence of other microorganisms, application of fertilizers and soil salinity (Bationo et al. 2000) may affect population diversity and size.

Mycorrhizae regulate plant communities by affecting competition, composition and succession (Kumar et al. 1999). In competition between plants, mycorrhizae in the soil favor the growth of one species and are detrimental to other competing species. AM may regulate competition between plants by making available to mycorrhizal plants, the resources that are not available to nonmycorrhizal neighbors. AM symbiosis may also increase intraspecific competition (Facelli et al. 1999). As a result, density of individuals of a single species would be reduced, thereby allowing the co-existence of individuals of different species. This would lead to an increase in species diversity.

Mycorrhizae govern species composition in communities by influencing plant fitness at the establishment phase and preventing nonmycorrhizal plants from growing in soils colonized by them. This has a selective advantage for the fungus. Maintaining a high proportion of compatible host species at the expense of noncompatible species provides the fungus with an undisturbed carbon supply (Francis and Read 1994). Owing to their role in nutrient uptake, mycorrhizae may play an important part in determining the rate and direction of the process by influencing either the outcome of succession or by affecting the composition and diversity of species (Smith and Read 1997).

The above pattern of succession seems to be true in temperate regions. In tropical countries like India, mycorrhizal plants act as pioneer species. It has been reported that mycorrhizal species like *Adhatoda vasica*, *Solanum xanthocarpum*, *Sporobolus* sp. and *Desmostachya* sp. form the pioneer vegetation in alkaline wastelands (Janardhanan et al. 1994).

The functioning of plant communities depends to a large extent on decomposition, which makes nutrient elements available to the plants. Decomposition is essentially carried out by the soil biota (bacteria, fungi, nematodes, arthropods, annelids), which breaks down the litter and organic matter of the soil (Zhu and Ehrenfeld 1996). The external mycelium of both ectomycorrhiza and AM interact with these organisms. Some soil organisms have been found to feed on AM spores. By bringing about changes in the abundance and activity of decomposers, mycorrhizal fungi are believed to hasten the process of decomposition and thereby the nutrient cycling.

An important role played by the fungal component in plant growth is the absorption of nutrients from the soil, making them available to the plants (Hooker and Black 1995; Goicoechea et al. 2000). Nitrogen, phosphorous and potassium are the important nutrient elements required by plants for their growth. AM assist in nutrient uptake by exploring the soil beyond the range of roots (Torrisi et al. 1999). Extra radical AM hyphae augment the uptake of nutrients from up to 12 cm away from the root surface (Cui and Caldwell 1996).

The network of hyphae may increase the availability of nutrients like N or P from locked sources by decomposing large organic molecules (George et al. 1995). Mycorrhizal fungi are also known to develop bridges connecting the root with the surrounding soil particles to improve both nutrient acquisitions by the plant and soil structure (Varma 1995; Hodge 2000). Unlike N_2-fixing bacteria that function as biological fertilizers, AM fungi do not add P to the soil. They only improve its availability to the plant. There is evidence that phosphatase activity is higher in the rhizosphere around AM than in nonmycorrhizal roots. P uptake is enhanced with the increase in root colonization by mycorrhizae. A system of barter operates, the colonized plant provides photosynthate to the fungus, in return, its extraradical hypha makes more P available to the host (Merryweather and Fitter 1995). Plants rely more on AM when growing in soils deficient in P (Bationo et al. 2000). Depriving a plant in its natural environment of mycorrhizae on a long-term basis can also reduce P acquisition. Plants that are nonmycorrhizal invest more in their vegetative tissues like shoots and roots. In contrast, in mycorrhizal plants, the functions of the roots are taken over by the AM hyphae, thereby permitting the host plant to invest its resources in reproductive organs.

Nitrogen occurs in the soil predominantly in the form of nitrate and ammonia, which is water-soluble and readily available for absorption. Studies with labelled N have revealed that the AM increases N uptake by plants (Bijbijen et al. 1996; Faure et al. 1998; Mädder et al. 2000). AM fungal hyphae have been credited with the uptake and transfer of large amounts of N from the soil to the host (Johansen et al. 1996; Hodge et al. 2000). However, there is little reciprocal transfer of N from the plant to the fungi, which makes uptake and assimilation of N by the symbiont essential for its growth (Bijbijen et al. 1996). Since AM form underground hyphal links between plants, N transfer between plants by means of such links is possible. Using labelled ^{15}N, Frey and Schüepp (1993) demonstrated that N flows from *Trifolium alexandrium* to *Zea mays* via AM fungal network. AM are believed to enhance N_2-fixation by symbiotic legumes by increasing root and nodule biomass, N_2-fixation rates, root N absorption rates, and plant N and P content (Olesniewicz and Thomas 1999). Mycorrhizae have also been reported to be involved in the uptake of other micro- and macro-nutrients like K, S, Mg, Zn, Cu, Ca and Na (Díaz et al. 1996; Hodge 2000).

Soil microorganisms, particularly saprophytic fungi affect the development and function of AM symbiosis. Fracchia et al. (2000) investigated the effect of the saprophytic fungus *Fusarium oxysporum* on AM colonization and plant dry matter was studied in greenhouse and field experiments using host plants, maize, sorghum, lettuce, tomato, wheat, lentil and pea and AM fungi, *Glomus mosseae*, *G. fasciculatum*, *G. intraradices*, *G. clarum* and *G.*

deserticola. The greatest plant growth and AM colonization responses in sterilized and nonsterilized soils was observed with pea, *G. deserticola* and sodium alginate pellets as carrier for *F. oxysporum* inoculum. Application of *F. oxysporum* increased shoot dry matter, N and P concentrations of pea and sorghum plants and the level of AM fungi colonization.

Piriformospora indica, a newly described axenically cultivable phytopromotional endosymbiont, which mimics the capabilities of AM fungi, was recently described by Varma et al. (1999) and Singh et al. (2000). The fungus has a broad host spectrum and inoculation with the fungus and application of culture filtrate promotes plant growth and biomass production. It mobilizes the insoluble phosphate and translocates the phosphorus to the host in an energy-dependent process. As a biological hardening agent of micropropagated plants, it renders more than 90 % survival rate for laboratory to field transferred plantlets. Regenerative protoplasts of *P. indica* have been successfully isolated, which opens up the possibility of improving symbiosis by transgenic manipulation of the fungal component in a symbiosis-specific manner.

In the ectomycorrhizal (EM) symbiosis between fungi and trees, the fungus completely ensheaths the tree roots and takes over water and mineral nutrient supply, while the plant supplies photosynthate (Wiemken and Boller 2002). N and P are the main elements limiting plant growth in terrestrial ecosystems. One of the great assets of the ectomycorrhizal symbiosis is its capability to short-circuit nutrient uptake from organic material to the symbiotic partner. In addition to mobilizing mineral nutrients from organic sources, EM fungi may also link plants to rock directly though secretion of organic acids and solubilizing nutrients from the mineral part of soil. Many EM fungi also retain considerable saprotrophic potential, for example, production of lignin-degrading enzymes, a quality that benefits the symbionts in the acquisition of nutrients from lignin-rich organic material.

Sulfur nutrition of plants is largely determined by sulfate uptake of the roots, the allocation of sulfate to the sites of sulfate reduction and assimilation, the reduction of sulfate to sulfide and its assimilation into reduced sulfur-containing amino acids and peptides and the allocation of reduced sulfur to growing tissues (Rennenberg 1999). EM colonization of oak and beech tree roots can alter the response of sulfate uptake to sulfate availability in the soil and enhance xylem transport of sulfate to the leaves. Simultaneously, sulfate reduction in the roots seems to be stimulated by EM association. These interactions between EM association and the processes involved in sulfur nutrition are required to provide sufficient amounts of reduced sulfur for increased protein synthesis that is used to enhance tree growth.

Information on the diversity of ericoid mycorrhizal endophytes in the Ericaeae and Epacridaceae has been compiled over the years by several authors (Varma and Bonfante 1994; Read 1996; Bergero et al. 2000; Berch 2001; Perotto et al. 2002). *Hymenoscyphus ericae* and *Oidiodendron* sp. appear to be the dominant fungi in the diverse assemblages of symbionts colonizing the plants. Unlike other mycorrhizal symbionts, where the fungal partner produces an extensive mycelial phase that grows from the host roots and act as an efficient nutrient collecting system, ericoid fungi produce little mycelial growth external to the root. It is now widely accepted that the major benefit conferred upon the ericaceous host plant by mycorrhizal infection is enzymatic degradation of organic nutrient sources in soil and transfer of much of the resulting products across the fungus–root interface (Cairney and Burke 1998). Ericoid mycorrhizal fungi produce a range of extracellular enzymes including cellulases, hemicellulases, ligninases, pectinases, phosphatases, proteases and polyphenol oxidases which not only have the potential to mediate utilization of organic sources of nitrogen and phosphorus in soil, but also allow them to decompose the plant cell wall, facilitating access to mineral nutrients sequestered within the walls of moribund plant cells.

Ericoid mycorrhizal fungi can interact with metals in the surrounding environment by releasing extracellular metabolites that can modify heavy metal bioavailability. Ericoid mycorrhizal symbiosis can reduce metal toxicity to the host, allowing plants to survive in soils with potentially toxic amounts of soluble and insoluble metal species. In addition to metabolites, fungi can also respond to the presence of metals with the release of specific proteins in the surrounding medium. The mechanism of arsenic tolerance in ericoid mycorrhizal fungi has been investigated by Sharples et al. (2000). Arsenic enters the cell through the phosphate transporter, causing the fungi to enhance both phosphate and arsenate uptake. Active and specific efflux mechanisms are adopted by ericoid fungi from polluted sites to decrease cellular concentrations of arsenic while retaining phosphate.

3.2 Actinorhiza

Actinorhiza is the symbiotic association between the actinomycete *Frankia* and the roots of several nonleguminous woody angiosperms. The symbiosis is established when Frankiae infect roots and lead to the development of nodules that are active in N_2 fixation. Actinorhizal plants are distributed among 24 genera of 8 angiosperm families (Verghese et al. 1998). These plants are neither related, nor do they share characters that would identify them as uniquely symbiotic. The large phylogenetic disparity in comparison to the symbiotic legumes suggests that relationship between angiosperms and *Frankia* occurred early in evolutionary time resulting in significant divergence since then.

Morphological, physiological and cytochemical criteria are employed to assign strains to the genus *Frankia* (Lechevalier 1994; Maunuksela 2001). The morphological features used for taxonomic purposes include the formation of septate, branching hyphae, production of multilocular sporangia, presence of nonmotile spores in multilocular sporangia and the production of thick-walled, lipid encapsulated structures called vesicles – the seat of nitrogen fixation. On the basis of host specificity, *Frankia* isolates have been classified into four major groups: (1) *Alnus–Myrica*; (2) *Casuarina–Myrica*; (3) *Myrica-Eleagnus*; (4) members of Elagenceae.

Actinorhizal genera have a worldwide distribution with a few exemptions. Africa, with the exception of *Myrica* species, is lacking in native actinorhiza. Actinorhizal genera can be characterized as inhabiting nutrient-poor sites in temperate regions. The *Frankia-Alnus* symbiosis is the most extensively studied actinorhiza. *Alnus*, *Casuarina* and *Elaeagnus* are the most widely distributed actinorhizal plants largely due to the introduction by man to all the continents. Although the N_2-fixing potential of Frankia-*Alnus* symbiosis may be high, the amount of nitrogen actually fixed is low because of unfavorable environmental conditions. Therefore, proper management practices that optimize efficiency of the nitrogen-fixing system are required (Dommergues 1997).

Frankia populations occur in three niches, the root nodules, the rhizosphere and the soil. In the soil, *Frankia* can be (1) a symbiont of actinorhizal plants, (2) an associate of nonhost plants or (3) a saprophyte. Although the biochemical and molecular events of the *Frankia*-actinorhizal plant symbiosis are not as well understood as the *Rhizobium-legume* symbiosis, there is a regulated series of events leading to this close association between *Frankia*, the compatible host plant and the subsequent formation of root nodules.

Frankia infection can be through (1) root hair (Casuarinaceae and Myricaceae) or (2) through intercellular spaces of the root epidermis and root cortex (Elaeagnceae and *Ceanothus*). In *Alnus*, infection is initiated via root hairs, which become branched in response to *Frankia* contact (Maunuksela 2001). The host cell produces wall-like material containing pectin, hemicellulose and encapsulates the *Frankia* hyphae within the host cells. Division of root cortical cells results in the formation of a prenodule. The actual nodule lobe originates in the pericycle and becomes infected by penetrating *Frankia* hyphae.

Actinorhizal plants are pioneer species that have the ability to colonize low-nitrogen and disturbed sites such as fires, volcanic eruptions and flooding. They facilitate succession in the sites by soil solubilization and augmenting N_2-content. A well-developed actinorhizal plant root system favors soil-binding capacity, which improves the quality of impoverished soils and strongly supports the use of these plants in land reclamation. Many actinorhizal plants are also mycorrhizal and possess the ability to absorb other nutrients. As succession progresses, non N_2-fixing plants are able to replace the original actinorhizal pioneers. *Myrica faya* growing at a volcanic site in Hawaii was able to fix 18.5 kg N/ha/year and significantly increased the amount of available N_2 in soils under the plants. Non N_2-fixing plants growing in the vicinity of *M. faya* accumulated greater biomass in comparison to plants growing at sites away from *Myrica*. This is indicative of the importance of actinorhizal plants in the ecosystem development. Actinorhizal plants are also used as intercrops for other tree species (Dommergues 1997).

3.3 Plant Growth-Promoting Rhizobacteria

The rhizosphere is the region of soil surrounding the roots that is subject to influence by the root and rhizobacteria are plant-associated bacteria that are able to colonize and persist on roots (Subba Rao 1999). Several genera of bacteria such as *Arthrobacter*, *Agrobacterium*, *Azotobacter*, *Burkholderia*, *Cellulomonas*, *Micrococcus*, *Flavobacterium*, *Mycobacterium*, *Pseudomonas* and others have been reported to be present in the rhizosphere (see chap. 12, this vol.). It has been demonstrated that the metabolic activities of bacteria associated with the rhizosphere are different from those of the nonrhizosphere soils. Electron and direct microscopy has revealed that up to 10 % of the root surface is colonized by microorganisms in a random fashion depending on the presence of soil organic matter. Some strains of plant growth-promoting rhizobacteria (PGPR) can effectively colonize plant roots and protect plants from diseases caused by a variety of root pathogens and growth promotion of plants through formation of plant growth hormones. Considerable progress has been made using molecular techniques to elucidate the important microbial factors or genetic traits involved in the PGPR-stimulated plant growth and in the suppression of fungal root diseases (Glick and Bashan 1997; Kumari and Srivastava 1999; Bloemberg and Lugtenberg 2001; Zehnder et al. 2001). Several genera of allelopathic nonpathogenic bacteria have been identified and characterized which produce plant growth-inhibiting allelochemicals (Barazani and Friedman 2001). Allelochemicals like phytoxins, geldanamycin, nigericin and hydanthocidin have been isolated from *Streptomyces viridochromogenes*.

PGPR can affect plant growth either directly or indirectly. The direct effect of PGPR include providing the host plants with fixed nitrogen, P and Fe solubilized from the soil and phytohormones that are synthesized by the bacteria (Glick 1995). The indirect effect on plant growth occurs when PGPR reduces or prevents the harmful effects of one or more phytopathogenic organisms. PGPR effective in biocontrol produce a variety of substances including antibiotics, siderophores and a variety of enzymes (chitinase, protease, lipase, β-1,3-glucanase etc.) to limit the damage to plants by phytopathogens. PGPR have also been reported to reduce heavy metal toxicity in plants (Burd et al. 2000).

Symbiotic nitrogen fixation has long been considered to be an excellent replacement of N fertilization. The most efficient nitrogen fixers are strains of *Rhizobium*, *Sinorhizobium*, *Mesorhizobium*, *Bradirhizobium* and *Azorhizobium*, which form a host-specific symbiosis with leguminous plants (Paul and

Clark 1998; Subba Rao 1999). The genes involved in nitrogen fixation, nitrogen assimilation and regulation in various bacteria have been studied extensively (Glick and Bashan 1997; Bloemberg and Lugtenberg 2001; Rengel 2002). Several of the *nif* and *fix* genes, involved in N_2-fixation, have been characterized in different nitrogen fixers. Most of the organism contains similar nitrogenase complexes. Increased efficacy of N_2-fixation can be achieved by selecting and manipulating the best combination of host genotype and bacteria. Improvement in the symbiotic relationship in suboptimal environmental situations related to soil-borne or environmental stress is also important to improve N_2-fixation.

Free-living N_2-fixing rhizobacteria are capable of fixing atmospheric nitrogen. The aerobic, free-living bacteria that utilize organic substrates as a source of energy include *Azotobacter*, found in neutral and alkaline soils. Members of the same family *Beijerinckia* and *Derxia* have a broader pH range and are more often found in acidic soils in the tropics. *Azospirillum*, *Acetobacter*, *Herbaspirillum* and *Azoarcus* have frequently been found associated with grasses (Steenhoudt and Vanderleyden 2000). *Azotobacter* and *Beijerinckia* require aerobic conditions for the production of energy required for N_2 fixation. However, in these organisms as well as other diazotrophs, the activity of nitrogenase is inhibited by O_2 and special mechanisms for the protection of nitrogenase are present. Facultative microaerophilic organisms such as *Azospirillum*, *Klebsiella* and *Bacillus* produce energy in the form of ATP by oxidative pathways in an environment where nitrogenase does not need to be as well protected from O_2. The amount of N_2 fixed by free-living diazotrophs such as *Azotobacter* and *Pseudomonas* is generally a few kilograms per hectare (Paul and Clark 1998). Nitrogen-fixing microorganisms in the waterlogged rice fields may contribute 40–50 kg per hectare which is a cumulative effect of free-living as well as symbiotic organisms such as blue-green algae, *Azotobacter*, *Azospirillum*, *Rhizobium*, *Beizerinckia*, *Clostridium*, *Desulfovibrio* and *Pseudomonas* (Subba Rao 1999).

Soil amendments and artificial inoculation of beneficial rhizobacteria can induce changes in rhizosphere microflora (Bashan 1998; Bai et al. 2002). Rhizosphere nitrogen fixation could be enhanced by incorporation of N_2-fixing capacity into common rhizosphere. The large scale application of PGPR in agriculture is attractive as it substantially reduces the use of chemical fertilizers and pesticides. A growing number of PGPR are being marketed, and at present, biofertilizer application accounts for approximately 65% of the N supply to crops worldwide (Bloemberg and Lugtenberg 2001). Integrated approaches have been applied with a combination of AM fungi or biocontrol fungi like *Trichoderma* and PGPR for the beneficial plant growth and disease control effects (Valdenegro et al. 2001; Elliot and Broschat 2002). Recently focus has also been directed towards the development and use of rhizobacteria as biocontrol agents to combat fungal diseases (Naseby et al. 2001; Unge and Jansson 2001).

3.4 Phosphate-Solubilizing Microorganisms

After nitrogen, phosphorus is the major plant growth-limiting nutrient, though P is abundant in soils in both inorganic and organic forms. Most of the mineral nutrients in soil solution are present in millimolar amounts, however, P is present only in micromolar (up to 10 µm) amounts. Low level of availability of P is due to high reactivity of soluble P with Ca, Fe and Al (Gyaneshwar et al. 2002). Calcium phosphates are the predominant form of P in calcareous soils, whereas inorganic P in acidic soil is associated with Fe and Al compounds. In soils with high organic matter, organic P may make up as much as 50 % of the total soluble P available in soil. Phosphate-solubilizing microorganisms (PSM) are ubiquitous in soils and play an important role in supplying P to plants in a sustainable manner. Although a lot of laboratory work on phosphate solubilization has been done, the results of field trials were highly variable (Nahas 1996).

In spite of the importance of PSM in agriculture, the detailed biochemical and molecular mechanisms of P solubilization is not known. Mineral P solubilizing ability of microbes could be linked to specific genes which may be present in even non P-solubilizing bacteria (Goldstein 1995). The ability to solubilize the mineral–phosphate complexes has been attributed to the ability of PSM to reduce the pH of the surroundings by releasing organic acids such as acetate, lactate, oxalate, tartarate, succinate, citrate, gluconate etc. (Kim et al. 1998; Ezawa et al. 2002). These organic acids can either dissolve the mineral phosphate as a result of anion exchange or can chelate Fe or Al ions associated with the phosphate. However, acidification does not seem to be the only mechanism of P solubilization, as the ability to reduce pH in some cases does not correlate with the ability to solubilize mineral phosphates (Jones 1998; Gyaneshwar et al. 2002).

Plants have been shown to benefit from the association with microorganisms under P-deficient conditions, either resulting from a better uptake of the available P or by accession of the nonavailable form of P-source. Various kinds of bacteria and fungi have been isolated and characterized for their ability to solubilize mineral phosphate complexes. Although P-solubilizing bacteria outnumber P-solubilizing fungi in soil, fungal isolates generally exhibit greater P-solubilizing ability than bacteria in both liquid and solid media (Goldstein 1995; Gyaneshwar et al. 2002). Phosphate-solubilizing strains of bacteria *Enterobacter agglomerans* (Kim et al. 1998) and *Azotobacter chroococcum* (Kumar and Narula 1999) have been isolated from wheat rhizosphere and characterized for solubilization of hydroxyapetite, tricalcium phosphate and Mussoorie rock phosphate in laboratory experiments. Nautiyal et al. (2000) described the isolation and characterization of four unidentified bacterial strains from the chickpea rhizosphere in alkaline soil. NBRI 2601 was the most efficient strain in terms of its capability to solubilize phosphorus in the presence of 10 % salt, pH 12 and 45 °C. Seed inoculation with an acid-tolerant strain of *Bacillus* sp. significantly increased the vegetative and grain yield of fingermillet, maize, amaranth, buckwheat and french bean (Pal 1998). Although plants inoculated with PSM exhibit increased growth and P contents in laboratory studies, large variations have been found in the effectiveness of inoculations in field conditions.

Phosphate solubilizing fungi and their role in plant nutrition and growth have been extensively studied. Among the known fungal genera are *Aspergillus* (Goenadi et al. 2000; Narsian and Patel 2000), *Penicillium* (Whitelaw et al. 1999; Reyes et al. 2001), *Rhizoctonia* (Jacobs et al. 2002) and *Cyathus* (Singal et al. 1991). Supplementation of *A. niger* cultivated on sugar beet waste material to soil significantly improved the growth rate and shoot P concentration of *Trifolium repens* (Vassilev et al. 1996). Reddy et al. (2002) reported the biosolubilization of different rock phosphates by three isolates of *A. tubingensis* for the first time. Altomare et al. (1999) investigated the capability of biocontrol fungus *Trichoderma harzianum* to solubilize MnO_2, metallic zinc and rock phosphate and discussed its possible role in plant growth. Application of encapsulated fungal or bacterial cell systems for effective use as soil microbial inoculants in P solubilization and plant nutrition has been discussed in detail by Vassilev et al. (2001).

Nodule formation in legumes is often limited by the availability of P (Subba Rao 1999). While there are only a few reports on P solubilization by *Rhizobium* (Chabot et al. 1996), the improvement in the efficiency of N_2-fixation in legumes has been demonstrated by supplementation of P in alfalfa, clover, cow pea and pigeon pea (Al-Niemi et al. 1997). In chickpea and barley growing in soils treated with insoluble phosphate and inoculated with *Mesorhizobium mediterraneum*, the P content increased by 100 and 125 %, respectively (Peix et al. 2001). The dry matter, N, K, Ca and Mg contents in both plants also increased significantly. A coculture inoculum of *Rhizobium meliloti* and a phosphate-solubilizing fungus, *Penicilium bilalii* increased the P uptake of several field crops (Rice et al. 1995). Co-inoculations of AM fungi with PSM have shown positive effects on plant growth and crop yield (Toro et al. 1997; Ezawa et al. 2002). Beneficial effects of enriching vermicompost by nitrogen-fixing and phosphate-solubilizing bacteria have also been demonstrated (Kumar and Singh 2001).

3.5 Lignocellulolytic Microorganisms

The high cellulose and lignin contents of plant residue incorporated into soil emphasize the importance of lignocellulolytic microorganisms in the mineralization processes in soil (Kuzyakov and Domanski 2000). The chemical composition of the entire plant residues, their decomposition and biochemical transformations in the soil during humification has been investigated in detail (Paul and Clark 1998). The importance of microbial biomass and extracellular lignocellulolytic enzyme activity in the assessment of soil quality is established by the essential role of soil microbes in nutrient cycling within agricultural ecosystems (Christensen and Johnston 1997). During the microbial degradation and humification of plant residues, about 80 % of the residual carbon is released to the atmosphere as CO_2 (Omar 1994). The amendment of infertile or saline soils with plant residues and their subsequent degradation by cellulolytic soil microflora with a concomitant increase in CO_2 could increase soil aeration, improve its structure and also increase soil fertility. The activities of cellulolytic microbes affect the availability of energy and specific nutrients to a group of organisms deficient in hydrolytic enzyme activities (Jensen and Nybroe 1999).

Soils managed with organic inputs generally have larger and more active microbial populations than those managed with mineral fertilizers (Badr El-Din et al. 2000). Reincorporation of organic matter into the soil improves soil fertility, enhances microbial growth and buffers the soil environment from sudden changes. There are many types of agroindustrial organic refuse which can be transformed and applied to soil as crop amendments, such as compost, thus reducing the need for chemical fertilizers. During the composting process, the organic substrate present in the agricultural wastes is mainly transformed oxidatively into a stabilized organic matter. The slow transformation of lignocellulosic material results in the formation of humic substances. Several researchers have established a positive correlation between the amount of humic substances and promotion of plant growth. Application of different combinations of coir with peat and vermiculate significantly increased the growth of tomato transplants with respect to root dry weight, stem diameter and leaf area (Arenas et al. 2002).

Straw incorporation could also be beneficial in enhancing symbiotic nitrogen fixation and crop growth (Abd-Alla and Omar 1998). In nonsymbiotic nitrogen fixation studies in the laboratory and in the field, a significant increase in nitrogenase activity associated with the breakdown of straw after inoculation with various combinations of cellulolytic fungi and bacteria has been reported (Halsall and Gibson 1991; Chapman et al. 1992). Application of wheat straw with cellulolytic fungi, *Trichoderma harzianum* significantly enhanced growth, nodulation, nodule efficiency and increased the concentration of Ca, Mg and K in the shoots and roots of fenugreek plants grown in saline soil (Abd-Alla and Omar 1998). The increase in dry matter production and nitrogen content was due to improved N_2 fixation reflected by enhanced formation and growth of nodules as well as nitrogenase activity.

Inoculation of straw with lignocellulolytic organisms offers potential for manipulating and improving the composting of cellulosic waste (Verstraete and Top 1999; Hart et al. 2002). Composts produced using this method provide a more sustainable approach to agriculture, enabling subsistence farmers to utilize their agricultural waste products as a means to improve soil quality. Saprophytic lignin-decomposing basidiomycetes isolated from plant litter were found to play an important role in soil aggregation and stabilization (Caesar-Ton That and Cochran 2000). The basidiomycete produced large quantities of extracellular water-insoluble and heat-resistant materials that bind soil particles into aggregates.

Differences in the chemical properties of the organic matter from highly lignocellulosic compost after incubation with two lignocellulolytic microorganisms were studied by Requena et al. (1996). Inoculation with *Trichoderma viride* and *Bacillus* sp. enhanced degradation processes and the degree of organic matter humification. Both degradation-humification pathways beneficially affected the lettuce growth demonstrating that inoculation with lignocellulolytic microbes may be a useful tool to improve agronomic properties of lignocellulosic wastes by modifying the chemical structure and properties of their organic matter. Rajbanshi et al. (1998) found significant positive effects of seeding material (substrates with a high number of degradative microbes) on total organic carbon and organic matter contents of grass straw-leaf mix.

Temporal changes in soil moisture, soil temperature, and carbon input from crop roots, rhizosphere products (root exudate, mucilage, sloughed cells etc.), and crop residues can have a large effect on soil microbial activity (Jensen et al. 1997; Ritz et al. 1997). Crop growth often stimulates an increase in the size of microbial biomass during the growing season and after harvest. Enzyme activity displays different temporal patterns of the various soil enzymes. Some cellulases are closely related to inputs of fresh organic materials, plant growth and plant residues, while others appear to be more sensitive to soil temperature and moisture.

Due to their dynamic nature, soil microbial biomass and soil enzymes respond quickly to changes in organic matter input. In a field experiment after 8 years of cultivation with low- or high-organic matter input, pronounced and constant increase in endocellulase and β-glucosidase activities and variable increase in microbial biomass carbon and cellobiohydrolase activity was observed over the sampling period (Debosz et al. 1999). Temporal variations in endocellulase activity showed a different pattern from those for β-glucosidase activity, with highest activity in the autumn/winter and early summer samplings. On all sampling dates, endocellulase activity in the higher organic matter was about 30 % higher than in the low organic matter treatments.

Specific organic amendments such as mulched straw has been reported to influence soil suppression of plant diseases (Knudsen et al. 1999). Many fungi, known as antagonists to plant pathogens, e.g., *Trichoderma* sp., produce a wide range of cellulolytic enzymes which are believed to be associated with their antagonistic abilities. Rasmussen et al. (2002) investigated the relationship between soil cellulolytic activity and suppression of seedling blight of barley caused by *Fusarium culmorum* in arable soils. A bioassay for disease suppression in test soils indicated that the samples from 6 to 13-cm depth exhibited positive correlation between soil suppressiveness and the activities of β-glucosidase and cellobiohydrolase, where soil representing the highest disease suppression had the highest activity. Furthermore, soil suppressiveness, as well as the enzyme activity significantly correlated with the soil content of total C and N.

4 Plant Growth Promoting Substances Produced by Soil Microbes

The ability of soil microorganisms to produce various metabolites stimulating plant growth is considered to be one of the most important factors in soil fertility (Frankenberger and Arshad 1995; Paul and Clark 1998; Subba Rao 1999). Some PGPR control the damage to plants from plant pathogens by a number of different mechanisms including physical displacement and outcompeting the phytopathogen, secretion of siderophores to prevent pathogens in the immediate vicinity from proliferating, production of enzymes, antibiotics and a variety of small molecules that inhibit the phytopathogen and stimulation of systemic resistance in plants (Glick and Bashan 1997). Microbially produced antibiotics have a potential role in indirectly promoting plant growth by controlling plant diseases (Kumari and Srivastava 1999). Two prominent antifungal antibiotics are griseofulvin, a metabolic product of *Penicillium griseofulvum* and aureofungin, a metabolic product of *Streptoverticillium cinnamomeum*.

Soil microorganisms produce a variety of phytohormones such as auxins, gibberellins, cytokinins, ethylene and abscisic acid. Auxin production is widespread among many soil and rhizosphere microorganisms (fungi, bacteria and actinomycetes) and algae (Martens and Frankenberger 1993). Tryptophan is considered the physiological precursor of auxin for both plant and soil microbes. A number of indole compounds and phenylacetic derivatives have been reported with auxin activity. Indole-3-acetic acid (IAA) is considered the most physiologically active auxin in plants. Auxins are known to affect cell enlargement involving cell wall extensibility. Plant growth responses also include root and shoot dry weights, root/stem elongation and root/shoot ratios. Species of *Agrobacterium*, *Azospirillum*, *Pseudomonas*, *Rhizobium*, *Ustilago*, *Gymnosporangium*, *Rhizopus* and *Synchytrium* produce IAA in pure cultures or in association with higher plants (Subba Rao 1999).

Gibberellins (GA) are an important group of plant hormones that are diterpenoid acids. The involvement of GA in almost all phases of plant growth and development, starting from germination to senescence is well known. However, the most prominent physiological effect of GA is in shoot elongation. Some other plant growth related functions of GA include overcoming dormancy and dwarfism in plants, inducing flowering in some photoperiodically sensitive and other low temperature-dependent plants, and contributing to fruit setting. Several soil microbes are known to produce gibberellins or gibberellin-like substances (Kumar and Lonsane 1989; Steenhoudt and Vanderleyden 2000). The common bacterial genera are *Arthrobacter*, *Azotobacter*, *Azospirillum*, *Pseudomonas*, *Rhizobium*, *Bacillus*, *Brevibacterium* and *Flavobacterium*. *Actinomyces* and *Nocardia* are the important actinomycetes and *Fusarium*, *Gibberella*, *Aletrnaria*, *Aspergillus*, *Penicillium* and *Rhizopus* are known fungi.

Cytokinins, N^6-substituted aminopurines, regulate cell division and differentiation in certain plant tissues. Cytokinins play an important role in nodule development and formation. Along with auxins, cytokinins stimulate mature root cells to undergo polyploid mitosis. Symbiotic N_2-fixing bacteria, *Rhizobium*, free-living N_2-fixing bacteria *Azospirillum* and *Azotobacter*, and mycorrhizal fungus, *Rhizopogon roseolus* are known to produce cytokinins in the rhizosphere along with other growth-promoting substances (Nieto and Frankenberger 1989). Other bacteria that produce cytokinins or cytokinin-like substances include *Agrobacterium*, *Bacillus*, *Paenibacillus* and *Pseudomonas* (Timmusk et al. 1999).

Ethylene (C_2H_4) is the only phytohormone that is a gas under physiological temperature and pressure. Ethylene is considered to be a promoter of senescence and an inhibitor of growth and elongation. It can also promote flowering, fruit ripening and stimulate cell elongation in certain plants (Elsgaard 2001). Bacterial species of *Aeromonas*, *Citrobacter*, *Arthrobacter*, *Erwinia*, *Serratia*, *Klebsiela and Streptomyces*, and fungal species of *Acremonium*, *Alternaria*, *Mucor*, *Fusarium*, *Pythium*, *Neurospora* and *Candida* are capable of producing ethylene (Subba Rao 1999).

Abscisic acid (ABA) is generally involved in deceleration or cessation of plant growth. ABA is active in regulating abscission of young leaves and fruits, dormancy of buds and seeds, and ripening of fruit. ABA production in two bacterial species, *Azospirillum brasilense* and *Rhizobium* spp. and several phytopathogenic fungi such as *Cercospora*, *Fusarium*, *Cladsporium*, *Monilia*, *Pestatoria* and *Verticillium* has been demonstrated (Frankenberger and Arshad 1995; Paul and Clark 1998).

Siderophores are low molecular weight (<1000 D), virtually Fe (III)-specific ligands, produced by microorganisms to combat low iron stress (Glick and Bashan 1997). The function of siderophores in microbial nutrition is to solubilize the insoluble iron in the external environment and transport it into the cell. The siderophores are produced as free ligands that become complexed with iron in the ambient medium to form a ferric complex (ferrated siderophore) which is then transported into the cell via specific transport receptors. Inside the cell, the siderophore is freed from the transport receptor protein and again released outside as free ligand (desferriform), to repeat the cycle. Siderophore production is a major strategy used by fungi for iron-acquisition (Suneja and Lakshaminarayan 1999). Selective utilization of siderophores in mixed microbial populations is a key mechanism for preventing the emergence of unwanted competing microorganisms. Besides their role in solubility and transport of iron, hydroxamate siderophores are also involved in iron storage.

Based on the chemical nature of their coordination sites, microbial siderophores are classified as hydroxamates, catecholates, carboxylates and mixed type. Hydroxamates are produced both by bacteria and fungi. In most fungi, a mixture of siderophores is produced which varies depending on cultivation conditions. Aspergilli produce ferricrocin accompanied by fusarinines, while certain penicillia produce ferrichrome accompanied by coprogen. Similar observations have been made with *Neurospora*, *Gliocladium*, *Trichoderma* and *Agaricus bisporus* (Neilands and Leong 1986). Fusarinines (fusigens) produced by species of *Fusarium* and *Penicillium* are linear and cyclic hydroxamic acids joined by ester bonds. Ericoid mycorrhizal fungi produce ferrichrome and fusarinines.

Varieties of bacterial hydroxamates are known. Ferrioxamine, produced by actinomycetes, *Nocardia* and *Pseudomonas stutzeri*, is a cyclic trihydroxamate. Citrate hydroxamates are characterized by the presence of two hydroxamates and one citrate group as ligand, it is a linear citratehydroxamic acid obtained from *Klebsiella pneumonia* and several enteric bacteria. Catecholate siderophores are generally less diverse than the hydroxamates, and are conjugated to amino acids or polyamine backbones. Species of *Bacillus*, *Aeromaonas* and *Erwinia* are known to produce catecholate siderophores. Carboxylate (complexone) siderophores are produced by *Rhizopus microsporus*, *Rhizobium meliloti* and *Staphylococcus hycius*. Pyoverdines, the mixed types form a wide class of mixed siderophores showing a great variety of structures. Some strains of fluorescent pseudomonads produce hydroxamate siderophores (ferribactin) in addition to pyoverdine siderophores.

The plant growth-promoting rhizobacteria (PGPR) owe their plant growth promoting activity to their stronger siderophores with higher stability constants that outgrow the other bacterial population in competition for iron and finally displace them from the root surface. The siderophore-producing PGPR have become important in the biological control of plant pathogens (Glick and Bashan 1997).

5 Conclusions

Microorganisms play an essential role in the functioning and sustainability of soil ecosystems including biogeochemical cycling of nutrients and biodegradation. Recent advances in soil community analysis using molecular and biochemical approaches have helped us understand the enormous microbial diversity and their functional significance in nutrient recycling in soil and plant development. Soil diversity exceeds that of aquatic environments and provides a great resource for the biological exploitation of novel organisms, processes and products. Microbes isolated from soil and developed as biofertilizers or inoculants play an important role in enhancing plant growth enhancing efficiency of biological nitrogen fixation, availability of P, trace elements such as Fe and Zn, and production of plant growth substances. The development of better screening procedures and understanding the genetic basis of rhizosphere competence will help in developing novel microbial inoculants that will be better suited to survive and perform their desirable function in a natural environment. As we explore the soil microbial diversity more, we must remember that the microbes evolve more quickly than we can study them, providing an ever-increasing diversity of function, not only in agriculture, but also for industrial applications.

Acknowledgements. The authors thank Mr. Manoj Kumar for the preparation of the manuscript.

References and Selected Reading

Abd-Alla MH, Omar SA (1998) Wheat straw and cellulolytic fungi application increases nodulation, nodule efficiency and growth of fenugreek (*Trigonella foenum-graceum* L.) grown in saline soil. Biol Fertil Soil 26:58–65

Al-Niemi TS, Kahn ML, McDermott TR (1997) P metabolism in the *Rhizobium tropici*-bean symbiosis. Plant Physiol 113:1233–1242

Altomare C, Norvell WA, Bjoerkman T, Harman GE (1999) Solubilization of phosphates and micronutrients by the plant growth-promoting and biocontrol fungus *Trichoderma harzianum* Rifai 1295–22. Appl Environ Microbiol 65:2926–2933

Arenas M, Vavrina CS, Cornell JA, Hanlon EA, Hochmuth GJ (2002) Coir as an alternative to peat in media for tomato transplant production. Hort Science 37:309–312

Badr El-Din SMS, Attia M, Abo-Sedera SA (2000) Field assessment of composts produced by highly effective cellulolytic microorganisms. Biol Fertil Soils 32:35–40

Bagyaraj DJ (1984) Biological interactions with VA mycorrhizal fungi. In: Powell CL, Bagyaraj DJ (eds) VA Mycorrhiza. CRC Press, Boca Raton, pp 131–153

Bagyaraj DJ, Varma A (1995) Interaction between arbuscular mycorrhizal fungi and plants. Their importance in sustainable agriculture in arid and semiarid tropics. Adv Microb Ecol 14:119–142

Bai Y, D'Aoust F, Smith DL, Driscoll BT (2002) Isolation of plant growth promoting *Bacillus* strains from soybean root nodules. Can J Microbiol 48:230–238

Barazani O, Friedman J (2001) Allelopathic bacteria and their impact on higher plants. Crit Rev Microbiol 27:41–55

Bashan Y (1998) Inoculants of plant growth-promoting bacteria for use in agriculture. Biotechnol Adv 16:729–770

Bationo A, Wani SP, Bielders CL, Vlek PLG, Mokwunye AU (2000) Crop residue and fertilizer management to improve soil organic carbon content, soil quality and productivity in the desert margins of West Africa. In: Lal R, Kimble JM, Steward BA (eds) Advances in soil science. Global climate change and tropical ecosystems. CRC Press, Washington, DC, pp 117–145

Bending GD, Turner MK, Jones JE (2002) Interactions between crop residue and soil organic matter quality and the functional diversity of soil microbial communities. Soil Biol Biochem 34:1073–1082

Berch SM (2001) Molecular diversity and phylogeny of ericoid mycorrhizal fungi. Plant Soil 244:55–66

Berch SM, Allen TR, Berbee ML (2002) Molecular detection, community structure and phylogeny of ericoid mycorrhizal fungi. Plant Soil 244:55–66

Bergero R, Perotto S, Girlanda M, Vidano G, Luppi AM (2000) Ericoid mycorrhizal fungi are common root associates of a Mediterranean ectomycorrhizal plant *Quercus ilex*. Mol Ecol 9:1639–1650

Bijbijen JN, Urquiaga S, Ismaili M, Alves BJR, Boddey RM (1996) Effect of arbuscular mycorrhizae on uptake of nitrogen by *Brachiaria arrecta* and *Sorghum vulgare* from soils labelled for several years with 15 N. New Phytol 133:487–494

Bloemberg GV, Lugtenberg BJJ (2001) Molecular basis of plant growth promotion and biocontrol by rhizobacteria. Curr Opin Plant Biol 4:343–350

Boddington CL, Dodd JC (1999) Evidence that differences in phosphate metabolism in mycorrhizae formed by species of *Glomus* and *Gigaspora* might be related to their life cycle strategies. New Phytol 142:531–538

Bridge P, Spooner B (2001) Soil fungi: diversity and detection. Plant Soil 232:147–154

Burd GI, Dixon DG, Glick BR (2000) Plant growth-promoting bacteria that decrease heavy metal toxicity in plants. Can J Microbiol 46:237–245

Caesar-TonThat TC, Cochran VL (2000) Soil aggregate stabilization by a saprophytic lignin-decomposing basidiomycete fungus. I. Microbiological aspects. Biol Fertil Soil 32:374–380

Cairney JWG, Burke RM (1998) Extracellular enzyme activities of the ericoid mycorrhizal endophyte *Hymenoscyphus ericae* (Read) Korf and Kernen: their likely roles in decomposition of dead plant tissue in soil. Plant Soil 205:181–192

Chabot R, Antoun H, Cesas MP (1996) Growth promotion of maize and lettuce by phosphate solubilizing *Rhizobium leguminosarum* biovar *phaseoli*. Plant Soil 184:311–321

Chapman SJ, Veal DA, Lynch JM (1992) Effect of oxygen concentration on dinitrogen fixation and volatile fatty acid production by *Clostridium butyricum* growing in association with fungi on cellulose and on wheat straw. J Appl Bacteriol 72:9–15

Christensen BT, Johnston AE (1997) Soil organic matter and soil quality lessons learned from long-term experiments at Askov and Rothamsted. In: Gregorich EG, Carter MR (eds) Soil quality for crop production and ecosystem health. Elsevier, Amsterdam

Cui M, Caldwell MM (1996) Facilitation of plant phosphate acquisition by arbuscular mycorrhizae from enriched to soil patches. I. Roots and hyphae exploiting the same volume. New Phytol 133:453–460

Dahllöf I (2002) Molecular community analysis of microbial diversity. Curr Opin Biotechnol 13:213–217

Debosz K, Rasmussen PH, Pedersen AR (1999) Temporal variations in microbial biomass C and cellulolytic enzyme activity in arable soils: effects of organic matter input. Appl Soil Ecol 13:209–218

De Fede KL, Pannaccione DG, Sexstone AJ (2001) Characterisation of dilution enrichment cultures obtained from size-fractionated soil bacteria by BIOLOG (R) community level physiological profiles and restriction analysis of 16S rRNA genes. Soil Biol Biochem 33:1555–1562

Degens BP (1998) Microbial functional diversity can be influenced by the addition of simple organic substrates to soil. Soil Biol Biochem 30:1981–1988

Degens BP, Scipper LA. Sparling GP, Duncan LC (2001) Is the microbial community in a soil with reduced catabolic diversity less resistant to stress or disturbance? Soil Biol Biochem 33:1143–1153

Diaz G, Azcon-Aguilar C, Honrubia M (1996) Influence of arbuscular mycorrhizae on heavy metal (Zn and Pb) uptake and growth of *Lygeum spartum* and *Anthyllis cytisoides*. Plant Soil 180:241–249

Dommergues YR (1997) Contribution of actinorhizal plants to tropical soil productivity and rehabilitation. Soil Biol Biochem 29:931–941

Elliott ML, Broschat TK (2002) Effects of a microbial inoculant on plant growth and rhizosphere bacterial populations of container-grown plants. Hort Technol 12:222–225

Elsgaard L (2001) Ethylene turnover in soil, litter and sediment. Soil Biol Biochem 33:249–252

Ezawa T, Smith SE, Smith FA (2002) P metabolism and transport in AM fungi. Plant Soil 244:221–230

Facelli E, Facelli JM, Smith SE, Mclaughlin MJ (1999) Interactive effects of arbuscular mycorrhizal symbiosis intraspecific competition and resource availability on *Trifolium subterraneum* cv. Mt. Barker. New Phytol 141:535–547

Faure S, Cliquet J-B, Thephany G, Boucaud J (1998) Nitrogen assimilation in *Lolium perenne* colonized by the arbuscular mycorrhizal fungus *Glomus fasciculatum*. New Phytol 138:411–417

Fitter AH, Graves JD, Watkins NK, Robinson D, Scrimgeour C (1998) Carbon transfer between plants and its control in networks of arbuscular mycorrhizas. Funct Ecol 12:406–412

Fortin JA, Becard G, Declerck S, Dalpe Y, St-Arnaud M, Coughlan AP, Piche Y (2002) Arbuscular mycorrhiza on root-organ cultures. Can J Bot 80:1–20

Fracchia S, Garcia-Romera I, Godeas A, Ocampo JA (2000) Effect of the saprophytic fungus *Fusarium oxysporum* on arbuscular mycorrhizal colonization and growth of plants in greenhouse and field trials. Plant Soil 223:175–184

Francis R, Read DJ (1994) The contributions of mycorrhizal fungi to the determination of plant community structure. Plant Soil 159:11–25

Frankenberger Jr WT, Arshad M (1995) Phytohormones in soils. Microbial production and function. Marcel Dekker, New York

Frey B, Schüepp H (1993) A role of vesicular arbuscular (VA) mycorrhizal fungi in facilitating inter-plant N transfer. Soil Biol Biochem 25:651–658

Glick BR (1995) The enhancement of plant growth by free-living bacteria. Can J Microbiol 41:109–117

Goenadi D, Siswanto H, Sugiarto Y (2000) Bioactivation of poorly soluble phosphate rocks with a phosphorus-solubilizing fungus. Soil Sci Soc Am J 64:927–932

Goldstein AH (1995) Recent progress understanding the molecular genetics and biochemistry of calcium phosphate solubilization by gram negative bacteria. Biol Agric Hort 12:185–193

George E, Marschner H, Jakobsen I (1995) Role for arbuscular mycorrhizal fungi in uptake of phosphorous and nitrogen from soil. Crit Rev Biotechnol 15:257–270

Glick BR, Bashan Y (1997) Genetic manipulation of plant growth-promoting bacteria to enhance biocontrol of phytopathogens. Biotechnol Adv 15:353–378

Goicoechea N, Antolin MC, Sanchez-Diaz M (2000) The role of plant size and nutrient concentrations in associations between *Medicago*, and *Rhizobium* and/or *Glomus*. Biol Plant 43:221–226

Gyaneshwar P, Naresh Kumar G, Parekh LJ, Poole PS (2002) Role of soil microorganisms in improving P nutrition of plants. Plant Soil 245:83–93

Halsall DM, Gibson AH (1991) Nitrogenase activity (C_2H_2 reduction) in straw-amended wheat belt soils in response to diazotroph inoculation. Soil Biol Biochem 23:987–998

Hart TD, De Leij FAAM, Kinsey G, Kinsey G, Kelley J, Lynch JM (2002) Strategies for the isolation of cellulolytic fungi for composting of wheat straw. World J Microbiol Biotechnol 18:471–480

Hattori T, Mitsui H, Haga H, Wakao N, Shikano S, Gorlach K, Kasahara Y, El-Beltagy A, Hattori R (1997) Advances in soil microbial ecology and the biodiversity Ant van Leeuwen Hoek 72:21–28

Hawksworth DL (1997) The fascination of fungi: exploring fungal diversity. Mycologist 11:18–22
Helgason T, Daniell TJ, Husband R, Fitter AH, Young JPW (1998) Ploughing up the wood-wide web. Nature 394:431
Hodge A (2000) Microbial ecology of arbuscular mycorrhiza. FEMS Microbiol Ecol 32:91–96
Hodge A, Robinson D, Fitter AH (2000) An arbuscular mycorrhizal inoculum enhances root proliferation in, but not nitrogen capture from nutrient rich patches in soil. New Phytol 145:575–584
Hooker JE, Black KE (1995) Arbuscular mycorrhizal fungi as a component of sustainable soil plant systems. Crit Rev Biotechnol 15:201–212
Hunter-Cevera JC (1998) The value of microbial diversity. Curr Opin Microbiol 1:278–285
Jacobs H, Boswell GP, Ritz K, Davidson FA, Gadd GM (2002) Solubilization of calcium phosphate as a consequence of carbon translocation by *Rhizobium solani*. FEMS Microbiol Ecol 40:65–71
Janardhanan KK, Abdul-Khaliq Naushin F, Ramaswamy K (1994) Vesicular arbusicular mycorrhiza in an alkaline usar land ecosystem. Curr Sci 67:465–469
Jenesen SL, Mueller T, Magid J, Nielsen NE (1997) Temporal variations of C and N mineralization, microbial biomass and extractable organic pools in soil after oilseed rape straw incorporation in the field. Soil Biol Biochem 29:1043–1055
Jensen LE, Nybroe O (1999) Nitrogen availability to *Pseudomonas fluorescens* DF57 is limited during decomposition of barley straw in bulk soil and in barley rhizosphere. Appl Environ Microbiol 65:4320–4328
Johansen A, Finaly RD, Olsson PA (1996) Nitrogen metabolism of external hyphae of the arbuscular mycorrhizal fungus *Glomus intaradices*. New Phytol 133:705–712
Jones DL (1998) Organic acids in the rhizosphere – a critical review. Plant Soil 205:25–44
Kim KY, Jordan D, McDonald GA (1998) *Enterobacter agglomerans*, phosphate solubilizing bacteria and microbial activity in soil: effect of carbon sources. Soil Biol Biochem 30:995–1003
Knudsen IMB, Debosz K, Hockenhull J, Jensen DF, Emholt S (1999) Suppressiveness of organically and conventionally managed soils towards brown foot rot of barley. Appl Soil Ecol 12:61–72
Koide RT, Dickie IA (2002) Effects of mycorrhizal fungi on plant populations. Plant Soil 244:307–317
Kozdroj J, van Elsas JD (2001) Structural diversity of microorganisms in chemically perturbed soil assessed by molecular and cytochemical approaches. J Microbiol Meth 43:197–212
Kumar PKR, Lonsane BK (1989) Microbial production of gibberellins: state of the art. Adv Appl Microbiol 34:29–139
Kumar V, Narula N (1999) Solubilization of inorganic phosphates and growth emergence of wheat as affected by *Azotobacter chroococcum* mutants. Biol Fertil Soils 28:301–305
Kumar A, Nivedita P, Upadhyaya RS (1999) VA mycorrhizae and revegetation of coal mine spoils: a review. Trop Ecol 40:1–10
Kumar V, Singh KP (2001) Enriching vermicompost by nitrogen fixing and phosphate solubilizing bacteria. Biores Technol 76:173–175
Kumari V, Srivastava JS (1999) Molecular and biochemical aspects of rhizobacterial ecology with emphasis on biological control. World J Microbiol Biotechnol 15:535–543
Kuzyakov Y, Domanski G (2002) Carbon input by plants into the soil. J Plant Nutr Soil Sci 163:421–431
Lechevalier MP (1994) Minireview: taxonomy of the genus *Frankia* (Actinomycetales). Int J Syst Bacteriol 44:1–8

Loreau M (2001) Microbial diversity, producer-decomposer interactions and ecosystem processes: a theoretical model. Proc R Soc Biol Sci B 268:303–309

Madder P, Vierheilig H, Boller T, Streitwolf-Engel B, Frey B, Christie P, Wiemken A. (2000) Transport of 15 N from a soil compartment separated by a polytetraflouroethylene membrane to plant roots via the hyphae of arbuscular mycorrhizal fungi. New Phytol 146:155–161

Martens DA, Frankenberger Jr WT (1993) Metabolism of tryptophan in soil. Soil Biol Biochem 25:1679–1687

Maunuksela L (2001) Molecular and physiological characterization of rhizosphere bacteria and *Frankia* in forest soils devoid of actinorhizal plants. Ph.D thesis. University of Helsinki, Helsinki, Finland

Merryweather J, Fitter A (1995) Phosphorus and carbon budgets mycorrhizal contributions in *Hyacinthoides nonscripta* (L.) Chouard ex Rothm under natural conditions. New Phytol 129:619–627

Miller RM, Jastrow JD (1994) Vesicular arbuscular mycorrhizae and plant biogeochemical cycling. In: Pfleger FL, Linderman RG (eds) Mycorrhizae and plant health, APS Press, St. Paul, USA, pp 189–212

Myrold DD (2000) Microorganisms. In: Alexander DE, Fairbridge RW (eds) Encyclopedia of Environmental Science. Kluwer, Amsterdam, pp 409

Nahas E (1996) Factors determining rock phosphate solubilization by microorganisms. World J Microbiol Biotechnol 112:567–572

Narsian V, Patel HH (2000) *Aspergillus aculeatus* as a rock phosphate solubilizer. Soil Biol Biochem 32:559–565

Naseby DC, Way JA, Bainton NJ, Lynch JM (2001) Biocontrol of Pythium in the pea rhizosphere by antifungal metabolite producing and non-producing *Pseudomonas* strains. J Appl Microbiol 90:421–429

Nautiyal CS, Bhadauria S, Kumar P, Lal H, Mondal R, Verma D (2000) Stress induced phosphate solubilization in bacteria isolated from alkaline soils. FEMS Microbiol Lett 182:291–296

Neilands JB, Leong SA (1986) Siderophores in relation to plant growth and disease. Annu Rev Plant Physiol 37:187–208

O'Donnell AG, Görres HE (1999) 16S rDNA methods in soil microbiology. Curr Opin Biotechnol 10:225–229

Olesniewics KS, Thomas RB (1999) Effects of mycorrhizal colonization on biomass production and nitrogen fixation of black locust (*Roina pseudoacacia*) seedlings grown under elevated atmospheric carbon dioxide. New Phytol 142:133–140

Omar SA (1994) Degradation and mineralization of wheat straw by some cellulolytic fungi in pure culture. Microbiol Res 149:157–161

Pal SS (1998) Interaction of an acid tolerant strain of phosphate solubilizing bacteria with a few acid tolerant crops. Plant Soil 198:169–177

Pankhurst CE, Pierret A, Hawke BG, Kirby JM (2002) Microbiological and chemical properties of soil associated with macrospores at different depths in a red-duplex soil in NSW Australia. Plant Soil 238:11–20

Paul EA, Clark FE (1998) Soil microbiology and biochemistry, 2nd edn. Academic Press, San Diego

Peix A, Rivas-Boyero AA, Mateos PF, Rodriguez-Barrueco C, Martinez-Molina E, Velazquez E (2001) Growth promotion of chickpea and barley by a phosphate solubilizing strain of *Mesorhizobium* and mediterraneum under growth chamber conditions. Soil Biol Biochem 33:103–110

Perotto S, Girlanda M, Martino M (2002) Ericoid mycorrhizal fungi: some new perspectives on old acquaintances. Plant Soil 244:41–53

Prosser JI (2002) Molecular and functional diversity in soil microorganisms. Plant Soil 244:9–17

Rajbanshi SS, Endo H, Sakamoto K, Inubushi K (1998) Stabilisation of chemical and biochemical characteristics of grass straw and leaf mix during in-vessel composting with and without seeding material. Soil Sci Plant Nutr 44:485–495

Ranjard L, Richaume A (2001) Quantitative and qualitative microscale distribution of bacteria in soil. Res Microbiol 152:707–716

Rasmussen PH, Knudsen IMB, Elmholt S, Jensen DF (2002) Relationship between soil cellulolytic activity and suppression of seedling blight of barley in arable soils. Appl Soil Ecol 19:91–96

Read DJ (1996) The structure and function of the ericoid mycorrhizal root. Ann Bot 77:365–374

Reddy MS, Kumar S, Babita K, Reddy MS (2002) Biosolubilization of poorly soluble rock phosphates by *Aspergillus tubingensis* and *Aspergillus niger*. Biores Technol 84:187–189

Rennenberg H (1999) The significance of ectomycorrhizal fungi for sulfur nutrition of trees. Plant Soil 215:115–122

Rengel Z (2002) Breeding for better symbiosis. Plant Soil 245:147–162

Requena N, Azcon R, Baca MT (1996) Chemical changes in humic substances from compost due to incubation with lignocellulolytic microorganisms and effects on lettuce growth. Appl Microbiol Biotechnol 45:857–863

Reyes I, Baziramakenga R, Bernier L, Antoun H (2001) Solubilization of phosphate rocks and minerals by a wild type strain and two UV-induced mutants *of Penicillium rugulosum*. Soil Biol Biochem 33:1741–1747

Rice WA, Olsen PE, Legget ME (1995) Co-culture of *Rhizobium meliloti* and a phosphorus solubilizing fungus (*Penicillium bilaii*) in sterile peat. Soil Biol Biochem 27: 703–705

Ritz K, Wheatley RE, Griffiths BS (1997) Effects of animal manure application and crop plants upon size and activity of soil microbial biomass under organically grown spring barley. Biol Fert Soils 24:372–377

Schipper LA, Dehgens BP, Sparling GP, Duncan LC (2001) Changes in microbial heterotrophic diversity along five plant successional sequences. Soil Biol Biochem 33:2093–2103

Sen R (2000) Budgeting for the wood – wide web. New Phytol 145:161–165

Sharples JM, Mehrag A, Chambers SM, Cairney JWG (2000) Mechanism of arsenate resistance in the ericoid mycorrhizal fungus *Hymenoscyphus ericae*. Plant Physiol 124:1327–1334

Shen D (1997) Microbial diversity and application of microbial products for agricultural purposes in China. Agric Ecosyst Environ 62:237–245

Simon L (1996) Phylogeny of the Glomales: Deciphering the past to understand the present. New Phytol 133:95–101

Singal R, Gupta R, Kuhad RC, Saxena RK (1991) Solubilization of inorganic phosphates by a Basidiomycetous fungus *Cyathus*. Ind J Microbiol 31:397–401

Singh A, Sharma J, Rexer K-H, Varma A (2000) Plant productivity determinants beyond minerals, water and light: *Piriformospora indica* – a revolutionary plant growth-promoting fungus. Curr Sci 79:1548–1554

Smit E, Leeflang P, Gommans S, van der Broek J, van MS, Werners K (2001) Diversity and seasonal fluctuations of the dominant members of the bacterial soil community in a wheat field as determined by cultivation and molecular methods. Appl Environ Microbiol 67:2284–2291

Smith SE, Read DJ (1997) Growth and carbon economy of VA mycorrhizal plants. In: Mycorrhizal symbiosis. 2nd edn. Academic Press, London, pp 105–125

Staley JT (1997) Biodiversity: are microbial species threatened? Curr Opin Biotechnol 8:340–345

Steenhoudt O, Vanderleyden J (2000) *Azospirrilum*, a freeliving nitrogen fixing bacterium closely associated with grasses: genetic, biochemical and ecological aspects. FEMS Microbiol Rev 24:487–506

Stepanov AL, Korpelal TK (1997) Microbial basis for the biotechnological removal of nitrogen oxides from flue gases. Biotechnol Appl Biochem 25:97–104

Subba Rao NS (1999) Soil microbiology, 4th edn, Science Publishers, New Hampshire

Suneja S, Lakshminarayana K (1999) Siderophore production of *Azotobacter*. In: Narula N (ed) *Azotobacter* in sustainable agriculture, CBS Publishers, New Delhi, pp 64–73

Timmusk S, Nicander B, Granhall U, Tillberg E (1999) Cytokinin production by *Paenibacillus polymyxa*. Soil Biol Biochem 31:1847–1852

Tiedje JM, Cho JC, Murray A, Travers D, Xia B, Zhou J (2001) Soil teeming with life: new frontiers for soil science. In: Rees RM, Ball BC, Campbell CD, Watson CA (eds) Sustainable management of soil organic matter, CAB International, London, pp 393–412

Toro M, Azcon R, Barea J (1997) Improvement of arbuscular mycorrhiza development by inoculation of soil with phosphate solubilizing rhizobacteria to improve rock phosphate bioavailability (32P) and nutrient cycling. Appl Environ Microbiol 63:4408–4412

Torrisi V, Pattinson GS, McGee PA (1999) Localized elongation of roots of cotton follows establishment of arbuscular mycorrhizas. New Phytol 142:103–112

Torsvik V, Øvreås L (2002) Microbial diversity and function in soil: from genes to ecosystems. Curr Opin Microbiol 5:240–245

Torsvik V, Sorheim R, Goksoyr J (1996) Total bacterial diversity in soil and sediment communities – a review. J Ind Mcirobiol 17:170–178

Unge A, Jansson J (2001) Monitoring population size, activity, and distribution of *gfp-luxAB*-tagged *Pseudomonas fluorescence* SBW25 during colonization of wheat. Microb Ecol 41:290–300

Valdenegro M, Barea JM, Azcon R (2001) Influence of arbuscular-mycorrhizal fungi, *Rhizobium meliloti* strains and PGPR inoculation on the growth of *Medicago arborea* used as nodel legume for revegetation and biological reactivation in a semi-arid mediterranean area. Plant Growth Regul 34:233–240

Varma A (1995) Ecophysiology of mycorrhizal fungi. In: Varma A, Hock B (eds) Mycorrhizae: structure, function, molecular biology and biotechnology, Springer-Verlag, Berlin Heidelberg New York, pp 561–592

Varma A, Bonfante P (1994) Utilization of cell wall related carbohydrates by ericoid mycorrhizal endophytes. Symbiosis 16:301–313

Varma A, Verma S, Sudha A, Sahay N, Butehorn B, Franken P (1999) *Piriformospora indica*, a cultivable plant-growth-promoting root endophyte. Appl Environ Microbiol 65:2741–2744

Vassilev N, Franco I, Vassileva M, Azcon R (1996) Improved plant growth with rock phosphate solubilized by *Aspergillus niger* grown on sugar beet waste. Biores Technol 55:237–241

Vassilev N, Vassileva M, Fenice M, Fedirici F (2001) Immobilized cell technology applied in solubilization of insoluble inorganic (rock) phosphates and P plant acquisition. Biores Technol 79:263–271

Verghese SK, Sarma G, Chauhan V, Misra AK (1998) Optimising actinorhizal symbiosis using molecular markers. In: Rai B, Dkahr MS (eds) New trends in microbial ecology. NEH University, Shilong and International Society for Conservation of Natural Resources, BHU, Varanasi, India, pp 350–354

Verstaete W, Top EM (1999) Soil clean up: lessons to remember. Int Biodeter Biodeg 43:147–153

Vierheilig H, Bennett R, Kiddle G, Kaldorf M, Ludwig-Muller J (2000) Differences in glucosionolate patterns and arbuscular mycorrhizal status of glucosionolate containing species. New Phytol 146:343–352

Watkins NK, Fitter AH, Graves JD, Robinson D (1996) Carbon transfer between C3 and C4 plants linked by a common mycorrhizal network, quantified using stable carbon isotopes. Soil Biol Biochem 28:471–477

Whitelaw MA, Harden TJ, Helyar KR (1999) Phosphate solubilization in solution culture by the soil fungus *Penicillium radicum*. Soil Biol Biochem 31:655–665

Wiemken V, Boller T (2002) Ectomycorrhiza: gene expression, metabolism and the wood-wide web. Curr Opin Plant Biol 5:1–7

Yin B, Crowley D, Sparovek G, De Melo WJ, Borneman J (2000) Bacterial functional redundancy along a soil reclamation gradient. Appl Environ Microbiol 66:4361–4365

Zehnder GW, Murphy JF, Sikora EJ, Kloepper JW (2001) Application of rhizobacteria for induced resistance. Eur J Plant Pathol 107:39–50

Zhu W, Ehrenfeld JG (1996) The effects of mycorrhizal roots on litter decomposition, soil biota and nutrients in a spodsolic soil. Plant Soil 179: 109–118

6 Signalling in the Rhizobia–Legumes Symbiosis

Dietrich Werner

1 Introduction

During the last few years, significant progress has been made in the understanding of signal production, signal perception and signal regulation in plants, microorganisms and animals. In mammalian systems, the notch signal regulation has been studied intensively with a number of proteins involved in the signal transport and the proteolytic modifications of the notch signals (Fig. 1). The involvement of several organelles such as lysosomes, endoplasmic reticulum and the Golgi network points to interesting similarities and differences to the signalling across and through the symbiosome membrane (Werner 1992). Four different mammalian notch homologues have been identified (Baron et al. 2002). The integrin-adhesion-receptor signalling is another surface related crosstalk in multicellular organisms (Schwartz and Ginsberg 2002). The cell adhesion involving integrins leads to a phosphorylation of different growth-factor receptors, including those for the fibroblast growth factor, the hepatocyte growth factor and the epidermal growth factor (Giancotti and Ruoslahti 1999). Very recently, a plant receptor-like kinase has been identified in the laboratory of Martin Parniske, The Sainsbury Laboratory, UK, which is required for the rhizobial legume symbiosis as well as for the arbuscular mycorrhiza symbiosis (Stracke et al. 2002). The SYMRK (symbiosis receptor-like kinase) genes have been studied and characterized in *Lotus* and in pea. The protein has a signal peptide, a transmembrane and an extracellular protein kinase domain. The SYMRK is part of a symbiotic signal transduction pathway with the perception of a microbial signal molecule, leading to a rapid symbiosis-related gene activation. In *Medicago sativa*, a "nodulation receptor kinase" NORK was identified with a predicted function in Nod-factor perception/transduction (Endre et al. 2002).

Besides the short-distance signalling between microorganisms and plant surfaces, long-distance signalling also affects the plant partner of the interaction. Using mutants of *Arabidopsis*, the role for long-chain fatty acids in cell-to-cell communication has been established and the plant hormones abscisic

Plant Surface Microbiology
A. Varma, L. Abbott, D. Werner, R. Hampp (Eds.)

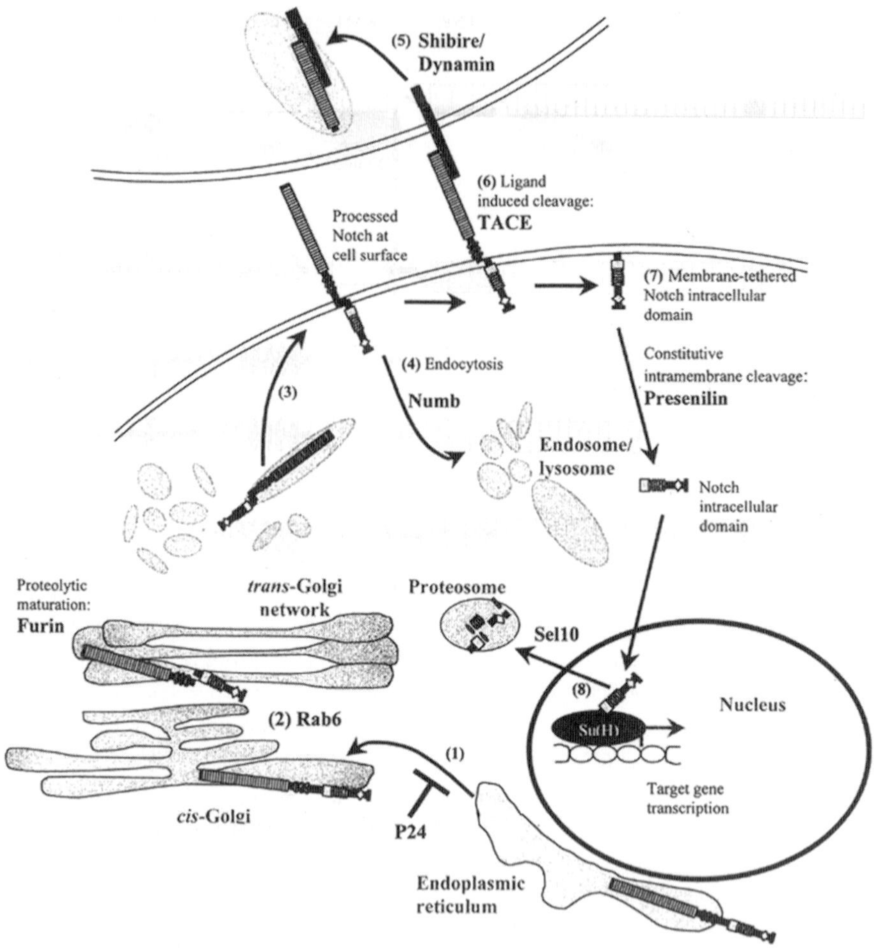

Fig. 1. Notch trafficking and signal regulation. Trafficking of Notch through the cell in conjunction with covalent modifications of Notch may confer potential points of regulation of Notch-signalling levels. Proteins known or suspected to affect particular steps in the transport process or the proteolytic modification of Notch are indicated in *bold*. *Numbers* indicate the potential decision points that could determine signal levels. *1* Notch trafficking to the Golgi body may be subject to quality control and cargo selection by P24 family proteins. *2* Small GTPases such as Rab6 may regulate transport to *trans*-Golgi where Furin-dependent proteolysis occurs. *3* Notch transported to the cell surface may undergo endocytosis (*4*), or *5* ligand-dependent activation at the cell surface. The latter is accompanied at least in some tissues by *6 trans*-endocytosis of the Notch extracellular domain into the adjacent ligand-bearing cell. *7* The remaining membrane-tethered intracellular domain undergoes Presenilin-dependent cleavage releasing Notch-intra for translocation to the nucleus leading to regulation to target gene expression. *8* Accumulation of Notch-intra in the nucleus may be regulated by its ubiquitination and proteosome-dependent degradation (Baron et al. 2002)

acid, ethylene and jasmonates are involved in the long-distance signalling, regulating, e.g., the stomata number by environmental signals such as the CO_2 partial pressure (Lake et al. 2002). Other long-distance signals, e.g., from roots to shoots, are transported in the xylem. This roots to shoots signalling has a lag time of only a few hours, persisting for several days (Jackson 2002). Abscisic acid (ABA) plays an important role in the signalling from the roots and the rhizosphere, e.g., in water stress for the regulation of stomata behavior (Wilkinson and Davies 2002). The signalling process includes the following steps: ABA sequestration in the root, ABA synthesis and catabolism in the root, the transfer of ABA across the root and into the xylem, the exchange of ABA from the xylem lumen to the xylem parenchyma in the shoot, the concentration of ABA in the leaf symplastic reservoir, the cleavage of ABA conjugates, the transfer of ABA from the leaf into the phloem and an assumed interaction between nitrate stress and the ABA signal.

Moreover, in unicellular eukaryotic model organisms specific signal molecules have been identified, such as phosphatidylinositol (3,5) bisphosphate modified by the phosphatidylinositol 3-phosphate 5-kinase in *Schizosacchromyces pombe* and which are required to respond to nutritional starvation (Morishita et al. 2002). This pathway is also necessary for mating the pheromone signal in this yeast. The question arises if these compounds also play a role in microbes – plant surface signalling. The costs for biological signalling are an important aspect in evolution. New results indicate that there is a shift to signals with high cost and an underutilization of signals with low costs (de Polavieja 2002). The signal structure follows a generalized Boltzmann form, penalizing signals with high costs and a high sensitivity to errors. In this respect, costs are defined in metabolic costs, time costs or risk costs.

In bacteria key components for major regulatory pathways have been identified such as the protein H-NS (Schröder and Wagner 2002). It is a small DNA binding protein, regulating a diverse range of genes such as for anaerobic growth phase activation, endochitinase, nitrate reductase, leucine responsive regulatory proteins, a proline/glycine betaine transport system, an activator for capsular polysaccharide synthesis and an invasion regulatory gene.

2 The Signals from the Host Plants

In principle, molecules released from plant surfaces can be substrates and signals for microorganisms. Types and functions of root exudates in the rhizosphere have recently been reviewed by Brimecombe et al. (2001), Neumann and Römheld (2001) and Uren (2001). Their functions as signal molecules have been reviewed by Werner (2001) and Werner and Müller (2002).

2.1 Phenylpropanoids: Simple Phenolics, Flavonoids and Isoflavonoids

The basis of this biochemical pathway is the shikimate pathway, producing aromatic amino acids and several vitamins and co-factors. The phenylalanine-ammonia-lyase (PAL) produces cinnamate from phenylalanine, which is a precursor of a large number of phenolics, phenylpropanoids and flavonoids (Paiva 2000). Major cinnamate derivatives are *para*-coumaric acid, caffeic acid, ferulic acid and sinapic acid, leading to plant surface polymers such as suberin and lignin. The ortho-hydroxylation of cinnamate gives coumarate, a precursor of coumarin, which has a strong antimicrobial activity. Another derivate is salicylic acid, which has also antimicrobial activity and a function in signal transduction during plant pathogenic interactions. Especially well studied is the function of acetosyringone, inducing *vir* gene expression in *Agrobacterium tumefaciens*, with a broad range of host plants. Dimers of phenylpropanoids form lignans such as the antifungal compound magnolol and the toxic compound podophyllotoxin. The biosynthesis of major classes of flavonoids and isoflavonoids are summarized in Fig. 2, according to Dixon and Steele (1999). Major enzymes involved are chalcone synthase, chalcone reductase, chalcone isomerase, flavone synthase I and II, flavonol synthase, isoflavone synthase and flavonoid 3′-hydroxylase. The major *nod* genes inducing compounds in legumes are isoflavones such as daidzein (R1=H) and genistein (R1=OH). The pterocarpan at the end of this pathway includes phytoalexins such as medicarpin from alfalfa and glyceollin I from soybeans. In addition to the intensively studied effects of flavonoids and isoflavonoids in the interaction of plants with microorganisms, the health-promoting effects in medical sciences are another intensively studied area. Genistein, for instance, has been shown in human cell lines to inhibit prostate tumor growth, stomach tumor growth and anti-angiogenesis (Rice-Evans and Miller 1996; Hollman and Katan 1998). Genistein has also an effect in preventing bone-loss caused by deficiency of estrogens in female mice (Ishimi et al. 1999).

Fig. 2. The biosynthesis of the major classes of flavonoid derivates. The enzymes are: *CHS* chalcone synthase, *CHR* chalcone reductase, *CHI* chalcone isomerase, *FSI* flavone synthase I, *FSII* flavone synthase II, *'IFS'* isoflavone synthase, consisting of 2-hydroxyisoflavanone synthase (*2HIS*) and 2-hydroxyisoflavanone dehydratase (*2-HID*), *F3bH* flavanone 3b hydroxylase, F3′H flavonoid 3′hydroxylase, *F3′5H* flavonoid 3′,5′-hydroxy-

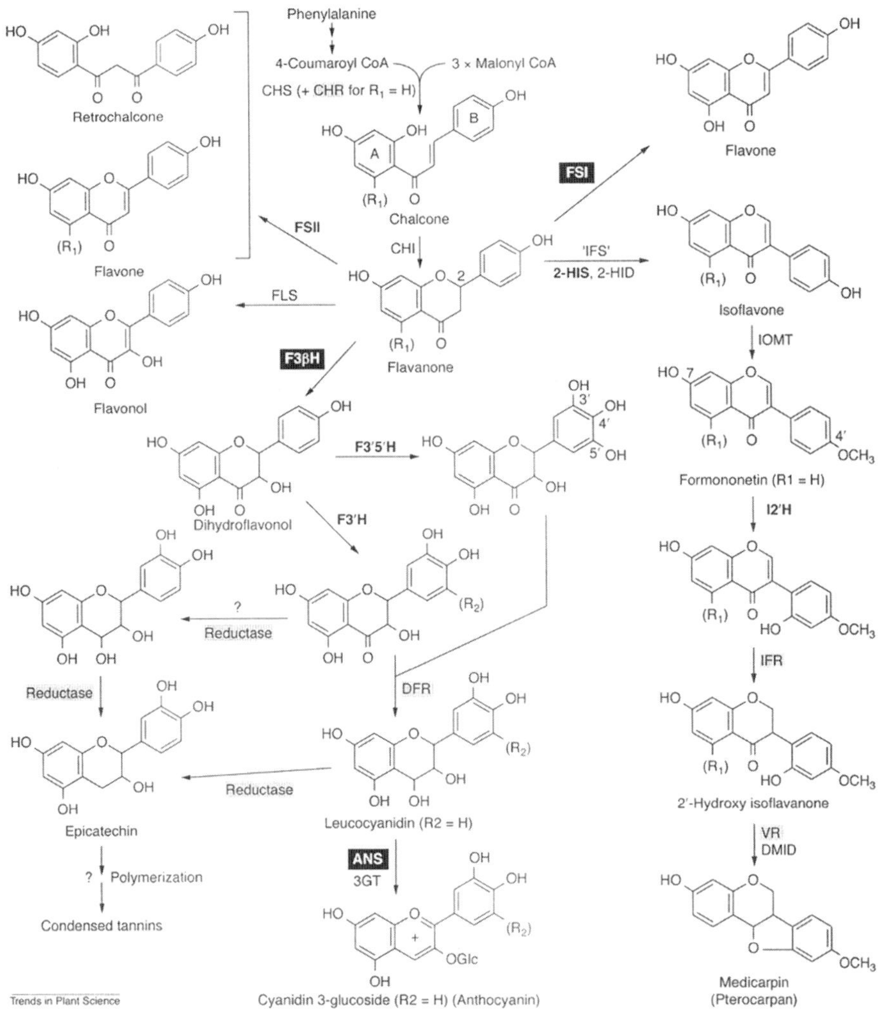

lase, *DFR* dihydroflavonol reductase, *ANS* anthocyanidin synthase, *3GT* anthocyanidin 3-glucosyltransferase, *IOMT* isoflavone *O*-methyltransferase, *IFR* isoflavone reductase, *VR* vestitone reductase, *DMID* 7,2′-dihydroxy, 4′-methoxyisoflavanol dehydratase. Enzymes in *white* are 2-oxoglutarate-dependent dioxygenases, in *black bold* are cytochrome P450s, and highlighted in *grey* are NADPH-dependent reductases. Simplifications include not discriminating between the 5-hydroxy (R1=OH) and 5-deoxy (R1=H) flavonoids and isoflavonoids, for which the loss of the 5-hydroxyl occurs because of the co-action of CHR with CHS, showing only the anthocyanin pathway leading to the compounds with a di-substituted B-ring (cyaniding derivatives). Parallel pathways function in the formation of anthocyanins with mono- and tri-substituted B-rings. In the latter, F3′5′H can act at the level of the dihydroflavonol with a mono-or di-substituted B-ring. The pathway to epicatechin from a dihydroflavonol is shown to follow two routes, both via leucocyanidin. It is unclear whether there is a specific form of DFR that functions only in condensed tannin biosynthesis. The 4′-*O*-methylation of the B-ring of isolflavones occurs in alfalfa, pea and other legumes, but not in bean or soybean (Dixon and Steele 1999)

2.2 Metabolization of Flavonoids and Isoflavonoids

A degradation pathway for luteolin by *Sinorhizobium meliloti* and for daidzein by *Bradyrhizobium japonicum* is shown in Fig. 3. Identified metabolites from luteolin were caffeic acid, phoroglucinol, protocatechuic acid and phenylacetic acid. Daidzein, p-coumaric acid, p-hydroxybenzoic acid, phenylacetic acid and resorcinol were major metabolites. Umbeliferone has also been found to be produced from coumestrol (Cooper et al. 1995).

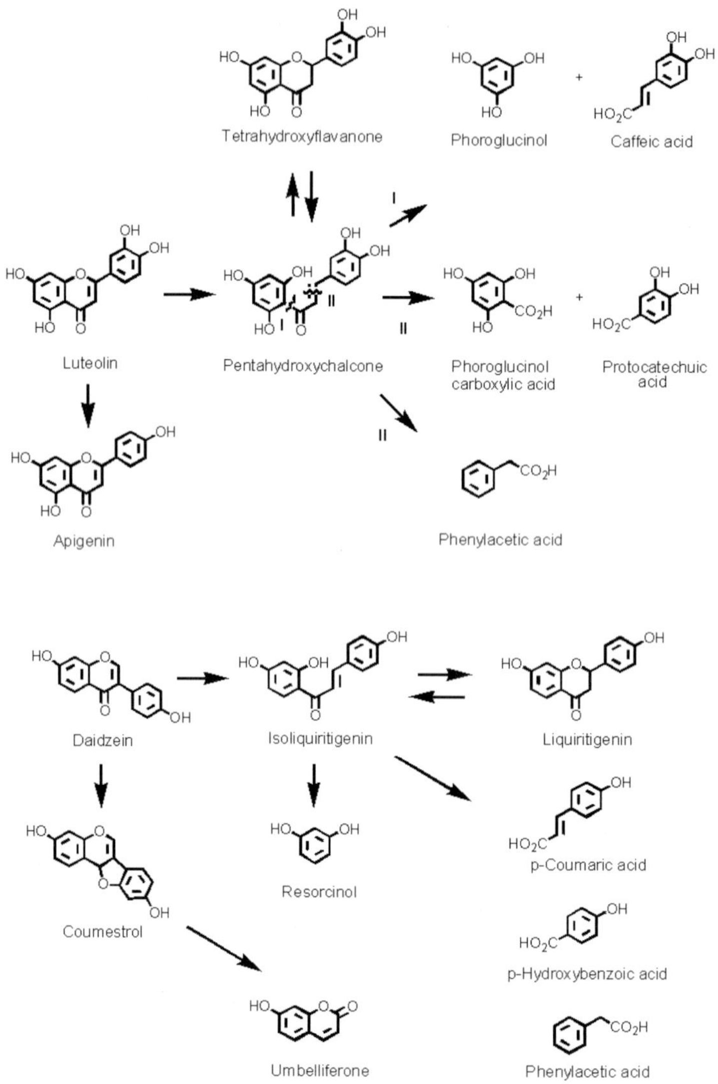

Fig. 3. Proposed degradation pathway for luteolin by *Sinorhizobium meliloti* and for daidzein by *Bradyrhizobium japonicum* (Cooper et al. 1995)

Flavanone metabolites from animals and humans are summarized in Table 1 (Heilmann and Merfort 1998). A number of flavanes, flavanoles, *trans*-3-hydroxyflavanes, 6-hydroxyflavanones, 6-hydroxyflavanes, 4-hydroxyflavanes, 3,6-dihydroxy-flavanes, 3,4-dihydroxy-flavanes and methoxyflavanes have been identified. The best studied flavane is catechin. The major metabolites of this compound in humans are 3-hydroxybenzoic acid, 3-hydroxyphenylpropionic acid, 3-hydroxyhippuric acid, 3,4-dihydroxyphenylbenzoic acid, 5-(3,4-dihydroxy)-valerianic acid and δ-(3-hydroxy-4-methoxyphenyl)-γ-valerolactone. Metabolites excreted from mammalians can, in certain locations of the soil, therefore also significantly increase the concentrations of flavonoid and isoflavonoid metabolites which may affect the plant-microbe interactions. Sulfation of flavonoids and phenolic dietary compounds by cytosolic sulfotransferases has been studied in detail by Pai et al. (2001). The mechanisms for a chemoprotective action of these compounds is the inhibition of the bioactivation of carcinogens by the human cytosolic sulfotransferases. The ten known cytosolic sulfotransferases have a different substrate specificity, e.g., the isoform PST has a high activity with flavonoids, but no activity with isoflavonoids (Pai et al. 2001).

A fluoroimmunoassay has been developed to detect small concentrations of daidzein and genistein as phytoestrogens in blood plasma. After synthesis of 4′-*O*-carboxymethyl-daidzein and 4′-*O*-carboxymethyl-genistein, these compounds were linked to bovine serum albumin and used to immunize rabbits. The antisera were cross-reactive with some isoflavonoids, but not with flavonoids (Wang et al. 2000). The assays could detect daidzein and genistein in the range between 1 and 370 nMol/l. The actual concentrations in the blood plasma were in the range between 4 and 7 nMol/l. The correlation coefficient between this fluoroimmunoassay and a reference method, using an isotope dilution gaschromatography mass spectrometry, was in the range of 0.95–0.99. Another method to detect low concentrations of estrogenic flavonoids was the development of a recombinant yeast strain in which the human estrogen receptor was stably integrated into the genome. The most active flavonoids in this assay were naringenin, apigenin, kaempferol, phloretin,

Table 1. Metabolites from flavanones (Heilmann and Merfort 1998)

Flavane-4-α-ol	4′-hydroxyflavane
Flavane-4-β-ol	4′-hydroxyflavane-4-α-ol
Trans-3-hydroxyflavanone	4′- hydroxyflavane-4-β-ol
Trans-3-hydroxyflavane-4-α-ol	3,6-dihydroxy-flavane-4-α-ol
Trans-3-hydroxyflavane-4-β-ol	3,6-dihydroxy-flavane-4-β-ol
Flavone-3-ol	3,4-dihydroxy-flavane-4-α-ol
6-Hydroxyflavanone	3,4-dihydroxy-flavane-4-β-ol
6-Hydroxyflavane-4-α-ol	4′-hydroxy-3′-methoxyflavane-4-α-ol
6-Hydroxyflavane-4-β-ol	4′-hydroxy-3′-methoxyflavane-4-β-ol

equol, genistein, daidzein and biochanin A. The main feature for the estrogenic activity in these compounds is a single hydroxyl group at the 4′-position of the B-ring of the flavan nucleus. It must be pointed out that the estrogenic activity of these flavonoids is 4,000–4,000,000 times lower than that of 17β-estradiol (Breinholt and Larsen 1998). In the *Handbook of Flavonoids* (Harborne 1994), about 900 different isoflavones, chalcones, pterocarpans, B-flavonoids, rotenoids, isoflavanes and coumestrols are listed. Nevertheless, new flavonoids can be found by studying less known legumes. In *Ulex airensis* and *Ulex europaeus* three new isoflavonoids called ulexin C, ulexin D and 7-*O*-methylisolupalbigenin could be isolated and characterized (Maximo et al. 2002).

Another important group of plant signals involved in symbiosis and defense reactions are fatty acid-derived signals (Weber 2002). The best studied compound in this category is jasmonic acid and its volatile methyl ester. Almost 20 different jasmonate signalling mutants in *Arabidopsis* are known (Staswick et al. 1998; Thomma et al. 1998; Hilpert et al. 2001). With *Arabidopsis* it has also been shown that keto, hydroxy and hydroperoxy fatty acids such as ketodienoic fatty acid can accumulate in plants after infection with *Pseudomonas syringae*. By infiltration of these compounds into *Arabidopsis* leaves, the gene encoding glutathione-S-transferase (GST1) has been demonstrated (Weber 2002). From a large number of flavonoids and isoflavonoids tested for their ability to inhibit the ascorbate-induced microsomal lipid peroxidation, kaempferol has been showed to have the highest activity of all flavonoids tested (Cos et al. 2001).

2.3 Vitamins as Growth Factors and Signal Molecules

On a molecular basis the best-studied system is the effect of biotin on growth and survival of *Sinorhizobium meliloti*. Already, nanomolar concentrations of biotin increase the colonization of alfalfa roots sevenfold and addition of avidin, a biotin-binding protein from eggs, reduces the colonization by a factor of 7. Biotin is a co-factor of carboxylation reactions and biotinylated carboxylases have been demonstrated in *Rhizobium etli* (Dunn 1996). From the biotin biosynthesis pathway, several genes such as *bioA*, *bioC* and *bioH* are apparently not functional and also for *bioD*, no homology genes have been found (Entcheva et al. 2001). On the other hand, components of a prokaryotic biotin transporter have been identified with bioMN, which are activated under biotin deficiency. Rhizobia have a very efficient uptake system for biotin. The kM values are 40 times lower than those for the transport system for *E. coli*. With biotin limitation the synthesis of PHB is significantly increased in *Sinorhizobium meliloti* and *Rhizobium etli*. Under biotin limitation the transcription rate of the gene, responsible for proteins of PHB degradation is down-regulated. Addition of biotin at nanomolar concentrations increases the activity of the *bdhA* gene more than fivefold, as demonstrated for strains with a lac-Z-fusion (Streit et al. 1996; Hofman et al. 2000).

Another very interesting signal molecule derived from a vitamin is lumichrome, which is produced from riboflavin and functions as a signal molecule in the rhizosphere (Philipps et al. 1999). Nanomolar concentrations of lumichrome around the roots increase the root respiration and also photosynthesis. This may be of general significance since a number of riboflavin-producing bacteria have been identified on roots in a significantly higher number than in the bulk soil (Strzelczyk and Rozycki 1985).

3 Signals from the Microsymbionts

3.1 Nod Factors

A breakthrough in understanding infection, nodulation and host specificity in the *Rhizobium*-legume symbiosis was the identification of Nod factors such as lipochitooligosaccharides (LCOs) produced and excreted by more than 30 different *nod, nol* and *noe* genes and their corresponding proteins from the microsymbionts. The first identified Nod factors were from *Rhizobium meliloti* (Lerouge et al. 1990) and from *Rhizobium leguminosarum* bv. *viciae* (Spaink et al. 1991). The general structure of Nod factors are *N*-acetylglucosamine backbone with four or five GlcNAc residues with different substituents at nine different positions such as *N*-methyl, O-carbamyl, *O*-acetyl, *O*-sulfyl, α-linked fucosyl, 2-*O*-methylfucosyl, 4-*O*-acetyl-2-*O*-methylfucosyl, 3-*O*-sulfate-2-*O*-methylfucosyl, ethyl, glyceryl, mannosyl and *N*-glycosyl groups. Another major residue variable is the fatty acid group attached to the nitrogen of the nonreducing end of the Nod factor. Fatty acids with 16–18 carbons and a different degree of unsaturation in different positions of the double bonds are mainly present. In addition, C18–C22 (ω-1)-hydroxy fatty acids,

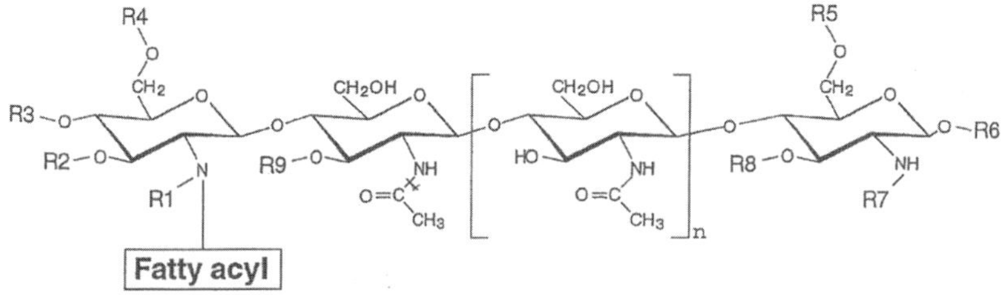

Fig. 4. General structure of the Nod factors produced by rhizobia. The presence of substituents numbered R1–R9 is variable within various strains of rhizobia. In the absence of specific substituents, the R groups stand for hydrogen (*R1*), hydroxy (*R2, R3, R4, R5, R6, R8*, and *R9*), and acetyl (*R7*) (Spaink 2000)

Table 2. Modifications of Nod factors (Modified from Spaink 2000 and Pacios-Bras et al. 2002)[a]

Bacterial strain	Nodulated plant tribes	GlcNAc residues (n)[b]	Special substituents[c]
S. meliloti	Galegeae	3,4,5	R4:Ac, R5:S, C16:2, C16:3, C26(ω-1)OH
R. leguminosarum			
bv. *Viciae* RBL5560	Galegeae	3,4,5	R4:Ac, C18:4
bv. *Viciae* TOM	Galegeae	3,4,5	R4:Ac, R5:Ac, C18:4
bv. *Viciae* A1	Galegeae	3,4,5	R4:Ac, R5:Ac, C18:4, C18:3
bv. *Trifolii* ANU842	Galegeae	3,4,5	R4:AC, R5:Ac, R6:Et, C20:4, C20:3, C18:3
R. galegae	Galegeae	4,5	R4:Cb, R9:Ac, C18:2, C18:3, C20:2, C20:3
M. huakuii	Galegeae	3,4,5	R5:S, R7:G, C18:4
M. loti			
E1R, NZP2235	Loteae	4,5	R1:Me, R3:Cb, R5:AcFuc
NZP2037	Loteae Genisteae	4,5	R1:Me, R2:Cb, R3:Cb, R5:AcFuc
NZP2213	Loteae	2,3,4,5	R1:Me, R3:Cb, R5:AcFuc, R9:Fuc
B. aspalati bv. carnosa	Crotalarieae	3,4,5	R1:Me, R3:Cb, R4:Cb
B. japonicum USDA110	Phaseoleae	5	R5:MeFuc
B. japonicum USDA135	Phaseoleae	5	R4:Ac, R5:MeFuc
B. elkanii USDA61	Phaseoleae	4,5	R1:Me, R4:Ac, R3:Cb, R5:MeFuc, R6:Gro
R. etli	Phaseoleae	4,5	R1:Me, R3:Cb, R5:AcFuc
R. etli KIM5S	Phaseoleae	6	R1:Me, R2-R6:H, R7:acetyl, R8:H, ring 5:acetyl or H
R. tropici	Phaseoleae, Mimoseae	4,5	R1:Me, R5:S, R6:Man
S. fredii			
USDA257	23 Tribes	3, 4,5	R5:MeFuc
NGR234	26 Tribes	4,5	R1:Me, R3:Cb, R4:Cb, R5:MeFuc/AcMe-Fuc/SmeFuc
Rhizobium sp. GRH2	Acacieae	3,5,6	R1:Me, R5:S
S. teranga bv. *acaciae*	Acacieae	5	R1:Me, R3/4:Cb, R5:S
Mesorhizobium ORS1001	Acacieae	5	R1:Me, R3/4:Cb, R5:S
A. caulinodans	Robinieae	4,5	R1:Me, R4:Cb, R5:Fuc, R8:Ara
S. sahelii	Robinieae	4,5	R1:Me, R3/4:Cb, R5:Fuc, R8:Ara
S. teranga bv. *sesbaniae*	Robinieae	4,5	R1:Me, R3/4:Cb, R5:Fuc, R8:Ara

[a] For backbone structure, see Fig. 4

[b] The underlined numbers of *N*-acetylglucosamine (GlcNAc) residues indicate the most abundant species

[c] The indicated substituents do not always occur in all lipochitin oligosaccharides (LCOs) produced, leading to a mixture of LCOs, which do or do not contain all possible substituents. Abbreviations: Me, *N*-methyl; Cb, *O*-carbamyl; Ac, *O*-acetyl; S, *O*-sulfyl; Fuc, α-linked fucosyl; MeFuc, 2-*O*-methylfucosyl, AcMeFuc, 4-*O*-acetyl-2- *O*-methylfucosyl; SmeFuc, 3-*O*-sulfate-2-*O*-methylfucosyl; Et, ethyl, Gro, glyceryl, Man, mannosyl; G, *N*-glycolyl; FA, fatty acyl

which are perhaps intermediates in the synthesis of C23 (ω-1) hydroxy fatty acyl groups in lipopolysaccharides, can be present in Nod factors of *Sinorhizobium meliloti* (Demont et al. 1994). Figure 4 and Table 2 summarize the large variations of Nod factors identified so far. A novel lipochitin oligosaccharide has recently been found in *Rhizobium etli* KIM5S (Pacios-Bras et al. 2002). This is the first case where the major LCO contains six oligosaccharide residues and differs by this point from all other rhizobia analyzed so far. An additional specificity was that the chitin backbone was deacetylated in one or two of the GlcNAc moieties, although these were only minor compounds. The fatty acids of these Nod factors were C16:0, C16:1, C18:0, C18:1 and C17:1. In this respect, the fatty acids are much more variable than those of *Rhizobium etli* strain CE3. Moreover, the host range of strain KIM5S of *Rhizobium* etli was different from the *Rhizobium* type strain CE3, since it could not nodulate *Lotus japonicus*, although it did nodulate *Siratro*. In *Sinorhizobium meliloti* it has been shown that an enzymatic *N*-deacetylation of the Nod factors decreases their biological activity, but increases the stability in the rhizosphere (Staehelin et al. 2000).

In all rhizobia the *nodABC* genes are essential for the synthesis of the core LCO: NodC synthesizes the chito-oligosaccharide backbone and nodB removes N-acetyl groups from the sugar at its nonreducing end. All other nod, nol and noe genes are responsible for the modifications of this general structure, as indicated in Table 2. NodD is a positive transcription regulator from the Lysr family and present in all rhizobia. In some rhizobial species such as *Sinorhizobium meliloti*, nodD genes are present in multiple forms and their proteins respond to different groups of flavonoids. NodG has the enzymatic activity of an 3-oxoacyl-acyl carrier protein reductase and is thereby homologous to FabG involved generally in fatty acid elongation (López-Lara and Geiger 2001).

3.2 Cyclic Glucans

Cyclic glucans in rhizobia are small molecules linked either by β-(1,2) glycosidic bonds with 17–40 units in *Rhizobium* and *Sinorhizobium* or by β-(1,3) and β-(1,6) glycosidic bonds in *Bradyrhizobium japonicum*. Dominant substituents can be either sn-1-phosphoglycerol (Breedveld and Miller 1998) or phosphocholine (Rolin et al. 1992). The function of the cyclic glucans in *Rhizobium*, *Sinorhizobium* and *Bradyrhizobium* is to protect against hypoosmotic conditions. Rhizobia also produce, however, large quantities of cyclic glucans in the endosymbiotic stage. A specific function during this stage is assumed to be an increase in the solubility of flavonoids and Nod factors (Morris et al. 1991; Schlaman et al. 1997). Another hypothesis is, that β-glucans play a decisive role in the suppression of the host plant defense response with rhizobia, compared to phytopathogenic bacteria.

3.3 Lipopolysaccharides

Lipopolysaccharides (LPS) of rhizobia have been studied in only a few species such as *Rhizobium etli* and *Rhizobium trifolii*. The structure contains three parts, the lipid A, the core chain and the repeat unit of the O-antigen chain. All three parts are very variable. Typical features of rhizobial LPS are the very long chain hydroxy fatty acids (Hollingsworth and Carlson 1989). The genes of the LPS core and O-antigen synthesis have been localized on a plasmid (Vinuesa et al. 1999). A mutation in a glycosyltransferase produced a rough colony phenotype with a disruption of the O-antigen biosynthesis.

The LPS in rhizobia may be involved in the infection process (Brewin 1998; Kannenberg et al. 1998). Their function is perhaps not in the first stages of symbiosis development, but in the release of the bacteria from the infection thread, and the first steps of the symbiosome membrane development. For the LPS moreover, a function in the suppression of the host plant defense response has been assumed, comparable to the LPS functions in plant pathogens (Schoonejans et al. 1987).

3.4 Exopolysaccharides

The exopolysaccharides (EPS) have been studied in detail by a large number of rhizobial strains (Becker and Pühler 1998; Becker et al. 1998; Van Workum and Kijne 1998). In *Sinorhizobium meliloti* two types of EPS forms could be discriminated, EPS I as a succinoglucan and EPS II as a galactoglucan with two size classes in each form, one with thousands of saccharide units and a low-molecular-weight class with only 8–40 saccharide units. All genes involved in the biosynthesis of the repeating units have been identified, especially in the laboratory of Alf Pühler.

Exopolysaccharides play a major role in the primary stage of the infection of the host plants. It has been suggested that EPS are involved in the suppression of a defense response by the host plants and EPS mutants are eliciting a pronounced plant defense response (Parniske et al. 1994). There are linkages between the lipopolysaccharide and extracellular polysaccharide synthesis. A knockout of the dTDP-L-rhamnose synthase affects lipopolysaccharide and extracellular polysaccharide production, as shown for *Azorhizobium caulinodans* (Gao et al. 2001). The mutation affecting this gene induced only ineffective nodular structures on the host *Sesbania rostrata*, with no bacteroids and leghemoglobin present in the nodules. The bacteria were trapped in thick-walled infection threads.

4 Signal Perception and Molecular Biology of Nodule Initiation

On the molecular and cellular level a large number of responses of legume roots to Nod factors (LCOs) are known (Table 3). In the epidermis and the root hairs, ion fluxes, plasma membrane depolarization and accumulation of calcium in the root hair tips have been observed within seconds. In the range of minutes; cytoskeleton modifications, root hair deformation and specific gene expression, e.g., for *ENOD 12* are found as well as calcium^{2+} spiking and phospholipase C and D activation. In the range of hours to days, formation of pre-infection threads and cell divisions in the nodule primordium can be observed, together with the expression of other early nodulins such as *ENOD 20*. At the same time, in the vascular tissue, an inhibition of polar auxin transport and a specific gene expression of *ENOD 40* follow. Several of these reactions are triggered by nanomolar concentrations of Nod factors.

Table 3. Responses of legume roots to Nod factors (modified from Cullimore et al. 2000; Hartog et al. 2001; Hogg et al. 2002)

Tissue	Responses	Rapidity of response	Concentration of Nod factors applied	Tested plant genera and species
Epidermis and root hairs	Ion fluxes	Seconds	nM	*Medicago*
	Plasma membrane depolarization	Seconds	nM	*Medicago*
	Increase in intracellular pH	Seconds	nM	*Medicago*
	Accumulation of Ca^{2+} in root hair tip	Seconds	nM	*Medicago, Vigna*
	Ca^{2+} spiking	10 min	nM	*Medicago, Pisum*
	Gene expression (e.g., *ENOD12, RIP1*)	Minutes–hours	fM–pM	*Medicago*
	Root hair deformation	Minutes–hours	nM–μM	Many
	Cytoskeleton modification	Minutes–hours	fM–pM	*Phaseolus, Vicia*
	Phospholipase C and D activation	Minutes–hours	nM	*Vicia sativa*
Cortex	Gene expression (e.g., *ENOD20*)	Hours–days	pM	*Medicago*
	Formation of pre-infection threads	Days	nM–μM	*Vicia*
	Cell division leading to nodule primordium formation	Days	nM–μM	Many
	Competitive nodulation blocking (Cnb)	Days	nM	*Pisum*
Vasculature	Inhibition of polar auxin transport	Minutes		*Trifolium*
	Gene expression (e.g., *ENOD40*)	24 h-days	nM–μM	*Glycine, Vicia, Medicago*

There is increasing evidence that there is more than one LCO receptor responsible for these very different biochemical and structural phenotypes of Nod factor responses. In *Medicago sativa* a number of responses such as root hair deformation, membrane depolarization and ion fluxes require a sulfate group on the reducing sugar, whereas nonsulfated factors can elicit an increase in the cytosolic pH in root hairs (Felle et al. 1996). The presence of more than one LCO receptor is supported by different affinities with 4 nM for the Nod factor binding site NFBS2 and 86 nM for the binding site NFBS1 (Gressent et al. 1999). Both receptors have a more than 100-fold higher affinity for Nod factors compared to chitin fragments. This means that they are different from the chitin fragment receptors in legumes and grasses (Stacey and Shibuya 1997). From *Glycine* and *Dolichos* a lectin nucleotype phosphohydrolase has been shown to have Nod factor-binding activity. It is plasma membrane located and may also have some functions in phosphate transport (Etzler et al. 1999; Thomas et al. 1999). With *Medicago truncatula* ENOD 12-gene activation, it has been shown that heterotrimeric G proteins may be involved in the LCO signal transduction pathway (Pingret et al. 1998). This indicates an interesting relationship to signalling concepts in animal cells (see Sect. 1). The involvement of Nod factors in the different signalling pathways with G protein involvement was found by studying the phospholipase C activity in *Medicago* roots, by using the G protein activator mastoparan (Hartog et al. 2001). Similar to Nod factors, mastoparan produces root hair deformation in zone 1. It also increases the concentration of phosphatidic acid and diacylglycerol pyrophosphate four- to sixfold. The concentration of Nod factors also plays an important role. Addition of Nod factors to the cultivar Afghanistan in pea roots strongly inhibits nodulation (Hogg et al. 2002). The most obvious phenotype was the inhibition of infection thread initiation. The gene involved had been identified as *sym2*^{A} in this pea cultivar. Nod factors (LCOs) also have effects in nonlegume cells such as tobacco protoplasts (Röhrig et al. 1996). They activate the expression of the *AX11* gene involved in auxin signalling. Auxin and LCOs are transduced in tobacco cells by different pathways at, or before, the *AX11* transcription. The biochemical study on Nod factor integration into membranes revealed that they are rapidly transferred between membranes and from membrane vesicles to root hair cell walls (Goedhart et al. 1999). It was also shown that the Nod factors did not flip-flop between different membrane leaflets. Nod factors are present in buffers as monomers at a concentration effective in biological systems of around 10 nM, but when dioleoylphosphatidylcholine (DOPC) vesicles are added, the Nod factors associate with these vesicles. Our limited knowledge of the involvement of calcium spiking in the Ca^{2+} signal pathway is obvious in the present models of calcium oscillations (Schuster et al. 2002). A general model involves six types of concentration variables: inositol-1,4,5-trisphosphate, cytoplasmic calcium, endoplasmic reticulum calcium and mitochondrial calcium, the occupied binding site of calcium buffers and the active IP_3-receptor calcium released channel. The long search for the molecular identification of the gene responsible for the regulation of nodule number on the host plants (supernodulation) was successful. Characterized were a receptor like kinase with the *HAR* 1 gene (Krusell et al. 2002; Nishimura et al. 2002) and the *GmNARK* gene from soybeans, a CLAVATA 1 like receptor kinase (Searle et al. 2002). *HAR* 1 and *NARK* are the same genes from different species, as summarized by Downie and Parniske (2002).

Besides Nod factors, rhizobia also excrete other components relevant for the symbiosis development, e.g., type III secretion systems (TTSSs). Rhizobial TTSS clusters contain an open reading frame, homologous to *ysc* and *hrc* genes (Bogdanove et al. 1996; Hueck 1998; Marie et al. 2001). Two proteins have been identified in *Rhizobium* NGR234, secreted in a TTSS dependent way: *nolX* and *y4xL*. In *Bradyrhizobium japonicum* a new two-component system, *ElmS* and *ElmR*, has recently been identified, coding for a putative regulator protein and a putative sensor histidine kinase with unknown functions (Mühlencoert and Müller 2002). In agreement with the results from cytological observations in the chapter of F. Dazzo (see Chap 27, this Vol.), the population density-dependent expression of *Bradyrhizobium japonicum* nodulation genes has been reported (Loh et al. 2002). Induction of nod genes is high at low culture densities and is repressed at high population densities. The expression of NolA and NodD2 was mediated by an extracellular factor excreted into the medium. Two rhizobia species with a very broad host range are *Rhizobium* strain NGR234 and *Rhizobium fredii* USDA257 (Pueppke and Broughton 1999). They nodulate a wide range of mimosoid legumes, especially *Acacia* species and also the nonlegume *Parasponia andersonii*. In a few cases, only *Rhizobium fredii* USDA257 could nodulate some host plants such as *Glycine max* and *Glycine soja*. The most important result was that there is no relationship between the origin of the host plants and the ability of the strains to nodulate specific host plant species. The strain NGR234 shows significant dynamics of the genome architecture (Mavingui et al. 2002) with a large-scale DNA rearrangement, cointegration and excision exist between the three replicons, the symbiotic plasmids, the megaplasmid and the chromosome.

Going from the laboratory to the field under natural conditions, especially in agricultural soils, we must realize that there are not only a large number of soil, microbial and plant factors involved, but nowadays also a large number of agrochemicals present in small quantities on seeds and also on emerging roots (Johnsen et al. 2001). Pesticide effects on specific populations of soil bacteria have been demonstrated for *Rhizobium* species (Ramos and Ribeiro 1993) and with nitrifying bacteria (Martinez-Toledo et al. 1992). The degradation of these pesticides by the microbial communities in soils is another relevant aspect, contributing to the complexity of effects of molecules on the plant surface microbial interaction (Soulas and Lagacherie 2001). A large number of resistance genes in *Rhizobium* species is a strategy to deal with many antimicrobial factors concentrated around plant roots. Several multidrug efflux pumps have been identified, e.g., in *Rhizobium etli* (González-Pasayo and Martinez-Romero 2000).

Every specific symbiotic interaction has also to take a look to other strategies of the symbiotic communication. There is increasing evidence that the genetic requirements in the symbiotic interaction, e.g., between different *Rhizobium* species, pathogens and arbuscular mycorrhiza fungi species with their respective host plants, partially overlap (Parniske 2000). Different symbiotic and pathogenic interactions finally branch to very specific functions and nutrient exchanges. Common pathways and different aspects of symbiosis and defense developments are fascinating aspects of future research (Werner et al. 2002).

Acknowledgements. I thank the European Union for support in the INCO-DEV Project ICA-CT-2001-10057, the JSPS, Japan, and Mrs Lucette Claudet for the excellent work for this article.

References and Selected Reading

Baron M, Aslam H, Flasza M, Fostier M, Higgs JE, Mazaleyrat SL, Wilkin MB (2002) Multiple levels of Notch signal regulation. Mol Membr Biol 19:27–38

Becker A, Pühler A (1998) Specific amino acid substitutions in the proline-rich motif of the *Rhizobium meliloti* ExoP protein result in enhanced production of low-molecular-weight succinoglycan at the expense of high-molecular-weight succinoglycan. J Bacteriol 180:395–399

Becker BU, Kosch K, Parniske M, Müller P (1998) Exopolysaccharide (EPS) synthesis in *Bradyrhizobium japonicum*: sequence, operon structure and mutational analysis of an *exo* gene cluster. Mol Gen Genet 259:161–171

Bogdanove AJ, Beer SV, Bonas U, Boucher CA, Collmer A, Coplin DL, Cornelis GR, Huang H-C, Hutcheson SW, Panopoulos NJ et al. (1996) Unified nomenclature for broadly conserved *hrp* genes of phytopathogenic bacteria. Mol Microbiol 20:681–683

Breedveld MW, Miller KJ (1998) Cell-surface β-glucans. In: Spaink HP, Kondorosi A, Hooykaas PJJ (eds) The *Rhizobiaceae*: molecular biology of model plant-associated bacteria. Kluwer, Dordrecht, pp 81–96

Breinholt V, Larsen JC (1998) Detection of weak estrogenic flavonoids using a recombinant yeast strain and a modified MCF7 cell proliferation assay. Chem Res Toxicol 11:622–629

Brewin NJ (1998) Tissue and cell invasion by *Rhizobium*: the structure and development of infection threads and symbiosomes. In: Spaink HP, Kondorosi A, Hooykaas PJJ (eds) The *Rhizobiaceae*: molecular biology of model plant-associated bacteria. Kluwer, Dordrecht, pp 347–360

Brimecombe MJ, De Leij FA, Lynch JM (2001) The effect of root exudates on rhizosphere microbial populations. In: Pinton R, Varanini Z, Nannipieri P (eds) The rhizosphere. Marcel Dekker, New York, pp 95–140

Cooper JE, Rao JR, Eveaert E, De Cooman L (1995) Metabolism of flavonoids by rhizobia. In: Tikhonovich IA, Provorov NA, Romanov VI, Newton WE (eds) Nitrogen fixation: fundamentals and applications. Kluwer, Dordrecht, pp 287–292

Cos P, Calomme M, Sindambiwe JB, De Bruyne T, Cimanga K, Pieters L, Vanden Berghe D (2001) Cytotoxicity and lipid peroxidation-inhibiting activity of flavonoids. Planta Med 67:515–519

Cullimore JV, Ranjeva R, Bono J-J (2001) Perception of lipo-chitooligosaccharidic Nod factors in legumes. Trends Plant Sci 6:24–30

Demont N, Ardourel M, Maillet F, Promé D, Ferro M et al. (1994) The *Rhizobium meliloti* regulatory *nodD3* and *syrM* genes control the synthesis of a particular class of nodulation factors *N*-acylated by (omega-1)-hydroxylated fatty acids. EMBO J 13:2139–2149

Dixon RA, Steele CL (1999) Flavonoids and isoflavonoids – a gold mine for metabolic engineering. Trends Plant Sci 4:394–400

Downie JA, Parniske M (2002) Fixation with regulation. Nature 420:369–370

Dunn MF, Encarnación S, Araíza G, Vargas MC, Dávalos A, Peralta H, Mora Y, Mora J (1996) Pyruvate carboxylase from *Rhizobium etli*: Mutant characterization, nucleotide sequence, and physiological role. J Bacteriol 178:5960–5970

Endre G, Kereszt A, Kevei Z, Mihacea S, Kalo P, Kiss GB (2002) A receptor kinase gene regulating symbiotic nodule development. Nature 417:962–966

Entcheva P, Liebl W, Johann A, Hartsch T, Streit WR (2001) Direct cloning from enrichment cultures, a reliable strategy for the isolation of complete operons and genes from microbial consortia. Appl Environ Microbiol 67:89–99

Etzler ME, Kalsi G, Ewing NN, Roberts NJ, Day RB, Murphy JB (1999) A Nod factor binding lection with apyrase activity from legume roots. Proc Natl Acad Sci USA 96:5856–5861

Felle HH, Kondorosi E, Kondorosi A, Schultze M (1996) Rapid alkalinization in alfalfa root hairs in response to rhizobial lipochitooligosaccharide signals. Plant J 10:295–301

Gao M, D'Haeze W, dee Rycke R, Wolucka B, Holsters M (2001) Knockout of an azorhizobial dTDP-L-rhamnose synthase affects lipopolysaccharide and extracellular polysaccharide production and disables symbiosis with *Sesbania rostrata*. Mol Plant Microbe Interact 7:857–866

Giancotti FG, Ruoslahti E (1999) Integrin signaling. Science 285:1028–1033

Goedhart J, Röhrig H, Hink MA, van Hoek A, Visser AJWG, Bisseling T, Gadella Jr TWJ (1999) Nod factors integrate spontaneously in biomembranes and transfer rapidly between membranes and to root hairs, but transbilayer flip-flop does not occur. Biochemistry 38:10898–10907

González-Pasayo R, Martinez-Romero E (2000) Multiresistance genes of *Rhizobium etli* CFN42. Mol Plant Microbe Interact 13:572–577

Gressent F, Drouillard S, Mantegazza N, Samain E, Geremia RA, Canut H, Niebel A, Driguez H, Ranjeva R, Cullimore J, Bono JJ (1999) Ligand specificity of a high-affinity binding site for lipochitooligosaccharidic Nod factors in *Medicago* cell suspension cultures. Proc Natl Acad Sci USA 96:4704–4709

Harborne JB (ed) (1994) The flavonoids. Chapman and Hall, London, 621 pp

Hartog M den, Musgrave A, Munnik T (2001) Nod factor-induced phosphatidic acid and diacylglycerol pyrophosphate formation: a role for phospholipase C and D in root hair deformation. Plant J 25:55–65

Heilmann J, Merfort I (1998) Aktueller Kenntnisstand zum Metabolismus von Flavonoiden II. Resorption und Metabolismus von Flavonen, Flavanonen, Flavanen, Proanthocyanidinen und Isoflavonoiden. Pharmazie in unserer Zeit 27:173–183

Hilpert B, Bohlmann H, op den Camp RO, Przybyla D, Miersch O, Buchala A, Apel K (2001) Isolation and characterization of signal transduction mutants of *Arabidopsis thaliana* that constitutively activate the octadecanoid pathway and form necrotic microlesions. Plant J 26:435–446

Hofmann K, Heinz EB, Charles TC, Hoppert M, Liebl W, Streit WR (2000) *Sinorhizobium meliloti* strain 1021 bioS and bdhA gene transcription are both affected by biotin available in defined medium. FEMS Microbiol Lett 182:41–44

Hogg B, Davies AE, Wilson KE, Bisseling T, Downie JA (2002) Competitive nodulation blocking of cv. Afghanistan pea is related to high levels of nodulation factors made by some strains of *Rhizobium leguminosarum* bv. viciae. Mol Plant Microbe Interact 15:60–68

Hollingsworth RI, Carlson RW (1989) 27-hydroxyoctacosanoic acid is a major structural fatty acyl component of the lipopolysaccharide of *Rhizobium trifolii* ANU 843. J Biol Chem 264:9300–9303

Hollman PC, Katan MB (1998) Bioavailability and health effects of dietary flavonols in man. Arch Toxicol (Suppl) 20:237–248

Hueck CJ (1998) Type III in protein secretion systems in bacterial pathogens of animals and plants. Microbiol Mol Biol Rev 62:379–433

Ishimi Y, Miyaura C, Ohmura M, Onoe Y, Sato T, Uchiyama Y, Ito M, Wang X, Suda T, Ikegami S (1999) Selective effects of genistein, a soybean isoflavone on β-lymphopoiesis and bone loss caused by estrogen deficiency. Endocrinology 140:1893–1900

Jackson MB (2002) Long-distance signalling from roots to shoots assessed: the flooding story. J Exp Bot 53:175–181

Johnsen K, Jacobsen CS, Torswik V, Sorensen J (2001) Pesticide effects on bacterial diversity in agricultural soils – a review. Biol Fertil Soils 33:443–453

Kannenberg EL, Reuhs BL, Forsberg LS, Carlson RW (1998) Lipopolysaccharides and K-antigens. Their structures, biosynthesis and functions. In: Spaink HP, Kondorosi A, Hooykaas PJJ (eds) The Rhizobiaceae: molecular biology of model plant-associated bacteria. Kluwer Acad, Dordrecht, pp 120–154

Krusell L, Madsen LH, Sato S, Aubert G, Genua A, Szczyglowski K, Duc G, Kaneko T, Tabata S, de Bruijn F, Pajuelo E, Sandal N, Stougaard J (2002) Shoot control of root development and nodulation is mediated by a receptor-like kinase. Nature 420:422–426

Lake JA, Woodwart FI, Quick WP (2002) Long-distance CO_2 signalling in plants. J Exp Biol 53:183–193

Lerouge P, Roche P, Faucher C et al. (1990) Symbiotic host-specificity of *Rhizobium meliloti* is determined by a sulphated and acylated glucosamine oligosaccharide signal. Nature 344:781–784

Loh J, Lohar DP, Andersen B, Stacey G (2002) A two-component regulator mediates population-density-dependent expression of the *Bradyrhizobium japonicum* nodulation genes. J Bacteriol 184:1759–1766

López-Lara IM, Geiger O (2001) The nodulation protein NodG shows the enzymatic activity of an 3-Oxoacyl-acyl carrier protein reductase. Mol Plant-Microbe Interact 14:349–357

Marie C, Broughton WJ, Deakin WJ (2001) Rhizobium type III secretion systems: legume charmers or alarmers? Curr Opin Plant Biol 4:336–342

Martinez-Toledo MV, Salmeron V, Gonzalez-Lopez J (1992) Effect of the insecticides pyrimifosmethyl and chlorpyrifos on soil microflora in an agricultural loam. Plant Soil 147:25–30

Mavingui P, Flores M, Guo X, Dávila G, Perret X, Broughton WJ, Palacios R (2002) Dynamics of genome architecture in *Rhizobium* sp. strain NGR 234. J Bacteriol 184:171–176

Maximo P, Lourenço A, Feio SS, Roseiro JC (2002) Flavonoids from *Ulex airensis* and *Ulex europaeus* spp. *europaeus*. J Nat Prod 65:175–178

Morishita M, Morimoto F, Kitamura K, Koga T, Fukui Y, Maekawa H, Yamashita I, Shimoda C (2002) Phosphatidylinositol 3-phosphate 5-kinase is required for the cellular response to nutritional starvation and mating pheromone signals in *Schizosaccharomyces pombe*. Genes Cells 7:199–215

Morris VJ, Brownsey GJ, Chilvers GR, Harris JE, Gunning AP, Stevens BJH (1991) Possible biological roles for *Rhizobium leguminosarum* extracellular polysaccharide and cyclic glucans in bacteria-plant interactions for nitrogen-fixing bacteria. Food Hydrocoll 5:185–188

Mühlencoert E, Müller P (2002) A novel two-component system of *Bradyrhizobium japonicum*: *ElmS* and *ElmR* are encoded in diverse orientations. DNA Sequence 13:93–102

Neumann G, Römheld V (2001) The release of root exudates as affected by the plant's physiological status. In: Pinton R, Varanini Z, Nannipieri P (eds) The Rhizosphere. Marcel Dekker, New York, pp 41–94

Nishimura R, Hayashi M, Wu G-J, Kouchi H, Imaizumi-Anraku H, Murakami Y, Kawasaki S, Akao S, Ohmori M, Nagasawa M, Harada K, Kawaguchi M (2002) HAR1 mediates systemic regulation of symbiotic organ development. Nature 420:426–429

Pacios-Bras C, van der Burgt YEM, Deelder AM, Vinuesa P, Werner D, Spaink HP (2002) Novel lipochitin oligosaccharide structures produced by *Rhizobium etli* KIM5s. Carbohydr Res 337:1193–1202

Pai TG, Suiko M, Sakakibara Y, Liu M-C (2001) Sulfation of flavonoids and other phenolic dietary compounds by the human cytosolic sulfotransferases. Biochem Biophys Res Com 285:1175–1179

Paiva NL (2000) An introduction to the biosynthesis of chemicals used in plant-microbe communication. J Plant Growth Regul 19:131–143

Parniske M (2000) Intracellular accommodation of microbes by plants: a common developmental program for symbiosis and disease? Curr Opin Plant Biol 3:320–328

Parniske M, Schmidt PE, Kosch K, Müller P (1994) Plant defense responses of host plants with determinate nodules induced by EPS-defective *exoB* mutants of *Bradyrhizobium japonicum*. Mol Plant Microbe Interact 7:631–638

Phillips DA, Joseph CM, Yang GP, Martínez-Romero E, Sanborn JR, Volpin H (1999) Identification of lumichrome as a *Sinorhizobium* enhancer of alfalfa root respiration and shoot growth. Proc Natl Acad Sci USA 96:12275–12280

Pingret JL, Journet EP, Barker DG (1998) *Rhizobium* Nod factor signalling: evidence for a G protein-mediated transduction mechanism. Plant Cell 10:659–672

Polavieja GG de (2002) Errors drive the evolution of biological signalling to costly codes. J Theor Biol 214:657–664

Pueppke SG, Broughton WJ (1999) *Rhizobium* sp. Strain NGR234 and *R. fredii* USDA257 share exceptionally broad, nested host ranges. Mol Plant Microbe Interact 12:293–318

Ramos MLG, Ribeiro WOJ (1993) Effect of fungicides on survival of *Rhizobium* on seeds and the nodulation of bean (*Phaseolus vulgaris* L.) Plant Soil 152:145–150

Rice-Evans CA, Miller NJ (1996) Antioxidant activities of flavonoids as bioactive components of food. Biochem Soc Trans 24:790–794

Röhrig H, Schmidt J, Walden R, Czaja I, Lubenow H, Wieneke U, Schell J, John M (1996) Convergent pathways for lipochitooligosaccharide and auxin signaling in tobacco cells. Proc Natl Acad Sci USA 93:13389–13392

Rolin DB, Pfeffer PE, Osman SF, Szwergold BS, Kappler F, Benesi AJ (1992) Structural studies of a phosphocholine substituted β-(1,3); (1,6) macrocyclic glucan from *Bradyrhizobium japonicum* USDA 110. Biochim Biophys Acta 1116:215–225

Schlaman HRM, Gisel AA, Quaedvlieg NEM, Bloemberg GV, Lugtenberg BJJ et al. (1997) Chitin oligosaccharides can induce cortical cell division in roots of *Vicia sativa* when delivered by ballistic microtargeting. Development 124:4887–4893

Schoonejans E, Expert D, Toussaint A (1987) Characterization and virulence properties of *Erwinia chrysanthemi* lipopolysaccharide-defective. øEC2-resistant mutants. J Bacteriol 169:4011–4017

Schröder O, Wagner R (2002) The bacterial regulatory protein H-NS – a versatile modulator of nucleic acid structures. Biol Chem 383:945–960

Schuster S, Marhl M, Höfer T (2002) Modelling of simple and complex calcium oscillations. From single-cell responses to intercellular signalling. Eur J Biochem 269:1333–355

Schwartz MA, Ginsberg MH (2002) Networks and crosstalk: integrin signalling spreads. Nature Cell Biol 4:E65-E68

Searle IR, Men AT, Laniya TS, Buzas DM, Searle IR, Men AE, Laniya TS, Buzas DM, Iturbe-Ormaetxe I, Carroll BJ, Gresshoff PM (2003) Long-distance signaling in nodulation directed by a CLAVATA1-like receptor kinase. Science 3299:109–112

Soulas G, Lagacherie B (2001) Modelling of microbial degradation of pesticides in soils. Biol Fertil Soils 33:551–557

Spaink HP (2000) Root nodulation and infection factors produced by rhizobial bacteria. Ann Rev Microbiol. 54:257–288

Spaink HP, Sheeley DM, van Brussel AAN et al. (1991) A novel highly unsaturated fatty acid moiety of lipooligosaccharide signals determines host specificity of *Rhizobium*. Nature 354:125–130

Stacey G, Shibuya N (1997) Chitin recognition in rice and legumes. Plant Soil 194:161–169

Staehelin C, Schultze M, Tokuyasu K, Poinsot V, Promé J-C, Kondorosi E, Kondorosi A (2000) N-deacetylation of *Sinorhizobium meliloti* Nod factors increases their stability in the *Medicago sativa* rhizosphere and decreases their biological activity. Mol Plant Microbe Interact 13:72–79

Staswick PE, Yuen GY, Lehman CC (1998) Jasmonate signaling mutants of *Arabidopsis* are susceptible to the soil fungus *Pythium irregulare*. Plant J 15:747–754

Stracke S, Kistner C, Yoshida S, Mulder L, Sato S, Kaneko T, Tabata S, Sandal N, Stougaard J, Szczyglowski K, Parniske M (2002) A plant receptor-like kinase required for both bacterial and fungal symbiosis. Nature 417:959–962

Streit WR, Joseph CM, Phillips DA (1996) Biotin and other water-soluble vitamins are key growth factors for alfalfa rhizosphere colonization by *Rhizobium meliloti* 1021. Molec Plant-Microbe Interact 9:330–338

Strzelczyk E, Rozycki H (1985) Production of B-group vitamins by bacteria isolated from soil, rhizosphere, and mycorrhizosphere of pine (*Pinus sylvestris* L.). Zbl Mikrobiol 140:293–301

Thomas C, Sun Y, Naus K, Lloyd A, Roux S (1999) Apyrase functions in plant phosphate nutrition and mobilizes phosphate from extracellular ATP. Plant Physiol 119:543–551

Thomma B, Eggermont K, Penninckx IAMA, Mauch-Mani B, Vogelsang R, Cammue BPA, Broekaert WF (1998) Separate jasmonate-dependent and salicylate-dependent defense-response pathways in *Arabidopsis* are essential for resistance to distinct microbial pathogens. Proc Natl Acad Sci USA 95:15107–15111

Uren NC (2001) Types, amounts, and possible functions of compounds released into the rhizosphere by soil-grown plants. In: Pinton R, Varanini Z, Nannipieri P (eds) The rhizosphere. Marcel Dekker, New York, pp 19–40

Van Workum WAT, Kijne JW (1998) Biosynthesis of rhizobial exopolysaccharides and their role in the root nodule symbiosis of leguminous plants. In: Romeo JT, Downum KR, Verpoorte R (eds.) Phytochemical signals and plant-microbe interactions. Plenum, New York, pp 139–166

Vinuesa P, Reuhs BL, Breton C, Werner D (1999) Identification of a plasmid-borne locus in *Rhizobium. etli* KIM5s involved in lipopolysaccharide O-chain synthesis and nodulation of *Phaseolus vulgaris*. J Bacteriol 181:5606–5614

Wang GJ, Lapcik O, Hampl R, Uehara M, Al-Maharik N, Stumpf K, Mikola H, Wähälä K, Adlercreutz H (2000) Time-resolved fluoroimmunoassay of plasma daidzein and genistein. Steroids 65:339–348

Weber H (2002) Fatty acid-derived signals in plants. Trends Plant Sci 7:217–224
Werner D (1992) Symbiosis of plants and microbes. Chapman and Hall, London, 389 pp
Werner D (2001) Organic signals between plants and microorganisms. In: Pinton R, Varanini Z, Nannipieri P (eds) The rhizosphere. Marcel Dekker, New York, pp 197–222
Werner D, Müller P (2003) Communication and efficiency in the symbiotic signal exchange. In: Heldmaier G, Werner D (eds) Environmental signal processing and adaptation. Springer, Berlin Heidelberg New York, pp 9–38
Werner D, Barea JM, Brewin NJ, Cooper JE, Katinakis P, Lindström K, O'Gara F, Spaink HP, Truchet G, Müller P (2002) Symbiosis and defence in the interaction of plants with microorganisms. Symbiosis 32:83–104
Wilkinson S, Davies WJ (2002) ABA-based chemical signalling: the co-ordination of responses to stress in plants. Plant Cell Environ 25:195–210

7 The Functional Groups of Micro-organisms Used as Bio-indicator on Soil Disturbance Caused by Biotech Products such as *Bacillus thuringiensis* and Bt Transgenic Plants

GALDINO ANDRADE

1 Introduction

Insects are usually controlled with insecticides. Of the insecticides 5 % are biological, and more than 90 % of biological insecticides are based on *Bacillus thuringiensis* (Bt; Sanchis 2000). The use of bio-insecticides has increased because of the growing need to obtain better quality food and to protect the environment, but very little is known about the impact these organisms have on the environment and mainly on the soil functional microorganism groups.

Due to the efficiency of bio-insecticides based on *B. thuringiensis*, the gene which produces the bio-insecticide crystal was introduced into plants to produce Bt-transgenic plants. Transformed tobacco using the Ti plasmodium from *Agrobacterium tumefasciens* was obtained in the 1980s. Later, the electroporation and bombardment or bio-ballistic of embryos method, which is more efficient for transformation of a greater number of plant species with the *cry B. thuringiensis* gene, was used (Peferoen 1997). The second generation of Bt-transgenic plants is presently obtained with the introduction of at least two *cry* genes in the plant genome, and there are already more than 20 species of transgenic plants of economic importance being used in a few countries (Sanchis 2000).

Although transgenic plants have been produced and sown for two decades, there is little information about their environmental impact. Currently proposed plant gene products will probably have less impact on soil ecosystems than some familiar and accepted practices. However, some transgenic plant products may have measurable adverse effects on soil organisms that will have to be monitored for some years after widespread introduction (Tomlin 1994).

Plant Surface Microbiology
A. Varma, L. Abbott, D. Werner, R. Hampp (Eds.)

Some studies have assessed *B. thuringiensis* spore and vegetative cell survival in the soil. The soil permanence of the protein crystal released by Bt-transgenic maize root exudates has also been assessed (Saxena et al. 1999). The analysis of the soil stability of *B. thuringiensis* indicated that the bacterium was not active, and that the number of cells inoculated either in the vegetative or in the spore form decreased rapidly a few hours after inoculation. However, the soil can be considered its natural deposit, as the spores are released into the soil after the death of an insect and persist in this form for several years until they find another host insect. The *B. thuringiensis* toxin produced by Bt-maize is transferred to the soil by root exudates, pollen and other plant parts. Tapp et al. (1995a) reported that the *B. thuringiensis* crystal is adsorbed by the clay minerals in the soil, and thus may be protected from biodegradation action of hydrolytic enzymes such as proteases, and may remain active in the soil for several months.

Pesticides have protocols to evaluate non-target effects where many organisms are used as biological indicators. Similar test protocols could be extended to plant bio-insecticide manufacturers. The soil biological process is poorly understood, and care should be taken to prevent further negative impacts on soil ecosystems from genetically modified organisms. Although these might be non-target events, all effects of Bt-plants on the soil environment must be well understood.

2 General Aspects of *Bacillus thuringiensis*

The *B. thuringiensis* bacterium is a Gram positive rod, aerobic, chemoheterotrophic, with perithiquious flagella that sporulates when the environmental conditions are not favourable. The bio-insecticide protein is formed when the sporulation event is activated and cells should be isolated from soil or infected insects.

Meadows (1993) suggested three hypotheses for the natural habitat of *B. thuringiensis* based on isolation studies. The bacterium could be an insect pathogen, a component of the normal flora of the phylosphere tree species, or a soil microorganism.

According to the first hypothesis, the release of the crystal would be a strategy to kill the insect larvae, which would permit spore germination and vegetative cell multiplication.

The second hypothesis was suggested by a study where great quantities of *B. thuringiensis* were found in tree species, and could be disseminated by the wind or rain, and the soil would only be a deposit where *B. thuringiensis* did not multiply (Smith and Couche 1991).

These studies have shown that *B. thuringiensis* is widely distributed in soils throughout the world. Wide distribution even in localities which cannot be correlated with the presence of insects reinforces the hypothesis that the bacterium uses the soil as a natural habitat (Martin and Travers 1989). Soil survival studies showed that the number of inoculated vegetative cells decreases rapidly and only viable spores are found after 5 days. Vegetative cell multiplication was not observed (Villas-Bôas et al. 2000).

Lereclus et al. (2000) suggested that during the sporulation phase the *cry* regulation gene has the function of producing high quantities of *Cry* protein to kill the insect, allowing the bacterium to complete its biological cycle, (spore germination – multiplication – sporulation – dispersion). Virulence factors, acting in unfavourable environmental conditions, enable the bacterium to damage and invade host tissues, obtaining ideal conditions for spore germination and cell multiplication.

3 Survival in the Soil

Knowledge of *B. thuringiensis* survival in the soil is important in the context of its use for biological control. It is also relevant in the study of the interactions with other soil microorganisms. The commercial *B. thuringiensis* formulas are composed of mixtures of spores and crystals, which are released in great quantities into the environment each year. The behaviour of these spores and crystals in the soil has been studied in the field, greenhouse and in sterilised and natural soils (Thomas et al. 2000; Villas Bôas et al. 2000). In these assessments, the spores were inoculated into the soil and their recovery was monitored.

Some reports demonstrated that, initially, the number of colony spore forming units of *B. thuringiensis* declines rapidly. Vilas-Boas et al. (2000) observed that after 24 h only 20 % of the spores survived in sterilised soils.

Addison (1993) observed that several factors could affect the survival of the spores, such as pH, moisture and nutrient availability. Spore viability seemed to be little influenced by soil type or temperature. The results on nutrient availability are controversial. Cell remains could be an extra source of nutrients for the native microbiota and inoculated bacteria, but *B. thuringiensis* is unable to use nutrients released by the lysis of inoculated dead cells (West et al. 1985).

Competition with soil microbiota is one of the factors that most affects spore viability in the soil. In the soil microcosm, *B. thuringiensis* spores have a greater mortality index because of competition from other microorganisms (Pruett et al. 1980).

The initial decline in the number of colony forming units after a 24-h permanence in the soil is about 80 %. The surviving cells rapidly produce spores, increasing the number of viable spores until the number of vegetative cells is matched. This means that at a given moment, the number of spores will equal that of the vegetative cells, and will remain stable for several months.

Nutrient limitation is one of the main characteristics of soils (Edwards 1993). Microorganisms develop strategies in this oligotrophic environment to capture nutrients and survive the environmental stress. Spore formation is one of the strategies of bacterium survival when the environmental conditions are unfavourable for growth. The spore can endure stress conditions for long periods of time.

When inoculated into the soil, the spore number falls, after this initial fall, the number of cells stabilised due to vegetative cell sporulation. Petras and Casida (1985) observed an exceptional fall in the spore number during the first 2 weeks, but the number of viable spores stabilised in the third week. However, West et al. (1985) reported that spores of the *B. thuringiensis* var aizawai HD137 presented low mortality when inoculated into sterilised soil and persisted with little decrease in the initial number for 135 days.

4 History of Bacillus thuringiensis-Transgenic Plants

At the end of the 1980s, tobacco plants were the first to receive the *B. thuringiensis cry* gene using the Agrobacterium transformation system. The *Agrobacterium tumefasciens* system was used in the transformation of several dicotyledon plants. However, the electroporation and particle bombardment methods are more efficient for the transformation of monocotyledon and other dicotyledon plants (Peferoen 1997).

The first transgenic plants obtained showed low expression of the complete gene for *Cry* protein production. From then onwards, only truncated genes were introduced which codify the toxic nucleus of the Cry protein, thereby increasing the expression in several plants such as tobacco (Mazier et al. 1997), sugar cane (Arencibia et al. 1997) and peanuts (Singsit et al. 1997). Bt-transgenic potato, cotton and maize cultivation (Schnepf et al. 1998) began in 1996 and, today, there are more than 20 transgenic plant species of agronomic interest on the market.

Promoters, such as CaMV 35 s of the cauliflower mosaic virus and ubiquitinine-1 from maize are being used to increase the expression of the *cry* gene to the required levels. In addition to this strategy, greater *cry* gene expression levels were obtained by altering the sequence of the gene to increase the cytosine and guanine content and enhance the expression level from 0.02 to 0.5 % of the plant soluble protein. However, the most pronounced expression level of the *cry* gene in plants (3–5 % soluble protein) was obtained with the introduction of the unmodified gene in chloroplasts, a cell organelle which has the transcription and translation apparatus similar to that of prokaryotes (McBride et al. 1995).

New generations of transgenic plants have been developed to express other types of *cry* genes or genes expressing proteins that could have insecticide action, such as protease inhibitors, lectin, kinases, cholesterol, oxidases, inhibitors of the α-amylase and Vip proteins (Sanchis 2000).

5 Persistence of the Protein Crystal in the Soil

The release of the use of transgenic plants with *cry* genes for agricultural pest control in the 1990s has raised a lot of controversy and concern in the scientific community. It is common sense that four factors must be very carefully assessed: (1) the potential of selection for insects resistant to the Cry proteins, (2) the persistence of the crystal released by the root exudates and lysis in the soil, (3) the non-selectivity towards other non-pathogenic insects, (4) the impact of the bio-insecticide crystal protein on the functional groups of soil microorganisms.

Several studies on the persistence of the *B. thuringiensis* toxin released by transgenic plants into the soil have shown that the toxin degradation is relatively quick during the first 45 days, and less than 25 % of the initial bio-activity is maintained after 120 days (Palm et al. 1994, 1996; Sims and Holden 1996; Sims and Ream 1997). On the other hand, Tapp et al. (1995a) showed that part of the insecticide activity of *B. thuringiensis* may be maintained because of the rapid adsorption and binding of the toxin in the soil clay minerals. Other authors reported that a substantial proportion of the *Cry1Ab* toxin released in the Bt-maize roots exudates could be detected and maintained their bio-insecticide activity for 234 days after release into the soil. This indicates that the protein crystal is very stable in the soil and is protected from microbial action due to adsorption by the soil clay (Saxena et al. 1999). Experimental results on adsorption of the protein to soil particles (Saxena and Stotzky 2000) indicated that the *cry1Ab* gene coded toxin released by Bt-maize root exudates in sandy soil supplemented with montmorillonite and caullinite bound preferentially to clay minerals. This confirmed that adsorption to clay minerals is one of the main factors in the permanence and activity maintenance of the bio-insecticide crystals in the soil.

Crystals were detected by the ELISA method (enzyme-linked immunosorbent assay). The permanence of the toxin was also determined in other soil types, with predominance of those with low organic matter content. Palm et al. (1994, 1996) observed a fall in the purified toxin concentration in the soil in the first 14 days after inoculation, with stabilisation after this period. In the case of transgenic plant-produced toxin, the fall in the soil concentration occurred for about 10 days and then remained stable. However, toxin production was continuous throughout the plant lifecycle with a consequent accumulation of bio-insecticide crystals in the soil. The lowest rate of recuperation after extraction was obtained in soil with a high quantity of organic matter, indicating that much of this protein may also be adsorbed by the soil organic matter.

Saxena and Stotzky (2000) observed that the *B. thuringiensis* toxin expressed in Bt-transgenic maize was released into the rhizosphere through exudates and lysates and that much of the released crystal remained active for several months. Although Bt-transgenic plants produce and accumulate significant quantities of bio-insecticide crystals in pollen, leaves and roots, their effect on the functional groups of soil microorganisms is little understood.

The influence of the mineralogical composition of the soil on the stability of the bio-insecticide crystals has been reported by several authors. Tapp et al. (1995a) showed that the toxin is rapidly adsorbed or linked to clay minerals in the soil, remaining protected from degradation by soil microorganisms. In another study, Tapp et al. (1995b) showed that the toxin adsorbed by clay minerals becomes resistant to hydrolytic action of the enzymes produced by the soil microbiota. It has also been shown that soils with high levels of organic matter have a high protein crystal adsorption capacity (Palm et al. 1994). Sims et al. (1996) observed that *B. thuringiensis var kurstaki* Cry1Ab toxin present in Bt-transgenic maize tissues incorporated in the soil can be detected by bioassays with insects susceptible to the bio-insecticide action of the crystal. According to the results obtained by these authors, the bioassay allows the detection of smaller quantities of the protein (around 0.5 ng/ml in the diet) compared to the ELISA test (50.0 ng/g of soil). Sims et al. (1997), working with bioassays on *B. thuringiensis var kurstaki* transgenic cotton toxin inactivation in soils, observed that the toxin mean life in the soil ranges from 15 to 32 days, and less than 25 % of the initial activity remains after 120 days.

6 Effect of Bacillus thuringiensis and Its Bio-insecticide Protein on Functional Soil Microorganism Assemblage

Plant roots and their surfaces constitute dynamic habitats densely colonised by soil-borne microbiota. The high microbial activity in these habitats is due to a flow of organic substances from the photosynthetic parts of the plants to the roots (Olsson and Person 1999). This flow consists of low molecular weight organic substances (e.g. sugars, fatty acids and amino acids), as well as more complex substances (e.g. starch, cellulose and proteins). The chemical composition of this organic matter (the rhizodeposition) varies among plant species and growth stages, and is affected by plant growth conditions (Curl and Truelove 1986). The functional groups of microorganisms of nitrogen, phosphorus and carbon cycling are important to the maintenance of nutrient turnover. These microorganisms interact with the plant roots, supply nutrients and participate actively in plant nutrition and growth (Andrade 1999).

Mycorrhizal fungi are ubiquitous soil inhabitants and form a symbiotic relationship with the roots of most terrestrial plants. When arbuscular mycorrhizae (AM) form, there are significant changes in the plant and root physiology. Photosynthetic rates increase and the nutritional status of the host tissues changes and thus, the quality and quantity of root exudates (Linderman 1992). Altered exudation induces changes in the composition of microbial communities in the rhizosphere soil (Andrade et al. 1997) that may influence formation and behaviour of rhizobia nodules. Such changes could influence the competition between rhizobia and other rhizobacteria. If bacteria selectively favoured in the rhizoplane enhanced rhizobia competitiveness, then nodulation would be favoured (Linderman 1992).

In general, soil productivity and nutrient cycling are influenced by soil microbial populations. The relationships among functional groups of microorganisms of C, N and P cycling, and their influence on the plant growth, are potential indicators to evaluate disturbance in the soil environment.

A corresponding rise in the input of Bt and its toxins into soil systems can be expected with the increased use of *B. thuringiensis*-based insecticides, whether by direct spraying, in insect cadavers, or in transgenic plant material or microorganisms (Addison 1993). Very little attention has been paid to the effects that *B. thuringiensis* might have on the indigenous soil assemblages, and the information that is available is often confusing. Petras and Casida (1985) reported that endogenous soil bacteria, actinomycetes, fungi and nematodes increased moderately compared with the control when using a spore and crystal suspension of *B. thurigiensis* subsp. *kurstaki* isolated from Dipelr, a commercial preparation. Pruett et al. (1980) inoculated *B. thuringiensis* subsp. *galleriae* into clay soil and reported that bacterial populations increased 2 weeks after inoculation and were still increasing at the end of the 135-day experiment.

In contrast to the above studies, Atlavinyté et al. (1982) reported a decrease in indigenous soil microbiota when *B. thuringiensis* subsp. *galleriae* was inoculated into the soil. Bacterial numbers had decreased 50 % and actinomycetes by 90 % after 45 days, and in contrast, fungal populations had increased by 300–500 % compared with the control.

The influence of *B. thuringiensis* subsp. *kurstaki* and its protein on functional groups of soil assemblages was assessed for the first time in our laboratory, and we discuss our findings as follows.

In non-sterile soil *B. thuringiensis* vegetative cells seemed to be unable to compete with the indigenous microorganisms in non-sterile soil. Under these conditions, the number of cells decreased drastically, sporulation occurred quickly and the number of spores was stable, approximately four log unit for at least 45 days. The cell number decrease was greater in non-sterile soil than in sterile soil conditions (Villas Bôas et al. 2000). The same results were found by Thomas et al. (2000). Their results suggested that, although the soils used were of different types and composition, *B. thuringiensis* apparently did not show biological activity after spores had been released into the environment and could persist for several years (Pruett et al. 1980; Pedersen et al. 1995). However, some species of *Bacillus* genera such as *B. megaterium*, *B. subtilis*, *B cereus* suppressed pathogen fungi and/or bacteria and saprophyte fungi populations in microcosm soil (Reddy and Rhae 1989; Halverson et al. 1993; Young et al. 1995; Kim et al. 1997). In many cases, the results showed a great decrease in the viable vegetative or spore form *Bacillus* units in the soil. Reddy and Rhae (1989) reported that a strain of *B. subtilis* introduced into an onion rhizosphere at a concentration of 7.2×10^5 seed^{-1} could be recovered at only 7×10^3 plant^{-1} after 30 days, despite the decrease in numbers, the *B. subtilis* was effective in suppressing indigenous soil microbiota in the rhizosphere. Young et al. (1995) also observed that *B. cereus* survival was not influenced by developing wheat roots and the absence of a rhizosphere effect may be due to the fact that *B. cereus* was isolated originally from non-rhizosphere soil. The *Bacillus* spp. are often reported to be present in low numbers in the rhizosphere compared with other bacteria, such as fluorescent pseudomonas (Elliot Juhnke et al. 1987).

Populations of C, P-cycling microorganisms and formed nodules changed during the plant growth period and were influenced by *B. thuringiensis* inoculation in soybean plants. No differences were found on assemblages of bacteria and fungi in soil inoculated with *B. thuringiensis*, but time influenced the populations. The time corresponded to plant growth, AM root colonisation and nodule formation. Some physiological changes during plant growth, including C compounds released to the medium, influenced bacteria growth (Amora-Lazcano and Azcón 1997). AM colonisation and *Bradyrhizobium japonicum* nodulation normally decreased the amount of plant root-derived and organic matter available for heterotrophic bacteria and other soil microorganism growth by altering the root cell permeability, thus affecting exudation (Schwab et al. 1983). The carbon cycling microbiota populations also decreased their number of cells, possibly because of changes in C concentration in the rhizosphere. Negative correlation between symbiotic and cellulolytic, amylolytic and proteolytic microorganisms shows that carbon compounds from the root are important factors for their proliferation. Deleterious effects of AM roots on soil bacteria have also been observed, suggesting C competition (Marschner and Crowley 1996), although AM fungi and rhizobia do not consume C from the rhizosphere due to their symbiotic condition (Secilia and Bagyaraj 1987; Paulitz and Linderman 1989). Cellulolytic and amylolytic microorganisms decreased their cell number during the experiment, whereas proteolytic microorganisms increased their population the first time. This result suggested that this group had an extra supply of nutrients from inoculated crystal protein. The faster decrease in the proteolytic cell number after day 15 could be explained by the small amount of ICP free in the soil. Saxena et al. (1999) suggested that ICP binds rapidly and tightly to clays and humic acids and is protected against microbial degradation by being bound to soil particles. AM infection and nodule number increased in the time following the plant growth. The saprophyte fungi population decreased when the soil was inoculated with Cry- strain, and the same effect was observed in AM infection. In another experiment carried out under axenic conditions in Petri dishes, Cry– and Cry+ strains showed an inhibitory effect against the growth of some saprophytes fungi. This fungistasis effect might be explained by degrading enzymes produced (Cody 1989) or another compound that would attack the fungus cell wall or inhibit the fungal growth. Probably, these enzymes and other metabolites are produced at a vegetative phase when *B. thuringiensis* multiplies, nevertheless, this does not happen in the soil (Thomas et al. 2000; Vilas Bôas et al. 2000). However, much of the cell contents were released during the cell/spores lysis after inoculation and could have a suppressor effect on soil microbiota.

The *B. thuringiensis* effect on plant growth was observed only in plants inoculated with Cry+ strain and ICP. Although the Cry+ strain inhibited mycorrhizal colonisation, plant growth was not affected, possibly because soil fertility status and nitrogen fixation were not affected by *B. thuringiensis* inoculum. The same results were found by Reddy and Rhae (1989) with *B. subtilis* and other rhizobacteria.

AM fungi were suppressed by *B. thuringiensis* inoculum due to the use of spores as inoculum, but due to the fact that *B. thuringiensis* is found in low numbers in the rhizosphere, it is difficult to explain the inhibitory effect mechanism. Other authors (Andrade et al. 1995; Bethlenfalvay et al. 1997) found the same inhibitory effect by *Bacillus* spp. on mycorrhizae fungi colonising pea plants. Some strains of *Bacillus* spp. can probably suppress the release of AM fungi and other soil microorganism cellular contents, but this subject needs more investigation to conclude the mechanisms involved.

The present data provide evidence that *B. thuringiensis* inoculum does not produce an effect on plant growth when soil fertility is involved. However, *B. thuringiensis* var. Kurstaki HD1 demonstrated inhibitory effects on some functional groups of microorganisms that could be involved in deleterious effects in the field when the nutritional condition is oligotrophic. However, the cumulative effect of protein crystal was not evaluated. It should also be emphasised that the accumulative effect of the protein crystal due to successive cultivation of Bt-transgenic plants has not yet been assessed, but some authors have suggested that there may be a deleterious effect on the microbiota (Saxena et al. 1999) and macrofauna (Donegan et al. 1997).

The groups of soil functional microorganisms may be either positively or negatively affected by *B. thuringiensis* products, whether produced by bacteria or transgenic plants. Up to now, the results obtained by microbial ecologists are still preliminary, and it is clear that exhaustive studies should be carried out before releasing these plants into the environment. The dynamic of the functional groups of microorganisms in the presence of these plants must be understood. In addition, the accumulative effect of the crystal on these microorganism groups should be assessed together with their subsequent effects on the bio-geochemical cycles. Confidence that Bt-plants will not damage the environment when released for intense cultivation will be obtained after the positive or negative effects they may have on the environment are established.

References and Selected Reading

Addison JA (1993) Persistence and nontarget effects of *Bacillus thuringiensis* in soil: a review. Can J For Res 23:2329–2342

Amora-Lazcano E, Azcón R (1997) Response of sulphur cycling microorganisms to arbuscular mycorrhizal fungi in the rhizosphere of maize. Appl Soil Ecol 6:217–222

Andrade G (1999) Interacciones microbianas en la rizosfera. In: Siqueira JO, Moreira FMS, Lopes AS, Guilherme LR, Faquin V, Furtinni AE, Carvalho JG (eds) Soil fertility, soil biology and plant nutrition interrelationships. Brazilian Soil Science Society/ Federal University of Lavras/Soil Science Department (SBCS/UFLA/DCS), Lavras, Brazil, pp 551–575

Andrade G, Azcón R, Bethlenfalvay GJ (1995) Mycorrhizae in sustainable agriculture 1. An agrosystem affecting rhizobacterium modifies plant soil responses to a mycorrhizal fungus. Appl Soil Ecol 2:195–202

Andrade G, Mihara KL, Linderman RG, Bethlenfalvay GJ (1997) Bacteria from rhizosphere and hyphosphere soils of different arbuscular mycorrhizal fungi. Plant Soil 192;71–79

Arencibia A, Vázquez RI, Prieto D, Téllez P, Carmona ER, Coego A, Hernández L, Selman-Housein G, De La Riva GA (1997) Transgenic sugarcane plants resistant to stem borer attack. Mol Breed 3:247–255

Bethlenfalvay GJ, Andrade G, Azcón-Aguilar C (1997) Mycorrhizae in sustainable agriculture. 2. Plant and soil microorganisms in nodulated and nitrate fertilized peas. Biol Fertil Soils 24:164–168

Cody RM (1989) Distribution of chitinase and chitibiose in *Bacillus*. Curr Microbiol 19:201–205

Curl EA, Truelove B (1986) The rhizosphere. Advances series in agricultural sciences, vol 15, Springer, Berlin Heidelberg New York, pp 288

Donegan KK, Seidler RJ, Fieland VJ, Schaller DL, Palm CJ, Ganio LM, Cardwell DM, Steinbergers Y (1997) Decomposition of genetically engineered tobacco under field conditions: persistence of the proteinase inhibitor I product and effects on soil microbial respiration and protozoa, nematode and microarthropod populations. J Appl Ecol 34:767–777

Elliot Juhnke M, Mathre DE, Sands DC (1987) Identification and characterization of rhizosphere-competent bacteria of wheat. Appl Environ Microbiol 53:2793–2799

Halverson LJ, Clayton MK, Handelsman J (1993) Population biology of *Bacillus cereus* UW85 in the rhizosphere of field-grown soybeans. Soil Biol Biochem 25:485–493

Kim DS, Cook RJ, Weller DM (1997) *Bacillus* sp. L324–92 for biological control of three root diseases of wheat grown with reduced tillage. Phytopathology 87:551–558

Lereclus D, Agaisse H, Grandvalet C, Salamitou S, Gominet M (2000) Regulation of toxin and virulence gene transcription in *Bacillus thuringiensis*. Int J Med Microbiol 290:295–299

Linderman RG (1992) Vesicular-arbuscular mycorrhizae and soil microbial interactions. In: Bethlenfalvay GJ, Linderman RG (eds) Mycorrhizae in sustainable agriculture. ASA Special Publication, Madison, WI, pp 45–70

Marschner P, Crowley DE (1996) Physiological activity of a bioluminescent *Pseudomas fluorescens* (strain 2–79) in the rhizosphere of mycorrhizal and non-mycorrhizal pepper (*Capsicum annum* L.). Soil Biol Biochem 18:191–196

Martin PAW, Travers RS (1989) Worldwide abundance and distribution of *Bacillus thuringiensis* isolates. Appl Environ Microbiol 55:2437–2442

Mazier M, Chaufaux J, Sanchis V, Lereclus D, Giband M, Tourneur J (1997) The *cry*IC gene from *Bacillus thuringiensis* provides protection against *Spodoptera littoralis* in young transgenic plants. Plant Sci 127:179–190

McBride KE, Svab Z, Schaaf D J (1995) Amplification of a chimeric gene in chloroplasts leads to an extraordinary level of an insecticidal protein in tobacco. Bio/technology. 13:362–365

Meadows MP (1993) *Bacillus thuringiensis* in the environment: ecology and risk assessment (1993) In: Entwistle PF, Cory JS, Bailey MJ, Higgs S (eds) *Bacillus thuringiensis* an environmental biopesticide: theory and practice. Wiley, Chichester, pp193–220

Olsson S, Person P (1999) The composition of bacterial population in soil fractions differing in their degree of adherence to barley roots. Appl Soil Ecol 12:205–215

Palm CJ, Donegan K, Harris D, Seidler RJ (1994) Quantification in soil of *Bacillus thuringiensis* var *kurstaki* δ-endotoxin from transgenic plants. Mol Ecol 3:145–151

Palm CJ, Schaller DL, Donegan KK, Seidler RJ (1996) Persistence in soil of transgenic plant produced *Bacillus thuringiensis* var *kurstaki* δ-endotoxin. Can J Microbiol 42:1258–1262

Paulitz TC, Linderman RG (1989) Interactions between fluorescent pseudomonads and VA mycorrhizal fungi. New Phytol 113:37–45

Pedersen JC, Damgaard PH, Eilenberg J, Hansen BM (1995) Dispersal of *Bacillus thuringiensis var. kurstaki* in an experimental cabbage field. Can J Microbiol 41:118–125

Peferoen M (1997) Progress and prospects for field use of Bt genes in crops. Trends Biotechnol 15:173–177

Petras SF, Casida Jr LE (1985) Survival of *Bacillus thuringiensis* spores in soil. Appl Environ Microbiol 50:1496–1501

Pruett CJH, Burges HD, Wyborn CH (1980) Effect of exposure to soil on potency and spore viability of *Bacillus thuringiensis.* J Invert Pathol 35:168–174

Reddy MS, Rhae JE (1989) *Bacillus subtilis* B-2 and selected onion rhizobacteria in onion seedling rhizospheres: effects on seedling growth and indigenous rhizosphere microflora. Soil Biol Biochem 21:379–383

Sanchis V (2000) Biotechnological improvement of *Bacillus thuringiensis* for agricultural control of insect pests: benefits and ecological implications. In: Charles JF, Delecluse A, Nielsen-Leroux C (eds) Entomophatogenic bacteria: from laboratory to field application. Kluwer Academic, Berlin

Saxena D, Stotzky G (2000) Insecticidal toxin from *Bacillus thuringiensis* is released from roots of transgenic Bt corn in vitro and in situ. FEMS Microbiol Ecol 33:35–39

Saxena D, Flores S, Stotzky G (1999) Transgenic plants; insecticidal toxin in root exudates from Bt corn. Nature 402:480

Schnepf E, Crickmore N, Van Rie J, Lereclus D, Baum J, Feitelson J, Zeigler DR, Dean DH (1998) *Bacillus thuringiensis* and its pesticidal crystal proteins. Microbiol Mol Biol Rev 62:775–780

Schwab SM, Menge JA, Leonard RT (1983) Quantitative and qualitative effects of phosphorus on extracts and exudates of sundangrass roots in relation to vesicular-arbuscular mycorrhiza formation. Plant Physiol 73:761–765

Secilia J, Bagyaraj DJ (1987) Bacteria and actinomycetes associated with pot cultures of vesicular-arbuscular mycorrhizas. Can J Microbiol 33:1067–1073

Sims SR, Holden LR (1996) Insect bioassay for determining soil degradation of *Bacillus thuringiensis* subsp. *kurstaki* CryIA(b) protein in corn tissue. Environ Entomol 25: 659–664

Sims SR, Ream JE (1997) Soil inactivation of the *Bacillus thuringiensis* subsp. *kurstaki* CryIIA insecticidal protein within transgenic cotton tissue: laboratory microcosm and field studies. J Agric Food Chem 45:1502–1505

Singsit C, Adang MJ, Lynch RE, Anderson WF, Wang A, Cardineau G, Ozias-Akins P (1997) Expression of a *Bacillus thuringiensis cry*IA(c) gene in transgenic peanut plants and its efficacy against lesser cornstalk borer. Transg Res 6:169–176

Smith RA, Couche GA (1991) The philloplane as a source of *Bacillus thuringiensis* variants. Appl Environ Microbiol 57:311–331

Tapp H, Stotzky G (1995a) Insecticidal activity of the toxins from *Bacillus thuringiensis* subspecies *kurstaki* and *tenebrionis* adsorbed and bound on pure and soil clays. Appl Environ Microbiol 61:1786–1790

Tapp H, Stotzky G (1995b) Dot blot enzyme-linked immunosorbent assay for monitoring the fate of insecticidal toxins from *Bacillus thuringiensis* in soil. Appl Environ Microbiol 61:602–609

Thomas DJI, Alun J, Morgan W, Whipps JM, Saunders JR (2000) Plasmid transfer between the *Bacillus thuringiensis* subspecies *kurstaki* and *tenebrionis* in laboratory culture and soil and in Lepidopteran and Coleopteran larvae. Appl Environ Microbiol 118–124

Tomlin AD (1994) Transgenic plant release: comments on the comparative effects of agriculture and foresty practices on soil fauna. Mol Biol 3:51–52

Villas-Bôas LA, Villas-Bôas GFLT, Saridakis HO, Lemos MVF, Lereclus D, Arantes OMN (2000) Survival and conjugation of *Bacillus thuringiensis* in a soil microcosm. FEMS Microbiol Ecol 31:255–259

West AW, Burges HD, Dixon TJ, Wyborn CH (1985) Survival of *Bacillus thuringiensis* and *Bacillus cereus* spore inocula in soil: effects of pH, moisture, nutrient availability and indigenous microorganisms. Soil Biol Biochem 17:657–665

Young CS, Lethbridge G, Shaw LJ, Burns RG (1995) Survival of inoculated *Bacillus cereus* spores and vegetative cells in non-planted and rhizosphere soil. Soil Biol Biochem 27:1017–1026

8 The Use of ACC Deaminase-Containing Plant Growth-Promoting Bacteria to Protect Plants Against the Deleterious Effects of Ethylene

Bernard R. Glick and Donna M. Penrose

1 Introduction

Plant growth-promoting bacteria can affect plant growth and development in two different ways: indirectly or directly (Glick 1995; Glick et al. 1999). Indirect promotion of plant growth occurs when these bacteria decrease or prevent some of the deleterious effects of a phytopathogenic organism by any one or more of several different mechanisms. In general, bacteria can directly promote plant growth by providing the plant with a compound that is synthesized by the bacterium or facilitating the uptake of nutrients.

There are several ways in which plant growth-promoting bacteria can directly facilitate the proliferation of their plant hosts. They may fix atmospheric nitrogen; produce siderophores which can solubilize and sequester iron and provide it to plants; synthesize phytohormones, including auxins, cytokinins, and gibberellins which can enhance various stages of plant growth; solubilize minerals such as phosphorus; and synthesize enzymes that can modulate plant growth and development (Brown 1974; Kloepper et al. 1986, 1989; Davison 1988; Lambert and Joos 1989; Patten and Glick 1996; Glick et al. 1999). A particular bacterium may affect plant growth and development using any one, or more, of these mechanisms. Moreover, many plant growth-promoting bacteria possess several properties that enable them to facilitate plant growth and, of these, may utilize different ones at various times during the life cycle of the plant.

The mechanism most often invoked to explain the various effects of plant growth-promoting bacteria on plants is the production of phytohormones, most notably auxin (Brown 1974; Tien et al. 1979; Patten and Glick 1996). Since plants as well as plant growth-promoting bacteria can synthesize auxin, it is important when assessing the consequences of treating a plant with a plant growth-promoting bacterium, to distinguish between the bacterial stimulation of plant auxin synthesis and bacterial auxin synthesis (Gaudin et al. 1994). To complicate matters, the response of plants to auxin-producing

Plant Surface Microbiology
A. Varma, L. Abbott, D. Werner, R. Hampp (Eds.)

bacteria may vary from one species of plant to another, as well as according to the age of the plant.

2 Ethylene

In higher plants ethylene is produced from L-methionine via the intermediates, S-adenosyl-L-methionine (SAM) and 1-aminocyclopropane-1-carboxylic acid (ACC; Yang and Hoffman 1984). The enzymes involved in this metabolic sequence are SAM synthetase, which catalyzes the conversion of methionine to SAM (Giovanelli et al. 1980); ACC synthase, which is responsible for the hydrolysis of SAM to ACC and 5′–methylthioadenosine (Kende 1989) and ACC oxidase which further metabolizes ACC to ethylene, carbon dioxide, and cyanide (John 1991).

Ethylene, which is produced in almost all plants, mediates a range of plant responses and developmental steps. Ethylene is involved in seed germination, tissue differentiation, formation of root and shoot primordia, root elongation, lateral bud development, flowering initiation, anthocyanin synthesis, flower opening and senescence, fruit ripening and degreening, production of volatile organic compounds responsible for aroma formation in fruits, storage product hydrolysis, leaf and fruit abscission, and the response of plants to biotic and abiotic stress (Matoo and Suttle 1991; Abeles et al. 1992; Frankenberger and Arshad 1995). In some instances, ethylene is stimulatory while in others it is inhibitory.

The term "stress ethylene" was coined by Abeles (1973) to describe the acceleration of ethylene biosynthesis associated with biological and environmental stresses, and pathogen attack (Morgan and Drew 1997). The increased level of ethylene formed in response to trauma inflicted by chemicals, temperature extremes, water stress, ultraviolet light, insect damage, disease, and mechanical wounding (Bestwick and Ferro 1998) can be both the cause of some of the symptoms of stress (e.g., onset of epinastic curvature and formation of arenchyma), and the inducer of responses which will enhance survival of the plant under adverse conditions (e.g., production of antibiotic enzymes and phytoalexins).

Chemicals have been used to control ethylene levels in plants. The application of compounds such as rhizobitoxin, an amino acid secreted by several strains of bacteria, and its synthetic analog, aminoethoxyvinylglycine (AVG), can inhibit ethylene biosynthesis; silver thiosulfate can inhibit ethylene action, and 2-chloroethylphosphoric acid (ethephon), regarded by some researchers as "liquid ethylene", can release ethylene (Abeles et al. 1992). Sisler and Serek (1997) discovered that cyclopropenes can block ethylene perception and are potentially useful for extending the life span of cut flowers and the display life of potted plants. In addition, tropolone compounds were isolated from wood by Mizutani et al. (1998). These compounds, which can inhibit the growth of wood-rotting fungi, were shown to inhibit the biosynthesis of ethylene in excised peach pits. Many of these chemicals are potentially harmful to the environment: AVG and silver thiosulfate are highly toxic in food, and silver thiosulfate causes blackspotting in flowers (Bestwick and Ferro 1998).

3 ACC Deaminase

In 1978, an enzyme capable of degrading ACC was isolated from *Pseudomonas* sp. strain ACP, and from the yeast, *Hansenula saturnus* (Honma and Shimomura 1978; Minami et al. 1998). Since then, ACC deaminase has been detected in the fungus, *Penicillium citrinum* (Honma 1993) and in a number of other bacterial strains (Klee and Kishore 1992; Jacobson et al. 1994, Glick et al. 1995; Campbell and Thomson 1996) all of which originated in the soil. Many of these microorganisms were identified by their ability to grow on minimal media containing ACC as its sole nitrogen source (Honma and Shimomura 1978; Klee et al. 1991; Honma 1993; Jacobson et al. 1994; Glick et al. 1995; Campbell and Thomson 1996; Burd et al. 1998; Belimov et al. 2001).

Enzymatic activity of ACC deaminase is assayed by monitoring the production of either ammonia or α-ketobutyrate, the products of ACC hydrolysis (Honma and Shimomura 1978). ACC deaminase has been found only in microorganisms, and there are no microorganisms that synthesize ethylene via ACC (Fukuda et al. 1993). However, there is strong evidence that the fungus, *Penicillium citrinum*, produces ACC from SAM via ACC synthase, one of the enzymes of plant ethylene biosynthesis, and degrades the ACC by ACC deaminase. It appears that the ACC, which accumulates in the intracellular spaces, can induce ACC deaminase (Jia et al. 2000).

ACC deaminase has been purified to homogeneity from *Pseudomonas* sp. strain ACP (Honma and Shimomura 1978), *Hansenula saturnus* (Minami et al. 1998), *Penicillium citrinum* (Jia et al. 1999) and partially purified from *Pseudomonas* sp. strain 6G5 (Klee et al. 1991) and *Pseudomonas putida* GR12-2 (Jacobson et al. 1994); enzyme activity is localized exclusively in the cytoplasm (Jacobson et al. 1994). The molecular mass and form is similar for the bacterial ACC deaminases. The enzyme is a trimer (Honma 1985); the size of the holoenzyme is approximately 104–105 kDa (Honma and Shimomura 1978; Honma 1985; Jacobson et al. 1994) and the subunit mass is approximately 36.5 kDa (Honma and Shimomura 1978; Jacobson et al. 1994). Similar subunit sizes were predicted from nucleotide sequences of cloned ACC deaminase genes from *Pseudomonas* sp. strains ACP (Sheehy et al. 1991) and 6G5 (Klee et al. 1991), and *Enterobacter cloacae* UW4 (Shah et al. 1997).

The molecular mass of the holoenzymes and subunits from *Hansenula saturnus* (69 and 40 kDa, respectively) and *Penicillium citrinum* (68 and 41 kDa, respectively) suggests that these ACC deaminases are dimers (Minami et al. 1998; Jia et al. 1999).

K_m values for the binding of ACC by ACC deaminase have been estimated for enzyme extracts from 12 microorganisms at pH 8.5. These values ranged from 1.5 to 17.4 mM (Honma and Shimomura 1978; Klee and Kishore 1992; Honma 1993) indicating that the enzyme does not have a particularly high affinity for ACC (Glick et al. 1999).

ACC deaminase activity has been induced in both *Pseudomonas* sp. strain ACP and *Pseudomonas putida* GR12–2 by ACC, at levels as low as 100 nM, (Honma and Shimomura 1978; Jacobson et al. 1994); both bacterial strains were grown on a rich medium and then transferred to a minimal medium containing ACC as its sole nitrogen source. The rate of induction, similar for the enzyme from the two bacterial sources, was relatively slow: complete induction required 8–10 h. Enzyme activity increased only approximately tenfold over the basal level of activity even when the concentration of ACC increased up to 10,000-fold.

Pyridoxal phosphate is a tightly bound cofactor of ACC deaminase in the amount of approximately three moles of enzyme-bound pyridoxal phosphate per mole of enzyme, or one mole per subunit (Honma 1985).

Genes encoding ACC deaminase have been cloned from a number of different soil bacteria including *Pseudomonas* sp. strains 6G5 and 3F2 (Klee et al. 1991; Klee and Kishore 1992), *Pseudomonas* sp. strain 17 (Campbell and Thomson 1996) *Pseudomonas* sp. strain ACP (Sheehy et al. 1991) and *Enterobacter cloacae* strains CAL2 and UW4 (Glick et al. 1995; Shah et al. 1998); yeast, *Hansenula saturnus* (Minami et al. 1998); and fungus, *Penicillium citrinum* (Jia et al. 1999).

The ACC deaminase genes from *Pseudomonas* sp. strains 6G5 and F17, and *Enterobacter cloacae* strains UW4 and CAL2 all have an ORF of 1014 nucleotides that encodes a protein containing 338 amino acids with a calculated molecular weight of approximately 36.8 kDa (Klee et al. 1991; Campbell and Thomson 1996; Shah et al. 1998). The genes from these strains are highly homologous to each other: at the nucleotide level 6G5, F17, UW4 and CAL2 are 85–95 % identical to each other (Campbell and Thomson 1996; Shah et al. 1998) and most of the dissimilarities are in the wobble position (Shah et al. 1998). However, the DNA sequences from strains UW4 and CAL2 show only about 74 % homology with the sequence of the ACC deaminase gene from *Pseudomonas* sp. strain ACP (Sheehy et al. 1991; Shah et al. 1998).

Sequence data indicate that strain UW4 contains a DNA region similar to that of the anaerobic transcription regulator, FNR, (fumarate and nitrate regulator) at positions –39 to –49 (Grichko and Glick 2000). Moreover, the ACC deaminase gene promoter in strain UW4 is under the transcriptional control of a nearby gene that has a DNA sequence similar to a leucine-responsive regulatory protein (LRP) and the LRP-like protein is transcriptionally regulated by ACC (Grichko and Glick 2000; Li and Glick 2001).

When a broad host range plasmid containing the ACC deaminase gene from *Enterobacter cloacae* UW4 was introduced into two nonplant growth-promoting bacteria, *Pseudomonas putida* ATCC 17399 and *Pseudomonas fluorescens* ATCC 17400, by conjugational transfer, the transconjugants acquired the ability to grow on minimal media using ACC as the sole source of nitrogen, and to promote the elongation of canola roots (Shah et al. 1998).

In 1998, Glick et al. proposed a model in which plant growth-promoting bacteria can lower plant ethylene levels and in turn stimulate plant growth. In this model, the plant growth-promoting bacteria bind to the surface of either the seed or root of a developing plant; in response to tryptophan and other small molecules in the seed or root exudates (Whipps 1990), the plant growth-promoting bacteria synthesize and secrete indole acetic acid (IAA; Fallik et al. 1994; Patten and Glick 1996), some of which is taken up by the plant. This IAA together with endogenous plant IAA, can stimulate plant cell proliferation, plant cell elongation or induce the activity of ACC synthase to convert SAM to ACC (Kende 1993).

Much of the ACC produced by this latter reaction is exuded from seeds or plant roots along with other small molecules normally present in seed or root exudates (Penrose and Glick 2001). The ACC in the exudates may be taken up by the bacteria and subsequently hydrolyzed by the enzyme, ACC deaminase, to ammonia and α-ketobutyrate. The uptake and cleavage of ACC by plant growth-promoting bacteria decreases the amount of ACC outside the plant. Increasing amounts of ACC are exuded by the plant in order to maintain the equilibrium between internal and external ACC levels. As a result of the activity of ACC deaminase, the presence of the bacteria induces the plant to synthesize more ACC than it would otherwise need and as well, stimulates the exudation of ACC from the plant.

Thus, plant growth-promoting bacteria are supplied with a unique source of nitrogen in the form of ACC that enables them to proliferate under conditions in which other soil bacteria may not flourish. As a result of lowering the ACC level within the plant, either the endogenous level or the IAA-stimulated level, the amount of ethylene in the plant is also reduced.

Plant growth-promoting bacteria that possess the enzyme ACC deaminase and are bound to seeds or roots of seedlings, can reduce the amount of plant ethylene and the extent of its inhibition on root elongation. Thus, these plants should have longer roots and possibly longer shoots as well, inasmuch as stem elongation is also inhibited by ethylene, except in flooding-resistant plants (Abeles et al. 1992).

3.1 Treatment of Plants with ACC Deaminase Containing Bacteria

Consistent with the above mentioned model, ACC deaminase activity was completely lost and the ability to promote the elongation of canola roots under gnotobiotic conditions was greatly diminished when the ACC deaminase gene (*acdS*) from *Enterobacter cloacae* UW4 was replaced, by homologous recombination, with a version of the same gene that contained a tetracycline resistance gene inserted within the coding region (Li et al. 2000). Results of an earlier study showed that ACC deaminase mutants of *Pseudomonas putida* GR12–2 did not promote the elongation of canola roots (Glick et al. 1994). However, in those experiments, the mutants were created by chemical mutagenesis, and as a result, one could never be certain that the mutations were within the ACC deaminase structural gene per se. In the experiments by Li et al. (2000), ACC deaminase function was specifically eliminated by replacing the functional gene with an inactive version in order to demonstrate that there is no ambiguity as to the nature of the ACC deaminase minus mutants.

It has been observed that both *Escherichia coli* and two different nonplant growth-promoting pseudomonads acquired the ability to significantly promote root elongation after they were transformed with a broad-host-range plasmid carrying the *Enterobacter cloacae* UW4 ACC deaminase gene and its upstream transcriptional regulatory region (Shah et al. 1998). Moreover, elongation of canola roots following treatment of seeds with an ACC deaminase-containing bacterium is invariably accompanied by a decrease in the level of ACC found inside the root (Penrose et al. 2001). These observations confirm the effectiveness of ACC deaminase in lowering ACC levels.

As mentioned earlier, many plants respond to biotic and abiotic stresses by synthesizing ethylene. Among these stresses is the presence of heavy metals in the environment. It has been reasoned that at least some of the inhibitory effect of heavy metals on plant growth is the consequence of the plant synthesizing excessive amounts of stress ethylene in response to the presence of the metal, especially during early seedling development. Prior to being planted in metal-contaminated soil, canola and tomato seeds were treated with a heavy metal-resistant bacterium that also contained ACC deaminase. Seeds inoculated with the bacterium, *Kluyvera ascorbata*, and then grown in the presence of high concentrations of nickel chloride were partially protected against nickel toxicity (Burd et al. 1998). The presence of this bacterium had no measurable influence on the amount of nickel accumulated per mg dry weight in either roots or shoots of canola plants. Therefore, the bacterial plant growth-promoting effect in the presence of nickel was not attributable to a reduction of nickel uptake by seedlings. Rather, it reflects the ability of the bacterium to lower the level of stress ethylene caused by the nickel.

Transgenic canola plants that express *Enterobacter cloacae* UW4 ACC deaminase were tested for the ability to proliferate and accumulate metal in the presence of high levels of arsenate in the soil. In both the presence and absence of the plant growth-promoting bacterium, *Enterobacter cloacae* CAL2, the transgenic plants grew significantly larger than nontransformed plants (Nie et al. 2002).

Flooding is a common biotic stress that affects many plants, often several times during the same growing season. Plant roots suffer a lack of oxygen as a consequence of flooding, and this in turn causes deleterious effects such as epinasty, leaf chlorosis, necrosis, and reduced fruit yield. Two of the ACC synthase genes, *LE-ACS7* and *LE-ACS2*, are rapidly induced in the roots of flooded tomato plants. Of these two genes, *LE-ACS7* is expressed earliest after flooding and *LE-ACS2* is expressed approximately 8 h after flooding; the gene, *LE-ACS7* is also involved in the early wound response of tomato leaves (Shiu et al. 1998). Since ACC oxidase-catalyzed ethylene synthesis cannot occur in the anaerobic environment of flooded roots, ACC is transported into the aerobic shoots where is converted to ethylene (Bradford and Yang 1980; Else and Jackson 1998). Treatment of tomato plants with ACC deaminase-containing plant growth-promoting bacteria significantly decreases the damage suffered by these plants – damage that is caused by the deleterious effects of ethylene which normally occurs as a consequence of flooding (Grichko and Glick 2001). These ACC deaminase-containing plant growth-promoting bacteria can act as a sink for ACC, thereby lowering the level of ethylene that can be formed in the shoots. The tomato plants are thus "protected" against flooding.

Many of the symptoms of a diseased plant arise as a direct result of the stress imposed by the infection. That is, much of the damage sustained by plants infected with fungal phytopathogens occurs as a result of the response of the plant to the increased levels of stress ethylene (Van Loon 1984). It has also been observed that exogenous ethylene often increases the severity of a fungal infection and, as well, ethylene synthesis inhibitors significantly decrease the severity of a fungal infection. In a study with over 60 different cultivars and breeding lines of wheat, ethylene production increased as a result of infection with the fungal phytopathogen, *Septoria nodorum*, and was correlated with increased plant disease susceptibility (Hyodo 1991). The damage caused by the fungal phytopathogen, *Alternaria*, decreased in cotton plants by treating them with chemical inhibitors of ethylene synthesis (Bashan 1994). The levels of both ethylene and disease severity decreased in melon plants infected by the fungal phytopathogen, *Fusarium oxysporum*, following treatment of the plants with ethylene inhibitors (Cohen et al. 1986). Fungal disease development increased in both cucumber plants infected with *Colletotrichum lagenarium* (Biles et al. 1990) and in tomato plants infected with *Verticillium dahliae* (Cronshaw and Pegg 1976) when the plants were pretreated with ethylene. Treatment with ethylene inhibitors decreased disease severity in roses, carnations, tomato, pepper, French-bean and cucumber infected with the fungus, *Botrytis cinerea* (Elad 1988 and 1990).

Several biocontrol strains were transformed with the *Enterobacter cloacae* UW4 ACC deaminase gene and the effect of the transformation was assessed by using the cucumber-*Pythium ultimum* system (Wang et al. 2000). The results of the experiments indicated that ACC deaminase-containing biocontrol bacterial strains were significantly more effective than biocontrol strains that lacked this enzyme. Moreover, transgenic tomato plants that express ACC deaminase are also protected, to a significant extent, against phytopathogen-mediated damage from several different phytopathogens (Lund et al. 1998; Robison et al. 2001). In effect, ACC deaminase acts synergistically with other mechanisms of biocontrol, such as the production of antibiotics or pathogenesis-related proteins, to prevent phytopathogens from damaging plants. As with other types of stress, it is assumed that ACC deaminase can act to prevent the accumulation of ACC that would otherwise occur as a result of environmental stress.

Ethylene is also a key signal in the initiation of senescence of flowers in most plants. For example, carnation flowers produce minute amounts of ethylene until there is an endogenous rise (climacteric burst) in the level of this phytohormone. This rise in endogenous ethylene concentration is responsible for flower senescence (Mol et al. 1995), which in carnations is characterized by in-rolling of their petals.

However, ethylene does not cause senescence in all flower families, and even the features of senescence that are caused by ethylene differ from plant to plant (Woltering and Van Doorn 1988): for example, Caryophyllaceae (e.g., carnations) show ethylene-mediated wilting of their petals, whereas ethylene causes petal abscission in Rosaceae (e.g., roses), but does not cause any senescence of petals in Compositae (e.g., sunflowers).

Since ACC is a key element in the senescence of flower petals, a reduction in endogenous ACC would lower the amount of ethylene synthesized by the flower and delay the senescence of the petals. Many cut flowers (e.g., carnations and lilies), sold commercially, are routinely treated with the ethylene inhibitor, silver thiosulfate, which in high concentrations is potentially phytotoxic and environmentally hazardous. However, the use of ACC deaminase-containing plant growth-promoting bacteria could be an environmentally friendly method of lowering ACC levels in cut flowers. As a first step toward determining the feasibility of this suggestion, carnation petals were treated with ACC deaminase-containing plant growth-promoting bacteria; petal senescence was delayed by several days when compared with untreated flower petals (Nayani et al. 1998).

4 Conclusions

There are a large number of situations in which the manipulation of ACC deaminase genes could be used to improve agricultural/horticultural/silvicultural practice. Organisms containing these genes may find use in, among other things, promoting early root development from either seeds or cuttings, increasing the life of cut flowers, protecting plants against a wide range of environmental stresses, facilitating the production of volatile organic compounds responsible for aroma formation and phytoremediation of contaminated soils.

Currently, many consumers worldwide are reluctant to embrace the use of genetically modified plants as sources of foods. Thus, for the foreseeable future it may be advantageous to use either natural or genetically engineered plant growth-promoting bacteria as a means of lowering plant ethylene levels rather than genetically modifying the plant itself to achieve the same end. Moreover, given the large number of different plants, the various cultivars of those plants and the multiplicity of genes that would need to be introduced into plants, it is not feasible to genetically engineer all plants to be resistant to all types of pathogens and environmental stresses. Rather, it makes a lot of sense to engineer plant growth-promoting bacteria to do this job, and the first step in this direction could well be the introduction of appropriately regulated ACC deaminase genes.

Acknowledgements. The work from our laboratory that is described here was supported by grants from the Natural Science and Engineering Research Council of Canada. We wish to acknowledge the role of numerous collaborators and students in the work described here including: Chunxia Wang, Geneviève Défago, Shimon Mayak, Varvara Grichko, Jiping Li, Mary Robison, Peter Pauls, Saleh Shah, Barbara Moffatt, Genrich Burd, Seema Nayani, Gina Holguin, Cheryl Patten, Chris Jacobson and Daniel Ovakim. Thanks are also due to Andrei Belimov for sharing his results prior to their publication.

References and Selected Reading

Abeles FB (1973) Ethylene in plant biology. Academic Press, New York, 302 pp

Abeles FB, Morgan PW, Saltveit ME Jr (1992) Ethylene in plant biology, 2nd edn. Academic Press, San Diego

Bashan Y (1994) Symptom expression and ethylene production in leaf blight of cotton caused by *Alternaria macrospora* and *Alternaria alternata* alone and combined. Can J Bot 72:1574–1579

Belimov AI, Safronova, VI, Sergeyeva TA, Egorova TN, Matveyeva VA, Tsyganov VE, Borisov AY, Tikhonovich IA, Kluge C, Preisfeld A, Dietz K-J, Stepanok VV (2001) Characterization of plant growth promoting rhizobacteria isolated from polluted soils and containing 1-aminocyclopropane-1-carboxylate deaminase. Can J Microbiol 27:642–652

Bestwick RK, Ferro AJ (1998) Reduced ethylene synthesis and delayed fruit ripening in transgenic tomatoes expressing S-adenosylmethionine hydrolase. US Patent No: 5,723,746

Biles CL, Abeles FB, Wilson CL (1990) The role of ethylene in anthracnose of cucumber, *Cucumis sativus*, caused by *Colletotrichum lagenarium*. Phytopathology 80732–736

Bradford KJ, Yang SF (1980) Xylem transport of 1-aminocyclopropane-1-carboxylic acid, an ethylene precursor, in waterlogged tomato plants. Plant Physiol 65:322–326

Brown ME (1974) Seed and root bacterization. Annu Rev Phytopathol 12:181–197

Burd GI, Dixon DG, Glick BR (1998) A plant growth-promoting bacterium that decreases nickel toxicity in seedlings. Appl Environ Microbiol 64:3663–3668

Campbell BG, Thomson JA (1996) 1-Aminocyclopropane-1-carboxylate deaminase genes from *Pseudomonas* strains. FEMS Microbiol Lett 138:207–210

Cohen R, Riov J, Lisker N, Katan J (1986) Involvement of ethylene in herbicide-induced resistance to *Fusarium oxysporum* f. sp. *melonis*. Phytopathology 76:1281–1285

Cronshaw DK, Pegg GF (1976) Ethylene as a toxin synergist in *Verticillium* wilt of tomato. Physiol Plant Pathol 9:33–38

Davison J (1988) Plant beneficial bacteria. Bio/Technology 6:282–286

Elad Y (1988) Involvement of ethylene in the disease caused by *Botrytis cinerea* on rose and carnation flowers and the possibility of control. Ann Appl Biol 113:589–598

Elad Y (1990) Production of ethylene in tissues of tomato, pepper, French-bean and cucumber in response to infection by *Botrytis cinerea*. Physiol Mol Plant Pathol 36:277–287

Else MA, Jackson MB (1998) Transport of 1-aminocyclopropane-1-carboxylic acid (ACC) in the transpiration stream of tomato (*Lycopersicon esculentum*) in relation to foliar ethylene production and petiole epinasty. Aust J Plant Physiol 25:453–458

Fallik E, Sarig S, Okon Y (1994) Morphology and physiology of plant roots associated with *Azospirillum*. In: Okon Y (ed) *Azospirillum*/plant associations. CRC Press, Boca Raton, pp 77–85

Frankenberger WT Jr, Arshad M (1995) Phytohormones in soil. Marcel Dekker, New York

Fukuda H, Ogawa T, Tanase S (1993) Ethylene production by microorganisms. Adv Microb Physiol 35:275–306

Gaudin V, Vrain T, Jouanin L (1994) Bacterial genes modifying hormonal balances in plants. Plant Physiol Biochem 32:11–29

Giovanelli J, Mudd SH, Datko AH (1980) Sulfur amino acids in plants. In: Miflin BJ (ed) Amino acids and derivatives, the biochemistry of plants: a comprehensive treatise. vol 5. Academic Press, New York, pp 453–505

Glick BR (1995) The enhancement of plant growth by free-living bacteria. Can J Microbiol 41:109–117

Glick BR, Jacobson CB, Schwarze MK, Pasternak JJ (1994) 1-Aminocyclopropane-1-carboxylic acid deaminase mutants of the plant growth promoting rhizobacterium *Pseudomonas putida* GR12-2 do not stimulate canola root elongation. Can J Microbiol 40:911–915

Glick BR, Karaturovíc DM, Newell PC (1995) A novel procedure for rapid isolation of plant growth promoting pseudomonads. Can J Microbiol 41:533–536

Glick BR, Penrose DM, Li J (1998) A model for the lowering of plant ethylene concentrations by plant growth-promoting bacteria. J Theor Biol 190:63–68

Glick BR, Patten CL, Holguin G, Penrose DM (1999) Biochemical and genetic mechanisms used by plant growth-promoting bacteria. Imperial College Press, London

Grichko VP, Glick BR (2000) Identification of DNA sequences that regulate the expression of the *Enterobacter cloacae* UW4 1-aminocyclopropane-1-carboxylate deaminase gene. Can J Microbiol 46:1159–1165

Grichko VP, Glick BR (2001) Amelioration of flooding stress by ACC deaminase-containing plant growth-promoting bacteria. Plant Physiol Biochem 39:11–17

Honma M (1985) Chemically reactive sulfhydryl groups of 1-aminocyclopropane-1-carboxylate deaminase. Agric Biol Chem 49:567–571

Honma M (1993) Stereospecific reaction of 1-aminocyclopropane-1-carboxylate deaminase. In: Pech JC, Latché A, Balagué C (eds) Cellular and molecular aspects of the plant hormone ethylene. Kluwer, Dordrecht, pp 111–116

Honma M, Shimomura T (1978) Metabolism of 1-aminocyclopropane-1-carboxylic acid. Agric Biol Chem 42:1825–1831

Hyodo H (1991) Stress/wound ethylene. In: Mattoo AK, Suttle JC (eds) The plant hormone ethylene. CRC Press, Boca Raton, pp 65–80

Jacobson CB, Pasternak JJ, Glick BR (1994) Partial purification and characterization of 1-aminocyclopropane-1-carboxylate deaminase from the plant growth promoting rhizobacterium *Pseudomonas putida* GR12-2. Can J Microbiol 40:1019–1025

Jia Y-J, Kakuta Y, Sugawara M, Igarashi T, Oki N, Kisaki M, Shoji T, Kanetuna Y, Horita T, Matsui H, Honma M (1999) Synthesis and degradation of 1-aminocyclopropane-1-carboxylic acid by *Penicillium citrinum*. Biosci Biotechnol Biochem 63:542–549

Jia Y-J, Ito H, Matsui H, Honma M (2000) 1-Aminocyclopropane-1-carboxylate (ACC) deaminase induced by ACC synthesized and accumulated in *Penicillium citrinum* intracellular spaces. Biosci Biotechnol Biochem 64:299–305

John P (1991) How plant molecular biologists revealed a surprising relationship between two enzymes, which took an enzyme out of a membrane where it was not located, and put it into the soluble phase where it could be studied. Plant Mol Biol Rep 9:192–194

Kende H (1989) Enzymes of ethylene biosynthesis. Plant Physiol 91:1–4

Kende H (1993) Ethylene biosynthesis. Annu Rev Plant Physiol Plant Mol Biol 44:283–307

Klee HJ, Kishore GM (1992) Control of fruit ripening and senescence in plants. US Patent No: 5,702,933

Klee HJ, Hayford MB, Kretzmer KA, Barry GF, Kishore GM (1991) Control of ethylene synthesis by expression of a bacterial enzyme in transgenic tomato plants. Plant Cell 3:1187–1193

Kloepper JW, Scher FM, Laliberté M, Tipping B (1986) Emergence-promoting rhizobacteria: description and implications for agriculture. In: Swinburne TR (ed) Iron, siderophores, and plant disease. Plenum, New York, pp 155–164

Kloepper JW, Lifshitz R, Zablotowicz RM (1989) Free-living bacterial inocula for enhancing crop productivity. Trends Biotechnol 7:39–43

Lambert B, Joos H (1989) Fundamental aspects of rhizobacterial plant growth promotion research. Trends Biotechnol 7:215–219

Li J, Glick BR (2001) Transcriptional regulation of the *Enterobacter cloacae* UW4 1-aminocyclopropane-1-carboxylate (ACC) deaminase gene (*acdS*). Antonie van Leewenhoek 80:255–261

Li J, Ovakim DH, Charles TC, Glick BR (2000) An ACC deaminase minus mutant of *Enterobacter cloacae* UW4 no longer promotes root elongation. Curr Microbiol 41:101–105

Lund ST, Stall, RE, Klee HJ (1998) Ethylene regulates the susceptible response to pathogen infection in tomato. Plant Cell 10:371–382

Mattoo AK, Suttle JC (1991) The plant hormone ethylene. CRC Press, Boca Raton

Minami R, Uchiyama K, Murakami T, Kawai J, Mikami K, Yamada T, Yokoi D, Ito H, Matsui H, Honma M (1998) Properties, sequence, and synthesis in *Escherichia coli* of 1-aminocyclopropane-1-carboxylate deaminase from *Hansenula saturnus*. J Biochem 123:1112–1118

Mizutani F, Golam Rabbany ABM, Akiyoshi H (1998) Inhibition of ethylene production by tropolone compounds in young excised peach pits. J Jpn Soc Hortic Sci 67:166–169

Mol JNM, Holton TA, Koes RE (1995) Floriculture: genetic engineering of commercial traits. Trends Biotechnol 13:350–355

Morgan PW, Drew CD (1997) Ethylene and plant responses to stress. Physiol Plant 100:620–630

Nayani S, Mayak S, Glick BR (1998) The effect of plant growth promoting rhizobacteria on the senescence of flower petals. Ind J Exp Biol 36:836–839

Nie L, Shah S, Rashid A, Burd GI, Dixon GD, Glick BR (2002) Phytoremediation of arsenate contaminated soil by transgenic canola and the plant growth-promoting bacterium *Enterobacter cloacae* CAL2. Plant Physiol Biochem 40:355–361

Patten CL, Glick BR (1996) Bacterial biosynthesis of indole-3-acetic acid. Can J Microbiol 42:207–220

Penrose DM, Glick BR (2001) Levels of 1-aminocyclopropane-1-carboxylic acid (ACC) in exudates and extracts of canola seeds treated with plant growth-promoting bacteria. Can J Microbiol 47:368–372

Penrose DM, Moffatt BM, Glick BR (2001) Determination of 1-aminocyclopropane-1-carboxylic acid (ACC) to assess the effects of ACC deaminase-containing bacteria on roots of canola seedlings. Can J Microbiol 47:77–80

Robinson MM, Shah S, Tamot B, Pauls PK, Moffatt BA, Glick BR (2001) Reduced symptoms of Verticillium wilt in tomato plants transformed with ACC deaminase to control ethylene biosynthesis. Mol Plant Pathol 2:135–145

Shah S, Li J, Moffatt BA, Glick BR (1997) ACC deaminase genes from plant growth promoting bacteria. In: Ogoshi A, Kobayashi K, Homma Y, Kodama F, Kondo N, Akino S (eds) Plant growth-promoting rhizobacteria: present status and future prospects. OECD, Paris, pp 320–324

Shah S, Li J, Moffatt BA, Glick BR (1998) Isolation and characterization of ACC deaminase genes from two different plant growth promoting rhizobacteria. Can J Microbiol 44:833–843

Sheehy RE, Honma M, Yamada M, Sasaki T, Martineau B, Hiatt WR (1991) Isolation, sequence, and expression in *Escherichia coli* of the *Pseudomonas* sp. strain ACP gene encoding 1-aminocyclopropane-1-carboxylate deaminase. J Bacteriol 173:5260–5265

Shiu OY, Oetiker JH, Yip WK, Yang SF (1998) The promoter of *LE-ACS7*, an early flooding-induced 1-aminocyclopropane carboxylate synthase gene of the tomato, is tagged by a Sol3 transposon. Proc Natl Acad Sci USA 95:10334–10339

Sisler EC, Serek M (1997) Inhibitors of ethylene responses in plants at the receptor level: recent developments. Physiol Plant 100:577–582

Tien TM, Gaskins MH, Hubell DH (1979) Plant growth substances produced by *Azospirillum brasilense* and their effect on the growth of pearl millet (*Pennisetum americanum* L). Appl Environ Microbiol 37:1016–1024

Van Loon LC (1984) Regulation of pathogenesis and symptom expression in diseased plants by ethylene. In: Fuchs Y, Chalutz E (eds) Ethylene: biochemical, physiological and applied aspects. Martinus Nijhoff/Dr. W. Junk, The Hague, pp 171–180

Wang C, Knill E, Glick BR, Défago G (2000) Effect of transferring 1–aminocyclopropane-1-carboxylic acid (ACC) deaminase genes into *Pseudomonas fluorescens* strain CHA0 and its *gacA* derivative CHA96 on their growth-promoting and disease-suppressive capacities. Can J Microbiol 46:898–907

Whipps JM (1990) Carbon utilization. In: Lynch JM (ed) The rhizosphere. Wiley Interscience, Chichester, pp 59–97

Woltering EJ, Van Doorn WG (1988) Role of ethylene in senescence of petals – morphological and taxonomical relationships. J Exp Bot 39:1605–1616

Yang SF, Hoffman NE (1984) Ethylene biosynthesis and its regulation in higher plants. Annu Rev Plant Physiol 35:155–189

9 Interactions Between Epiphyllic Microorganisms and Leaf Cuticles

Lukas Schreiber, Ursula Krimm and Daniel Knoll

1 Introduction

Leaves of higher plants are exposed to the atmosphere. Due to the pronounced two-dimensional structure of leaves, the surface area of plants is significantly enlarged. This allows an efficient absorption of visible light used in photosynthesis and it supports the rapid gas exchange of carbon dioxide and oxygen, occurring across stomates. With most leaves, stomates representing small pores, cover only between 0.5 to 1 % of the total leaf surface area (Larcher 1996), whereas the largest part of the leaf surface is covered by the plant cuticle forming the major interface between the leaves and the atmosphere (Kerstiens 1996). The cuticle developed during evolution when plants moved from their aqueous habitats to the dry land. It protects land living plants from desiccation. The water potential in the atmosphere is nearly always lower than the water potential of plants, which causes a constant driving force for the flow of water from the plant body to the atmosphere (Nobel 1991). Without the cuticle forming a very efficient transport barrier for the passive diffusion of water from the turgescent plant to the atmosphere, most of the land-living higher plants would never be able to survive.

Besides this major function as a watertight barrier, the plant cuticle also limits the leaching of ions and nutrients from the leaf interior (Tukey 1970), and it forms a mechanical barrier for most microorganisms trying to infect the living leaf tissues (Mendgen 1996; Schafer 1998). Looking at the surfaces of healthy, green leaves collected in the environment in their natural habitats using different microscopical techniques (fluorescence microscopy, confocal laser scanning microscopy or scanning electron microscopy), it becomes obvious that leaf surfaces are always covered by epiphyllic microorganisms to a certain degree (Fig. 1). This epiphyllic flora is composed of bacteria, yeasts and filamentous fungi belonging to different systematic categories (Morris et al. 1996). The degree of coverage strongly depends on a series of parameters like the plants species, the structure of the leaf surface, the habitat of the plant and the age of the leaf (Preece and Dickinson 1971; Dickinson and Preece

Plant Surface Microbiology
A. Varma, L. Abbott, D. Werner, R. Hampp (Eds.)

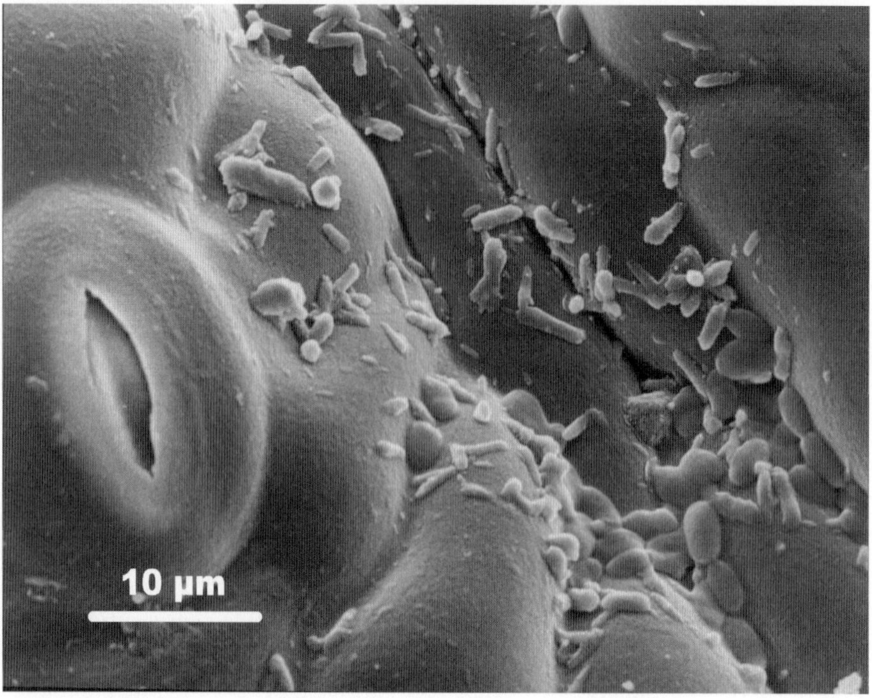

Fig. 1. Micrograph (SEM) of the lower stomatous leaf side of walnut (*Juglans regia* L.). A dense epiphyllic microflora of bacterial (*smaller rod-like cells*) and yeast cells (*larger spherical cells*) in a depression of the lower leaf surface can be seen

1976). Population densities of leaf surface microorganisms are characterised by large fluctuations, since they are strongly dependent on environmental conditions (Fokkema and van den Heuvel 1986). Rapid changes from favourable environmental conditions (e.g., high humidity and low irradiance) to unfavourable conditions (e.g., low humidity, high irradiance and high temperatures) as they can naturally occur within hours or days are followed by rapid changes in the density and the number of epiphyllic microorganisms (Leben 1988).

Thus, it can be concluded that the phyllosphere forms a characteristic habitat for microorganisms, a fact which has largely been neglected in the past (Beattie and Lindow 1995). Plant biologists normally investigate the structure and the function of the cuticle as a barrier for water and organic compounds (Schönherr and Riederer 1989), whereas plant pathologists are mostly interested in the interaction between pathogens and the living epidermal cells or the cell walls (Dixon and Lamb 1990). Environmental microbiologists are mostly interested in describing the epiphyllic population dynamics, the species composition and their potential as antagonists

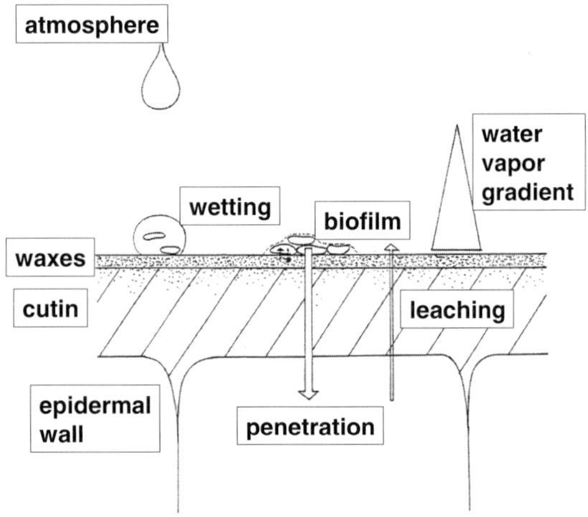

Fig. 2. A scheme of the phyllosphere as a habitat for microorganisms showing most of the relevant climatic and plant parameters determining the living conditions of the leaf surface

(Andrews 1992; Jacques and Morris 1995; Fiss et al. 2000). However, looking at the leaf surface as a microhabitat with very specific boundary conditions, investigations of the parameters, limitations and interactions between the lipophilic leaf surface and the microorganisms have rarely been carried out (Fig. 2). This habitat is characterised by an extreme microclimate due to large variations in climatic parameters like light intensity and temperature (Andrews and Harris 2000). Due to specific physical, chemical and biological properties of leaf surfaces, the phyllosphere is also dominated by a low availability of water and nutrients (Schönherr and Baur 1996; Beattie and Lindow 1999). Investigating the microbial ecology of the phyllosphere will be a combined approach including plant ecophysiological and microbiological tools. In the following, several important aspects of the microbial ecology of the phyllosphere will be discussed and selected examples for the interactions occurring between epiphyllic microorganisms and the leaf surface will be given.

2 Physical and Chemical Parameters of the Phyllosphere

The plant cuticle covering the leaf surface is a lipophilic, extracellular biopolymer. It is composed of the cutin polymer (Kolattukudy 2001), which is a polyester of esterified hydroxy fatty acids, and of cuticular waxes (Walton 1990), deposited as monomeric compounds to the cutin polymer (intracuticular waxes) and to the cutin surface (epicuticular waxes). Cuticular waxes are basically linear long chain aliphatic compounds of different chain length and different substance classes. Typical wax constituents are alkanes, aldehydes, primary and secondary alcohols, acids and esters composed of the respective acids and alcohols (Bianchi 1995). Besides these linear long-chain aliphatics, cuticular waxes of some plant species are dominated by a large degree by triterpenoic acids and triterpenols (Gülz 1994; Markstädter et al. 2000). The chemical environment which will be sensed by epiphyllic microorganisms living on leaf surfaces will be the outermost layer of wax compounds forming the true interface between the leaf and the atmosphere. For this reason, analytical chemistry such as gas chromatography coupled to different detector systems (FID and MS) is an important tool for describing the chemical environment of the leaf surface (Riederer and Markstädter 1996).

Epicuticular waxes often form characteristic three-dimensional structures like platelets (Fig. 3), rods, ribbons or filaments (Jeffree 1986). This significantly increases leaf surface roughness. As a consequence, water drops, small particles, as well as spores and bacterial cells located on the tips of these crystals strongly reduce the attachment of particles to the leaf surface. Thus, rain or water can simply wash off these loosely attached particles on rough leaf surfaces (Barthlott and Neinhuis 1997). Both parameters, the very hydrophobic nature of cutin and wax and the often very pronounced roughness of the leaf surface, are responsible for the fact that leaf surfaces are a very dry habitat since water is very efficiently rejected (Holloway 1970). Nevertheless, with increasing leaf age in most cases epicuticular wax crystals tend to disappear, probably due to erosion, and the factor roughness will become less significant

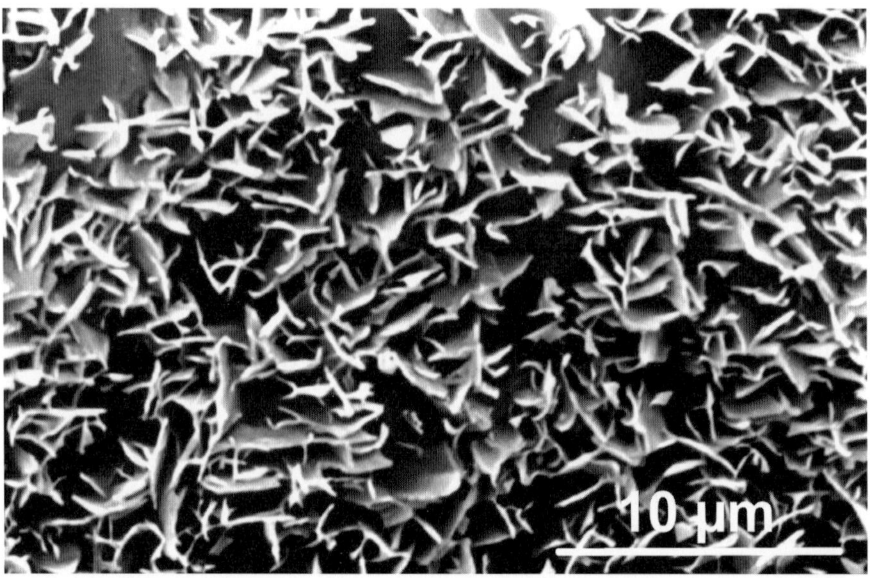

Fig. 3. Micrograph (SEM) of the upper astomatous leaf side of oak (*Quercus robur* L.) showing the dense accumulation of epicuticular wax crystals. The crystals, having the shape of small platelets, are oriented in a rectangular angle to the leaf surface leading to a pronounced surface roughness

for epiphyllic microorganisms trying to colonise older leaves (Neinhuis and Barthlott 1998).

In addition, cuticular waxes are responsible for establishing the transport barrier of the plant cuticle (Riederer and Schreiber 1995). Extraction of cuticular waxes with appropriate solvents such as chloroform increased cuticular transport for water and dissolved compounds by two to three orders of magnitude (Schönherr and Riederer 1989). At room temperature, waxes form solid partially crystalline aggregates with a high degree of order (Reynhardt and Riederer 1994; Schreiber et al. 1997) and, thus, they efficiently seal the amorphous cutin polymer, which itself is fairly permeable for water and dissolved compounds (Schönherr and Riederer 1989). The structure of the cuticle can best be compared to the technical principle realised in wax-coated papers, where the wax establishes a transpiration barrier, whereas the cellulose polymer forms a stable matrix for deposition of the wax (Fox 1958). Thus, although epiphyllic microorganisms live on a substrate, below which the best conditions in terms of water supply and nutrient concentrations exist, the leaf surface is an environment with extremely unfavourable conditions, because this reservoir below the cuticle is rarely accessible to leaf surface microorganisms under normal conditions (Schönherr and Baur 1996).

3 Leaf Surface Colonisation and Species Composition

Freshly emerging leaves are basically clean, unwettable and they often have a pronounced roughness due to epicuticular waxes crystals (Neinhuis and Barthlott 1998). Pronounced succession in leaf surface colonisation has been described by several authors (Ercolani 1991; Kinkel 1991; Blakeman 1993). Normally, the first detectable microorganisms are bacteria starting to colonise the leaf surface (Blakeman 1991). Later in the season, yeasts become more and more abundant in the phyllosphere due to additional nutrients like pollen and high amounts of sugars becoming available by the activity of aphids (Stadler and Müller 1996). Towards the end of the season, especially with deciduous trees, leaf surfaces are often densely covered with filamentous fungi. This might be related to decreasing barrier properties of the cuticle due to leaf ageing.

Once epiphyllic microorganisms have succeeded in colonising the leaf surface they are strongly attached to the surface (Romantschuk 1992) and can rarely be removed even after excessive washing (Schreiber and Schönherr 1993). They often tend to protect themselves in an extracellular matrix (Beattie and Lindow 1999), and it has also been shown that biofilms, containing different bacterial species, may develop in the phyllosphere (Morris et al. 1997, 1998). The species living in the leaf surface belong to diverse taxonomic groups. Most abundant bacterial species which have been described belonged to the genera *Corynebacterium*, *Erwinia*, *Pseudomonas*, *Xan-*

thomonas and *Bacillus* (Ercolani 1991; Morris et al. 1998). *Cladosporium*, *Alternaria* and *Aureobasidium* have been described as being abundant filamentous fungal species and *Cryptococcus* and *Sporobolomyces* were described as abundant yeast species in the phyllosphere (Andrews and Harris 2000; Blakeman 1993).

However, it must be mentioned that the description of the epiphyllic microflora up to now is exclusively based on an identification of the species after cultivation on standard media. However, it is well known today from environmental microbiology that many bacterial species cannot be cultivated with standard techniques (Amann et al. 1995). PCR-based approaches showed that the bacterial species composition of aquatic environments, but also of soil and rhizosphere communities, is much more complex and diverse as it was originally concluded from cultivation-based approaches (Marilley et al. 1998; Tiedje et al. 1999; Ogram 2000). A similar approach has rarely been carried out in the leaf surface and it is absolutely necessary in leaf surface microbiology in future in order to obtain a more realistic and complete picture of the species composition in the phyllosphere. Results of one of the first approaches, comparing the bacterial species identified using PCR versus cultivation-based techniques (Yang et al. 2001), in fact, yielded two quite different pictures of the species composition of the phyllosphere. This proves that our knowledge of the species composition on leaf surfaces obtained from cultivation-based techniques is still rather limited and needs further research.

4 Alteration of Leaf Surface Wetting

Investigations of the seasonal development of leaf surface wetting have shown several times that leaf surfaces become more and more wettable with increasing leaf age (Cape 1983; Turunen and Huttunen 1989; Cape and Percy 1993; Neinhuis and Barthlott 1998). This was normally attributed to chemical changes in the physico-chemical properties of the waxy leaf surface at the leaf/atmosphere interface caused by environmental pollution. In addition, it was shown that wax erosion due to the constant exposure of the leaf surface to wind, rain and the deposition of dust particles from the atmosphere to the leaf surface also occurs (van Gardingen et al. 1991), and may be further contributed to these observed increases in wetting. However, epiphyllic microorganisms as a further parameter contributing to an increased wetting of the leaf surface may not be neglected here.

In simple model experiments, silanised glass surfaces, which are rarely wetted by water due to their high hydrophobicity, were colonised by bacteria and wetting properties were quantified by measuring contact angles (Knoll and Schreiber 1998, 2000). From these experiments, it became obvious that already at a coverage of 10% of the total surface, contact angles decreased by 25°

(Fig. 4A). Maximum effects were a decrease of the contact angle from about 95° to 30° at a coverage of 70 %. Similar results were obtained when clean ivy leaf surfaces were colonised by bacteria. A bacterial coverage of 10 % of the leaf surface resulted in a decrease in the contact angle by 25° and only a 25 % coverage resulted in decrease from 90° to 40° (Fig. 4B). These experiments clearly proved that leaf surface wetting properties can be altered to a large degree by the presence of epiphyllic microorganisms.

Using scanning electron microscopy, gas chromatography and contact angle measurements in parallel, investigation of needle (*Abies grandis* Lindl.) and leaf surfaces (*Juglans regia* L.) during one season supported this observation (Schreiber 1996; Knoll and Schreiber 1998). The pronounced increase in

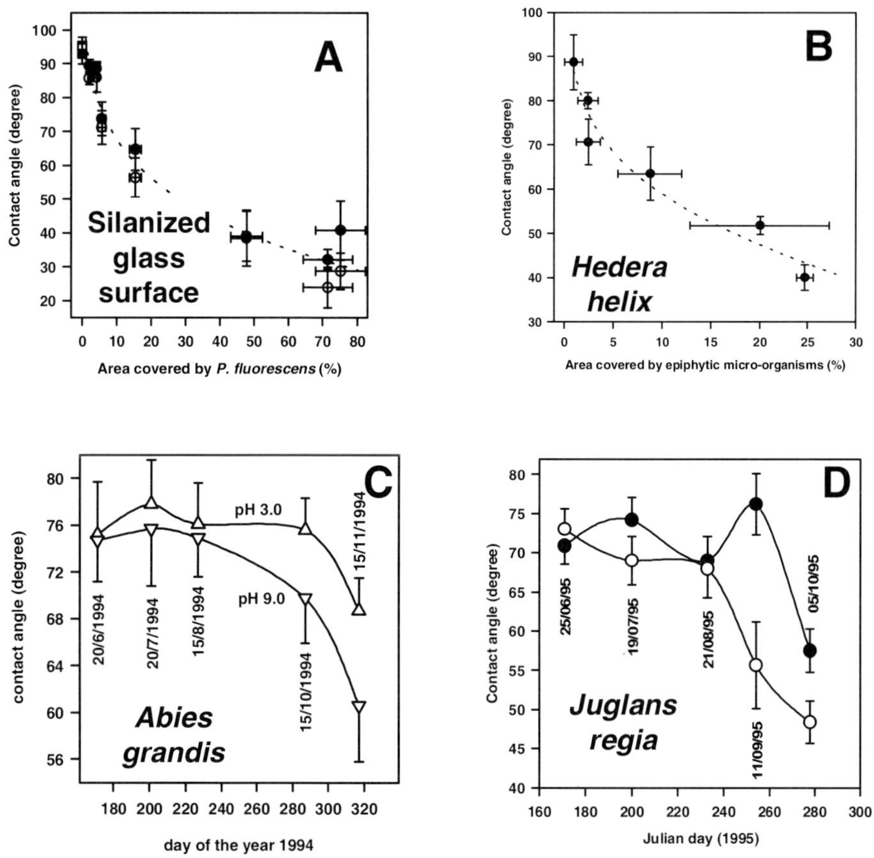

Fig. 4. Degree of wetting of a silanised glass surface (**A**) and an ivy (*Hedera helix* L.) leaf surface (**B**) as a function of the coverage of the leaf surface with epiphyllic microorganisms. Seasonal increase of needle (*Abies grandis* Lindl.) surface (**C**) and leaf (*Juglans regia* L.) surface (**D**) wetting due to increasing amounts of microorganisms growing in the phyllosphere

the needle and the leaf surface wetting properties quantified by contact angle measurements (Fig. 4C, D) was always in parallel with a significant increase in the colonisation of the needle and leaf surfaces with epiphyllic microorganisms as seen in scanning electron microscopy. However, changes in the qualitative and quantitative wax composition, measured by gas chromatography, were not at all correlated with the changes in the wetting properties of the leaf surfaces (Schreiber 1996; Knoll and Schreiber 1998).

From this, it is evident that leaf surface microorganisms have the ability to significantly change leaf wettability by altering the physico-chemical properties of leaf surfaces. This is probably an important ecological strategy of epiphyllic microorganisms improving the living conditions in their environment. Increased wetting will increase the water availability in the leaf surface, which in turn is highly favourable for the microorganisms living there. Furthermore, increased wetting will also more easily lead to the formation of thin water films, which is necessary in order to dissolve substances leaching from the apoplast to the leaf surface. As a consequence, the availability and the amount of nutrients in the phyllosphere will increase as well, which again is favourable for epiphyllic microorganisms.

5 Interaction of Bacteria with Isolated Plant Cuticles

It is generally believed that plant cuticles form more or less impermeable mechanical barriers for bacteria (Agrios 1995). Whereas fungi may have the ability to penetrate the cuticle using extracellular enzymes (Schäfer 1998), for bacteria an infection of the leaf tissue only seems to be possible via stomates or hydathodes forming natural openings or via artificial openings like cracks caused by injuries. In order to test this hypothesis, isolated cuticular membranes from different plant species were mounted in transpiration chambers and cuticular water permeability was quantified as a measure of the effect of microorganisms on leaf surface barrier properties.

Cuticular water permeability of selected species (*Vinca major* L., *Hedera helix canariensis* L. and *Prunus laurocerasus* L.) was measured before and after inoculation with *Pseudomonas fluorescens*, which was chosen as a characteristic and representative epiphyllic microorganism. With all three investigated species, cuticular water permeability significantly increased by factors between 40 to 60 % after inoculation with *P. fluorescens* for 10–12 days (Fig. 5). In parallel to the observed increase in cuticular water permeability, it was always observed that the bacteria had successfully penetrated the cuticle, since bacteria were growing on the inner side of the isolated cuticle, which was sterile at the beginning of the experiment. From this observation, it must be concluded that the bacteria had induced additional defects to the transport barrier of the cuticle, leading to increased rates of water permeability as well as paths for penetrating the cuticle (Knoll 1998).

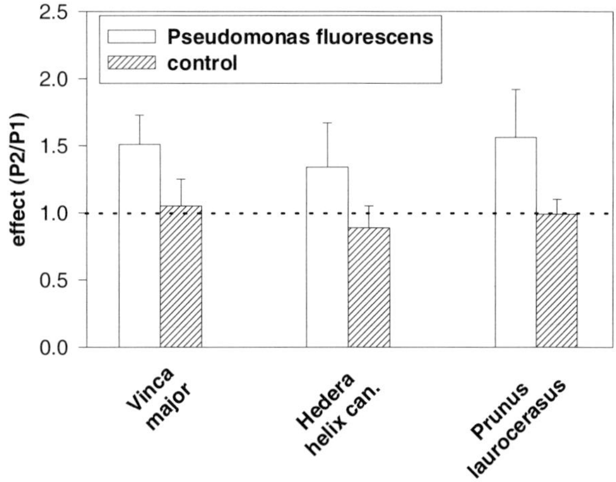

Fig. 5. Interaction between *Pseudomonas fluorescens* growing on isolated cuticles of different plant species and cuticular water permeability. The effect, which was calculated from the ratio of cuticular water permeability after inoculation divided by cuticular transpiration before inoculation, indicates the relative increase in cuticular water permeability after inoculation with bacteria

However, at the moment, the mechanism as to how this was achieved by the bacteria is not clear. One possibility might be a dissolution of the cutin polymer by extracellular bacterial enzymes. Alternatively, one could also image a pure physical basis. It was shown in the past that cuticular permeability for water and many dissolved compounds can be increased by surfactants (Riederer and Schönherr 1990). A similar mechanism might be used by microorganisms, since for many of them it has been shown that they are able to synthesise biosurfactants (Persson et al. 1988; Bunster et al. 1989; Karanth et al. 1999). Moreover, it also may not be forgotten that in reality there are living epidermal cells below the cuticle. They probably will significantly contribute to inhibiting leaf surface microorganisms from penetrating the cuticle, which is not the case in the artificial system using isolated cuticular membranes. Future work will have to concentrate on this important question of the interaction between epiphyllic microorganisms and the plant cuticle.

6 Conclusions

In conclusion, it must be stated that lipophilic surfaces of leaves form microhabitats for many microorganisms, although living conditions in terms of water and nutrient availability and climatic conditions in the phyllosphere are far from optimal. Specific interactions between epiphyllic microorganisms and the plant cuticle, leading to increased leaf surface wetting and elevated rates of cuticular permeability, have been shown to occur. Nevertheless, there is still a series of questions which deserves further attention in future research. Using molecular biological tools, a more realistic description of the diversity of the species composition in the phyllosphere must be achieved. Physiological experiments will have to analyse in more detail the mechanisms forming the basis for the different interactions occurring between epiphyllic microorganisms and the plant cuticle. Furthermore, an important question is to what extent the aggregation of different epiphyllic species forming biofilms increases their ecological fitness in the phyllosphere. Answering these questions in the future will significantly help to improve our knowledge of the microbial ecology of the phyllosphere.

Acknowledgements. The authors gratefully acknowledge financial support of this work by the Deutsche Forschungsgemeinschaft and the FCI.

References and Selected Reading

Agrios GN (1995) Plant pathology. San Diego, Academic Press

Amann R, Ludwig W, Schleifer KH (1995) Phylogenetic identification and in situ detection of individual microbial cells without cultivation. Microbiol Rev 59:143–169

Andrews JH (1992) Biological control in the phyllosphere. Annu Rev Phytopathol 30:603–635

Andrews JH, Harris RF (2000) The ecology and biogeography of microorganisms of plant surfaces. Annu Rev Phytopathol 38:145–180

Barthlott W, Neinhuis C (1997) Purity of the sacred lotus, or escape from contamination in biological surfaces. Planta 202:1–8

Beattie GA, Lindow SE (1995) The secret life of foliar bacterial pathogens on leaves. Annu Rev Phytopathol 33:145–172

Beattie GA, Lindow SE (1999) Bacterial colonization of leaves: a spectrum of strategies. Phytopathology 89:353–359

Bianchi G (1995) Plant waxes. In: Hamilton RJ (ed) Waxes: chemistry, molecular biology and functions. The Oily Press, Dundee, pp 175–222

Blakeman JP (1991) Foliar bacterial pathogens: epiphytic growth and interactions on leaves. J Appl Bacteriol 70:49–59

Blakeman JP (1993) Pathogens in the foliar environment. Plant Path 42:479–493

Bunster L, Fokkema NJ, Schippers B (1989) Effect of surface-active *Pseudomonas* sp. on leaf wettability. Appl Environ Microbiol 55:1340–1345

Cape JN (1983) Contact angles of water droplets on needles of Scots pine (*Pinus sylvestris*) growing in polluted atmospheres. New Phytol 93:293–299

Cape JN, Percy KE (1993) Environmental influences on the development of spruce needle cuticles. New Phytol 125:787–799

Dickinson CH, Preece TF (1976) Microbiology of aerial plant surfaces. Academic Press, London

Dixon RA, Lamb CJ (1990) Molecular communication in interactions between plants and microbial pathogens. Annu Rev Plant Phys Plant Mol Biol 41:339–367

Ercolani GL (1991) Distribution of epiphytic bacteria on olive leaves and the influence of leaf age and sampling time. Microb Ecol 21:35–48

Fiss M, Kucheryava N, Schonherr J, Kollar A, Arnold G, Auling G. (2000) Isolation and characterization of epiphytic fungi from the phyllosphere of apple as potential biocontrol agents against apple scab (*Venturia inaequalis*). J Plant Dis Prot 107:1–11

Fokkema NJ, van den Heuvel J (1986) Microbiology of the phyllosphere. Academic Press, New York

Fox RC (1958) The relationship of wax crystal structure to the water vapor transmission rate of wax films. Tech Assoc Pulp Paper Ind 41:283–289

Gülz PG (1994) Epicuticular leaf waxes in the evolution of the plant kingdom. J Plant Physiol 143:453–464

Holloway PJ (1970) Surface factors affecting the wetting of leaves. Pest Sci 1:156–163

Jacques MA, Morris CE (1995) A review of issues related to the quantification of bacteria from the phyllosphere. FEMS Microbiol Ecol 18:1–14

Jeffree CE (1986) The cuticle, epicuticular waxes and trichomes of plants, with reference to their structure, functions and evolution. In: Juniper BE, Southwood R (eds) Insects and plant surfaces. Edward Arnold, London, pp 23–64

Karanth NGK, Deo PG, Veenanadig NK (1999) Microbial production of biosurfactants and their importance. Curr Sci 77:116–126

Kerstiens G (1996) Signalling across the divide: a wider perspective of cuticular structure–function relationships. Trends Plant Sci 1: 125–129

Kinkel LL (1991) Microbial population dynamics on leaves. Annu Rev Phytopathol 35:327–347

Knoll D (1998) Die Bedeutung der Kutikula bei der Interaktion zwischen epiphyllen Mikroorganismen und Blattoberflächen. PhD Thesis, University of Würzburg, Germany

Knoll D, Schreiber L (1998) Influence of epiphytic micro-organisms on leaf wettability: wetting of the upper leaf surface of *Juglans regia* and of model surfaces in relation to colonization by microorganisms. New Phytol 140:271–282

Knoll D, Schreiber L (2000) Plant-microbe interactions: wetting of ivy (*Hedera helix* L.) leaf surfaces in relation to colonization by epiphytic microorganisms. Microb Ecol 41:33–42

Kolattukudy PE (2001) Polyesters in higher plants. Adv Biochem Engin Biotech 71:1–49

Larcher W (1996) Physiological plant ecology: ecophysiology and stress physiology of functional groups. Springer, Berlin Heidelberg New York

Leben C (1988) Relative humidity and the survival of epiphytic bacteria with buds and leaves of cucumber plants. Phytopathology 78:179–185

Marilley L, Vogt G, Blanc M, Aragno M (1998) Bacterial diversity in the bulk soil and rhizosphere fractions of *Lolium perenne* and *Trifolium repens* as revealed by PCR restriction analysis of 16S rDNA. Plant Soil 198:219–224

Markstädter C, Federle W, Jetter R, Riederer M, Hölldobler B (2000) Chemical composition of the slippery epicuticular wax blooms on Macaranga (Euphorbiaceae) ant-plants. Chemoecology 10:33–40

Mendgen K (1996) Fungal attachment and penetration. In: Kerstiens G (ed) Plant cuticles: an integrated functional approach. BIOS Scientific Publishers, Oxford, pp 175–188

Morris CE, Monier JM, Jacques MA (1997) Methods for observing microbial biofilms directly on leaf surfaces and recovering them for isolation of culturable microorganisms. Appl Environ Microbiol 63:1570–1576

Morris CE, Monier JM, Jacques MA (1998) A technique to quantify the population size and composition of the biofilm component in communities of bacteria in the phyllosphere. Appl Environ Microbiol 64:4789–4795

Morris CE, Nicot PC, Nguyen CN (1996) Aerial plant surface microbiology. Plenum Press, New York

Neinhuis C, Barthlott W (1998) Seasonal changes of leaf surface contamination in beech, oak, and gingko in relation to leaf micromorphology and wettability. New Phytol 138:91–98

Nobel PS (1991) Physicochemical and environmental plant physiology. Academic Press, San Diego

Ogram A (2000) Soil molecular microbial ecology at age 20: methodological challenges for the future. Soil Biol Biochem 32:1499–1504

Persson A, Oesterberg E, Dostalek M (1988) Biosurfactant production by *Pseudomonas fluorescens* 378: growth and product characteristics. Appl Microbiol Biotech 29:1–4

Preece TF, Dickinson CH (1971) Ecology of leaf surface microorganisms. Academic Press, London

Reynhardt EC Riederer M (1994) Structures and molecular dynamics of plant waxes. II Cuticular waxes from leaves of *Fagus sylvatica* L. and *Hordeum vulgare* L. Eur Biophys J 23:59–70

Riederer M, Markstädter C (1996) Cuticular waxes: a critical assessment of current knowledge. In: Kerstiens G (ed) Plant cuticles: an integrated functional approach. BIOS Scientific Publishers, Oxford, pp 189–200

Riederer M, Schönherr J (1990) Effects of surfactants on water permeability of isolated plant cuticles and on the composition of their cuticular waxes. Pest Sci 29:85–94

Riederer M, Schreiber L (1995) Waxes: the transport barriers of plant cuticles. Plant waxes. In: Hamilton RJ (ed) Waxes: chemistry, molecular biology and functions. The Oily Press, Dundee, pp 131–156

Romantschuk M (1992) Attachment of plant pathogenic bacteria to plant surfaces. Annu Rev Phytopathol 30:225–243

Schäfer W (1998) The involvement of fungal cutinase in early processes of plant infection. Mol Genet Host-Specific Toxins Plant Dis 13:273–280

Schönherr J, Baur P (1996) Cuticle permeability studies: a model for estimating leaching of plant metabolites to leaf surfaces. In: Morris CE, Nicot PC, Nguyen CN (eds) Aerial plant surface microbiology. Plenum Press, New York, pp 1–23

Schönherr J, Riederer M (1989) Foliar penetration and accumulation of organic chemicals in plant cuticles. Rev Environ Cont Toxicol 108:1–70

Schreiber L (1996) Wetting of the upper needle surface of *Abies grandis*: influence of pH, wax chemistry and epiphyllic microflora on contact angles. Plant Cell Environ 19:455–463

Schreiber L, Schönherr J (1993) Determination of foliar uptake of chemicals: influence of leaf surface microflora. Plant Cell Environ 16:743–748

Schreiber L, Schorn K, Heimburg T (1997) ^{2}H NMR study of cuticular wax isolated from *Hordeum vulgare* L. leaves: identification of amorphous and crystalline wax phases. Eur Biophys J 26:371–380

Stadler B, Müller T (1996) Aphid honeydew and its effect on the phyllosphere microflora of *Picea abies* (L.) Karst. Oecologia 108:771–776

Tiedje JM, Asuming Brempong S, Nusslein K, Marsh TL, Flynn SJ (1999) Opening the black box of soil microbial diversity. Appl Soil Ecol 13:109–122

Tukey HB (1970) The leaching of substances from plants. Annu Rev Plant Phys 21: 305–324

Turunen M, Huttunen S (1989) A review of the response of epicuticular wax of conifer needles to air pollution. J Environ Qual 19:35–45

van Gardingen PR, Grace J, Jeffree CE (1991) Abrasive damage by wind to the needle surfaces of *Picea sitchensis* (Bong) Carr and *Pinus sylvestris* L. Plant Cell Environ 14:185–193

Walton TJ (1990) Waxes, cutin and suberin. Meth Plant Biochem 4:105–158

Yang CH, Crowley DE, Borneman J, Keen NT (2001) Microbial phyllosphere populations are more complex than previously realized. Proc Natl Acad Sci USA 98:3889–3894

10 Developmental Interactions Between Clavicipitaleans and Their Host Plants

James F. White Jr., Faith Belanger, Raymond Sullivan, Elizabeth Lewis, Melinda Moy, William Meyer and Charles W. Bacon

1 Introduction

Clavicipitalean fungi have evolved to survive as saprophytes, degrading organic material, as well as biotrophs of plants, fungi, nematodes, and insects. They have become particularly successful as epibionts and endophytes of grasses. We believe that the associations between clavicipitalean fungi and their hosts constitute unique biotrophic symbioses where the stages of physiological adaptation to the plant host may be examined to gain an understanding of how evolution among these fungi has progressed.

2 Endophyte/Epibiont Niche

In recent years, awareness has developed that many microbes colonize and inhabit interior and exterior surfaces of plants. Many microbes may colonize plants without eliciting defense responses from host plants or causing disease symptoms (Bacon and White 2000). The benefits to plants of hosting beneficial microbes are numerous. Diazotrophic bacterial endophytes in sugarcane have been shown to fix atmospheric nitrogen that enables hosts to grow indefinitely in soils low in available nitrogen. *Bacillus subtilis*-infected seedlings of many plants have been shown to have an enhanced growth rate and survival in pathogen-laden soils. Tall fescue seedlings infected by the endophyte *Neotyphodium coenophialum* show enhanced resistance to "damping off" disease caused by *Rhizoctonia solani* (Gwinn and Gavin 1992). Mature plants of *F. arundinacea* show increased drought tolerance and resistance to above ground and below ground insect and nematode pests (Gwinn et al. 1991). Similarly, several grasses infected by the endophytes *Epichloë typhina*, *E. festucae*, and *E. clarkii* were found to deter the feeding of migratory locusts; while endophyte-free plants were readily consumed by the locusts (Lewis et al. 1993). Arizona fescue (*Festuca arizonica*) infected by a *Neotyphodium*

Plant Surface Microbiology
A. Varma, L. Abbott, D. Werner, R. Hampp (Eds.)

endophyte has not been found to possess insect deterrent properties, instead growth enhancements have been proposed (Faeth et al. 2000). Finally several species of fine fescues infected by *Epichloë festucae* were found to have an increased resistance to the dollar spot disease caused by *Sclerotinia homeocarpa*.

It seems evident that plants benefit tremendously from the colonization of symbiotic microbes. The benefits to hosting mutualistic microbes likely outweigh losses in terms of nutrient use by the microbes. The widespread level of infection of grasses by *Epichloë* and asexual forms in *Neotyphodium* are evidence that these associations also have evolutionary value. In one study of endophytes in grasses, infection levels in some hosts (e.g., *Achnatherum robustum* and *Festuca versuta*) were estimated to be greater than 90 % in populations throughout the ranges of the grasses (White 1987).

3 Coevolution of Clavicipitalean Fungi with Grass Hosts

It is evident that *Epichloë* (Clavicipitaceae; Ascomycetes) and related asexual endophytes have co-evolved with the cool-season (C-3) grasses in which they perennate (White 1988). Species of these endophytes are unknown from warm-season (C-4) grasses (White 1987). On the other hand, endophytic species in genus *Balansia*, also in family Clavicipitaceae, appear to have co-evolved with warm-season grasses and are rarely or never found on cool-season hosts (White and Owens 1992). In co-evolving with grasses, it is logical to expect that their interactions with hosts became more sophisticated.

4 The Jump from Insects to Plants

4.1 Trans-Kingdom Jump

Analysis of rDNA 26S sequence data indicates that the predominantly insect-infecting subfamily Cordycipitoideae (Clavicipitaceae) is the most deeply rooted group and is, therefore likely ancestral to grass-infecting species (Sullivan et al. 2000). A trans-kingdom host jump is postulated to have occurred to plants. Such a jump could have occurred gradually through intermediate forms that were parasitic on both insects and plants.

4.2 Intermediate Stages in the Transition to Plants

Several Cordycipitoideae exhibit stages of such a transition. Most of the Cordycipitoideae (e.g., *Cordyceps militaris* and *C. sinensis*) infect insect hosts and mummify them, using their necrotrophied bodies as energy to fuel fungal development (Tzean et al. 1997). In these associations, there is no association with plants and no opportunity for the fungi to adapt to plants as hosts. However, some species of Cordycipitoideae infect scale insects that are sedentary and parasitic on plant hosts by use of stylets with which they penetrate and suck sugars from host vascular tissues. In these species some simple adaptations for parasitism of plants are evident. In *Hypocrella africana*, *H. gaertneriana*, and *H. schizostachyi*, infection of the scale insect is biotrophic with the fungus obtaining nutrients from the plant through the living body of the insect (Hywel-Jones and Samuels 1998). Here, the parasitized scale insect is a bridge to obtain plant nutrients; however, the fungus does not interface directly with the plant in any way. This is an indirect adaptation to parasitism on plants. The quantity of nutrients available to *Hypocrella* in this type of association far exceeds that available in the body of the insect. Hywel-Jones and Samuels (1998) estimated that the stroma attained some 1000 times or more the mass of the body of the insect. *Hyperdermium bertonii* exhibits another step toward direct parasitism of plants. This species infects scale insects, necrophytizes them, then develops epibiotically on the surface of the plant, nourishing itself on sugars that continue to flow from the stylet wound left by the scale insect (Sullivan et al. 2000). Although *H. bertonii* relies on scale insects to prepare its parasitism site on plants, it directly absorbs and utilizes plant sugars. It is also possible that *H. bertonii* produces compounds that interfere with scar tissue development to prevent the stylet wound from sealing. This possibility should be further evaluated. However, at present we have no evidence that wound retardant compounds or growth regulator compounds are being produced by *H. bertonii*. Regardless, it is evident that *H. bertonii* has taken physiological steps in adapting to growth on plant sugars. In experiments, conducted in vitro where *H. bertonii* is grown on a minimal medium containing minerals and combinations of simple sugars glucose and fructose, we have demonstrated that mycelium and conidial production are stimulated by equal ratios of glucose to fructose; while higher levels of fructose in media induce the fungus to differentiate pigmentation and its mature stromal morphology. *Hyperdermium bertonii* has adapted to utilize changes in host sugar content on which it nourishes itself to guide its development. Sucrose, leaking directly from the stylet wound, is cleaved to its component monomers glucose and fructose. The glucose is likely preferentially absorbed. As a result, fructose is left behind to accumulate in the liquid film of sugars on which the fungus grows. Increasing concentrations of fructose, or the fructan polymers of it, are the probable cues employed by the epiphyte to shift its growth from early stroma development to differentiation and maturation. The possession of invertases by *H. bertonii* may also be evidence of adaptation to plants. Sucrose is only available in plant tissues. It is a short step from the condition of *Hyperdermium* to infection of plants without the use of insects.

4.3 Parasitism of Grass Meristematic Tissues

On the meristematic tissues of grasses, wounds created by insects are unnecessary since the tissues of meristems, such as the inflorescence primordia, are bathed in sucrose. *Atkinsonella hypoxylon* illustrates this point. *Atkinsonella hypoxylon* grows superficially on young leaves of grasses as an epiphyte, perhaps degrading wax in the cuticle to obtain nutrients for epiphytic growth (White et al. 1991). When the grass begins to produce an inflorescence primordium, sucrose is mobilized into the primordium to provide energy for its development. The primordium is surrounded by nutrients in liquid, which in turn is surrounded by layers of developing leaves. It is believed that the sudden increase in sucrose availability and concentration triggers rapid mycelial growth that eventually results in formation of the stroma (White and Chambless 1991). It is interesting that leaves and inflorescence primordia within the stroma never develop a cuticular layer that would impede flow of nutrients and moisture to the fungus. Prevention of cuticular development and prevention of maturation of the inflorescence primordial tissues may be a growth regulator effect, although they have not yet been identified for this species.

5 Developmental Differentiation of Endophytic and Epiphyllous Mycelium

5.1 Plant Cell Wall Alteration

Epichloë sp. illustrate many of the physiological capacities needed by Clavicipitaceae to colonize grasses (White et al. 1991). *Epichloë* inhabits leaf sheaths and growing tillers of grass plants. Endophytic mycelium is largely nonbranched and exclusively intercellular (Fig. 1) and often seen to adhere closely to parenchyma cell walls as if attached by glue. This may be due to the partial degradation of cell wall components by the endophytic mycelium. While it is possible that cell walls are modified by endophytes, they remain largely intact as evidenced by electron microscopic studies. It is notable that during stroma development, profound changes have been observed in cell walls of the grass epidermis. Walls of epidermal cells appear to lose structural integrity with mycelium of the endophyte frequently penetrating the wall (Fig. 2).

5.2 Endophytic Mycelial Growth

Endophytic mycelium in young leaves or elongating tillers is frequently narrow (1 µm across), straight, oriented parallel to the axis of expansion of the cells and plant organ (Fig. 3). Sometimes in very young tissues, the hyphae

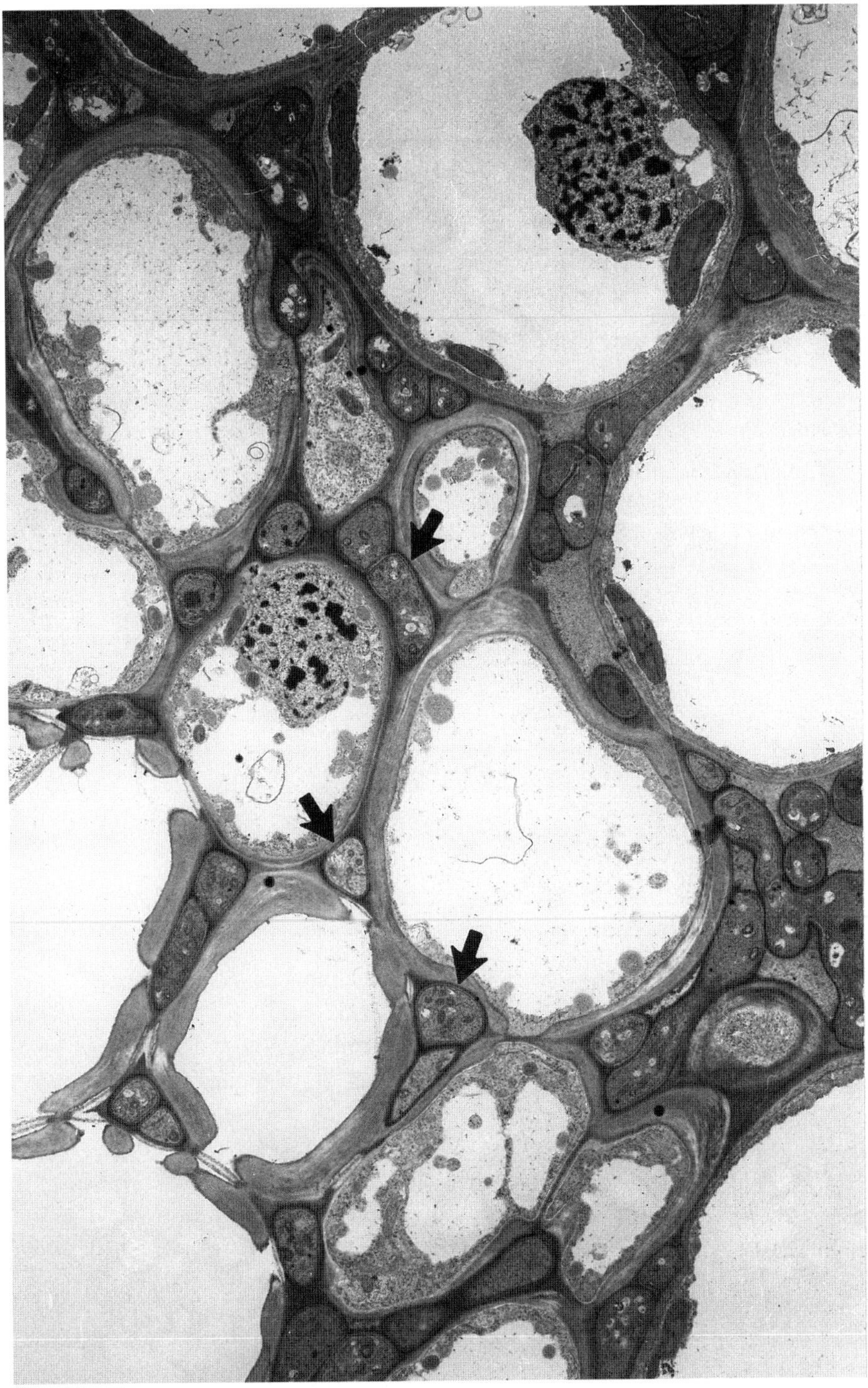

Fig. 1. Electron micrograph (TEM) showing hyphae (*arrows*) of *Epichloë amarillans* in intercellular spaces of vascular tissues of the grass *Agrostis hiemalis* (×10,000)

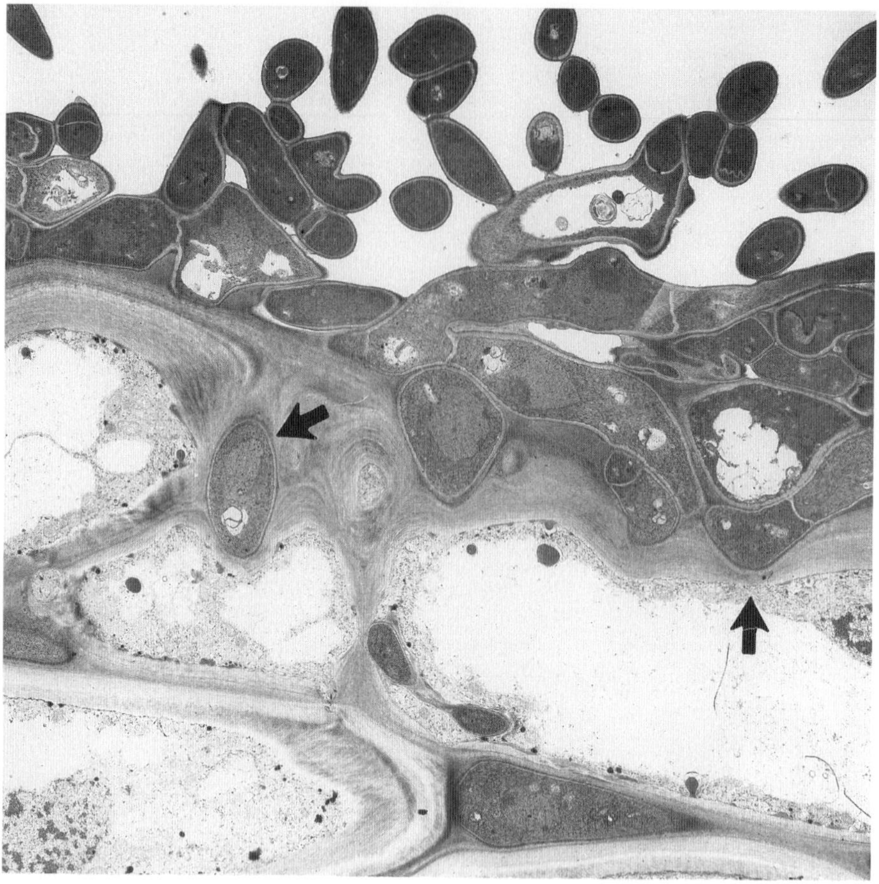

Fig. 2. Electron micrograph (TEM) showing hyphae (*arrows*) of *Epichloë amarillans* penetrating epidermal tissues of the grass *Agrostis hiemalis* (×10,000)

may be observed to taper to a fine point on the ends where it has been stretched and sheered during elongation of the plant tissues. Sheered hyphae are seen to recover rapidly with elongation of the sheared ends of the endophytic hyphae. In later stages of growth, the endophytic hyphae fully elongate, then become convoluted, apparently due to excessive elongation. This has the effect of increasing the surface area of the cell wall that an individual hypha may come into contact with. Whether an increased contact surface area results in increased nutrients leaking from the parenchyma cells to the endophytic mycelium is yet to be determined. Such endophytic mycelium is abundant in leaf sheaths where many nutrients are stored, but they are rare in the leaf blades where photosynthesis is occurring. It is the abundant presence of photosynthate within the cells of the leaf sheath that likely accounts for the abundance of mycelium in this tissue.

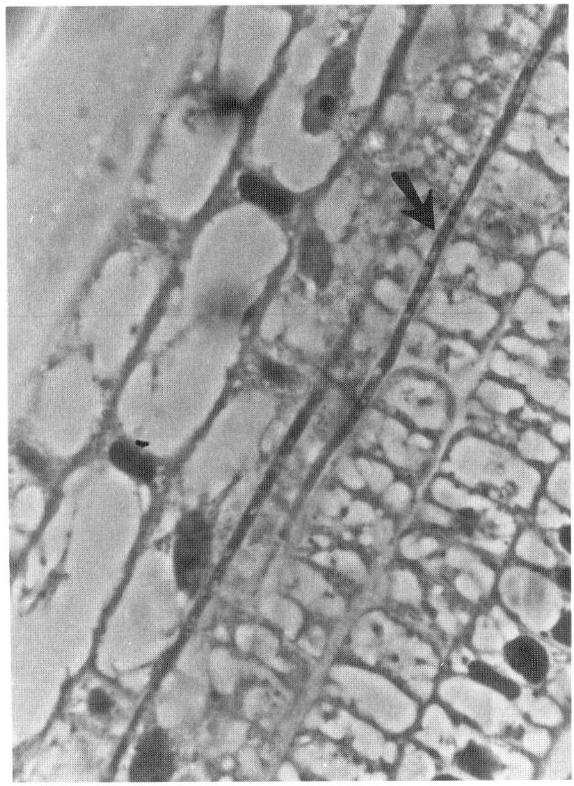

Fig. 3. Section of developing *Agrostis hiemalis* culm showing endophytic hypha (*arrow*) oriented parallel to the direction of culm elongation (×1000)

5.3 Control of Endophytic Mycelial Development

Endophytic mycelium is never observed to produce conidia within tissues of the plant. One experiment suggests that conidial development and branching are suppressed by unknown factors present within tissues of the leaf sheath parenchyma. Media containing basal salts (Murashige and Skoogs), agar (1 %), ground leaf sheath tissues of *Agrostis hiemalis* (0.5 % dry wt.), and low concentrations of glucose (0.5 %) produced mycelium of *E. amarillans* that sparsely branched and rarely produced conidiogenous cells, while controls that lacked only the ground leaf sheath tissues, branched and produced conidia abundantly.

5.4 Epiphyllous Mycelial Development

Some species of *Epichloë* and their asexual derivatives have been found to produce an epiphyllous stage where they grow superficially on the surface of leaf blades. Epiphyllous mycelium tends to be present in the groves at cell junctions of the epidermis, may adhere closely to the cuticular surface, is frequently branched, and produces abundant wind-disseminated conidia (White et al. 1996; Moy et al. 2000). The epiphyllous network of mycelium and conidia are frequently connected to internal sources of nutrients by intercellular bridges, but are shielded from direct association with the interior leaf substances by the waxy cuticle layer on which it spreads. There is no evidence that species of *Epichloë* have the capacity to degrade cuticular waxes. Previous studies of the capacity of various Clavicipitaceae to colonize and degrade paraffin showed that *Epichloë* does not colonize paraffin beads in agar culture and apparently cannot degrade waxes to gain nutrients, although other Clavicipitaceae, such as the predominantly epiphytic *Atkinsonella hypoxylon*, do posses that capacity (White et al. 1991).

5.5 Expression of Fungal Secreted Hydrolytic Enzymes in Infected Plants

All fungi, whether saprophytic, pathogenic, or mutualistic, acquire their carbon and nitrogen by absorption of small molecules from their surroundings. Fungi typically secrete numerous enzymes that function in degradation of polymeric substances in the environment to their monomeric constituents that can then be absorbed by the fungal cells.

Endophytic fungi are exclusively intercellular and do not invade the plant cells. They must, therefore obtain all their carbon and nitrogen compounds from the nutrient-poor apoplastic space. Endophytic fungal-secreted proteins are likely to be important components of the mutualistic interaction as they are located at the interface of the two species. Fungal secreted proteins are expected to be synthesized for growth and nutrient acquisition and perhaps for defense.

We have detected expression of several fungal-secreted enzymes in *Poa ampla* infected with a *Neotyphodium* sp. endophyte. A fungal subtilisin-like proteinase was purified from infected leaf sheaths and cDNA and genomic clones for the gene were characterized (Lindstrom and Belanger 1994; Reddy et al. 1996). The fungal proteinase was found to be expressed at surprisingly high levels in the infected plant tissues. It was estimated to be 1–2 % of the total leaf sheath protein, suggesting it was a major fungal protein. The amino acid sequence of the proteinase is homologous to proteinases believed to be important in pathogenicity of entomopathogenic, nematophagous, and mycoparasitic fungi (Geremia et al. 1993; Bonants et al. 1995; St. Leger 1995).

A fungal secreted endochitinase and an endo-β-1,6-glucanase are also expressed in the infected *P. ampla* plants. Sequencing of cDNA clones for the chitinase and glucanase revealed they are 38 and 74 % identical, respectively, to the homologous enzymes from *Trichoderma harzianum*. *T. harzianum* is a potent mycoparasite of many plant pathogenic fungi (Papavizas 1985; Chet 1987). Because of this property, it is being investigated as a potential biocontrol agent in crop production.

The physiological roles of the endophytic proteinase, chitinase, and endo-β-1,6-glucanase are not yet known. The endophytic chitinase transcript is very abundant as determined from a blot of total RNA isolated from the infected plants. This is similar to the situation with the endophytic proteinase (Reddy et al. 1996). The endo-β-1,6-glucanase appears to be expressed at lower levels. Several roles have been proposed for chitinases and endoglucanases from filamentous fungi. Roles in hyphal growth, branching, and autolysis have been proposed (Bartnicki-Garcia 1973; Gooday and Gow 1990; Peberdy 1990) as well as roles in mycoparasitism. Functions in fungal growth and/or in mycoparasitism would be relevant to endophytic infection. Interestingly, the homologous proteinase, chitinase, and endo-β-1,6-glucanase from *T. harzianum* are believed to be synergistic components of its mycoparasitic activity (Geremia et al. 1993; Garcia et al. 1994; Lora et al. 1995). These hydrolytic enzymes function together to break down the cell walls of the fungal hosts allowing entry of the *T. harzianum* hyphae.

Expression of these hydrolytic enzymes in endophyte-infected plants raises the possibility that they may also function as a mycolytic system for the endophyte. Such a system could provide the endophyte with a source of nutrients in addition to plant derived nutrients found in the apoplast. With a mycolytic system, the endopytic hyphae located on the surface of the plant (Moy et al. 2000) would have access to additional sources of nutrients from other surface-located fungi. By attacking invading fungi, an endophytic mycolytic system could also protect the plants from pathogenic fungi, perhaps resulting in enhanced disease resistance. Current research is aimed at determining the roles of these enzymes in endophyte infection.

6 Modifications of Plant Tissues for Nutrient Acquisition

6.1 Development of the Stroma in *Epichloë*

The development of sexual reproductive structures in plants poses some special problems for Clavicipitaceae. Larger quantities of nutrients are needed to provide the fuel for construction of the external mycelial stroma on which are produced first spermatia, then perithecia and ascospores (White and Bultman 1987; Bultman et al. 1995). To obtain large quantities of nutrients from hosts, many other groups of biotrophic fungi, e.g., powdery mildews, downy mildews, and rusts may produce haustoria to suck nutrients from individual host cells (Alexopoulos et al. 1996). However, clavicipitalean plant biotrophs have another strategy. They grow on meristematic tissues before the cuticle has been formed and by some unknown mechanism prevent development of the waxy cuticle and alter the epidermis itself, effectively removing a key barrier to the flow of nutrients to the stromal mycelium. The following two examples will illustrate this method of nutrient acquisition by these clavicipitaleans. In *Epichloë* the abundance of sucrose in the developing inflorescence primordium triggers the fungus to proliferate rapidly and permeate the young inflorescence and the leaf sheath of a leaf that surrounds it. This process is comparable to that already suggested for stroma development in *Atkinsonella hypoxylon*, except that in *Epichloë* the mycelium is endophytic and frequently permeates vascular tissues as well as nonvascular tissues (White et al. 1991). The stroma is composed of a mix of plant tissues and fungal mycelium. These stromata are much like those of the scale insect parasites *Hypocrella africana*, *H. gaertneriana*, and *H. schizostachyi*, in that the host tissues embedded within the stromata remain alive, but are modified so that nutrients will flow freely into the developing stromata. Plant tissues embedded within the stroma are not only permeated by mycelium, but also possess epidermal cells that are hypertrophied, often collapsed, and lack waxy cuticles (White et al. 1997). Through these modifications of the host tissues, the endophyte removes all barriers to nutrient flow into the stromal mycelium. The development of mycelium within the vascular bundle enhances the transfer of nutrients to the fungal stroma. By mummifying the living inflorescence primordium and the sheath of the leaf that surrounds it, the fungus can intercept all nutrients that are transported into the flowering tiller. Mature stromata of *Epichloë* always possess the stromal leaf blade emergent from the top of the stroma (Fig. 4). The reason for this emergent leaf blade is unknown, but may be a source of plant hormones that are needed as a signal to the plant to continue to send nutrients into the culm. Experimental work is needed to evaluate this hypothesis.

6.2 Stroma Development in Myriogenospora

A second clavicipitalean biotroph that modifies host tissues for nutrient acquisition during stroma development is the epiphytic fungus *Myriogenospora atramentosa*. *Myriogenospora atramentosa* grows superficially on the epidermis of young leaves at the crown of many warm-season grasses and sedges. As the leaves develop, conidia of *M. atramentosa* proliferate on the folded leaves of the grass. The leaves continue to expand and the conidial stroma develops into a linear black perithecial stroma, composed of a single line of perithecia (Figs. 5, 6). The plant leaf tissues beneath the stroma are modified with hypertrophied epidermal cells that lack a cuticular layer (Rykard et al. 1985; White and Glenn 1994). The absence of a cuticle layer on the leaf epidermis and modification of the epidermal cells by the fungus permits *M. atramentosa* to absorb nutrients directly through the epidermis of the leaf blades to provide energy for stroma development.

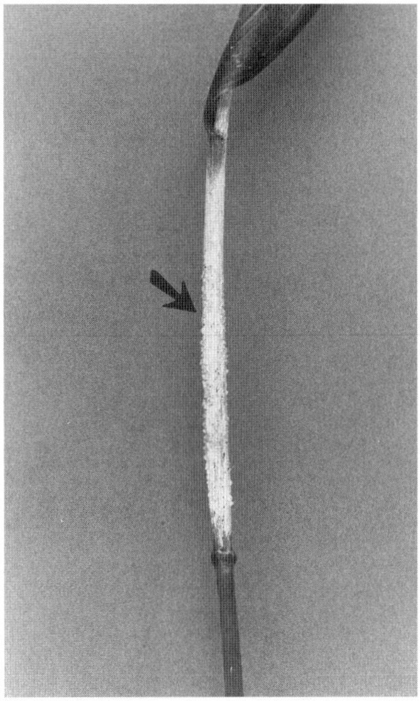

Fig. 4. Stroma (*arrow*) of *Epichloë amarillans* showing white stromal mycelium and apical stromal leaf (×3)

Fig. 5. Black, linear, stroma (*arrow*) of *Myriogenospora atramentosa* on upper surface of leaf of *Andropogon* sp. (×2)

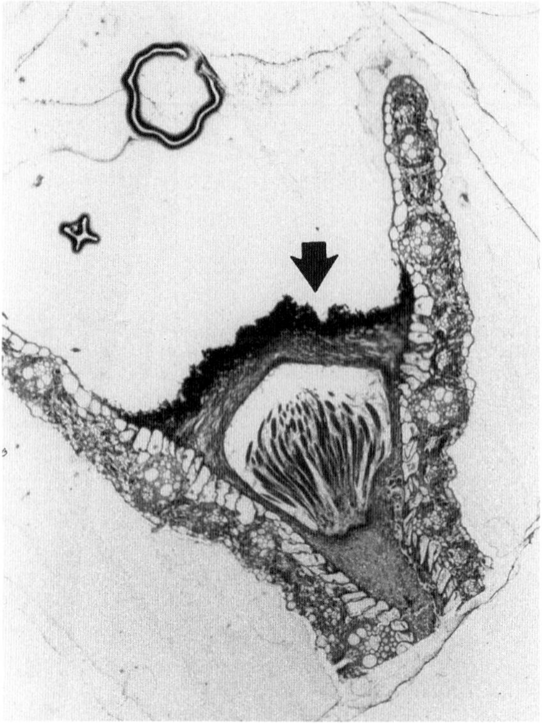

Fig. 6. Cross-section of stroma of *Myriogenospora atramentosa* showing a single perithecium (*arrow*) bordered by the leaf blades on either side (×500)

6.3 Mechanisms for Modifying Plant Tissues

The mechanisms whereby the Clavicipitaceae alter development of plant tissues is unknown. One hypothesis is that at least some of their secondary products may have growth regulator effects. In this respect, it is notable that several Clavicipitaceae, including *Epichloë festucae* and *Balansia epichloë* have been shown to produce the plant auxin indole acetic acid (IAA; Porter et al. 1985; Yue et al. 2000). Indeed, other indole derivatives such as the ergot alkaloids may also possess auxin-like effects. One effect that auxin has is to loosen cell wall fibers, allowing cells to expand.

Moubarak et al. (1993) demonstrated that ergovaline, an ergot alkaloid commonly produced by *Epichloë/Neotyphodium* endophytes, interferes with cell membrane polarization and ATPase activities in animal tissues. These data suggest a potential mechanism by which ergot alkaloids may alter physiology and structure of plant tissues and acquire nutrients from those tissues. If ergot alkaloids, such as ergovaline, inhibit ATPases in grass cells, they may enhance leakage of nutrients from cells adjacent to mycelium. Without use of ATPases, plant cells would be incapable of utilizing active transport proteins to reacquire leaking nutrients. Almost no research has been pursued to evaluate impacts of clavicipitalean-produced secondary metabolites on plant tissues themselves. The current scientific wisdom holds that clavicipitalean secondary metabolites have impacts on animal tissues as feeding deterrents and other defensive compounds. Whether ergot alkaloids, auxin-like compounds or other secondary metabolites of these fungi are involved in effecting changes in plant tissues embedded within or adjacent to stromal mycelium must be further evaluated. It seems likely that this will be a fruitful area for future investigation.

6.4 Evaporative-Flow Mechanism for Nutrient Acquisition

The stroma of *Epichloë* maintains a constant flow of water and nutrients into its mycelium through an evaporation-driven process (White and Camp 1996; White et al. 1997). Water evaporates rapidly from the surface of stromata. As water evaporates from the stroma it is replaced by water from the plant. This process establishes a flow of water and dissolved nutrients into the stroma from mycelium interfacing with the vascular bundles and other tissues embedded within the stroma. Evaporative flow mimics the enhanced transpiration that occurs in developing inflorescences during elongation of the flowering tillers of uninfected grasses, but here the stroma is the recipient of the nutrients.

6.5 The Cytokinin Induction Hypothesis

In *Atkinsonella hypoxylon* stromata form on the inflorescence primordium and include parts of several leaves as well (Fig. 7). The stromata are gray (sometimes with areas of a yellow pigment) and produce several different spore states, including cup-shaped sporodochia that produce moist masses of ephelidial conidia, and a layer of neotyphodial conidia borne on tips of elongate conidiogenous cells. It is reasonable to expect that the fungus would coordinate its development with that of its host grass. For example, the fungus mycelium must be able to detect when it is growing on an inflorescence primordium rather than on the tiller meristems. On the tiller meristems it will produce a low biomass of nonpigmented mycelium and ephelidial conidia, but no neotyphodial conidia or other structures; while on the inflorescence primordium the entire suite of morphological structures is produced. One way for the fungus to coordinate its development to that of the host plant would be to use compounds present in the host during different stages of development as 'cues' to initiate developmental stages in the fungus. Our approach to the search for host compounds that may serve as cues for fungal development has been a trial and error approach. Over several years, we have screened hundreds of compounds that might be present in grass tissues to determine how they affect differentiation of *A. hypoxylon*.

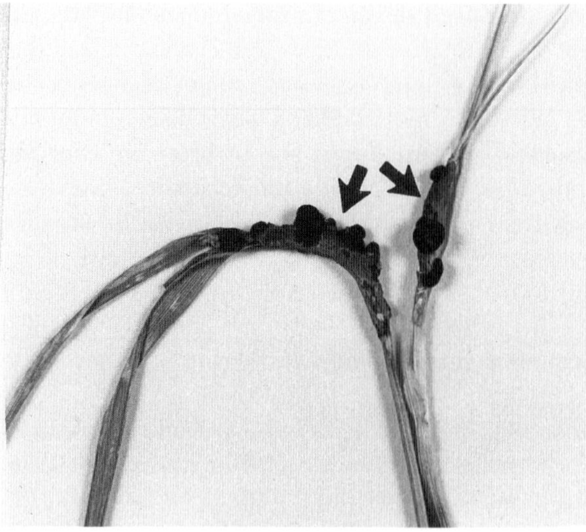

Fig. 7. Two stromata (*arrows*) of *Atkinsonella hypoxylon* on culms of *Danthonia spicata* (×4)

Studies on *Atkinsonella hypoxylon* and *A. texensis* in vitro have demonstrated that certain media additives will induce the fungus to develop in a way comparable to that seen on the host grass inflorescence primordia (Bacon and White 1994). When these claviciptaleans are grown on media containing agar (1 %), basal salts (Murashige and Skoog; Sigma Chemical Company, Inc.), and glucose (3 %), colonies are white, with no aerial mycelium or conidia of any type. This is an undifferentiated mycelium, the fungus equivalent of 'callus tissue'. Stroma-like colonies with gray pigmentation, sporodochia producing ephelidial conidia, and a layer of neotyphodial conidiogenous cells and conidia can be induced by inclusion of 100 ppm of the cytokinin zeatin or kinetin (Research Organics, Inc. Cleveland, Ohio) in the medium. The stroma-like states in culture are most striking when the grass cytokinin zeatin is employed. Partial induction of stroma-like states may be induced through use of 1 % citrate (sodium or potassium salt) and 0.1–0.5 % acetate (sodium or potassium salt). With acetate in the medium the gray pigmentation is seen to develop, but differentiated reproductive cells do not form. With citrate in the medium, pigmentation, sporodochia and ephelidial conidia form, but the neotyphodial conidia do not form. Because induction of differentiation is incomplete with the use of acetate and citrate, we believe that these compounds are not the primary cues for stroma differentiation, but instead may be indirectly causing differentiation by turning on secondary metabolism pathways. On the other hand, cytokinins are plant hormones and are expected to be present in the developing ovary tissues embedded within the fungal stroma since ovaries produce cytokinins for regulation of their own development (Miller 1961; Mauseth 2003). Thus the presence of cytokinins may be a key signal for the fungus to begin a sequence of developmental events that end in production of the mature stroma. Some preliminary differential display studies were also conducted to examine genes that may be upregulated and downregulated when *A. hypoxylon* was exposed to cytokinin. The results of these differential display studies showed that several genes were turned on while several others were turned off, however, none of the genes were identified. One preliminary study on another clavicipitalean-producing stromata on inflorescence primordia, *Epichloë festucae*, employed a monoclonal antibody-based cytokinin detection kit (Phytodetek-t-ZR, Sigma, St. Louis, Missouri) to compare levels of cytokinin in stromata and other tissues of the grass. The result of this test suggested that cytokinin was present in high concentrations within the stromata. However, this test must be considered preliminary because of the possibility for cross-reactivity of the antibody with other compounds. More precise tests for the presence of cytokinins must be employed to evaluate levels in the stromata. Presently, the hypothesis that cytokinins are a key cue for development of stromata on the grass inflorescence primordium for the grass inflorescence-colonizing clavicipitaleans is an interesting hypothesis. However, additional work must be done to evaluate this hypothesis.

7 Evolution of Asexual Derivatives of *Epichloë*

7.1 Reproduction and Loss of Sexual Reproduction

One notable feature of genus *Epichloë* is the abundance of asexual species, often classified in form genus *Neotyphodium*. Formation of asexual derivatives is apparently a relatively frequent phenomenon based on how common these asexual forms are in grasses (White 1987). In the sexual cycle of *Epichloë* stromata are produced on grasses, and on stromata spermatia develop. In a heterothallic mating process symbiotic flies in genus *Botanophila* (Anthomyidae) vector spermatia between stromata of the opposite mating type (Bultman et al. 1995). Following deposition of spermatia on a compatible stroma, an ascogenous (dikaryotic) mycelium develops in which perithecia and ascospores form. Meiosis takes place within the asci to result in the haploid ascospores that are ejected from asci onto surrounding vegetation, where they may germinate to form wind-disseminated conidia (White and Bultman 1987). Precisely how primary infections of grasses occur is still unknown, but may involve a period of epiphyllous growth prior to penetration of plant tissues (Moy et al. 2000). Other investigators (Diehl 1950; Chung and Schardl 1997) have suggested that ovules may be the site of entry into plants. However, definitive data that will answer this question are still lacking. The asexual forms of *Epichloë* are seed-transmitted and stromata do not form on grass inflorescences. Since these asexual forms do not form stromata, sexual recombination does not occur. Seed transmission is the result of growth of the endophyte in inflorescence primordia. When ovules differentiate in the primordia, the fungus is incorporated into tissues of the nucellus. When the embryo differentiates within the nucellus, it is invaded by endophytic mycelium and the next generation of host has been effectively colonized (White and Cole 1986).

7.2 The Hypotheses

Two hypotheses have been proposed to explain loss of the sexual cycles by species of *Epichloë*. In the 'hybridization hypothesis' it is suggested that hybridization between two different species of *Epichloë* results in 'hybrids' that cannot undergo sexual reproduction due to meiotic incompatibility of the two sets of chromosomes (Schardl and Wilkerson 2000). The frequent occurrence of multiple sets of genes in some asexual endophytes supports this hypothesis (Leuchtmann and Clay 1990; Tsai et al. 1994; Cabral et al. 1999). The occurrence of asexual forms such as the endophyte of *Lolium rigidifolium* that do not show multiple copies of genes is problematic for the hybridization hypothesis (Moon et al. 2000). The second problem with the hybridization hypothesis is that it suggests a very unlikely scenario. It suggests that haploid spermatia of one species fuse with haploid mycelium on a stroma of an opposite species to produce a dikaryotic mycelium. The next steps would involve formation of perithecia, asci, and ascospores. Within the asci the two nuclei from different species of *Epichloë* would fuse to become a diploid which would be immediately followed by meiosis to result in production of the haploid ascospores. Without formation of ascospores, the hybrid would be unable to spread. If the two genomes were meiotically incompatible as the 'hybridization hypothesis' suggests that first meiosis would not occur and ascospores could not be produced. This hypothesis invokes meiotic incompatibility, yet demands that meiosis occurred at least once following hybridization. It seems unlikely that hybridization and meiotic incompatibility account for the origins of asexual *Epichloë* endophytes. It should be noted that speciation by hybridization does work in plants. However, in plants meiosis does not occur immediately after hybridization, instead, a diploid forms. The diploid may reproduce clonally for a time (Grant 1977).

The plurality of gene copies present within many asexual endophytes may be an indication of a parasexual process that is acting in asexual endophytes to produce variation. To evaluate whether multiple gene copies reflect parasexual recombinations within populations of asexual endophytes, it is necessary to conduct populational studies on gene variation. To this point studies examining gene variation in asexual endophytes have involved only a few isolates. It will be important to determine whether this parasexual recombination (hybridization) is a populational phenomenon and occurring relatively frequently within populations of the fungi or is a rare event, resulting in the origins of new species. Until we understand how frequent the asexual recombinatorial events are, their importance and significance will be speculation.

The alternative hypothesis to explain loss of stromata and the sexual cycle invokes ecological factors and is termed the 'environmental selection hypothesis'. This hypothesis suggests that stroma development reduces fitness of the symbiotic unit (grass and endophyte) and is selected against under certain environmental conditions. It is supported by work demonstrating that stromata increase the losses of water from plant tissues (White and Camp 1996), and decrease the fecundity of hosts by replacing inflorescences with stromata (White and Chambless 1991). It has been observed that stromata tend to form on plants in soils that contain high levels of moisture, whereas asexual forms occur in plants that live in soils that range from very dry to moist. Additional ecological studies are needed to confirm the association between soil moisture and stromata occurrence.

7.3 The Process of Stroma Development and its Loss

To understand the mechanism of loss of stroma-forming ability in *Epichloë*, it is necessary to understand the mechanism of stroma development and the interactions between endophyte and grass during stroma development. Kirby (1961) proposed that the capacity to form stromata on grasses was a function of the growth rate of fungal mycelium in the inflorescence primordium versus the growth rate of the inflorescence primordium. That is, endophytes that grow rapidly in the inflorescence primordium tissues can outgrow the inflorescence primordium, surround it, and trap it in a stromal mycelium, thus successfully forming a stroma. If an endophyte cannot grow rapidly enough to trap the inflorescence primordium in a stromal mycelium, the inflorescence emerges, and develops flowers and seeds that may contain the endophyte.

Central to the issue of stroma development is the question of which nutrients provide the energy for stroma formation. Lam et al. (1995) demonstrated that *Epichloë festucae* possesses the sucrose degrading enzyme invertase and suggested that this enzyme may play a role in stroma development. Earlier work by White et al. (1991) suggests another mechanism. White et al. (1991) examined the growth rate of a range of endophytes producing stromata of different sizes on several different sugars likely to be found in grass inflorescence primordia. These sugars included glucose, fructose, xylose, and arabinose. Glucose and fructose result from the cleavage of sucrose that is abundant in and around the inflorescence primordium tissues. Xylose and arabinose are sugars present in the cell wall polysaccharides of grasses and may be available in meristems of the primordium. In this study it was found that there is a positive correlation between the size of stromata and the growth rate on a selection of sugars. Apparently, the larger the stroma formed on a particular host, the faster an endophyte must grow to develop that stroma. It was further found that endophytes that failed to reach the critical growth rate on any of the sugars, tended to produce fewer stromata per plant than endophytes that grew rapidly on all of the sugars. From an evolutionary perspective, selection against stroma development may be selection for endophytes that grow more slowly on nutrients available in host tissues. This hypothesis is consistent with at least one important observation. Many asexual endophytes (e.g., *Neotyphodium coenophialum* and *N. lolii*) are slow growing in culture while stroma-forming endophytes grow comparably faster.

7.4 The Shift from Pathogen to Mutualist

Much is known of the biochemistry and genetics of the interactions between plants and pathogenic organisms and how these interactions result in disease or in plant resistance (Oliver and Osbourne 1995; Hammond-Kosack and Jones 1996). Mutualistic associations, such as those between the fungal endophytes and their grass hosts, are believed to have evolved from pathogenic associations (Clay 1988). Little is known regarding the genetic changes that result in a change from a pathogenic to a mutualistic lifestyle.

Plant fungal pathogens typically secrete a number of plant cell wall degrading enzymes such as cellulases, glucanases, xylanases, and polygalacturonases. It is likely that expression of these cell wall degrading enzymes plays some role in pathogenicity (Oliver and Osbourne 1995; Mendgen et al. 1996), although disruption of individual genes has not resulted is reduced virulence (Scott-Craig et al. 1990; Apel et al. 1993; Schaeffer et al. 1994; Bowen et al. 1995; Sposato et al. 1995). The presence of other genes encoding the same enzyme activity and synergistic activity of different cell wall degrading enzymes in pathogenicity may explain these results.

Claviceps purpurea, a plant pathogen closely related to the *Epichloë* and *Neotyphodium* endophytes, secretes a polygalacturonase during infection of rye ovaries (Tenberge et al. 1996). Polygalacturonase activity is believed to be important in splitting the host middle lamellae allowing intercellular growth of the fungus (Tenberge et al. 1996). Since the fungal endophytes also have an intercellular mode of growth, we have investigated the possibility of endophytic polygalacturonase expression in the *Neotyphodium* sp. endophyte that infects the grass *Poa ampla*. No hybridization was detected in a DNA blot using the cloned *C. purpurea* gene as a probe. Also, nothing was detected in PCR reactions using degenerate primers based on conserved amino acid regions of polygalacturonase genes from diverse organisms. It appears that this endophytic fungus may have lost the gene(s) for polygalacturonase. Perhaps loss of this cell wall degrading activity is a factor in the evolution of pathogen to mutualist. Since the fungal endophytes are exclusively intercellular and do not invade the plant cells, it is likely that genes for other cell wall degrading enzymes have also been lost. Ultimately, genome sequencing of an endophytic fungus will reveal the differences in enzyme coding capacity between fungal pathogens and fungal endophytes.

8 Conclusions

Our understanding of the range of physiological interactions between clavicipitalean mycosymbionts and grasses is virtually nonexistent. The majority of the research to date has focused on the agronomic aspects of the toxicity problem or on ecology of the hosts as modified by these fungi. As a consequence, physiology of clavicipitalean – plant interactions is a fertile and potentially important area of research.

References and Selected Reading

Alexopoulos CJ, Mims CW, Blackwell M (1996) Introductory mycology. Wiley, New York

Apel PC, Panaccione DG, Holden FR, Walton JD (1993) Cloning and targeted gene disruption of XYL1, a 1,4-xylanase gene from the maize pathogen *Cochliobolus carbonum*. Mol Plant-Microbe Interact 6:457–473

Bacon CW, White JF Jr (1994) Stains, media, and procedures for analyzing endophytes. In: Bacon CW, White JF Jr (eds) Biotechnology of endophytic fungi of grasses. CRC Press, Boca Raton, Florida, pp 47–58

Bacon CW, White JF Jr (2000) Microbial endophytes. Marcel-Dekker, New York, pp 341–388

Bartnicki-Garcia S (1973) Fundamental aspects of hyphal morphogenesis. In: Ashworth JM, Smith JE (eds) Microbial differentiation. Cambridge University Press, Cambridge

Bonants PJM, Fitters PFL, Thijs H, den Belder E, Waalwijk C, Henfling JWDM (1995) A basic serine protease from *Paecilomyces lilacinus* with biological activity against *Meloidogyne hapla* eggs. Microbiology 141:775–784

Bowen JK, Templeton MD, Sharrock KR, Crowhurst RN, Rikkerink EHA (1995) Gene inactivation in the plant pathogen *Glomerella cingulata*: three strategies for the disruption of the pectin lyase gene *pnlA*. Mol Gen Genet 246:196–205

Bultman TL, White JF Jr, Bowdish TI, Welch AM, Johnston J (1995) Mutualistic transfer of *Epichloë* spermatia by *Phorbia* flies. Mycologia 87:182–189

Cabral D, Cafaro M, Saidman B, Lugo M, Reddy PV, White JF Jr (1999) Evidence supporting the occurrence of a new species of endophyte in some South American grasses. Mycologia 91:315–325

Chet I (1987) *Trichoderma* – application, mode of action, and potential as a biocontrol agent of soil borne plant pathogenic fungi. In: Chet I (ed) Innovative approaches to plant disease control. Wiley, New York

Chung KR, Schardl CL (1997) Sexual cycle and horizontal transmission of the grass symbiont, *Epichloë typhina*. Mycological Res 101:295–301

Clarke BB, White, JF Jr, Funk CR Jr, Sun S, Huff DR, Hurley RH (2003) Enhanced resistance to dollar spot in endophyte-infected fine fescues. Plant Dis (in press)

Clay K (1988) Clavicipitaceous fungal endophytes of grasses: coevolution and the change from parasitism to mutualism. In: Pirozynski KA, Hawksworth DL (eds) Coevolution of fungi with plants and animals. Academic Press, London, pp 79–105
Diehl (1950) Balansia and Balansiae in America, USDA Monograph, US Govt Printing Office, Washington, DC, 99 pp
Faeth S, Sullivan H, Hamilton CE (2000) What maintains high levels of *Neotyphodium* endophytes in native grasses? A dissenting view and alternative hypotheses, In: Paul VH, Krohn K, Dapprich PD, Gutter B (eds) Proceedings Fourth International *Neotyphodium*/Grass Interactions Symposium. The University of Paderborn, Soest, p 14
Garcia I, Lora JM, de la Cruz J, Benitez T, Llobell A, Pintor-Toro JA (1994) Cloning and characterization of a chitinase (CHIT42) cDNA from the mycoparasitic fungus *Trichoderma harzianum*. Curr Genet 27:83–89
Geremia RA, Goldman GH, Jacobs D, Ardiles W, Vila SB, Van Montagu M, Herrera-Estrella (1993) A. Molecular characterization of the proteinase-encoding gene, prb1, related to mycoparasitism by *Trichoderma harzianum*. Mol Microbiol 8:603–613
Gooday GW, Gow NAR (1990) Enzymology of tip growth in fungi. In: Heath IB (ed) Tip growth in plant and fungal cells. Academic Press, New York, pp 31–58
Grant V (1977) Organismic Evolution. WH Freeman, San Francisco
Gwinn KD, Gavin AM (1992) Relationship between endophyte infestation level of tall fescue seed lots and *Rhizoctonia zeae* seedling disease. Plant Dis 76:911–914
Gwinn KD, Blank CA, Cole AM, Pless CD (1991) Resistance of endophyte-infected tall fescue seedlings to pathogens and pests. Tenn Farm Home Sci 160:72
Hammond-Kosack KE, Jones JDG (1996) Resistance gene-dependent plant defense responses. Plant Cell 8:1773–1791
Hywel-Jones NL, Samuels GJ (1998) Three species of Hypocrella with large stromata pathogenic on scale insects. Mycologia 90:36–46
Kirby EJM (1961) Host-parasite relations in the choke disease of grasses. Trans Br Mycol Soc 44:493–503
Lam CK, Belanger FC, White JF Jr, Daie J (1995) Invertase activity in *Epichloë/Acremonium* fungal endophytes and its possible role in choke disease. Mycol Res 99:867–873
Lane GA, Christensen MJ, Miles CO (2000) Coevolution of fungal endophytes with grasses: the significance of secondary metabolites. In: Bacon CW, White JF Jr (eds) Microbial Endophytes. Marcel-Dekker, New York, pp 341–388
Leuchtmann A, Clay K (1990)Isozyme variation in the *Acremonium/Epichloë* fungal endophyte complex. Phytopathology 80:1133–1139
Lewis GC, White JF Jr, Bonnefont J (1993) Evaluation of grasses infected with fungal endophytes against locusts. Ann Appl Biol; Tests Agrochem Cultivars 14:142–143
Lindstrom JT, Belanger FC (1994) A novel fungal protease expressed in endophytic infection of *Poa* species. Plant Physiol 102:645–650
Lora JM, de la Cruz J, Llobell A, Benitez T, Pintor-Toro JA (1995) Molecular characterization and heterologous expression of an endo-β-1,6-glucanase gene from the mycoparasitic fungus *Trichoderma harzianum*. Mol Gen Genet 247:639–645
Mauseth JD (2003) Botany: An introduction to plant biology. Jones and Bartlett, Boston
Mendgen K, Hahn M, Deising H (1996) Morphogenesis and mechanisms of penetration by plant pathogenic fungi. Ann Rev Phytopathol 34:367–386
Miller CO (1961) A kinetin-like compound in maize. Proc Natl Acad Sci USA 47:170–174
Moon CD, Scott B, Schardl CL, Christensen MJ (2000) The evolutionary origins of *Epichloë* endophytes from annual ryegrasses. Mycologia 92:1103–1118
Moubarak AS, Piper EL, West CP, Johnson ZB (1993) Interaction of purified ergovaline from endophyte-infected tall fescue with synaptosomal ATPase enzyme system. J Agric Food Chem 41:407–409

Moy M, Belanger F, Duncan R, Freehof A, Leary C, Meyer W, Sullivan R, White JF Jr (2000) Identification of epiphyllous mycelial nets on leaves of grasses infected by clavicipitaceous endophytes. Symbiosis 28:291–302

Oliver R, Osbourn A (1995) Molecular dissection of fungal phytopathogenicity. Microbiology 141:1–9

Papavizas GC (1985) *Trichoderma* and *Gliocladium*: biology, ecology, and potential for biocontrol. Annu Rev Phytopathol 23:23–54

Peberdy JF (1990) Fungal cell walls – a review. In: Kuhn PJ, Trinci APJ, Jung MJ, Goosey MW, Copping LG (eds) Biochemistry of cell walls and membranes in fungi. Springer, Berlin Heidelberg New York

Porter JK, Bacon CW, Cutler HG, Arrendale RF, Robbins JD (1985) *In vitro* auxin production by *Balansia epichloë*. Phytochemistry 24:1429–1431

Reddy PV, Lam CK, Belanger FC (1996) Mutualistic fungal endophytes express a proteinase which is homologous to proteases suspected to be important in fungal pathogenicity. Plant Physiol 111:1209–1218

Rykard DM, Bacon CW, Luttrell ES (1985) Host relations of *Myriogenospora atramentosa* and *Balansia epichloë* (Clavicipitaceae). Phytopathology 75:950–956

Schaeffer JH, Leykam J, Walton JD (1994) Cloning and targeted gene disruption of EXG1, encoding exo-1,3-glucanase, in the phytopathogenic fungus Cochliobolus *carbonum*. Appl Environ Microbiol 60:594–598

Schardl CL, Wilkinson HH (2000) Hybridization and cospeciation hypotheses for the evolution of grass endophytes. In: Bacon CW, White JF Jr (eds) Microbial endophytes. Marcel-Dekker, New York, pp 63–83

Scott-Craig JS, Panaccione DG, Cervone F, Walton JD (1990) Endopolygalacturonase is not required for pathogenicity of *Cochliobolus carbonum* on maize. Plant Cell 2:1191–1200

Sposato P, Ahn J-H, Walton JD (1995) Characterization and disruption of a gene in the maize pathogen *Cochliobolus carbonum* encoding a cellulose binding domain and hinge region. Mol Plant-Microbe Interact 8:602–609

St. Leger RJ (1995) The role of cuticle-degrading proteases in fungal pathogenesis of insects. Can J Bot 73:1119–1125

Sullivan RF, Bills GF, Hywel-Jones NL, White JF Jr (2000) *Hyperdermium*: a new clavicipitalean genus for some tropical epibionts of dicotyledonous plants. Mycologia 92:908–919

Tenberge KB, Homann V, Oeser B, Tudzynski P (1996) Structure and expression of two polygalacturonase genes of *Claviceps purpurea* oriented in tandem and cytological evidence for pectinolytic enzyme activity during infection of rye. Phytopathology 86:1084–1097

Tsai H-F, Liu J-S, Staben C, Christensen MJ, Latch GCM, Siegel MR, Schardl CL (1994) Evolutionary diversification of fungal endophytes of tall fescue grass by hybridization with *Epichloë* species. Proc Natl Acad Sci USA 91:2542–2546

Tzean SS, Hsieh LS, Wu WJ (1997) Atlas of entomopathogenic fungi from taiwan. Council of Agriculture, Yuan, Taiwan, Republic of China

White JF Jr (1987) Widespread distribution of endophytes in the Poaceae. Plant Dis 71:340–342

White JF Jr (1988) Endophyte-host associations in forage grasses. XI. A proposal concerning origin and evolution. Mycologia 80:442–446

White JF Jr, Cole GT (1986) Endophyte-host associations in forage grasses. IV. The endophyte of *Festuca versuta*. Mycologia 78:102–107

White JF Jr, Bultman TL (1987) Endophyte-host associations in forage grasses. VIII. Heterothallism in *Epichloë typhina*. Am J Bot 74:1716–1721

White JF Jr, Chambless DA (1991) Endophyte-host associations in forage grasses. XV. Clustering of stromata-bearing individuals of *Agrostis hiemalis* infected by *Epichloë typhina*. Am J Bot 78:527–533

White JF Jr, Owens JR (1992) Stromal development and mating system of *Balansia epichloë*, a leaf-colonizing endophyte of warm-season grasses. Appl Environ Microbiol 58:513–519

White JF Jr, Glenn AE (1994) A study of two fungal epibionts of grasses: structural features, host relationships, and classification in genus *Myriogenospora* Atk. (Clavicipitales). Am J Bot 81:216–223

White JF Jr, Camp CR (1996) A study of water relations of *Epichloë amarillans* White, an endophyte of the grass *Agrostis hiemalis* (Walt.) B.S.P. Symbiosis 18:15–25

White JF Jr, Breen JP, Morgan-Jones G (1991) Substrate utilization in selected *Acremonium*, *Atkinsonella*, and *Balansia* species. Mycologia 83:601–610

White JF Jr, Morrow AC, Morgan-Jones G, Chambless DA (1991) Endophyte-host associations in forage grasses. XIV. Primary stromata formation and seed transmission in *Epichloë typhina*: developmental and regulatory aspects. Mycologia 83:72–81

White JF Jr, Martin TI, Cabral D (1996) Endophyte-host associations in grasses. XXIII. Conidia formation by *Acremonium* endophytes in the phylloplanes of *Agrostis hiemalis* and *Poa rigidifolia*. Mycologia 88:174–178

White JF Jr, Bacon CW, Hinton DM (1997) Modifications of host cells and tissues by the biotrophic endophyte *Epichloë amarillans* (Clavicipitaceae; Ascomycotina). Canadian J Bot 75:1061–1069

Yue C, Miller CJ, White JF, Richardson M (2000) Isolation and characterization of fungal inhibitors from *Epichloë festucae*. J Agric Food Chem 48:4687–4692

11 Interactions of Microbes with Genetically Modified Plants

Michael Kaldorf, Chi Zhang, Uwe Nehls, Rüdiger Hampp and François Buscot

1 Introduction

The introduction of molecular biological methods into plant breeding has offered the possibility to construct genetically modified plants (GMPs) with new qualities. Major goals of genetic engineering are the improvement of product quality as well as the enhancement of resistance or tolerance to pathogen infections, herbicides and abiotic stress factors.

Attempts to improve the quality of agricultural products include the manipulation of the softening of fruits like strawberry (Jimenez-Bermudez et al. 2002) and tomato (Quiroga and Fraschina 1997) in order to allow longer storage after harvesting, the modification of oil composition of oilseed crops (Thelen and Ohlrogge 2002), or the elevation of the provitamin A content of rice (Ye et al. 2000) and tomato (Romer et al. 2000). Even in forestry, increased wood production and quality are of great commercial interest (Mullin and Bertrand 1998). For example, the lignin content of transgenic aspen, in which the lignin biosynthesis pathway was downregulated by antisense inhibition, was greatly reduced (Hu et al. 1999), indicating that some technical limitations for the use of these fast growing trees for cellulose fiber production (e.g., in paper industry) might be reduced by genetic engineering.

In contrast to the examples given above, the basic target of constructing GMPs with enhanced resistance to biotic or abiotic stress factors is not a modified product quality, but an enhanced productivity and reduction of the production costs in agriculture and forestry. The possibility to overcome different types of abiotic stress in GMPs has been demonstrated in several experiments [e.g., drought-resistant sugar beet (Pilon-Smits et al. 1999), salt-tolerant tomato plants (Zhang and Blumwald 2001), or aluminium-resistant *Brassica napus* plants (Basu et al. 2001)], but until now, none of these GMPs is being used for commercial production.

All GMPs introduced on a large scale into agriculture in the 1990s possess resistance genes either against herbicides or against plant pathogens. Many different herbicide-resistant transgenic plants like corn, cotton, lettuce,

Plant Surface Microbiology
A. Varma, L. Abbott, D. Werner, R. Hampp (Eds.)

poplar, potato, rapeseed, soybean, sugar beet, tobacco, tomato, and wheat have been developed and field-tested (Saroha et al. 1998). Especially soybean, corn and oilseed resistant to the herbicide glyphosate were planted on at least 15 million ha in the USA, Argentina, Canada and other countries since 1998, occupying about 70 % of the total release area for GMPs in 1998 (Warwick et al. 1999; Owen 2000). Among the pathogen-resistant GMPs, transgenic corn (Owen 2000), cotton (Perlak et al. 2001) and other plants producing insecticidal proteins from *Bacillus thuringiensis* (Bt) are grown on a similar scale (11.4 million ha worldwide in 2000, Shelton et al. 2002) as herbicide-resistant GMPs. Transgenic plants with different resistances to many other viral, bacterial and fungal pathogens have been described (e.g., Düring et al. 1993; Punja 2001; Solomon-Blackburn and Barker 2001, and references cited therein), which are not yet used commercially.

Since the first field experiment in 1986, more than 25,000 field trials with GMPs have been performed worldwide (Warwick et al. 1999), and for all GMPs, some common features have to be demonstrated in field experiments prior to commercial production. First, the new quality of the transgenic plant must be stable under field conditions. Second, all other characteristics important for the agricultural use of a plant should remain unaffected in the GMPs compared to their parental breeds. Third, negative effects on the environment and particularly on nontarget organisms have to be low or missing. Such nontarget effects include cases like the negative and – in the worst case – lethal impact of Bt transgenic corn pollen on larvae of the monarch butterfly (Losey et al. 1999; Hansen Jesse and Obrycki 2000), which correspond to environmental risks without direct influence on the performance of the GMPs in the field. A further category of nontarget effects includes reduced compatibility of GMPs in symbiotic interactions or damage of plant growth promoting bacteria, which are not only environmental risks, but might also reduce the productivity of GMPs in the field. Especially in the case of pathogen-resistant GMPs, a negative impact on nontarget organisms is likely and has to be investigated thoroughly prior to the decision to use a GMP commercially.

The aim of this review is to summarize the effects of transgenic plants on nontarget microorganisms. Depending on the specific characters of a GMP, these effects might be positive (e.g., enrichment of plant growth-promoting bacteria), neutral, or even negative (e.g., increase in plant pathogenic bacteria or fungi). Experimental work in this field can be grouped into three categories: (1) analysis of effects of GMPs on changes in microorganism communities at the root surface and in the rhizosphere; (2) investigations that focus on positive interactions between plants and microorganisms like mycorrhizal or *Rhizobium* symbioses that are important factors for plant nutrition and health; (3) analysis of horizontal gene transfer (HGT) events as a result of tight interactions between GMPs and microorganisms.

2 Changes in Microbial Communities Induced by Genetically Modified Plants

Many studies dealing with the impact of GMPs on microbial communities have been published since the urgent call of Morra (1994) for this kind of research (Table 1). At present, transgenic potatoes producing T4 lysozyme to obtain resistance to *Erwinia carotovora* and other bacterial pathogens (Düring et al. 1993) are the best characterized GMP system with respect to microbial interactions. In addition, some of the commercially most important GMPs like corn and cotton producing the insecticidal *Bacillus thuringiensi* (Bt) toxin, or glyphosate-tolerant *Brassica* sp., as well as different other GMPs have been tested.

Selective advantages of specific bacteria in the rhizosphere of GMPs were demonstrated in two studies. A T4 lysozyme-tolerant *Pseudomonas putida* strain with antibacterial activity to the bacterial pathogen *Erwinia carotovora* was introduced into the rhizosphere of T4 lysozyme-producing potatoes. During flowering of the plants, when the highest lysozyme level was detected in planta, significantly more colonies of the introduced *P. putida* strain could be reisolated from transgenic potatoes compared to controls (Lottmann et al. 2000). In the second experiment, the culture of transgenic *Lotus corniculatus* plants producing different opines led to a significant increase in opine-utilizing bacteria in the rhizosphere. When the opine-producing plants were replaced by nontransgenic plants, the concentration of different fractions of opine utilizers in the soil slowly decreased over 22 weeks. However, even then, the density of the bacterial fraction specifically using mannopine was still five times higher than in soil which had not been planted with the transgenic *Lotus* plants (Oger et al. 2000). This demonstrates that alterations in the soil microflora induced by the cultivation of GMPs may be persistent.

Nontarget effects of GMPs on bacterial communities seem to be common. A broad spectrum of both physiological and molecular biological methods like community-level physiological profile (CLPP; Griffiths et al. 2000; Dunfield and Germida 2001), BIOLOG substrate utilization test (Siciliano et al. 1998; Di Giovanni et al. 1999), fatty acid methyl ester (FAME) patterns (Siciliano et al. 1998; Dunfield and Germida 2001), plate-counting of bacterial groups (e.g., spore-forming or cellulose-utilizing bacteria, Donegan et al. 1999) and T-RFLP (Lukow et al. 2000) have been used to characterize bacterial communities associated with GMPs. In addition, population analyses of protozoa, nematodes and microarthropods have been included in some studies (e.g., Donegan et al. 1997; Saxena and Stotzky 2001). Significant changes in the rhizosphere of different GMPs have been shown (see Table 1), which are not necessarily linked directly to the presence of new gene product(s). For example, changes in the rhizospheric bacterial community of transgenic cotton plants producing the *Bt* toxin were significant, but the purified *Bt* toxin itself displayed no detectable effect on soil microorganisms (Donegan et al.

Table 1: Studies assessing the impact of GMPs on plant-associated and soil microorganisms

GMP species, new characteristics	Group(s) of organisms investigated	Observations	Reference
Brassica sp., tolerance to the herbicide glyphosate	Soil bacterial communities	Indications for changes in the soil bacterial communities	Siciliano et al. (1998)
Brassica sp., tolerance to the herbicide glyphosate	Soil bacterial communities	Significant changes in the soil bacterial communities	Dunfield and Germida (2001)
Gossypium hirsutum, Bt toxin production resistance to insects	Leaf material decomposing bacterial, fungal, and protozoan populations	Two of three transgenic cotton lines caused significant stimulation and qualitative changes of bacterial and fungal populations	Donegan et al. (1995)
Lotus corniculatus, opine production	Opine-utilizing bacteria, soil bacterial community	Increase of opine-utilizing bacteria in the rhizosphere of the GMPs; effect persistent for 22 weeks	Oger et al. (2000)
Medicago sativa, α-amylase or lignin peroxidase production	Soil bacteria, fungi, protozoa, nematodes and micro-arthropods	Significant changes in bacterial populations of lignin peroxidase plants; population levels of fungi, protozoa, nematodes and microarthropods not affected	Donegan et al. (1999)
Medicago sativa, α-amylase or lignin peroxidase production	Soil bacterial communities	Significant changes in bacterial populations associated with lignin peroxidase plants	Di Giovanni et al. (1999)
Nicotiana tabaccum, proteinase inhibitor I, resistance to insects	Protozoa, nematodes, and microarthropods	Changes in nematode populations, reduced Collembola populations on litter from GMPs	Donegan et al. (1997)
Solanum tuberosum, T4 lysozyme production, resistance to bacteria	Plant-associated bacteria	Minor effects on community structure	Lottmann et al. (1999)
Solanum tuberosum, T4 lysozyme production	Introduced, pathogen-antagonistic bacteria with high lysozyme tolerance	Significant increase of introduced lysozyme-tolerant bacteria on GMPs	Lottmann et al. (2000)
Solanum tuberosum, T4 lysozyme production	Pseudomonads and enterics from the rhizosphere	No correlation between phenotypic or genotypic profile and transgenic character	Lottmann and Berg (2001)

GMP species, new characteristics	Group(s) of organisms investigated	Observations	Reference
Solanum tuberosum, T4 lysozyme production	Rhizosphere bacterial community	Only minor effects of lysozyme	Heuer and Smalla (1999)
Solanum tuberosum, T4 lysozyme production	Rhizosphere bacterial community	No detectable effects of lysozyme production	Heuer et al. (2002)
Solanum tuberosum, T4 lysozyme production	Pathogenic *Erwinia carotovara* strains; soil bacteria	Similar lysozyme sensitivity of *Erwinia* and most other soil bacteria	de Vries et al. (1999)
Solanum tuberosum, T4 lysozyme production	*Bacillus subtilis*	Increased killing of *B. subtilis* on the hairy roots of lysozyme-producing plants	Ahrenholtz et al. (2000)
Solanum tuberosum, Bt toxin production	Total bacterial and fungal populations colonizing leaves	Only minor effects of the GMPs	Donegan et al. (1996)
Solanum tuberosum, production of anti-feedant lectines	Soil bacterial communities; protozoa	No effect on protozoa, but significant changes in the physiological profiles of bacterial communities	Griffiths et al. (2000)
Solanum tuberosum, Barstar/ Barnase genes, fungal pathogen resistance	Soil bacterial community structure	Some significant differences between GMPs and control plants	Lukow et al. (2000)
Zea mays, Bt toxin production	Soil bacteria, fungi, protozoa, and nematodes	No significant differences detected	Saxena and Stotzky (2001)

1995). Similarly, the effects of transgenic potatoes producing the lectin GNA on nontarget soil organisms could not be attributed to the formation of the lectin GNA itself (Griffiths et al. 2000). So far, only one GMP producing an antibacterial substance, namely T4 lysozyme-producing potatoes with enhanced resistance to *Erwinia carotovora*, has been investigated in detail for nontarget effects on soil bacteria. Most soil bacteria were lysozyme-sensitive when tested in laboratory experiments with pure cultures (de Vries et al. 1999). In addition, increased killing of *Bacillus subtilis* was observed on the root surface of T4 lysozyme-producing potatoes from the field, and this effect was ascribed directly to the lysozyme release by the roots (Ahrenholtz et al. 2000). Nevertheless, the production of lysozyme had only a minor influence on the bacterial phyllo- and rhizosphere communities (Heuer and Smalla 1999; Heuer et al. 2002), which was considered negligible relative to natural factors by the authors. Further studies with the same system which focused on potentially beneficial plant-associated microbes like auxin-producing bacteria or bacteria antagonistic to the pathogenic *E. carotovora* did not reveal correlations between the transgenic character of plants and the pheno- or genotypic features of bacterial isolates (Lottmann and Berg 2001). Thus, up to now, there is no direct evidence from field experiments that the primary product of transgene expression is responsible for significant changes in the soil microbial community in any GMP. Instead, secondary effects of GMP generation, like somaclonal variation and changes in general plant metabolism induced by the transgene insertion or expression, may contribute to a major part of the effects described above.

3 Impact of Genetically Modified Plants on Symbiotic Interactions

The question whether the genetical transformation of plants might reduce their ability to form mutualistic symbioses with microorganisms has been addressed in a surprisingly small number of studies.

Biological nitrogen fixation accounts for about 65 % of the nitrogen utilized in agriculture worldwide (Vance and Graham 1995). The ability to reduce atmospheric nitrogen to ammonia (nitrogen fixation) is restricted to prokaryotes. Beside free-living and plant-associated bacteria, members of the Rhizobiaceae living symbiotically in the typical root nodules of legumes such as alfalfa, clover, pea, and soybean are the agriculturally most important group of nitrogen fixing organisms. The symbiotic interaction between rhizobia and legumes requires a sequential signal exchange between both partners, and therefore, exhibits a high degree of host specificity (Bothe 1993). Transgenic plants have been used as a tool to investigate the host recognition of rhizobia (Diaz et al. 1989, 2000). For example, the transfer of lectin genes between different legumes has been shown as a possible way to modify host specificity

(Diaz et al. 2000; van Rhijn et al. 2001). While the GMPs used in these studies have been modified specifically to change the plant–rhizobia interactions, possible alterations in the rhizobia symbiosis would be untargeted in pathogen- or herbicide-resistant legumes. The constitutive expression of a rice basic chitinase gene with putative antifungal effects in alfalfa had no negative influence on the interaction with *Rhizobium* (Masoud et al. 1996). No further information is available about interference between genetical modifications of plants and nitrogen fixing bacteria, particularly about the herbicide-resistant soybean cultivars planted on a large scale since 1996.

The second important group of plant-microbe symbioses is the mycorrhiza. Under natural conditions, the roots of most plants are colonized by mycorrhizal fungi, which increase their uptake of water and nutrients, as well as their resistance against pathogens and abiotic stress (Smith and Read 1997). Ectomycorrhiza (EM) is the dominating type of mycorrhiza in gymnosperms and many woody angiosperms. EM formation is accompanied by morphological changes of both the fungal hyphae and the plant fine roots. Typically, hyphae form a mantle of varying thickness around the fine roots. From there they extend into the apoplast of the root cortex, forming a highly branched network and thus establishing a large surface area for solute exchange, the Hartig net (Kottke and Oberwinkler 1986). Arbuscular mycorrhiza (AM) can be found in mosses, ferns and many angiosperms, including agronomically important plants like barley, corn, potato, rice, soybean, and wheat (Smith and Read 1997). The successful use of transgenic host plants in basic research on mycorrhiza has demonstrated that the use of common molecular biological methods, like the introduction of antibiotic resistance genes [e.g., transgenic aspen carrying a hygromycin resistance gene in addition to indoleacetic acid-biosynthetic genes, (Tuominen et al. 1995; Hampp et al. 1996)] or reporter gene constructs (e.g., the *gus* reporter gene system, Gianinazzi-Pearson et al. 2000) into plants, has normally no impact on mycorrhiza formation. Only in the case of the symbiosis-related gene enod40 from *Medicago truncatula*, overexpression of the gene accelerated AM colonization, while transgenic lines with suppressed enod40 transcript levels exhibited reduced mycorrhization (Staehelin et al. 2001).

Negative nontarget effects of GMPs on mycorrhizal fungi seem to be most likely in GMPs constitutively expressing antifungal proteins in order to obtain resistance against fungal pathogens (Glandorf et al. 1997). Transgenic *Nicotiana sylvestris* plants with more than tenfold enhanced chitinase activity were significantly less colonized by the fungal pathogen *Rhizoctonia solani* compared to control plants. However, neither the quantity of AM colonization nor the anatomy of AM hyphae, arbuscules or vesicles were significantly affected in the chitinase overproducing plants (Vierheilig et al. 1993). In a further study, several pathogenesis-related (PR) proteins were constitutively expressed in transgenic tobacco plants to investigate their influence on the AM fungus *Glomus mosseae* (Vierheilig et al. 1995). Two acidic and two basic chitinases, one acidic and two basic glucanases, as well as three PR proteins of unknown function had no detectable influence on AM colonization. Only one of the PR proteins tested, an acidic, extracellular β-1,3-glucanase of class II, reduced the mycorrhization of transgenic tobacco roots by *G. mosseae* significantly. This observation demonstrates that mycorrhiza formation could be affected in GMPs expressing antifungal PR proteins. Therefore, a case-to-case investigation of GMPs with increased fungal pathogen resistance seems to be necessary to exclude negative effects on AM formation.

In addition to the quantity of mycorrhizal colonization, the structural and functional diversity of mycorrhizal fungi colonizing the root system might be influenced in GMPs. Probably due to methodical difficulties, this question has not been addressed for AM fungi. Compared to the rather uniform morphology of AM fungi, which makes their morphological identification quite difficult, EM fungi exhibit many morphological and anatomical characters that could be used for their characterization (Agerer 1991). In combination with PCR-RFLP and sequence analysis of the ITS region within the fungal rDNA (Buscot et al. 2000), EM communities can be described with a sufficient resolution to compare the mycorrhization of transgenic and nontransgenic trees, even in the field. A release experiment with transgenic aspen carrying the *rolC* gene from *Agrobacterium rhizogenes* (Fladung and Muhs 2000) was accompanied by a detailed analysis of the EM status of the trees. Although *rolC* modified the hormonal balance in the trees, and therefore, might have affected their mycorrhization ability, no significant difference in the degree of mycorrhization was observed in the transformed aspen. The structure of the EM community of the different aspen lines was similar in the first two years of the experiment (Kaldorf et al. 2002), but in the third and fourth years, a significantly reduced EM diversity was observed on the *rolC* transgenic aspen compared to controls (Kaldorf et al. 2001). In addition, one EM morphotype formed by *Phialocephala fortinii* appeared to be significantly less represented on the transgenic line "E2/5" compared to all other transgenic and control lines (Kaldorf et al. 2002). This reduced compatibility for one mycobiont represents the first example of a clone-specific effect concerning mycorrhization of transgenic plants.

4 Horizontal Gene Transfer

Three potential pathways have been proposed for the spread of GMPs or the transgenes introduced into these plants. Two of these pathways, the establishment of self-sustaining GMP populations and the introgression of genes into wild populations, regarded as the major risks of GMPs for natural plant communities (Wolfenbarger and Phifer 2000), do not involve plant/microorganism interactions. The third possibility is the horizontal transfer of genes from GMPs to microorganisms, which might lead to bacterial or fungal strains carrying genes from GMPs. The exchange of genetic information between different bacterial species by transformation, transduction or conjugation seems to be common in nature (Krishnapillai 1996; Wöstemeyer et al. 1997 and references therein). The detailed analysis of DNA and amino acid sequence data has indicated that horizontal transmission of genes, even between bacteria and eukaryotes or between eukaryotes from different systematic kingdoms, probably occurred in rare cases during evolution (Dröge et al. 1998 and references therein), but there is no experimental access to further investigation or verification of such horizontal gene transfer (HGT) events.

The focus of the experimental work on HGT has been the question whether antibiotic resistance genes, used as selectable markers in GMPs, can be transferred to bacteria, enhancing the frequency of antibiotic-resistant bacteria of medical importance (Nielsen et al. 1998). Beside the relevance of this question for the risk assessment of GMPs, the transfer of antibiotic-resistance genes is easy to detect compared to a possible horizontal transfer of genes which cannot be used as a selectable marker for the isolation of transformed bacteria. Therefore, experimental data about the possible HGT of other genes are scarce.

The transformation of different bacterial species has been demonstrated under optimized laboratory conditions using isolated plasmid DNA, total DNA from GMPs and even homogenized plant material from GMPs as the source for antibiotic-resistance genes (Schlüter et al. 1995; Gebhard and Smalla 1998). The efficiency of the integration of the *npt*II gene, causing resistance to kanamycin, into the genome of *Acinetobacter* sp. strongly depended on the presence of homologous sequences in the bacterial DNA (Nielsen et al. 1997). This observation was confirmed by de Vries et al. (2001) using *Acinetobacter* sp. and *Pseudomonas stutzeri* as well as by Bertolla et al. (2000) using the plant pathogenic bacterium *Ralstonia solanacearum* as recipient for antibiotic-resistance genes.

While transformation of bacteria is common under optimized laboratory conditions, all experiments under natural conditions indicated that the frequency of HGT is drastically reduced compared to optimized conditions. Although DNA from transgenic plants can persist in soil for up to 2 years (Gebhard and Smalla 1999), the availability of DNA from GMPs could be a limiting factor for HGT. Even under otherwise optimized conditions (e.g., use of purified DNA from transgenic sugar beet as source for the npt*II* gen, construction of an *Acinetobacter* strain carrying a deleted *npt*II gene to allow homologous recombination in the recipient bacteria), the frequency of HGT was low in sterilized soil microcosms and below the detection limit in non-sterilized soil (Nielsen et al. 2000). In a field release experiment with *npt*II-transgenic sugar beet, a total of 4000 kanamycin-resistant colonies of soil bacteria isolated from the field release site was checked for the presence of the *npt*II gene from the transgenic plants by dot blot hybridization and PCR. None of the isolates carried the *npt*II gene, indicating a natural kanamycin resistance of all strains tested, which was not acquired by HGT (Gebhard and Smalla 1999). Thus, the conclusion of Bertolla and Simonet (1999) that we are a long way from demonstrating that plant–bacterium gene transfer does occur under natural conditions, is still valid.

Approaches to detect HGT from transgenic plants to eukaryotic microorganism are sparse. Particular fungi often grow in intimate contact with plants or – in the case of endoparasitic and mycorrhizal symbioses – even within plants. In these cases, uptake of plant DNA by fungi might be more likely compared to soil bacteria, as the plant DNA does not come in contact with soil. Indeed, evidence has been presented that the phytopathogenic fungus *Plasmodiophora brassicae* takes up host plant DNA during each infection cycle (Bryngelsson et al. 1988), but interactions between *Plasmodiophora* and transgenic *Brassica* sp. have not been investigated. HGT from plants to saprophytic fungi has also been reported (Hoffmann et al. 1994). Transgenic plants expressing the hygromycin gene (*hph*) as selection marker under control of a fungal promoter were generated. After cocultivation of dead plant material together with *Aspergillus*, fungal progenies were isolated that revealed resistance to hygromycin B on selective agar plates (Hoffmann et al. 1994). The *hph* gene and other foreign DNA sequences could be detected in some of these hygromycin B-resistant fungal strains. Nevertheless, in most cases the foreign DNA was not stably integrated into the *Aspergillus* genome.

In the following, we present some unpublished data evaluating the possibility of HGT in mycorrhizal symbioses. Ectomycorrhizas are of special interest in this aspect, due to the long life time of the host trees. Therefore, there is a need to investigate the possibility of HGT from transformed forest trees to EM fungi. Plant cells frequently die during EM interaction and thus, fungal hyphae of the Hartig net come in close contact with plant DNA. Filamentous fungi are naturally not very competent in the uptake of large DNA fragments. In EM however, hyphae of the Hartig net are coenocytic and have a highly enlarged plasma membrane surface area due to extensive invaginations (Kottke and Oberwinkler 1987). Therefore, they might be more competent for DNA uptake than normal hyphae.

Two different approaches were used to study HGT between plant and fungal cells in ectomycorrhizas. In the first approach, transgenic aspen carrying the *rolC* gene from *Agrobacterium rhizogenes* under control of the light-inducible *rbcS* promoter from potato (Fladung et al. 1997) were grown in vitro together with the ectomycorrhizal ascomycete *Phialocephala fortinii* strain 5B, isolated from mycorrhizal aspen roots collected in the field (Fladung et al. 2000). After 12 weeks of cocultivation, *P. fortinii* was reisolated from colonized aspen roots. Fungal hyphae growing out from mycorrhizas were transferred to fresh medium for further growth to avoid contamination with plant material. Genomic DNA was isolated from the fungal mycelium and analyzed for the presence of the *rolC* gene. To enhance the sensitivity of the PCR assay, a "nested" PCR strategy was followed. The first *rolC* specific primer pair should amplify a 950-bp DNA fragment. In a second PCR step, a 500-bp fragment of *rolC* should be amplified from the products of the first PCR using a second inner primer pair, again specific for *rolC*. Isolated DNA from transgenic aspen leaves was used as positive control for the nested PCR, while the quality of the fungal DNA was checked with the primer pair ITS1/ITS4 (White et al. 1990), specific for a part of fungal rDNA clusters. The *rolC* gene was not detected in any of the 24 *Phialocephala* colonies analyzed. The number of 24 samples was sufficient to demonstrate that the uptake of plant DNA in *Phialocephala* EM does not occur on a regular basis, as suggested for the *Plasmodiophora–Brassica* interaction (Bryngelsson et al. 1988).

The second approach was with transgenic plants that contained a small marker gene, which could confer herbicide resistance into the target organism after HGT. The advantage of this strategy is that a large number of samples can be simultaneously screened, but only a small number of the samples able to grow on the selection medium have to be investigated in detail. In order to monitor HGT in ectomycorrhizas formed between poplar and *Amanita muscaria*, a 1250-bp *EcoRI/XbaI* fragment of pBG (Straubinger et al. 1992) containing the *Streptomyces hygroscopicus* bar gene under the control of the *Cochlibolus heterostrophus GPD1* promoter was inserted into the *Agrobacterium* vector pBI121 (Clontech, Palo Alto, CA, USA; Fig. 1). The function of

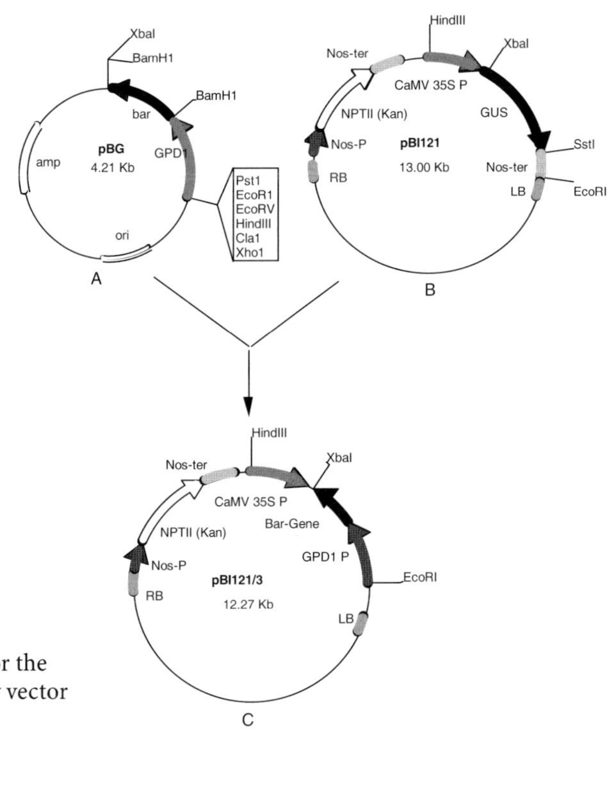

Fig. 1. Cloning strategy for the construction of the binary vector pBI121/3

the GPD/bar construct in the ectomycorrhizal fungus *A. muscaria* was previously verified by PEG-mediated protoplast transformation. Transgenic poplars containing the GPD/bar construct were generated by *Agrobacterium*-mediated transformation. Twenty plants were isolated that originate from different calli. PCR amplification was carried out with genomic DNA from transgenic poplars using *bar*-specific primers. PCR products of the expected size were obtained from 19 out of a total of 20 putative transgenic poplar plants (Fig. 2). Isolated PCR-fragments of three clones were sequenced and revealed the introduced bar gene.

For the investigation of a HGT event, 35,000 ectomycorrhizas formed between transgenic poplars and *A. muscaria* were isolated and transferred to selective agar plates. After the first round of selection, 102 putative Basta resistant fungal colonies were obtained. However, none of these colonies was able to grow after transfer to a fresh selection medium. Genomic DNA isolated from fungal hyphae initially growing on the selection medium was investigated for the presence of the *bar*-gene using PCR. No bar-fragment could be obtained from any of these investigated clones. The utilization of primers for

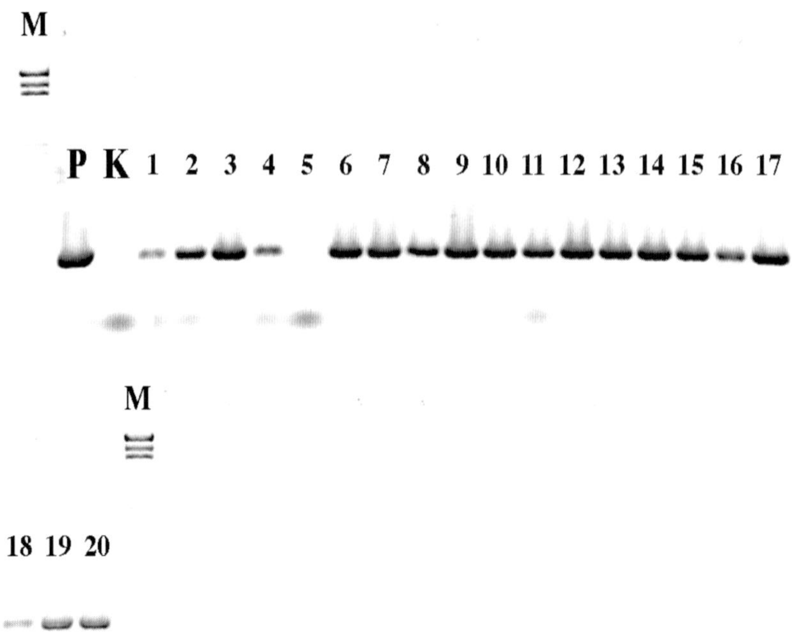

Fig. 2. Analysis of genomic DNA isolated from putative kanamycin-resistant poplar transformants. PCR amplification was performed on genomic DNA using primers specific for the *bar* gene that amplifies an internal fragment of 550-bp length. *Lanes 1 to 20* Isolated DNA from putative transformants. *P* Positive control with diluted DNA of pBI121/3, *K* DNA isolated from a nontransformed poplar plant, *M* molecular size marker (λ/*Hind*III DNA marker)

a single copy gene of *A. muscaria* (SCIV038, Nehls et al. 2001) revealed PCR fragments in any case, indicating that no inhibitors of the PCR reaction were present in the genomic DNA preparation. The reason for false positives was most probably the low herbicide concentration in the growth medium. A concentration of 200 µg/ml Basta (as used in this study) results in fungal background growth. This relatively low Basta-concentration was chosen to recognize also lateral transfer of the resistance gene lacking its heterologous promoter. In this case, the bar-gene might integrate behind a weak *A. muscaria* promoter, resulting in only a weak herbicide resistance.

The 35,000 mycorrhizas investigated in this study represent, of course, only a limited sample number. Nevertheless, since each mycorrhiza does contain a large number of competent fungal hyphae in direct contact with plant epidermal cells that die under the selection conditions, this sample number is large enough to reveal that HGT from the tree to the fungal partner is a quite rare or maybe completely missing event in EM symbiosis, at least under axenic conditions.

5 Conclusions

Taken together, many of the studies cited above demonstrate that transgenic plants can induce changes in soil microorganism communities. Nevertheless, the importance of these findings is unclear as in most studies, the modifications in the rhizosphere of GMPs were not compared to the natural variance in the rhizosphere of different plant breeds generated by conventional methods. For example, the mycorrhization capacity of modern wheat varieties with high pathogen resistance has been shown to be reduced (Hetrick et al. 1992). Such potentially negative effects would be considered unacceptable in the case of any GMP introduced into agriculture.

Concerning the investigations on HGT, there is some evidence for the possibility of HGT not only between bacteria, but also between plants and microorganisms. In soil, HGT must be a rare event, as several attempts to detect HGT in field experiments failed. Despite the missing evidence for HGT in the field, the possibility of HGT should be kept in mind for risk assessment. The question to be answered in a case-to-case consideration is whether a possible rare HGT of the introduced genes from GMPs to microorganisms might cause specific problems. This is unlikely if the transgene itself is common in nature. For example, a natural transfer of the *rolC* gene from *Agrobacterium* to other bacteria seems much more likely than a HGT from the *rolC* transgenic aspen mentioned above to microorganisms. On the other hand, artificial genes generated by genetic engineering might have a high risk potential when released into nature.

References and Selected Reading

Agerer R (1991) Characterization of ectomycorrhiza. In: Norris JR, Read DJ, Varma AK (eds) Methods in microbiology, vol 23. Academic Press, London, pp 25–73

Ahrenholtz I, Harms K, de Vries J, Wackernagel W (2000) Increased killing of *Bacillus subtilis* on the hair roots of transgenic T4 lysozyme-producing potatoes. Appl Environ Microbiol 66:1862–1865

Basu U, Good AG, Taylor GJ (2001) Transgenic *Brassica napus* plants overexpressing aluminium-induced mitochondrial manganese superoxide dismutase cDNA are resistant to aluminium. Plant Cell Environ 24:1269–1278

Bertolla F, Simonet P (1999) Horizontal gene transfers in the environment: natural transformation as a putative process for gene transfers between transgenic plants and microorganisms. Res Microbiol 150:375–384

Bertolla F, Pepin R, Passelegue-Robe E, Paget E, Simkin A, Nesme X, Simonet P (2000) Plant genome complexity may be a factor limiting in situ the transfer of transgenic plant genes to the phytopathogen *Ralstonia solanacearum*. Appl Environ Microbiol 66:4161–4167

Bothe H (1993) Metabolism of inorganic nitrogen compounds. Progress in Botany 54. Springer, Berlin Heidelberg New York, pp 201–217

Bryngelsson T, Gustafsson M, Green B, Lind C (1988) Uptake of host DNA by the parasitic fungus *Plasmodiophora brassicae*. Physiol Mol Plant Pathol 33:163–171

Buscot F, Munch JC, Charcosset JY, Gardes M, Nehls U, Hampp R (2000) Recent advances in exploring physiology and biodiversity of ectomycorrhizas highlight the functioning of these symbioses in ecosystems. FEMS Microbiol Rev 699:1–14

de Vries J, Harms K, Broer I, Kriete G, Mahn A, Düring K, Wackernagel W (1999) The bacteriolytic activity of transgenic potatoes expressing a chimeric T4 lysozyme gene and the effect of T4 lysozyme on soil- and phytopathogenic bacteria. Syst Appl Microbiol 22:280–286

de Vries J, Meier P, Wackernagel W (2001) The natural transformation of the soil bacteria *Pseudomonas stutzeri* and *Acinetobacter* sp. by transgenic plant DNA strictly depends on homologous sequences in the recipient cells. FEMS Microbiol Lett 195: 211–215

Diaz CL, Melchers LS, Hooykaas PJJ, Lugtenberg EJJ, Kijne JW (1989) Root lectin as a determinant of host plant specificity in the *Rhizobium*-legume symbiosis. Nature 338:579–581

Diaz CL, Spaink HP, Kijne JW (2000) Heterologous rhizobial lipochitin oligosaccharides and chitin oligomers induce cortical cell divisions in red clover roots, transformed with the pea lectin gene. Mol Plant-Microbe Int 13:268–276

Di Giovanni GD, Wartrud LS, Seidler RJ, Widmer F (1999) Comparison of parental and transgenic alfalfa rhizosphere bacterial communities using Biolog GN metabolic fingerprinting and enterobacterial repetitive intergenic consensus sequence-PCR (ERIC-PCR). Microb Ecol 37:129–139

Donegan KK, Palm CJ, Fieland VJ, Porteous LA, Ganio LM, Schaller DL, Bucao LQ, Seidler RJ (1995) Changes in levels, species and DNA fingerprints of soil microorganisms associated with cotton expressing the *Bacillus thuringiensis* var. *kurstaki* endotoxin. Appl Soil Ecol 2:111–124

Donegan KK, Schaller DL, Stone JK, Ganio LM, Reed G, Hamm PB, Seidler RJ (1996) Microbial populations, fungal species diversity and plant pathogen levels in field plots of potato plants expressing the *Bacillus thuringiensis* var. *tenebrionis* endotoxin. Transgen Res 5:25–35

Donegan KK, Seidler RJ, Fieland VJ, Schaller DL, Palm CJ, Ganio LM, Cardwell DM, Steinberger Y (1997) Decomposition of genetically engineered tobacco under field conditions: persistence of the proteinase inhibitor I product and effects on soil microbial respiration and protozoa, nematode and microarthropod populations. J Appl Ecol 34:767–777

Donegan KK, Seidler RJ, Doyle JD, Porteus LA, di Giovanni G, Widmers F, Wartrud LS (1999) A field study with genetically engineered alfalfa with recombinant *Sinorhizonium meliloti*: effects on the soil ecosystem. J Appl Ecol 36:920–936

Dröge M, Pühler A, Selbitschka W (1998) Horizontal gene transfer as a biosafety issue: a natural phenomenon of public concern. J Biotechnol 64:75–90

Dunfield KE, Germida JJ (2001) Diversity of bacterial communities in the rhizosphere and root interior of field-grown genetically modified *Brassica napus*. FEMS Microbiol Ecol 38:1–9

Düring K, Porsch P, Fladung M, Lörz H (1993) Transgenic potato plants resistant to the phytopathogenic bacterium *Erwinia carotovora*. Plant J 3:587–598

Fladung M, Muhs H-J (2000) Field release with *Populus tremula* (*rolC*-gene) in Großhansdorf. In: Umweltbundesamt (ed) Release of transgenic trees – present achievements, problems, future prospects. Humboldt-Universität, Berlin, pp 40–45

Fladung M, Kumar S, Ahuja MR (1997) Genetic transformation of *Populus* genotypes with different chimaeric gene constructs: transformation efficiency and molecular analysis. Transgen Res 6:111–121

Fladung M, Kaldorf M, Buscot F, Muhs H-J (2000) Untersuchungen zur Stabilität und Expressivität fremder Gene in Aspenklonen (*Populus tremula* und *P. tremula* x *P. tremuloides*) unter Freilandbedingungen. In: Schiemann J (ed) Biologische Sicherheitsforschung bei Freilandversuchen mit transgenen Organismen und anbaubegleitendes Monitoring. BEO (Projektträger Biologie, Energie, Umwelt des BMBF), Braunschweig-Jülich, Braunschweig, pp 77–83

Gebhard F, Smalla K (1998) Transformation of *Acinetobacter* sp. strain BD413 by transgenic sugar beet DNA. Appl Environ Microbiol 64:1550–1554

Gebhard F, Smalla K (1999) Monitoring field release of genetically modified sugar beets for persistence of transgenic plant DNA and horizontal gene transfer. FEMS Microbiol Ecol 28:261–272

Gianinazzi-Pearson V, Arnould C, Oufattole M, Arango M, Gianinazzi S (2000) Differential activation of H^+-ATPase genes by an arbuscular mycorrhizal fungus in root cells of transgenic tobacco. Planta 211:609–613

Glandorf DCM, Bakker PAHM, van Loon LC (1997) Influence of the production of antibacterial and antifungal proteins by transgenic plants on the saprophytic soil microflora. Acta Bot Neerl 46:85–104

Griffiths BS, Geoghegan IE, Robertson WM (2000) Testing genetically engineered potato, producing the lectins GNA and Con A, on non-target soil organisms and processes. J Appl Ecol 37:159–170

Hampp R, Ecke M, Schaeffer C, Wallenda T, Wingler A, Kottke I, Sundberg B (1996) Axenic mycorrhization of wild type and transgenic hybrid aspen expressing T-DNA indoleacetic acid-biosynthetic genes. Tree 11:59–64

Hansen Jesse LC, Obrycki JJ (2000) Field deposition of Bt transgenic corn pollen: lethal effects on the monarch butterfly. Oecologia 125:241–248

Hetrick BAD, Wilson GWT, Cox TS (1992) Mycorrhizal dependence of modern wheat varieties, landraces, and ancestors. Can J Bot 70:2032–2040

Heuer H, Smalla K (1999) Bacterial phyllosphere communities of *Solanum tuberosum* L. and T4-lysozyme-producing transgenic variants. FEMS Microbiol Ecol 28:357–371

Heuer H, Kroppenstedt RM, Lottmann J, Berg G, Smalla K (2002) Effects of T4 lysozyme release from transgenic potato roots on bacterial rhizosphere communities are negligible relative to natural factors. Appl Environ Microbiol 68: 1325–1335

Hoffmann T, Golz C, Schieder O (1994) Foreign DNA sequences are received by a wild-type strain of *Aspergillus niger* after co-culture with transgenic higher plants. Curr Genet 27:70–76

Hu WJ, Harding SA, Lung J, Popko JL, Ralph J, Stokke DD, Tsai CJ, Chiang VL (1999) Repression of lignin biosynthesis promotes cellulose accumulation and growth in transgenic trees. Nat Biotechnol 17:808–812

Jimenez-Bermudez S, Redondo-Nevado J, Munoz-Blanco J, Caballero JL, Lopez-Aranda JM, Valpuesta V, Pliego-Alfaro F, Quesada MA, Mercado JA (2002) Manipulation of strawberry fruit softening by antisense expression of a pectate lyase gene. Plant Physiol 128:751–759

Kaldorf M, Buscot F, Fladung M, Muhs H-J (2001) Establishment of mycorrhizas on *rolC*-transgenic aspen in a field trial. Abstract of the conference "Molekularbiologie der Pilze", Jena, 30.9.-3.10.2002, pp 50

Kaldorf M, Fladung M, Muhs H-J, Buscot F (2002) Mycorrhizal colonization of transgenic aspen in a field trial. Planta 214: 653–660

Kottke I, Oberwinkler F (1986) Mycorrhiza of forest trees – structure and function. Tree 1:1–24

Kottke I, Oberwinkler F (1987) The cellular structure of the Hartig net: coenocytic and transfer cell-like organization. Nord J Bot 7:85–95

Krishnapillai V (1996) Horizontal gene transfer. J Genet 75:219–232

Losey JE, Rayor LS, Carter ME (1999) Transgenic pollen harms monarch larvae. Nature 399:214

Lottmann J, Berg G (2001) Phenotypic and genotypic characterization of antagonistic bacteria associated with roots of transgenic and non-transgenic potato plants. Microbiol Res 156:75–82

Lottmann J, Heuer H, Smalla K, Berg G (1999) Influence of transgenic T4-lysozyme-producing potato plants on potentially beneficial plant-associated bacteria. FEMS Microbiol Ecol 29:365–377

Lottmann J, Heuer H, de Vries J, Mahn A, Düring K, Wackernagel W, Smalla K, Berg G (2000) Establishment of introduced antagonistic bacteria in the rhizosphere of transgenic potatoes and their effect on bacterial community. FEMS Microbiol Ecol 33:41–49

Lukow T, Dunfield PF, Liesack W (2000) Use of the T-RFLP technique to assess spatial and temporal changes in the bacterial community structure within an agricultural soil planted with transgenic and non-transgenic potato plants. FEMS Microbiol Ecol 32:241–247

Masoud SA, Zhu Q, Lamb C, Dixon RA (1996) Constitutive expression of an inducible β-1,3-glucanase in alfalfa reduces disease severity caused by the oomycete pathogen *Phytophthora megasperma* f. sp. *medicagini*, but does not reduce disease severity of chitin-containing fungi. Transgen Res 5:313–323

Morra MJ (1994) Assessing the impact of transgenic plant products on soil organisms. Mol Ecol 3:53–55

Mullin TJ, Bertrand S (1998) Environmental release of transgenic trees in Canada – potential benefits and assesment of biosafety. Forestry Chron 74:203–219

Nehls U, Bock A, Ecke M, Hampp R (2001) Differential expression of the hexose-regulated fungal genes *AmPAL* and *AmMst1* within *Amanita/Populus* ectomycorrhizas. New Phytol 150:583–589

Nielsen KM, Gebhard F, Smalla K, Bones AM, van Elsas JD (1997) Evaluation of possible horizontal gene transfer from transgenic plants to the soil bacterium *Acinetobacter calcoaceticus* BD413. Theor Appl Genet 95:815–821

Nielsen KM, Bones AM, Smalla K, van Elsas JD (1998) Horizontal gene transfer from transgenic plants to terrestrial bacteria – a rare event? FEMS Microbiol Rev 22:79–103

Nielsen KM, van Elsas JD, Smalla K (2000) Transformation of *Acinetobacter* sp. strain BD413 (pFG4Δ*nptII*) with transgenic plant DNA in soil microcosms and effects of kanamycin on selection of transformants. Appl Environ Microbiol 66:1237–1242

Oger P, Mansouri H, Dessaux Y (2000) Effect of crop rotation and soil cover on alteration of the soil microflora generated by the culture of transgenic plants producing opines. Mol Ecol 9:881–890

Owen MDK (2000) Current use of transgenic herbicide-resistant soybean and corn in the USA. Crop Prot 19:765–771

Perlak FJ, Oppenhuizen M, Gustafson K, Voth R, Sivasupraniam S, Heering D, Carey B, Ihrig RA, Roberts JK (2001) Development and commercial use of Bollgard® cotton in the USA – early promises versus today's reality. Plant J 27:489–501

Pilon-Smits EAH, Terry N, Sears T, van Dun K (1999) Enhanced drought resistance in fructan-producing sugar beet. Plant Physiol Bioch 37:313–317

Punja ZK (2001) Genetic engineering of plants to enhance resistance to fungal pathogens – a review of progress and future prospects. Can J Plant Pathol 23:216–235

Quiroga GOS, Fraschina AA (1997) Evaluation of sensory attributes and biochemical parameters in transgenic tomato fruit with reduced polygalacturonase activity. Food Sci Technol Int 3:93–102

Romer S, Fraser PD, Kiano JW, Shipton CA, Misawa N, Schuch W, Bramley PM (2000) Elevation of the provitamin A content of transgenic tomato plants. Nat Biotechnol 18:666–669

Saroha MK, Sridhar P, Malik VS (1998) Glyphosate-tolerant crops: genes and enzymes. J Plant Biochem Biotechnol 7:65–72

Saxena D, Stotzky G (2001) *Bacillus thuringiensis* (Bt) toxin released from root exudates and biomass of Bt corn has no apparent effect on earthworms, nematodes, protozoa, bacteria, and fungi in soil. Soil Biol Biochem 33:1225–1230

Schlüter K, Fütterer J, Potrykus I (1995) "Horizontal" gene transfer from a transgenic potato line to a bacterial pathogen (*Erwinia chrysanthemi*) occurs – if at all – at an extremely low frequency. Biotech 13:1094–1098

Shelton AM, Zhao JZ, Roush RT (2002) Economic, ecological, food safety, and social consequences of the development of Bt transgenic plants. Annu Rev Entomol 47:845–881

Siciliano SD, Theoret CM, de Freitas JR, Hucl PJ, Germida JJ (1998) Differences in the microbial communities associated with the roots of different cultivar of canola and wheat. Can J Microbiol 44:844–851

Smith SE, Read DJ (1997) Mycorrhizal symbiosis, 2nd edn. Academic Press, San Diego

Solomon-Blackburn RM, Barker H (2001) Breeding virus resistant potatoes (*Solanum tuberosum*): a review of traditional and molecular approaches. Heredity 86:17–35

Staehelin C, Charon C, Boller T, Crespi M, Kondorosi A (2001) *Medicago truncatula* plants overexpressing the early nodulin gene enod40 exhibit accelerated mycorrhizal colonization and enhanced formation of arbuscules. Proc Natl Acad Sci USA 98:15366–15371

Straubinger B, Straubinger E, Wirsel S, Turgeon G, Yoder O (1992) Versatile fungal transformation vectors carrying the selectable bar gene of *Streptomyces hygroscopicus*. Fungal Genet Newslet 39:82–83

Thelen JJ, Ohlrogge JB (2002) Metabolic engineering of fatty acid biosynthesis in plants. Metab Eng 4:12–21

Tuominen H, Sitbon F, Jacobsson C, Sandberg G, Olsson O, Sundberg B (1995) Altered growth and wood characteristics in transgenic hybrid aspen expressing *Agrobacterium tumefaciens* T-DNA indoleacetic acid-biosynthetic genes. Plant Physiol 109:1179–1189

Vance CP, Graham PH (1995) Nitrogen fixation in agriculture: application and perspectives. In: Tikhonovich IA, Provorov NA, Romanov VI, Newton WE (eds) Nitrogen fixation: fundamentals and applications. Kluwer, St. Petersburg, pp 77–86 van Rhijn P, Fujishige NA, Lim PO, Hirsch AM (2001) Sugar-binding activity of pea lectin enhances heterologous infection of transgenic alfalfa plants by *Rhizobium leguminosarum* biovar *viciae*. Plant Physiol 126:133–144

Vierheilig H, Alt M, Neuhaus J-M, Boller T, Wiemken A (1993) Colonization of transgenic *Nicotiana sylvestris* plants, expressing different forms of *Nicotiana tabacum* chitinase, by the root pathogen *Rhizoctonia solani* and by the mycorrhizal symbiont *Glomus mosseae*. Mol Plant-Microbe Int 6:261–264

Vierheilig H, Alt M, Lange J, Gut-Rella M, Wiemken A, Boller T (1995) Colonization of transgenic tobacco constitutively expressing pathogenesis-related proteins by the vesicular arbuscular mycorrhizal fungus *Glomus mosseae*. Appl Environ Microbiol 61:3031–3034

Warwick SI, Beckie HJ, Small E (1999) Transgenic crops: new weed problems for Canada? Phytoprotection 80:71–84

White TJ, Bruns T, Lee S, Taylor J (1990) Amplification and direct sequencing of fungal ribosomal RNA genes for phylogenetics. In: Innis MA, Gelfand DH, Sninsky JJ, White TJ (eds) PCR protocols: a guide to methods and applications. Academic Press, San Diego, pp 315–322

Wolfenbarger LL, Phifer PR (2000) The ecological risks and benefits of genetically engineered plants. Science 290:2088–2093

Wöstemeyer J, Wöstemeyer A, Voigt K (1997) Horizontal gene transfer in the rhizosphere: a curiosity or a driving force in evolution? Adv Bot Res 24:399–429

Ye X, Al-Babili S, Klöti A, Zhang J, Lucca P, Beyer P, Potrykus I (2000) Engineering the provitamin A (β-carotene) biosynthetic pathway into (carotenoid-free) rice endosperm. Science 287:303–305

Zhang HX, Blumwald E (2001) Transgenic salt-tolerant tomato plants accumulate salt in foliage but not in fruit. Nat Biotechnol 19:765–768

12 Interaction Between Soil Bacteria and Ectomycorrhiza-Forming Fungi

Rüdiger Hampp and Andreas Maier

1 Introduction

Roots constitute important plant organs for water and nutrient uptake. However, they also release a wide range of carbon compounds of low molecular weight which are called exudates. These compounds form the basis for an environment rich in diversified microbiological populations, the rhizosphere (Hiltner 1904; the rhizosphere has been defined as a narrow zone of soil which is influenced by living roots). Bacteria are an important part of these populations. In addition, roots of most terrestrial plants develop symbiotic structures (mycorrhiza) with soil-borne fungi. In these interactions, the fungal partner provides the plant with improved access to water and nutrients in the soil due to more or less complex hyphal structures that emanate from the root surface and extend far into the soil. The plant, in return, supplies carbohydrates for fungal growth and maintenance (Smith and Read 1997; Hampp and Schaeffer 1998). Due to leakage and the turnover of mycorrhizal structures, these solutes are also released into the rhizosphere where they can be accessed by other microorganisms. The term "rhizosphere" has, therefore been extended to "mycorrhizosphere" (Oswald and Ferchau 1968). In the latter, two different zones can be distinguished: the surface of the mycorrhizal structure, affected by both root and fungus, and that occupied by fungal hyphae only. The latter has been termed "hyphosphere" (Marschner 1995). Soil free of plant and fungal components has been referred to as "bulk soil" (Andrade et al. 1997). It is reasonable to believe that these different spheres may differ in their microbial activities, and it has been shown that microbial communities within the rhizosphere are distinct from those of nonrhizosphere soil (Curl and Truelove 1986; Whipps and Lynch 1986).

Interactions between soil bacteria and symbiotic fungi can be both negative and positive. Mycorrhiza-forming fungi have been shown to reduce bacterial viability (Meyer and Linderman 1986). Due to the transfer and exudation of plant-derived organic compounds to soil microsites not accessible to roots, fungi can promote bacterial growth and survival (Hobbie 1992; Söder-

Plant Surface Microbiology
A. Varma, L. Abbott, D. Werner, R. Hampp (Eds.)

ström 1992; Frey-Klett et al. 1997). Furthermore, there is evidence that soil bacteria can also enhance the formation of mycorrhizal structures, either by promoting growth (helper bacteria; Garbaye 1994) or by protecting them from pathogenic micro-organisms.

2 Bacteria

Free-living soil bacteria which are beneficial for plant growth are named plant-growth-promoting rhizobacteria (PGPR; Kloepper et al. 1989). These include species and strains which belong to the genera *Azotobacter, Pseudomonas, Burkholderia, Acetobacter, Herbaspirillum* and *Bacillus* (Glick 1995; Probanza et al. 1996; see also Barea, Chap. 20, this Vol.).

In contrast to agricultural soils where bacteria dominate, fungi constitute the major fraction of the microbial flora of forest soils especially in acidic, organic soils under cold climates (Söderström et al. 1983; Nohrstedt et al. 1989).

3 Bacterial Community Structure

Abundance and micro-stratification of bacteria and fungi inhabiting the organic layers of a Scots pine forest were analyzed by Berg et al. (1998). They counted approx. $5x10^9$ bacteria/g dry wt. of organic matter. The mean bacterial biomass was between 0.34 and 0.25 mg C/g dry wt. of organic matter. This compared to a fungal biomass of between 0.05 and 0.009 mg C /g dry wt..

Abundance of bacteria and fungi is influenced by the soil water content, and clear seasonal patterns with a peak of microbial biomass in winter were reported. The ratio of carbon due to bacteria biomass/fungal biomass was 2:1 in fresh litter and 28:1 in humus. This is in contrast to reports which give evidence that in acid forest soils the fungal biomass exceeds that of bacteria (Söderström et al. 1983; Nohrstedt et al. 1989; see also citations in Berg et al. 1998). The ratio may, however, be altered by increasing N input (Verhoef and Brussaard 1990), which may interfere with existing soil food webs (Moore and Hunt 1988).

Generally, bacterial densities in forest soils are an order of magnitude lower than those determined from nursery peat (Timonen et al. 1998). In forest humus, the common soil species, *Pseudomonas fluorescens*, a potential mycorrhiza helper bacterium in pot cultures, could not be identified in the acidic environment. In contrast, spore-forming bacteria such as *Bacillus* ssp., which are also classified as helper bacteria (Garbaye and Duponnois 1992), could be identified in mycorrhizospheres of pine.

Following colonization with ectomycorrhiza (ECM)-forming fungi, changes in root exudates result in greater numbers of microbes in the rhizosphere and a change in the species types found (Oswald and Ferchau 1968; Malajczuk 1979;

Linderman 1988). Different ectomycorrhizospheres indicate an overall bacterial-enrichment gradient from bulk soil to rhizosphere to mycorrhizosphere (Frey et al. 1997).

Active exudation of readily usable carbon-rich substrates into the mycorrhizosphere results in enhanced catabolic community development in natural lignin-rich forest humus (Heinonsalo et al. 2000). The driving force for community development/diversity is obviously the continuous supply of carbohydrates from the host plant. A substrate-utilization analysis showed that simple carbohydrates are readily used by all inner and outer mycorrhizosphere bacteria (Timonen et al. 1998). Mannitol, an important intermediate storage form of carbohydrate in many fungi, was preferred by bacterial populations from all types of mycorrhizospheres. Bacteria from bulk soil, in contrast, show a preference for organic and amino acids (Timonen et al. 1998).

Bacterial communities from mycorrhizospheres of *Pinus sylvestris* are characterized by a preferential utilization of carbohydrates and organic and amino acids (Frey et al. 1997; Timonen et al. 1998; Frey-Klett et al. 2000). Bacteria associated with *Suillus bovinus* ectomycorrhiza favored mannitol, while those co-occurring with *Paxillus involutus* preferred fructose as carbon source. Additional carbon sources used by the bacteria (trehalose, glycogen, mannitol, N-acetyl-D-glucosamine; Heinonsalo et al. 2000) suggest a limited saprophytic turnover in acidic forest soils.

Rhizosphere bacteria can also make use of contaminating hydrocarbons as shown by a decrease in nonpolar hydrocarbons in the mycorrhizosphere (Heinonsalo et al. 2000).

4 Association of Bacteria with Fungal/Ectomycorrhizal Structures

Symbiotic interactions between roots and soil fungi comprise different types, the most important ones being endo- and ectomycorrhizas. Endomycorrhiza is the most abundant form in soils of most ecosystems. Typical for this mycorrhiza is the presence in roots of a series of structures which facilitate solute exchange between the partners of symbiosis. These comprise arbuscules, vesicles, coiled hyphae etc. (Smith and Read 1997). Ectomycorrhizas, the focus of this chapter, are mainly formed with roots of forest trees belonging to temperate and boreal regions. They are characterized by defined morphological structures. Extraradical mycelia which exploit the soil, form a mantle structure around fine roots of their host plant. From there hyphae emanate into the cell wall of cortex cells, forming a large surface area (Hartig net) which facilitates solute exchange (Smith and Read 1997; see also Kottke Chap. 13, this Vol.).

In contrast to endomycorrhiza-forming fungi (compare Barea, Chap. 20, this Vol.), information about the interaction between bacteria and ectomycorrhiza-forming fungi is still rather limited.

Pioneering work in this field has been carried out by Garbaye (for a review see Garbaye 1994). Experiments carried out with *Picea abies*, *Pinus nigra*, *Pinus sylvestris*, *Pseudotsuga menziesii*, and *Quercus robur* (Garbaye et al. 1992) indicated that soil bacteria can stimulate the inoculation of roots with ectomycorrhiza-forming fungi, thereby also reducing the adverse effect of pathogens. Both effects resulted in a better seedling growth, and thus the term "helper bacteria" was coined. For more recent work, see Dunstan et al. (1998) and Poole et al. (2001).

5 Bacteria Associated with Sporocarps and Ectomycorrhiza

Twenty seven bacterial species were isolated from both the sporocarps of *Suillus grevillei* and the ECMs of *S. grevillei*/*Larix decidua* (Varese et al. 1996). The genera *Pseudomonas*, *Bacillus*, and *Streptomyces* were predominant. From sporocarps of white truffles (*Tuber* sp.), bacterial strains such as *Micrococcus*, *Moraxella*, *Staphylococcus* and *Pseudomonas* could be isolated (Citterio et al. 1995). Gram-positive bacteria seldom stimulated in vitro fungal growth. Among gram-negative bacteria, *Pseudomonas* strains enhanced growth. *Streptomyces* significantly inhibited the fungus. Bacterial supernatants were not effective. Volatiles enhanced fungal growth to some extent, but not significantly. Most of the bacteria isolated produced siderophores.

A distinction between the different structures of ECM showed that bacteria primarily occurred on the surface of the mantle and in the interhyphal spaces (Schelkle et al. 1996), but also deep within the mantle (Foster and Marks 1967).

Bacteria of subclasses of proteobacteria (containing plant-growth-promoting rhizobacteria such as *Burkholderia*, *Azospirillum*, *Acetobacter* and *Herbaspirillum*) were detected in high numbers on mantle surfaces (Mogge et al. 2000). The two most common fungi on beech, *Lactarius vellereus* and *Lactarius subdulcis*, were associated with members of the α- and β-subclasses of the proteobacteria. These bacteria have been shown to be abundant in winter and early spring (Timonen et al. 1998).

Electron microscopy of ECM with *Pinus sylvestris* and *S. bovinus* and *Paxillus involutus* (Nurmiaho-Lassila et al. 1997) also revealed bacteria on the mantle surface and at inter- and intracellular locations in the mantle and the Hartig net (*S. bovinus*). Fungal strands were colonized only by a few bacteria, while the outermost external fine hyphae had extensive monolayers of bacteria attached.

ECM with *P. involutus* were mostly devoid of bacteria, while the external mycelium supported bacteria (Nurmiaho-Lassila et al. 1997). From their observations, the authors conclude that single ECM fungi create defined mycorrhizosphere habitats with distinct populations of bacteria. Knowing that several different types of ECM can be formed on the same root in close vicinity, a large local biodiversity of ECM-specific bacterial populations could be postulated.

Intracellular bacteria as detected in certain endomycorrhiza-forming fungi (Bianciotto et al. 1996) are also described for *P. sylvestris*/*Paxillus involutus* mycorrhizospheres; Nurmiaho-Lassila et al. (1997) identified *Burkholderia cepacia* in extracts from the respective mycorrhizas. Intracellular bacteria were also detected in the mycelium of the ectomycorrhizal fungus *Laccaria bicolor* S238 N (Bertaux, Frey-Klett, Hartmann, Schmidt, Garbaye, pers. comm.).

6 Benefits from Bacteria/Ectomycorrhiza Interactions

Bacteria are producers of antibiotics. Newer studies show that a variety of genera, species and strains of these bacteria (e.g., *Bacillus subtilis*, *Pseudomonas fluorescens*) can inhibit the growth of pathogenic fungi (*Fusarium oxysporum*; *Cylindrocarpon* sp.) in co-culture with ECM fungi such as *Laccaria bicolor*, *L. proxima* and *Suillus granulatus* (Schelkle and Peterson 1996). They can, however, also affect ECM fungi. *Burkholderia cepacia* significantly reduced the in vitro growth of mycelia of *Paxillus involutus*. *B. cepacia*, *Pseudomonas chlororaphis*, *Ps. fluorescens*, and *P. involutus* reduced the mycelial growth of the root pathogens *Fusarium moniliforme*, *F. oxysporum*, and *Rhizoctonia solani* (Pedersen et al. 1999). *Burkholderia cepacia* also reduced the formation of ECM short roots by *P. involutus* on lodgepole pine and white spruce seedlings in the short term (2 months), but not upon longer incubation (4 months). *Pseudomonas chlororaphis* and *Ps. fluorescens* did not reduce mycelial growth and mycorrhiza formation. Treatment of the seedlings with either *B. cepacia* or *P. involutus* increased their survival in the presence of some of the root pathogens investigated. From the data given by Pedersen et al. (1999) it can thus be concluded that the simplest protective system exists when bacteria do not inhibit fungal growth/mycorrhiza formation, but affect potential root pathogens (see also Frey-Klett et al. 2000). There are obviously also synergistic effects between these bacteria and ECM fungi such as *L. proxima* in inhibiting pathogens (Schelkle and Peterson 1996).

In addition to preventing pathogen attacks, bacteria can also support ECM development directly. This has been shown for different host/fungus combinations (Garbaye 1994; Frey-Klett et al. 1997). In general, the effect ascribed to the presence of bacteria consists of a significantly increased number of infected root tips (Dunstan et al. 1998; Poole et al. 2001).

This should also have an impact on the respective host plant. Probanza et al. (2001) investigated the effect of a co-cultivation with *P. tinctorius* and PGPR belonging to the genus *Bacillus* in enhancing growth of *Pinus pinea*. Although the bacterial strains promoted seedling growth, this effect could not be related to a synergistic interaction with the fungus. A stimulation of shoot and root biomass production was also observed for *Acacia holoserica* seedlings, mycorrhizal with *Pisolithus alba* and after co-cultivation with two fluorescent pseudomonad strains (Founoune et al. 2002). After 3 months of co-culture, the bacterial inoculants disappeared, showing how difficult such experiments are to interpret. Obviously, the amount of inoculum supplied can also play an important role (Frey-Klett et al. 1999).

7 Possible Mechanisms of Interaction

As pointed out by Schelkle and Peterson (1996), and in addition to the production of antibiotics, protective or "helper" effects could be due to competition for nutrients in the rhizosphere. The formation of siderophores, for example could be such a mechanism. Siderophores are iron chelators which make iron available for uptake by the bacteria. As these compounds are species-specific, Fe-chelates can only be taken up by those bacteria that are able to produce them. Protective bacteria synthesizing siderophores could thus out-compete pathogens with regard to Fe (Neidhardt et al. 1990). Similar mechanisms are known for ECM fungi (Watteau and Berthelin 1990). Siderophore release from bacteria into the mycorrhizosphere could also improve absorption of Fe by the mycorrhizal fungus.

Protection from pathogens could, however, also be a mass effect, simply due to the large number of nonpathogenic bacteria that accumulate in the rhizosphere due to the high nutrient supply. However, as outlined by Garbaye (1994), there can be many more mechanisms, such as an improved receptivity of the root for mycorrhizal infection, a modification of the rhizospheric soil, improvement of the root-fungus recognition, stimulation of germination of fungal propagules, as well as an enhancement of fungal growth in the rhizosphere (see also Brule et al. 2001) which would increase the probability of contact between fungus and root (compare Dunstan et al. 1998).

In nutrient-poor acidic forest soils modification by micro-organisms should be an important factor; the C-rich environment provided by the plant is attractive for soil micro-organisms, leading to the formation of functionally compatible microbial communities. These are jointly able to co-mobilize soil nutrients such as P and N in and around the vegetative mycelium. In addition, N-fixing bacterial species including *Bacillus* spp. are possibly present in the mycorrhizosphere of forest trees (Li et al. 1992) as the vegetative mycelium represents a niche that is ideally suited for the selection and enrichment of associative N-fixing bacteria (Sen 2000).

In many of the possible mechanisms, phytohormones such as IAA could play an important role. A study on the rooting of derooted shoot hypocotyls of spruce showed that *Laccaria bicolor* and *Pseudomonas fluorescens* BBc6 (MHB) both increased the number of roots formed per rooted hypocotyl (Karabaghli et al. 1998). The same effect was caused by the addition of IAA alone (control). Both organisms produced IAA in pure culture.

8 Biochemical Evidence for Interaction

Streptomycetes are widely distributed saprobic soil bacteria which produce a wide range of compounds affecting other organisms. Becker et al. (1999) studied mycorrhiza-associated *Streptomyces* strains with regard to their effect on the protein pattern of ECM-forming *Laccaria bicolor* and *Cenococcum geophilum*, and on two pathogenic fungi (*Armillaria ostoyae* and *A. gallica*). One of the strains improved the growth of ECM fungi while inhibiting that of the pathogens. The effects could be related to differences in fungal gene expression (mRNA) and the protein profile obtained after in vitro translation; new proteins were induced by the strain supporting the growth of ECM-fungi, while the *Streptomyces* strain leading to adverse effects caused the disappearance of bands.

New techniques allow for the annotation of such protein spots. Only this way is it possible to obtain information about the function of the respective protein. In the following, we give an example for such an approach for *Amanita muscaria.*

A. muscaria is a fungus which develops ECM with a wide range of forest trees. Grown in dual culture with bacterial isolates obtained from soil samples in close vicinity to spruce roots, this fungus exhibited distinct changes in protein pattern. Most effective were isolates which were members of the *Actinomycetes.*

After 10 weeks of dual inoculation of *A. muscaria* with a respective soil bacterium in Petri dishes, the hyphae of the fungal mycelium changed their phenotype in comparison to controls. The hyphal diameter decreased, while cell length and the extent of hyphal branching increased. To investigate the molecular mechanisms behind these morphological changes, the proteome of *A. muscaria* was screened for differentially expressed polypeptides (two-dimensional SDS-PAGE electrophoresis). In Fig. 1, the protein patterns for mycelium from *A. muscaria* in pure culture (A) and after dual culture with the bacterium (B) are compared. The pattern reveals about 100 well-separated protein spots of which about 20 polypeptides were recognized as differentially expressed. Twelve spots were excised from the gels for sequence analysis by MALDI-TOF (matrix-assisted laser desorption/ionization time of flight) mass spectrometry.

Reliable matches to known protein sequences with the peptide mass fingerprints were obtained for 7 of 12 selected spots. As an example, Table 1 gives the peptide masses obtained from protein spot no. 78. They show identity with several predicted peptide masses of actin 1 from the saprophytic fungus *Schizophyllum commune* and for actin 2 from the ectomycorrhizal fungus *Suillus bovinus* (Tarkka et al. 2000).

Actins are highly conserved cytoskeletal proteins that are present in all eucaryotic cells. They are probably involved in various processes such as cytoplasmic streaming, cell shape determination, tip growth, cell wall deposition,

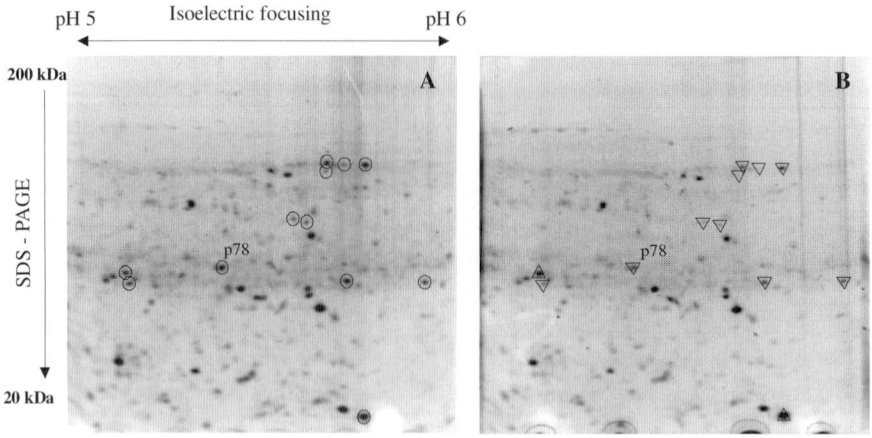

Fig. 1. Two-dimensional maps of mycelial proteins from *Amanita muscaria.* Protein (300 μg) was loaded onto IEF gels. **A** *A. muscaria* pure culture; **B** dual culture of *A. muscaria* with a soil bacterium. Differentially expressed proteins are indicated in **A** (*open circles*). Downregulation (*closed inverted triangles*) or upregulation (*closed triangles*) of the analogous spots is indicated in **B.** (Gels were stained with SYPRO Ruby fluorescent dye; molecular probes, Eugene, OR, USA.). Results obtained from the computer-aided evaluation were rigorously compared by visual analysis of the original gels. Stained protein spots were excised and digested in-gel with modified trypsin (Promega) according to Williams and Stone (1997)

etc. (Sheterline et al. 1992). At first sight, it looks surprising that a protein which is important in cell shape determination is downregulated when hyphal morphology changes. However, the decrease in the amount of actin coincides with the decrease of the fungal diameter. Interestingly, the amount of actin protein increases during fruiting body formation of *A. muscaria.* The hyphae of the fruiting body are swollen and branched (Manachére et al. 1983) and the results thus indicate differential regulation of actin during changes in *A. muscaria* hyphal growth pattern.

These results emphasize the usefulness of proteome analysis in identifying molecular events occurring in fungus bacteria interactions.

Table 1. Protein features and data from the peptide mass fingerprint of the protein in spot no. 78. Comparison with the computer-generated peptides from *Schizophyllum commune* and *Suillus bovinus* indicates identity with fungal actins. Mass spectra of peptide mixtures were obtained by MALDI-TOF (matrix-assisted laser desorption/ionization time of flight) mass spectrometer (Dr. C. Niehaus, University of Bielefeld, Germany). The database search using the proteolytic peptide masses was performed with the Mascot program developed by Perkins et al. (1999)

Protein identity	p78	Actin 1	Actin 2
Organism	*Amanita muscaria*	*Schizophyllum commune*	*Suillus bovinus*
MW (Da)	Approx. 42,000	41,876	41,979
pI	5.35	5.30	5.31
	Observed peptides MW (Da)	Calculated peptides MW (Da)/sequence	Calculated peptides MW (Da)/sequence
	780.59	780.45/IVAPPER	780.45/IVAPPER
	908.69	908.54/IVAPPERK	908.54/IVAPPERK
	1141.69	1141.54/GYPFTTTAER	-
	1485.89	1485.68/QEYDESGPGIVHR	1485.68/QEYDESGPGIVHR
	1589.19	1588.88/LDLAGRDLTDFLIK	1589.81/DLTDCLIKNLTER
	1790.19	1789.88/SYELPDGQVITIGNER	1789.88/SYELPDGQVITIGNER

9 Impacts of Environmental Pollution

Microcosms with *S. bovinus, P. involutus/Pinus sylvestris* in forest humus amended with petroleum hydrocarbon were investigated with regard to fungus/bacteria responses (Sarand et al. 1998). Hyphae emanating from mycorrhizas formed patches around contaminations with a microbial biofilm at the hydrocarbon/fungus interface. Bacteria consisted of isolates of *Ps. fluorescens*. This opens possibilities for the bioremediation of environmental pollutants by the use of degradative micro-organisms. Sarand et al. (1999) tested m-toluate as a model compound for petrol-contaminated sites. Fungal survival (*Suillus bovinus*) on medium containing this compound was increased in co-culture on agar plates with degradative bacterial strains of *Ps. fluorescens*. The activity of the bacterium was not affected in a tripartite system containing *S. bovinus/Pinus sylvestris* mycorrhizas. The fungus was not able to degrade m-toluate, although mycorrhizal fungi are able to produce enzymes capable of degrading complex organic compounds (see Sarand et al. 1999).

10 Conclusions

Many of the experiments carried out in order to investigate a possible interaction between bacteria and ECM-forming fungi have been carried out under sterile conditions or in pot cultures. Experience shows that, when transferred to field conditions the respective bacteria will not thrive, but will soon be substituted by other genera, species or strains. Thus, a more promising approach is to collect bacteria from mycorrhizas obtained from natural sites and introduce these into laboratory experiments, with dual cultures being the easiest way to investigate molecular interaction. In our experience, Gram-positive bacteria such as Actinomycetes, although largely neglected, are abundant at least in mycorrhizospheres of spruce stands, and are thus important candidates for future approaches.

Acknowledgements. We gratefully acknowledge critical reading and helpful suggestions by Dr. Garbaye (INRA, Nancy, France). As far as our own work is concerned, we are indebted to the Deutsche Forschungsgemeinschaft for financial support (Graduate School "Infection Biology")

References and Selected Reading

Andrade G, Mihara KL, Linderman RG, Bethlenfalvay GJ (1997) Bacteria from rhizosphere and hyphosphere soils of different arbuscular-mycorrhizal fungi. Plant Soil 192:71–79

Becker DM, Bagley ST, Podila GK (1999) Effects of mycorrhizal-associated streptomycetes on growth of *Laccaria bicolor*, *Cenococcum geophilum*, und Armillaria species and on gene expression in *Laccaria bicolor*. Mycologia 91:33–40

Berg MP, Kniese JP, Verhoef HA (1998) Dynamics and stratification of bacteria and fungi in the organic layers of a Scots pine forest soil. Biol Fertil Soils 26:313–322

Bianciotto V, Bandi C, Minerdi D, Sironi M, Tichy HV, Bonfante P (1996) An obligately endosymbiotic mycorrhizal fungus itself harbors obligately intracellular bacteria. Appl Environ Microbiol 62:3005–3010

Brule C, Frey-Klett P, Pierrat JC, Courrier S, Gerard F, Lemoine MC, Rousselet JL, Sommer J, Garbaye J (2001) Survival in the soil of the ectomycorrhizal fungus *Laccaria bicolor* and the effects of a mycorrhiza helper *Pseudomonas fluorescens*. Soil Biol Biochem 33:1683–1694

Citterio B, Cardoni P, Potenza L, Amicucci A, Atocchi V, Gola G, Nuti M (1995) Isolation of bacteria from sporocarps of *Tuber magnatum* pico, *tuber borchii* vitt. and *Tuber maculatum* vitt.: identification and biochemical characterization. In: Stocchi V, Bonfante P, Nuti M (eds) Biotechnology of ectomycorrhizae. Plenum Press, New-York, pp 241–248

Curl EA, Truelove B (1986) The rhizosphere. Springer, Berlin Heidelberg New York, pp 1–8

Dunstan WA, Malajczuk N, Dell, B (1998) Effects of bacteria on mycorrhizal development and growth of container-grown *Eucalyptus diversicolor* F. Muell. seedlings. Plant Soil 201:241–249

Foster RC, Marks GC (1967) The fine structure of the mycorrhizas of *Pinus radiata*. Aust J Biol Sci 19:1027–1038

Founoune H, Duponnois R, Ba AM, Sall S, Branget I, Lorquin J, Neyra M, Chote JL (2002) Mycorrhiza helper bacteria stimulate ectomycorrhizal symbiosis of *Acacia holosericea* with *Pisolithus alba*. New Phytol 153:81–89

Frey-Klett P, Brule C, Garbaye J (1997) A proposed model for the mycorrhiza-helper effect of *Pseudomonas fluorescens* BBc6 on the ectomycorrhizal system *Laccaria bicolor*S238N-Douglas fir. 4th International Workshop on Plant Growth Promoting Rhizobacteria, Sapporo, Japan

Frey-Klett P, Churin JL, Pierrat JC, Garbaye J (1999) Dose effect in the dual inoculation of an ectomycorrhizal fungus and a mycorrhiza helper bacterium in two forest nurseries. Soil Biol Biochem 31:1555–1562

Frey-Klett P, Chavatte M, Courriers S, Martinotii G, Pierrat JC, Garbaye J (2000) Ectomycorrhizosphere effect of the Douglas fir-*Laccaria bicolor* symbiosis on the functional diversity of fluorescent pseudomonads in a forest nursery. 5th International Workshop on PGPR, Cordoba, Argentine. www.ag.auburn.edu/argentina/pdfmanuscripts/freyklett.pdf

Frey P, Frey-Klett P, Garbaye J, Berge O, Heulin (1997) Metabolic and genotyping fingerprinting of fluorescent pseudomonads associated with the Douglas fir-*Laccaria bicolor* mycorrhizosphere. Appl Environ Microbiol 63:1852–1860

Frey-Klett P, Pierrat JC, Garbaye J (1997) Location and survival of Mycorrhiza helper *Pseudomonas fluorescens* during establishment of ectomycorrhizal symbiosis between *Laccaria bicolor* and Douglas Fir. Appl Environ Microbiol 63:139–144

Garbaye J (1994) Helper bacteria: a new dimension to the mycorrhizal symbiosis. New Phytol 128:197–210

Garbaye J, Churin JL, Duponnois R (1992) Effects of substrate disinfection, fungicide treatment and mycorrhiza helper bacteria (MHB) on ectomycorrhiza formation of pedunculate oak inoculated with *Laccaria laccata* in two bare-root nurseries. Biol Fertil Soils 13:55–47

Garbaye J, Duponnois R (1992) Specificity and function of mycorrhiza helper bacteria (MHB) associated with the *Pseudotsuga menziesii–Laccaria laccata symbiosis*. Symbiosis 14:335–344

Glick BR (1995) The enhancement of plant growth by free living bacteria. Can J Microbiol 41:109–117

Hampp R, Schaeffer C (1998) Mycorrhiza – Carbohydrate and energy metabolism. In: Varma A, Hock B (eds) Mycorrhiza – structure, function, molecular biology and biotechnology. Springer, Berlin Heidelberg New York, pp 273–303

Heinonsalo J, Jorgensen KS, Haahtela K, Sen R (2000) Effects of *Pinus sylvestris* root growth and mycorrhizosphere development on bacterial carbon source utilization and hydrocarbon oxidation in forest and petroleum-contaminated soils. Can J Microbiol 46:451–464

Hiltner L (1904) Über neuere Erfahrungen und Probleme auf dem gebiet der Bodenbakteriologie und unter besonderer Berücksichtigung der Gründüngung und Brache. Arb Dtsch Landwirt Ges 98:59–78

Hobbie SE (1992) Effects of plant species on nutrient cycling. Trends Ecol Evol 7:336–339

Karabaghli C, Frey-Klett P, Sotta B, Bonnet M, Le Tacon F (1998) *In vitro* effects of *Laccaria bicolor* S238 N and *Pseudomonas fluorescens* strain BBc6 on rooting of derooted shoot hypocotyls of Norway spruce. Tree Physiol 18:103–111

Kloepper JW, Lifshitz R, Zablotowicz RM (1989) Free-living bacterial inocula for enhancing crop productivity. Trends Biotechnol 7:39–43

Li CY, Massicotte HB, Moore LVH (1992) Nitrogen-fixing *Bacillus* sp. associated with Douglas-fir tuberculate ectomycorrhizae. Plant Soil 140:35–40

Linderman RG (1988) Mycorrhizal interactions with the rhizosphere microflora: the mycorrhizosphere effect. Phytopathology 78:366–371

Malajczuk N (1979) The microflora of unsuberized roots of *Eucalyptus calophylla* R. Br. and *Ecalyptus marginata* Donn ex Sm seedlings grown in soils suppressive and conductive to *Phytophthora cinnamomi* Rands. II. Mycorrhizal roots and associated microflora. Aust J Bot 27:235–254

Manachére G, Rober JC, Durand R, Bret JP, Févre M (1983) Differentiation in the basidiomycetes. In: Smith JE (ed) Mycology series; vol 4. Fungal differentiation: A contemporary synthesis. Marcel Dekker, New York, pp 481–514

Marschner H (1995) Mineral nutrition of higher plants. Academic Press, London, 571 pp

Meyer JR Linderman RG (1986) Response of subterranean clover to dual inoculation with vesicular-arbuscular mycorrhizal fungi and a plant growth-promoting bacterium, *Pseudomonas putida*. Soil Biol Biochem 18:185–190

Mogge B, Loferer C, Agerer R, Hutzler P, Hartmann A (2000) Bacterial community structure and colonialization patterns of *Fagus sylvatica* L. ectomycorrhizospheres as determined by fluorescence in situ hybridization and confocal laser scanning microscopy. Mycorrhiza 9:271–278

Moore JC, Hunt HW (1988) Resource compartmentation and the stability of real ecosystems. Nature 333:261–263

Neidhardt FC, Ingraham JL, Schaechter M (eds) (1990) Physiology of the bacterial cell. Sinauer, MA, USA

Nohrstedt H-Ö, Arnebrandt K, Bååth E, Söderström B (1989) Changes in carbon content, respiration rate, ATP content, and microbial biomass in nitrogen-fertilized pine forest soils in Sweden. Can. J For Res 19:323–328

Nurmiaho-Lassila E-L, Timonen S, Haahtela K, Sen R (1997) Bacterial colonialization patterns of intact *Pinus sylvestris* mycorrhizospheres in dry pine forest soil: an electron microscopy study. Can J Microbiol 43:1017–1035

Oswald ET, Ferchau HA (1968) Bacterial associations of coniferous mycorrhizae. Plant Soil 28:187–192

Pedersen EA, Reddy MS, Chakravarty P (1999) Effect of three species of bacteria on damping-off, root rot development, and ectomycorrhizal colonialization of lodgepole pine and white seedlings. Eur J For Pathol 29:123–134

Perkins DN, Pappin DJ, Creasy D, Cotrell JS (1999) Probability-based protein identification by searching sequence databases using mass spectrometry data. Electrophoresis 20:3551–3567

Poole EJ, Bending GD, Whipps JM, Read DJ (2001) Bacteria associated with *Pinus sylvestris-Lactarius* rufus ectomycorrhizas and their effects on mycorrhiza formation *in vitro.* New Phytol 151:743–751

Probanza A, Lucas JA, Guiterrez Mañero JF (1996) The influence of native rhizobacteria on European alder (*Alnus glutinosa* (L.) Gaertn.) growth. I. Characterization of growth promoting and growth inhibiting bacterial strains. Plant Soil 182:59–66

Probanza A, Mateos JL, Lucas Garcia JA, Ramos B, de Felipe MR, Guiterrez Manero JF (2001) Effects of inoculation with PGPR *Bacillus* and *Pisolithus tinctorius* on *Pinus pinea* L. growth, bacterial rhizosphere colonialization, and mycorrhizal infection. Microb Ecol 41:140–148

Sarand I, Timonen S, Nurmiaho-Lassila E-L, Koivula T, Haahtela K, Romantschuk M, Sen R (1998) Microbial biofilms and catabolic plasmid harbouring degradative fluorescent pseudomonads in Scots pine mycorrhizospheres developed on petroleum contaminated soil. FEMS Microbiol Ecol 27:115–126

Sarand I, Timonen S, Koivula T, Peltola R, Haahtela K, Sen R, Romantschuk M (1999) Tolerance and biodegradation of m-toluate by Scots pine, a mycorrhizal fungus and fluorescent pseudomonads individually and under associative conditions. J Appl Microbiol 86:817–826

Schelkle M, Peterson RL (1996) Suppression of common root pathogens by helper bacteria and ectomycorrhizal fungi in vitro. Mycorrhiza 6:481–485

Schelkle M, Ursic M, Farquhar M, Peterson RL (1996) The use of laser scanning confocal microscopy to characterize mycorrhizas of *Pinus strobus* L. and to localize associated bacteria. Mycorrhiza 6:431–440

Sen R (2000) Budgeting for the wood-wide web. New Phytol 145:161–165

Sheterline P, Handel SE, Molloy C, Hendry KAK (1992) The nature and regulation of actin filament turnover in cells. Acta Histochem 41:303–309

Smith SE, Read DJ (1997) Mycorrhizal symbiosis, 2nd edn. Academic Press, Cambridge

Söderström B (1992) Ecological potential of ectomycorrhizal mycelium. In: Read DJ, Lewis DH, Fitter AH, Alexander IJ (eds) Mycorrhizas in ecosystems. Cambridge University Press, Cambridge, pp 77–83

Söderström B, Bååth E, Lundgren B (1983) Decrease in soil microbial activity and biomass owing to nitrogen amendments. Can J Microbiol 29:1500–1506

Tarkka MT, Vasara R, Gorfer M, Raudaskoski M (2000) Molecular characterization of actin genes from homobasidiomycetes: two different actin genes from *Schizophyllum commune* and *Suillus bovinus*. Gene 251:27–35.

Timonen S, Jorgensen KS, Haahtela K, Sen R (1998) Bacterial community structure and defined locations of *Pinus sylvestris–Paxillus involutus* mycorrhizospheres in dry pine forest humus and nursery peat. Can J Microbiol 44:499–513

Varese GC, Portinario S, Trotta A, Scannerini S, Luppi-Mosca AM, Martinotti MG (1996) Bacteria associated with *Suillus grevillei* sporocarps and ectomycorrhizae and their effects on in vitro growth of the Microbiont. Symbiosis 21:129–147

Verhoef HA, Brussaard L (1990) Decomposition and nitrogen mineralization in natural and agro-ecosystems: the contribution of soil animals. Biogeochemistry 11:175–211

Watteau F, Berthelin J (1990) Iron solubilization by mycorrhizal fungi producing siderophores. Symbiosis 9:59–67

Whipps JM, Lynch JM (1986) The influence of rhizosphere on crop productivity. Adv Microb Ecol 9:187–244

Williams KR, Stone KL (1997) Enzymatic cleavage and HPLC peptide mapping of proteins. In: Walker J (ed) Molecular biotechnology. Humana Press, Totowa, pp 155–167

13 The Surface of Ectomycorrhizal Roots and the Interaction with Ectomycorrhizal Fungi

INGRID KOTTKE

1 Introduction

Most of the trees in the temperate and alpine regions live in symbiosis with root fungi forming ectomycorrhizas. Ectomycorrhizas (ECM) display a very specified cellular organization. A fungal sheath covers the root surface and the hyphae invade intercellularly between the root cortical cells establishing the so-called Hartig net. The hyphal sheath is formed in a species-specific manner (Agerer 1998), but the architecture of the Hartig net is similar in all the ectomycorrhizas, independent of plant and fungal species (Blasius et al. 1986, Kottke and Oberwinkler 1987, 1989). Establishing the Hartig net, hyphal growth undergoes important changes. The hyphae invade as multi-branched, fan-like lobes in intimate juxtaposition, starting at the root surface and finally covering the root cortical cells in a dense mono-layer (Fig. 1; Jacobs et al. 1989; Brunner and Scheidegger 1992; Kottke et al. 1996).

The Hartig net structure is only established in so-called short roots, a special root type of the ectomycorrhiza-forming plants (Marks and Foster 1973; Wong et al. 1990). It was hypothesized that the surface of these rootlets might trigger the attachment of hyphae and the change of their growth characters (Jacobs et al. 1989; Brunner and Scheidegger 1992; Kottke 1997; Bonfante et al. 1998). Cysteine-rich, moderately hydrophobic proteins ("hydrophobins") in the walls of the ectomycorrhizal fungus *Pisolithus tinctorius* (Pers.) Coker & Couch were shown to be highly expressed in the early stage of mycorrhiza formation and were considered to attach the hyphae to the root surface of *Eucalyptus globulus* ssp. *bicostata* Kirkp. (Tagu et al. 1996, 2000, 2001; Martin et al. 1999). A hydrophobic root surface was, therefore postulated and a cuticle-like layer on the surface of short roots may be the substrate for adhesion of ectomycorrhizal fungi (Kottke 1997). Recent ultrastructural studies comparing long and short roots have supported this hypothesis by revealing the origin of the cuticle-like layer on short, but not on long roots. Differences were also detected in the amounts of methyl-esterified pectins in the cortical cell walls of both the root types. Furthermore, when establishing the Har-

Plant Surface Microbiology
A. Varma, L. Abbott, D. Werner, R. Hampp (Eds.)

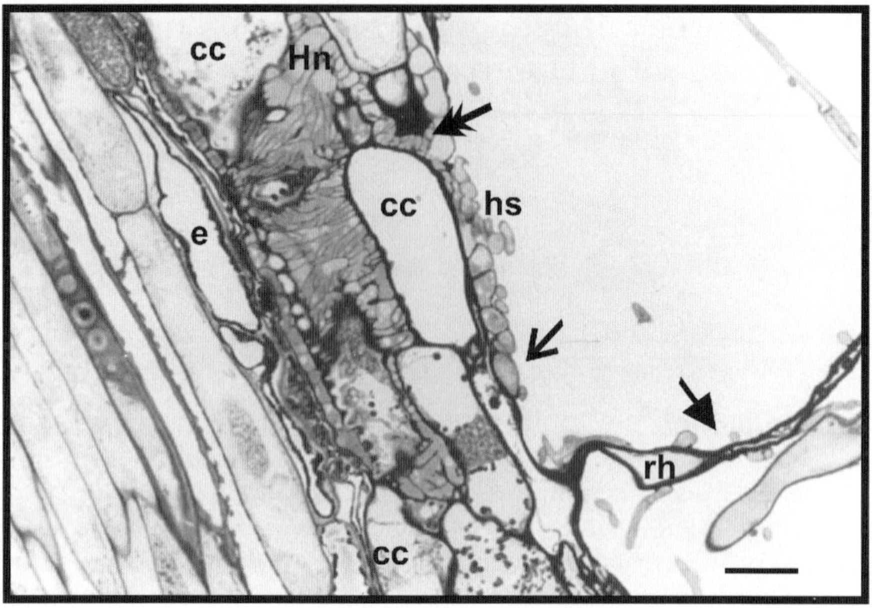

Fig. 1. Ectomycorrhiza formation by *Laccaria amethystea* and *Picea abies*. Longitudinal section through an early state mycorrhiza. Hyphal attachment to root hairs (*arrowhead*) without changes of hyphal morphology; hyphal attachment to root surface followed by hyphal enlargement (*arrow*) and lobe-like growth of hyphae (*double arrow*); typical Hartig net structure established between root cortical cells down to the endodermis (scale 15 µm). *cc* Cortical cell, *e* endodermis, *hs* hyphal sheath, *Hn* Hartig net, *rh* root hair

tig net, the cuticle-like layer has to be penetrated by the hyphae. This process has not been shown before and may be considered as a locally restricted aggressive or saprophytic phase during ECM formation.

2 Long and Short Roots of Ectomycorrhiza-Forming Plants

Ectomycorrhizas are exclusively formed by perennial, woody plants belonging to Pinaceae or to distinct families within the Rosidae (sensu "Angiosperm Phylogenetic Group", Bremer et al. 1998). The root system of these ECM-forming plants is divided into main, or "long roots" of relatively fast and unlimited growth and secondary "short roots" of slow and limited growth (Noelle 1910; Clowes 1951; Marks and Foster 1973). Ectomycorrhizae are typically formed on short roots. However, long roots may become mycorrhizal after turning into a resting stage (Wilcox 1968b). It was speculated that the growth rate of hyphae might not compete with the growth rate of long roots, thus preventing mycorrhiza formation (Marks and Foster 1973). However, sig-

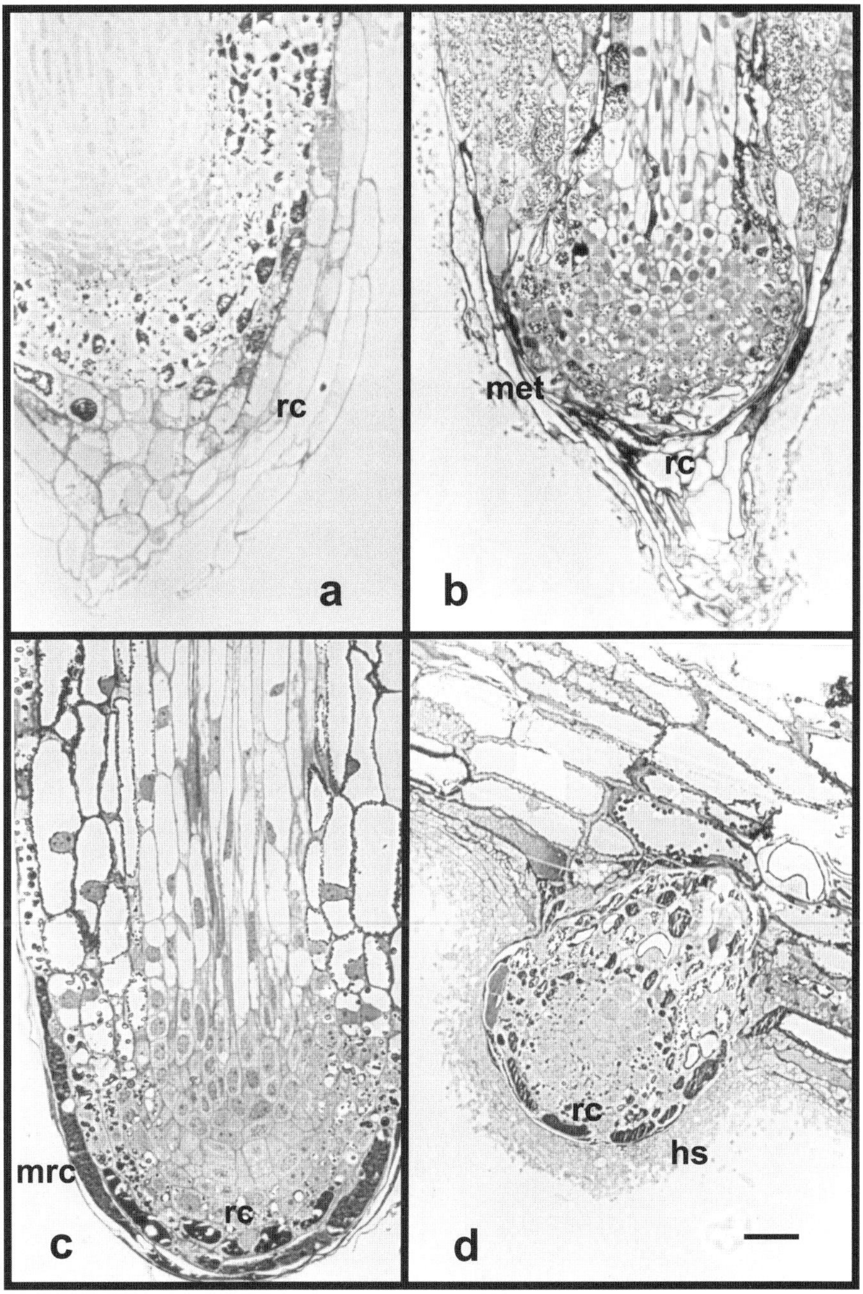

Fig. 2a–d. Root cap formation of several Pinaceae. **a** Long root of *Picea abies* with non-suberized, decaying root cap cells; **b** long root of *Larix decidua* at resting stage with metacutization of root cap; **c** short root of *Pinus sylvestris* with suberized root cap cells containing phenols; **d** root budding in *Picea abies* from below a hyphal sheath, bud covered by suberized root cap cells (scale 15 μm). *hs* Hyphal sheath, *met* metacutization, *mrc* moribund root cap cell, *rc* root cap

nificant structural differences exist between the two root types and may be even more important for mycorrhiza formation or failure.

Fast growing long root tips are covered by a conspicuous root cap consisting of non-suberized, rapidly decaying cells (Fig. 2a). Dormant long roots and short roots have in common that the root cap cells are few and become suberized (Fig. 2b, c). This type of root cap cells is also found in root buds even when emerging from below a hyphal sheath (Fig. 2d). The process was termed metacutization ("Metakutinisierung" Müller 1906) and was found to occur in a multitude of gymnosperm and angiosperm perennial species irrespectively of the epidermal or cortical cell type on the root surface of these two taxonomic groups (Plaut 1918). It was described in detail from light microscope studies of *Fagus sylvatica* L. (Clowes 1954), *Betula alleghaniensis* Britt., *Alnus crispa* (Ait.) Pursh, *Eucalyptus pilularis* Smith (Massicotte et al. 1986, 1987a,b), *Abies procera* Rehder (Wilcox 1954), *Picea abies* [L.] Karst. (Kottke et al. 1986) and *Pinus* spp. (Hatch and Doak 1933).

3 A Cuticle-Like Layer on the Surface of Short Roots

Ultrastructural investigations yielded further details on the fate of the suberized root cap cells of short roots. Young root cap cells become suberized by a lamellar layer imposed on the inner side of the cell walls (Figs. 3a, b, 4a). As suberin is only weakly stained by osmium and lead, a suberin layer appears electron-translucent (Sitte 1975; Kottke and Oberwinkler 1990). Lamellae are visible in the suberin layer if waxes are present additionally (Fig. 4a; Sitte 1975). The suberin layer progressively increases with ageing of the cells. Finally, these cells accumulate phenolic substances, become impermeable and moribund (Fig. 3a, b). Short roots proliferate slowly under the root cap cells (Fig. 2 c) and remain covered by their residues (Fig. 4b, c). The dead root cap cells progressively detach from the root (Figs. 3b, 4b, c), but the innermost, suberized root cap cell walls remain tightly connected to the root cortical cell layer (Figs. 3 c, 4b, c). Thus, the suberin layer of the innermost root cap cells covers the whole surface of short roots, similar to a fine cuticle. During the elongation of the root the suberin layer is thinned out (Fig. 4e). It fades away on the root hairs (Fig. 3d) covering only the root hair base (Fig. 4d). At the most proximal parts of the rootlets, the suberin layer may also fade away on the surface of cortical cells (Fig. 5a, c), but the cell junctions remain tightly covered by the suberin layer all along the rootlet (Fig. 5a, b). The whole situation is illustrated by a scheme (Fig. 6).

The suberin layer is covered by a thin layer of electron-dense material. Phenols are strongly stained by osmium and lead and thus appear electron-dense. It is not always easy to discern if this material originates from insoluble phenolic residues of the former vacuole or from cell walls of the deceased root cap cells as vacuoles and moribund cell walls of root cap cells may contain high

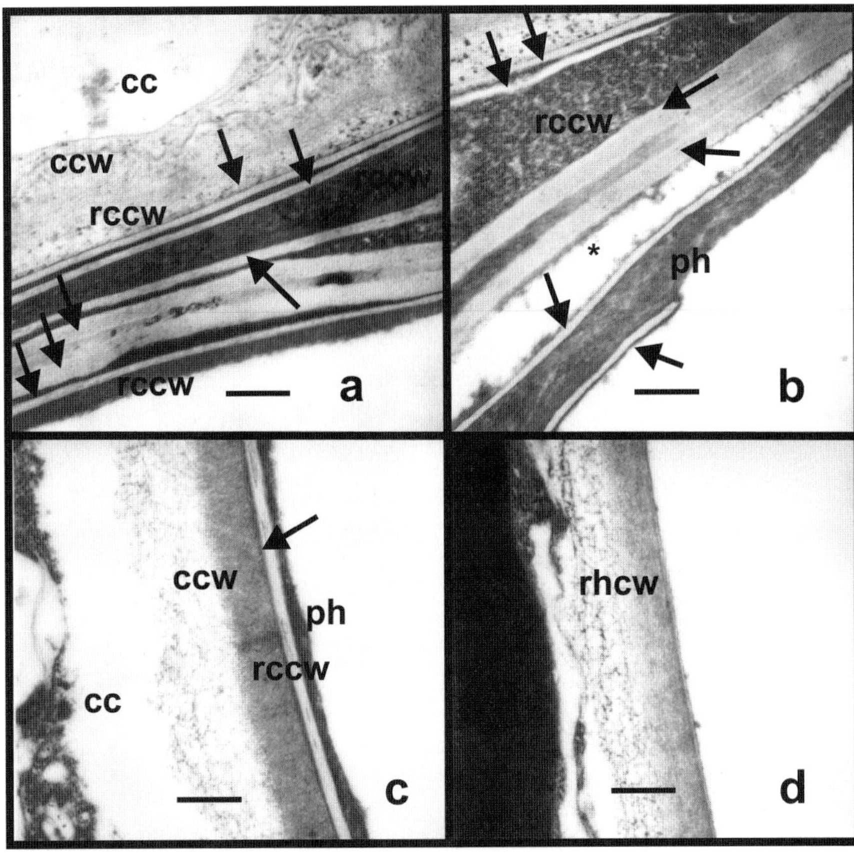

Fig. 3. a, b Suberized (*arrows*) root cap cell layers close to the root apex of *Picea abies*. The outer layers detaching (*) and partly decomposed. Cell wall or vacuolar, phenolic residues form the superficial layer. Scale 0.5 µm. **c** Superficial layer on short root cortical cell formed by the suberized root cap cell wall lined by phenolic residues (*arrow*). Scale 0.3 µm. **d** Cell wall of root hair with no suberin layer (scale 0.3 µm). *cc* Cortical cell, *ccw* cortical cell wall, *ph* phenolic residues, *rccw* root cap cell wall, *rhcw* root hair cell wall

amounts of phenolic residues (Fig. 2 c). An attempt was undertaken to clarify the situation by carefully studying the cell layers (Fig. 3a, b). Additional hints for recognition of cell wall material were obtained by immunogold labelling (see below). At the final stage of root development, when only the innermost root cap cell wall and its suberin layer are preserved on the root cortical cell wall (Fig. 3 c), the thin, electron-dense layer on top of the suberin layer can only be interpreted as the phenolic residues of the former vacuole. Dehydration of mycorrhizas in alcohol and embedding in LRWhite resin may obscure the suberin layer (Kottke 1997; Bonfante et al. 1998), but high pressure cryofixation, dehydration by acetone and embedding in Araldite/Epon or embed-

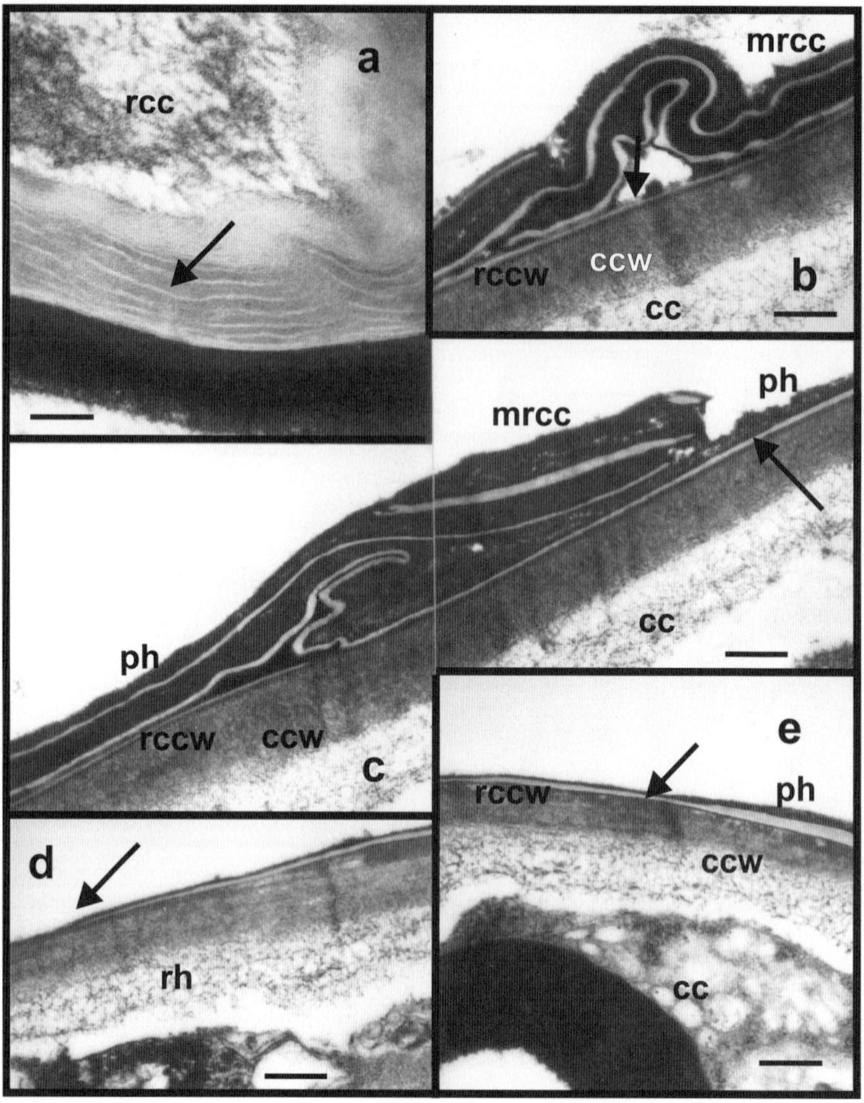

Fig. 4a–e. Cuticle-like layer on surface of short roots of *Picea abies*. **a** Lamellate structure of the suberin layer (*arrowhead*). Scale 0.1 µm. **b, c** Moribund root cap cells detaching from root cortical cell, suberized innermost root cap cell wall preserved in tight contact to cortical cell wall (*arrowheads*). Scale 0.5 µm. **d** Suberin layer fading away at root hair basis (*arrowhead*). Scale 0.5 µm. **e** Thinning of suberin layer (*arrowhead*) during elongation of cortical cells (scale 0.5 µm). *cc* Cortical cell, *ccw* cortical cell wall, *ph* phenolic residues, *rcc* root cap cell, *rccw* root cap cell wall, *rh* root hair

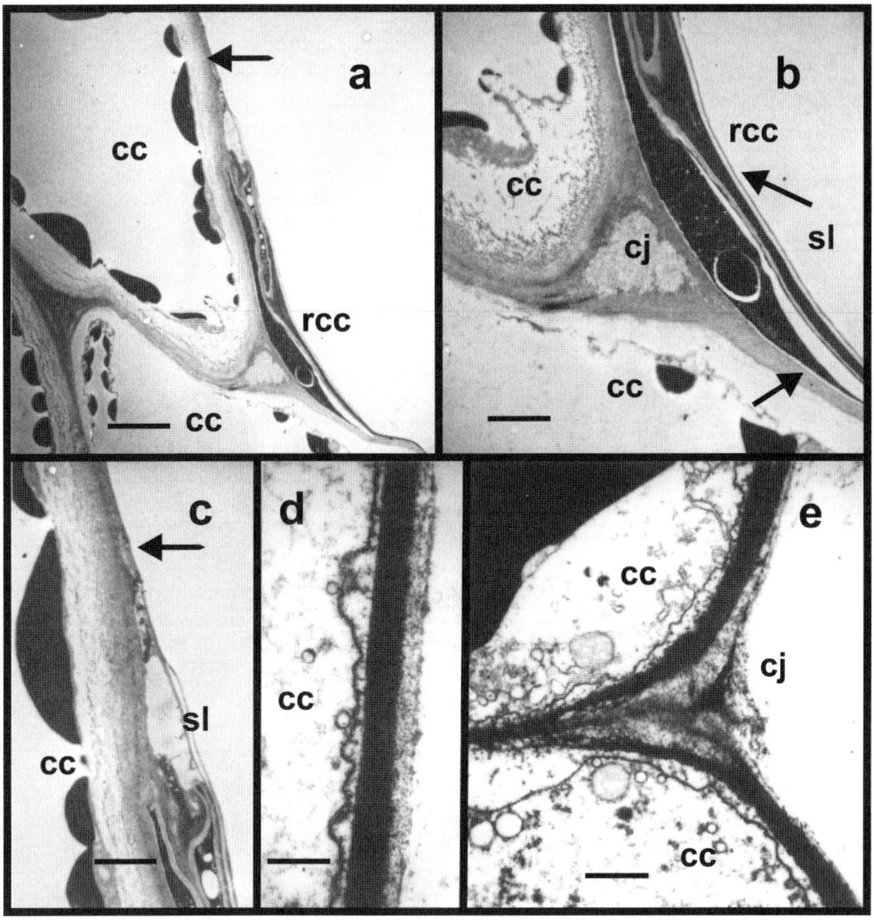

Fig. 5a–e. Suberized root cap cell layer covering cell junctions of short roots, but not of long roots (*Picea abies*). **a** Residues of suberized root cap cells covering the cell junction at the proximal part of a short root, fading away on the cortical cell (*arrow*). Scale 3 µm. **b** Enlargement of cell junction displaying several suberin layers of moribund root cap cells. Scale 1 µm. **c** Enlargement of cortical cell displaying fading away of the suberin layer (*arrow*). Scale 1 µm. **d** No suberin layer on the cortical cell wall of a long root. Scale 1 µm. **e** No suberin layer on top of the cell junction of a long root (scale 1 µm). *cc* Cortical cell, *cj* cell junction, *rcc* root cap cell, *sl* suberin layer

ding in Spurr's resin after fixation in glutaraldehyde yields clear results. The electron-dense layer on the root hair surface is no longer considered as a cuticle as was erroneously given in Kottke (1997).

The root cap cells of long roots are not suberized and no cuticle-like layer exists on the surface of long roots and their cell junctions (Fig. 5d, e).

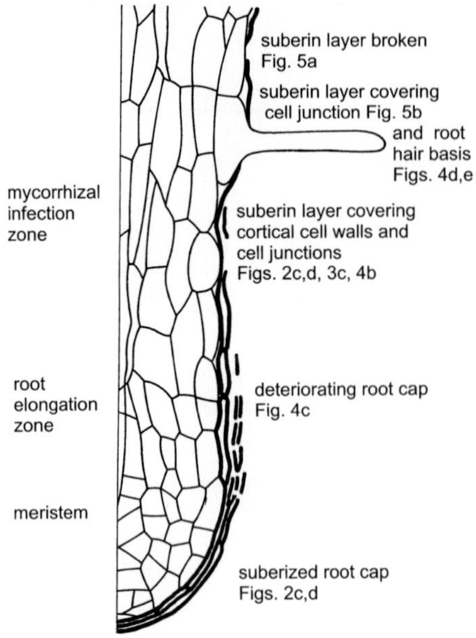

Fig. 6. Scheme illustrating the fate of the suberized root cap cells of short roots presumably in most ectomycorrhiza forming tree species. Figures refer to given micrographs

4 Involvement of the Cuticle-Like Layer in Mycorrhiza Formation

The cuticle-like, suberin layer covering short roots, by displaying a hydrophobic surface, appears to be involved in hyphal attachment at the beginning of ECM development. The suberin layer on the cell junctions is a barrier, however, that has to be penetrated when the hyphae invade between the root cortical cells establishing the Hartig net.

5 Involvement of the Cuticle-Like Layer in Hyphal Attachment

Using scanning electron microscopy, hyphae of *P. tinctorius* and *Paxillus involutus* (Batsch) Fr. attached to rootlets of *Quercus acutissima* Carruth or *Betula* spp., respectively, were found to be embedded in a mucilaginous material (Massicotte et al. 1987a, b; Brunner and Scheidegger 1992; Oh et al. 1995). Transmission electron microscopy revealed that adhesion of hyphae to the root surface was aided by polysaccharide fibrils and binding sites of mannose (Piché et al. 1983a; Thomson et al. 1989; Wong et al. 1990; Lei et al. 1991; Tagu et al. 2000). Laurent et al. (1999) identified cell-adhesion proteins in the cell

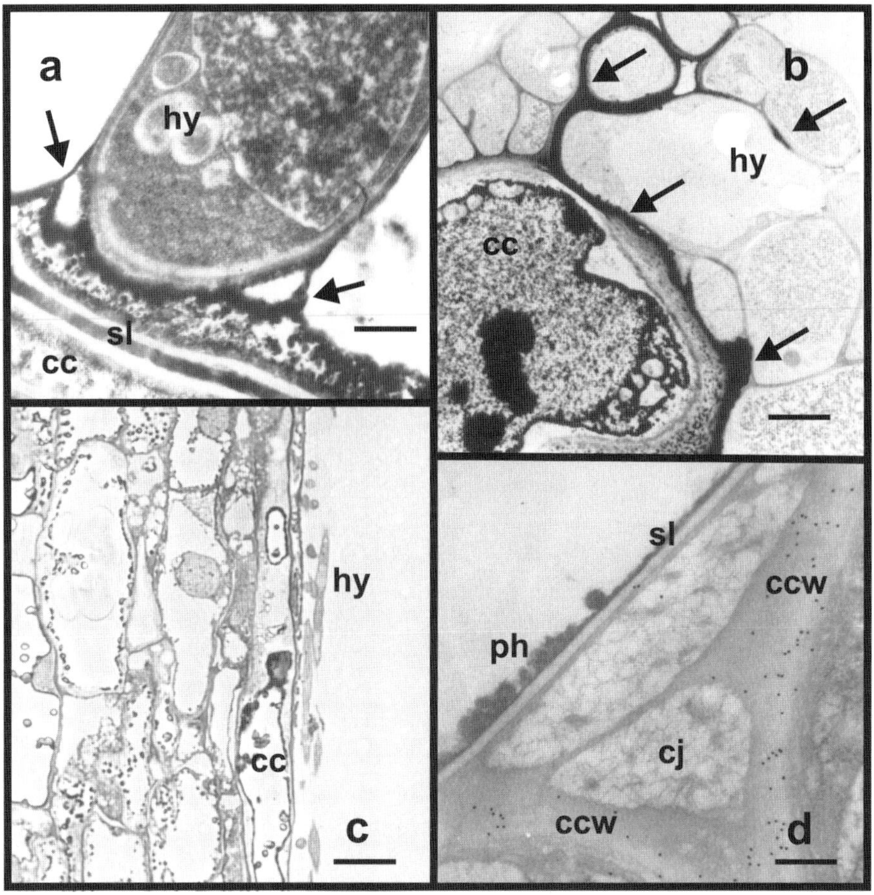

Fig. 7. a Adhesion pad of *Laccaria amethystea* hyphae in contact with the suberin layer of a root cap cell on top of the cortical cell. Scale 0.5 µm. **b** Swift positive reaction of cysteine-rich proteins in the cell wall of hyphae in contact with the root and each other (*Laccaria amethystea–Picea abies*). Scale 1.5 µm. **c** Long root of *Picea abies* displaying no attachment of *Laccaria amethystea* hyphae. Scale 15 µm. **d** Immunogold labelling of methyl-esterified pectins by the monoclonal antibody JIM7. The cell junction is covered by a suberized root cap cell wall lined by phenolic residues (short root of *Picea abies*). Scale 0.5 µm. *cc* Cortical cell, *ccw* cortical cell wall, *cj* cell junction, *ph* phenolic residues, *hy* hypha, *sl* suberin layer

walls of the ectomycorrhiza-forming fungus *P. tinctorius*. Investigation of the attachment of *Laccaria amethystea* (Bull.) Murrill to *Picea abies* short roots showed formation of an adhesion pad (Fig. 7a) which was strongly stained by the Swift reaction for cysteine-rich proteins (Lewis and Knight 1977; Kottke 1997). The surface of the hyphae in contact to the suberin layer and to each other is stained similarly (Fig. 7b). Attachment of hyphae to the basis of root hairs by Swift-positive material was found previously for *P. tinctorius* and

Picea mariana Mill. B.S.P. (Thomson et al. 1989). The superficial layer of the fungal wall may contain hydrophobins, cysteine-rich proteins, self-assembling at the wall/air interface (Wösten et al. 1994; Wessels 1997; Wösten and Vocht 2000). Hydrophobins were localized using antibodies in mycorrhizas formed by *P. tinctorius* and *E. globulus* and in mycorrhizas of *Tricholoma terreum* (Schaeff.) Quél. with the compatible host *Pinus sylvestris* L. (Mankel et al. 2000; Tagu et al. 2001). The cuticle-like layer may thus be considered as the hydrophobic surface appropriate for hyphal attachment by hydrophobins.

Attachment to the tips of root hairs was observed (Kottke 1997), but occurred only within a defined, susceptible zone (Thomson et al. 1989). Hyphae may also attach to the surface of root cap cells (Bonfante et al. 1998). This kind of attachment differs from that to the short root surface. Staining for cysteine-rich proteins was found to be negative (Kottke 1997). Attachment was neither followed by enlargement of hyphae or lobed ramification, nor by any digesting process (Thomson et al. 1989; Kottke 1997). Instead, thickening of fungal wall has been observed and the appearance of β-1,3-glucans in the root cell wall was shown (Bonfante et al. 1998).

No attachment to the surface of long roots was found. Hyphae grow along the long roots in acropetal direction without apparent changes (Fig. 7 c).

6 Digestion of the Suberin Layer and the Cell Wall of the Root Cap

The cuticle-like layer covers all the cell junctions of short roots (Figs. 5a, 6). The hyphae, therefore, must penetrate the suberin layer and the wall of the moribund root cap cell when establishing the Hartig net. Vesicles, probably containing a cutinase-like enzyme were frequently observed in hyphae dissolving the suberin (Fig. 8a). The hyphae split away the suberized root cap cell wall and proliferate below, on top of the surface of the cortical cell (Fig. 8b, c, d). This process may explain why finally, when the hyphal sheath covers the rootlet, the cuticle-like layer is no longer found. The suberin layer became integrated into the hyphal sheath (Fig. 8d).

The hyphae digest the suberin layer locally and disrupt the root cap cell wall, but do not attack the wall of the live cortical cell (Fig. 8a, b). While the enzyme activity remains to be proven in situ, there are many indications for a controlled cell wall hydrolyzing activity of ECM fungi (for review, see Cairney and Burke 1994). ECM fungi digest cell wall material, including the suberin layer of the moribund root cap, but not material of live cells during mycorrhiza formation (Chilvers 1968; Piché et al. 1983b; Kottke and Oberwinkler 1986). A strict spatial and temporal regulation of enzyme activity has, thus, to be expected when the hyphae contact alive cells.

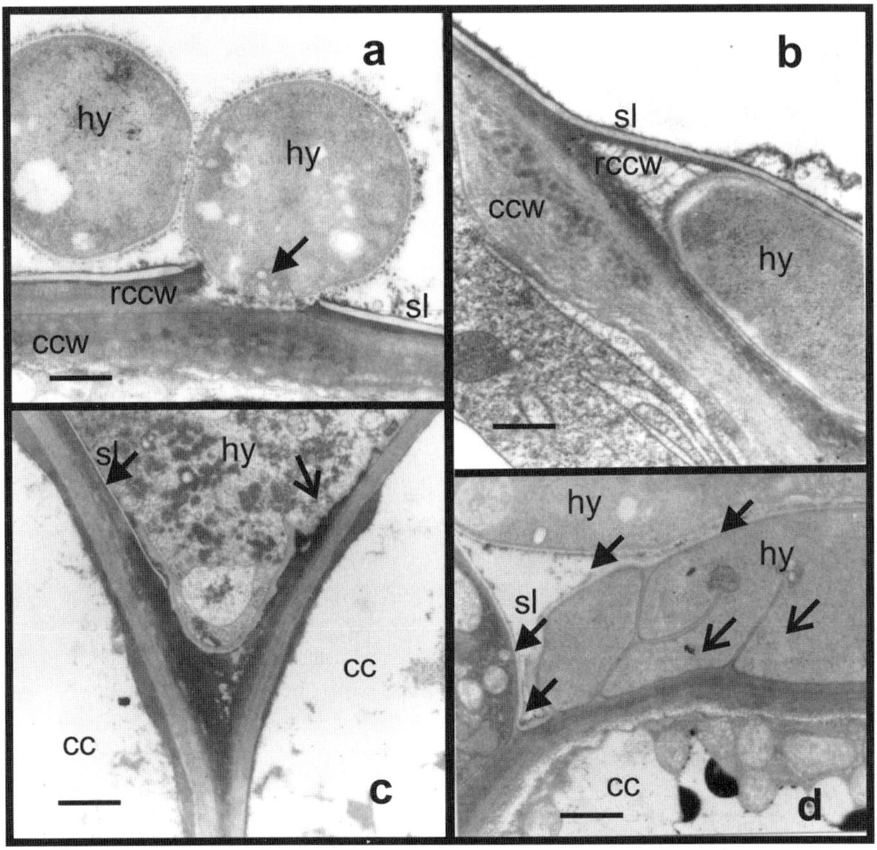

Fig. 8a–d. Cuticle-like suberin layer involved in mycorrhiza formation (*Laccaria amethystea-Picea abies*). **a** Local digestion of the suberin layer and root cap cell wall, no disturbance of cortical cell wall, vesicles probably containing enzymes (*arrowhead*, scale 1 µm). **b** Hypha splitting off the suberin layer and proliferating beneath, on the surface of the cortical cell wall. Scale 0.5 µm. **c** Hypha penetrating between cell junction, suberin layer partly digested (*arrow*) and partly preserved (*arrowhead*), lobed growth of hyphae visible (*arrow*). Scale 1 µm. **d** Hyphae proliferating under the suberin layer (*arrowheads*) show lobed branching typical of Hartig net structure (*arrows*). Scale 1 µm. *cc* Cortical cell, *ccw* cortical cell wall, *hy* hypha, *rccw* root cap cell wall, *sl* suberin layer

7 Hartig Net Formation

Lobed growth of hyphae indicating the initialization of the Hartig net was found in connection with the digestion of the cuticle-like layer (Fig. 8 c, d). Thomson et al. (1989) described the formation of hyphal lobes at the base of root hairs. Lobed hyphal growth on the root surface has been observed by SEM (Jacobs et al. 1989; Brunner and Scheidegger 1992), but the connection to suberin digestion is shown here for the first time. It remains to be elucidated if there is a direct signalling link between the digestion of the suberin and the change of hyphal growth characters. After digesting the suberin layer and disrupting the root cap cell wall, the hyphae come into direct contact with the live cortical cells. There may then be additional signals involved in triggering the hyphal growth changes at the root surface (Salzer et al. 1997, 2000).

8 Pectins in the Cortical Cell Walls of Nonmycorrhizal Long and Mycorrhizal Short Roots

Methyl-esterified pectins were localized in the root cell walls of *Picea abies* using the monoclonal antibody JIM7 (K. Roberts, John Innes Institute, Norwich UK; Fig. 7d). No difference in the amount of pectin was found between cortical cells in contact to hyphae and those lacking hyphal contact when short roots and mycorrhizas were compared (Fig. 9). The cortical cells of noncolonized, long roots, however, were significantly more densely marked by the antibody (Fig. 9). There was no difference between both root types in the amounts of methyl-esterified pectins in the cell walls of the meristems (Fig. 9). During differentiation, the amounts of methyl-esterified pectins obviously increase in cortical cell walls of long roots, but are reduced in cortical cell walls of short roots. There is no indication for a digestion of pectins by the hyphae as no changes in the amounts of pectins was found during early stages of *Laccaria amethystea- Picea abies* mycorrhiza formation. Balestrini et al. (1996) could not find any indication for polygalacturonase activity during ECM development between *Coryllus avellana* and *Tuber magnatum* either. The authors supposed de-esterification of the pectins according to increased labelling of de-esterified pectins after mycorrhiza formation. In the case of *P. abies*, however, immunogold labelling by the monoclonal antibody JIM5 showed low amounts of de-esterified pectins and labelling decreased from inner cortical cells to outer cortical cells (not shown). High labelling of methyl-esterified pectins was detected in roots of *Daucus carota* L. and *Avena sativa* L. (Knox et al. 1990). This finding would support the view that fast growing roots contain high amounts of methyl-esterified pectins in cortical cells.

It is unclear whether the amounts of methyl-esterified pectins have any influence on mycorrhiza formation. There is too little knowledge on the importance of methyl-esterified pectins for stability or plasticity of cell walls and cell-to-cell adhesion (Liners et al. 1994). Previously, reduction of the cell wall-bound ferulic acid, linking pectic substances in the cell wall matrix, was found to occur during mycorrhiza formation of *Picea abies*, *Larix decidua* and *Arbutus menziesii* (Münzenberger et al. 1990, 1995, Weiss et al. 1999). Less rigid cortical cell walls were considered a prerequisite for intercellular hyphal penetration during Hartig net establishment.

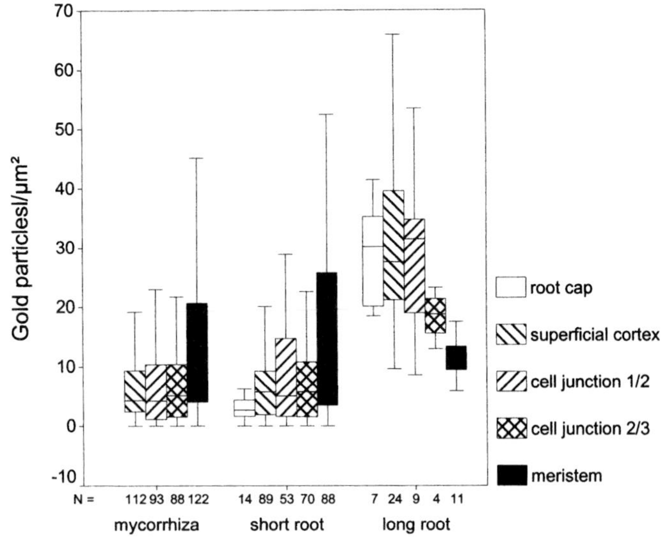

Fig. 9. Amount of immunogold labelling by the monoclonal antibody JIM7 against methyl-esterified pectins. Counting of gold granules was carried out by means of image analysis in different compartments of ectomycorrhizas, short roots, and long roots. Material collected from in vitro cultures of *Picea abies* inoculated by *Laccaria amethystea*

9 Conclusions

The surface of short roots appears to be important in ECM initialization. So far we have only started to understand the process. Further research is needed to clarify changes of cell wall components and signal exchanges involved. Some progress was, however, obtained by structural and molecular investigations during the last few years. Tight attachment of hyphae to the root surface is established between hydrophobins on the hyphal surface and the hydrophobic root surface. The hydrophobic root surface derives from the residue of the suberized root cap of short roots. The lack of a suberized root cap might be involved in the lack of a Hartig net in long roots. Digestion of the suberin layer and the root cap cell wall may mean the occurrence of a slight, transient and locally restricted aggressive phase during ectomycorrhiza formation and may explain the slight, transient defence reactions in the early phase (Salzer et al. 1997, 2000). The lack of pectin digestion by the fungus might avoid severe defence reactions of the root. The locally restricted digestion of the moribund, suberized root cap cell wall may, however, alternatively be looked upon as a saprophytic phase of interaction. Ectomycorrhizal fungi phylogenetically derive from saprophytes and not from parasites (Bresinsky et al. 1999) and many species have preserved saprophytic growth facilities.

When dissolving the suberin layer locally, the hyphae start lobed growth typical of Hartig net structure. It is unclear so far if the digestion of the cuticle-like layer has itself an inductive effect on hyphal growth. Signals obtained from the live cortical cells, reached after digestion of the suberized root cap layer, may be more decisive in change of hyphal growth characters. The described phenomena are unique in ECM formation and, as far is known, do not occur in any other plant-fungus interaction system.

Acknowledgements. The valuable comments on the manuscript and the introduction to immunogold-labelling by Paola Bonfante is greatly appreciated. I also express my gratitude to Bettina Grüninger and Esther Strasdas for carrying out the JIM7/JIM5 labelling studies.

References and Selected Reading

Agerer R (1998) Colour atlas of ectomycorrhizae. Einhorn-Verlag, Schwäbisch Gmünd, 140 pp

Balestrini R, Hahn MG, Bonfante P (1996) Location of cell-wall components in ectomycorrhizae of *Coryllus avellana* and *Tuber magnatum*. Protoplasma 191: 55–69

Blasius D, Feil W, Kottke I, Oberwinkler F (1986) Hartig net structure and formation in fully ensheathed ectomycorrhizas. Nordic J Bot 6: 837–842

Bonfante P, Balestrini R, Martino E, Perotto S, Plassard C, Mousain D (1998) Morphological analysis of early contacts between pine roots and two ectomycorrhizal *Suillus* strains. Mycorrhiza 8:1–10

Bremer K, Chase MW, Stevens PF (1998) An ordinal classification for the families of flowering plants. Ann Mo Bot Gard 85:531–553

Bresinsky A, Jarosch M, Fischer M, Schönberg I, Wittmann-Bresinsky B (1999) Phylogenetic relationship within Paxillus s. l. (Basidiomycetes, Boletales): Separation of a southern hemisphere genus. Plant Biol 1:327–333

Brunner I, Scheidegger C (1992) Ontogeny of synthesized *Picea abies*(L.) Karst.- *Hebeloma crustuliniforme* (Bull. ex St Amans) Quél. ectomycorrhizas. New Phytol 120:359–369

Cairney JW, Burke RM (1994) Fungal enzymes degrading plant cell walls: their possible significance in the ectomycorrhizal symbiosis. Mycol Res 98:1345–1356

Chilvers GA (1968) Low power electron microscopy of the root cap region of eucalypt mycorrhizas. New Phytol 67:663

Clowes FAL (1951) The structure of mycorrhizal roots of *Fagus sylvatica*. New Phytol 50:1–16

Clowes FAL (1954) The root-cap of ectotrophic mycorrhizas. New Phytol 53:525–9

Hatch AB, Doak KD (1933) Mycorrhizal and other features of the root system of *Pinus*. J Arnold Arbor 14:85–99

Jacobs PF, Peterson RL, Massicotte HB (1989) Altered fungal morphogenesis during early stage ectomycorrhiza formation in *Eucalyptus pilularis*. Scann Microsc 3:249–255

Knox JP, Linstead PJ, King J, Cooper C, Roberts K (1990) Pectin esterification in spatially regulated both within cell walls and between developing tissues of root apices. Planta 181:512–521

Kottke I (1997) Fungal adhesion pad formation and penetration of root cuticle in early stage *Picea abies-Laccaria amethystea* mycorrhizas. Protoplasma 196:55–64

Kottke I, Oberwinkler F (1986) Root-fungus interactions observed on initial stages of mantle formation and Hartig net establishment in mycorrhizas of *Amanita muscaria* (L. ex Fr.) Hooker on *Picea abies* (L.) Karst. in pure culture. Can J Bot 64: 2348–2354

Kottke I, Oberwinkler F (1987) Cellular structure and function of the Hartig net: coenocytic and transfer cell-like organization. Nordic J Bot 7:85–95

Kottke I, Oberwinkler F (1989) Amplification of root-fungus interface in ectomycorrhizae by Hartig net architecture. Ann Sci For 46 Suppl:737s–740s

Kottke I, Oberwinkler F (1990) Comparative investigations on the differentiation of the endodermis and the development of the Hartig net in mycorrhizae of *Picea abies* and *Larix decidua*. Trees 4:41–48

Kottke I, Rapp C, Oberwinkler F (1986) Zur Anatomie gesunder und "kranker" Feinstwurzeln von Fichten: Meristem und Differenzierungen in Wurzelspitzen und Mykorrhizen. Eur J For Pathol 16:159–171

Kottke I, Münzenberger B, Oberwinkler F (1996) Structural approach to function in ectomycorrhizas. In: Rennenberg H, Eschrich W, Ziegler H (eds) Trees – contribution to modern tree physiology. SPB Academic, The Hague, pp 3–22

Laurent P, Voiblet C, Tagu D, de Carvalho D, Nehls U, De Bellis R, Ballestrini R, Bauw G, Bonfante P, Martin F (1999) A novel class of ectomycorrhiza-regulated cell wall polypeptides in *Pisolithus tinctorius*. Mol Plant Microbe Interact 12:862–871

Lei J, Wong KK, Piché Y (1991) Extracellular Concanavalin A-binding sites during early interaction between *Pinus banksiana* and two closely related genotypes of the ectomycorrhizal basidiomycete *Laccaria bicolor*. Mycol Res 95:357–363

Lewis PR, Knight DP (1977) Staining methods for sectioned material. North-Holland Publishing Company, Amsterdam, 311 pp

Liners F, Gaspar T, Van Cutsem P (1994) Acetyl- and methyl-esterification of pectins of friable and compact sugar-beet calli: consequences for intercellular adhesion. Planta 192:545–556

Mankel A, Krause K, Genenger M, Kost G, Kothe E (2000) A hydrophobin accumulated in the Hartig net of ectomycorrhiza formed between *Tricholoma terreum* and its compatible host tree is missing in an incompatible association. J Appl Bot 74:95–99

Marks GC, Foster RC (1973) Structure, morphogenesis and ultrastructure of ectomycorrhizae. In: Marks GC, Kozlowski TT (eds) Ectomycorrhizae. Their ecology and physiology. Academic Press, New York, London, pp 1–41

Martin F, Laurent P, de Carvalho D, Voiblet C, Balestrini R, Bonfante P, Tagu D (1999) Cell wall proteins of the ectomycorrhizal basidiomycete *Pisolithus tinctorius*: identification, function, and expression in symbiosis. Fungal Gen Biol 27:161–174

Massicotte HB, Peterson RL, Ackerley CA, Piche Y (1986) Structure and ontogeny of *Alnus crispa-Alpova diplophloeus* ectomycorrhizae. Can J Bot 64:177–192

Massicotte HB, Peterson RL, Ackerly CA (1987a) Ontogeny of *Eucalyptus pilularis-Pisolithus tinctorius* ectomycorrhizae. I. Light microscopy and scanning electron microscopy. Can J Bot 65:1927–1939

Massicotte HB, Peterson RL, Ackerly CA (1987b) Ontogeny of *Eucalyptus pilularis-Pisolithus tinctorius* ectomycorrhizae. II. Transmission electron microscopy. Can J Bot 65:1940–1947

Müller H (1906) Über die Metakutinisierung der Wurzelspitze und über die verkorkten Scheiden in den Achsen der Monokotyledonen. Bot Z 4:54–64

Münzenberger B, Heilemann J, Strack D, Kottke I, Oberwinkler F (1990) Phenolics of mycorrhizas and non-mycorrhizal roots of Norway spruce. Planta 182:142–148

Münzenberger B, Kottke I, Oberwinkler F (1995) Reduction of phenolics in mycorrhizas of *Larix decidua* Mill. Tree Physiol 15:191–196

Noelle W (1910) Studien zur vergleichenden Anatomie und Morphologie der Koniferenwurzeln mit Rücksicht auf die Systematik. Bot Z 68:169–266

Oh KI, Melville LH, Peterson RL (1995) Comparative structural study of *Quercus serrata* and *Q. acutissima* formed by *Pisolithus tinctorius* and *Hebeloma cylindrosporum*. Trees 9:171–179

Piché Y, Peterson RL, Ackerley KA (1983a) Early development of ectomycorrhizal short roots of pine (*Pinus strobus*). Scann Electron Microsc 111:1467–1474

Piché Y, Peterson RL, Howarth MJ, Fortin JA (1983b) A structural study of the interaction between the ectomycorrhizal fungus *Pisolithus tinctorius* and *Pinus strobus* roots. Can J Bot 61:1185–119

Plaut M (1918) Über die morphologischen und mikroskopischen Merkmale der Periodizität der Wurzel, sowie über die Verbreitung der Metakutisierung der Wurzelhaube im Pflanzenreich. Festschr. 100. j. Best. K. Württ. Landw. Hochsch. Hohenheim. Verlag Ulmer, Stuttgart, S 129–151

Salzer P, Boller T (2000) Elicitor induced reactions in mycorrhizae and their suppression. In: Podila GK, Douds DD Jr (eds) Current advances in mycorrhizae research. Symposium Series, APS Press, The American Phytopathological Society, St. Paul, Minnesota, pp 1–10

Salzer P, Münzenberger B, Schwacke R, Kottke I, Hager A (1997) Signalling in ectomycorrhizal fungus-root interactions. Trees – contribution to modern tree physiology. SPB Academic, The Hague, pp 339–356

Sitte, P (1975) Die Bedeutung der molekularen Lamellenbauweise von Korkzellwänden. Biochem Physiol Pflanzen 168:287–297

Tagu D, Martin T (1996) Molecular analysis of cell wall proteins expressed during the early steps of ectomycorrhiza development. New Phytol 133:73–85

Tagu D, Lapeyrie F, Ditengou F, Lagrange H, Laurent P, Missoum N, Nehls U, Martin F (2000) Molecular aspects of ectomycorrhiza development. In: Podila GK, Douds DD Jr (eds) Current advances in mycorrhizae research. Symposium Series, APS Press, The American Phytopathological Society, St. Paul, Minnesota, pp 69–90

Tagu D, De Bellis R, Balestrini R, De Vries OM, Piccoli G, Stocchi V, Bonfante P, Martin F (2001) Immunolocalization of hydrophobin HYDPt-1 from the ectomycorrhizal basidiomycete *Pisolithus tinctorius* during colonization of *Eucalyptus globulus* roots. New Phytol 149:127–135

Thomson J, Melville IH, Peterson RL (1989) Interaction between the ectomycorrhizal fungus *Pisolithus tinctorius* and root hairs of *Picea mariana* (Pinaceae). Am J Bot 76:632–636

Weiss M, Schmidt J, Neumann D, Wray V, Christ R, Strack D (1999) Phenylpropanoids in mycorrhizas of Pinaceae. Planta 208:491–502

Wessels JG (1997) Hydrophobins: Proteins that change the nature of the fungal surface. Adv Microbial Physiol 38:1–45

Wilcox HE (1968) Morphological studies of the roots of Redpine, *Pinus resinosa*. II. Fungal colonisation of roots and the development of mycorrhizae. Am J Bot 55:688–700

Wilcox HE (1954) Primary organization of active and dormant roots of noble-fir, *Abies procera*. Am J Bot 41:812–821

Wösten HA, Asgeirdottir SA, Krook JH, Drenth JH, Wessels JG (1994) The fungal hydrophobin Sc3p self-assembles at the surface of aerial hyphae as a protein membrane constituting the hydrophobic rodlet layer. Eur J Cell Biol 63:122–129

Wösten HA, de Vocht ML (2000) Hydrophobins, the fungal coat unravelled. Biochim Biophys Acta Rev Biomembr 1469:79–86

Wong K, Montpetit D, Piché Y, Lei J (1990) Root colonization by four closely related genotypes of ectomycorrhizal basidiomycete *Laccaria bicolor* (Maire) Orton – comparative studies using electron microscopy. New Phytol 116:669–679

14 Cellular Ustilaginomycete – Plant Interactions

ROBERT BAUER and FRANZ OBERWINKLER

1 Introduction

The Ustilaginomycetes comprises more than 1300 species in ca. 80 genera of basidiomycetous plant parasites. They occur throughout the world, although many species are restricted to tropical, temperate or arctic regions. Some species of *Ustilago* and *Tilletia*, e.g., the barley, wheat or maize smut fungi, are well known because they are of economic importance (Trione 1982; Thomas 1989). For example, from 1983 to 1988 the barley smut fungi reduced annual yields from 0.7 to 1.6% in the prairie provinces in central Canada, causing annual losses of about US $8,000,000 (Thomas 1989). *Tilletia contraversa* is important in the international wheat trade (Trione 1982) and 2–5% in a corn field are generally infected by *Ustilago maydis*, while up to 80% of a field can be infected if conditions are good for the fungus. On the other hand, the galls of *U. maydis* are regarded as a delicacy in the Mesoamerican tradition. They are known in Mexico as "Huitlacoche" and in parts of the USA. as "maize mushroom", "Mexican truffles" or "caviar azteca".

This chapter focuses predominantly on the cellular interaction of the Ustilaginomycetes that represents one of the three classes of the Basidiomycota (Begerow et al. 1997).

2 The Term Smut Fungus

Like the terms agaric, polypore, lichen etc., the term smut fungus circumscribes the organization and life strategy of a fungus, but it is not a taxonomic term. Smut fungi evolved in different fungal groups. Most smut fungi are in the Ustilaginomycetes. Other smut fungi, in the Microbotryales, are members of the Urediniomycetes (Bauer et al. 1997; Begerow et al. 1997). There are significant convergences between the urediniomycetous and the ustilaginomycetous phragmobasidiate smut fungi. Certain taxa of both groups are similar with

Plant Surface Microbiology
A. Varma, L. Abbott, D. Werner, R. Hampp (Eds.)

respect to soral morphology, teliosporogenesis, life cycle, basidial morphology and host range.

3 Life Cycle

The species of the Ustilaginomycetes share an essentially similar life cycle with a saprobic haploid phase and a parasitic dikaryophase (e.g., Sampson 1939). The haploid phase usually commences with the formation of basidiospores after meiosis of the diploid nucleus in the basidium and ends with the conjugation of compatible haploid cells to produce dikaryotic, parasitic mycelia. The dikaryotic phase ends with the production of basidia.

Almost all Ustilaginomycetes sporulate on or in parenchymatic tissues of the hosts. In the ustilaginomycetous smut fungi, the young basidium becomes thick-walled and at maturity separates from the sorus, thus functioning as a dispersal agent, the teliospore. The teliospores are usually the most conspicuous stage in the smut's life cycle. Most of the Ustilaginomycetes are dimorphic, producing a yeast or yeast-like phase in the haploid state.

4 Hosts

The Ustilaginomycetes are ecologically well characterized by their plant parasitism. Two species occur on spike mosses (Bauer et al. 1999), one on ferns, two on conifers, whereas all other Ustilaginomycetes parasitize angiosperms with a high proportion of species on monocots, especially on Poaceae and Cyperaceae. Thus, of the ca. 1300 species, ca. 42 % occurs on Poaceae and ca. 15 % on Cyperaceae. Concerning the hosts two points are remarkable: (1) with a few exceptions the teliospore-forming species of the Ustilaginomycetes parasitize nonwoody herbs, whereas those without teliospores prefer woody trees or bushes. However, almost all species sporulate on parenchymatic tissues of the hosts. (2) Two of the angiosperm families with the highest number of species, the Orchidaceae with about 20,000 species and the Poaceae with about 9000 species, play quite a different role for the Ustilaginomycetes. There are no known species on Orchidaceae while the Poaceae are the most important host family of the Ustilaginomycetes. This can be tentatively explained by the completely different ecological strategies of the two families. Orchid species subsist with a few isolated individuals and are highly specialized for insect pollination. The Poaceae, however, disperse their dusty pollen by the wind and cover about a third of the land surface with numerous individuals. The ecology of the Ustilaginomycetes, with dusty teliospores or basidiospores dispersed by the wind and with the requirement of extensive host populations for successful infection, corresponds well to the ecology of the Poaceae.

5 Cellular Interactions

Information concerning the cellular interaction of the Ustilaginomycetes has come from only a few studies (Mims 1982, 1991; Mims and Nickerson 1986; Luttrell 1987; Nagler 1989; Nagler et al. 1990; Snetselaar and Tiffany 1990; Mims et al. 1992; Snetselaar and Mims 1994; Bauer et al. 1995, 1997; Martinez et al. 1999). Hyphae of the Ustilaginomycetes in contact with host plant cells possess zones of host–parasite interaction with fungal deposits resulting from exocytosis of primary interactive vesicles. These zones provide ultrastructural characters diagnostic for higher groups in the Ustilaginomycetes (Bauer et al. 1997). Initially, primary interactive vesicles with electron-opaque contents accumulate in the fungal cell (Fig. 1). Depending on the fungal species, the primary interactive vesicles may fuse with one another before being exocytosed from the fungal cytoplasm. Electron-opaque deposits also

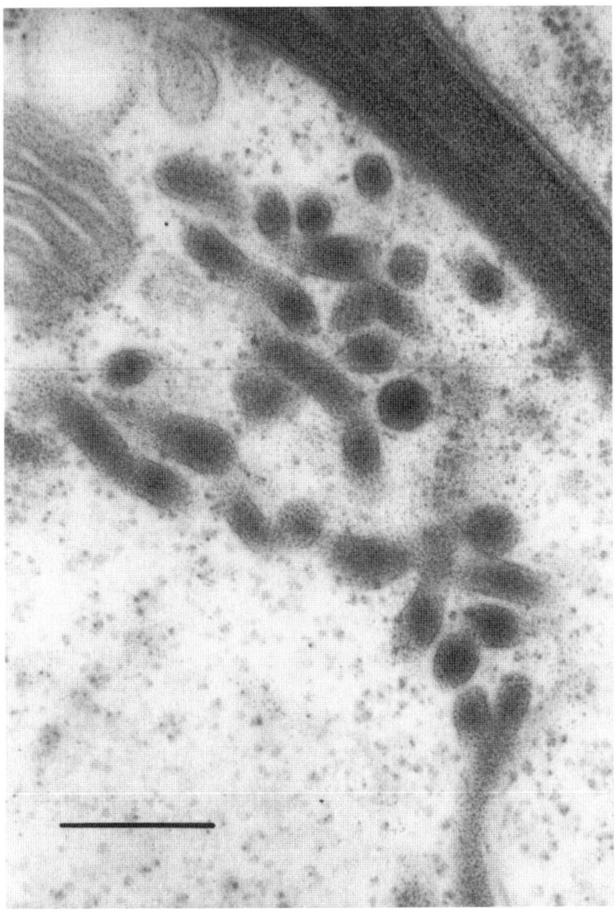

Fig. 1. Interactive vesicles in a hypha of *Exobasidium pachysporum. Scale bar* 0.2 μm

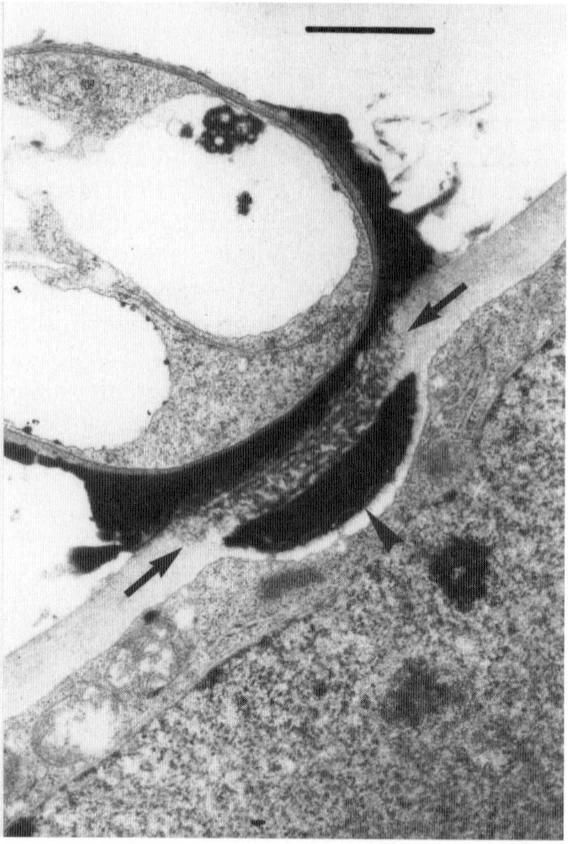

Fig. 2. Transfer stage between *Mycosyrinx cissi* (*upper cell*) and its host (*lower cell*). Note the infiltrated host cell wall (*between the arrows*) and the electron-opaque deposit at the host side (*arrrowhead*). *Scale bar* 0.1 µm

appear at the host side, opposite the point of contact with the fungus. Detailed studies indicate that these deposits at the host side originate from the exocytosed fungal material by transfer towards the host plasma membrane (Fig. 2; Bauer et al. 1995, 1997).

The following major types, minor types and variations were recognized by Bauer et al. (1995, 1997).

5.1 Local Interaction Zones

Short-term production of many primary interactive vesicles per interaction site results in local interaction zones. This type of cellular interaction characterizes the Entorrhizomycetidae and Exobasidiomycetidae (Bauer et al. 1997).

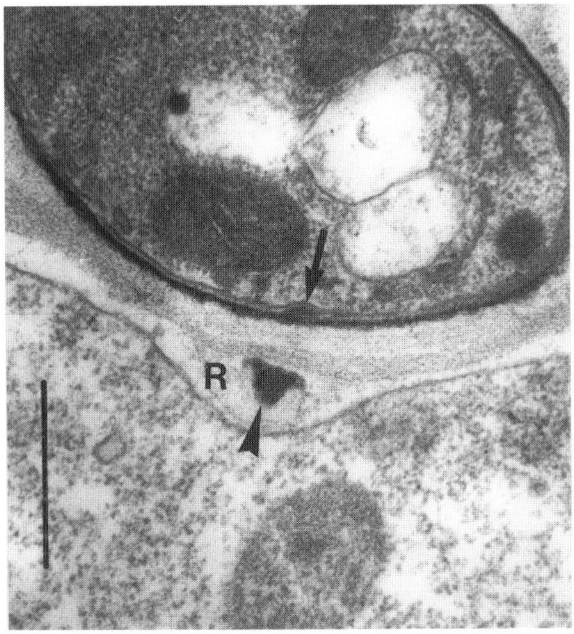

Fig. 3. Local interaction zone without interaction apparatus between *Conidiosporomyces ayresii* (*upper cell*) and its host (*lower cell*) showing the secretion profile of one interactive vesicle (*arrow*). Note the electron-opaque deposit at the host side (*arrowhead*). Host response to infection is visible at *R*. *Scale bar* 0.5 µm

5.1.1 Local Interaction Zones without Interaction Apparatus

Primary interactive vesicles fuse individually with the fungal plasma membrane (Fig. 3). Depending upon the species, local interaction zones without interaction apparatus are present in intercellular hyphae or in haustoria. This type of cellular interaction characterizes the Entorrhizomycetidae, Georgefischeriales, Tilletiales and Microstromatales (Bauer et al. 1997).

5.1.2 Local Interaction Zones with Interaction Apparatus

Fusion of the primary interactive vesicles precedes exocytosis. This type of cellular interaction characterizes the Exobasidianae (Bauer et al. 1997).

5.1.2.1 Local Interaction Zones with Simple Interaction Apparatus

Primary interactive vesicles fuse to form one large secondary interactive vesicle per interaction site (Fig. 4). Interaction zones of this type are only located in intercellular hyphae. This type of cellular interaction characterizes the Entylomatales (Bauer et al. 1997).

5.1.2.2 Local Interaction Zones with Complex Interaction Apparatus

Numerous primary interactive vesicles fuse to form several secondary interactive vesicles per interaction site. Fusion of the secondary interactive vesicles

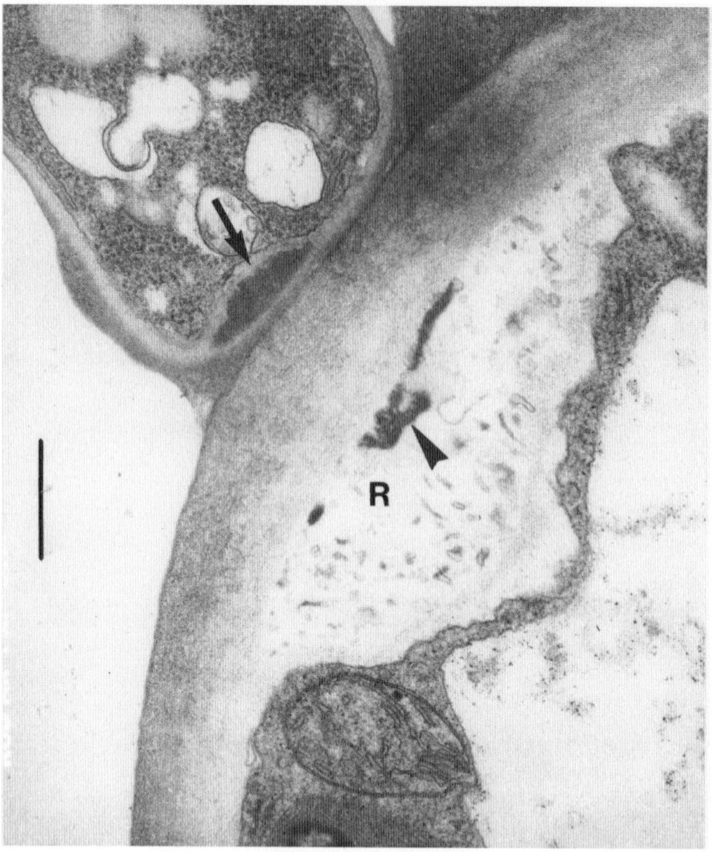

Fig. 4. Local interaction zone between *Entyloma hieracii* (*upper cell*) and its host (*lower cell*) showing the exocytosis profile of a simple interaction apparatus (*arrow*). Note the electron-opaque deposit at the host side (*arrowhead*). Host response to infection is visible at *R*. *Scale bar* 0.5 µm

then results in the formation of a complex cisternal net. This type of cellular interaction characterizes the Doassansiales and Exobasidiales (Bauer et al. 1997).

1. Local interaction zones with complex interaction apparatus containing cytoplasmic compartments (Fig. 5)
   The intercisternal space of the cisternal net finally becomes integrated in the interaction apparatus. Depending upon the species, interaction zones of this type are formed by intercellular hyphae or haustoria. This type of cellular interaction characterizes the Doassansiales (Bauer et al. 1997).
2. Local interaction zones with complex interaction apparatus producing interaction rings (Fig. 6)

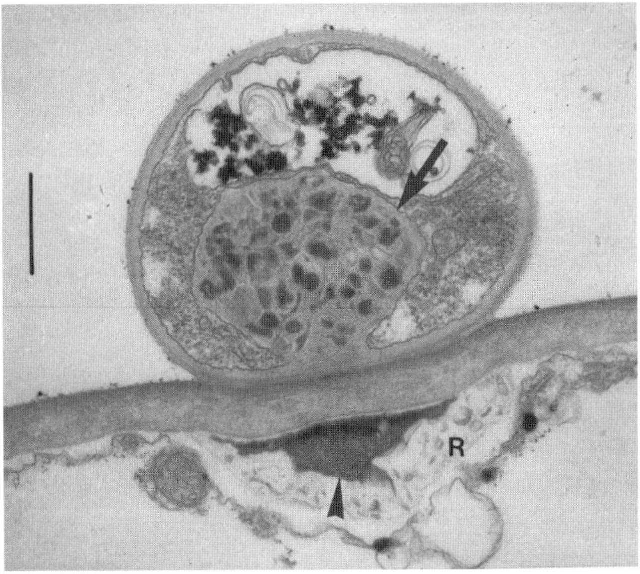

Fig. 5. Local interaction zone between *Doassinga callitrichis* (*upper cell*) and its host (*lower cell*) showing the exocytosis profile of a complex intercisternal interaction apparatus (*arrow*). The interaction apparatus and its intercisternal space is excluded from the cytoplasm. Note the electron-opaque deposit at the host side (*arrowhead*). Host response to infection is visible at *R*. *Scale bar* 0.5 µm

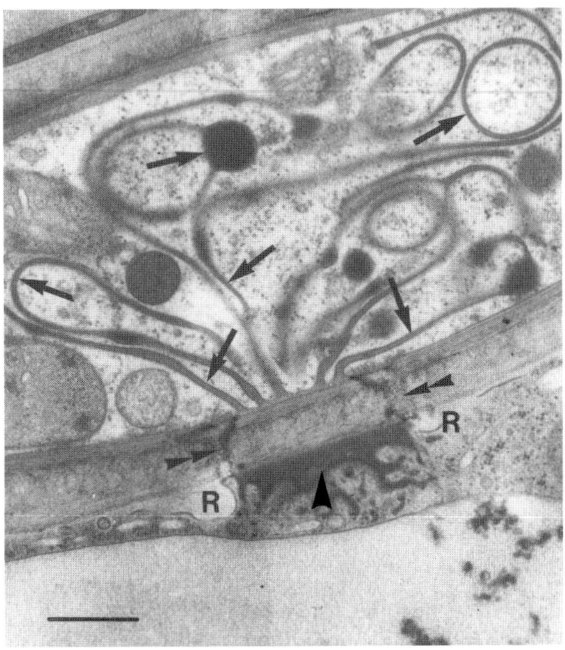

Fig. 6. Local interaction zone between *Exobasidium pachysporum* (*upper cell*) and its host (*lower cell*) showing the exocytose profile of a complex interaction apparatus (*arrows*) and the sectioned interaction ring (*double arrowheads*). Note the electron-opaque deposit at the host side (*arrow*). Initial host response to infection is visible at *R*. *Scale bar* 0.5 µm

The intercisternal space does not become integrated in the interaction apparatus. The transfer of fungal material towards the host plasma membrane occurs in two or three steps. The first transfer results in the deposition of a ring at the host plasma membrane. Depending upon the species, interaction zones of this type are located in intercellular hyphae or haustoria. This type of cellular interaction characterizes the Exobasidiales (Bauer et al. 1997).

5.2 Enlarged Interaction Zones

Continuous production and exocytosis of primary interactive vesicles results in the continuous deposition of fungal material at the whole contact area with the host cell. Depending upon the species, this type of interaction zone is located in intercellular hyphae (Fig. 2), intracellular hyphae or haustoria (Fig. 7). This type of cellular interaction characterizes the Ustilaginomycetidae (Bauer et al. 1997).

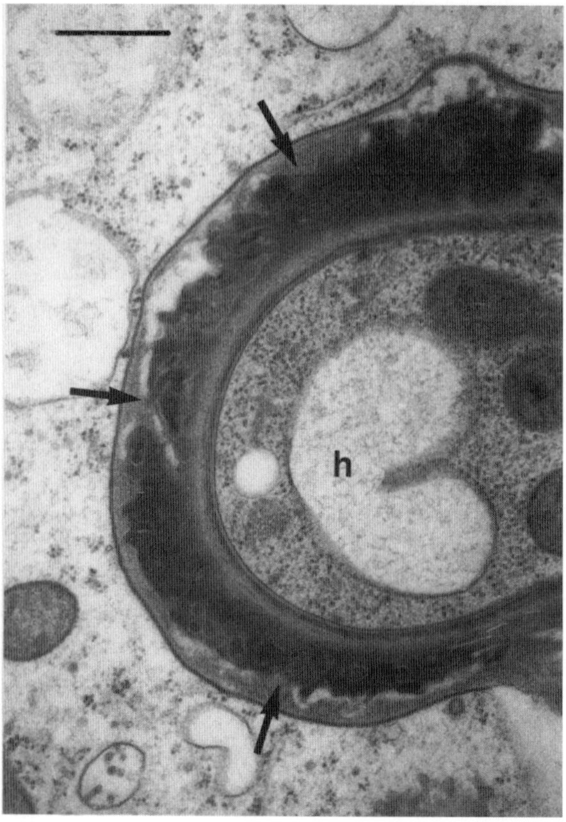

Fig. 7. Haustorial apex (*h*) of *Ustacystis waldsteiniae* encased by an electron-opaque vesicular matrix (*arrows*). *Scale bar* 0.5 µm

6 Conclusions

Similar development of the different interaction types occurring in the Ustilaginomycetes reveals that these interaction types are homologous to one another, thus reflecting variations of a common ancestral type. Accordingly, during the phylogenetic history the cellular interactions gradually specialized and optimized. An apomorphy for the Ustilaginomycetes is the presence of interaction zones with fungal deposits resulting from exocytosis of primary interactive vesicles. The contents of the primary interactive vesicles are transferred towards the host plasma membrane by different mechanisms in the various taxa. This parasitic process is unique among the basidiomycetes (e.g., see Littlefield and Heath 1979). Interestingly, a similar parasitic process may occur in the downy mildews (Hickey and Coffey 1977; Coffey and Wilson 1983; Wetherbee et al. 1985). The similarities include the presence of densely stained vesicles at the penetration region, the localized increase in the electron opacity of the host cell, and the deposition of electron-opaque material between host cell wall and host plasma membrane. Because of numerous fundamental differences between the downy mildews and the Ustilaginomycetes, these similarities must be interpreted as a result of convergent evolution.

The transfer of fungal material towards the host plasma membrane appears to be unusual and its function is basically unknown. Bauer et al. (1995) studied the cellular interaction of the ustilaginomycete *Ustacystis waldsteiniae* in detail and hypothesized the following scenario for this fungus: the transferred fungal material stabilizes and binds the associated host plasma membrane and, thus prevents on the one hand, membrane recycling via endocytosis. On the other hand, exocytosis of Golgi products of the host cell at this point results in the formation of coralloid vesicular buds extending into the fungal deposit. Finally, the vesicular buds separate from the host cytoplasm. Bauer et al. (1995) assumed that in this interaction scenario the following three characteristics are advantageous for the parasite: (1) the Golgi products extruded via exocytosis could serve as direct nutriment for the parasite, (2) the formation of the coralloid vesicular buds extending into the fungal deposits results in a greatly increased transfer-like host – parasite contact surface, and (3) the content of the vesicular buds could also serve as direct nutriment for the parasite.

Acknowledgements. We thank Uwe Simon for critically reading the manuscript, and the Deutsche Forschungsgemeinschaft for financial support.

References and Selected Reading

Bauer R, Mendgen K, Oberwinkler F (1995) Cellular interaction of the smut fungus *Ustacystis waldsteiniae*. Can J Bot 73:867–883

Bauer R, Oberwinkler F, Vánky K (1997) Ultrastructural markers and systematics in smut fungi and allied taxa. Can J Bot 75:1273–1314

Bauer R, Vánky K, Begerow D, Oberwinkler F (1999) Ustilaginomycetes on *Selaginella*. Mycologia 91:475–484

Begerow D, Bauer R, Oberwinkler F (1997) Phylogenetic studies on large subunit ribosomal DNA sequences of smut fungi and related taxa. Can J Bot 75:2045–2056

Coffey MC, Wilson UE (1983) An ultrastructural study of the late-blight fungus *Phytophthora infestans* and its interaction with the foliage of two potato cultivars possessing different levels of general (field) resistance. Can J Bot 61:2669–2685

Hickey EL, Coffey MD (1977) A fine-structural study of the pea downy mildew fungus *Peronospora pisi* in its host *Pisum sativum*. Can J Bot 55:2845–2858

Littlefield LJ, Heath MC (1979) Ultrastructure of rust fungi. Academic Press, New York

Luttrell ES (1987) Relations of hyphae to host cells in smut galls caused by species of *Tilletia*, *Tolyposporium*, and *Ustilago*. Can J Bot 65:2581–2591

Martinez C, Roux C, Dargent R (1999) Biotrophic development of *Sporisorium reilianum* f. sp. *zeae* in vegetative shoot apex of maize. Phytopathology 89:247–253

Mims CW (1982) Ultrastructure of the haustorial apparatus of *Exobasidium camelliae*. Mycologia 74:188–200

Mims CW (1991) Using electron microscopy to study plant pathogenic fungi. Mycologia 83:1–19

Mims CW, Nickerson NL (1986) Ultrastructure of the host-pathogen relationship in the red leaf disease of lowbush blueberry caused by the fungus *Exobasidium vaccinii*. Can J Bot 64:1338–1343

Mims CW, Snetselaar KM, Richardson EA (1992) Ultrastructure of the leaf stripe smut fungus *Ustilago striiformis*: host-pathogen relationship and teliospore development. Int J Plant Sci. 153:289–290

Nagler A, Oberwinkler F (1989) Haustoria in *Urocystis* (Tilletiales). Plant Syst Evol 165:17–28

Nagler A, Bauer R, Oberwinkler F, Tschen J (1990) Basidial development, spindle pole body, septal pore, and host-parasite-interaction in *Ustilago esculenta*. Nordic J Bot 10:457–464

Sampson K (1939) Life cycles of smut fungi. Trans Br Mycolog Soc 23:1–23

Snetselaar KM, Tiffany LH (1990) Light and electron microscopy of sorus development in *Sorosporium provinciale*, a smut of big bluestem. Mycologia 82:480–492

Snetselaar KM, Mims CW (1994) Light and electron microscopy of *Ustilago maydis* hyphae in maize. Mycolog Res 98:347–355

Thomas PL (1989) Barley smuts in the prairie provinces of Canada, 1983–1988. Can J Phytopathol 11:133–136

Trione EJ (1982) Dwarf bunt of wheat and its importance in international wheat trade. Plant Dis 66:1083–1088

Wetherbee R, Hinch JM, Clarke AE (1985) Response of *Zea mays* roots to infection with *Phytophthora cinnamomi* II. The cortex and stele. Protoplasma 126:188–197

15 Interaction of *Piriformospora indica* with Diverse Microorganisms and Plants

Giang Huong Pham, Anjana Singh, Rajani Malla, Rina Kumari, Ram Prasad, Minu Sachdev, Karl-Heinz Rexer, Gerhard Kost, Patricia Luis, Michael Kaldorf, François Buscot, Sylvie Herrmann, Tanja Peskan, Ralf Oelmüller, Anil Kumar Saxena, Stephané Declerck, Maria Mittag, Edith Stabentheiner, Solveig Hehl, and Ajit Varma

1 Introduction

An axenically cultivable Mycorrhiza-like-fungus has been described by Varma and his collaborators. The fungus was named *Piriformospora indica* based on its characteristic pear-shaped chlamydospores (Verma et al. 1998). *P. indica* tremendously improves the growth and overall biomass production of diverse hosts, including legumes (Varma et al. 1999, 2001; Singh et al. 2002a), medicinal and other plants of economic importance (Rai et al. 2001; Singh et al. 2003a, b). Interestingly, the host spectrum of *P. indica* is very much like arbuscular mycorrhizal fungi (AMF). In addition, a pronounced growth promotional effect was seen with terrestrial orchids (Blechert et al. 1999; Singh and Varma 2000; Singh et al. 2000, 2002b). The fungus also provides protection when inoculated into the tissue culture-raised plantlets by overcoming the 'transient transplant shock' on transfer to the field and renders almost 100 % survival (Sahay and Varma 1999, 2000). The fungus has great potential in forestry, horticulture, agriculture, viticulture and especially for better establishment of tissue culture-raised plants much needed in the plant industry (Singh et al. 2003). This would open up numerous opportunities for the optimization of plant productivity in both managed and natural ecosystems, while minimizing the risk of environmental damage. The properties of the fungus, *Piriformospora indica*, have been patented (Varma and Franken 1997, European Patent Office, Muenchen, Germany. Patent No. 97121440.8–2105, Nov. 1998). The culture has been deposited at Braunschweig, Germany (DMS No.11827). An 18S rDNA fragment was deposited at EMBL under the accession number AF 014929.

Plant Surface Microbiology
A. Varma, L. Abbott, D. Werner, R. Hampp (Eds.)

The fungus forms inter- and intracellular hyphae in the root cortex, often differentiating into dense hyphal coils and chlamydospores. Like AM fungi, hyphae multiply within the host cortical tissues and never traverse through the endodermis. Likewise, they also do not invade the aerial portion of the plant (stem and leaves).

This chapter details the interaction of *P. indica* with various groups of microorganisms and higher plants.

2 Interaction with Microorganisms

2.1 Rhizobacteria

Piriformospora indica and the respective bacteria *Pseudomonas fluorescence* and *Azotobacter chroococcum* were placed on defined modified Aspergillus medium (see Chap. 30). After 7-day incubation at 25 °C, it was found that *Ps. fluorescence* completely blocked the growth of the fungus. *P. indica* acquired immense, but tiny chlamydospores, perhaps to overcome the stress (Fig. 1). Plausible reasons for the inhibition could be the production of ammonia, HCN, siderophores, antibiotics or chitinase. In contrast, *Az. chroococcum* promoted the growth of the fungus which produced extensive mycelium with low and delayed sporulation. The strains of *Pseudomonas* sp. and *Ps. putrida* also

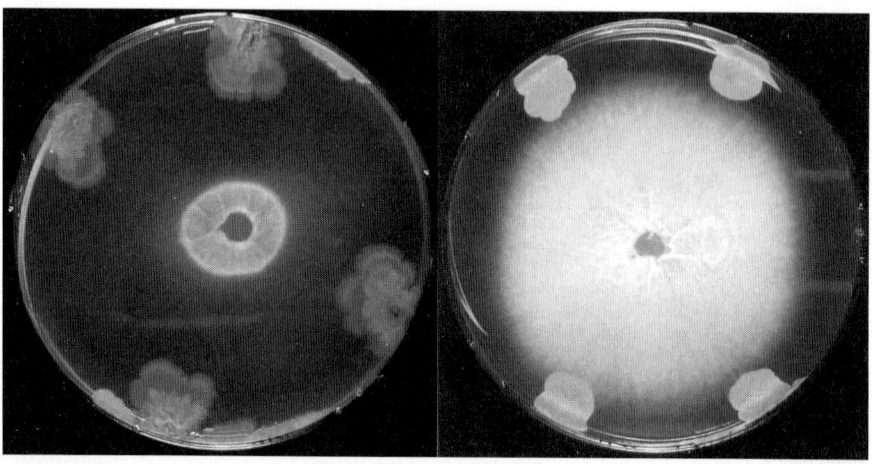

Fig. 1. Interactions with *Pseudomonas fluorescens* (*left*) and *Bradyrhizobium* sp. (*right*). *P. indica* was grown in the center of plates with modified *Aspergillus* medium for 48 h. Then freshly grown (early log phase) bacteria were inoculated four times at an equal distance close to the margin of the plate. Incubation was done at 25±2 °C. Photographed after 5 days. The growth of *P. indica* was strongly suppressed by *Ps. fluorescens* and promoted by *Bradyrhizobium* sp.

initially blocked the growth of *P. indica*, but the fungus rapidly recovered. *Bacillus subtilis* had neutral effects, whereas strains of *Azospirillum* and *Bradyrhizobium* promoted the growth of the fungus.

2.2 *Chlamydomonas reinhardtii*

Ch. reinhardtii and *P. indica* were allowed to grow on MMN 1/10 medium. The alga was inoculated as a streak either only on one side or on both sides of a fungal disc. Both microorganisms grew well in both experiments, but the growth of the alga was more intense and the color of the colony was a much darker green if there was only one streak (Fig. 2). At this stage, to our knowledge, no suitable explanation can be offered for these phenomena.

2.3 *Sebacina vermifera*

One disc each of *P. indica* and *S. vermifera* were placed on Aspergillus medium. The distance between the two inocula was 4 cm. Both fungi grew normally without inhibiting the growth of each other. The most interesting part was that after 7 days at the intersection of two colonies, hyphae turned highly intertwined, inflated and produced a large number of chlamydospores. Therefore, both strains were able to block each other with a typical deadlock

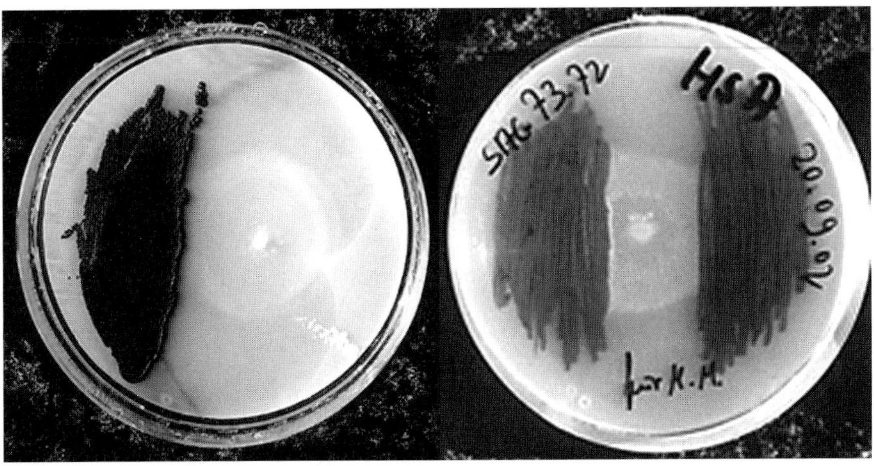

Fig. 2. Interaction with *Chlamydomonas reinhardtii. P. indica* was grown on MS medium for 48 h. Thereafter, the green alga was streaked on one side (*left*) or on both sides (*right*) of the mycelium. Incubation was carried out under 52 µmol/m light and 24±2 °C temperature. *Left* Dark green strongly grown algal colonies

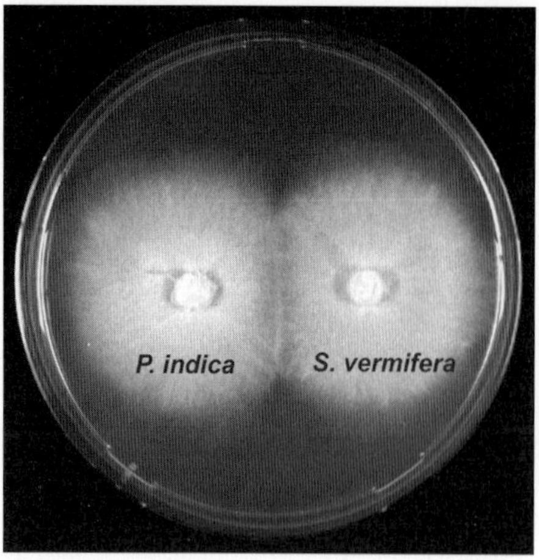

Fig. 3. Interaction with *Sebacina vermifera*. The fungal inocula were placed on *Aspergillus* medium about 3 cm apart and incubated for 5 days. The mycelia formed a sharp demarcation line where they touched

reaction. No evidence for the production of basidia and basidiospores was recorded (Fig. 3).

2.4 Other Soil Fungi

Several commonly occurring soil fungi were tested for the interaction with *P. indica*. The results were highly diverse (Fig. 4). The growth of *Aspergillus sydowii*, *Rhizopus stolonifer*, and *Aspergillus niger* was completely blocked by *P. indica*. The growth of *Cunninghamella echinulata* was reduced, whereas *Rhizopus oryzae*, *As. flavus* and *Aspergillus* sp. completely blocked the growth of *P. indica*. The data indicated that *P. indica* divulges a wide range of interaction types with diverse soil fungi.

2.5 *Gaeumannomyces graminis*

In his pioneering work, Dehne (1982) was able to show that AM fungi are able to reduce soil-borne diseases and/or the severity of diseases caused by root pathogens. *P. indica* was challenged with a virulent root and seed pathogen *G. graminis* (Fig. 5). In a confrontational experiment, initially the mycelia were not able to overcome each other, resulting in sharp borderlines between the colonies. After prolonged incubation, *P. indica* started to invade into the area of *G. gramins* and caused a lysis of the root pathogen hyphae.

In another experiment, when the *P. indica* was allowed to grow earlier and the pathogen was inoculated later in the center of the solidified Aspergillus medium, the pathogen growth was completely blocked. A culture filtrate of *P. indica* also completely inhibited the growth of the pathogen. These experi-

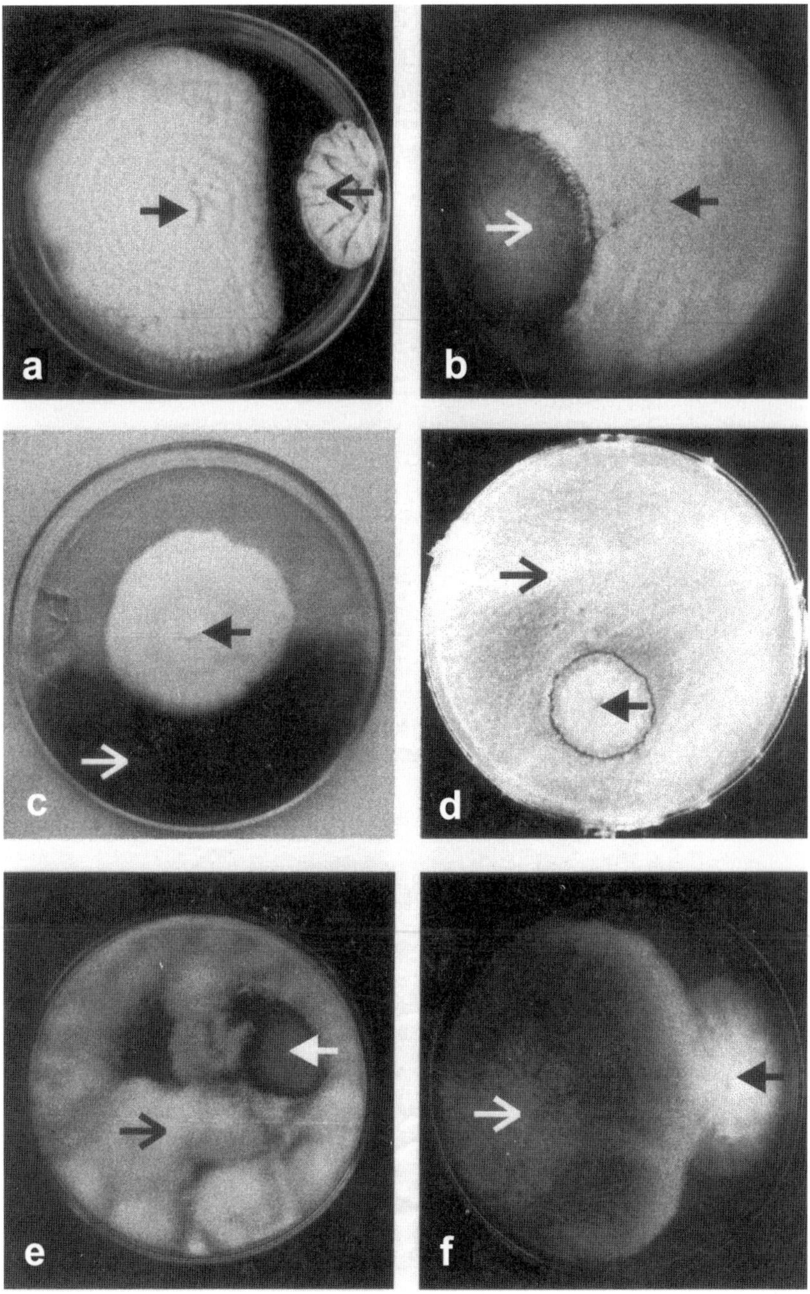

Fig. 4a–f. Interactions with soil fungi. One disc of *P. indica* and a soil fungus each were placed on the solidified *Aspergillus* medium. *Closed arrows* indicate where *P. indica* was inoculated, *open arrows* indicate the inoculation placement of the respective fungus. **a** *Aspergillus sydowii*, **b** *Rhizopus stolonifer*, **c** *Aspergillus niger*, **d** *Rhizopus oryzae*, **e** *Aspergillus flavus*, **f** *Aspergillus* sp. *P. indica* strongly suppressed the growth of some fungi studied (**a–c**) or was itself suppressed (**f**)

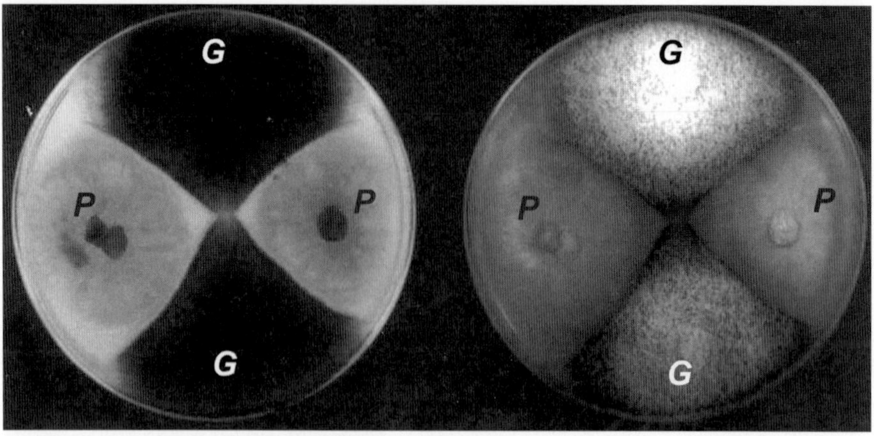

Fig. 5. Interaction with *Gaeumannomyces graminis* (*G*). Two discs of *P. indica* (*P*) and the root pathogen were each placed at equal distances. Incubation was done for 7 days on *Aspergillus* medium in the dark at 25±2 °C. *Right* Top view of the mycelia, *left* bottom view of the mycelia shining through the agar. The mycelia formed a sharp demarcation line where they made contact; after prolonged incubation, *P. indica* invaded the hyphal mat of *G. graminis*

ments demonstrated that *P. indica* is able to act as a potential agent for biological control of root diseases; however, the chemical nature of the inhibitory factor is still unknown.

3 Interaction with Bryophyte

To test the ability of *P. indica* to interact with different kinds of moss and liverwort, the plants were first grown in axenic culture. In co-culture with the fungus, *Eurhynchium praelongum* and *Cephalozia bicuspidata* were weakly colonized without causing severe symptoms to the gametophyte. In *Riccardia incurvata* heavy colonization took place, but the growth promotional effect was hardly significant. In this liverwort the interaction was intense and the fungus entered deeply into the thallus. In a further study, the interaction found in this species will be compared to the interaction found in *Aneura pinguis*, a liverwort of the same family Aneuraceae.

4 Interaction with Higher Plants

A large number of diverse higher plants (mono- and dicots) interacted with *P. indica* (Table 1). This included terrestrial, annual and perennial herbs, and woody plants. Interestingly, *P. indica* mimics a number of symbiotic proper-

Table 1. Host spectrum tested for *P. indica*

Hosts
Abrus precatorius L.
Acacia catechu (L.) Willd.
A. nilotica (L.) Del.
Adhatoda vasica Nees
Aneura pinguis (L.) Dumort.
Arabidopsis thaliana (L.) Heynh.
Artemisia annua L.
Azadirachta indica A. Juss.
Bacopa monnieria (L.) Wett.
Cassia angustifolia Vahl
Chlorophytum borivilianum Santapau & R.R. Fernandez
Ch. tuberosum Baker
Cicer arietinum L.
Coffea arabica L.
Cymbopogon martini (Roxb.) W. Wats.
Dactylorhiza fuchsii (Druce) Soo'
D. incarnata (L.) Soo'
D. maculata (L.) Verm.
D majalis (Rchb.) P.F. Hunt & Summerh.
D. purpurella (Steph's) Soo'
Daucus carota L.
Delbergia sissoo Roxb.
Glycine max (L.) Merr.
Lycopersicon esculentum Mill.
Nicotiana attenuata Torr. ex S. Wats L.
N. tabaccum L.
Oryza sativa L.
Petroselinum crispum (Mill.) A. W. Hill
Pisum sativum L.
Populus tremula L.
P. tremuloides Michx. (clone Esch5)
Prosopis chilensis (Mol.) Stuntz
P. juliflora (Sw.) DC.
Quercus robur L. (clone oak DF 159)
Setaria italica (L.) P. Beauv.
Solanum melongena L.
Sorghum vulgare Pers.
Spilanthes calva DC.
Tagetes erecta L.
Tectona grandis L.
Terminalia arjuna Wight & Arn.
Tephrosia purpurea (L.) Pers.
Vigna mungo (L.) Hepper
V. radiata (L.) R. Wilczek
Withania somnifera (L.) Dunal
Zea mays L.
Zizyphus nummularia Burm. fil.

Data are based on the root colonization analysis in vivo and in vitro (cf. Varma et al. 2001; Singh et al. 2003a, b)

Fig. 6. Interactions with *Zea mays* and *Setaria italica*. The substratum was sterilized and filled into pots (1 kg). Fungal inoculum (1 % w/v) was thoroughly mixed with the soil. Plants were irrigated with tap water on alternate days to maintain about 70 % soil moisture. They were grown under greenhouse conditions maintained at 25±2 °C, 16 h light/8 h dark with fluorescent light intensity 1000 lux and relative humidity 70 %. *P. indica* promoted the growth of both monocots

ties, which are characteristics of AM fungi (Singh et al. 2002a, b; Singh et al. 2003a, b; Varma et al. 2001). It colonizes the root cortex in a variety of host plants and improves their overall biomass production.

4.1 Monocots

Recently, we have selected the model plants, *Zea mays* L. and *Setaria italica* (L.) Beauv. for in-depth studies. The roots were colonized and the growth of the plants was highly promoted as a result of interaction with the fungus (Fig. 6). The phytopromotional influence was evident from early stages of the interaction.

4.2 Legumes

P. indica promotes the growth and survival of tissue culture raised-tropical legumes like *Cicer arietinum* L., *Vigna radiata* (L.) R. Wilczek, *Pisum sativum* L., *Vigna mungo* L. Hepper (Fig. 7) and *Glycine max* (L.) Merr. The fungal colonization resulted in 100 % survival of the in vitro raised plants, whereas it was less than 50 % in uninoculated plants.

A dramatic increase in the plant growth was observed in *C. arietinum* and *V. mungo* as compared to their corresponding controls. The percent increase in plant height was 35.7 and 14.2 %, respectively, and the increase in fresh weight was 90 and 11 %, respectively, as compared to corresponding controls.

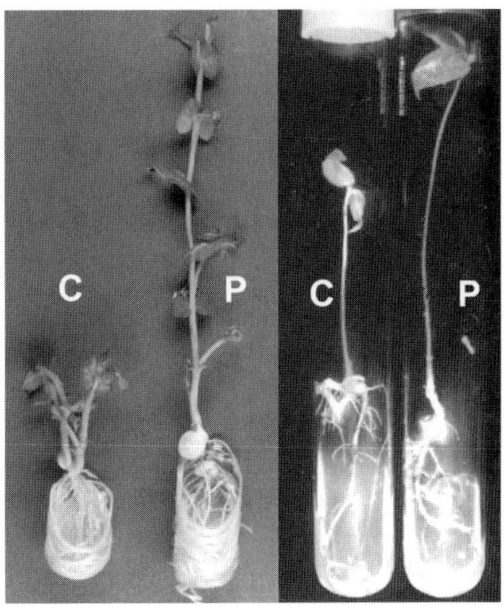

Fig. 7. Interactions with *Pisum sativum* (*left*) and *Vigna mungo* (*right*). Surface-sterilized seeds of the legumes were germinated on water agar. Approximately 3-cm-long young seedlings were placed on MS agar slants and incubated with *P. indica* (*P*). Control plants (*C*) did not receive any fungus. Tubes were incubated at 25±2 °C and 1000 lux. Photographs were taken after 6 days. The fungus promoted the growth of both legumes

Treated roots were colonized by the fungus and produced extramatrical, inter- and intracellular hyphae. Chlamydospores were observed at maturity.

4.3 Orchids

Seeds of *Dactylorhiza purpurella* (Steph's.) Soó and *D. majalis* (Rchb. F.) Hunt and Summerh. were surface-sterilized and inoculated with *P. indica* (Fig. 8). After 2 weeks, seeds of *D. purpurella* started germination. After the appear-

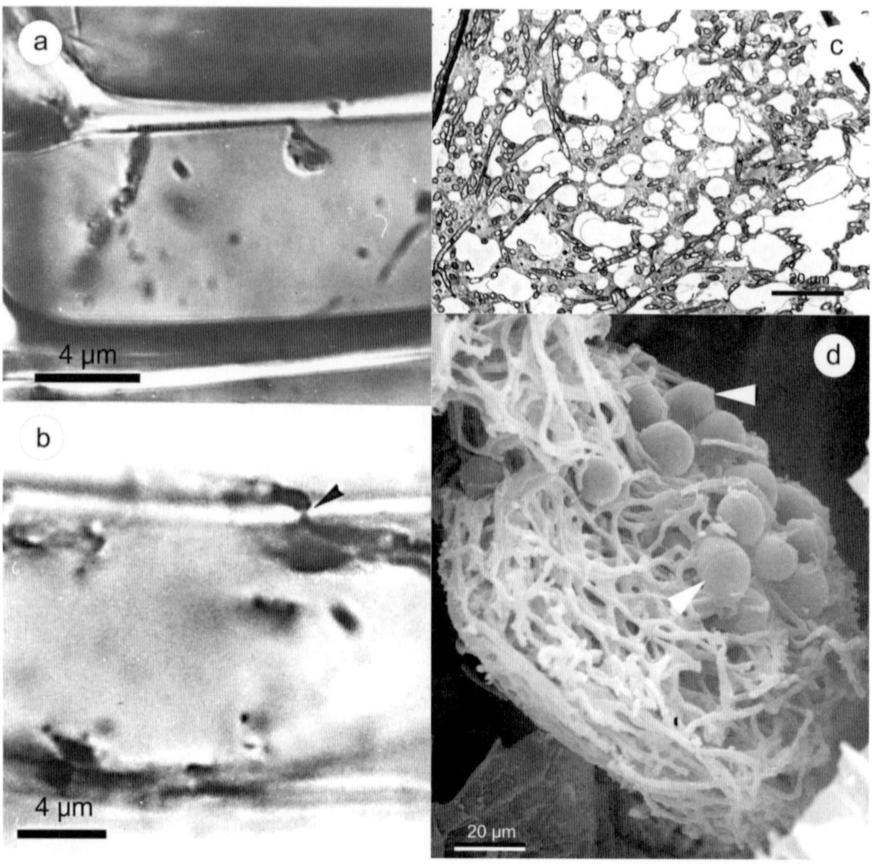

Fig. 8a–d. Interaction with *Dactylorhiza majalis*. Seeds of the orchid were surface-sterilized and germinated on oat agar. When some of the seeds started to swell, *P. indica* was added. The plates were incubated in the dark at room temperature. **a** Hyphae penetrating into the protocorm testa without traversing into the epidermis. **b** Hypha penetrating into a rhizoid (*arrowhead*), growing towards the protocorm and then entering into the cortex. **c** Semithin section of a peloton formed in a living cortical cell. **d** SEM picture of a peloton formed in a cortical cell, *arrowhead* pointing at starch grana (Blechert et al. 1999; Varma et al. 2001)

ance of the rhizoids the fungus penetrated into them, growing towards the protocorm. Inter- and intracellular hyphae spread from the basal swelling of the rhizoids and typical pelotons were formed in living cortical cells. Digestion of the pelotons started 17 days after inoculation. Differences in the growth of the protocorms inoculated with *P. indica* and the corresponding controls were obvious after the intracellular interactions began. *P. indica* was found to be a typical orchid mycorrhizal fungus in vitro, promoting growth in all tested *Dactylorhiza* species (Blechert et al.1999; Singh et al. 2000, 2002b; Varma et al. 2001).

4.4 Medicinal Plants

The tissue culture-raised plants and seedlings from surface-sterilized seeds of medicinal plants, like *Spilanthes calva* DC, *Withania somnifera* (L.) Dunal, *Bacopa monnieria* (L.) Wett., *Adhatoda vasica* L., *Azadirachta indica* A.Juss. (neem), *Artemisia annua* L., *Chlorophytum tuberosum* Baker, *C. borivilianum* Santapau & R. R. Fernandez (musli), and *Termnalia arjuna* L. were inoculated with *P. indica* in mist chambers and nurseries before being transferred to the field (Fig. 9).

Significant increases in growth and yield of the plant species were recorded relative to uninoculated controls. Shoot and root length, biomass, basal stem, leaf area, overall size, inflorescence number, flower and seed production were all enhanced in the presence of the fungus. Net primary productivity was also higher than in control plants. The results clearly indicate the commercial potential of *P. indica* for large-scale cultivation of medicinal plants. The differences in growth observed may have been caused by a greater absorption of water and mineral nutrients due to extensive colonization of roots and the proliferation of the mycelium into the soil.

In another pot trial experiment, neem seedlings were inoculated with *Glomus mosseae*, *Scutellospora gilmorei*, and *P. indica*. The treatment was conducted using pots with sterile and natural soils. Plant growth of *P. indica*-treated plants was found to be drastically improved compared to those plants treated with AM fungi and controls. *P. indica*-treated plants attained maximum height, healthier foliage and a well developed subterranean.

Bacopa monnieria (L.) Wett. is considered to be important because the whole plant has medicinal value. *P. indica* colonizes the roots of tissue culture-raised plants and promoted the overall plant biomass. A biological hardening rendered almost 100 % survival on transfer from the laboratory to the field.

S. calva and *W. somnifera* were treated with *P. indica* in a field trial. A pronounced growth response following the *P. indica* inoculation was observed. The basal stem and leaf areas of treated plants were enhanced. Interestingly, large kidney-shaped inflorescences were observed on inoculated *S. calva*

Fig. 9a–f. Interactions with medical plants. *P. indica* was used for the inoculation of young seedlings of **a** *Adhatoda vasica*, **b** *Azadirachta indica*, **c** *Terminalia arjuna*, **d** *Spilanthus calva*, **e** *Withania somnifera* and **f** *Chlorophytum borivillianum* growing in pots. After the establishment of the interaction, the latter three plant species were tranferred to the field where the pictures were taken. In **a–c**, **e**, the control is on the *left*, inoculated plants are on the *right*. In all experiments growth and flowering of the treated plants were obviously promoted

plants, however, these kidney-shaped inflorescences were never observed in control plants. The length of the inflorescence and the number of flowers on inoculated *S. calva* plants also increased compared to the controls. Similarly, the number of flowers of the inoculated *W. somnifera* was higher than in the controls. For both inoculated medicinal plant species, the number of seeds was higher than for the controls (Rai et al. 2001), and the overall root biomass of the inoculated plants was higher than that of the corresponding controls. The fresh and dry weights of both underground and aboveground parts of *S. calva* and *W. somnifera*-inoculated plants were higher than in the controls.

4.5 Economically Important Plants

Plants of economic importance tested in vivo and in vitro were *Tagetes erecta* L. (marigold), *Nicotiana tabaccum* L. (tobacco), *Lycopersicon esculentum* Mill. (tomato) and *Solanum melongena* L. (bringal). In a pot trial marigold inoculated with *P. indica* showed healthier plants with early bud formation and enlarged flowers compared to the control (Fig. 10).

Hypocotyl of germinated seeds of tobacco were taken as an explant for callus development. Callusing regeneration of the shoot was established on MS medium (Murashige and Skoog 1962). Biological hardening of the regenerated plantlets with *P. indica* recorded the maximum capacity for regaining the tensile strength of the stem (Fig. 11). Plants treated with *G. mosseae* possessed less tensile strength, but more than the control plantlets.

Early root induction was recorded in the bringal root organ culture interacting with *P. indica* (Tables 2, 3). This study indicated that *P. indica* is a potent

Fig. 10. Interaction with *Tagetes erecta*. The experiment was conducted as described in Fig. 6. *P. indica*-inoculated plants (*P*) were extensive in growth and had bigger flowers compared to the control (*C*)

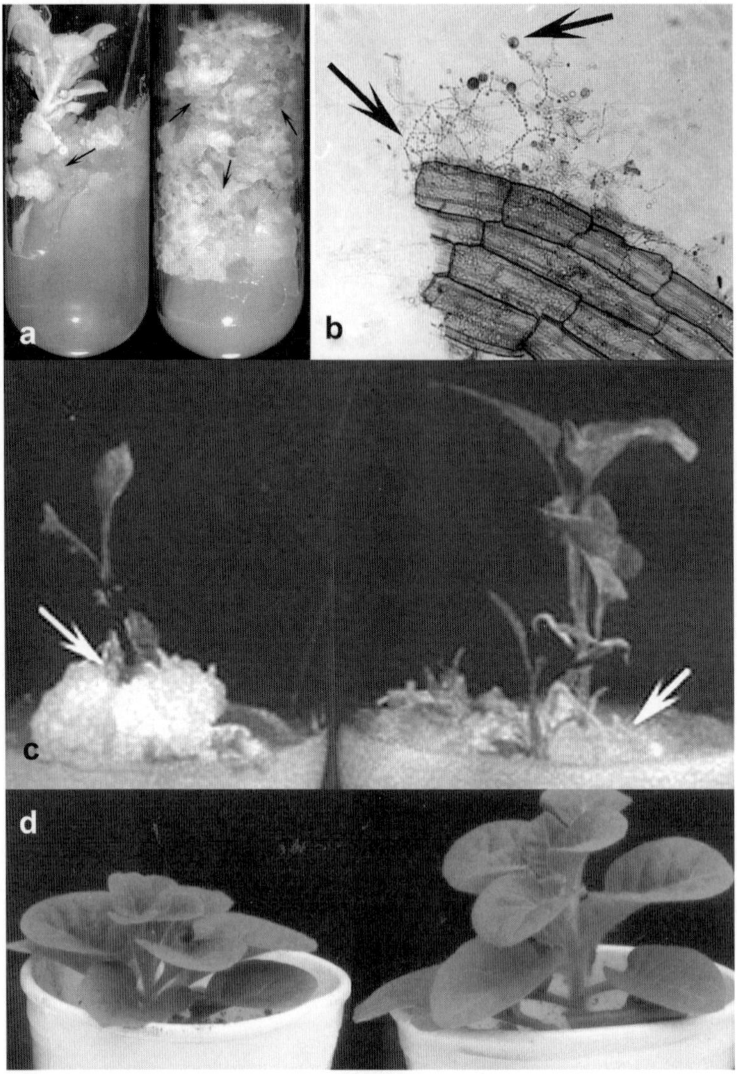

Fig. 11a–d. Interaction with tissue culture-raised *Nicotiana tabacum*. Surface-sterilized seeds of tobacco were germinated on 1/2 strength MS medium. Fifteen days after germination, hypocotyl was transferred for callus formation on MS media supplemented with NAA 2 mg/l and BAP 0.5 mg/l. Cultures were grown in a controlled tissue culture laboratory. Mycelia of *P. indica* were grown in culture bottles on minimal medium. Regenerated shoots were transferred to these bottles. Observations were made after 15 days of treatment; root fragments were stained with Trypan blue. After 3 weeks the plants were transferred to pots with sterile substratum and grown in a mist chamber. **a** Massive callus formation on 1/2 MS medium of the *P. indica*-treated plants (*right*) compared to the control (*left*). **b** Root fragment of inoculated tobacco plantlet colonized by the fungus. **c** Differentiation of the callus cultures after 15 days of inoculation on regeneration medium; *left* control, *right* treated plantlet. **d** Tobacco plants after 8 weeks in a mist chamber, *left* control, *right* treated plant

plant growth-promoting fungus. It is not only the mycelium in association with the root which exerts this effect, fungus culture filtrate containing fungal exudates (these may be hormones, proteins, enzymes, polyamines, amino acids, etc.) also exhibited almost the same effect.

P. indica-inoculated tomato seedlings grown in glass tubes on MS medium were about three times higher than the control and proliferate root biomass was observed. Soon after inoculation adventitious roots appeared high above the surface of the substrate. As a result of prolonged incubation, the subapical region of the shoot became chlorotic and finally the leaves wilted, probably due to the restricted nutrition. However, adventitious roots started developing again from the apex, thus maintaining a green tip.

Table 2. Morphological features of tobacco plantlets after treatment with different mycobionts (biological hardening)

Mycobiont	2 Weeks	4 Weeks
Control	Leaves lost turgor pressure and stem the tensile strength	Some leaves turned yellow, stem turned brown
G. mosseae	Leaves lost turgor pressure and stem the tensile strength	Some leaves dried, stem regained tensile strength
P. indica	Leaves lost turgor pressure and stem the tensile strength	Plants were healthy, stem regained tensile strength

cf. Sahay and Varma (1999)
Morphological features of micropropagated tobacco plantlets subjected to biological hardening with *Glomus mosseae* and *Piriformospora indica*. Soil substrata were sterilized soil:sand mixture (3:1). Inocula (spores, hyphae, colonized root pieces, etc.) were included at 1 % (w/v) to each pot (10×6 cm)

Table 3. Plant biomass and percent-colonization as a result of interaction of the plantlets with mycobionts (biological hardening)

Mycobionts	Fresh weight (g/plant)	Colonization (%)
Control	2.18±0.199	nd
Glomus mosseae	2.69±0.145	64±15.16
Piriformospora indica	3.08±0.266	76±25.10

Plants were harvested and the total fresh weight was recorded. Fungal root colonization was estimated after staining with Trypan blue. RM ANOVA ON RANKS test shows $\chi^2=8.40$ with 3 degrees of freedom. $P_{(est)}=0.0384$, $P_{(exact)}=0.0190$. The differences in the median value among the treatment groups are greater than would be expected by chance, i.e., there is a statistically significant difference ($P=0.0190$). Data represent mean ±SD, nd, not detected

4.6 Timber Plants

Young seedlings of *Populus tremula* L., *Quercus robur* L. and *Dalbergia sissoo* Roxb. ex DC. were tested in the green house and followed by field trials. In a pot culture experiment, *P. indica*-inoculated plants of *Po. tremula* were strongly promoted according to their biomass production compared to the controls (Fig. 12). *Qu. robur* (clone oak DF 159) was micro-propagated and rooted as described by Herrmann et al. (1998). *P. indica* was pre-cultivated for 7 days on *Aspergillus* medium (Kaefer 1977). Co-culture was performed on MMN 1/10 medium (Marx 1969) in Petri dishes. The oak shoots grew out of the dish and to avoid rapid wilting, the inoculated dishes were grown in large Petri dishes (radius of 145 mm) with moistened absorbent paper sheets. Co-cultivation was carried out at 25 °C and a photoperiod of 16 h illumination (97 W m^{-2}; OSRAM L 115 W/20SA cool white) for 10 weeks. At the end of the experiment, the roots were colonized by the fungus (Fig. 12).

Precedent results (Herrmann et al. 1998) showed that an ectomycorrhizal fungus *Piloderma croceum* Erikss. & Hjortst. was able to enhance plant growth of oaks before any mycorrhizal formation occurred. Hence, *P. indica* and *P. croceum* were able to enhance root development. In further investigations it would be of interest to compare the effects of both fungi separately and in combination on root initiation and elongation, and analyse in which

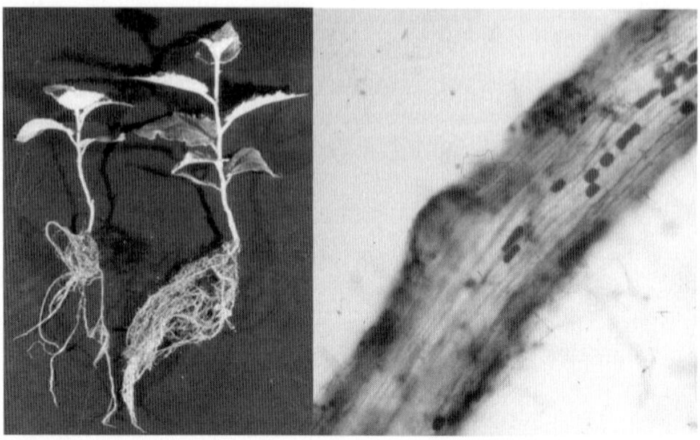

Fig. 12. Interactions with ectomycorrhizal plants. *Left* Seedlings of *Populus tremula* (obtained from Köln, Germany) were treated with *P. indica*. The experimental design was the same as described in Fig. 6. The larger plant was treated with the fungus, the smaller plant without fungus. *Right Quercus robur* (clone oak DF 159) seedlings were treated with *P. indica* under laboratory conditions. It was micro-propagated and rooted as described by Herrmann et al. (1998). *P. indica* was pre-cultivated for 7 days on *Aspergillus* medium. Co-culture was performed on 1/10 MMN medium in Petri dishes. Co-cultivation was conducted at 25 °C and a photoperiod of 16 h illumination (97 W/m) over a duration of 10 weeks. Picture shows hyphae and spores formed within and around the roots

manner diffusing substances may be involved in the phenomena. A similar positive response was recorded for *P. tremula* and *D. sissoo.*

4.7 Unexpected Interactions with Wild-Type and Genetically Modified Populus Plants

Hybrid aspen *Populus tremula* x *P. tremuloides* Michx. (clone Esch5, kindly supplied by Dr. M. Fladung, Federal Research Centre for Forestry and Forest Products, Grosshansdorf, Germany) were cultivated in vitro at 25 °C in permanent light. Four plants were grown in each Magent GA7 vessel containing 80 ml of WPM medium (see Chap. 30). When 2-cm cuttings of the shoot including the tip were transferred to fresh medium, rooting was initiated at the cutting site in most transplants after 5–6 days. When *Populus* was inoculated with *P. indica* after rooting of transplants had commenced, stimulation in the root growth was observed after 5 days. A clear stimulation of root branching and an apparent increase in root length were observed (Fig. 13). This result confirms the observations made for other plants. No significant changes were observed in plant shoot growth.

In a second experiment, the fungus was allowed to grow in the WPM medium 1 week prior to the *Populus* transplants. Interestingly, plant growth and rooting pattern changed under these conditions. The salient changes recorded were:

- inhibition of root formation at the cutting site where rooting normally occurs without inoculation with fungus.
- aerial root formation was induced.
- deformations occurred in aerial roots when they came into contact with the surface of the fungus-inoculated medium, and they failed to grow into the medium.

Shoot growth of inoculated plants was suppressed compared to the control. After 6 weeks of cultivation, a profuse fungal mat appeared on the surface of the medium, however, plants were not killed. Observed under the light microscope, fungal infection was not detected in aerial roots.

We speculated that one or more chemical compounds were produced by the mycelium which were responsible for the changes mentioned above. To prepare a crude extract, plant material was removed from the cultures. The remaining medium (with or without fungus) was autoclaved for 30 min at 121 °C. These extracts were mixed with the same volume of double strength WPM medium and filled into culture vessels. After solidification, four transplants without roots were transferred into these media. After 5 days in the medium prepared with the plant extract, the rooting was initiated at the cutting site. In contrast, the plants incubated with the extract prepared from both types of media did not form any roots (neither in the medium nor in the air).

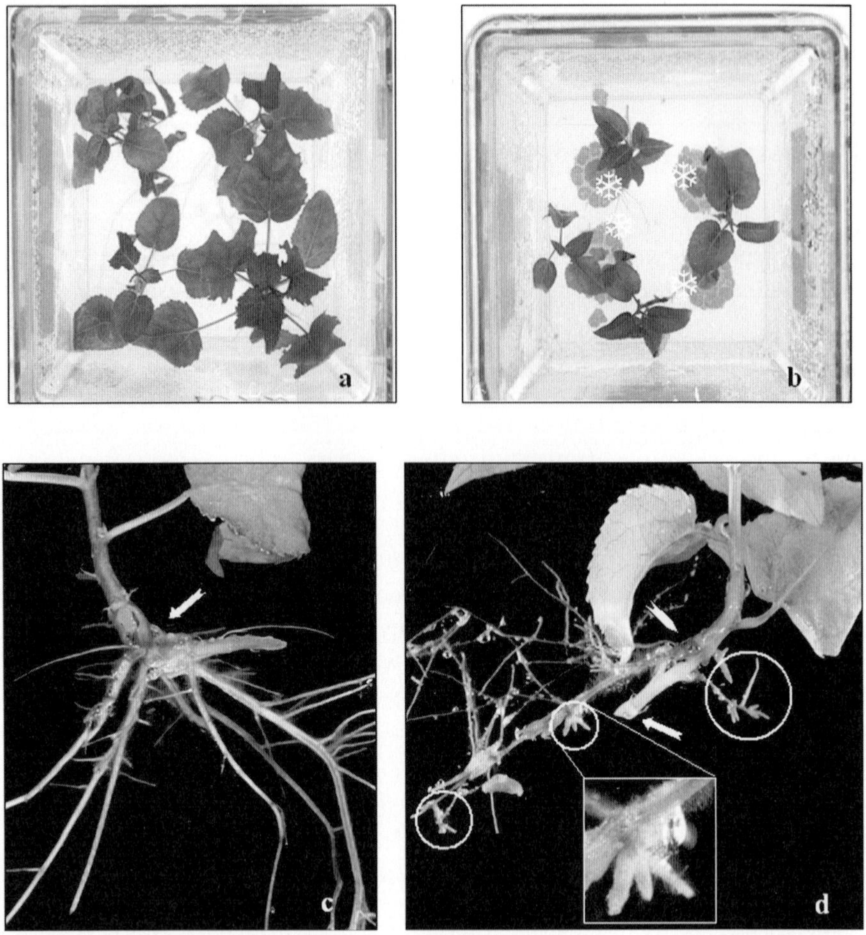

Fig. 13a–d. Interaction with in vitro grown *Populus* clone Ech5 on WPM. Explants with a length of 2 cm, including one terminal and 5–6 side buds were transferred to Magenta GA7 vessels containing 80 ml of WPM medium. **a** Top view of a noninoculated control after 4 weeks of cultivation at 25 °C in constant light; **b** four discs of *P. indica* culture (*snowflake*) were placed onto the medium and incubated for 7 days at 25 °C prior to the introduction of the plant cuttings; **c** normal rooting started at the cutting site (*arrow*) after 1 week. The photo shows the typical rooting pattern after 4 weeks; **d** rooting pattern of an inoculated plantlet after 4 weeks of co-cultivation with *P. indica*; rooting at the cutting site was completely blocked (*arrow*). Instead, aerial rooting appeared above the medium, ca. 1 cm apart from the cutting site (*arrow*). Root tips, when in contact with the medium, showed a modified morphology (see *circles*). *Inset* shows a magnified view of the modified root tips

Water extracts of fungus and plant-grown medium prepared at room temperature had no influence on rooting.

4.8 Non Mycorrhizal Plants

Most species of the plants are normally infected by mycorrhizal fungi, but some plant taxa do not usually form generally recognizable mycorrhizas (Tester et al. 1987). Members of, e.g., Brassicaceae, Chenopodiaceae and Amaranthaceae (Read 1999; Singh et al. 2003a, b; Varma 1998, 1999; Varma et al. 2001), belong to this exceptional group of nonmycorrhizal plants. The mechanism which determines the nonhost nature of plant species preventing the establishment of a functional symbiosis is not known. Present knowledge of the sequence of fungal development leading to the establishment of functional mycorrhiza suggests that the nonhost nature of plants lies in their inability to trigger expression of fungal genes involved in hyphal commitment to the symbiotic status.

It would be useful to assess these plants with respect to their interaction with *P. indica*. In vitro studies with *P. indica* and *S. vermifera* recorded that these two symbiotic fungi profusely interacted with the root system of the crucifer plants, *Brassica juncea* (L.) Czern. et Coss. (mustard), *Brassica oleracea* var. *capitata* L. (cabbage), and the Chenopodiaceae *Spinacia oleracea* L. (spinach). Although some of these plants were said to be able to form AM (Tester et al. 1987), no mycorrhizal interactions with *Glomus mosseae* were found in the pot trials we conducted. Instead, all the plants inoculated with *P. indica* were colonized by the fungus and recorded phytopromotional effects in comparison to the control. However, a high degree of variation was recorded in with respect to biomass and their length. Cabbage responded most positive with *P. indica*. Different results were recorded in root systems. In cabbage, the fungi profusely colonized inter- and intracellularly the root cortex cells. Colonization in mustard was less in comparison to cabbage and followed by spinach.

Further experiments indicated that *P. indica* did not invade the root of myc$^-$ mutants of pea (*Pisum sativum* L.) and soyabean (*Glycine max* (L.) Merr.). When the fungus was confronted with these mutants, plant growth was suppressed and the fungal morphology was severely affected. The sporulation of P. *indica* interacting with wild types was homogenous, while it was heterogeneous in myc$^-$ mutants (Fig. 14). During the co-cultivation the mycelia turned brown and produced a copious amount of mucilage.

Fig. 14. Interaction with myc$^-$ mutant of *Pisum sativum*. Wild-type and myc$^-$ mutant of pea were inoculated with *P. indica*. Growth conditions were as described in Fig. 7. *Left* SEM-picture of the surface of colonized roots (wild-type) and mycelial mats of *P. indica* showing dense masses of well differentiated chlamydospores. *Right* SEM-picture of the root surface and the mycelial mat after interaction with myc$^-$ mutant and 14 days of incubation with a low amount of morphologically heterogeneous chlamydospores

4.9 *Arabidopsis thaliana*

A. thaliana (L.) Heynh. seedlings were pre-germinated for 2 weeks on MS and then transferred to Aspergillus medium. Plants were further cultivated with or without the fungus *P. indica* at 22 °C and under short day conditions. Control plants (without fungus) remained small with limited root growth and branching. In contrast, the co-cultivation with *P. indica* resulted in promotion of the plant growth and extensive root proliferation and elongation. An observation of the inoculated roots under the light microscope revealed that the fungus colonized the root surface and the cortical zone. Chlamydospores were produced by external hyphae (extramatrical) and within the root cortex and root hairs (intracellular). The fungal colonization reduced the root hair formation (Fig. 15).

In another independent study, *A. thaliana* was cultivated on MYP-agar. A 1-month-old culture was flooded with 10 ml sterilized water to gain a chlamydospore suspension for inoculation. Three-day-old *A. thaliana* seedlings were inoculated each with 10 µl of chlamydospore suspension. After 5, 10, 17, and 31 days of co-cultivation, plants were harvested and the roots examined with the light microscope and SEM. Moreover, FDA (fluoresceindiacetate) was used to discriminate between living and dead cells. At the latest, 17 days after inoculation, the whole root surface of *A. thaliana* was covered with mycelium. Most of the hyphae were growing between root hairs and some were closely attached to the rhizodermis. Several of these closely attached hyphae were following the anticlinal, axial cell walls of the rhizodermal cells. Keijer (1996) found that this is an indication of the beginning of an interaction between *Rhizoctonia solani* and its hosts.

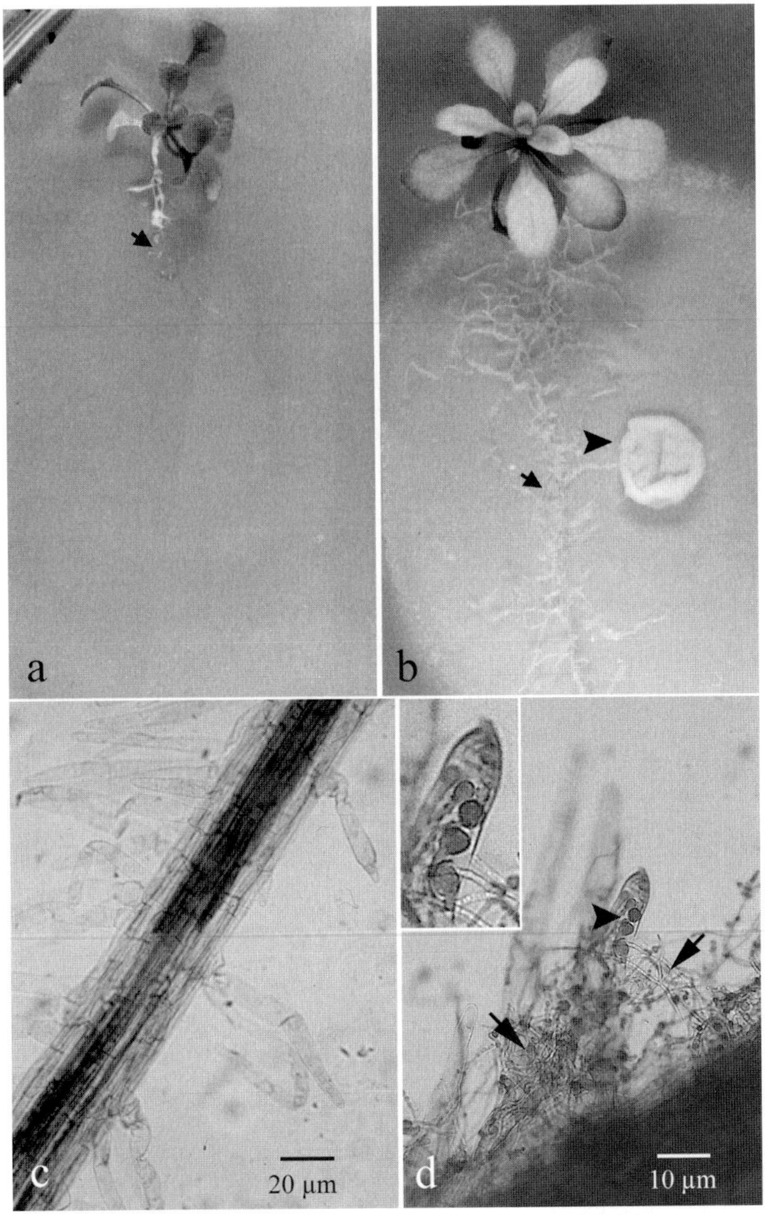

Fig. 15a–d. Interaction with *Arabidopsis thaliana*. Seedlings were germinated for 2 weeks on MS and then transferred to Aspergillus medium. Plants were cultivated at 22 °C and under short day conditions. **a** Control plants remained small with limited root growth and branching; **b** in contrast, the co-cultivaton with *P. indica* resulted in the promotion of plant growth and extensive root proliferation; **c** control roots produced a large number of root hairs growing uniformly from the top to the base; **d** inoculated roots as seen under the light microscope after staining with cotton blue: *P. indica* colonized the root surface and the cortical zone, spores were produced by external hyphae and in the roots and root hairs (*inset*)

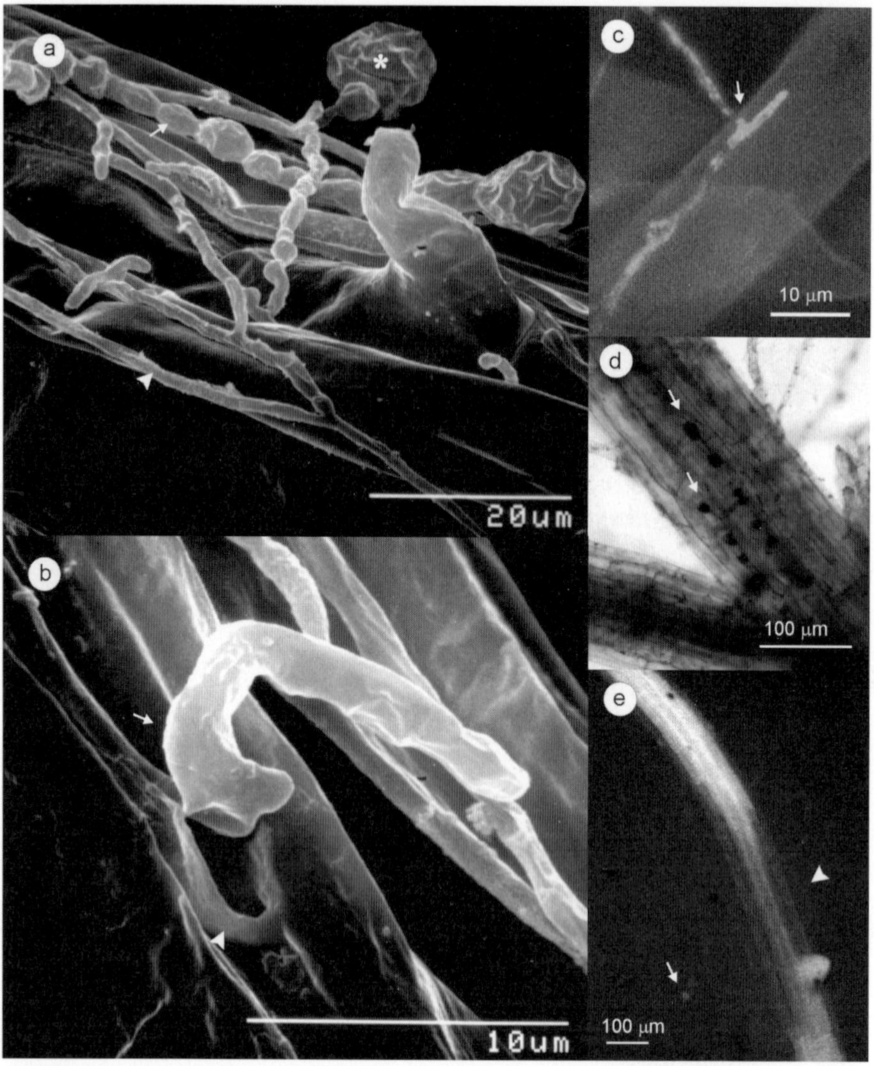

Fig. 16a–e. Interaction with *Arabidopsis thaliana.* **a** SEM picture of hyphae closely attached to the root surface. Besides normal hyphae (*arrowhead*) *P. indica* also forms coralloid hyphae (*arrow*) and chlamydospores (*asterisk*, collapsed due to preparation); **b** SEM picture of a hypha probably forming an appressorial swelling (*arrow*), which has caused an imprint (*arrowhead*) on the surface of a rhizodermal cell; **c** epifluorescence LM-picture of a hypha entering a root hair (*arrow*). Staining: aniline blue; **d** root stained with cotton blue shows intracellular chlamydospores in the rhizodermis (*arrows*); **e** a segment of the root stained with FDA and observed in epifluorescence showing lower FDA-fluorescence (*arrowhead*) than adjacent regions, indicating less vitality. In bright field it was clearly visible that this region was covered with hyphae (*arrow* chlamydospore)

From 10 days after inoculation on, some hyphae formed irregular, globular swellings, which were called coralloid hyphae (Fig. 16). These hyphae preceded the production of chlamydospores, which were produced terminally. Mature spores could be found from 17 days after inoculation on. On the surface of rhizodermal cells occasionally slightly swollen hyphal tips could be observed. In some cases an imprint in the host cell wall was visible, probably caused by mechanical pressure in combination with enzymes, indicating an appressorial function of these hyphal tips (Fig. 16). In some host cells intracellular hyphae could be found and at the penetration point, the hyphae had generally reduced their diameter. Host cell wall thickenings as an answer to the penetration could never be observed. Like hyphae of the external mycelium, intracellular hyphae formed coralloid swellings and sometimes also produced chlamydospores. A necrotrophic potential of *P. indica* could be demonstrated by vitality tests. Control roots and regions, which were free from mycelium were always stained bright green, which indicated the vitality of the corresponding cells. Root areas that were covered with hyphae often showed weaker or no fluorescence (Fig. 16). This indicates that the fungus is able to cause local damage to the root cortex of this host.

4.10 Root Organ Culture

Root organ culture of *Daucus carota* L. (carrot) was prepared as described by Bécard and Piche (1992). *P. indica* interacted with the root organ culture of carrot in the same way as was found in other plants tested. The infection rate, as a portion of infected root length, has been calculated to be 17% 9 weeks after inoculation, 50% in the most successful culture and 40% after pro-

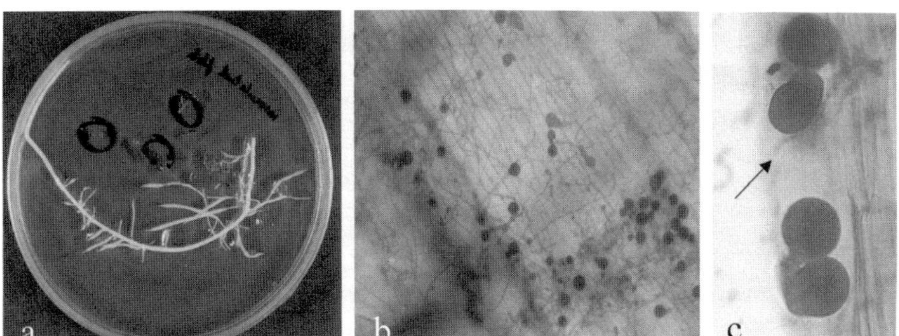

Fig. 17a–c. Interaction with transformed *Daucus carota* (Queen Anne's-lace) root. Root organ cultures were inoculated with *P. indica* and grown for 20 days. **a** Dark circles represent the place for inocula; **b** hyphae and chlamydospores on the surface of the roots; **c** intracellular sporulation as seen with the LM

longed incubation. Newly developed lateral roots were preferred infection sites and most infections started 0.5–1 cm behind the root tips. Bécard and Fortin (1988) observed a similar pattern for AMF. The preferential site for primary infection by the germ tubes of germinating AM spores was the elongation zone of the main root, where lateral root primordia formed. (Fig. 17).

5 Cell Wall Degrading Enzymes

The exo-oxidative enzyme laccase has been detected in a large number of basidiomycete ectomycorrhizal fungi, in a few ectomycorrhizal ascomycetes and only in one endomycorrhizal species (Gramss et al. 1998). All the ascomycetes tested showed the presence of laccase, but in basidiomycetes only 33 out of 44 species were found to be active (Table 4). However, to our knowl-

Table 4. Laccase activity in mycorrhizal fungi

Fungal species	Systematic positions	Laccase activities	References
Ectomycorrhizal basidiomycetes			
Amanita gemmata	Agaricales, Amanitaceae	(+)	Gramss et al. (1998)
Amanita muscaria	Agaricales, Amanitaceae	(+)	Gramss et al. (1998)
Amanita rubescens	Agaricales, Amanitaceae	(+)	Gramss et al. (1998)
Amanita spissa	Agaricales, Amanitaceae	(+)	Gramss et al. (1998)
Amanita strobiliformis	Agaricales, Amanitaceae	(+)	Gramss et al. (1998)
Boletinus cavipes	Boletales, Gyrodontaceae	(–)	Gramss et al. (1998)
Boletus edulis	Boletales, Boletaceae	(–)	Gramss et al. (1998)
Boletus erythropus	Boletales, Boletaceae	(–)	Gramss et al. (1998)
Boletus luridus	Boletales, Boletaceae	(–)	Gramss et al. (1998)
Boletus piperatus	Boletales, Boletaceae	(–)	Gramss et al. (1998)
Cortinarius varius	Agaricales, Cortinariaceae	(+)	Gramss et al. (1998)
Hebeloma crustuliniforme	Agaricales, Cortinariaceae	(+)	Gramss et al. (1998)
Hebeloma edurum	Agaricales, Cortinariaceae	(–)	Gramss et al. (1998)
Hebeloma hiemale	Agaricales, Cortinariaceae	(+)	Gramss et al. (1998)
Hebeloma sinapizans	Agaricales, Cortinariaceae	(+)	Gramss et al. (1998)
Laccaria amethystina	Agaricales, Tricholomataceae	(+)	Muenzenberger et al. (1997)
Lactarius deliciosus	Agaricales, Russulaceae	(+)	Gramss et al. (1998)
Lactarius deterrimus	Agaricales, Russulaceae	(+)	Gramss et al. (1998)
Lactarius necator	Agaricales, Russulaceae	(–)	Gramss et al. (1998)
Lactarius rufus	Agaricales, Russulaceae	(+)	Gramss et al. (1998)
Lactarius torminosus	Agaricales, Russulaceae	(+)	Gramss et al. (1998)
Leccinum scabrum	Boletales, Boletaceae	(+)	Gramss et al. (1998)
Leccinum versipelle	Boletales, Boletaceae	(+)	Gramss et al. (1998)
Paxillus involutus	Boletales, Paxillaceae	(+)	Gramss et al. (1999)

edge, the presence of laccase genes in these fungi was shown only in *Lactarius rufus* (Chen et al. 2003). Most fungi tested showed a strong laccase activity. *P. indica* and *S. vermifera* ss. Warcup and Talbot also showed a positive reaction to the ABTS test (oxidation of 2,2′- azino-bis (3-ethylthiazoline-6-sulfonate), i.e., presence of laccase activity (Fig. 18). However, the reaction was faster in the former fungus than in the latter. Laccase activity was observed in the very young culture (5 days old), whereas in *S. vermifera* the reaction appeared stronger after 10–12 days.

Other cell wall-degrading enzymes detected in *P. indica* are given in Table 5. All major cell wall degrading enzymes were present in *P. indica* except monoxygenase and phenoloxidase. Table 6 shows the activities of CMC-ase, xylanase and polygalacturonase in the plants incubated with *P. indica*. In the case of polygalacturonase, the enzyme activity was higher at the initial stage

Table 4. (*Continued*)

Fungal species	Systematic positions	Laccase activities	References
Pisolithus tinctorius	Boletales, Sclerodermataceae	(–)	Gramss et al. (1998)
Russula aeruginea	Agaricales, Russulaceae	(+)	Gramss et al. (1998)
Russula foetens	Agaricales, Russulaceae	(+)	Gramss et al. (1998)
Russula violeipes	Agaricales, Russulaceae	(+)	Gramss et al. (1998)
Scleroderma citrinum	Boletales, Sclerodermataceae	(–)	Gramss et al. (1998)
Suillus aeruginascens	Boletales, Boletaceae	(+)	Gramss et al. (1998)
Suillus granulatus	Boletales, Boletaceae	(low)	Gramss et al. (1998)
Suillus grevillei	Boletales, Boletaceae	(+)	Gramss et al. (1998)
Suillus luteus	Boletales, Boletaceae	(+)	Gramss et al. (1998)
Suillus variegatus	Boletales, Boletaceae	(low)	Gramss et al. (1998)
Tricholoma fulvum	Agaricales, Tricholomataceae	(+)	Gramss et al. (1998)
Tricholoma imbricatum	Agaricales, Tricholomataceae	(+)	Gramss et al. (1998)
Tricholoma lascivum	Agaricales, Tricholomataceae	(+)	Gramss et al. (1998)
Tricholoma scalpturatum	Agaricales, Tricholomataceae	(+)	Gramss et al. (1998)
Tricholoma subannulatum	Agaricales, Tricholomataceae	(+)	Gramss et al. (1998)
Tricholoma terreum	Agaricales, Tricholomataceae	(+)	Gramss et al. (1998)
Tricholoma ustaloides	Agaricales, Tricholomataceae	(+)	Gramss et al. (1998)
Xerocomus badius	Boletales, Boletaceae	(low)	Gramss et al. (1999)
Xerocomus chrysenteron	Boletales, Boletaceae	(–)	Gramss et al. (1998)
Xerocomus subtomentosus	Boletales, Boletaceae	(–)	Gramss et al. (1998)
Ectomycorrhizal ascomycetes			
Morchella conica	Pezizales	(+)	Gramss et al. 1998
Morchella elata	Pezizales	(+)	Gramss et al. 1998
Morchella esculenta	Pezizales	(+)	Gramss et al. 1998
Tuber sp.	Pezizales	(+)	Miranda et al. 1992
Endomycorrhizal fungi			
Glomus etunicatus	Glomales	(+)	Nemec, 1981

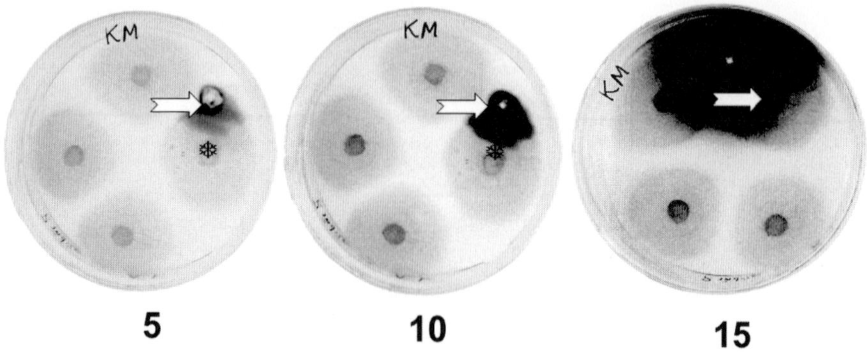

Fig. 18. Laccase activity in *P. indica*. The fungus was grown for 5 days on *Aspergillus* medium, then a small hole (*arrow*) was cut into the agar at the border of the growing mycelial mat. Three drops (30 µl) of ABTS were added. Green coloration appeared and was monitored at different time intervals. The evolution of the coloration is given with time, but the coloration was immediately obtained after a few seconds. *Snowflake* indicates the fungal growth

Table 5. Cell wall degrading enzymes from *P. indica*

Compound	Detection	Possible enzymes	Important for the degradation of
ABTS	+	Laccase	Lignin
Ferulic acid	+	Ferulase	Lignin
Vanillin	–	Monoxygenase	Lignin
Tannin	–	Phenoloxidase	Phenols
Starch	+	Amylase	Plant storage polysaccharides
Cellulose	+	Cellulase	Cellulose
Gelatine	+	Protease	Proteins
Pectin	+	Pectinase	Pectin
Lipid	+	Lipase	Fat
Xylan	+	Xylanase	Hemicellulose
Chitin	+	Chitinase	Chitin

cf. Bütehorn (1999) and Varma et al. (2001)

(after 20 min) of incubation, i.e., up to 0.058 µmol ml^{-1} min^{-1}, and declined to 0.041 µmol ml^{-1} min^{-1} after 80 min of incubation.

Fungal hyphae entered the cells randomly through the cell wall. It seems the entry was facilitated by the combined action of wall degrading enzymes and mechanical pressure.

Under normal conditions AMF do not invade vascular systems and the aerial parts of their hosts. Despite heavy root colonization this is also true for *P. indica* and *S. vermifera*, although these fungi were able to produce heavy amounts of cell wall-degrading enzymes.

Table 6. Extracellular hydrolytic enzymes produced in vitro by *P. indica*

	IU[a]
CMCase	0.013
Xylanase	0.062
Polygalacturonase	0.017

cf. Varma et al. (2001)

[a] 0.5 % Na-polypectate (Sigma) was used as the substrate. One unit activity (IU) of Pgase is defined as the amount of enzyme which releases 1 mol of carboxyl group as equivalent to the amount of Na-thiosulphate added to neutralize the residual iodine. Polymethylgalacturonase (PMG) was completely absent

6 Conclusions

Piriformospora indica is a wide host range organism. It interacts with plant growth-promoting rhizobacteria (PGPRs), terrestrial mycobionts, green algae, lower and higher plants. Among the PGPRs, *Pseudomonas fluorescence* inhibited the fungus, while strains of *Azotobacter*, *Bradyrhizobium* and *Azospirillum* over all enhanced the fungal growth. In vitro and in vivo studies as well as field trials have proved phytopromotional effects on most plants tested. Exceptions were myc$^-$ mutants of pea and soyabean, where the hyphae did not invade and the plant growth was negatively influenced. Results obtained from the interaction with *Arabidopsis thaliana* and Hybrid aspen (*Populus tremula* x *P. tremuloides)* were interesting as they opened new vistas to understand the mechanism and molecular basis of plant-fungus symbiosis.

There are still lots of unanswered questions: how does the fungus promote the growth of the plants and why is the growth of nonhosts reduced? what is the mechanism of root pathogen suppression? these are only two examples of such questions to be answered in the future.

Acknowledgments. The Indian authors are thankful to DBT, DST, CSIR, UGC, and the Government of India for partial financial assistance. We are thankful to Dr. Michael Weiss, Germany for providing 28 s rDNA analysis of *P. indica*.

References and Selected Reading

Bécard G, Fortin JA (1988) Early events of vesicular-arbuscular mycorrhiza formation on Ri T-DNA transformed roots. New Phytol 108:211–218

Bécard G, Piche Y (1992) Establishment of vesicular – arbuscular mycorrhiza in root organ culture: Review and proposed methodology. Methods Microbiol 24:89–108

Blechert O, Kost G, Hassel A, Rexer R-H, Varma A (1999) First remarks on the symbiotic interactions between *Piriformospora indica* and terrestrial orchids. In: Varma A, Hock B (eds) Mycorrhizae 2nd edn. Springer, Berlin Heidelberg New York, pp 683–688

Bütehorn B (1999) Erste Zytologische und molekulare Untersuchungen zu *Piriformospora indica*, einem pflanzenwachstumsfördernden Endophyten. PhD Thesis, Marburg, Germany

Chen DM, Bastias BA, Taylor AFS, Cairney JWG (2003) Identification of laccase-like genes in ectomycorrhizal basidiomycetes and transcriptional regulation by nitrogen in *Piloderma byssinum*. New Phythol 157:547–554

Dehne HW (1982) Interaction between VAM fungi and plant pathogens. Phytopathology 72:1115–1119

Gramss G, Kirsche B, Voigt K-D, Günther T, Fritsche W (1998) Conversion rates of five polycyclic aromatic hydrocarbons in liquid cultures of fifty-eight fungi and the concomitant production of oxidative enzymes. Mycol Res 103:1009–1018

Gramss G, Günther T, Fritsche W (1999) Spot tests for oxidative enzymes in ectomycorrhizal, wood-, and litter decaying fungi. Mycol Res 102:67–72

Herrmann S, Munch J-C, Buscot F (1998) A gnotobiotic system with oak micro-cuttings to study specific effects of mycobionts on plant morphology before, and in the early phase of ectomycorrhiza formation by *Paxillus involutus* and *Piloderma croceum.* New Phytol 138:203–212

Kaefer E (1977) Meiotic and mitotic recombination in *Aspergillus* and its chromosomal aberrations. Adv Genet 19:33–131

Keijer J (1996) The initial steps of the interaction process in *Rhizoctonia solani.* In: Sneh B, Jabaji-Hare S, Neate S, Dijst G (eds) *Rhizoctonia* species: taxonomy, molecular biology, ecology, pathology and disease control. Kluwer, Dordrecht, pp 149–162

Marx DH (1969) The influence of ectotrophic mycorrhizal fungi on the resistance of pine roots to pathogenic infections. I. Antagonism of mycorrhizal fungi to root pathogenic fungi and soil bacteria. Phytopathology 59:153–163

Miranda M, Bonfigli A, Zarivi O, Ragnelli AM, Pacioni G, Botti D (1992) Truffle tyrosinase: properties and activity. Plant Sci 81:175–182

Münzenberger B, Otter T, Wustrich D, Polle A (1997) Peroxidase and laccase activities in mycorrhizal and non-mycorrhizal fine roots of Norway spruce (*Picea abies*) and larch (*Larix decidua*). Can J Bot 75:932–938

Murashige T, Skoog F (1962) A revised medium for rapid growth and bioassays with tobacco tissue cultures. Physiol Plant 15:473–497

Nemec S (1981) Histochemical characteristics of *Glomus etunicatus* infection on *Citrus limon* fibrous roots. Can J Bot 59:609–617

Rai M, Acharya D, Singh A, Varma A (2001) Positive growth responses of the medicinal plants *Spilanthes calva* and *Withania somnifera* to inoculation by *Piriformospora indica* in a field trial. Mycorrhiza 11:123–128

Read DJ (1999) Mycorrhiza – The state of art. In: Varma A, Hock B (eds) Mycorrhizae 2nd edn. Springer, Berlin Heidelberg New York, pp 3–34

Sahay NS, Varma A (1999) *Piriformospora indica*; a new biological hardening tool for micropropagated plants. FEMS Microbiol Lett 181:297–302

Sahay NS, Varma A (2000) Biological approach towards increasing the survival rates of the micropropagated plants. Curr Sci 78:126–129

Singh A, Varma A (2000) Orchidaceous mycorrhizal fungi. In: Mukerji KG, Chamola BP, Singh J (eds) Mycorrhizal biology. Kluwer Academic/Plenum Publishers, New York, pp 265–288

Singh A, Sharma J, Rexer K-H, Varma A (2000) Plant productivity determinants beyond minerals, water and light: *Piriformospora indica* – A revolutionary plants promoting fungus. Curr Sci 79:101–106

Singh Ar, Singh An, Varma A (2002a) *Piriformospora indica* – in vitro raised leguminous plants: a new dimension in establishment and phytopromotion. Ind J Biotechnol 1:371–376

Singh An, Singh Ar, Rexer K-H, Kost G, Varma A (2002b) Root endosymbiont: *Piriformospora indica* – a boon for orchids. J Orchid Soc India 15:89–102

Singh An, Singh Ar, Kumari M, Rai MK, Varma A (2003a) Biotechnological importance of *Piriformospora indica* Verma et al. – a novel symbiotic mycorrhiza-like fungus: an overview. Indian J Biotechnol 2:65–75

Singh An, Singh Ar, Kumari M, Kumar S, Rai MK, Sharma AP, Varma A (2003b) Unmassing the accessible treasures of the hidden unexplored microbial world. In: Prasad BN (ed) Biotechnology in sustainable biodiversity and food security. Science Publishers, Enfield, NH, pp 101–124

Tester M, Smith SE, Smith FA (1987) The phenomenon of "nonmycorrhizal" plants. Can J Bot 65:419–431

Varma A (1998) Mycorrhizae, the friendly fungi: what we know and how do we know? In: Varma A (ed) Mycorrhiza manual. Springer, Berlin Heidelberg New York, pp 1–24

Varma A (1999) Functions and applications of arbuscular mycorrhizal fungi in arid and semi-arid soils. In: Varma A, Hock B (eds) Mycorrhiza. Springer, Berlin Heidelberg New York, pp 521–556

Varma A, Verma S, Sudha, Sahay NS, Franken P (1999) *Piriformospora indica*, a cultivable plant growth promoting root endophyte with similarities to arbuscular mycorrhizal fungi. Appl Environ Microbiol 65:2741–2744

Varma A, Singh A, Sudha, Sahay N, Sharma J, Roy A, Kumari M, Rana D, Thakran S, Deka D, Bharati K, Franken P, Hurek T, Blechert O, Rexer K-H, Kost G., Hahn A, Hock B, Maier W, Walter M, Strack D, Kranner I (2001) *Piriformospora indica*: A cultivable mycorrhiza-like endosymbiotic fungus. In: Hock B (ed) Mycota IX. Springer, Berlin Heidelberg New York, pp 123–150

Verma S, Varma A, Rexer K-H, Hassel A, Kost G, Sarbhoy A, Bisen P, Buetehorn B, Franken P (1998) *Piriformospora indica* gen. nov; a new root-colonizing fungus. Mycologia 90:895–909

16 Cellular Basidiomycete–Fungus Interactions

ROBERT BAUER and FRANZ OBERWINKLER

1 Introduction

While basidiomycetes are well known as saprobes, ectomycorrhizal symbionts or parasites of plants (e.g., Bauer et al. 2001; Hibbett and Thorn 2001), their role as parasites of other fungi has received scant attention. Thus, the ultrastructure of the host–parasite interaction in basidiomycetous mycoparasites has been studied only in a few species (Bauer and Oberwinkler 1990a, b, 1991; Oberwinkler and Bauer 1990; Oberwinkler et al. 1990a, c, 1999; Zugmaier et al. 1994; Kirschner et al. 2001a).

In this chapter, our data concerning the interfungal cellular interaction of basidiomycetes are summarized.

2 Occurrence of Mycoparasites Within the Basidiomycota

The division Basidiomycota comprises the classes Urediniomycetes, Ustilaginomycetes and Hymenomycetes (Swann and Taylor 1993; Begerow et al. 1997). Mycoparasites occur in two of these groups: while scattered throughout the Urediniomycetes mycoparasites form one of the basal lineages of the Hymenomycetes (Swann et al. 2001; Weiß and Oberwinkler 2001). Urediniomycetous mycoparasites include the genera *Colacogloea*, *Colacosiphon*, *Cryptomycocolax*, *Cystobasidium*, *Heterogastridium*, *Mycogloea*, *Naohidea*, *Occultifur*, *Spiculogloea*, *Zygogloea*, and some species of *Platygloea* (Bandoni 1956, 1984; Oberwinkler 1990; Oberwinkler and Bauer 1990; Oberwinkler et al. 1990a, b; Roberts 1994, 1996, 1997; Kirschner et al. 2001a). However, the phenomenon of mycoparasitism may be more widespread among Urediniomycetes than is currently suspected. Many species of Urediniomycetes (e.g., members of *Agaricostilbum*, *Atractogloea*, *Camptobasidium*, *Chionosphaera*, *Leucosporidium*, *Naiadella*, *Rhodosporidium* or *Sporidiobolus*), currently thought to be saprobes may be capable of parasitizing fungi (Oberwinkler and Bandoni 1982, 1989; Marvanová and Bandoni 1987; Marvanová and

Plant Surface Microbiology
A. Varma, L. Abbott, D. Werner, R. Hampp (Eds.)

Suberkropp 1990; Bauer et al. 1997; Roberts 1997; Kirschner et al. 2001b). Within the Hymenomycetes, mycoparasitism is common among the Tremellales sensu Bandoni (1984). In addition, some lichen parasites (Diederich 1996) also may belong in the Tremellales.

3 Hosts

Hosts of the basidiomycetous mycoparasites are either ascomycetes or basidiomycetes. Mycoparasitism on chytrids or zygomycetes is unknown. There appears to be no phylogenetic correlation between the basidiomycetous mycoparasites and their respective host fungi. In other words, a distinct basidiomycetous group of mycoparasites usually occurs on both ascomycetes and basidiomycetes.

4 Cellular Interactions

The basidiomycetous interfungal cellular interactions can be divided into two main types at the structural level.

4.1 Colacosome-Interactions

Mycoparasites in the Microbotryomycetidae (Bauer et al. 1997; Swann et al. 1999) are generally characterized by their interaction via a unique organelle, the colacosome, formed at the interface between the parasite and its fungal host. This mycoparasitic organelle was firstly described in detail from the interaction of the parasite *Colacogloea peniophorae* and its host *Hyphoderma praetermissum* (Oberwinkler et al. 1990a; Bauer and Oberwinkler 1991). Colacosomes develop in the contact area between the parasite and its host as illustrated in Fig. 1. They are positioned at the inner surface of the parasite cell outside the cytoplasm, but inside the cell wall. Their shape is globular, subglobular or beaked. The central part of the colacosome is electron-opaque, 0.3–0.4 µm diameter, enclosed by a membrane and surrounded by an electron-transparent, unstructured sheath of approximately 0.05 µm diameter. The colacosome is covered by the plasmalemma. A thin secondary cell wall layer is often present along the plasmalemma covering the colacosome.

During formation (Fig. 2), the plasma membrane of the parasite is folded into the cytoplasm, then recurves, and finally fuses with itself at a distance of 0.2–0.3 µm from the original outgrowth. Consequently, it is surrounded by a membrane as a derivative from the plasma membrane. The globose compartment is now separated from the cytoplasm by an electron-transparent, intermembranaceous space. After separation from the cytoplasm, the vesicular

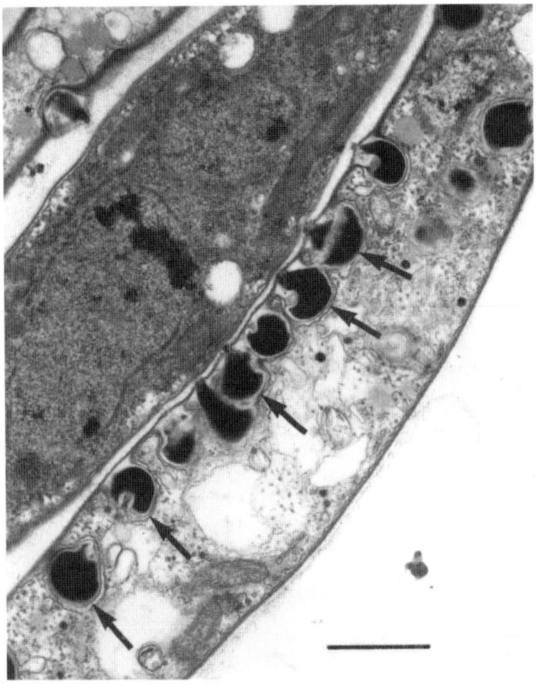

Fig. 1. Hypha of *Colacogloea peniophorae* with colacosomes (*arrows*) contacting host hypha. *Bar* 1 μm

core becomes homogeneous and finally more and more electron-opaque. Simultaneously, the intermembranaceous space between the central part of the colacosome and the cytoplasm increases slightly in thickness. Interaction starts with intrusion of electron-opaque core material of the colacosome into the cell wall of the parasite (Fig. 3). The cell wall close to the intrusion peg becomes electron-transparent and indistinct in substructure. Intrusion then continues through the closely attached cell wall of the host into an electron-transparent protuberance formed between the cell wall and the plasmalemma of the host (Fig. 4).

In *Colacogloea peniophorae* colacosomes develop in great numbers close together (Fig. 1). It is evident that in most cases hyphae possessing colacosomes and their host hyphae lie for a relatively long distance side by side closely attached to one another (Fig. 1). Furthermore, the host hyphae often form one or two spirals around the colacosome-possessing hyphae (Bauer and Oberwinkler 1991). As discussed by Bauer and Oberwinkler (1991), this situation may be explained as follows: in the beginning, the parasite hypha grows loosely in the host fructifications. After a first, probably accidental, contact of the hypha with a host hypha, colacosomes develop rapidly and in great number. The electron-opaque content of the colacosomes penetrates the host cell wall. Thus, the colacosomes combine both cells and the first contact

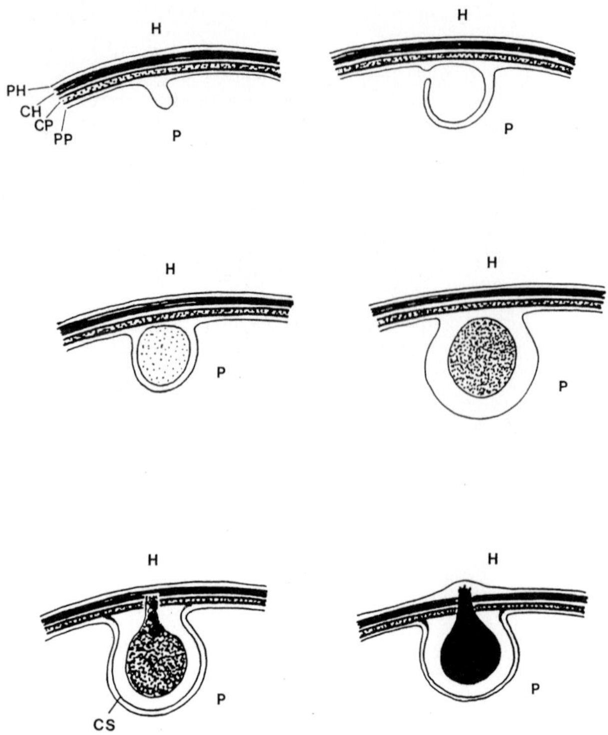

Fig. 2. Diagram of colacosome development, modified from Bauer and Oberwinkler (1991). Abbreviations and symbols: *CH* cell wall of the host *Hyphoderma praetermissum*, *CP* cell wall of the parasite *Colacogloea peniophorae*, *CS* secondary cell wall layer, *H* cell of the host *Hyphoderma praetermissum*, *P* cell of the parasite *Platygloea peniophorae*, *PH* plasma membrane of the host *Hyphoderma praetermissum*, *PP* plasma membrane of the parasite *Colacogloea peniophorae*. (*top left*) Initial stage of invagination of the plasma membrane of the parasite. (*top right*) The plasmalemma of the parasite recurves. (*middle left*) Delimitation of the young colacosome from the cytoplasm. (*middle right*) The central part of the colacosome becomes homogeneous and more and more electron-opaque. The electron-transparent sheath of the colacosome increases in thickness. (*lower left*) The electron-opaque core material penetrates the cell wall of the parasite and begins to intrude the cell wall of the host. (*lower left*) Final developmental stage with colacosome penetration through host cell wall

remains stable. Furthermore, if the parasite and/or the host hypha continue to grow, additional colacosomes are rapidly developed. Consequently, the number of connections between both organisms is continually increased and both are forced to grow in close contact to each other. The development of colacosomes is, therefore accompanied by an increase of the host – parasite interface. In this sense, the colacosomes could serve as connecting agents. It is unclear from the present data, however, whether or not the colacosomes are involved in host–parasite metabolism functions as no specific attempts to

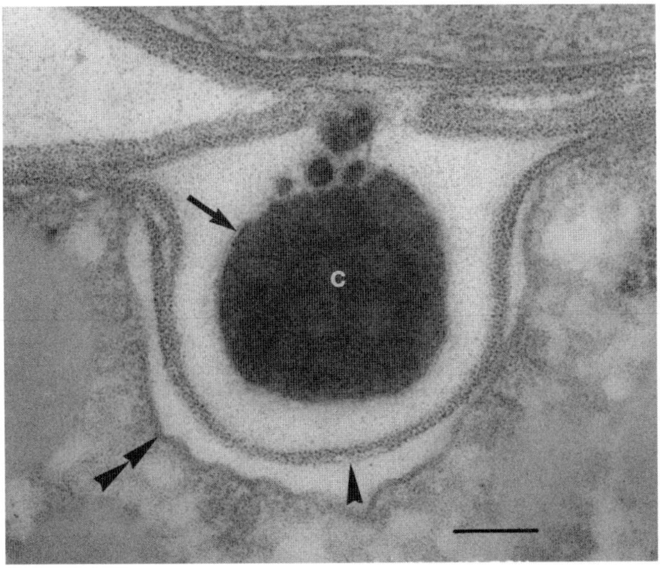

Fig. 3. Hypha of *Colacogloea peniophorae* (*lower cell*) in contact with a host hypha (*upper cell*). The electron-opaque core (*c*) of the colacosome intrudes into the cell wall of the parasite. Note the tripartite membrane (*arrow*) around the core of the colacosome (*c*). The colacosome is covered by a thin secondary cell wall layer (*arrowhead*) and the plasma membrane (*double arrowhead*) of the parasite. *Bar* 0.1 μm

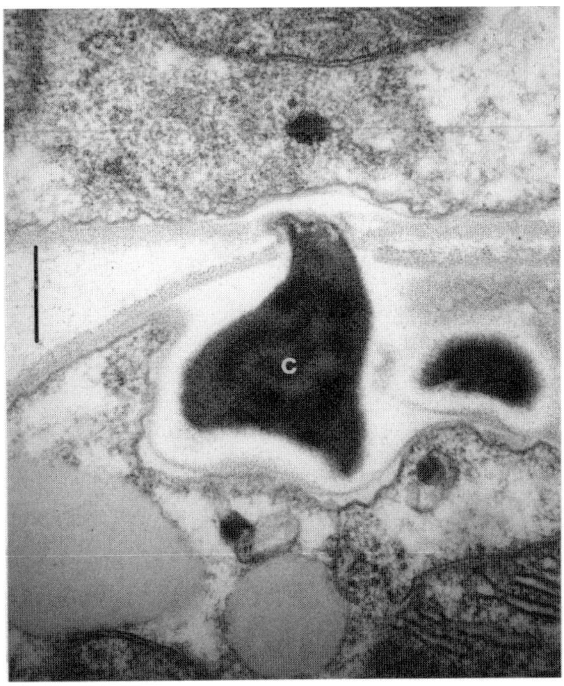

Fig. 4. Hypha of *Colacogloea peniophorae* (*lower cell*) in contact with a host hypha (*upper cell*). Final stage of host – parasite interaction with the content of the electron-opaque core of the colacosome (*c*) penetrating the host cell wall. *Bar* 0.2 μm

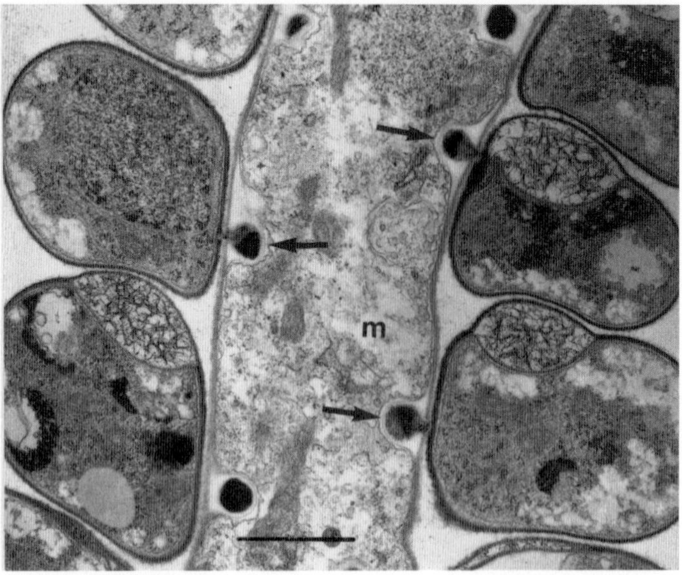

Fig. 5. Longitudinally sectioned hypha of "*Mycospira*" (*m*) surrounded by a hyphal spiral of the host fungus (three-dimensional configuration reconstructed from serial sections). Note the colacosomes (*arrows*) at the contact area. *Bar* 1 μm

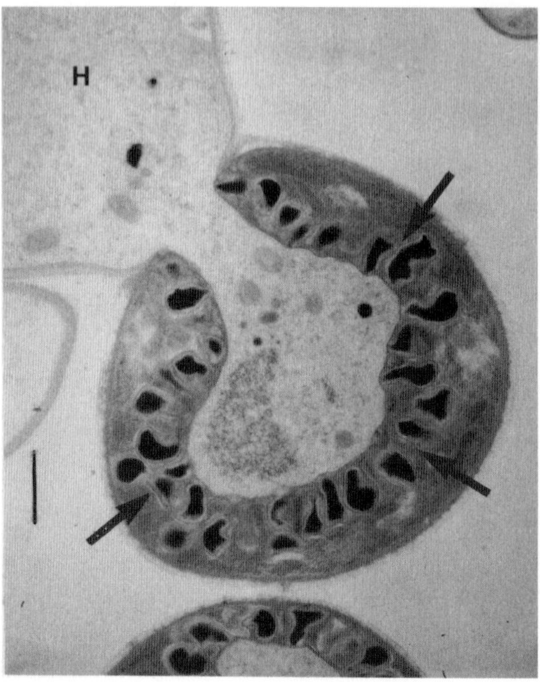

Fig. 6. Host cell (*H*) intruding into a hyphal cell of *Colacogloea* sp. Colacosomes (*arrows*) surround the intracellular part of the host cell which lacks a cell wall. *Bar* 1 μm

identify them have been made. However, the change in the electron density of the core material of the colacosome suggests that an alteration of the chemical composition occurs after separation of the colacosome from the cytoplasm. Furthermore, the penetration of the parasite and host cell wall appears to be enzymatic since both cell walls are not distorted at the site of penetration.

This interpretation is reinforced by the mycoparasitic behavior of a currently undescribed basidiomycete, called here "*Mycospira*". In contact with its host, a member of *Tulasnella*, this fungus develops colacosomes in exact spirals. As a consequence, the host fungus grows in spirals around the colacosome-possessing hyphae (Fig. 5). Thus, the formation of colacosomes results

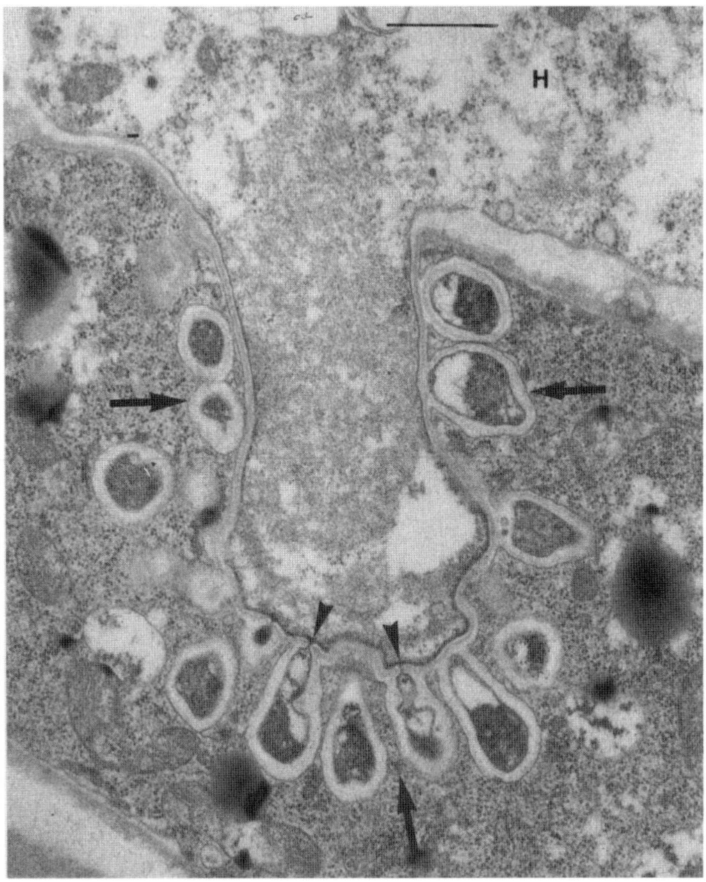

Fig. 7. Host cell (*H*) intruding into a hyphal cell of *Cryptomycocolax abnorme*. Colacosomes (*arrows*) surround the intracellular part of the host cell which lacks a cell wall. Two fusion pores are visible at *arrowheads*. Note that the colacosomes are more electron-transparent than in Fig. 6. *Bar* 1 μm

in a greatly increased contact zone between the parasite and its host. This is also the case in a third type of colacosome-arrangement. In *Colacogloea bispora* (Oberwinkler et al. 1999), *Colacogloea* sp. (Fig. 6), *Colacosiphon* (Kirschner et al. 2001a), *Cryptomycocolax* (Oberwinkler and Bauer 1990), *Krieglsteinera*, and *Heterogastridium*, the appearance of colacosomes is associated with curious interaction structures: filamentous outgrowths of the host cells are intimately enclosed by galloid parasite cells. Numerous colacosomes are present along the contact area between the host intrusion and the parasite cell (Fig. 6). These host intrusions always terminate in the parasite cell. They are unseptate, often branched and, astonishingly, lack cell walls, thus giving the impression of haustoria (of the host into the parasite!!!). In *Cryptomycocolax* a second type of colacosome was found along the cytoplasmic intrusions of the host formed into the hyphae of the parasite (Fig. 7; Oberwinkler and Bauer 1990). These colacosomes have a more electron-transparent core and they fuse with the host cell via a pore of approximately 7–14 nm in diameter (Figs. 7, 8). It is clear that the cellular interaction of *Cryptomycocolax* is complex and currently misunderstood.

Colacosomes have also been found in *Atractocolax*, *Leucosporidium*, *Mastigobasidium*, *Rhodosporidium* and *Sporidiobolus*, indicating a potential for mycoparasitism in these genera that have been assumed to be saprobic (Kreger van Rij and Veenhuis 1971; Bauer et al. 1997; Kirschner et al. 1999).

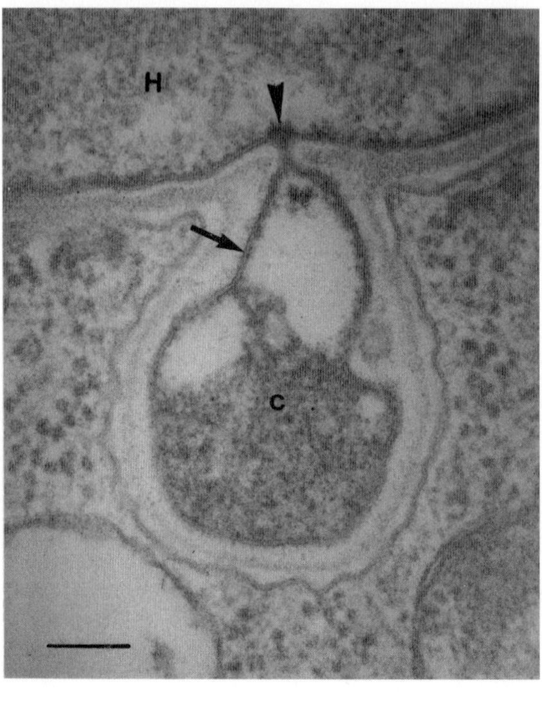

Fig. 8. Colacosome of *Cryptomycocolax abnorme* in contact with the host cytoplasm (*H*) showing the fusion pore (*arrowhead*) in detail. Note that the pore membranes are continuous with both the host plasma membrane and the membrane surrounding the core (*c*) of the colacosome (*arrow*). *Bar* 0.1 μm

4.2 Fusion-Interaction

Typical fusion mycoparasites (Bauer and Oberwinkler 1990a, b) are the Tremellales (including the Filobasidiales) of the Hymenomycetes (Bandoni 1984, 1995). Astonishingly, however, fusion mycoparasites are also scattered throughout the Urediniomycetes. For example, the members of *Cystobasidium*, *Mycogloea*, *Naohidea*, *Occultifur*, *Spiculogloea* and *Zygogloea* are fusion mycoparasites (unpubl. data). Usually, basidiomycetous fusion mycoparasites interact with their respective hosts by specialized interactive cells, designated often as "tremelloid haustorial cells". These cells were first described and designated as "haustoria" by Olive (1947). Each tremelloid haustorial cell is subtended by a clamp and consists of a subglobose basal part with one or more thread-like filaments (e.g., see Oberwinkler et al. 1984) that are capable of fusing with host cells via a pore of approximately 14–19 nm (Figs. 9, 10; Bauer and Oberwinkler 1990a, b). Thus, a direct cytoplasm – cytoplasm connection between the parasites and their respective hosts occurs. As discussed by Bauer and Oberwinkler (1990a), the following stages in the development of the cel-

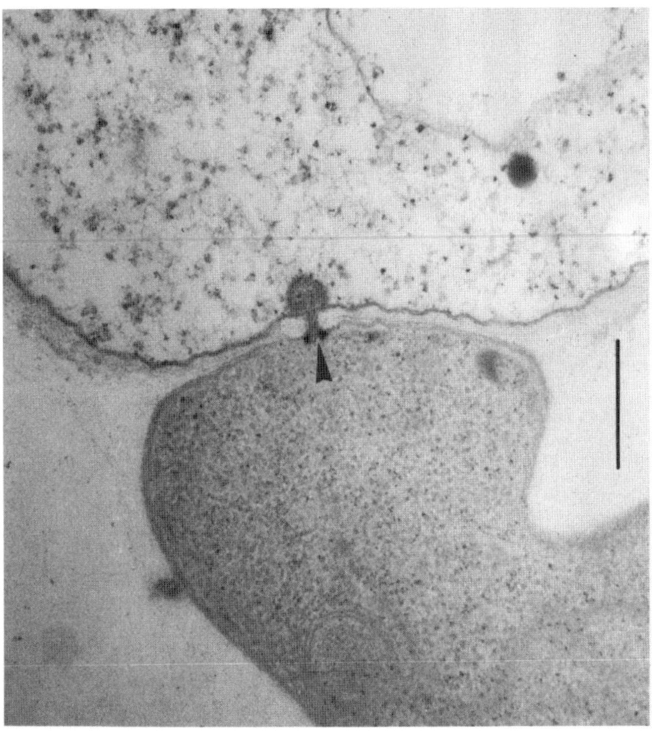

Fig. 9. Haustorial filament of *Tetragoniomyces uliginosus* (*lower cell*) in contact with a host hypha (*upper cell*) demonstrating the fusion pore (*arrowhead*). *Bar* 0.5 μm

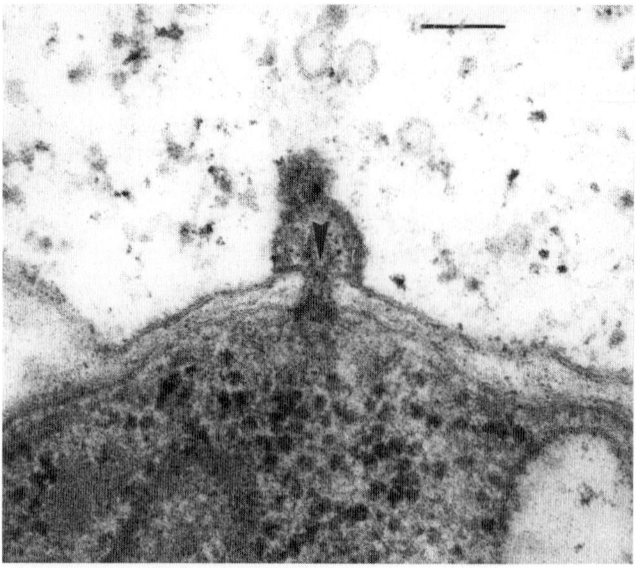

Fig. 10. Haustorial filament of *Tetragoniomyces uliginosus* (*lower cell*) in contact with a host hypha (*upper cell*) illustrated to show the fusion pore (*arrowhead*) in detail. Note that the pore membranes are continuous with the plasma membranes of both cells. *Bar* 0.1 μm

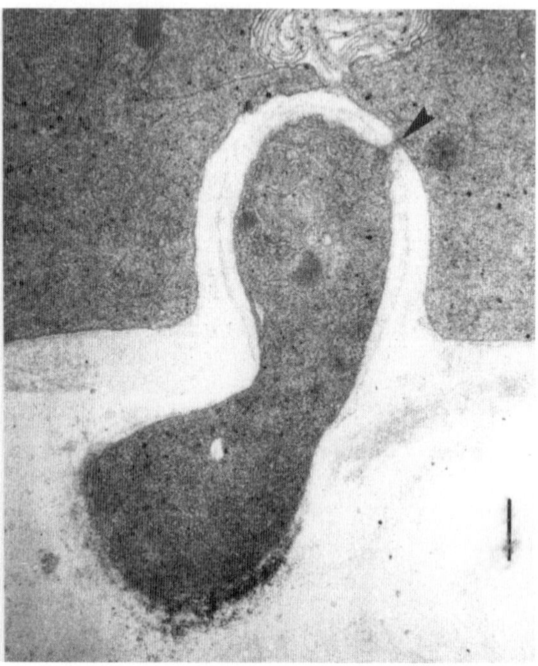

Fig. 11. Haustorial filament of *Christiansenia pallida* (*lower cell*) penetrating a host cell (*upper cell*). One fusion pore is visible at *arrowhead*. *Bar* 0.2 μm

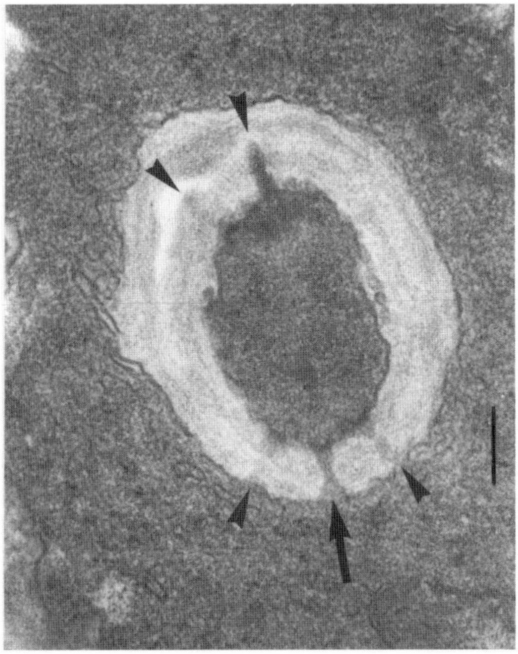

Fig. 12. Transverse section through a penetrating haustorial filament of *Christiansenia pallida*. One fusion pore medianly sectioned (*arrow*) and four pores nonmedianly sectioned (*arrowheads*). *Bar* 0.2 μm

lular interaction can be recognized: (1) contact of the haustorial filament with the host cell, (2) nesting of the haustorial filament into the host cell wall, and (3) fusion of the haustorial and host cell protoplasts via a pore.

In the interaction between *Christiansenia pallida* and its host (Bauer and Oberwinkler 1990b), the haustorial filament forms one or more protrusions into the host cells where a lot of fusion pores develop (Figs. 11, 12).

Direct cytoplasm – cytoplasm connections between mycoparasites and their respective hosts represent an unusual type of cellular interaction. As discussed by Bauer and Oberwinkler (1990a), this type of interaction may be considered as most effective. Substances required by the parasite do not need to cross membranes or cell walls. Thus, the fusion pores could serve as direct avenues for nutrients (Hoch 1977).

5 Basidiomycetous Mycoparasitism, a Result of Convergent Evolution?

The different mode of mycoparasitism occurring in the basidiomycetes, as discussed above, suggests that mycoparasitism may have evolved at least twice (or more) in the basidiomycetous history. Thus, it appears that the Microbotryomycetidae (for the subclass, see Swann et al. 1999) arose independently from the fusion mycoparasites as colacosome-mycoparasites (Bauer et al. 1997). For the occurrence of fusion mycoparasites in the Urediniomycetes on the one hand, and in the Hymenomycetes on the other, two explanations are possible: (1) the fusion mycoparasitism occurring in these two groups is a result of convergent evolution, and, alternatively (2), the common ancestor of both groups or of the basidiomycetes in general was a fusion mycoparasite. Additional phylogenetic studies are necessary to clarify this situation.

6 Conclusions

The phenomenon of mycoparasitism may be more widespread among basidiomycetes than is currently suspected. Our data illustrate that mycoparasitic basidiomycetes evolved a fascinating array of different strategies at the cellular level to benefit from host metabolites.

Acknowledgements. We thank Uwe Simon for critically reading the manuscript, and the Deutsche Forschungsgemeinschaft for financial support.

References and Selected Reading

Bandoni RJ (1956) A preliminary survey of the genus *Platygloea*. Mycologia 48:821–840

Bandoni RJ (1984) The Tremellales and Auriculariales: an alternative classification. Trans Mycol Soc Jpn 25:489–530

Bandoni RJ (1995) Dimorphic heterobasidiomycetes: taxonomy and parasitism. Stud Mycologia 38:13–27

Bauer R, Oberwinkler F (1990a) Direct cytoplasm-cytoplasm connection: an unusual host-parasite interaction of the tremelloid mycoparasite *Tetragoniomyces uliginosus*. Protoplasma 154:157–160

Bauer R, Oberwinkler F (1990b) Haustoria of the mycoparasitic heterobasidiomycete *Christiansenia pallida*. Cytologia 55:419–424

Bauer R, Oberwinkler F (1991) The colacosomes: new structures at the host-parasite interface of a mycoparasitic basidiomycete. Bot Acta 104:53–57

Bauer R, Oberwinkler F, Vánky K (1997) Ultrastructural markers and systematics in smut fungi and allied taxa. Can J Bot 75:1273–1314

Bauer R, Begerow D, Oberwinkler F, Piepenbring M, Berbee ML (2001) Ustilaginomycetes. In: McLaughlin DJ, McLaughlin EG, Lemke PA (eds) Mycota VII Part B, Systematics and evolution. Springer, Berlin Heidelberg New York, pp 57–83

Begerow D, Bauer R, Oberwinkler F (1997) Phylogenetic studies on large subunit ribosomal DNA sequences of smut fungi and related taxa. Can J Bot 75:2045–2056

Diederich P (1996) The lichenicolous heterobasidiomycetes. Bibliotheca Lichenologica 61:1–198

Hibbett DS, Thorn RG (2001) Basidiomycota: Homobasidiomycetes. In: McLaughlin DJ, McLaughlin EG, Lemke PA (eds) Mycota VII. Part B, Systematics and evolution. Springer, Berlin Heidelberg New York, pp 122–168

Hoch HC (1977) Mycoparasitic relationships: *Gonatobotrys simplex* parasitic on *Alternaria tenuis*. Phytopathology 67:309–314

Kirschner R, Bauer R, Oberwinkler F (1999) *Atractocolax*, a new heterobasidiomycetous genus based on a species vectored by conifericolous bark beetles. Mycologia 91:538–543

Kirschner R, Bauer R, Oberwinkler F (2001a) *Colacosiphon*: a new genus described for a mycoparasitic fungus. Mycologia 93:634–644

Kirschner R, Begerow D, Oberwinkler F (2001b) A new *Chionosphaera* associated with conifer inhabiting bark beetles. Mycol Res 105:1403–1408

Kreger-van Rij NJW, Veenhuis M (1971) Some features of the genus *Sporidiobolus* observed by electron microscopy. Antonie van Leeuwenhoek 37:253–255

Marvanová L, Bandoni RJ (1987) *Naiadella fluitans* gen. et sp. nov.: a conidial basidiomycete. Mycologia 79:578–586

Marvanová L, Suberkropp K (1990) *Camptobasidium hydrophilum* and its anamorph, *Crucella subtilis*: a new heterobasidiomycete from streams. Mycologia 82:208–217

Oberwinkler F (1990) New genera of auricularioid heterobasidiomycetes. Rep Tottori Mycol Inst 28:113–127

Oberwinkler F, Bandoni RJ (1982) *Atractogloea*: a new genus in the Hoehnelomycetaceae (Heterobasidiomycetes). Mycologia 74:634–639

Oberwinkler F, Bauer R (1989) The systematics of gastroid, auricularioid heterobasidiomycetes. Sydowia 41:224–256

Oberwinkler F, Bauer R (1990) *Cryptomycocolax*: a new mycoparasitic heterobasidiomycete. Mycologia 82:671–692

Oberwinkler F, Bandoni RJ, Bauer R, Deml G, Kisimova-Horovitz L (1984) The life history of *Christiansenia pallida*, a dimorphic, mycoparasitic heterobasidiomycete. Mycologia 76:9–22

Oberwinker F, Bauer R, Bandoni RJ (1990a) *Colacogloea*: a new genus in the auricularioid heterobasidiomycetes. Can J Bot 68:2531–2536

Oberwinkler F, Bauer R, Bandoni RJ (1990b) Heterogastridiales: a new order of basidiomycetes. Mycologia 82:48–58

Oberwinkler F, Bauer R, Schneller J (1990 c) *Phragmoxenidium mycophilum* sp. nov., an unusual mycoparasitic heterobasidiomycete. Syst Appl Microbiol 13:186–191

Oberwinkler F, Bauer R, Tschen J (1999) The mycoparasitism of *Platygloea bispora*. Kew Bull 54:763–769

Olive LS (1947) Notes on the Tremellales on Georgia. Mycologia 39:90–108

Roberts P (1994) *Zygogloea gemellipara*: an auricularioid parasite of *Myxarium nucleatum*. Mycotaxon 52:241–246

Roberts P (1996) Heterobasidiomycetes from Majorca and Cabrera (Balearic Islands). Mycotaxon 60:111–123

Roberts P (1997) New heterobasidiomycetes from Great Britain. Mycotaxon 63:195–216

Swann EC, Taylor JW (1993) Higher taxa of basidiomycetes: an 18S rRNA gene perspective. Mycologia 85:923–936

Swann EC, Frieders EM, McLaughlin DJ (1999) *Microbotryum*, *Kriegeria* and the changing paradigm in basidiomycete classification. Mycologia 91:51–66

Swann EC, Frieders EM, McLaughlin DJ (2001) Urediniomycetes. In: McLaughlin DJ, McLaughlin EG, Lemke PA (eds) Mycota VII Part B, Systematics and evolution. Springer, Berlin Heidelberg New York, pp 37–56

Weiß M, Oberwinkler F (2001) Phylogenetic relationships in Auriculariales and related groups – hypotheses derived from nuclear ribosomal DNA sequences. Mycol Res 105:403–415

Zugmaier W, Bauer R, Oberwinkler F (1994) Mycoparasitism of some *Tremella* species. Mycologia 86:49–56

17 Fungal Endophytes

SITA R. GHIMIRE and KEVIN D. HYDE

1 Introduction

Fungal endophytes have been isolated from almost every vascular plant studied and much has been written about their role and ecology. In this paper we review these aspects, but also review the role of molecular techniques in endophyte identification, the possible relationship with host specificity of fungal saprobes and suggest future areas for study.

2 Definition of a Fungal Endophyte

The term endophyte was introduced by De Bary (1866) and was initially applied to any organism found within a plant (Wilson 1995). The meaning of the term endophyte has been refined over time with the addition of new information (Siegel et al. 1984; Carroll 1986; Petrini 1986). Petrini (1991) used the term endophyte to mean all organisms inhabiting plant organs that at some time in their life can colonize internal plant tissues without causing apparent harm to the host. This has been the most widely used definition of endophytes and also includes the organisms that have a more or less lengthy epiphytic phase and also latent pathogens (Petrini 1991; Schulz et al. 1998). There has however, been a certain level of disagreement expressed by some mycologists over the inclusion of plant pathogens as endophytes, since endophytes are nonaggressive, nonpathogenic and have developed a mutualistic role with their hosts (Freeman and Rodriguez 1993; Tyler 1993; Stone et al. 1994; Sinclair and Cerkauskas 1996).

Studies on the endophyte composition in different hosts have identified organisms with varying roles within their hosts. Organisms having weak parasitic associations, localized infection, quiescent infection, latent infection and aggressive parasitic relationships with their hosts have often been recovered (Jersch et al. 1989; Kehr 1992; Gotz et al. 1993; Kehr and Wulf 1993; Williamson 1994; Agrios 1997). Wilson (1995) provided a working definition

Plant Surface Microbiology
A. Varma, L. Abbott, D. Werner, R. Hampp (Eds.)

of the term by analyzing the different levels of endophytic association and stated that "endophytes are fungi or bacteria which, for all or part of their life cycle, invade the tissues of living plants and cause unapparent and asymptomatic infections entirely within plant tissues but causes no symptoms of the disease". The same organism may also be described as a saprobe or pathogen at other times (Boddy and Griffith 1989).

3 Role of Endophytes

Endophytes have previously been defined as mutualists and are closely related to virulent pathogens, but have limited pathogenicity, and have probably evolved directly from plant pathogenic fungi (Carroll 1988). The mutualistic symbiosis includes the lack of destruction of most cells or tissues, nutrient and chemical cycling between the fungus and hosts, enhanced longevity and photosynthetic capacity of cell and tissue under the influence of infection, enhanced survival of fungus, and a tendency towards greater host specificity than is seen in necrotrophic infections (Lewis 1973). It is often difficult to differentiate between an endophyte and pathogen as many plant pathogens undergo an extensive phase of asymptomatic latent infection before the appearance of disease symptoms and the mutation in a single genetic locus can change a pathogen to nonpathogenic endophytic organism with no effect on its host specificity (Freeman and Rodriguez 1993). Latent infection is the state in which a host is infected with a pathogen, but does not show any symptoms and persists until signs or symptoms are prompted to appear by environmental or nutritional conditions or by the state of maturity of the host or pathogen (Agrios 1997). The latent infection is considered as the highest level of parasitism because the host and parasite coexist for a period of time with minimal damage to the host. Hence, the relationship between plant pathogenic fungi and host is considered as parasitic.

Wilson (1995) argued that the term endophyte bears much affinity to the term pathogen and stated that it is often difficult to be able to classify a particular species. Sinclair and Cerkauskas (1996) compared endophyte colonization and latent infections by fungi and stated that they are distinctly different. Endophytic fungi are asymptomatic and considered mutualistic, whereas latent infecting fungi are parasitic and cannot be considered mutualistic. Rather, they are considered to be one of the most advanced stages of parasitism as the host and parasite co-exist for a period of time with minimal effect on the host (Sinclair and Cerkauskas 1996). Hammon and Faeth (1992) suggested that the disproportionate amount of attention that has been paid to the study of grass endophytes has lead to the impression that all endophytes must be mutualists. There seems to be a greater probability of mutualism in the fungal species that are transmitted through seeds, as transmission will increase directly as a result of host survival. The association where only one fungus is associated within the host plant is more likely to be mutualistic (Hammon and Faeth 1992).

The endophytes associated with grasses have received much attention, and many of these have been found to produce physiologically active alkaloids that cause their hosts to be toxic to mammals and increase their resistance to insect herbivores (Funk et al. 1983; Clay 1988; Cheplick and Clay 1988; Prestidge and Gallagher 1988). In the grasses and other plant hosts, endophytes have also been shown to enhance plant growth, reduce infection by nematodes, increase stress tolerance and increase nitrogen uptake in nitrogen deficit-soils (Latch et al. 1985; Clay 1987, 1990; Kimmons 1990; Bacon 1993; Gasoni and Stegman De Gurfinkel 1997; Rommert et al. 1998; Verma et al. 1999; Bultman and Murphy 2000;). Several reviews are available on secondary metabolite production by endophytes (Miller 1986; Clay 1991; Petrini et al. 1992). Endophytes in culture can produce biologically active compounds (Brunner and Petrini 1992) including several alkaloids, paxilline, lolitrems and tertraenone steroids (Dahlman et al. 1991), antibiotics (Fisher et al. 1984a, b) and plant growth promoting factors (Petrini et al. 1992). Endophytes are increasingly being identified as a group of organisms capable of providing a source of secondary metabolites for use in biotechnology and agriculture (Bills and Polishook 1992).

4 Modes of Endophytic Infection and Colonization

The colonization of plant tissues by endophytes, plant pathogens and mycorrhizae involves several steps involving host recognition, spore germination, penetration of the epidermis and tissue colonization (Petrini 1991, 1996). The inoculum source of fungal endophytes is widely considered to be the airborne spores, and also seed transmission and transmission of propagules by insect vectors (Petrini 1991). A high level of genetic diversity of endophyte isolates suggests that infection foci arise from different strains of fungi derived from constant new inoculum (Hammerli et al. 1992; Rodrigues et al. 1993). In terms of mechanical and enzymatic elements of penetration by endophytic fungi, it can be assumed that endophytes adopt the same strategy for penetration of host tissue as pathogens (Petrini et al. 1992). Fungi can invade plant tissues by direct cuticular penetration, via appressoria formed on the cuticle, after which penetration occurs through the cuticle and epidermal cell wall or via natural openings like stomata (O'Donnell and Dickinson 1980; Muirhead and Deverall 1981; Kulik 1988; Cabral et al. 1993; Viret et al. 1993; Viret and Petrini, 1994). Following penetration the infection may be inter-cellular or intra-cellular and may be limited to one cell or in a limited area around the penetration site. Limited cytological work on nonclavicipitaceous endophytes have shown that the infection of these endophytes in host plants may be inter- or intra-cellular and often localized in single cells (Stone 1988; Suske and Acker 1989, Cabral et al. 1993).

Some endophytic fungi (including those which are latent pathogens) are host-specific, whereas others seem to invade any available hosts (Carroll 1988; Petrini et al. 1992). When studying the infection of *Juncus* spp. with the endophytes *Stagnospora innumerosa* and *Drechslera* spp., Cabral et al. (1993) observed callose formation in the individual cells as a host defense response. Sieber et al. (1991) found that the synthesis of highly specialized enzymes associated the penetration of cuticular layers of the host by the endophyte *Melanconium* spp. Several other investigations have also reported a growth response of host calli to endophytes (Hendry et al. 1993; Peters et al. 1998).

Schulz et al. (1999) studied the secondary metabolites produced by endophytes and their host interactions in order to understand why endophytic infections are symptomless. The production of herbicidally active substances was three times that of soil isolates and twice that of phytopathogenic fungi, whereas the phenolic metabolites in the host were higher in the roots of plants infected with an endophyte than in those infected with pathogens. Their study hypothesized that both the pathogen–host and endophyte–host interaction involved constant mutual antagonisms, at least in part based on the secondary metabolites the partners produce. The pathogen – host interaction was thought to be imbalanced and resulted in disease while that of the endophytes and its host is a balanced antagonism.

5 Isolation of Endophytes

Techniques for endophyte isolation and culture have been developed gradually over time. Bacon and White (1994) have written an excellent review on staining, media and procedure for analyzing endophytes. Endophytes can be isolated from various plant parts such as seeds, leaf and stem and direct isolation of ascospores is also in practice. The plant and plant parts collected for studying endophytic communities should look apparently healthy, in order to minimize the compounding effect because of plant pathogenic and saprobic species. Young tissue is appropriate for isolation as older tissues often contain many additional fungi that make isolation of slow growing fungi difficult (Bacon and White 1994). The samples should be processed in the shortest time possible after collection. Plant parts for investigation should be cut into small pieces to facilitate sterilization and isolation processes. Bills (1996) discussed various surface sterilization techniques in detail. Any method can be used for surface sterilization provided that it can eliminate most of the epiphytic fungi from the exterior tissues and encourage the growth of the internal mycobiota. The method used by Petrini et al. (1992) has been used extensively and found very successful in studying endophytes (Rodrigues and Samuels 1990; Schulz et al. 1993). This method comprises dipping samples in 96% ethanol for 1 min, then in 65% commercial Chlorox (final concentration 3.25% aqueous sodium hypochlorite) for 10 min and finally in 96% ethanol for 30 s. Malt extract agar is considered the most suitable media for the growth and sporulation of endophytic fungi (Bills and Polishook 1992; Bills 1996). Amendment of medium with streptomycin sulfate is practised to prevent bacterial contamination. To prevent the fast growing fungi overgrowing the plate, a growth inhibitor, Ross Bengal is added to the agar.

Surface-sterilized plant tissues are plated in an appropriate medium amended with antibiotics and Rose Bengal and incubated at room temperature with periodic light and darkness. Incubated plates are checked after 1 week of incubation at regular intervals for fungal development. If the colony is very small and there is a risk of engulfment by other colonies, it needs to be subcultured. Subcultured isolates are generally maintained at room temperature for many weeks to study morphological and other characteristics. Some isolates may fail to produce reproductive structures even after several months. Subculture of these isolates onto medium with autoclaved host tissue strips can promote sporulation (Matsushima 1971). In general, sterile isolates should be checked regularly for fruiting bodies over a period of 3–4 months and the isolates failing to produce fruiting body are referred to as sterile "morphotypes" depending on the characteristics of culture.

Other methods to promote sporulation of morphospecies should also be tried. Guo et al. (1998) used twigs in conical flasks over a 3-month period to promote sporulation of endophytes. Other methods can be designed, but should try to mimic the situation in nature as closely as possible.

6 Molecular Characterization of Endophytes

Molecular approaches have been used to resolve the problems in fungal taxonomy and in the identification of fungi (Rollo et al. 1995; Ma et al. 1997; Zhang et al. 1997; Ranghoo et al. 1999). The use of molecular techniques for the direct detection and identification of fungi within natural habitats has been reviewed by Liew et al. (1998). Molecular techniques have mainly been used in the detection and identification of mycorrhizal fungi and phytopathogenic fungi directly from within plant tissues (Mills et al. 1992; Johanson and Jeger 1993; Beck and Ligon 1995; Bonito et al. 1995; Abbas et al. 1996; Chambers et al. 1998). Similarly, molecular techniques have been employed to detect and identify fungi from the grass clothing of Iceman, from bamboo leaves and glacial ice strata (Rollo et al. 1995; Ma et al. 1997; Zhang et al. 1997).

A most frequently encountered problem in endophyte study is the presence of mycelia sterilia, making their identification difficult (Guo et al. 2000). Variable proportions of mycelia sterilia have been reported ranging from 11% of isolates from palm (*Trachycarpus fortunei*) in China, 13% of endophytes obtained from two *Licuala* species in Brunei and Australia, 15% of isolates from evergreen shrubs in western Oregon, 16.5% of isolates from fronds of *Livistona chinensis* in Hong Kong, 27% of isolates from leaves of *Sequoia sempervirens* in central California and 54% of isolates obtained from twigs of *Quercus ilex* in Switzerland (Petrini et al. 1982; Espinosa-Garcia and Langenheim 1990; Fisher et al. 1994; Taylor et al. 1999; Frohlich et al. 2000; Gou et al. 2000). A high proportion of unidentified endophytic isolates resulting from traditional methodology has prompted various workers to develop methodology to improve sporulation in mycelia sterilia (Matsushima 1971; Guo et al. 1998; Taylor et al. 1999; Frohlich et al. 2000). The problem of having many nonidentifiable mycelia sterilia, however still remains. Hence, molecular techniques could be the best alternative to identify this taxa.

There have been only a small number of studies using molecular techniques to investigate endophytic fungal communities. Random amplified polymorphic DNA (RAPD) markers were used to study the genotypic diversity in the populations of *Rabdocline parkeri* from Douglas fir (McCutcheon and Carroll 1993). Specific PCR primers were used to amplify rDNA fragments of the endophyte *Acremonium coenophialum* from infected tall fescue tissues (Doss and Welty 1995). The genetic diversity of *Epichloe typhina*, an endophyte in *Bromus erectus*, was studied using a microsatellite-containing locus as a molecular marker (Groppe et al. 1995). Guo et al. (2000) performed phylogenetic analysis based on rDNA of 19 morphospecies from frond tissues of *Livistona chinensis* and found that they were filamentous Ascomycota belonging to the different taxonomic levels in the Loculoascomycetes and Pyrenomycetes. The 5.8S gene and flanking internal transcribed spacer of rDNA were used in detection and taxonomic placement of endophytic fungi within frond tissues of *Livistona chinensis* (Guo et al. 2001). Ribosomal DNA sequence analysis was used to validate the morphospecies concept used in endophyte study to group mycelia sterilia (Hyde et al. 2001). Therefore, rDNA sequence analysis is in frequent use to resolve the identification problem associated with endophytic fungi. These studies show increasing an use of molecular techniques in detection, identification, and population and ecological studies of endophytes.

7 Are Endophytes Responsible for Host Exclusivity/Recurrence in Saprobic Fungi?

There has been much debate as to whether saprobic fungi are host-specific as this has important implications for estimates of fungal numbers (Hawksworth 1991, 2001; Fröhlich and Hyde 1999; Hyde 2001). Zhou and Hyde (2001) explored the literature on host-specific saprobes and came to the conclusion that it was hard to prove that saprobic fungi were host-specific. They introduced the terms host exclusivity and host recurrence as more suitable for use with saprobic fungi. Host exclusivity is the exclusive occurrence of a strictly saprobic fungus on a particular host, while host recurrence is the frequent or predominant occurrence of a fungus on a particular host.

The basis for host recurrence in saprobic fungi is interesting and several factors may be responsible. Wong and Hyde (2001) thought that host exclusive saprobes may be responding to differences in physical structure or nutrient levels of the potential hosts. There is also the possibility that enzyme production capabilities may influence whether a certain fungus can decay a certain host. However, recent studies have shown that most fungi can produce a wide range of enzymes capable of degrading simple sugars and cellulose (Lumyong et al. 2002). Fewer fungi can produce enzymes capable of digesting lignins (Leung and Pointing 2002). This would however, restrict fungi to lignified versus nonlignified plant tissues and is unlikely to be responsible for host recurrence (restricting fungi to certain plant species) as a large range of host tissues incorporate lignins into their tissues.

Wong and Hyde (2001) studied the saprobes on six grass and one sedge species in Hong Kong and found that certain fungi showed host exclusivity or specificity. They hypothesized that these fungi may be host-specific endophytes that later become saprobes. There is much circumstantial evidence supporting this hypothesis and this has been discussed by Zhou and Hyde (2001) and Hyde (2001).

8 Conclusions

There have now been many studies on the diversity and ecology of endophytes of grass and nongrass hosts in both tropical and temperate regions (Viret and Petrini 1994; Bussaban et al. 2001; Photita et al. 2001). The problem in most of these studies is that two uninformative groups of fungi are generally isolated; the first major group being typical endophytic genera such as *Colletotrichum*, *Phomopsis* and *Phyllosticta* while the second are mycelia sterilia. The first group are rarely recorded as saprobes on the host, although some may be pathogens. Therefore, the role of these fungi is puzzling and they may actually have no function. It is possible that the spores have landed on the plant surface and produced a germ tube which has penetrated the plant stoma, but then cannot progress further due to plant defense. Future studies should, therefore concentrate on the role of these common endophytes, rather than provide uninformative lists with ecological data that have little consequence.

The mycelia sterilia may be a more important group, but until we can find some way to identify more of them, it is impossible to elucidate their function. Methods need to be developed to stimulate these fungi to sporulate, or at least molecular techniques need to be refined in order to make identification simpler. Future studies should, therefore concentrate on developing these methods, rather than providing uninformative data on mycelia sterilia with ecological data that again have little consequence.

Studies on endophytes have shown the beneficial roles of endophytic associations to the host as protection against mammals, resistance to insect herbivores and other pathogenic fungi, increased growth and development, nutrient uptake and stress tolerance in plants including agriculturally important crops. The actual mechanisms involved for such phenomenon, however are poorly understood. The current understanding of secondary metabolites of host and endophyte origin and their interactions is limited. This area demands further studies which could lead to the discovery of novel compounds of biotechnological and agricultural importance. The mechanism by which a latent form of a pathogen turns pathogenic and vice-versa could be an interesting area of research of the highest plant pathological significance.

References and Selected Reading

Abbas JD, Hertrick BAD, Jurgenson JE (1996) Isolate specific detection of mycorrhizal fungi using genome specific primer pairs. Mycologia 88:939–946

Agrios GN (1997) Plant pathology. Academic Press, London

Bacon CW (1993) Abiotic stress tolerances (moisture and nutrients) and photosynthesis in endophyte-infected tall fescue. Agric Ecosyst Environ 44:123–141

Bacon CW, White JF (1994) Stain, media and procedure for analyzing endophytes. In: Bacon CW, White JF (eds) Biotechnology of endophytic fungi of grasses, CRC Press, Boca Raton, pp 47–56

Beck JJ, Ligon JM (1995) Polymerase chain reaction assays for the detection of *Stagonospora nodorum* and *Septoria tritici* in wheat. Phytopathol 85:319–324

Bills GF (1996) Isolation and analysis of endophytic fungal communities from woody plants. In: Redlin SC, Carris LM (eds) Endophytic fungi in grasses and woody plants: systematic, ecology and evolution, APS Press, St. Paul, MN, pp 31–65

Bills GF, Polishook JD (1992) Recovery of endophytic fungi from *Chamaecyparis thyroides*. Sydowia 44:1–12

Boddy L, Griffith GS (1989) Role of endophytes and latent invasion in the development of decay communities in sapwood of angiospermous trees. Sydowia 41:41–73

Bonito RD, Elliott ML, Jardin EAD (1995) Detection of the arbuscular mycorrhizal fungus in roots of different plants species with the PCR. Appl Environ Microbiol 61:2809–2810

Brunner F, Petrini O (1992) Taxonomic studies of *Xylaria* species and xylariaceous endophytes by isozymeelectrophoresis. Mycol Res 96:723–733

Bultman TL, Murphy JC (2000) Do fungal endophytes mediate wound-induced resistance? In: Bacon CW, White JF (eds) Microbial endophytes, Marcel Dekker, New York

Bussaban B, Lumyong, S, Lumyong P, McKenzie EHC, Hyde KD (2001) Endophytic fungi from *Amomum siamense*. Can J Microbiol 47:943–948

Cabral D, Stone JK, Carroll G (1993) The internal mycobiota of *Juncas* sp.: microscopic and cultural observations of infection pattern. Mycol Res 97:367–376

Carroll G (1986) The biology of endophytism in plants with particular references to woody perennials. In: Fokkema NJ, Van den Heuvel J (eds) Microbiology of phyllosphere, Cambridge University Press, Cambridge, pp 205–222

Carroll G (1988) Fungal endophytes in stems and leaves: from latent pathogen to mutualistic symbiont. Ecology 69:2–9

Chambers SM, Sharples JM, Cairney JWG (1998) Towards a molecular identification of the *Pisonia* mycobiont. Mycorrhiza 7:319–321

Cheplick GP, Clay K (1988) Acquired chemical defenses in grasses: The role of fungal endophytes. Oikos 52:309–318

Clay K (1987) Effect of fungal endophytes on the seed and seedling biology of *Lolium perenne* and *Festuca arundinaceae*. Oecologia (Berlin) 73:358–362

Clay K (1988) Clavicipitaceous fungal endophytes of grasses: co-evolution and change from parasitism to mutualism. In: Pyrozynski KA, Hawksworth DL (eds) Co-evolution of fungi with plants and animals. Academic Press, New York, pp 79–105

Clay K (1990) Fungal endophytes of grasses. Annu Rev Ecol Syst 21:275–297

Clay K (1991) Fungal endophytes, grasses and herbivores. In: Barbosa P, Krischik VA, Jones CG (eds) Microbial mediation of plant herbivore interactions. Wiley, New York, pp 199–226

Dahlman DL, Eichenseer H, Siegel MR (1991) Chemical perspective on endophyte grass interactions and their implications to insect herbivory. In: Barbosa P, Krischik VA, Jones CG (eds) Microbial mediation of plant herbivore interactions. Wiley, New York, pp 227–252

De Bary A (1866) Morphologie und Physiologie der Pilze, Flechten, und Myxomyceten. Hofmeister's Handbook of Physiological Botany. vol 2. Leipzig

Doss PR, Welty RE (1995) A polymerase chain reaction based procedure for detection of *Acremonium coenophialum* in tall fescue. Phytopathol 85:913–917

Espinosa-Garcia FJ, Langenheim JH (1990) The leaf fungal endophyte community of a coastal red wood population diversity and spatial patterns. New Phytol 116:89–97

Fisher PJ, Anson AE, Pertini O (1984a) Antibiotic activity of some endophytic fungi from ericaceous plants. Bot Helv 94:249–253

Fisher PJ, Anson AE, Pertini O (1984b) Novel antibiotic activity of an endophyte *Cryptosporiopsis* sp. isolated from *Vaccinium myrtillus*. Trans Br Mycol Soc 83:145–148

Fisher PJ, Pertini O, Petrini LE, Sutton, BC (1994) Fungal endophytes from leaves and twigs of *Quercus ilex* L. from England, Majorca and Switzerland. New Phytol 127: 133–137

Freeman S, Rodriguez RJ (1993) Genetic conversion of a fungal plant pathogen to a nonpathogenic, endophytic mutualist. Science 260:75

Fröhlich J, Hyde KD (1999). Biodiversity of palm fungi in the tropics: are global fungal diversity estimates realistic? Biodivers Conserv 8:977–1004

Fröhlich J, Hyde KD, Petrini O (2000) Endophytic fungi associated with palm. Mycol Res 104:1202–1212

Funk CR, Halisky PM, Johnson MC, Siegel MR, Stewart AV, Ahamad S, Hurley RH, Harvey IC (1983) An endophytic fungi and resistance to Sod Webworms: association of *Lolium perenne* L. Bio/technology April:189–191

Gasoni L, Stegman De Gurfinkel B (1997) The endophyte *Cladorrhinum foecundissimum* in cotton roots: phosphorus uptake and host growth. Mycol Res 101:867–870

Gotz M, Zornabach W, Boyle C (1993) Life cycle of *Mycosphaerelle brassicicola* (Duby) Lindau and ascospore production in vitro. J Phytopathol 139:298–308

Guo LD, Hyde KD Liew ECY (1998) A method to promote sporulation in palm endophytic fungi. Fungal Divers 1:109–113

Guo LD, Hyde KD, Liew ECY (2000) Identification of endophytic fungi from *Livistona chinensis* based on morphology and rDNA sequences. New Phytol 147:617–630

Guo LD, Hyde KD, Liew ECY (2001) Detection and taxonomic placement of endophytic fungi within frond tissues of *Livistina chinensis* based on rDNA sequences. Mol Phylogenet Evol 20:1–13

Groppe K, Sanders I, Wiemken A, Boller T (1995) A microsatellite marker for studying the ecology and diversity of fungal endophytes (*Epichloe* spp.) in grasses. Appl Environ Microbiol 61:3943–3949

Hammerli UA, Brandle UE, Petrini O, McDermott JM (1992) Differentiation of isolates of *Discula umbrinella* (teleomorph: *Apiognomonia errabunda*) from beech, chestnut and oak using RAPD markers. Mol Plant-Microbe Interact 5:479–483

Hammon K.E, Faeth SH (1992) Ecology of plant-herbivore communities: a fungal component. Nat Toxins 1:197–208

Hawksworth DL (1991) The fungal dimension of biodiversity: magnitude, significance, and conservation. Mycolog Res 95:641–655

Hawksworth DL (2001) The magnitude of fungal diversity: the 1.5 million species estimate revisited. Mycol Res 105:1422–1432

Hendry SJ, Boddy L, Lonsdale D (1993) Interaction between callus culture of European beech, indigenous ascomycetes and derived fungal extracts. New Phytol 123:421–428

Hyde KD (2001) Where are the missing fungi? Does Hong Kong have the answers? Mycol Res 105:1514–1518

Hyde KD, Lacap DC, Liew ECY (2001) An evaluation of the fungal 'morphospecies' concept based on ribosomal DNA sequences (Abstract). Phytopathology 91:S113

Jersch S, Scherer C, Hutz G, Schlosser E (1989) Proanthocianidins as a basis for quiescence of *Botrytis cinerea* in immature strawberry fruits. Z Pflanzenkrankheiten Pflanzenschutz 96:365–378

Johanson A, Jeger MJ (1993) Use of PCR for detection of *Mycosphaerella fijiensis* and *M. musicola*, the causal agent of Sigayoka leaf spot in banana and plantain. Mycol Res 97:670–674

Kehr RD (1992) Pezicula cancer of *Quercus rubra* L., caused by *Pezicula cinnamomea* (DC) Sacc. II. Morphology and biology of the causal agent. Eur J For Pathol 22:29–40

Kehr RD, Wulf A (1993) Fungi associated with above ground portion of declined oaks (*Quercus rubra*) in Germany. Eur J For Pathol 23:18–27

Kimmons CA (1990) Nematode reproduction on endophyte infected and endophyte free tall fescue. Plant Dis 74:757–761

Kulik MM (1988) Observations by scanning electron and brightfield microscopy on the mode of penetration of soybean seedlings by *Phomopsis phaseoli*. Plant Dis 72:115–118

Latch GCM, Hunt WF, Musgrave DR (1985) Endophytic fungi affect growth of perennial rye grass. New Zealand J Agric Res 28:165–168

Leung PC, Pointing SB (2002) Effect of different carbon and nitrogen regimes on Poly R decolorization by white rot fungi. Mycolog Res 106:86–92

Lewis DH (1973) Concept in fungal nutrition and the origin of biotrophy. Biol Rev 48:261–278

Liew ECY, Guo LD, Ranghoo VM, Goh TK, Hyde KD (1998) Molecular approaches to assessing fungal diversity in the natural environment. Fungal Divers 1:1–17

Lumyong S, Lumyong P, McKenzie EH, Hyde KD (2002) Enzymatic activity of endophytic fungi of six native seedling species from Doi Suthep-Pui National Park, Thailand. Can J Microbiol 48(12):1109–1112

Ma LJ, Catramis CM, Rogers SO, Starmer WT (1997) Isolation and characterization fungi entrapped in glacial ice. Inoculum 48:23–24

Matsushima T (ed) (1971) Microfungi of the Solomon Islands and Papua New Guinea. Matsushima, Kobe, Japan, pp 15–20

McCutcheon TL, Carroll GC (1993) Genotypic diversity in populations of a fungal endophytes from Douglas fir. Mycologia 85:180–186

Miller JD (1986) Toxic metabolites of epiphytic and endophytic fungi of conifers needles. In: Fokkema, NJ, Heuvel JVD (eds) Microbiology of phyllosphere. Cambridge University Press, Cambridge, pp 223–231

Mills PR, Sreenivasaprasad S, Brown AE (1992) Detection and differentiation of *Colletotrichum gloeosporioides* isolates using PCR. FEMS Microbiol Lett 98:137–144

Muirhead IF Deverall BJ (1981) Role of appressoria in latent infection of banana fruits by *Colletotrichum musae*. Physiol Plant Pathol 19:77–84

O'Donnell J and Dickinson CH (1980) Pathogenicity of *Alternaria* and *Cladosporium* isolated on *Phaseolus*. Trans Br Mycol Soc 74:335–342

Peters S, Aust AJ, Draeger S, Schulz B (1998) Interaction in dual cultures of endophytic fungi with host and non-host plant calli. Mycologia 90:360–367

Petrini O (1986) Taxonomy of endophytic fungi of aerial plant tissue. In: Fokkema NJ, Van den Heuvel J (eds) Microbiology of phyllosphere. Cambridge University Press, Cambridge, pp 175–187

Petrini O (1991) Fungal endophytes of tree leaves. In: Andrews J, Hirano S (eds) Microbila ecology of leaves, Springer, Berlin Heidelberg New York, pp 179–197

Petrini O (1996) Ecological and physiological aspect of host specificity in endophytic fungi. In: Redlin SC, Carris LM (eds) Endophytic fungi in grasses and woody plants. APS Press, St. Paul, MN

Petrini O, Stone J, Carroll FE (1982) Endophytic fungi in evergreen shrubs in western Oregon: a preliminary study. Can J Botany 60:789–796

Petrini O, Fisher PJ, Petrini LE (1992) Fungal endophytes of bracken (*Pteridium aquilinum*) with some reflections on their use in biological control. Sydowia 44:282–293

Photita W, Lumyong S, Lumyong P, Hyde KD (2001) Endophytic fungi of wild banana (*Musa acuminata*) at Doi Suthep Pui National Park, Thailand. Mycol Res 105:1508–1514

Prestidge RA, Gallagher RT (1988) Endophyte fungus confers resistance to rye grass: Argentine stem weevil larval studies. Ecol Entomol 13:429–435

Ranghoo VM, Hyde KD, Liew ECY, Spatafora JW (1999) Family placement of *Ascotaiwanian* and *Ascolacicola* based on DNA sequences from the large subunit rRNA gene. Fungal Divers 2:159–168

Rodrigues KF, Samuels GJ (1990) Preliminary study of endophytic fungi in tropical palm. Mycol Res 94:827–830

Rodrigues KF, Leucthmann A, Petrini O (1993) Endophytic species of *Xylaria*: Cultural and isozymic studies. Sydowia 45:116–138

Rollo F, Sassaroli S, Ubaldi M (1995) Molecular phylogeny of the fungi of the Iceman's grass clothing. Curr Genet 28:289–297

Rommert AK, Strack D, Aust HJ, Schulz B (1998) Verhalten sich Endophyten unter N-stress als Schwachenparasiten? Bielefelder Okol Beitr 14:307–311

Schulz B, Wanke U, Draeger S and Aust HJ (1993) Endophytes from herbaceous plants and shrubs: effectiveness of surface sterilization methods. Mycol Res 97:1447–1450

Schulz B, Guske S, Dammann U, Boyle C (1998) Endophyte-host interactions. II. Defining symbiosis of the endophyte-host interaction. Symbiosis 25:213–227

Schulz B, Rommert AK, Dammann U, Aust HJ, Strack D (1999) The endophyte-host interaction: a balanced antagonism? Mycol Res 103:1275–1283

Sieber TN, Sieber-Canavesi F, Dorworth CE (1991) Endophytic fungi of red alder (*Alnus rubra*) leaves and twigs in British Colombia. Can J Bot 69:407–411

Siegel MR, Johnson M., Varney DR, Nesmith WC, Buckner RC, Bush LP, Burrus PB (1984) A fungal endophyte in tall fescue: incidence and dissemination. Phytopathology 74:932–937

Sinclair JB, Cerkauskas RF (1996) Latent infection vs. endophytic colonization by fungi. In: Redlin SC, Carris LM (eds) Endophytic fungi in grasses and woody plants: systematics, ecology and evolution. APS Press, St. Paul, pp 3–29

Stone JK (1988) Fine structure of latent infection by *Rhabdocline parkeri* on Douglas fir, with observation on uninfected epidermal cells. Can J Bot 66:45–54

Stone JK, Viret O, Petrini O, Chapela IH (1994) Histological studies of host penetration and colonization by endophytic fungi. In: Pertini O, Ouellette G (eds) Host wall alterations by parasitic fungi. APS Press, St. Paul, pp 115–126

Suske, J, Acker G (1989) Identification of endophytic hyphae of *Lophodermium piceae* in tissues of green, symptomless Norway spruce needles by immunoelectron microscopy. Can J Bot 67:1768–1774

Taylor JE, Hyde KD, Jones EBJ (1999) Endophytic fungi associated with the temperate palm *Trachycarpus fortunei* within and outside its natural geographical range. New Phytol 142:335–346

Tyler BM (1993) To kill or not to kill: the genetic relationship between a parasite and endophyte. Trends Microbiol 1:252–254

Verma A, Verma S, Sudha, Sahay N, Butehorn B, Franken P (1999) *Piriformospora indica*, a cultivable plat growth promoting root endophyte. Appl Environ Microbiol 65:2741–2744

Viret O, Petrini O (1994) Colonization of beech leaves (*Fagus sylcatica*) by the endophyte *Discula umbrinella* (teleomorph, *Apiognomonia errabunda*). Mycol Res 98:423–432

Viret O, Scheidegger C, Pertini O (1993) Infection of beech leaves (*Fagus sylcatica*) by the endophyte Discula umbrinella (teleomorph, Apiognomonia errabunda) – low temperature scanning electron microscopy studies. Can J Bot 71:1520–1527

Williamson B (1994) Latency and quiescence in survival and success of fungal plant pathogen. In: Blackman JP, Williamson B (eds) Ecology of plant pathogens. CAB International, London, pp 187–207

Wilson D (1995) Endophyte – the evolution of term, a classification of its use and definition. Oikos 73:274–276

Wong MKM, Hyde KD (2001) Diversity of fungi on six species of Gramineae and one species of Cyperaceae in Hong Kong. Mycol Res 105:1485–1491

Zhang W, Wildel JF, Clark LG (1997) Bamboozled again! Inadvertent isolation of fungal rDNA sequences from bamboos (Poaceae: Bambusoideae). Mol Phylogenet Evol 8:205–217

Zhou DQ, Hyde KD (2001) Host-specificity, host-exclusivity and host-recurrence in saprobic fungi. Mycol Res 105:1449–1457

18 Mycorrhizal Development and Cytoskeleton

MARJATTA RAUDASKOSKI, MIKA TARKKA and SARA NIINI

1 Introduction

The formation of mycorrhiza requires morphological changes, both in the plant cells and fungal hyphae, necessary for the development and maintenance of the signal and nutrient exchange at plant fungal interfaces. It is well known that cytoskeletal elements play a central role both in the morphogenesis of plant root cells (Barlow and Baluška 2000) and of fungal hyphae (Raudaskoski et al. 2001). In the present review, the plant and fungal genes encoding the structural proteins of main cytoskeletal elements, microtubules (MTs) and microfilaments (MFs), are described. Some speculations of the functional significance of cytoskeletal rearrangements observed in plant cells and fungal hyphae at the formation of endo- and ectomycorrhiza are presented. The reorganization of the cytoskeleton results from interactions with proteins that serve by themselves as targets for intra- and extracellular signal mediating pathways (Johnson 1999; Kost et al. 1999b). The presence of such pathways in mycorrhiza is discussed. The different phases in the cell cycle also requires rearrangements in the cytoskeleton (Mews et al. 1997; John et al. 2001). This aspect is shortly discussed in association with known effects of mycorrhiza on the plant cell cycle. Finally, the general methods used in visualization of cytoskeletal components are shortly introduced.

2 Cytoskeletal Components

The cytoskeleton is composed of filamentous structures whose arrangements are continuously changing in living cells in response to different developmental and environmental cues. In plant and fungal cells there are two main cytoskeletal proteins: actin and tubulin. Actin monomers polymerize to thin filaments known as MFs or actin filaments. Tubulin polymerizes to MTs. Both MFs and MTs are polarized structures with minus and plus ends. The minus end is often attached to some subcellular structure while the plus end is

Plant Surface Microbiology
A. Varma, L. Abbott, D. Werner, R. Hampp (Eds.)

mainly untouched and is able to polymerize and depolymerize more freely than the minus end. The polarity of the cytoskeletal elements is well studied in animal cells while in plant cells and fungal hyphae, their polarity is known only in a few cases (Euteneuer and McIntosh 1980).

In eukaryotic cells, MTs participate in cell division, cell shape changes, cell motility and intracellular organelle trafficking as well as in cell wall synthesis (Wasteneys 2000; Lloyd and Hussey 2001). Actin cytoskeleton is involved in many different developmental processes including the establishment of cell polarity and tip growth (Hepler et al. 2001; Raudaskoski et al. 2001), division plane determination, cell elongation, positioning of proteins on membranes and cytoplasmic streaming (Meagher et al. 1999). Recently, connections have been revealed between MT and actin cytoskeleton mediated by interaction between MF- and MT-associated motor molecules or scaffold proteins (Goode et al. 2000). Thus, the reorganization in one cytoskeletal component may also lead to the reorganization of the other.

2.1 Expression of Tubulin-Encoding Genes

MTs in eukaryotic cells are composed of polymerized α- and β-tubulin heterodimers. A less abundant form, γ-tubulin occurs also in eukaryotic cells (Liu et al. 1993). γ-Tubulin functions in MT organizing centers in centrosomes in animal cells (Joshi et al. 1992) and in spindle pole bodies in fungi (Oakley et al. 1990). Higher plants have no discrete MT organizing centers and γ-tubulin appears to be dispersed around the cells (Joshi and Palevitz 1996). Plant α- and β-tubulin gene families consist of five to ten genes while in fungi the number is much lower with only one or two family members (Joyce et al. 1992; Kopczak et al. 1992; Snustad et al. 1992; Villemur et al. 1992, public databases). Analysis of tubulin gene expression, mainly done in *Arabidopsis* and maize, has shown that transcripts of *tubulin genes* occur in all plant tissues, but their accumulation can be specific for different plant organs or developmental stages (Montoliu et al. 1989; Hussey et al. 1990; Joyce et al. 1992; Villemur et al. 1994) or induced by environmental factors (Kerr and Carter 1990a). Tubulin expression studies in mycorrhizal plants are of special interest, since the invading fungus alters the expression pattern of tubulins known from uninfected root (Bonfante et al. 1996).

In maize it has been shown that the accumulation of the transcripts from all six α-tubulin genes is relatively high in the root tip, but low in the root cortex (Joyce et al. 1992). More detailed analyses at cellular level have shown that in the maize root the preferential expression of *tua1* and *tua3* genes occurs in root meristem cells differentiating into cortex and vascular tissue, respectively. The transcripts of *tua2* gene accumulate in maize root epidermis (Uribe et al. 1998), while *tua4* transcripts are mainly expressed in root vascular tissue (Joyce et al. 1992). In agreement with the idea that a mycorrhizal fungus might affect the expression pattern of tubulin genes is the observation that *tua3* transcripts increased in the differentiated cortical cells at the invasion of an endomycorrhizal fungus. Similarly, in the transgenic tobacco, in which the Gus expression took place under the promoter of maize tubulin gene *tua3*, Gus activity occurred in differentiated cortical cells infected by endomycorrhizal fungus (Bonfante et al. 1996). As with α-tubulins the expression of *Arabidopsis* β-tubulins is relatively high in the root tip and vascular tissue, but low in root cortical cells (Villemur et al. 1994). The *tub6* and *tub8* genes are preferentially expressed in the root tip and vascular cylinder while the high expression of *tub4* gene seems to be a unique feature for vascular tissue (Villemur et al. 1994). It has not been investigated whether the formation of endomycorrhiza is affecting the expression pattern of β-tubulin genes.

Until now, *Eucalyptus globulus* is the only ectomycorrhiza forming plant in which tubulin expression during *ectomycorrhiza* formation has been studied at the RNA level. An α-tubulin gene was shown to be upregulated during formation of symbiosis, and the upregulation of α-tubulin expression paralleled the increased formation of lateral roots in *Eucalyptus* seedlings that were in contact with the fungal mycelium (Diaz et al. 1996). In *Pinus sylvestris- Suillus bovinus* and *P. contorta–S. variegatus* ectomycorrhiza the expression of tubulins has been analyzed at the protein level (Timonen et al. 1993, 1996; Niini et al. 1996). Different mobility of plant and fungal α-tubulin allowed their comparison in one-dimensional (1-D) immunoblots, which suggested that in mature ectomycorrhiza the fungal α-tubulin dominated (Timonen et al. 1996). The comparison of the amount of plant and fungal α-tubulin during the development of *P. contorta–S. variegates* ectomycorrhiza for 60 days also indicated that the amount of plant α-tubulin decreased gradually, probably due to the development of fungal sheath around the root (Timonen et al. 1996). No such comparisons could be made between plant and fungal β-tubulin or actin due to their similar mobility during the electrophoretic separation.

The immunoblots of two-dimensional gels from three root types of *P. sylvestris* radicles, main root and first order laterals and short roots as well as from different developmental stages of *P. sylvestris–S. bovinus* ectomycorrhiza revealed more differences in the tubulin protein patterns than 1-D immunoblots (Niini et al. 1996). Three plant α-tubulins were detected in all root types, but the pattern in the short roots differed from that in radicles and first-order laterals. This is an interesting observation, since the formation of ectomycorrhiza occurs in *Pinus* short roots probably due to their reduced growth rate that could be associated with the occurrence of the short-root-specific α-tubulin pattern. During the development of ectomycorrhiza the initial short root-specific α-tubulin pattern gradually changed and two new α-tubulins were distinguished in mature ectomycorrhiza. Whether the α-tubulin protein patterns in *P. sylvestris* short roots and ectomycorrhiza results from alterations in the expression of α-tubulin genes or post-transcriptional modifications of α-tubulin proteins, will be resolved when the number and the degree of transcript accumulation of α-tubulin encoding genes in *P. sylvestris* short roots have been clarified. In the immunoblots from the different root types and ectomycorrhiza, three different β-tubulins were identified. For the expression of *P. sylvestris* β-tubulins no alterations comparable to those in α-tubulin were observed.

The expression of the tubulin genes in endo- and ectomycorrhizal fungi has also obtained some attention. Fragments encoding 267 amino acids from the central part of the β-tubulin gene have been cloned from several endomycorrhizal fungi including *Glomus mossae*, *G. geosporum*, *G. coronatum*, *G. clarum*, *Gigaspora rosea*, *Acaulospora laevis*, and *Scutellospora castanea* (M. Stommel and P. Franken, public databases). The deduced amino acid sequences of the fragments suggest clearly that there are at least two β-tubulin encoding genes in the endomycorrhizal fungi. Comparison of the deduced amino acid sequences of the β-tubulin genes in and between different species indicates that in each species the β-tubulin amino acid sequences are more similar to β-tubulins of other species than to those within the same species. The *G. rosea* β-tubulin transcripts accumulate in dormant and germinated spores, in extraradical hyphae and in pea mycorrhiza (Franken et al. 1997; Bütehorn et al. 1999), while α- and β-tubulin protein was detected in the hyphae of *G. mossae* elicited by the host plant (Åström et al. 1994). None of these experiments have yet shown whether the expression of either the β-tubulin gene is associated with the formation of *endomycorrhiza* or some other specific stage in the life cycle of the endomycorrhizal fungi.

The immunoblots of α- and β-tubulins from the ectomycorrhizal fungus *S. bovinus* and from its ectomycorrhiza with *P. sylvestris* indicated the presence of three α- and two β-tubulin polypeptides (Niini et al. 1996) in nonsymbiotic hyphae and ectomycorrhiza. From the filamentous homobasidiomycete *Schizophyllum commune* that is closely related to *S. bovinus*, three α- and two β-tubulins have also been identified by 2-D gel electrophoresis. Until now, only two α- and one β-tubulin encoding genes have been isolated from *S. commune* in spite of several attempts (Russo et al. 1992; Raudaskoski unpublished data). The higher number of polypeptides than tubulin encoding genes suggests that fungal α- and β-tubulins are targets for posttranslational modifications. Recently, a β-tubulin encoding cDNA highly similar to that of *S. commune* has been isolated from *S. bovinus*. By using the encoding region of the *S. bovinus* β-tubulin gene as a probe, a high, but similar amount of β-tubulin mRNAs were detected both in nonsymbiotic and symbiotic hyphae (Lahdensalo et al., unpublished). Both in vegetative and ectomycorrhizal hyphae of *S. bovinus* the tubulin polypeptides occurred in doublet patterns (Niini et al. 1996), which are thought to be due to allelic differences between tubulins of the haploid genomes present in the *dikaryotic hyphae* of *S. bovinus*. This needs to be certified by cloning and further analysis of *S. bovinus* tubulin genes.

2.2 Expression of Actin-Encoding Genes

Arabidopsis is known to have ten actin genes, two of which are pseudogenes (Meagher et al. 1999). Out of the eight expressed actin genes *ACT2*, *ACT8*, and *ACT7* are named *Arabidopsis* vegetative actin genes due to the accumulation pattern of transcripts. *ACT2* is expressed in young and old vegetative tissue, but *ACT8* only in a subset of the organs and tissues expressing *ACT2* (An et al. 1996a). *ACT7* is expressed in young expanding vegetative tissues and is also involved in phytohormone responses (Kandasamy et al. 2001). The rest of the *Arabidopsis* actin genes appear to be associated with reproductive processes (An et al. 1996b; Huang et al. 1997; Meagher et al. 1999). In spite of the high number of actin genes in most plant species and the important cell biological functions of the actin cytoskeleton, only a few analyses about the expression of different actin genes in root tissue have been performed (McLean et al. 1990) and the effect of mycorrhiza on the expression of actin genes at mRNA level has not been investigated.

One-dimensional analyses of actin expression at the protein level have been performed in *Pinus* ectomycorrhiza. The similar mobility of plant and fungal actins in the immunoblots of 1-D gels from *P. sylvestris- S. bovinus* ectomycorrhiza made it difficult to record the contribution of each symbiotic partner to the actin signal (Timonen et al. 1993). The presence of the actin signal throughout the different developmental stages of ectomycorrhiza was suggested to be a reliable marker for still metabolically active symbiosis (Timonen et al. 1996).

In the immunoblots of 2-D gels four actin polypeptides were detected in *P. sylvestris* radicles, main roots and first order laterals and in short roots. The polypeptides were also detected in *P. sylvestris- S. bovinus* young and dichotomous mycorrhizal short roots. In fully mature coralloid mycorrhiza only two actin polypeptides occurred. Whether they represented plant or fungal actin or both was not possible to conclude. The presence of four actin polypeptides in *Pinus* root tissues is in agreement with the occurrence of a similar number of actin polypeptides in *Vicia faba* roots (Janssen et al. 1996), while in the roots of *Phaseolus vulgaris* one and two actin polypeptides were detected in symbiotic root nodules and in uninfected roots, respectively (Pérez et al. 1994).

In the immunoblots of 2-D gels of the vegetative hyphae of *S. bovinus* two actin polypeptides occurred that were also detected in ectomycorrhiza (Niini et al. 1996). Recently, two actin-encoding genes, *Sbact1* and *Sbact2*, were isolated from *S. bovinus* nonsymbiotic hyphae (Tarkka et al. 2000). Northern hybridization with specific probes for each actin gene of *S. bovinus* indicated that both actins are expressed in vegetative hyphae and ectomycorrhiza. In vegetative hyphae, the expression rate and protein level of Sbact1 was tenfold higher than that of Sbact2. A ten times higher accumulation of *Sbact1* than *Sbact2* mRNA was also observed in ectomycorrhizal hyphae, although the analyzed sample represented a pool of young and mature ectomycorrhizal roots. In the future, the analysis of different developmental stages of ectomycorrhiza might improve our understanding of the expression pattern of the *S. bovinus actin genes* in symbiosis. From *S. commune*, a nonectomycorrhizal homobasidiomycete closely related to *S. bovinus*, two actin-encoding genes were also isolated (Tarkka et al. 2000), which indicates that filamentous homobasidiomycetes differ from filamentous deutero- and ascomycetes, having only a single actin gene. Recently, two actin-encoding genes have also been identified in the genome sequence of *Schizosaccharomyces pombe* (Wood et al. 2002).

3 Organization of Cytoskeleton in Endomycorrhiza

3.1 Root Cells

Indirect immunofluorescence (IIF) microscopy and related methods have been used to study the structure of the cytoskeleton in nonmycorrhizal and endomycorrhizal root cells of tobacco (Genre and Bonfante 1997, 1998), *Asparagus* (Matsubara et al. 1999), *Medicago truncatula* (Blancaflor et al. 2001), and in protocorms of orchid seeds (Uetake et al. 1997; Uetake and Peterson 1997,1998). Protocorms develop at the base of germinating orchid seeds and invasion of the parenchyma cells by a symbiotic fungus is necessary for the further development of the embryo. Several common features were observed in the reorganization of MT cytoskeleton in roots and protocorms after invasion of the symbiotic fungus into the plant cells. In all three cases the fungus invades differentiated parenchyma cells containing mainly transversely orientated cortical (below the plasma membrane) MTs connected with a few cytoplasmic MTs to the nucleus. After hyphal penetration, the plasma membrane separating the fungal hyphae from the plant cell cytoplasm, the perifungal membrane (Uetake and Peterson 1998), expands to follow the branching of the fungal hyphae. The growth of the hyphae in the intracellular space leads to formation of an arbuscule in endomycorrhiza and a peloton of hyphal coils in orchid mycorrhiza. The hyphal growth is associated with profound reorganization of MT cytoskeleton in the plant cell (Uetake et al. 1997; Uetake and Peterson 1997; Genre and Bonfante 1997, 1998; Matsubara et al. 1999; Blancaflor et al. 2001). During fungal invasion the cortical MTs of the plant cell disappear probably through depolymerization, and new MTs, less well orientated, reappear at the plasma membrane surrounding the intracellular hyphae.

The signals and mechanisms behind the observed reorganization of MT cytoskeleton in plant cells colonized by endomycorrhizal fungi are not yet known. However, it can be speculated that the invasion and proliferation of fungal hyphae in the space between cell wall and plasma membrane of a differentiated plant parenchyma cell with full turgor pressure could cause mechanical stress (Ko and McCulloch 2000; Stamatas and McIntire 2001) at the plant cell membrane, perhaps associated with chemical signals from the fungus to the plant cell. These signals could cause the depolymerization of MTs at the plasma membrane surrounding the cell and direct MT repolymerization at the expanding plasma membrane surrounding the growing hyphae, together with altered gene expression of the invaded parenchyma cells. In line with this idea is the observation that in transgenic tobacco the GUS expression under the promoter of the maize tubulin gene *Tuba3* increased in differentiated cortical cells infected by endomycorrhizal fungus (Bonfante et al. 1996). The accumulation of *Tuba3* transcripts was similarly observed to increase in maize cortical cells during the invasion of the cells by an endomycorrhizal fungus. Cell wall material is deposited into the extracellular space between the plant cell membrane and fungal hyphae (Peterson et al. 1996), which could require the presence of MTs as these are required for cell plate formation in meristems (Lloyd and Hussey 2001), primary wall formation in elongating cells (Wasteneys 2000), and secondary wall thickenings in differentiating cells (Chaffey et al. 2000). In addition, proteins necessary for the nutrient exchange between the symbionts probably are synthesized and transported to the perifungal membrane (Rosewarne et al.1999; Hahn and Mendgen 2001), which could also require transport along the MTs.

The distribution of MFs has also been studied in endomycorrhiza formed in tobacco roots (Genre and Bonfante 1998) and in protocorm cells (Uetake and Peterson 1997). In noninfected cells, MFs appeared to have the distribution reported for parenchyma cells in a large number of plants investigated (Staiger 2000), with thin and thick MF cables crossing the cell cytoplasm. At fungal invasion, reorganization of MFs were observed in cortical cells of tobacco, in which the cables disappeared and MFs seemed to become tightly associated with the plasma membrane surrounding the arbuscular branches (Genre and Bonfante 1998). In protocorm cells, no clear reorganization of MFs was observed, but the distribution of MFs was comparable to that in uninfected cells (Uetake and Peterson 1997), which was a result quite different from the reorganization of MTs in the protocorm cells at fungal invasion. The different behavior of MFs in tobacco and protocorm cells at fungal invasion is suggested to be due to the physiological difference between the endomycorrhizal and orchid endosymbiotic fungus (Genre and Bonfante 1997), or it could result from differences in the processing of the cells for confocal microscopic investigation (Uetake and Peterson 1997). In tobacco root cells, the accumulation of MFs in close association with the plasma membrane surrounding the fungus is suggested to be due to the involvement of the actin cytoskeleton in localization of proteins necessary for membrane transport and signal transduction between symbionts (Genre and Bonfante 1997). Noteworthy is that at the invasion of the plant cell by the endomycorrhizal fungus, the plant cell nucleus moves from the periphery of the cell to the center and the central vacuole becomes fragmented (Bonfante and Perotto 1995). These processes could also be controlled by rearrangements of MTs or MFs, or by both cytoskeletal elements.

3.2 Fungal Hyphae

In the research of cytoskeleton in endomycorrhiza, the structure and function of fungal MTs and MFs have gained less attention than those of plant cells. In the IIF microscopical investigation of nonsymbiotic hyphae of *Glomus mossae*, the MTs were visualized with tubulin and MFs with actin antibodies in the multinucleate hyphae originating from germinated spores (Åström et al. 1994). MTs extended in the cortical and central parts of the hyphae to the extreme hyphal tip, continued from the main hypha into a branch and the position of nuclei appeared to follow the MT tracks. Long MFs were also visualized in the hyphae (Åström et al. 1994). MTs and MFs could be involved in the intra- and intercellular nuclear movements and cytoplasmic streaming, respectively, recorded in living nonsymbiotic hyphae of different endomycorrhizal fungi (Bago et al. 1998; Giovannetti et al. 1999). The presence of both MTs and MFs in nonsymbiotic hyphae suggests that these structures could also play a significant role in hyphal morphogenesis associated with the formation of endomycorrhiza, such as the differentiation of the appressorium at the root surface at the beginning of the symbiosis and the formation of vesicular and arbuscular structures in the plant cell after the establishment of the symbiosis.

4 Organization of Cytoskeleton in Ectomycorrhiza

4.1 Root Cells

The effect of ectomycorrhiza formation on the plant cell cytoskeleton is more difficult to investigate than that of endomycorrhiza or orchid mycorrhiza. In the ectomycorrhizal symbiosis, the fungal hyphae grow between the cortical cells of the host plant, forming a hyphal network for nutrient exchange called the Hartig net. The plant cells of the Hartig net have thick cell walls and accumulations of secondary metabolites such as phenols and starch. The thick cell walls inhibit rapid penetration of fixatives necessary for preservation of cytoskeletal elements, which is seen as lack of MTs or MFs from the published ultrastructural studies of ectomycorrhiza. Autofluorescence of secondary metabolites hampers the recording of cytoskeletal structures when they have been preserved during fixation. In spite of these difficulties, some knowledge of cytoskeletal structure has been obtained in *Pinus sylvestris–Suillus bovinus* ectomycorrhiza (Timonen et al. 1993; Niini and Raudaskoski 1998).

P. sylvestris has a root system consisting of three morphologically and anatomically different root types, which is also found in other pines as well as in eucalypts and beeches (Smith and Read 1997). The primary root or the main root has an undetermined capacity for continuous growth, the lateral roots have a somewhat limited ability to elongate, and the so-called short roots have a very limited ability to grow. In order to survive, they have to be colonized by a symbiotic fungus (Robertson 1954; Wilcox 1968). In some pines, i.e., *P. sylvestris* and *P. strobus*, the mycorrhiza only forms in the short roots (Piché et al. 1983). In Scots pine seedlings, two types of short roots are distinguished: one type consists of long and slender roots with a high number of root hairs, the other of truncated, robust roots with a round apex and only a few root hairs. The majority of the short roots of pine seedlings belong to the latter type and mycorrhiza is only formed in this root type (Niini and Raudaskoski 1998).

The IIF microscopical studies on the MT cytoskeleton in nonmycorrhizal and mycorrhizal short roots have indicated some common features. The most prominent MT cytoskeleton is detected in meristematic and vascular tissue. In the meristems of truncated short roots the MTs are often vertical in interphase cells and mitotic spindles are horizontally oriented. These features suggest that in the meristem the cells elongate and divide horizontally which probably is the reason for the blunt form of the tip in the truncated short roots (Niini and Raudaskoski 1998). Almost next to the meristem in short roots occur xylem elements with cell wall thickenings and cortical cells with amyloplasts, which indicates that cell differentiation takes place very close to the short root tip. In the topmost cell layer of the meristem the direction of cell divisions is no longer horizontal, but oblique or vertical. From the central part of this layer originate vertically elongated cells with horizontally orientated MTs, and from the borders cells with amyloplasts representing differentiating vascular and cortical tissue, respectively. Thus, in short roots divisions in only one cell layer seem to provide initials for the elongation and differentiation zones while in lateral roots the transition zone between the meristem and elongation zone appears to consist of four to five cell layers with vertical cell divisions producing initials for cell elongation and differentiation. This must affect the growth rate of the roots and could explain why the elongation of the short roots is retarded and tissue maturation occurs closer to the meristem in the short roots than in the laterals.

In nonmycorrhizal short roots, very few MTs are detected in the cortical cells for which a high number of amyloplasts is typical (Fig. 1A). In ectomycorrhizal roots, MT fluorescence is only observed in association with the nucleus in cortical cells (Fig. 1B). The low number or absence of MTs from cortical cells seems to be a specific feature of *Pinus* short roots. Although the tubulin expression in cortical cells of *Arabidopsis* (Villemur et al. 1994), and maize (Joyce et al. 1992), roots is low, MTs are always detected at the inner face of the plasma membrane in uninfected cortical cells of these plants. It

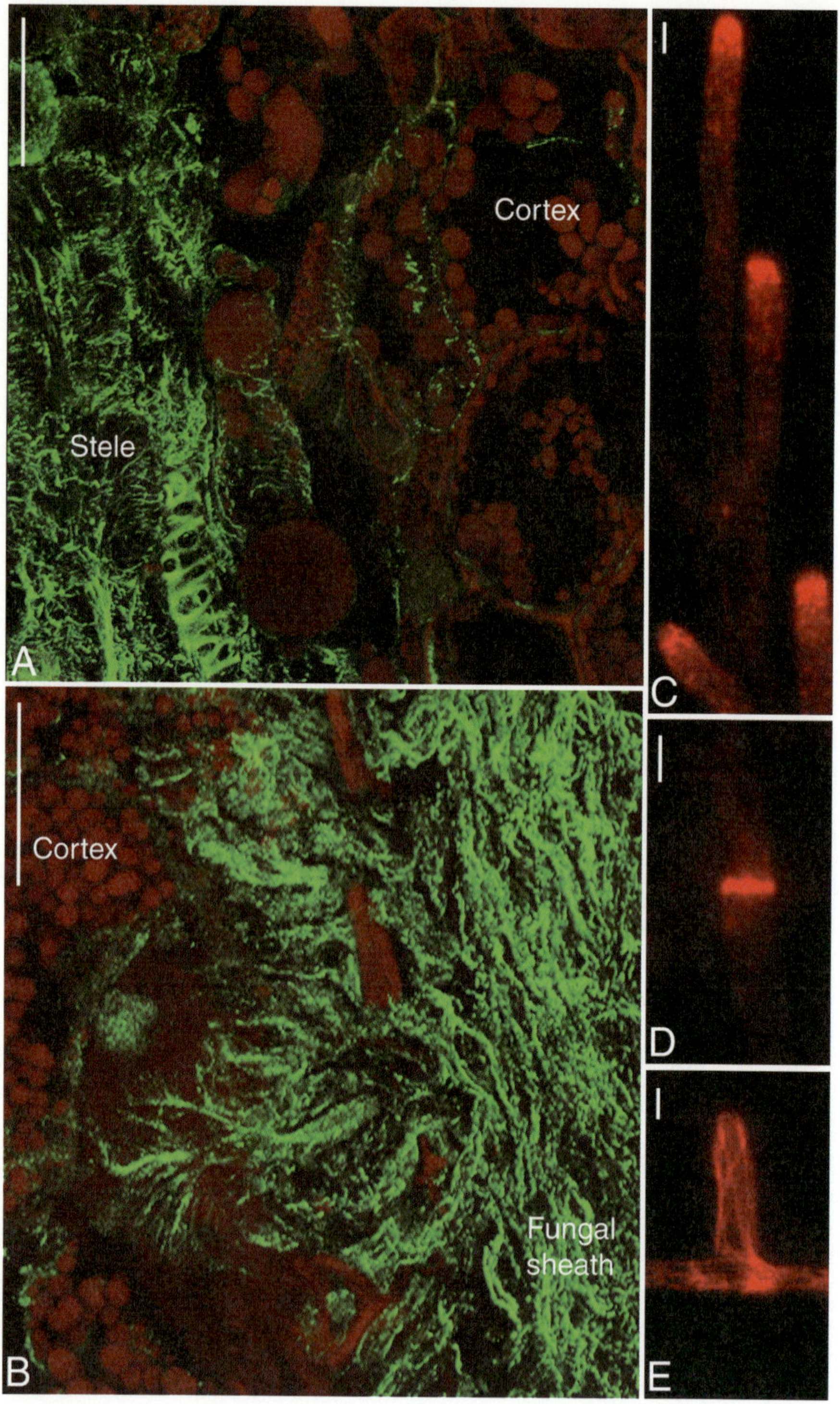

Cortex
Stele
A
C
Cortex
D
Fungal sheath
B
E

is possible that the antibody used for IIF microscopical detection of MTs in *Pinus* is not able to recognize the MTs of cortical cells although it visualizes well MTs in meristem and vascular tissue. The MT cytoskeleton of cortical cells of *Pinus* could also be more sensitive to the processing of the samples for IIF microscopy than the MTs in meristem and vascular tissue. Interestingly, it has not been possible to visualize MFs in the cortical cells of short roots although plenty of MFs are seen in vascular tissue, where they were already visualized in the early days of the study of MFs in plant cells (Pesacreta et al. 1982).

The structure of cell wall and cytoskeleton in differentiating cortical cells in short roots of pine is of special interest, since this is the region of the short root in which the ectomycorrhizal fungus invades and establishes the Hartig net. When root morphogenesis and ectomycorrhiza formation in Scots pine was studied (Niini et al. 1996; Tarkka et al. 1998), a group of polypeptides with molecular weight slightly above 43 kDa were observed to be short root-specific. By using peptide sequencing, it was shown that the polypeptides represented a group of peroxidases. By reverse genetics a full-length cDNA of one of the peroxidases, *Psyp1*, was cloned and sequenced (Tarkka et al. 2001). The signal sequence suggests that Psyp1 is secreted and could be involved in cell wall formation. In ectomycorrhiza *Psyp1* expression is downregulated, which agrees with the idea that the growth of the fungal hyphae in the intercellular space might inhibit the cortical cell wall differentiation (Niini 1998). There may be signalling or linkages between adjacent plant cells that can regulate the organization of their cytoskeletal structures. This exchange of information might be mediated through plasmodesmata, or alternatively, through the intervening cell wall (Canut et al. 1998; Overall et al. 2001).

Fig. 1. Microtubule cytoskeleton visualized with indirect immunofluorescence technique with α-tubulin antibody and viewed with laser scanning confocal microscopy in *Pinus sylvestris* short root (**A**) and ectomycorrhiza with *Suillus bovinus* (**B**). **A** In cortex only few microtubules are distinguished in the cortical cells with numerous round amyloplasts. In stele microtubules with mainly transverse orientation are abundant in elongating cells differentiating to vascular tissue. Strong vertical bands represent wall thickenings in a xylem cell. **B** Microtubules are hardly seen in pine cortical cells, but they are clearly distinguished as long tracks in hyphae forming the fungal sheath and penetrating into the root cortex. **A,B** *Bars* 20 µm. **C–E** Cytoskeletal elements in *Suillus bovinus* hyphae visualized with rhodamine-phalloidin staining of actin (**C**, **D**) and indirect immunofluorescence microscopy with α-tubulin antibody (**E**). **C** A strong actin signal at hyphal tip, **D** an actin ring at the site of the future septum. **E** Microtubule tracks in a hyphal branch. **C–E** *Bars* 10 µm

4.2 Fungal Hyphae

An extensive MT cytoskeleton with strictly axial orientation is observed in the long polarized apical cells of ectomycorrhiza-forming and other filamentous fungi (Salo et al. 1989; Raudaskoski et al. 1991; Niini and Raudaskoski 1998; Raudaskoski et al. 2001). In the dikaryotic hyphae of these fungi the longitudinally running MTs (Fig. 1E) appear to keep the two nuclei with different mating type genes close to each other by forming a cage of crossing MTs around the nuclear pair (Runeberg et al. 1986; Salo et al. 1989). Actin is visualized as a cap in the hyphal tips (Fig. 1C), as small plaques along the apical cell and as a ring (Fig. 1D) at the site where the cross wall will be formed (Salo et al. 1989; Raudaskoski et al. 1991; Gorfer et al. 2001).

Comparison of the structure/organization of actin cytoskeleton in the taxonomically closely related slow-growing ectomycorrhiza-forming *S. bovinus* and fast-growing wood-decaying *S. commune* reveals several differences. In the slow-growing hyphae of *S. bovinus* the actin cap is more extensive than in *S. commune,* and the cap can be easily visualized with fluorochrome-labeled phalloidin that binds only to filamentous actin (Barak et al. 1980). In contrast, the visualization of the actin cap at the hyphal tips of *S. commune* succeeds only with an actin antibody, which visualizes actin monomers in addition to actin filaments. These differences suggest that the structure of MFs is more stable and their number is higher at the hyphal tip in the slow-growing ectomycorrhizal than in the fast-growing wood-decay fungus. In the hyphae of *S. bovinus,* it is also possible to distinguish occasionally actin filaments (Gorfer et al. 2001) comparable to those seen at a specific growth phase in budding yeast cells of *S. cerevisi*ae (Kilmartin and Adams 1984). Actin filaments are not observed in the hyphae of fast-growing filamentous fungi, such as *S. commune* (Runeberg et al. 1986; Raudaskoski et al. 1991) or *Neurospora crassa* (Heath et al. 2000). These observations suggest that there probably are some basic differences in the structure of the actin cytoskeleton between slow-growing and fast-growing filamentous fungi. The question whether these differences are associated with the ability of *S. bovinus* to form ectomycorrhiza with *P. sylvestris* root cells and whether these specific features occur in all ectomycorrhiza-forming fungi has to be answered the in the future.

The use of drugs against polymerized tubulin and actin has given insights into their roles in hyphal growth. The depolymerization of the MT cytoskeleton with an anti-MT drug leads to strong branching of the hyphae in ectomycorrhizal fungi such as *Amanita muscaria*, *Hebeloma cylindrosporum*, *Paxillus involutus*, and *S. bovinus* (Niini and Raudaskoski 1993). In contrast, the depolymerization of actin filaments from the hyphae of *S. bovinus* with cytochalasin D leads to swelling of the hyphal tip cells and loss of the polarized growth pattern (Niini 1998; Raudaskoski et al. 2001). Nonpolarized growth and strong branching of hyphae are also observed when an ectomycorrhizal fungus is associated with the plant root cells (Kottke and Oberwinkler 1987; Timonen et al. 1993; Niini 1998; Raudaskoski et al. 2001). The similar morphology of nonsymbiotic hyphae treated with inhibitors and of the hyphae grown in association with the root cells have led us to the hypothesis that the change in the hyphal morphology is due to a signal from root cells, the recognition and transduction of which leads to a reorganization of the actin cytoskeleton.

5 Regulation of Actin Cytoskeleton Organization in Fungal Hyphae and Plant Cells

Recently, several GTPases known to play an important role in linking extracellular signals to reorganization of actin cytoskeleton in yeasts and mammalian cells (Johnson 1999) have been isolated from *S. bovinus* and *S. commune*. The GTPases are conserved molecular switches that are normally anchored to the plasma membrane by a C-terminally attached farnesyl tail. In the active form the protein is bound to GTP, in the inactive one to GDP (Nuoffer and Balch 1994).

Until now, a *Ga*, a *Cdc42* and a *rac* cDNA and two *ras* cDNAs have been isolated and characterized from *S. bovinus* and three *Ga* cDNAs, a *Cdc42,* a *ras* and a *rho3* cDNA from *S. commune* (Gorfer et al. 2001; Raudaskoski et al. 2001). Out of these genes only *ras* had been isolated before from an ectomycorrhizal fungus *Laccaria laccata* (Sundaram et al. 2001). The GTPases cloned from *S. bovinus* are expressed in vegetative hyphae, but also during ectomycorrhiza formation (Gorfer et al. 2001; Raudaskoski et al. 2001). *SbCdc42* cDNA is able to complement the temperature-sensitive *S. cerevisiae cdc42* mutation causing disruption of actin cytoskeleton (Johnson and Pringle 1990), which suggests that SbCdc42 is also involved in regulating the organization of actin cytoskeleton as it is in yeast and animal cells (Gorfer and Raudaskoski, unpubl. data).

The small GTPases were chosen as the target for research in the ectomycorrhizal fungus *S. bovinus* on the basis of the following previous results: (1) IIF microscope analyses show rearrangement of cytoskeletal elements in *S. bovinus* hyphae at the formation of ectomycorrhiza (Timonen et al. 1993; Niini 1998; Raudaskoski et al. 2001). (2) The reorganization of the cytoskeleton in fungal hyphae occurs without differential expression of fungal tubulin or actin genes (Niini et al. 1996; Tarkka et al. 2000). (3) In *S. commune*, closely related to *S. bovinus*, the mating interaction necessary for sexual reproduction is regulated by the signal transduction pathway starting from the pheromone-G-protein-coupled receptor (GPCR) interaction (Wendland et al. 1995; Vaillancourt et al. 1997; Raudaskoski 1998; Raudaskoski et al. 1998; Fowler et al. 1999). In animal cells (Schmidt and Hall 1998), and in the yeast *S. cerevisiae* (Johnson 1999), the effect of small GTPases (e.g., Cdc42, Rac and Rho) on the organization of the actin cytoskeleton is initiated by the binding of a signal molecule to a G-protein coupled receptor, which is the receptor-type shown to mediate the signalling between haploid hyphae at the mating interaction in *S. commune*.

A high number of different types of ligands/signals are recognized by the members of the GPCR superfamily in animal cells (Neer 1995). Recently, it has been shown that a set of GPCRs may function in fungi as well, since a distinct GPCR system from the one recognizing the pheromones senses glucose in yeasts (Versele et al. 2001). It could thus be speculated that in ectomycorrhiza-forming fungi a GPCR originally involved in sexual reproduction is able to interact with signal molecules produced by plant roots, due to a mutational change in its structure. The contact with the plant would be signalled into the hyphal cells where it might lead to a change in the organization of actin cytoskeleton from highly polarized to a more relaxed form via a pathway involving the small GTPases Cdc42 and Rac1 (Gorfer et al. 2001). This could then lead to the development of a hyphal morphology suitable for the symbiotic growth. Perhaps it is meaningful that the majority of ectomycorrhiza-forming fungi belong to homobasidiomycetes. This group of filamentous fungi includes species, such as *S. commune* and *Coprinus cinereus*, in which sexual reproduction is regulated by the signal transduction pathway starting from pheromone-GPCR interaction (Fowler et al. 1999; Olesnicky et al. 1999).

In endomycorrhizal fungi, the characteristic changes in hyphal morphology, the formation of the appressorium on the root surface at the beginning of colonization and the strong branching of hyphae at arbuscule formation, may require the reorganization of actin cytoskeleton. It may be speculated that the different hyphal morphologies are due to the perception of plant signals, which could be mediated to the actin cytoskeleton through the small GTPases belonging to the Rho subfamily.

Small GTPases of the Rho subfamily exist also in plants, where they are addressed as Rac or Rop proteins. In *Arabidopsis* 11 *Rac/Rop* genes have been identified (Kost et al. 1999a; Li et al. 2001). The expression analysis of the constitutively active mutant form of Rac/Rop proteins unable to hydrolyse GTP or the dominant negative mutant form unable to exchange GDP to GTP has shown that the GTPases are involved in the regulation of apical growth and organization of MFs in pollen tubes (Kost et al. 1999a; Zheng and Yang 2000; Fu et al. 2001), and in root hairs (Molendijk et al. 2001). Rac/Rop proteins appear also to be involved in the regulation of organization of the actin cytoskeleton during stomatal closure in response to abscisic acid (Lemichez et al. 2001). The expression of constitutively active and dominant negative forms of Rac/Rop under the 35S universal promoter indicated that these proteins participate in multiple distinct signalling pathways that control plant growth, development and responses to the environment (Li et al. 2001).

6 Actin Binding Proteins

In eukaryotic cells, a high number of actin binding proteins (ABPs) regulate polymerization and depolymerization, bundling and cross-linking of MFs, and movement of cargo along the MFs. In animal cells, many of the ABP encoding genes have been isolated and the function of the corresponding proteins has been characterized. In plants and filamentous fungi, the ABP research is just beginning. In *Arabidopsis*, the members of the gene families encoding ADF proteins (actin depolymerizing factor/cofilin; Dong et al. 2001), profilins (Ramachandran et al. 2000), Arp2 (Klahre and Chua 1999), villins (Klahre et al. 2000), and myosins (Reddy and Day 2001) have been cloned and their expression patterns in different plant organs characterized.

ADF protein family members interact with actin monomers and filaments in a pH-sensitive manner. When ADF/cofilin binds to filamentous (F) actin it accelerates the dissociation of subunits from the pointed ends of actin filaments (Bamburg 1999; Cooper and Schafer 2000). The properties of maize cofilins have been analysed by using recombinant ZmADF1 and ZmADF3 proteins (Hussey et al. 1998). ZmADF3 has the ability to bind monomeric (G) actin and F-actin and to decrease the viscosity of polymerized actin solutions, indicating an ability to depolymerize actin filaments (Lopez et al. 1996). ZmADF3 is phosphorylated on Ser6 by a calcium-stimulated protein kinase in plant extracts (Smertenko et al. 1998), which suggests that phosphorylation regulates cofilin's actin binding activity and affects the stability of the actin cytoskeleton in a manner shown in animal cells (Daniels and Bokoch 1999).

Profilin is a G-actin binding protein known to interact in animal and yeast cells with proline-rich motifs of other proteins and with polyphosphoinositides. The interaction of profilin with G-actin provides a mechanism to sequester actin monomers and promote actin depolymerization. It appears that profilin may also be involved in promoting actin polymerization. This might take place by binding to proline-rich motifs in proteins that convey intra- or extracellular signals to reorganization of actin cytoskeleton (Mullins 2000). In *Arabidopsis*, the *PFN-1* gene, from profilin gene family with eight to ten members, is expressed in root and root hairs, and in a ring of cells in the elongating zone of the root (Ramachandran et al. 2000), in which the profilin levels could be involved in the regulation of cell elongation as a rate-limiting factor.

Plants have also several genes with high homology to animal villin (Klahre et al. 2000). The first plant villin was isolated from pollen tubes as a 135-kDa actin-bundling protein (Yokota et al. 1998; Yokota and Shimmen 1999). Its identity as a villin-gelsolin family member (Vidali et al. 1999) was confirmed by partial amino acid sequencing and by isolating the corresponding cDNA from a pollen grain cDNA expression library. Immunodetection of villin revealed its co-localization with actin bundles in pollen tubes. Due to the gelsolin-like headpiece, villin may also act as an actin-severing protein, although this activity has not yet been demonstrated for plant villins.

In *Arabidopsis*, 17 myosin encoding genes have been identified, which on the basis of phylogenetic analysis, fall into plant-specific myosin classes VIII (4 genes) and XI (13 genes). The three cloned myosins from maize and four from *Helianthus annuus* also fall into these classes (Reddy and Day 2001). The true structure, enzymatic properties, intracellular localization and physiology of plant myosin are not yet known.

The structure and function of actin binding proteins in association with the reorganization of the actin cytoskeleton in plant cells at mycorrhiza formation has not yet been studied, but seems to require attention. In endomycorrhiza, rearrangement of actin cytoskeleton was observed at the colonization of tobacco cortical cells by endomycorrhizal fungus (Genre and Bonfante 1997). Actin cables, typical for noncolonized cortical root cells, disappeared and MF polymerized to a network at the plasma membrane surrounding the arbuscule branches. The observed changes may be hypothesized to require both the activation and function of proteins involved in the reorganization of the actin cytoskeleton as a consequence of the perception of signals from the intruding fungus.

The research on actin binding proteins in filamentous fungi seems to be restricted to myosin (McGoldrick et al. 1995), in *Aspergillus nidulans* and to actin-related proteins Arp1 (Plamann et al. 1994) in *Neurospora crassa*, although in yeast, *Saccharomyces cerevisiae*, most actin binding proteins previously described in plants are present and have been studied in detail (Ayscough 1998). In *A. nidulans*, the myosin I encoding gene *myoA* has been cloned and different mutational studies have indicated that MYOA is necessary for the maintenance of polarized growth, secretion and endocytosis (McGoldrick et al. 1995; Osherov et al. 1998; Yamashita and May 1998). The yeast *S. cerevisiae* and *Schizosaccharomyces pombe*, similar to animal cells, contain myosins from classes II and V in addition to those belonging to class I which predicts that additional myosins will be detected in future from filamentous fungi.

7 Microtubule-Associated Proteins

7.1 Plant Cells

Microtubule-associated proteins (MAPs) are divided into structural and motor proteins. Several genes encoding structural plant MAPs have been isolated in the last few years. One of the MAP encoding genes was isolated from *Arabidopsis*, where its heat-sensitive mutation resulted in the disintegration of cortical microtubules in leaf, hypocotyl and root cells. The gene was named

MOR1 (Whittington et al. 2001), its product is homologous to animal MAPs belonging to class TOGp-XMAP215, and in plant cells, the MOR1 protein is essential for cortical microtubule organization (Wasteneys 2002). Proteins with a MW of 60–68 kDa which associate in vitro with MTs, cause their polymerization and bundling and induce cross-bridges between them, have been biochemically purified from tobacco BY-2 cells (Jiang and Sonobe 1993) and carrot tissue culture cells (Chan et al. 1996, 1999). These proteins are called MAP-65 proteins and screening of a tobacco BY-2 cell cDNA library with an antiserum raised against them led to the isolation of a cDNA clone named *NtMAP65–1a*. The clone encodes a 580 amino acid polypeptide with no homology with any known animal MAPs. Further screening of the same library led to the isolation of two other similar cDNA clones, *NtMAP65–1b* and *NtMAP65–1c* (Smertenko et al. 2000). NtMAP65–1a protein induces polymerization, but not bundling of the MTs in vitro, and it is localized to areas of overlapping microtubules in spindle and phragmoplast and on a certain subpopulation of stable cytoplasmic MTs (Smertenko et al. 2000; Lloyd and Hussey 2001). The isolation of genes encoding NtMAP65 proteins and characterization of the properties of the proteins in vitro and in vivo have clearly shown that there are plant-specific MAPs whose ability to interact and bind to microtubules is different from that of animal MAPs.

The group of motor MAPs consists of kinesins and dyneins. Kinesins are mainly plus end-directed MT-dependent motor molecules, but minus end-directed kinesins are also known (Miki et al. 2001). The heavy chains of kinesins are formed of a motor domain with ATP and MT binding regions and ATP hydrolyzing activity; stalk domain with coiled-coil structure and a cargo-binding domain. Recently, it has become obvious that kinesin and myosin share a common core structure in the ATP-binding region that converts energy from ATP into protein motion, using a similar conformational strategy (Vale and Milligan 2000). Kinesins are monomeric, homodimeric or homotetrameric (Kim and Endow 2000). Several small polypeptides can be associated with kinesins and they regulate the activity and function of these motor proteins. The position of the ATP binding domain, whether on the N- or C-terminus of the polypeptide, makes the protein either a plus or a minus end-directed motor.

The kinesin superfamily is divided into various subfamilies including conventional kinesins and several kinesin-related proteins (KRPs; Kim and Endow 2000). Conventional kinesins are mainly involved in the (inter) intracellular transport of membranous organelles, whereas most KRPs function in nuclear division. Genes encoding KRPs have been isolated from tobacco (Mitsui et al. 1996; Asada et al. 1997), potato (Reddy et al. 1996b), and *Arabidopsis* (Mitsui et al. 1993; Liu et al. 1996; Reddy et al. 1996a). Of these, the best studied is TKRP125 (Asada et al. 1997), which is a plus end-directed motor molecule with an ATP hydrolyzing domain at the N-terminus. Structurally, it is a BimC-type kinesin that functions as a homotetramer (Kim and Endow 2000).

TKRP125 is located in anaphase spindle and phragmoplast. A similar location has been recently shown for a TKRP125 homologue and a KRP slightly deviating from TKRP125 isolated from carrot tissue culture (Barroso et al. 2000). In plants, a subfamily of kinesin-like genes encodes proteins with a calmodulin-binding domain. Genes encoding members of this subfamily have been isolated from *Arabidopsis* (AKCBP; Reddy et al. 1996a), potato (PKCBP; Reddy et al. 1996b), and tobacco (TCK1; Wang et al. 1996). The genes encode minus end-directed motor proteins with the calmodulin-binding region in the motor domain at the very C-terminus of the polypeptide, suggesting that calcium/calmodulin may be involved in the regulation of the microtubule-based movements in plants. Genes encoding other types of minus end-directed motor molecules are also present in *Arabidopsis* and they have been named *katA* to *katE* (Mitsui et al. 1993, 1994). The production of minus end-directed KatB and KatC kinesins increases during M-phase of the cell cycle in tobacco BY-2 cell cultures, suggesting that mitotic spindle and phragmoplast may also be their site of action (Mitsui et al. 1996). Dyneins are large minus end-directed motors. Until now, no dynein heavy chain encoding gene has been detected in expressed sequence tag (EST) databases from flowering plants or in the recently sequenced *Arabidopsis* genome (Lawrence et al. 2002).

In endomycorrhiza, the organization of MT cytoskeleton changes when the fungus colonizes the plant cell (Genre and Bonfante 1997). In addition, in *P. sylvestris* ectomycorrhiza cortical cells surrounded by the Hartig net seem to contain less cortical MTs than the noninfected cortical cells (Niini 1998). The reorganization of the MT cytoskeleton takes place not only in the colonized endomycorrhizal plant cells, but also in the cells adjacent to the colonized ones, and when the arbuscules have completely collapsed, the transversely oriented cortical MTs next to the plasma membrane reappear (Blancaflor et al. 2001). The reorganization of the MT cytoskeleton observed in connection with symbiosis requires destabilization (depolymerization) of existing cortical MTs, polymerization of new MT areas at the plasma membrane next to the colonizing fungus, and repolymerization of cortical MTs after the senescence of the arbuscular structure. The observed changes present a very dynamic behavior of MT cytoskeleton in endomycorrhiza, which probably involves the function of both structural and motor MAPs in plant cells.

7.2 Fungal Hyphae

In filamentous fungi, the genes encoding MT-associated motor proteins kinesins and dyneins have caught more attention than structural MAPs. The first kinesin-related protein was cloned from ascomycete *Aspergillus nidulans* (Enos and Morris 1990) as a temperature-sensitive mutation that blocked mitosis in germinating conidia. Later studies have shown that this *bimC* (blocked in mitosis) gene encodes a protein with amino-terminal motor and coiled-coil tail domains. The protein forms homotetramers and it is required for nuclear division. A motor protein has been biochemically characterized from *Neurospora crassa* (Steinberg and Schliwa 1996) and the zygomycete *Syncephalastrum racemosum* (Steinberg 1997), and the production of an antibody against the *N. crassa* protein helped to isolate the first fungal conventional kinesin encoding gene, *Nkin* (Steinberg and Schliwa 1995). Using probes derived from conserved regions of kinesins and screening existing genomic databases, has led to the isolation of conventional kinesins from *Ustilago maydis* (Lehmler et al. 1997), *S. racemosum* (Grummt et al. 1998), *Nectria haematococca* (Wu et al. 1998), and *A. nidulans* (Requena et al. 2001). Conventional kinesins belong to the fastest kinesins known in eukaryotic cells, moving as dimers on MTs with a velocity of 2.5 µm/s in in vitro gliding experiments, which is three- to fivefold faster than that of their animal counterparts (Steinberg 1997; Grummt et al. 1998). The structural features responsible for the high gliding velocity of fungal kinesins are not yet well understood, but under active investigation (Kallipolitou et al. 2001).

The deletion of the conventional kinesin-encoding gene from any of the filamentous fungi investigated leads to a reduction of polarized growth and the size of *Spitzenkörper*, the secretory vesicle aggregation at the hyphal tip (Lehmler et al. 1997; Seiler et al. 1997; Wu et al. 1998; Seiler et al. 1999; Requena et al. 2001). These observations support the idea that conventional kinesin is involved in the transportation of components necessary for hyphal growth along the MTs towards the hyphal tip. It is noteworthy that in no case the mutation of the conventional kinesin gene has led to a complete cessation of growth, which implies the existence of other hyphal transportation systems, either based on other less efficient kinesins, or on the actin – myosin system. In *A. nidulans*, the deletion of kinesin also caused disturbance in nuclear distribution in the germ tube and hyphae suggesting that conventional kinesin plays a role in nuclear migration (Requena et al. 2001). In addition, the functional analysis suggested that the conventional kinesin of *A. nidulans* is involved in destabilization of MTs (Requena et al. 2001).

The MT-associated motor cytoplasmic dynein is generally proposed to provide the motive force for nuclear movement in filamentous fungi (Morris et al. 1995; Fischer 1999; Suelmann and Fischer 2000). The heavy chain of cytoplasmic dynein is a large polypeptide with a region for dimerization at the N-terminal and globular motor domain with four ATP binding and MT-binding sites at the C-terminal portion of the polypeptide. In association with the N-terminus of the dynein heavy chain, smaller polypeptides called intermediate and light chains occur, which are involved in the regulation of dynein heavy chain activity and function (Steinberg 1998, 2000). A multi-subunit complex dynactin is also required for efficient MT-associated transport by cytoplasmic dynein. In the dynactin complex actin-related protein ARP1 is the most abundant and p150^{Glued} the largest subunit. The dynactin complex also includes several other polypeptides that play an important role both in the activation of the dynein motor molecule and in the binding of dynein to membranes (King and Schroer 2000).

Dynein heavy chain encoding genes have been characterized from *A. nidulans* with the help of a temperature-sensitive mutation *nudA* that affects nuclear distribution, and from *N. crassa* as a morphological mutation with curly hyphae named *ro-1*. In the temperature-sensitive mutation of the *nudA* gene the nuclear movement from conidia into the germ tube failed and small hyphal colonies were formed at restrictive temperature (Xiang et al. 1994). In the *ro-1* mutant of *N. crassa*, nuclear aggregates formed in the hyphae at some distance from the hyphal tip (Plamann et al. 1994). The cloning and sequencing of the *nudA* and *ro-1* genes revealed that they both encode the cytoplasmic dynein heavy chains with ATP-binding motor domain at the C-terminus. Later, the dynein heavy chains have been identified from *N. haematococca* (Inoue et al. 1998), and *U. maydis* (Straube et al. 2001), by using degenerate oligonucleotide primers. Interestingly, this approach has led to the identification of a dynein with split motor domain in *U. maydis*, in which the N-terminus of the motor domain including the four ATP binding regions is encoded by *dyn1* and the MT-binding part by *dyn2* gene. Some of the *ro-1*-like phenotypic *N. crassa* mutants carry mutations in genes encoding subunits of the dynactin complex (Bruno et al. 1996). This has facilitated the isolation of *ro-4* and *ro-3* genes encoding the actin-related protein, Arp1 (Plamann et al. 1994), and $p150^{Glued}$ (Tinsley et al. 1996). Several other *ro* genes, such as *ro-2*, *ro-7* and *ro-12*, encoding different subunits of the dynactin complex have been identified (Lee et al. 2001).

By comparing the effects of the deletion of kinesin (*Nkin*), or dynein (*ro-1*), and both proteins on nuclear distribution, vesicle transport, secretion and vacuole formation in *N. crassa* hyphae, it was concluded that conventional kinesin indeed is responsible for the apical transport of vesicles destined for secretion, whereas dynein is responsible for nuclear movements and transport of vacuole precursors in the opposite direction. The latter phenomenon is suggested to support the formation of vacuoles in the basal part of *N. crassa* hyphae (Seiler et al. 1999), while conventional kinesin encoded by *kin2* was shown to be responsible for the accumulation of vacuoles to the basal part of in *U. maydis* dikaryotic hyphae (Steinberg et al. 1998). These results indicate that mutations in different motor molecules may result in the same phenotype in fungi belonging to different taxa, the phenotype in this case being the vacuolation of the basal part of a hypha. Interestingly, in a dynein-deficient mutant of *N. haematococca*, astral-like arrays of cytoplasmic MTs radiating from nuclear spindle pole bodies were missing, which probably causes the clustered nuclear distribution in the mutant hyphae. In filamentous fungi, the astral MTs are suggested to be responsible for post-mitotic nuclear migration and anchoring of the interphase nuclei to membrane structures (Aist and Bayles 1988; Salo et al. 1989; Raudaskoski et al. 1991; Morris et al. 1995; Inoue et al. 1998; Raudaskoski 1998).

Recent observations on living nonsymbiotic hyphae of endomycorrhizal fungi (Bago et al. 1998) have revealed active nuclear movements. This phenomenon has been observed in anastomosing hyphae before the establishment of symbiosis (Giovannetti et al. 1999), and between hyphae growing out from infected roots (Giovannetti et al. 2001). The nuclear movements was shown to be accompanied by cytoplasmic flow. These observations have awoken interest in the mechanism responsible for the nuclear movements, e.g., whether it is MT- or actin cytoskeleton-dependent. In ectomycorrhizal hyphae mobility of vacuoles has been observed in and between adjacent cells of young dikaryotic hyphae of *Pisolithus tinctorius* (Shepherd et al. 1993), as well as in hyphae growing out from the *Eucalyptus pilularis–P. tinctorius* ectomycorrhiza (Allaway and Ashford 2001). The vacuole motility includes tubule extensions and retractions, undulating movements, projections of tubules from spherical vacuoles and fusions of tubules with spherical vacuoles and other tubules (Cole et al. 1998). The motile vacuolar system has a high phosphorus, nitrogen and potassium content. A comparable content of ions also occurs in the hyphal vacuoles of the mycorrhizal sheath and the Hartig net (Ashford et al. 1999). This suggests that the tubular vacuolar system is perhaps involved in the transfer of phosphorus and nitrogen, both in short distance transport from cell to cell and in long distance transport from hyphal tips to the plant/fungal interface (Cole et al. 1998). Interestingly, the motility of tubular vacuolar system is dependent on an intact MT cytoskeleton (Hyde et al. 1999).

8 Cell Cycle and Cytoskeleton in Mycorrhiza

In *P. sylvestris* ectomycorrhiza, an increase in the number of short roots was observed in the root region that was in contact with the fungus (Niini and Raudaskoski 1998). Similarly, in *Eucalyptus grandis- Pisolithus* ectomycorrhiza, the number of root tips increased in seedlings inoculated with the symbiotic fungus (Burgess et al. 1995). In the colonized short roots of *P. sylvestris* the fungus also activates the cell divisions in the root tip leading to the formation of dichotomous and coralloid mycorrhiza (Niini 1998), unique for pines. The increase in short root number and the formation of dichotomous and coralloid short roots suggests that the presence of fungal mycelium activates the cell cycle in the pericycle for short root production and in the root tip meristem at the formation of dichotomous and coralloid short roots. In contrast, in endomycorrhiza the plant cell cycle appears not to be activated, but the chromatin of the plant nucleus in the root cortical cells decondensates which leads to nuclear hypertrophy (Berta et al. 1990).

Activation of the cell cycle in the root cortex occurs when root nodules are formed in the legume plants associated with *Rhizobium* bacteria (Mylona et al. 1995), or in the roots of woody plants at actinorhizal nodule formation

(Berg 1999). The integration of T-DNA from an unmodified Ti-plasmid of *Agrobacterium tumefaciens* into the plant genome induces tumor formation in plants by activating the plant cell cycle. *Agrobacterium* T-DNA is known to encode enzymes necessary for cytokinin and auxin synthesis, which in turn promote tumor formation via reactivation of the cell cycle in *Agrobacterium*-infected plants (Sheng and Citovsky 1996). In *Rhizobium* it has been shown that the *Nod*-factors, the signal molecules produced by *Rhizobium*, are necessary for reactivation of the cell cycle. Nod factors are shown to elicit local reduction in plant auxin transport and auxin accumulation, which probably stimulates root cortical cell division (Mathesius et al. 1998; Boot et al. 1999). Formation of a nodule-like structure can also be induced by treatment with auxin transport inhibitors (Hirsch and Fang 1994).

Early experiments by Slankis (1949) using isolated *P. sylvestris* roots and exogenous auxin (IAA) treatment showed that it was possible to obtain mycorrhiza-like dichotomous branching without the fungus. Some ectomycorrhizal fungi including *S. bovinus* have been shown to be able to produce IAA or cytokinin (Beyrle 1995). The effect of auxin on root branching and on the formation of dichotomous and coralloid short roots has been recently reinvestigated in several pine species by using auxin and auxin transport inhibitors (Kaska et al. 1999). This research indicated that auxin induces a marked increase in the formation of lateral roots while the treatment with auxin transport inhibitors induced dichotomous and coralloid short roots.

The relationship between the hormone treatments or the effect of the fungal hyphae on the expression of cell cycle regulating genes in *Pinus* has not been investigated. In plants as in other eukaryotic organisms, the cyclins and cyclin-dependent kinases (CDKs) are key regulators of cell cycle (Mironov et al. 1999). Both A- and B-type cyclins are expressed during mitosis, the A-type cyclins being also active during S-phase progression. D-type cyclins have an important role in the G1 to S phase transition. Transcription of D-type cyclins can be induced by the phytohormone cytokinin or by sucrose, which means that mitogenic signals stimulate transcription of D-cyclins and modulate cell cycle activity. Recently, a new D-type cyclin has been identified in *Arabidopsis* that is expressed during lateral root formation and the expression of which is stimulated by sucrose (De Veylder et al. 1999).

In higher-plant cells, the MT organization is regulated during the cell cycle (Vantard et al. 2000). Especially G2-phase and mitosis are accompanied by changes in the distribution of MTs. In early G2 the cortical interphase MTs accumulate to form the preprophase band (PPB) which precisely marks the future site for the cell plate formation. The breakdown of the nuclear envelope leads to the PPB disassembly and polymerization of spindle MTs. After karyokinesis the phragmoplast is formed at the site marked at G2 phase by the PPB. The phragmoplast consists of a ring of anti-parallel, inter-digitating microtubules and actin filaments, which are thought to transport the vesicles containing the material for cell plate construction. A close relationship between the function of cell cycle and MT cytoskeleton is suggested by the association of cell cycle-regulating components with specific MT arrays in dividing plant cells. B-cyclins and cyclin-dependent kinase Cdc2a are immunolocalized to preprophase band, mitotic spindle and chromosomes, while A1 cyclin is detected at cytokinesis in association with phragmoplast MTs. At interphase, Cdc2a kinase and A cyclin are localized in the nucleus (John et al. 2001).

In ectomycorrhiza, the association of plant root cells with the fungus leads probably to mitogenic stimuli that activate the cell cycle in pericycle and root meristematic tissue, seen in *Pinus* as production of short roots with dichotomous and coralloid tips. Cell cycle activation involves the regulation of MT dynamics so that the transition of one MT array to another is achieved during different cell cycle phases and also at cellular differentiation.

The Nobel Prize in Physiology or Medicine 2001 was awarded to L.H. Hartwell, R.T. Hunt and Sir Paul M. Nurse for their discoveries of key regulators of the cell cycle. The yeasts *Schizosaccharomyces pombe* and *Saccharomyces cerevisiae* were important model organisms for the prize winning research. In spite of this, very little is known about cell cycle regulation in filamentous fungi. The central genes have only been isolated and their function studied in the filamentous fungus, *A. nidulans* (Ye et al. 1999). In future, the isolation and characterization of the genes involved in regulation of the cell cycle (Tarkka 2001) and the cytoskeleton from both symbiotic partners, tree roots and fungal hyphae, will provide us with new insight about the development of the ectomycorrhizal association.

9 Cytoskeletal Research Methods

In the previous sections the use of in situ hybridization (Uribe et al. 1998), and GUS-reporter gene constructs (An et al. 1996a, b; Chu et al. 1998; Kandasamy et al. 2001), were described in localization transcripts of actin and tubulin gene products or for visualization of their activity in different plant tissues, even in endomycorrhiza (Bonfante et al. 1996). The application of these methods requires that the genes encoding tubulin or actin, and a transformation system are available. When actins or tubulins are detected at the protein level by Western blotting (Raudaskoski et al. 1987), commercially produced antibodies are applied. When commercial antibodies are used, the possibility always exists that some of the tubulin or actin proteins are not recognized. Therefore, efforts have been made to produce antibodies specific for the different actin or tubulin proteins (McLean et al. 1990). The availability of such antibodies would also be helpful for investigating the distribution of the different cytoskeletal proteins in plant and fungal cells by immunocytochemical methods.

9.1 Indirect Immunofluorescence Microscopy

Most antibodies that lead to signal detection in immunoblots can also be used to visualize the intracellular structure of actin filaments or MTs by indirect immunofluorescence (IIF) microscopy (Fig. 1A, B, E) as discussed in previous sections. IIF is based on the use of two antibodies, primary and secondary. The primary antibody is raised against a cytoskeletal protein or its peptide while the fluorochrome-labeled secondary antibody is able to recognize the primary antibody. If the primary antibody is monoclonal, it binds to a single epitope of the target protein. A polyclonal primary antibody binds several epitopes of the target protein. The immunocytochemical methods can also be applied at the electron microscopic level, although then the secondary antibody has to be labeled with gold particles. MFs are often visualized by fluorochrome-labeled phallotoxins which bind specifically to filamentous actin (Barak et al. 1980).

The successful visualization of MFs and MTs in root cells and fungal hyphae requires a fixative that penetrates quickly into the plant tissue and stabilizes the cytoskeletal elements in the polymerized form. A method that stops the metabolism of the cell even more quickly than chemical fixative and preserves the structure of MTs and MFs in fungal and plant cells is cryo-fixation (Raudaskoski et al. 1987, 1991, 1994; Åström et al. 1994; Lancelle et al. 1997; Bourett et al. 1998). Pretreatment of plant cells with an MF-stabilizing agent MBS (3-maleimidobenzoiz acid N-hydroxysuccinimide) ester has also been used (Sonobe and Shibaoka 1989; Miller et al. 1999; de Ruijter et al. 2001). Although the method leads to well-preserved MFs, the MBS pretreatment may also cause redistribution of the target molecules and the pretreatment may lead to less flexible images of actin cytoskeleton.

The penetration of the antibody into the plant or fungal cell requires enzymatic digestion of the cell wall and permeabilization of the plasma membrane with a detergent after fixation. During the enzymatic digestion, the protease activity in the enzyme preparation has to be reduced by protease inhibitors and/or by adding 1 % bovine serum albumin in the digestion buffer. For enzymatic treatment and for labeling with the antibodies, cells or cell rows can be isolated or cut manually from the fixed material (Uetake et al. 1997; Uetake and Peterson 1997, 1998), which can also be embedded in 15 % agar (Genre and Bonfante 1997, 1998) or cyanoacrylate (Blancaflor et al. 2001) for sectioning. The visualization of cytoskeletal elements also succeeds well when the fixed plant roots are embedded in Steedsman's wax and then sectioned (Baluška et al. 1992, 1995, 1997, 2001; Olinevich et al. 2001). Before labeling with the primary antibody the sections are dewaxed in ethanol, passed through a graded ethanol series diluted with PBS and either treated with a cell wall-degrading enzyme (Baluška et al. 1992), or not (Balu_ka et al. 2001). With these methods good preservation and visualization of cytoskeletal components is achieved in the roots of herbaceous plants, even with endomycorrhiza. Cells with well-preserved cytoskeletal structures can then be examined either by a regular fluorescence microscope or by a laser scanning confocal microscope.

In ectomycorrhiza, the thick hydrophobic sheath around the root formed by fungal hyphae inhibits the penetration of the fixative or rapid freezing of the fungal and plant cells. The fixation of ectomycorrhizal root in situ under vacuum facilitates and speeds up the penetration of the fixative into the ectomycorrhizal root (Timonen et al. 1993; Niini and Raudaskoski 1998). In ectomycorrhiza it is necessary to prepare sections from the fixed roots and until now, the best results in immunolocalization of cytoskeletal elements have been achieved by using cryosections (Fig. 1A, B; Niini and Raudaskoski 1998). No success in the visualization of MTs or MFs has yet been achieved by embedding the *Pinus* short roots or ectomycorrhizal roots in wax.

9.2 Microinjection Method

Microinjection of fluorescent-labeled phalloidin or tubulin has been used to visualize MFs (Schmit and Lambert 1990; Zhang et al. 1993; Cleary 1995; Kim et al. 1995; Miller et al.1996) and MTs (Zhang et al. 1990; Yuan et al. 1994) in living plant cells. However, the visualization of cytoskeletal elements succeeds with microinjection only in a few plant cell types, such as stamen hair cells of *Tradescantia* (Zhang et al. 1993), or epidermal cells of leaves (Yuan et al. 1994), while cytoskeletal elements in narrow fungal hyphae are not easily studied with this method.

9.3 Green Fluorescence Protein Technique

The MFs and MTs in plant and fungal cells have been successfully visualized with the help of green fluorescence protein (GFP) fused to actin, tubulin or to a cytoskeleton-associated protein. In plant cells, tubulin-GFP fusion protein has been used to visualize MTs in different *Arabidopsis* tissues (Ueda et al. 1999; Whittington et al. 2001). Transient expression of the MT binding domain of mammalian MAP4 labeled with GFP and transgenic plants with the same construct have been used for visualization of the MT organization in epidermal cells of fava bean (Marc et al.1998) and in *Arabidopsis* root cells (Bao et al. 2001). Fusion of GFP to the actin-binding domain of talin has led to visualization of actin cytoskeleton transiently in tobacco BY-2 suspension cells (Kost et al. 1998) and pollen tubes (Kost et al. 1998, 1999a, b; Fu et al. 2001). Constitutive expression of the same construct visualized actin cytoskeleton in different tissues of *Arabidopsis* including root hairs (Kost et al. 1998; Baluška et al. 2000; Baluška and Volkmann 2002). Recently, both MTs and MFs were visualized in living onion epidermal cells by using the MT binding domain of MAP4 and actin binding domain of talin fused with different spectral variants of GFP (Blancaflor 2002).

MTs (Straight et al. 1997) and actin (Doyle and Botstein 1996) of yeast have also been successfully visualized by GFP-tubulin and -actin fusions as well as the MTs in the hyphae of *Aspergillus nidulans* (Han et al. 2001). Recently, it has been shown that GFP can be used as a reporter of a gene function in the hyphae of filamentous basidiomycetes (Lugones et al. 1999; Ma et al. 2001), although no fungal protein has yet been localized by fusion to GFP in these fungi. In future it seems possible, at least in endomycorrhiza forming plants, to visualize the effect of fungal growth on the plant cytoskeleton in living root cells by GFP-fusion proteins. The application of GFP for visualization of cytoskeletal elements in endo- and ectomycorrhizal fungi during vegetative or symbiotic growth requires the development of an efficient transformation system for these fungi (Pardo et al. 2002).

Acknowledgements. The authors thank Erja Laitiainen (M.Sc.) for technical help in preparing the manuscript. The work was supported by a grant from the Academy of Finland to M.R.

References and Selected Reading

Aist JR, Bayles CJ (1988) Video motion analysis of mitotic events in living cells of the fungus *Fusarium solani*. Cell Motil Cytoskel 9:325–336

Allaway WG, Ashford AE (2001) Motile tubular vacuoles in extramatrical mycelium and sheath hyphae of ectomycorrhizal systems. Protoplasma 215:218–225

An Y-Q, McDowell JM, Huang S, McKinney EC, Chambliss S, Meagher RB (1996a) Strong, constitutive expression of the *Arabidopsis ACT2/ACT8* actin subclass in vegetative tissues. Plant J 10:107–121

An Y-Q, Huang S, McDowell JM, McKinney EC, Meagher RB (1996b) Conserved expression of the *Arabidopsis ACT1* and *ACT3* actin subclass in organ primordia and mature pollen. Plant Cell 8:15–30

Asada T, Kuriyama R, Shibaoka H (1997) TKRP125, a kinesin-related protein involved in the centrosome-independent organization of the cytokinetic apparatus in tobacco BY-2 cells. J Cell Sci 110:179–189

Ashford AE, Vesk PA, Orlovich DA, Markovina A-L, Allaway WG (1999) Dispersed polyphosphate in fungal vacuoles in *Eucalyptus pilularis/Pisolithus tinctorius* ectomycorrhizas. Fungal Genet Biol 28:21–33

Åström H, Giovannetti M, Raudaskoski M (1994) Cytoskeletal components in the arbuscular mycorrhizal fungus *Glomus mosseae*. Mol Plant-Microb Interact 7:309–312

Ayscough KR (1998) In vivo functions of actin-binding proteins. Curr Opin Cell Biol 10:102–111

Bago B, Zipfel W, Williams RM, Chamberland H, Lafontaine JG, Webb WW, Piché Y (1998) In vivo studies on the nuclear behavior of the arbuscular mycorrhizal fungus *Gigaspora rosea* grown under axenic conditions. Protoplasma 203:1–15

Baluška F, Volkmann D (2002) Pictures in cell biology. Actin-driven polar growth of plant cells. Trends Cell Biol 12:14

Baluška F, Parker JS, Barlow PW (1992) Specific patterns of cortical and endoplasmic microtubules associated with cell growth and tissue differentiation in roots of maize (*Zea mays* L.). J Cell Biol 103:191–200

Baluška F, Barlow PW, Hauskrecht M, Kubica Š, Parker JS, Volkmann D (1995) Microtubule arrays in maize root cells. Interplay between the cytoskeleton, nuclear organization and post-mitotic cellular growth patterns. New Phytol 130:177–192

Baluška F, Kreibaum A, Vitha S, Parker JS, Barlow PW, Sievers A (1997) Central root cap cells are depleted of endoplasmic microtubules and actin microfilament bundles: implications for their role as gravity-sensing statocytes. Protoplasma 196:212–223

Baluška F, Salaj J, Mathur J, Braun M, Jasper F, Samaj J, Chua N-H, Barlow PW, Volkmann D (2000) Root hair formation: F-actin-dependent tip growth is initiated by local assembly of profilin-supported F-actin meshworks accumulated within expansin-enriched bulges. Dev Biol 227:618–632

Baluška F, Jasik J, Edelmann HG, Salajová T, Volkmann D (2001) Latrunculin B-induced plant dwarfism: plant cell elongation is F-actin-dependent. Dev Biol 231:113–124

Bamburg JR (1999) Proteins of the ADF/cofilin family: essential regulators of actin dynamics. Annu Rev Cell Dev Biol 15:185–230

Bao Y, Kost, B, Chua N-H (2001) Reduced expression of α-tubulin genes in *Arabidopsis thaliana* specifically affects root growth and morphology, root hair development and root gravitropism. Plant J 28:145–157

Barak LS, Yocum RR, Nothnagel EA, Webb WW (1980) Fluorescence staining of the actin cytoskeleton in living cells with 7-nitrobenz-2-oxa-1,3-diazolephallacidin. Proc Natl Acad Sci USA 77:980–984

Barlow PW, Baluška F (2000) Cytoskeletal perspectives on root growth and morphogenesis. Annu Rev Plant Physiol Plant Mol Biol 51:289–322

Barroso C, Chan J, Allan V, Doonan J, Hussey P, Lloyd C (2000) Two kinesin-related proteins associated with the cold-stable cytoskeleton of carrot cells: characterization of a novel kinesin, DcKRP120-2. Plant J 24:859–868

Berg HR (1999) Cytoplasmic bridge formation in the nodule apex of actinorhizal root nodules. Can J Bot 77:1351–1357

Berta G, Sgorbati S, Soler V, Fusconi A, Trotta A, Citterio A, Bottone MG, Sparvoli E, Scannerini S (1990) Variations in chromatin structure in host nuclei of a vesicular arbuscular mycorrhiza. New Phytol 114:199–203

Beyrle H (1995) The role of phytohormones in the function and biology of mycorrhizas. In: Varma A, Hock B (eds) Mycorrhiza. Structure, function, molecular biology and biotechnology. Springer, Berlin Heidelberg New York, pp 365–390

Blancaflor EB (2002) The cytoskeleton and gravitropism in higher plants. J Plant Growth Regul 21:120–135

Blancaflor EB, Zhao L, Harrison MJ (2001) Microtubule organization in root cells of *Medicago truncatula* during development of an arbuscular mycorrhizal symbiosis with *Glomus versiforme*. Protoplasma 217:154–165

Bonfante P, Perotto S (1995) Strategies of arbuscular mycorrhizal fungi when infecting host plants. New Phytol 130:3–21

Bonfante P, Bergero R, Uribe X, Romera C, Rigau J, Puigdomènech P (1996) Transcriptional activation of a maize α-tubulin gene in mycorrhizal maize and transgenic tobacco plants. Plant J 9:737–743

Boot KJM, Van Brussel AAN, Tak T, Spaink HP, Kijne JW (1999) Lipochitin oligosaccharides from *Rhizobium leguminosarum* bv.*viciae* reduce auxin transport capacity in *Vicia sativa* subsp. *nigra* roots. Mol Plant-Microb Interact 12:839–844

Bourett TM, Czymmek KJ, Howard RJ (1998) An improved method for affinity probe localization in whole cells of filamentous fungi. Fungal Gen Biol 24:3–13

Bruno KS, Tinsley JH, Minke PF, Plamann M (1996) Genetic interactions among cytoplasmic dynein, dynactin, and nuclear distribution mutants of *Neurospora crassa*. Proc Natl Acad Sci USA 93:4775–4780

Burgess T, Laurent P, Dell B, Malajczuk N, Martin F (1995) Effect of fungal-isolate aggressivity on the biosynthesis of symbiosis-related polypeptides in differentiating eucalypt ectomycorrhizas. Planta 195:408–417

Bütehorn B, Gianinazzi-Pearson V, Franken P (1999) Quantification of β-tubulin RNA expression during asymbiotic and symbiotic development of the arbuscular mycorrhizal fungus *Glomus mosseae*. Mycol Res 103:360–364

Canut H, Carrasco A, Galaud J-P, Cassan C, Bouyssou H, Vita N, Ferrara P, Pont-Lezica R (1998) High affinity RGD-binding sites at the plasma membrane of *Arabidopsis thaliana* links the cell wall. Plant J 16:63–71

Chaffey N, Barlow P, Barnett J (2000) A cytoskeletal basis for wood formation in angiosperm trees: the involvement of microfilaments. Planta 210:890–896

Chan J, Rutten T, Lloyd C (1996) Isolation of microtubule-associated proteins from carrot cytoskeletons: a 120 kDa map decorates all four microtubule arrays and the nucleus. Plant J 10:251–259

Chan J, Jensen CG, Jensen LCW, Bush M, Lloyd CW (1999) The 65-kDa carrot microtubule-associated protein forms regularly arranged filamentous cross-bridges between microtubules. Pro Natl Acad Sci USA 96:14931–14936

Chu B, Wilson TJ, McCune-Zierath C, Snustad DP, Carter JV (1998) Two β-tubulin genes, *TUB1* and *TUB8*, of *Arabidopsis* exhibit largely nonoverlapping patterns of expression. Plant Mol Biol 37:785–790

Cleary AL (1995) F-actin redistributions at the division site in living *Tradescantia* stomatal complexes as revealed by microinjection of rhodamine-phalloidin. Protoplasma 185:152–165

Cole L, Orlovich DA, Ashford AE (1998) Structure, function, and motility of vacuoles in filamentous fungi. Fungal Genet Biol 24:86–100

Cooper JA, Schafer DA (2000) Control of actin assembly and disassembly at filament ends. Curr Opin Cell Biol 12:97–103

Daniels RH, Bokoch GM (1999) p21-activated protein kinase: a crucial component of morphological signaling? TIBS 24:350–355

de Ruijter NCA, Esseling JJ, Emons AMC (2001) The roles of calcium and the actin cytoskeleton in regulation of root hair tip growth by rhizobial signal molecules. In: Geitmann A, Cresti M, Heath IB (eds) Cell biology of plant and fungal tip growth. IOS Press, NATO Sci Ser 328:55–67

De Veylder L, de Almeida Engler J, Burssens S, Manevski A, Lescure B, van Montagu M, Engler G, Inze D (1999) A new D-type cyclin of *Arabidopsis thaliana* expressed during lateral root primordial formation. Planta 208:453–462

Diaz EC, Martin F, Tagu D (1996) Eucalypt α-tubulin: cDNA cloning and increased level of transcripts in ectomycorrhizal root system. Plant Mol Biol 31:905–910

Dong C-H, Kost B, Xia G, Chua N-H (2001) Molecular identification and characterization of the *Arabidopsis AtADF1*, *AtADF5* and *AtADF6* genes. Plant Mol Biol 45:517–527

Doyle T, Botstein D (1996) Movement of yeast cortical actin cytoskeleton visualized in vivo. Proc Natl Acad Sci USA 93:3886–3891

Dutcher SK (2001) The tubulin fraternity: alpha to eta. Curr Opin Cell Biol 13:49–54

Enos AP, Morris NR (1990) Mutation of a gene that encodes a kinesin-like protein blocks nuclear division in *A. nidulans*. Cell 60:1019–1027

Euteneuer U, McIntosh JR (1980) Polarity of midbody and phragmoplast microtubules. J Cell Biol 87:509–515

Fischer R (1999) Nuclear movement in filamentous fungi. FEMS Microbiol Rev 23:38–69

Fowler TJ, Desimone SM, Mitton MF, Kurjan J, Raper CA (1999) Multiple sex pheromones and receptors of a mushroom-producing fungus elicit mating in yeast. Mol Biol Cell 10:2559–2572

Franken P, Lapopin L, Meyer-Gauen G, Gianinazzi-Pearson V (1997) RNA accumulation and genes expressed in spores of the arbuscular mycorrhizal fungus, *Gigaspora rosea.* Mycol 89:293–297

Fu Y, Wu G, Yang Z (2001) Rop GTPase-dependent dynamics of tip-localized F-actin controls tip growth in pollen tubes. J Cell Biol 152:1019–1032

Genre A, Bonfante P (1997) A mycorrhizal fungus changes microtubule orientation in tobacco root cells. Protoplasma 199:30–38

Genre A, Bonfante P (1998) Actin versus tubulin configuration in arbuscule-containing cells from mycorrhizal tobacco roots. New Phytol 140:745–752

Giovannetti M, Azzolini D, Citernesi AS (1999) Anastomosis formation and nuclear and protoplasmic exchange in arbuscular mycorrhizal fungi. Appl Environ Microbiol 65:5571–5575

Giovannetti M, Fortuna P, Citernesi AS, Morini S, Nuti MP (2001) The occurrence of anastomosis formation and nuclear exchange in intact arbuscular mycorrhizal networks. New Phytol 151:717–724

Goode BL, Drubin DG, Barnes G (2000) Functional cooperation between the microtubule and actin cytoskeletons. Curr Opin Cell Biol 12:63–71

Gorfer M, Tarkka MT, Hanif M, Pardo AG, Laitiainen E, Raudaskoski M (2001) Characterization of small GTPases Cdc42 and Rac and the relationship between Cdc42 and actin cytoskeleton in vegetative and ectomycorrhizal hyphae of *Suillus bovinus.* MPMI 14:135–144

Grummt M, Pistor S, Lottspeich F, Schliwa M (1998) Cloning and functional expression of a 'fast' fungal kinesin. FEBS Lett 427:79–84

Hahn M, Mendgen K (2001) Signal and nutrient exchange at biotrophic plant-fungus interfaces. Curr Opin Plant Biol 4:322–327

Han G, Liu B, Zhang J, Zuo W, Morris NR, Xiang X (2001) The *Aspergillus* cytoplasmic dynein heavy chain and NUDF localize to microtubule ends and affect microtubule dynamics. Curr Biol 11:719–724

Heath IB, Gupta G, Bai S (2000) Plasma membrane-adjacent actin filaments, but not microtubules, are essential for both polarization and hyphal tip morphogenesis in *Saprolegnia ferax* and *Neurospora crassa.* Fungal Genet Biol 30:45–62

Hepler PK, Vidali L, Cheung AY (2001) Polarized cell growth in higher plants. Annu Rev Cell Dev Biol 17:159–187

Hirsch AM, Fang Y (1994) Plant hormones and nodulation: what's the connection? Plant Mol Biol 26:5–9

Huang S, An Y-Q, McDowell JM, McKinney EC, Meagher RB (1997) The *Arabidopsis ACT11* actin gene is strongly expressed in tissues of the emerging inflorescence, pollen, and developing ovules. Plant Mol Biol 33:125–139

Hussey PJ, Haas N, Hunsperger J, Larkin J, Snustad DP, Silflow CD (1990) The β-tubulin gene family in *Zea mays*: two differentially expressed β-tubulin genes. Plant Mol Biol 15:957–972

Hussey PJ, Yuan M, Calder G, Khan S, Lloyd CW (1998) Microinjection of pollen-specific actin-depolymerizing factor, ZmADF1, reorientates F-actin strands in *Tradescantia* stamen hair cells. Plant J 14:353–357

Hyde GJ, Davies D, Perasso L, Cole L, Ashford AE (1999) Microtubules, but not actin microfilaments, regulate vacuole motility and morphology in hyphae of *Pisolithus tinctorius.* Cell Motil Cytoskel 42:114–124

Inoue S, Turgeon BG, Yoder OC, Aist JR (1998) Role of fungal dynein in hyphal growth, microtubule organization, spindle pole body motility and nuclear migration. J Cell Sci 111:1555–1566

Janssen M, Hunte C, Schulz M, Schnabl H (1996) Tissue specification and intracellular distribution of actin isoforms in *Vicia faba* L. Protoplasma 191:158–163

Jiang C-J, Sonobe S (1993) Identification and preliminary characterization of a 65 kDa higher-plant microtubule-associated protein. J Cell Sci 105:891–901

Jiang C-J, Weeds AG, Hussey PJ (1997) The maize actin-depolymerizing factor, ZmADF3, redistributes to the growing tip of elongating root hairs and can be induced to translocate into the nucleus with actin. Plant J 12:1035–1043

John PCL, Mews M, Moore R (2001) Cyclin/Cdk complexes: their involvement in cell cycle progression and mitotic division. Protoplasma 216:119–142

Johnson DJ (1999) Cdc42: an essential rho-type GTPase controlling eukaryotic cell polarity. Microbiol Mol Biol Rev 63:54–105

Johnson DJ, Pringle JR (1990) Molecular characterization of CDC42, a *Saccharomyces cerevisiae* gene involved in the development of cell polarity. J Cell Biol 111:143–152

Joshi HC, Palevitz, BA (1996) γ-Tubulin and microtubule organization in plants. Trends Cell Biol 6:41–44

Joshi HC, Palacios MJ, McNamara L, Cleveland DW (1992) γ-Tubulin is a centrosomal protein required for cell cycle-dependent microtubule nucleation. Nature 356:80–83

Joyce CM, Villemur R, Snustad DP, Silflow CD (1992) Tubulin gene expression in maize (*Zea mays* L.). Change in isotype expression along the developmental axis of seedling root. J Mol Biol 227:97–107

Kallipolitou A, Deluca D, Majdic U, Lakämper S, Cross R, Meyhöfer E, Moroder L, Schliwa M, Woehlke G (2001) Unusual properties of the fungal conventional kinesin neck domain from *Neurospora crassa*. EMBO J 20:6226–6235

Kandasamy MK, Gilliland LU, McKinney EC, Meagher RB (2001) One plant actin isovariant, ACT7, is induced by auxin and required for normal callus formation. Plant Cell 13:1541–1554

Kaska DD, Myllylä R, Cooper JB (1999) Auxin transport inhibitors act through ethylene to regulate dichotomous branching of lateral root meristems in pine. New Phytol 142:49–58

Kerr GP, Carter JV (1990a) Tubulin isotypes in rye roots are altered during cold acclimation. Plant Physiol 93:83–88

Kilmartin JV, Adams AEM (1984) Structural rearrangements of tubulin and actin during the cell cycle of the yeast *Saccharomyces*. J Cell Biol 98:922–933

Kim AJ, Endow SA (2000) A kinesin family tree. J Cell Sci 113:3681–3682

Kim M, Hepler PK, Eun S-O, Ha KS, Lee Y (1995) Actin filaments in mature guard cells are radially distributed and involved in stomatal movement. Plant Physiol 109:1077–1084

King SJ, Schroer TA (2000) Dynactin increases the processivity of the cytoplasmic dynein motor. Nat Cell Biol 2:20–24

Klahre U, Chua N (1999) The *Arabidopsis* actin-related protein 2 (AtARP2) promoter directs expression in xylem precursor cells and pollen. Plant Mol Biol 41:65–73

Klahre U, Friederich E, Kost B, Louvard D, Chua N-H (2000) Villin-like actin-binding proteins are expressed ubiquitously in *Arabidopsis*. Plant Physiol 122:35–47

Ko KS, McCulloch CA (2000) Partners in protection: interdependence of cytoskeleton and plasma membrane in adaptations to applied forces. J Membr Biol 174:85–95

Kopczak SD, Haas NA, Hussey PJ, Silflow CD, Snustad DP (1992) The small genome of Arabidopsis contains at least six expressed α-tubulin genes. Plant Cell 4:539–547

Kost B, Spielhofer P, Chua N-H (1998) A GFP-mouse talin fusion protein labels plant actin filaments in vivo and visualizes the actin cytoskeleton in growing pollen tubes. Plant J 16:393–401

Kost B, Lemichez E, Spielhofer P, Hong Y, Tolias K, Carpenter C, Chua N-H (1999a) Rac homologues and compartmentalized phosphatidylinositol 4,5-bisphosphate act in a common pathway to regulate polar pollen tube growth. J Cell Biol 145:317–330

Kost B, Mathur J, Chua N-H (1999b) Cytoskeleton in plant development. Curr Opin Cell Biol 2:462–470

Kottke I, Oberwinkler F (1987) The cellular structure of the Hartig net: coenocytic and transfer cell-like organization. Nord J Bot 7:85–95

Lancelle SA, Cresti M, Hepler PK (1997) Growth inhibition and recovery in freeze-substituted *Lilium longiflorum* pollen tubes: structural effects of caffeine. Protoplasma 196:21–33

Lawrence CJ, Malmberg RL, Muszynski MG, Dawe RK (2002) Maximum likelihood methods reveal conservation of function among closely related kinesin families. J Mol Evol 54:42–53

Lee IH, Kumar S, Plamann M (2001) Null mutants of the *Neurospora* actin-related protein 1 pointed-end complex show distinct phenotypes. Mol Biol Cell 12:2195–2206

Lehmler C, Steinberg G, Snetselaar KM, Schliwa M, Kahmann R, Bölker M (1997) Identification of a motor protein required for filamentous growth in *Ustilago maydis*. EMBO J 16:3464–3473

Lemichez E, Wu Y, Sanchez J-P, Mettouchi A, Mathur J, Chua N-H (2001) Inactivation of AtRac1 by abscisic acid is essential for stomatal closure. Genes Dev 15:1808–1816

Li H, Shen J-J, Zheng Z-L, Lin Y, Yang Z (2001) The Rop GTPase switch controls multiple developmental processes in *Arabidopsis*. Plant Physiol 126:670–684

Liu B, Marc J, Joshi HC, Palevitz BA (1993) A γ-tubulin-related protein associated with the microtubule arrays of higher plants in a cell cycle-dependent manner. J Cell Sci 104:1217–1228

Liu B, Cyr RJ, Palevitz BA (1996) A kinesin-like protein, KatAp, in the cells of *Arabidopsis* and other plants. Plant Cell 8:119–132

Lloyd C, Hussey P (2001) Microtubule-associated proteins in plants – why we need a map. Nat Rev Mol Cell Biol 2:40–47

Lopez I, Anthony RG, Maciver SK, Jiang C-J, Khan S, Weeds AG, Hussey PJ (1996) Pollen specific expression of maize genes encoding actin depolymerizing factor-like proteins. Proc Natl Acad Sci USA 93:7415–7420

Lugones LG, Scholtmeijer K, Klootwijk R, Wessels JGH (1999) Introns are necessary for mRNA accumulation in *Schizophyllum commune*. Mol Microbiol 32:681–689

Ma B, Mayfield MB, Gold MH (2001) The green fluorescent protein gene functions as a reporter of gene expression in *Phanerochaete chrysosporium*. Appl Environ Microbiol 67:948–955

Marc J, Granger CL, Brincat J, Fisher DD, Kao T-H, McCubbin AG, Cyr RJ (1998) A *GFP-MAP4* reporter gene for visualizing cortical microtubule rearrangements in living epidermal cells. Plant Cell 10:1927–1939

Mathesius U, Schlaman HRM, Spaink HP, Sautter C, Rolfe BG, Djordjevic MA (1998) Auxin transport inhibition precedes root nodule formation in white clover roots and is regulated by flavonoids and derivatives of chitin oligosaccharides. Plant J 14:23–34

Matsubara Y, Uetake Y, Peterson RL (1999) Entry and colonization of *Asparagus officinalis* roots by arbuscular mycorrhizal fungi with emphasis on changes in host microtubules. Can J Bot 77:1159–1167

McGoldrick CA, Gruver C, May GS (1995) *myoA* of *Aspergillus nidulans* encodes an essential myosin I required for secretion and polarized growth. J Cell Biol 128:577–587

McLean BG, Eubanks S, Meagher RB (1990) Tissue-specific expression of divergent actins in soybean root. Plant Cell 2:335–344

Meagher RB, McKinney EC, Vitale AV (1999) The evolution of new structures: clues from plant cytoskeletal genes. Trends Genet 15:278–284

Mews M, Sek FJ, Moore R, Volkmann D, Gunning BES, John PCL (1997) Mitotic cyclin distribution during maize cell division: implications for the sequence diversity and function of cyclins in plants. Protoplasma 200:128–145

Miki H, Setou M, Kaneshiro K, Hirokawa N (2001) All kinesin superfamily protein, KIF, genes in mouse and human. Pro Natl Acad Sci USA 98:7004–7011

Miller DD, Lancelle SA, Hepler PK (1996) Actin microfilaments do not form a dense network in *Lilium longiflorum* pollen tube tips. Protoplasma 195:123–132

Miller DD, de Ruijter NCA, Bisseling T, Emons AMC (1999) The role of actin in root hair morphogenesis: studies with lipochito-oligosaccharide as a growth stimulator and cytochalasin as an actin perturbing drug. Plant J 17:141–154

Mironov V, de Veylder L, van Montagu M, Inze D (1999) Cyclin-dependent kinases and cell division in plants – the nexus. Plant Cell 11:509–521

Mitsui H, Yamaguchi-Shinozaki K, Shinozaki K, Nishikawa K, Takahashi H (1993) Identification of a gene family (*kat*) encoding kinesin-like proteins in *Arabidopsis thaliana* and the characterization of secondary structure of KatA. Mol Gen Genet 238:362–368

Mitsui H, Nakatani K, Yamaguchi-Shinozaki K, Shinozaki K, Nishikawa K, Takahashi H (1994) Sequencing and characterization of the kinesin-related genes *katB* and *katC* of *Arabidopsis thaliana*. Plant Mol Biol 25:865–876

Mitsui H, Hasezawa S, Nagata T, Takahashi H (1996) Cell cycle-dependent accumulation of a kinesin-like protein, KatB/C, in synchronized tobacco BY-2 cells. Plant Mol Biol 30:177–181

Molendijk AJ, Bischoff F, Rajendrakumar CSV, Friml J, Braun M, Gilroy S, Palme K (2001) *Arabidopsis thaliana* Rop GTPases are localized to tips of root hairs and control polar growth. EMBO J 20:2779–2788

Montoliu L, Rigau J, Puigdomènech P (1989) A tandem of α-tubulin genes preferentially expressed in radicular tissues from *Zea mays*. Plant Mol Biol 14:1–15

Morris NR, Xiang X, Beckwith SM (1995) Nuclear migration advances in fungi. Trends Cell Biol 5:278–282

Mullins RD (2000) How WASP-family proteins and the Arp2/3 complex convert intracellular signals into cytoskeletal structures. Curr Opin Cell Biol 12:91–96

Mylona P, Pawlowski K, Bisseling T (1995) Symbiotic nitrogen fixation. Plant Cell 7:869–885

Neer EJ (1995) Heterotrimeric G proteins: organizers of transmembrane signals. Cell 80:249–257

Niini S (1998) Growth pattern and cytoskeleton in *Suillus bovinus* hyphae, *Pinus sylvestris* roots and in *Pinus sylvestris–Suillus bovinus* ectomycorrhiza. Dissertationes Biocentri Viikki Universitatis Helsingiensis 16

Niini S, Raudaskoski M (1998) Growth patterns in non-mycorrhizal and mycorrhizal short roots of *Pinus sylvestris*. Symbiosis 25:101–114

Niini SS, Tarkka MT, Raudaskoski M (1996) Tubulin and actin protein patterns in Scots pine (*Pinus sylvestris*) roots and developing ectomycorrhiza with *Suillus bovinus*. Physiol Plant 96:186–192

Nuoffer C, Balch WE (1994) GTPases: multifunctional molecular switches regulating vesicular traffic. Annu Rev Biochem 63:949–990

Oakley BR, Oakley CE, Yoon Y, Jung MK (1990) γ-Tubulin is a component of the spindle pole body that is essential for microtubule function in *Aspergillus nidulans*. Cell 61:1289–1301

Olesnicky NS, Brown AJ, Dowell SJ, Casselton LA (1999) A constitutively active G-protein-coupled receptor causes mating self-compatibility in the mushroom *Coprinus*. EMBO J 18:2756–2763

Olinevich OV, Khokhlova LP, Raudaskoski M (2001) Effect of abscisic acid and cold acclimation on the cytoskeletal and phosphorylated proteins in different cultivars of *Triticum aestivum* L. Cell Biol Int 24:365–373

Osherov N, Yamashita RA, Chung YS, May GS (1998) Structural requirements for in vivo myosin I function in *Aspergillus nidulans*. J Biol Chem. 273:27017–27025

Overall RL, Dibbayawan TP, Blackman LM (2001) Intercellular alignments of the plant cytoskeleton. J Plant Growth Regul 20:162–169

Pardo AG, Hanif M, Raudaskoski M, Gorfer M (2002) Genetic transformation of ectomycorrhizal fungi mediated by *Agrobacterium tumefaciens*. Mycol Res 106:132–137

Pérez HE, Sánchez N, Vidali L, Hernández JM, Lara M, Sánchez F (1994) Actin isoforms in non-infected roots and symbiotic root nodules of *Phaseolus vulgaris* L. Planta 193:51–56

Pesacreta TC, Carley WW, Webb WW, Parthasarathy MV (1982) F-actin in conifer roots. Proc Natl Acad Sci USA 79:2898–2901

Peterson RL, Bonfante P, Faccio A, Uetake Y (1996) The interface between fungal hyphae and orchid protocorm cells. Can J Bot 74:1861–1870

Piché Y, Peterson RL, Ackerley CA (1983) Early development of ectomycorrhizal short roots of pine. Scan Electron Microsc III:1467–1474

Plamann M, Minke PF, Tinsley JH, Bruno KS (1994) Cytoplasmic dynein and actin-related protein Arp1 are required for normal nuclear distribution in filamentous fungi. J Cell Biol 127:139–149

Ramachandran S, Christensen HEM, Ishimaru Y, Dong C-H, Chao-Ming W, Cleary AL, Chua N-H (2000) Profilin plays a role in cell elongation, cell shape maintenance, and flowering in *Arabidopsis*. Plant Physiol 124:1637–1647

Raudaskoski M (1998) The relationship between B-mating-type genes and nuclear migration in *Schizophyllum commune*. Fungal Genet Biol 24:207–227

Raudaskoski M, Åström H, Penttilä K, Virtanen I, Louhelainen J (1987) Role of microtubule cytoskeleton in pollen tubes: an immunocytochemical and ultrastructural approach. Biol Cell 61:177–188

Raudaskoski M, Rupe_ I, Timonen S (1991) Immunofluorescence microscopy of the cytoskeleton in filamentous fungi after quick-freezing and low-temperature fixation. Exp Mycol 15:167–173

Raudaskoski M, Mao W-Z, Yli-Mattila T (1994) Microtubule cytoskeleton in hyphal growth. Response to nocodazole in a sensitive and a tolerant strain of the homobasidiomycete *Schizophyllum commune*. Eur J Cell Biol 64:131–141

Raudaskoski M, Färdig M, Uuskallio M (1998) The structure of pheromone and receptor gene transcripts in *Bα1* and *Bβ1* mating-type loci of *Schizophyllum commune*. In: Van Griensven LJLD, Visser J (eds) Proceedings of the fourth meeting on the genetics and cellular biology of basidiomycetes. Mushroom Experimental Station, Horst, Netherlands, pp 119–124

Raudaskoski M, Pardo AG, Tarkka MT, Gorfer M, Hanif M, Laitiainen E (2001) Small GTPases, cytoskeleton and signal transduction in filamentous homobasidiomycetes. In: Geitmann A, Cresti M, Heath IB (eds) Cell biology of plant and fungal tip growth. IOS Press, NATO Sci Ser 328:123–136

Reddy ASN, Day IS (2001) Analysis of the myosins encoded in the recently completed *Arabidopsis thaliana* genome sequence. Genome Biol 2:research0024.1–0024.17

Reddy ASN, Safadi F, Narasimhulu SB, Golovkin M, Hu X (1996a) A novel plant calmodulin-binding protein with a kinesin heavy chain motor domain. J Biol Chem 271: 7052–7060

Reddy ASN, Narasimhulu SB, Safadi F, Golovkin M (1996b) A plant kinesin heavy chain-like protein is a calmodulin-binding protein. Plant J 10:9–21

Requena N, Alberti-Segui C, Winzenburg E, Horn C, Schliwa M, Philippsen P, Liese R, Fischer R (2001) Genetic evidence for a microtubule-destabilizing effect of conventional kinesin and analysis of its consequences for the control of nuclear distribution in *Aspergillus nidulans*. Mol Microbiol 42:121–132

Robertson NF (1954) Studies on the mycorrhiza of *Pinus sylvestris*. I. The pattern of development of mycorrhizal roots and its significance for experimental studies. New Phytol 53:253–283

Rosewarne GM, Barker SJ, Smith SE, Smith FA, Schachtman DP (1999) A *Lycopersicon esculentum* phosphate transporter (*LePT1*) involved in phosphorus uptake from a vesicular-arbuscular mycorrhizal fungus. New Phytol 144:507–516

Runeberg P, Virtanen I, Raudaskoski M (1986) Cytoskeletal elements in the hyphae of the homobasidiomycete *Schizophyllum commune* visualized with indirect immunofluorescence and NBD-phallacidin. Eur J Cell Biol 41:25–32

Russo P, Juuti JT, Raudaskoski M (1992) Cloning, sequence and expression of a β-tubulin-encoding gene in the homobasidiomycete *Schizophyllum commune*. Gene 119:175–182

Salo V, Niini SS, Virtanen I, Raudaskoski M (1989) Comparative immunocytochemistry of the cytoskeleton in filamentous fungi with dikaryotic and multinucleate hyphae. J Cell Sci 94:11–24

Schmit A-C, Lambert A-M (1990) Microinjected fluorescent phalloidin in vivo reveals the F-actin dynamics and assembly in higher plant mitotic cells. Plant Cell 2:129–138

Schmidt A, Hall MN (1998) Signaling to the actin cytoskeleton. Annu Rev Cell Dev Biol 14:305–338

Seiler S, Nargang FE, Steinberg G, Schliwa M (1997) Kinesin is essential for cell morphogenesis and polarized secretion in *Neurospora crassa*. EMBO J 16:3025–3034

Seiler S, Plamann M, Schliwa M (1999) Kinesin and dynein mutants provide novel insights into the roles of vesicle traffic during cell morphogenesis in *Neurospora*. Curr Biol 9:779–785

Sheng J, Citovsky V (1996) Agrobacterium-plant cell DNA transport: Have virulence proteins, will travel. Plant Cell 8:1699–1710

Shepherd VA, Orlovich DA, Ashford AE (1993) Cell-to-cell transport via motile tubules in growing hyphae of a fungus. J Cell Sci 105:1173–1178

Slankis V (1949) Wirkung von alpha-indolylessigsäure auf die dichotomische Verzweigung isolierter wurzeln von *Pinus sylvestris*. Sven Bot Tidskr 43:603–607

Smertenko AP, Jiang C-J, Simmons NJ, Weeds AG, Davies DR, Hussey PJ (1998) Ser6 in the maize actin-depolymerizing factor, ZmADF3, is phosphorylated by a calcium-stimulated protein kinase and is essential for the control of functional activity. Plant J 14:187–193

Smertenko A, Saleh N, Igarashi H, Mori H, Hauser-Hahn I, Jiang C-J, Sonobe S, Lloyd CW, Hussey P (2000) A new class of microtubule-associated proteins in plants. Nat Cell Biol 2:750–753

Smith SE, Read DJ (1997) Mycorrhizal symbiosis. Second edition. Academic Press, San Diego, CA

Snustad DP, Haas NA, Kopczak SD, Silflow CD (1992) The small genome of *Arabidopsis* contains at least nine expressed β-tubulin genes. Plant Cell 4:549–556

Sonobe S, Shibaoka H (1989) Cortical fine actin filaments in higher plant cells visualized by rhodamine-phalloidin after pretreatment with m-maleimidobenzoyl N-hydroxysuccinimide ester. Protoplasma 148:80–86

Staiger CJ (2000) Signaling to the actin cytoskeleton in plants. Annu Rev Plant Physiol Plant Mol Biol 51:257–263

Stamatas GN, McIntire LV (2001) Rapid flow-induced responses in endothelial cells. Biotechnol Prog 17:383–402

Steinberg G (1997) A kinesin-like mechanoenzyme from the zygomycete *Syncephalastrum racemosum* shares biochemical similarities with conventional kinesin from *Neurospora crassa*. Eur J Cell Biol 73:124–131

Steinberg G (1998) Organelle transport and molecular motors in fungi. Fungal Genet Biol 24:161–177

Steinberg G (2000) The cellular roles of molecular motors in fungi. Trends Microbiol 8:162–168

Steinberg G, Schliwa M (1995) The *Neurospora* organelle motor: a distant relative of conventional kinesin with unconventional properties. Mol Biol Cell 6:1605–1618

Steinberg G, Schliwa M (1996) Characterization of the biophysical and motility properties of kinesin from the fungus *Neurospora crassa*. J Biol Chem 271:7516–7521

Steinberg G, Schliwa M, Lehmler C, Bölker M, Kahmann R, McIntosh JR (1998) Kinesin from the plant pathogenic fungus *Ustilago maydis* is involved in vacuole formation and cytoplasmic migration. J Cell Sci 111:2235–2246

Straight AF, Marshall WF, Sedat JW, Murray AW (1997) Mitosis in living budding yeast: anaphase a but no metaphase plate. Science 277:574–578

Straube A, Enard W, Berner A, Wedlich-Söldner R, Kahmann R, Steinberg G (2001) A split motor domain in a cytoplasmic dynein. EMBO J 20:5091–5100

Suelmann R, Fischer R (2000) Nuclear migration in fungi – different motors at work. Res Microbiol 151:247–254

Sundaram S, Kim SJ, Suzuki H, Mcquattie CJ, Hiremah ST, Podila GK (2001) Isolation and characterization of a symbiosis-regulated *ras* from the ectomycorrhizal fungus *Laccaria bicolour*. MPMI 14:618–628

Tarkka MT (2001) Developmentally regulated proteins in Scots pine roots and ectomycorrhiza. PhD Thesis, University of Helsinki, Finland. Yliopistopaino, Helsinki, Finland. ISBN 951-45-9645-5

Tarkka MT, Niini SS, Raudaskoski M (1998) Developmentally regulated proteins during differentiation of root system and ectomycorrhiza in Scots pine (*Pinus sylvestris*) with *Suillus bovinus*. Physiol Plant 104:449–455

Tarkka MT, Vasara R, Gorfer M, Raudaskoski M (2000) Molecular characterization of actin genes from homobasidiomycetes: two different actin genes from *Schizophyllum commune* and *Suillus bovinus*. Gene 251:27–35

Tarkka MT, Nyman TA, Kalkkinen N, Raudaskoski M (2001) Scots pine expresses short-root-specific peroxidases during development. Eur J Biochem 268:86–92

Timonen S, Finlay RD, Söderström B, Raudaskoski M (1993) Identification of cytoskeletal components in pine ectomycorrhizas. New Phytol 124:83–92

Timonen S, Söderström B, Raudaskoski M (1996) Dynamics of cytoskeletal proteins in developing pine ectomycorrhiza. Mycorrhiza 6:423–429

Tinsley JH, Minke PF, Bruno KS, Plamann M (1996) p150^{Glued}, the largest subunit of the dynactin complex, is nonessential in *Neurospora* but required for nuclear distribution. Mol Biol Cell 7:731–742

Ueda K, Matsuyama T, Hashimoto T (1999) Visualization of microtubules in living cells of transgenic *Arabidopsis thaliana*. Protoplasma 206:201–206

Uetake Y, Peterson RI (1997) Changes in actin filament arrays in protocorm cells of the orchid species, *Spiranthes sinensis*, induced by the symbiotic fungus *Ceratobasidium cornigerum*. Can J Bot 75:1661–1669

Uetake Y, Peterson RI (1998) Association between microtubules and symbiotic fungal hyphae in protocorm cells of the orchid species, *Spiranthes sinensis*. New Phytol 140:715–722

Uetake Y, Farquhar ML, Peterson RL (1997) Changes in microtubule arrays in symbiotic orchid protocorms during fungal colonization and senescence. New Phytol 135:701–709

Uribe X, Torres MA, Capellades M, Puigdomènech P, Rigau J (1998) Maize α-tubulin genes are expressed according to specific patterns of cell differentiation. Plant Mol Biol 37:1069–1078

Vaillancourt LJ, Raudaskoski M, Specht CA, Raper CA (1997) Multiple genes encoding pheromones and a pheromone receptor define the Bβ1 mating-type specificity in *Schizophyllum commune*. Genetics 146:541–551

Vale RD, Milligan RA (2000) The way things move: looking under the hood of molecular motor proteins. Science 288:88–95

Vantard M, Cowling R, Delichere C (2000) Cell cycle regulation of the microtubular cytoskeleton. Plant Mol Biol 43:691–703

Versele M, Lemaire K, Thevelein JM (2001) Sex and sugar in yeast: two distinct GPCR systems. EMBO Rep 2:574–579

Vidali L, Yokota E, Cheung AY, Shimmen T, Hepler PK (1999) The 135 kDa actin-bundling protein from *Lilium longiflorum* pollen is the plant homologue of villin. Protoplasma 209:283–291

Villemur R, Joyce CM, Haas NA, Goddard RH, Kopczak SD, Hussey PJ, Snustad DP, Silflow CD (1992) Alpha-tubulin gene family of maize (*Zea mays* L.). Evidence for two ancient alpha-tubulin genes in plants. J Mol Biol 227:81–96

Villemur R, Haas NA, Joyce CM, Snustad DP, Silflow CD (1994) Characterization of four new β-tubulin genes and their expression during male flower development in maize (*Zea mays* L.). Plant Mol Biol 24:295–315

Wang W, Takezawa D, Narasimhulu SB, Reddy ASN, Poovaiah BW (1996) A novel kinesin-like protein with a calmodulin-binding domain. Plant Mol Biol 31:87–100

Wasteneys GO (2000) The cytoskeleton and growth polarity. Curr Opin Plant Biol 3:503–511

Wasteneys GO (2002) Microtubule organization in the green kingdom: chaos or self-order? J Cell Sci 115:1345–1354

Wendland J, Vaillancourt LJ, Hegner J, Lengeler KB, Laddison KJ, Specht CA, Raper CA, Kothe E (1995) The mating-type locus *Ba1*of *Schizophyllum commune* contains a pheromone receptor gene and putative pheromone genes. EMBO J 14:5271–5278

Whittington AT, Vugrek O, Wei KJ, Hasenbein NG, Sugimoto K, Rashbrooke MC, Wasteneys GO (2001) MOR1 is essential for organizing cortical microtubules in plants. Nature 411:610–613

Wilcox HE (1968) Morphological studies of the roots of red pine, *Pinus resinosa*. II. Fungal colonization of roots and the development of mycorrhizae. Am J Bot 55:688–700

Wood V, Gwilliam R, Rajandream MA, Lyne M, Lyne R, Stewart A, Sgouros J, Peat N, Hayles J, Baker S, Basham D, Bowman S, Brooks K, Brown D, Brown S, Chillingworth T, Churcher C, Collins M, Connor R, Cronin A, Davis P, Feltwell T, Fraser A, Gentles S, Goble A, Hamlin N, Harris D, Hidalgo J, Hodgson G, Holroyd S, Hornsby T, Howarth S, Huckle EJ, Hunt S, Jagels K, James K, Jones L, Jones M, Leather S, McDonald S, McLean J, Mooney P, Moule S, Mungall K, Murphy L, Niblett D, Odell C, Oliver K, O'Neil S, Pearson D, Quail MA, Rabbinowitsch E, Rutherford K, Rutter S, Saunders D, Seeger K, Sharp S, Skelton J, Simmonds M, Squares R, Squares S, Stevens K, Taylor K, Taylor RG, Tivey A, Walsh S, Warren T, Whitehead S, Woodward J, Volckaert G, Aert R, Robben J, Grymonprez B, Weltjens I, Vanstreels E, Rieger M, Schafer M, Muller-Auer S, Gabel C, Fuchs M, Fritzc C, Holzer E, Moestl D, Hilbert H, Borzym K, Langer I, Beck A, Lehrach H, Reinhardt R, Pohl TM, Eger P, Zimmermann W, Wedler H, Wambutt R, Purnelle B, Goffeau A, Cadieu E, Dreano S, Gloux S, Lelaure V, Mottier S, Galibert F, Aves SJ, Xiang Z, Hunt C, Moore K, Hurst SM, Lucas M, Rochet M, Gaillardin C, Tallada VA, Garzon A, Thode G, Daga RR, Cruzado L, Jimenez J, Sanchez M, del Rey F, Benito J, Dominguez A, Revuelta JL, Moreno S, Armstrong J, Forsburg SL, Cerrutti L, Lowe T, McCombie WR, Paulsen I, Potashkin J, Shpakovski GV, Ussery D, Barrell BG, Nurse P (2002) The genome sequence of *Schizosaccharomyces pombe*. Nature 415:871–880

Wu Q, Sandrock TM, Turgeon BG, Yoder OC, Wirsel SG, Aist JR (1998) A fungal kinesin required for organelle motility, hyphal growth, and morphogenesis. Mol Biol Cell 9:89–101

Xiang X, Beckwith SM, Morris NR (1994) Cytoplasmic dynein is involved in nuclear migration in *Aspergillus nidulans*. Proc Natl Acad Sci USA 91:2100–2104

Yamashita RA, May GS (1998) Constitutive activation of endocytosis by mutation of *myoA*, the myosin I gene of *Aspergillus nidulans*. J Biol Chem 273:14644–14648

Ye XS, McGuire SL, Wolkow T, Tang A, Fincher R, Hamer JE, Osmani SA (1999) Interaction between developmental and cell cycle regulators is required to morphogenesis in *Aspergillus nidulans*. EMBO J 18:6994–7001

Yokota E, Shimmen T (1999) The 135-kDa actin-bundling protein from lily pollen tubes arranges F-actin into bundles with uniform polarity. Planta 209:264–266

Yokota E, Takahara K, Shimmen T (1998) Actin-bundling protein isolated from pollen tubes of lily: biochemical and immunocytochemical characterization. Plant Physiol 116:1421–1429

Yuan M, Shaw PJ, Warn RM, Lloyd CW (1994) Dynamic reorientation of cortical microtubules, from transverse to longitudinal, in living plant cells. Proc Natl Acad Sci USA 91:6050–6053

Zhang DH, Callaham DA, Hepler PK (1990) Regulation of anaphase chromosome motion in *Tradescantia* stamen hair cells by calcium and related signalling agents. J Cell Biol 111:171–182

Zhang D, Wadsworth P, Hepler PK (1993) Dynamics of microfilaments are similar, but distinct from microtubules during cytokinesis in living, dividing plant cells. Cell Motil Cytoskeleton 24:151–155

Zheng Z-L, Yang Z (2000) The Rop GTPase switch turns on polar growth in pollen. Trends Plant Sci 5:298–303

19 Functional Diversity of Arbuscular Mycorrhizal Fungi on Root Surfaces

M. Zakaria Solaiman and Lynette K. Abbott

1 Introduction

Arbuscular mycorrhizal (AM) fungi can promote host plant growth by increasing phosphorus (P) uptake from soil while simultaneously obtaining carbon (C) from the photosynthate of the host plant. However, reductions in plant growth associated with AM fungi have also been recorded (e.g. Graham and Eissenstat 1998; Graham and Abbott 2000) which can be linked to carbon and phosphorus exchange (Koide and Elliott 1989). Both growth promotion and reduction depend upon the particular plant–fungal combination (Johnson et al. 1997) and soil conditions. P and C exchange between the host plant and mycorrhizal fungus also depends on environmental and biological variables (Jakobsen 1998). The combined effects of P uptake and transfer to the plant and C release to the fungus are important considerations for the functioning of arbuscular mycorrhizas. While the mechanisms of nutrient exchange between AM fungi and the host plant remain speculative (Schwab et al. 1991; Saito 2000; Smith et al. 2001), more is known about the plant genes involved in P transfer (Harrison 1999) than the fungal genes (Rausch et al. 2001). Symbiotic exchange of nutrients in arbuscular mycorrhizas, especially transport along hyphae and transfer to the host plant, has been reviewed (Saito 2000; Smith et al. 2001) and it has been pointed out that the mechanisms of symbiotic nutrient exchange may be more diverse than originally expected (Saito 2000).

AM fungi occur in soil and in association with roots as communities of organisms that may simultaneously interact with the roots of one or several co-existing plant species. Species of AM fungi differ in their mode of colonisation and their capacity to form hyphae in soil and within the root (Abbott et al. 1992).

Although hyphal characteristics may be distinctive for some fungi (Dodd et al. 2000), they are not usually present as discrete organisms and are difficult to distinguish from one another within and on the surface of roots. Although the fungi may have markedly different characteristics, they appear to function

Plant Surface Microbiology
A. Varma, L. Abbott, D. Werner, R. Hampp (Eds.)

in a similar manner, but with different levels of efficiency depending on their abundance as well as their intrinsic characteristics. Furthermore, their symbiotic response depends on environmental conditions and the relative abundance of other AM fungi associated with the roots of the same plant.

The purpose of this review is to discuss the functional diversity of AM fungi and its significance in the context of interactions at root surfaces and the potential consequences of this for plant growth and plant community structure.

2 Mycorrhiza Formation and Ecological Specificity

Arbuscular mycorrhizal fungi live symbiotically with the roots of approximately 60% of terrestrial plants (Brundrett and Abbott 2002). About 200 species of AM fungi have been described so far in the Glomales, but unequivocal evidence of their capacity to form mycorrhizas is not available for many species (Walker and Trappe 1993). A putative zygosporic stage has only been reported for the life cycle of *Gigaspora decipiens* (Tommerup and Sivasithamparam 1990). Knowledge of genetic diversity of AM fungi is poorly defined.

Associations between hosts and symbionts are usually non-specific in AM symbiosis (Mosse 1975). Most of the evidence for non-specificity in these associations has been demonstrated by inoculating roots with propagules of species of AM fungi in separate pot cultures (Smith and Read 1997). However, it may be possible for a plant grown in field soil to be preferentially colonised by one of the species of the AM fungi present. This could result from differences in the infectivity and/or quantity of propagules of each species, or from differences in the susceptibility of roots to colonisation by each fungus. This phenomenon has been defined as 'ecological specificity' by McGonigle and Fitter (1990).

Arbuscular mycorrhizal fungi are obligate symbionts and they depend on the formation of mycorrhizas to take up carbon from the root for completing their life cycles. Fungal growth does not continue much in axenic culture in the absence of the host plant. The obligate status of AM fungi, the coenocytic nature of their spores (Becard and Pfeffer 1993) and the lack of demonstration of recombination (Rosendahl and Taylor 1997) limit the opportunities for fundamental research on their interactions with plant roots. Molecular techniques based on DNA analysis provide a number of possibilities to develop specific probes for AM fungi for determining phylogenetic relationships and diversity and for their identification in soil and plant roots (Jacquot et al. 2000).

2.1 Establishment of the Symbiosis

The establishment of a functional symbiosis between AM fungi and host plants involves a sequence of recognition events between the fungus and the plant (Giovanetti et al. 1993a; Giovannetti and Sbrana 1998). Mycorrhizal colonisation has several phases (Tester et al. 1987; Gianinazzi-Pearson and Gianinazzi 1989) including spore germination, hyphal growth in soil, hyphal attachment to roots, appressorium formation, intraradical penetration and intraradical growth involving the formation of arbuscules and coils (Smith and Read 1997; Wegel et al. 1998). The developmental stages of fungal interaction with the plant are associated with plant signals inducing gene expression and recognition between the two partners of the symbiosis (Giovanetti and Sbrana 1998).

2.2 Spore Germination and Hyphal Growth

Spores of AM fungi can germinate under appropriate storage and environmental conditions. A host signal is not necessary for this step in the life cycle. The first response of a fungus to a host root is stimulation of hyphal growth. It is well documented that host roots can promote hyphal growth of AM fungi and induce changes in hyphal growth pattern and morphology by stimulating branching and inducing the formation of hyphal fans (Giovannetti et al. 1993b). Hyphal contact with the surface of the host root occurs at random and the increased branching of a fungus near the root surface would increase the probability of root interception. Root architecture and root density also influence the likelihood of root and hypha interception (Abbott and Robson 1984).

2.3 Role of Plant Root Exudates

Although hyphal growth can be increased in response to root exudates from host plants (Mosse 1962, 1988; Koske and Gemma 1992), there is no direct evidence for the release of inhibitory compounds from non-host roots. Indirect evidence has raised this as a possibility (Ocampo et al. 1980; Holliday 1989). The exudates from non-hosts appeared to lack factors which induced hyphal growth (Giovannetti and Sbrana 1998; Nagahashi 2000).

Hyphal elongation of *G. fasciculatus* was enhanced by exudates from *Trifolium repens* when the plants were grown under phosphate-deficient conditions (Elias and Safir 1987). This effect was reduced when phosphorus was added. Root exudates from both non-mycorrhizal and mycorrhizal peas inhibited hyphal growth of *Gigaspora margarita* (Balaji et al. 1994). In contrast, mycorrhizal *Pisum sativum* and its non-mycorrhizal isogenic mutant did not form root exudates that had different effects on *Glomus mossae* (Giovanetti et al. 1993a). Generally, the amount, type and form of root exudates have the potential to influence fungal growth before the fungus meets the root as well as after the hyphae contact the root. In addition, the formation of appressoria and hyphal penetration of the root surface may involve recognition processes (Koske and Gemma 1992).

3 Functioning of Arbuscular Mycorrhizas in Nutrient Exchange

Nutrient exchange in arbuscular mycorrhizas can occur between the hyphae and host root cells. External hyphae penetrate the root surface and proceed into the root cortex forming arbuscules and coils (Saito 2000). The arbuscule is a complex intracellular hyphal structure formed in Arum-type mycorrhizas (Smith and Smith 1997). It has been assumed that the arbuscule is a likely site for symbiotic nutrient transfer in Arum-type mycorrhizas (Bonfante-Fasolo 1987; Smith and Smith 1997), but carbon exchange may also occur across walls of non-arbuscular hyphae (Smith and Read 1997). Variation in arbuscule development in Arum-type mycorrhizas could reflect different characteristics of roots and fungi (Smith and Dickson 1991).

Arbuscular mycorrhizal fungi with a range in demand for phosphorus have been isolated from south-western Australia. In particular, an isolate of *S. calospora* either increased or decreased plant growth depending on the phosphate status of the soil (Thomson et al. 1986). This fungus has a high demand for carbon from the plant relative to P transfer (Pearson et al. 1994), and forms considerably more external hyphae than some other fungi that are highly effective at enhancing P uptake across a range of P supply (Abbott and Robson 1985). Indeed, under some circumstances, it can be inefficient in P transfer (Smith et al. 2000). In addition, *S. calospora* can interact with *Glomus invermaium* during colonisation of roots, restricting growth of *G. invermaium* in other parts of the same root system during some stages of its colonisation (Pearson et al. 1993). However, colonisation and activity of *S. calospora* can be stopped once sporulation has taken place (Pearson and Schweiger 1994), resulting in resumption of colonisation (and presumably P uptake and transfer) by *G. invermaium*. This example demonstrates the dynamics of activities of two AM fungi on root surfaces, but the extent to which this occurs generally for all isolates of these or other species is not known. Unfortunately, relatively few isolates have been investigated for most species of AM fungi in any environment so the extent to which generalisations can be made, even for species, is unknown (Morton and Bentivenga 1994).

In field soils, as several species of AM fungi would generally be involved in co-colonisation of roots, P uptake and transfer into the plant might be affected if one fungus interfered with colonisation by another. The outcome would depend on the functional diversity of the species present, i.e. their efficiency in P uptake and transfer. Circumstantial evidence of seasonal variation in the function of different species of AM fungi present in a woodland environment has been clearly demonstrated (Merryweather and Fitter 1998b). There have been few direct measurements of P uptake by communities of AM fungi (Jakobsen et al. 2001), but it is difficult to attribute P transferred to the activity of a particular fungus within a community.

The relative abundance of AM fungi in field-grown plants can be manipulated by environmental and management processes that occur in the field, where dual AM fungal occupancy of roots is the norm (Merryweather and Fitter 1998a). With manipulation of P supply, the important observations of Thomson et al. (1986) can be used to identify the P status of the plant at which there is a point of transition from growth enhancement to growth depression in plants colonised by *S. calospora*. In contrast, *G. invermaium* did not display such a physiological transition in this study.

3.1 Metabolic Activity During Mycorrhiza Formation

Changes in metabolic activities during mycorrhiza formation provide evidence for hypotheses related to biochemical mechanisms of C and P exchange between symbionts (Saito 2000). These have mostly been examined by comparing mycorrhizal and non-mycorrhizal roots, but without identifying the mechanisms of nutrient exchange. Changes in quantity and concentration of soluble carbohydrates in roots have not shown consistent trends with mycorrhizal colonisation (Pakovsky 1989; McArthur and Knowels 1993; Pearson et al. 1994; Solaiman and Saito 1997).

The localisation of alkaline phosphatase (ALPase) in arbuscular hyphae was observed by histochemical study (Saito 1995). ALPase has also been located in vacuoles in intraradical hyphae of AM fungi (Gianinazzi et al. 1979), and its activity varied with environmental conditions (Jabaji-Hare et al. 1990). A correlation between the number of ALPase-active arbuscules and P uptake by mycorrhizal plants indicates that arbuscular ALPase plays a significant role in phosphorus transformation from the AM fungus to the host plant (Tijssen et al. 1983). Histochemical observation of intraradical hyphae has located lipid and polyphosphate (polyP) granules in hyphae. Although polyP molecules are likely to play an important role in P translocation, their existence has been contradicted. Recently, a successive extraction method showed that a granular fraction of polyP was contained in hyphae of *Gigaspora margarita* (Solaiman et al. 1999). The contribution of polyP was calculated from these data and it has been concluded that the contribution is not significant (Smith et al. 2001).

3.2 Gene Expression During Mycorrhiza Formation

A phosphate (Pi) transporter *GvPT* was isolated from the AM fungus *Glomus versiforme* that resembles the yeast high-affinity Pi transporter *PHO84* and the plant *Pht1* transporter (Harrison and van Buuren 1995). This transporter gene is thought to be active in P uptake by fungal hyphae outside roots. The molecular mechanisms at the fungus–root interface involved in Pi efflux from the fungus into the apoplastic space and subsequently, into the cortical cells of the root are not well understood. It would be interesting to know whether these transporters comprise new protein families and to identify their site of expression. Observations so far are not conclusive. For example, Liu et al. (1998) concluded that the expression patterns of *MtPT1* and *MtPT2* cloned from *Medicago truncatula* roots are not consistent with their involvement at the symbiotic interface. Similarly, expression of the tomato Pi transporter gene *LePT1* has been observed in cortical cells having arbuscules and it was thought to be involved in the uptake of Pi by the plant from the fungus (Rosewarne et al. 1999). However, *LePT1* transcript levels were less in mycorrhizal compared to non-mycorrhizal plants. A Pi transporter (*MtPT1*) from *M. truncatula* roots was consistent with its role in P transport at the root/soil interface (Chiou et al. 2001).

The expression of a plant H^+-ATPase gene was increased in barley roots colonised by *G. intraradices* (Murphy et al. 1997). This demonstrated the presence of a high H^+-ATPase activity in the periarbuscular membrane of mycorrhizal roots. The isolated hexose transporter was of host origin, and in situ hybridisation showed it was expressed in cortical cells in the area colonised by the AM fungus (Harrison 1996). Therefore, the efflux of sugar from the host cell to the apoplast may be mediated by the transporter. Gene expression has been demonstrated for P transporters (Harrison and van Buuren 1995, Rosewarne et al. 1999), but the genes involved in carbon flows have not been identified. A model for the pathways and regulation of P uptake in mycorrhizal plants was proposed by Rosewarne et al. (1999) based upon expression of P transporters in uncolonised roots (Liu et al. 1998), mycorrhizal roots (Rosewarne et al. 1999) and mycorrhizal fungi (Harrison and van Buuren 1995). According to the proposed model, cloning of the arbuscule-specific Pi transporter genes is required to investigate the mechanisms of P transfer in arbuscules in comparison with other parts of the symbiosis (such as intraradical and external hyphae), but phosphate can alter expression of the Pi transporter gene (Maldonado-Mendoza et al. 2001).

3.3 Nutrient Exchange Mechanisms in Arbuscular Mycorrhizas

Isolation of the fungus from host tissue in mycorrhizas has assisted in identifying the location of biochemical activities in nutrient exchange (Saito 2000).

Intraradical hyphae were first isolated from host roots by enzymic digestion with cellulase and pectinase and hand-sorted under a dissecting microscope (Capaccio and Callow 1982). This is a time-consuming process which might reduce the metabolic activity of the hyphae (McGee and Smith 1990). Subsequently, a more suitable method was developed for isolating metabolically active intraradical hyphae from onion roots colonised by an AM fungus with only 1 h of enzymic digestion (Saito 1995). The metabolic activity of the isolated hyphae was not affected by this technique (Saito 1995; Solaiman and Saito 1997). The isolated arbuscules remained functional in membrane transport for at least 4 h. This corresponded with the in vivo NMR study by Shachar-Hill et al. (1995). Using this isolation technique, polyphosphate metabolism in isolated intraradical hyphae (Solaiman et al. 1999) and P efflux from the isolated intraradical hyphae have been observed (Solaiman and Saito 2001). P efflux from intraradical hyphae was coupled with polyP hydrolysis.

A hyperarbuscule-forming plant mutant was screened from a large collection of nodulation mutants in model legume *Lotus japonicus* (Senoo et al. 2000; Solaiman et al. 2000). The characteristic features of this mutant are: (1) it produces a higher number of arbuscules per unit length of roots compared to the wild-type plant, (2) the individual arbuscules formed in this mutant are well developed, and (3) metabolic activity of these arbuscules is higher than of those formed in the wild-type plant. Furthermore, a method of intact arbuscule isolation from the hyperarbuscule-forming mutant was developed without using enzymic digestion of colonised roots. The isolated arbuscules were metabolically active when tested with succinate dehydrogenase (SDH; Solaiman et al. 2000), alkaline phosphatase (ALP) and acid phosphatase (ACP) histochemical staining. This mutant is suitable for molecular genetic study and for further investigation of the exchange of P and C between the symbionts.

Phosphorus in soil solution is absorbed by external hyphae through the Pi transporter, and the absorbed phosphate is condensed into polyP and translocated into the intraradical hyphae by protoplasmic streaming (Cooper and Tinker 1981; Harrison and van Buuren 1995). The factors potentially regulating the uptake, transport and transfer of phosphate from the fungus have been summarised by Saito (2000) as: (1) expression and regulation of the Pi transporter, (2) protoplasmic streaming of motile vacuoles, (3) synthesis and decomposition of polyP, and (4) release of Pi across the fungal membrane in the arbuscule. There has been no other information on the fungal Pi transporter in arbuscular mycorrhizas since the reports of Harrison and van Buuren (1995) and Maldonado-Mendoza et al. (2001). Synthesis and degradation of polyP in extraradical and intraradical hyphae (Solaiman et al. 1999; Ezawa et al. 2001) and an efflux of Pi from the intraradical hyphae have been demonstrated (Solaiman and Saito 2001). In spite of these advances, current knowledge is still based upon a limited number of host – fungus combinations. Demonstrations of morphological and genetic diversity of AM fungi imply that the mechanism of symbiotic nutrient exchange might be diverse (Smith and Smith 1996), but the taxonomic diversity found might not necessarily be linked to the functional role of the AM fungi. Both inter- and intraspecific variation in the effectiveness of AM fungi has been reported (Franke-Snyder et al. 2001). Studies are required that show whether fungal diversity reflects quantitative rather than qualitative differences in functioning. The current understanding of functional diversity is that plants can respond differently to different AM fungi, not only at the level of colonisation, nutrient uptake, growth, but also at the level of gene expression (Burleigh et al. 2002).

4 Functional Diversity of Arbuscular Mycorrhizal Fungi in Root and Hyphal Interactions

Functional diversity of AM fungi associated with roots of plants in different ecosystems is not well understood. The dynamics of interactions between roots and hyphae provide a framework for predicting how diversity of AM fungi might be related to mycorrhiza function, but it is difficult to measure. It is almost impossible to predict the functional diversity of AM fungi at an ecosystem level based simply on what fungi are present in soil. On a theoretical basis, the functional diversity of AM fungi under field conditions cannot be assumed to be directly correlated with a measure of diversity of AM fungi, even if the fungi present differ in P uptake and transfer under controlled conditions. This is because of differences in processes such as rate of root colonisation, interactions between fungi during colonisation and other strategies that include shutting down of hyphal infectivity in association with sporulation, as can occur for both *S. calospora* and *A. laevis* (Jasper et al. 1993; Pearson and Schweiger 1993).

There is increasing interest in the potential role of AM fungi in influencing plant community structure (Read 1990; van der Heijden et al. 1998a, b; Klironomos 2002; Franke-Snyder et al. 2001). However, as yet there is little evidence to support the hypothesis that the diversity of AM fungi is an important factor influencing plant community structure under natural field conditions. This would require extensive quantification of AM fungi within roots for time-intervals that are of significance to plant and fungal growth cycles following an appropriate approach (Merryweather and Fitter 1998a, b). On the contrary, there is considerable potential for the plant communities to influence the fungal community structure through preferential effects on colonisation by particular fungi and influences on sporulation (Sanders and Fitter 1992).

4.1 Diversity of Arbuscular Mycorrhizal Fungi Inside Roots

The benefit to plants through enhanced P uptake is expected to be markedly altered according to the effectiveness of dominant AM fungi inside roots. However, species diversity of AM fungi has generally been examined from the perspective of presence or absence of fungi in soil, not in roots (van der Heijden et al. 1998a; Bever et al. 2001). There is little evidence of a simple relationship between the relative abundance of morphotypes of AM fungi inside roots and spores in soil (Scheltema et al. 1987; Merryweather and Fitter 1998a). The effectiveness of AM fungi in taking up P in these studies is generally not considered, but neither is it easy to determine, particularly as effectiveness can change for the same fungus depending on the P status of the soil and plant (Thomson et al. 1986). The diversity of AM fungi needs to be investigated further in relation to their relative abundance inside the roots (Abbott and Gazey 1994). In addition, the methods applied to selecting combinations of species of AM fungi in studies of the role of species diversity need to be considered carefully (Wardle 1999). Another consideration is that AM fungi form associations with plants that differ markedly in their growth habits and susceptibility to colonisation, and this makes assessment of the impact of diversity of AM fungi even more difficult to predict or measure at an ecosystem level.

The diversity of AM fungi could be relevant to communities of AM fungi in the same way that high plant species diversity may help stabilise plant community structure and ecosystem processes (Klironomos et al. 2000; Tilman 1996). However, high diversity can also lead to competitive exclusion and cause a reduction in the number of co-existing species (Huston 1994). For AM fungi, this may only reduce the abundance of some species below levels of detection (Bever et al. 2001). The competitive ability of species of AM fungi in roots is likely to be an important factor in determining the dominance of AM fungi inside roots as well as in soil. This is compounded by seasonal changes in infectivity of AM fungi (Merryweather and Fitter 1998b), which is influenced by fungal life cycles (Abbott and Gazey 1994) such as sporulation and associated changes in infectivity of the hyphae (Pearson and Schweiger 1993).

As AM fungi occur as communities in soil and in roots, the extent to which they are likely to collectively contribute to P uptake (Jakobsen et al. 2001; Solaiman and Abbott 2003) depends on the mycorrhiza dependency of the host plant (van der Heijden et al. 1998b). Although there have been intensive studies of single AM fungus function, there have been few investigations of the contributions of communities of AM fungi. Reconstructed communities of AM fungi in soil can promote plant growth (Daft 1983; Daft and Hogart 1983), or have no effect on plant growth (Sylvia et al. 1993). A correlation between the occurrence of AM fungal morphotypes and seasonal P uptake for AM fungi in a natural ecosystem was observed (Merryweather and Fitter 1998b), which may be related to differences among fungi in the functional characteristics of hyphae they form in soil (Smith et al. 2000).

4.2 Relationship Between Hyphae in the Root and in the Soil

The quantity of hyphae in the vicinity of roots associated with mycorrhizal roots can vary greatly (Sylvia 1986) and change with time (Bethlenfalvay et al. 1982). Hyphae of AM fungi play key roles in the formation and functioning of mycorrhizas (Abbott et al. 1992). Hyphae in soil, originating from either an established hyphal network or from other propagules (spores, vesicles and root fragments), lead to the recognition and subsequent colonisation of roots. The distribution of hyphae and associated sporulation will determine where propagules are located in relation to newly formed roots. The roles of the hyphae in both P uptake and soil stabilisation are dependent on their distribution within the soil matrix and their interaction with the root surface.

Abbott and Robson (1984) hypothesised for *G. invermaium* that low initial levels of infective hyphae in the soil would lead to small amounts of hyphae in soil relative to the amount formed inside the root. For high initial densities of infective hyphae of this fungus in soil, the exponential phase of colonisation of roots was expected to occur in parallel with extensive development of hyphae in the soil. In contrast, there was no similar relationship between the formation of hyphae in soil and within the root expected for *S. calospora* which consistently produced large amounts of hyphae in soil, irrespective of the density of hyphae within the root.

There is relatively little information about the longevity of hyphae in soil (Sylvia 1988; Hamel et al. 1990). This would be important for predicting the activities of hyphae for both colonisation and P uptake. The majority of studies of mycorrhizas measure the extent of colonisation of roots at one point in time. This measure is of little value for understanding functional diversity of AM fungi because the AM fungi within the root may either be highly active or have ceased activity for some time. Most routine techniques for assessing mycorrhizas do not assess any functional attribute of the fungus. Therefore, care is required in extrapolating from levels of mycorrhizal root colonised to functional diversity of communities of AM fungi present in soil.

5 Role of Arbuscular Mycorrhizal Fungi Associated with Roots in Soil Aggregation

The AM fungal hyphae can enhance soil aggregation by using more than one mechanism. In clayey soils, entanglement of soil particles by hyphae can occur (Tisdall and Oades 1982; Oades 1984). There is insufficient hyphal length to extend around particles of sand and a more likely mechanism is cross-linking of particles by hyphae in sandy soils (Degens 1997). AM fungal hyphae may bind soil aggregates by exuding a glycoprotein (Wright et al. 1996; Wright and Upadhyaya 1998). Some information on species differences in soil aggregation is available which indicates that AM fungi commonly produce glomalin, but the amount varies considerably for different species (Wright and Upadhaya 1998). If AM fungal communities differ in their sensitivity to disturbance, the capacity of the species present to form hyphae as well as their ability to produce glomalin should influence the degree of soil aggregation.

6 Environmental Influence on Functional Diversity of Arbuscular Mycorrhizal Fungi

The soil environment includes many physical and chemical properties that are continuously being modified by dynamic biological processes. Rhizosphere soil is influenced by plant root exudates and microbial activity and the AM fungi that live in this habitat have adapted to a wide range of environmental conditions (Stahl and Christensen 1991; Giovannetti and Gianinazzi-Pearson 1994). Environmental stresses on AM fungi could include: (1) high or low levels of nutrients, (2) waterlog and drought, (3) soil acidity, (4) salinity, (5) high levels of toxic metals, (6) biotic factors (e.g. fauna that feed on hyphae), and (7) absence of suitable host plants for long periods. AM fungi can adapt to both low and high levels of soil nutrients (Solaiman and Hirata 1997). As they are aerobic, waterlogging has a considerable impact on their diversity and on their functioning. One of the most important soil factors influencing the distribution of species of AM fungi is soil pH (Robson and Abbott 1989). AM fungi show considerable diversity in their response to soil pH and changes in soil pH can affect the relative abundance of species inside roots (Sano et al. 2002). This has potential to influence the structure of communities of AM fungi in soil. There is also evidence that agricultural practices such as pesticide applications, cropping sequences and soil disturbance can affect diversity of AM fungi in soil (Dodd and Jeffries 1989; Sieverding 1991; Johnson et al. 1997).

7 Role of Plant Mutants in Studying the Interactions Between Arbuscular Mycorrhizal Fungi and Roots

In symbiotic associations between AM fungi and plant roots, genetic control imposed by each symbiont is poorly understood. Understanding of the genetic and molecular basis of this symbiosis has been prevented by the obligate nature of the fungal symbiont and by the lack of mycorrhiza formation on the model plant *Arabidopsis thaliana*. Recently, *Medicago truncatula* and *Lotus japonicus* have been chosen as model plants for research of plant – microbe symbioses (Sagan et al. 1995; Jiang and Gresshoff 1997; Bonfante et al. 2000; Senoo et al. 2000). The use of molecular genetic approaches in model legumes will rapidly increase knowledge of host genetic determinants of arbuscular mycorrhizas. An essential step in this process has been the generation, screening and analysis of mycorrhizal mutants (Marsh and Schultze 2001).

Plant mutants are valuable tools in unravelling complex events that occur during cell and tissue differentiation in plants that show impaired formation of arbuscular mycorrhizas (Peterson and Guinel 2000). Since the first description of myc^{-} mutants (Duc et al. 1989), there has been increasing interest in using them to address questions related to various key steps in the colonisation process (Senoo et al. 2000; Wyss et al. 1990). As these fungi are obligate symbionts, it has been difficult to study the interaction between the symbionts during colonisation. The interaction in early colonisation phases of *Allium porrum* L. (leek) roots by the AM fungus *Glomus versiforme* have been described (Garriock et al. 1989) and reviewed (Giovanetti et al. 1994). Mutants of pea (*Pisum sativum* L.) and faba bean (*Vicia faba* L.) were not colonised by AM fungi (Duc et al. 1989). These mycorrhizal (myc^{-}) mutants were also unable to form functional root nodules (nod^{-}). The myc^{-} mutants have aborted infections (Gianinazzi-Pearson et al. 1991). In contrast, nod^{-} mutants of soybean were colonised by *Glomus mossae* to the same extent as wild-type (nod^{+}) soybean plants (Wess et al. 1990). The myc^{-} mutants should be screened against different AM fungi in a range of soils to see whether resistance is horizontal or if some fungi can overcome resistance as in the case of certain nod^{-} plants in the presence of different *Rhizobium* populations (Lie and Timmermans 1983). Recently, it has been shown that some AM fungi can colonise the mutant of tomato, *rmc*, demonstrating that the fungi can overcome resistance to successful colonisation (Gao et al. 2001). This new tool would help exploration of genetic variability in AM fungi. It would also open the possibility of controlling plant – fungus specificity in the presence of communities of AM fungi in field soils.

Mycorrhizal mutants were screened from the model plant *Lotus japonicus* (Senoo et al. 2000) after inoculation with *Glomus* sp. R-10. These mutants were characterized and categorized into *mcbep* (mycorrhizal colonisation blocked at epidermis) and *mcbex* (mycorrhizal colonisation blocked at exodermis) based on the detailed assessment of colonisation and microscopic observation (Senoo et al. 2000). Isolation and cloning of the gene will facilitate understanding of its function, and it could be used to probe a range of hosts to determine its distribution and expression. It is essential to expand the collection of mutants in order to build up a comprehensive description of the molecular genetic basis of successful mycorrhization.

8 Conclusion and Future Research Needs

Research is needed that integrates knowledge of ecological and genetic characteristics of arbuscular mycorrhizas for predicting P-related functions of AM fungi in soil communities. For example, the diversity among Arum-type AM fungi in arbuscule formation in roots in relation to the intensity of mycorrhizal colonisation and function is not known for fungi that differ in colonisation aggressiveness and response to P supply. Neither are the relationships understood between arbuscule formation and P/C exchange for AM fungi that differ in arbuscular, intraradical and external hyphal growth characteristics. It would be interesting to compare Pi transporter gene expression in arbuscules, intraradical and external hyphae in mycorrhizas formed by AM fungi that differ in (1) arbuscular and other colonisation characteristics, and (2) functional response to phosphate supply and the presence of co-colonising AM fungi that have different C demands. Finally, clarification is required of the actual roles of AM fungi in natural environments. The importance of AM fungal diversity is currently of considerable interest, but the limited evidence available is not yet sufficient to support claims that diversity of these potentially symbiotic organisms is of wide-scale significance in regulating the structure of plant communities.

References and Selected Reading

Abbott LK, Robson AD (1984) The effect of root density, inoculum placement and infectivity of inoculum on the development of vesicular-arbuscular mycorrhizas. New Phytol 97:285–299

Abbott LK, Robson AD (1985) Formation of external hyphae in soil by four species of vesicular-arbuscular mycorrhizal fungi. New Phytol 99:245–255

Abbott LK, Gazey C (1994) An ecological view of the formation of VA mycorrhizas. Plant Soil 159:69–78

Abbott LK, Robson AD, Jasper DA, Gazey C (1992) What is the role of VA mycorrhizal hyphae in soil? In: Read DJ, Lewis DH, Fitter AH, Alexander IJ (eds) Mycorrhizas in ecosystems. CAB International, Wallingford, pp 37–41

Balaji B, Ba AM, LaRue TA, Tepfer D, Piche Y (1994) *Pisum sativum* mutants insensitive to nodulation are also insensitive to invasion in vitro by the mycorrhizal fungus *Gigaspora margarita*. Plant Sci 102:195–203

Becard G, Pfeffer PE (1993) State of nuclear division in arbuscular mycorrhizal fungi during in vitro development. Protoplasma 194:62–68

Bethlenfalvay GJ, Brown MS, Pakovsky RS (1982) Relationships between host and endophyte development in mycorrhizal soybeans. New Phytol 90:537–543

Bever JD, Schultz PA, Pringle A, Morton JB (2001) Arbuscular mycorrhizal fungi: more diverse than meets the eye, and the ecological tale of why. BioSci 51:923–931

Bonfante-Fasolo P (1987) Vesicular arbuscular mycorrhizae: fungus-plant interactions at the cellular level. Symbiosis 3:249–268

Bonfante P, Genre A, Faccio A, Martin I, Schauser L, Stougaard J, Web J, Parniske M (2000) The *Lotus japonicus Ljsym4* gene is required for the successful symbiotic infection of root epidermis cells. Mol Plant-Microbe Interact 13:1109–1120

Brundrett MC, Abbott LK (2002) Arbuscular mycorrhizas in plant communities. In: Sivasithamparam K, Dixon KW, Barrett RL (eds) Microorganisms in Plant Conservation and Biodiversity. Kluwer, Dordrecht, pp 151–193

Burleigh SH, CavagnaroT, Jakobsen I (2002) Functional diversity of arbuscular mycorrhizas extends to the expression of plant genes involved in P nutrition. J Exp Bot 53:1593–1601

Capaccio LCM, Callow JA (1982) The enzymes of polyphosphate metabolism in vesicular-arbuscular mycorrhizas. New Phytol 91:81–91

Chiou TJ, Liu H, Harrison MJ (2001) The spatial expression patterns of a phosphate transporter (MtPT1) from *Medicago truncatula* indicate a role in phosphate transport at the root/soil interface. Plant J 25:281–293

Cooper KM, Tinker PB (1981) Translocation and transfer of nutrients in vesicular-arbuscular mycorrhizas. IV. Effect of environmental variables on movement of phosphorus. New Phytol 88:327–339

Daft MJ (1983) The influence of mixed inocula on endomycorrhizal development. Plant Soil 71:331–337

Daft MJ, Hogart BG (1983) Competitive interactions amongst four species of *Glomus* on maize and onion. Trans Br Mycol Soc 80:339–345

Degens BP (1997) Macroaggregation of soils by biological bonding and binding mechanisms and factors affecting this: a review. Aust J Soil Res 35:431–446

Dodd JC, Jeffries P (1989) Effects of herbicides on three vesicular-arbuscular mycorrhizal fungi associated with winter wheat (*Triticum aestivum* L.). Biol Fertil Soils 7:113–119

Dodd JC, Boddington CL, Rodriguez A, Gonzalez-Chavez C, Mansur I (2000) Mycelium of arbuscular mycorrhizal fungi (AMF) from different genera: form, function and detection. Plant Soil 226:131–151

Duc G, Trouvelot A, Gianinazzi-Pearson V, Gianinazzi S (1989) First report of non-mycorrhizal plant mutants (myc$^-$) obtained in pea (*Pisum sativum* L.) and faba bean (*Vicia faba* L.). Plant Sci 60:215–222

Elias KS, Safir GR (1987) Hyphal elongation of *Glomus fasciculatus* in response to root exudates. Appl Environ Microbiol 53:1928–1933

Ezawa T, Smith SE, Smith FA (2001) Differentiation of polyphosphate metabolism between the extra- and intraradical hyphae of arbuscular mycorrhizal fungi. New Phytol 149:555–563

Franke-Snyder M, Douds Jr DD, Galvez L, Phillips JG, Wagoner P, Drinkwater L, Morton JB (2001) Diversity of communities of arbuscular mycorrhizal (AM) fungi present in conventional versus low-input agricultural sites in eastern Pennsylvania, USA. Appl Soil Ecol 16:35–48

Gao L-L, Delp G, Smith SE (2001) Colonization patterns in a mycorrhiza-defective mutant tomato vary with different arbuscular-mycorrhizal fungi. New Phytol 151: 477–491

Garriock ML, Peterson RL, Ackerley CA (1989) Early stages in colonization of *Allium porrum* (leek) roots by the vesicular-arbuscular mycorrhizal fungus, *Glomus versiforme*. New Phytol 112:85–92

Gianinazzi S, Gianinazzi-Pearson V, Dexheimer J (1979) Enzyme studies on the metabolism of vesicular-arbuscular mycorrhiza. 3. Ultrastructural localization of acid and alkaline phosphatase in onion roots infected by *Glomus mosseae* (Nicol. and Gerd.). New Phytol 82:127–132

Gianinazzi-Pearson V, Gianinzzi S (1989) Cellular and genetical aspects of interactions between host and fungal symbionts in mycorrhizae. Genome 31:336–341

Gianinazzi-Pearson V, Gianinazzi S, Guillemin JP, Trouvelot A, Duc G (1991) Genetic and cellular analysis of resistance of vesicular-arbuscular mycorrhizal fungi in pea mutants. In: Hennecke, H, Varma DPS (eds) Advances in molecular genetics of plant – microbe interactions. Kluwer, Dordrecht, pp 336–342

Giovannetti M, Gianinazzi-Pearson V (1994) Biodiversity in arbuscular mycorrhizal fungi. Mycol Res 98:705–715

Giovannetti M, Sbrana S, Logi C (1994) Early processes involved in host recognition by arbuscular mycorrhizal fungi. New Phytol 127:703–709

Giovannetti M, Sbrana C (1998) Meeting a non-host: the behaviour of AM fungi. Mycorrhiza 8:123–130

Giovannetti M, Sbrana C, Avio L, Citernesi AS, Logi C (1993a) Differential hyphal morphogenesis in arbuscular mycorrhizal fungi during preinfection stages. New Phytol 125:587–593

Giovanetti M, Avio L, Sbrana L, Citernesi AS (1993b) Factors affecting appressorium development in the vesicular-arbuscular mycorrhizal fungus *Glomus mosseae* (Nicol. and Gerd.) Gerd. and Trape. New Phytol 123:114–122

Graham JH, Eissenstat DM (1998) Field evidence for the carbon cost of citrus mycorrhizas. New Phytol 140:103–110

Graham JH, Abbott LK (2000) Wheat responses to aggressive and non-aggressive arbuscular mycorrhizal fungi. Plant Soil 220:207–218

Hamel C, Fyles H, Smith DL (1990) measurement of development of endomycorrhizal mycelium using three different vital stains. New Phytol 115:297–302

Harrison MJ (1996) A sugar transporter from *Medicago truncatula*: altered expression pattern in roots during vesicular-arbuscular (VA) mycorrhizal associations. Plant J 9:491–503

Harrison MJ (1999) Molecular and cellular aspects of the arbuscular mycorrhizal symbiosis. Ann Rev Plant Physiol Plant Mol Biol 50:361–389

Harrison MJ, van Buuren ML (1995) A phosphate transporter from the mycorrhizal fungus *Glomus versiforme*. Nature 378:626–629

Holliday P (1989) A dictionary of plant pathology. Cambridge University Press, Cambridge, pp 369

Huston MA (1994) Biological diversity: the coexistence of species on changing landscapes. Cambridge University Press, Cambridge, pp 1–681

Jabaji-Hare SH, Therien J, Charest PM (1990) High resolution cytochemical study of the vesicular-arbuscular mycorrhizal association, *Glomus clarum-Allium porrum*. New Phytol 114:481–496

Jakobsen I (1998) Transport of phosphorus and carbon in arbuscular mycorrhizas. In: Varma A, Hock B (eds) Mycorrhiza: structure, function, molecular biology and biotechnology. Springer, Berlin Heidelberg New York, pp 305–332

Jakobsen I, Gazey C, Abbott LK (2001) Phosphate transport by communities of arbuscular mycorrhizal fungi in intact soil cores. New Phytol 149:95–103

Jacquot E, van Tuinen D, Gianinazzi S, Gianinazzi-Pearson V (2000) Monitoring species of arbuscular mycorrhizal fungi in planta and in soil by nested PCR: application to the study of the impact of sewage sludge. Plant Soil 226:179–188

Jasper DA, Abbott LK, Robson AD (1993) The survival of infective hyphae of vesicular-arbuscular mycorrhizal fungi in dry soil: an interaction with sporulation. New Phytol 124:473–479

Jiang Q, Gresshoff PM (1997) Classical and molecular genetics of the model legume *Lotus japonicus*. Mol Plant-Microbe Interact 10:59–68

Johnson NC, Graham JH, Smith FA (1997) Functioning of mycorrhizal associations along the mutualism-parasitism continuum. New Phytol 135:575–585

Klironomos JN (2002) Feedback with soil biota contributes to plant rarity and invasiveness in communities. Nature 417:67–70

Klironomos JN, McCune J, Hart M, Neville J (2000) The influence of arbuscular mycorrhizae on the relationship between plant diversity and productivity. Ecol Lett 3:137–141

Koide R, Elliott G (1989) Cost, benefit and efficiency of the vesicular-arbuscular mycorrhizal symbiosis. Funct Ecol 3:252–255

Koide RT, Schreiner RP (1992) Regulation of the vesicular-arbuscular mycorrhizal symbiosis. Ann Rev Plant Physiol Plant Mol Biol 43:557–581

Koske RE, Gemma JN (1992) Fungal reactions to plants prior to mycorrhiza formation. In: Allen MF (ed) Mycorrhizal functioning. Routledge, Chapman and Hall, New York, pp. 3–36

Lie TA, Timmermans PCJM (1983) Host-genetic control of nitrogen fixation in the legume-*Rhizobium* symbiosis: complication in the genetic analysis due to material effects. Plant Soil 75:449–453

Liu H, Trieu AT, Blaylock LA, Harrison MJ (1998) Cloning and characterization of two phosphate transporters from *Medicago truncatula* roots: regulation in response to phosphate and to colonisation by arbuscular mycorrhizal (AM) fungi. Mol Plant-Microbe Interact 11:14–22

Maldonado-Mendoza IE, Dewbre GR, Harrison MJ (2001) A phosphate transporter gene from the extra-radical mycelium of an arbuscular mycorrhizal fungus *Glomus intraradices* is regulated in respect to phosphate in the environment. Mol Plant-Microbe Interact 4:1140–1148

Marsh JF, Schultze M (2001) Analysis of arbuscular mycorrhizas using symbiosis-defective plant mutants. New Phytol 150:525–532

McArthur DAJ, Knowels NR (1993) Influence of vesicular-arbuscular mycorrhizal fungi on the response of potato to phosphorus deficiency. Plant Physiol 101:147–160

McGee PA, Smith SE (1990) Activity of succinate dehydrogenase in vesicular-arbuscular mycorrhizal fungi after enzymic digestion from roots of *Allium porrum*. Mycol Res 94:305–308

McGonigle T, Fitter AH (1990) Ecological specificity of vesicular-arbuscular mycorrhizal associations. Mycol Res 94:120–122

Merryweather JW, Fitter AH (1998a) The arbuscular mycorrhizal fungi of *Hyacinthoides non-scripta*-I. Diversity of fungal taxa. New Phytol 138:117–129

Merryweather JW, Fitter AH (1998b) The arbuscular mycorrhizal fungi of *Hyacinthoides non-scripta*-II. Seasonal and spatial patterns of fungal populations. New Phytol 138:131–142

Morton JB, Bentivenga SP (1994) Levels of diversity in endomycorrhizal fungi (Glomales, Zygomycetes) and their role in defining taxonomic and non-taxonomic groups. Plant Soil 159:47–59

Mosse B (1962) The establishment of vesicular-arbuscular mycorrhiza under aseptic conditions. J Gener Microbiol 27:509–520

Mosse B (1975) Specificity in VA mycorrhizas. In: Sanders FE, Mosse B, Tinker PB (eds) Endomycorrhizas. Academic Press, London, pp 469–484

Mosse B (1988) Some studies relating to "independent growth of vesicular-arbuscular endophytes". Can J Bot 66:2533–2540

Murphy PJ, Langridge P, Smith SE (1997) Cloning plant genes differentially expressed during colonisation of roots of *Hordeum vulgare* by the vesicular-arbuscular mycorrhizal fungus *Glomus intraradices*. New Phytol 135:291–301

Nagahashi G (2000) In vitro and in situ techniques to examine the role of roots and root exudates during AM fungus-host interactions. In: Kapulnik Y, Douds Jr DD (eds) Arbuscular mycorrhizas: physiology and function. Kluwer, Dordrecht, pp 287–300

Oades JM (1984) Soil organic matter and structural stability: Mechanisms and implications for management. Plant Soil 76:319–337

Ocampo JA, Martin J, Hayman DS (1980) Influence of plant interactions on vesicular-arbuscular mycorrhizal infections. I. Host and non-host plants grown together. New Phytol 84:27–35

Pakovsky RS (1989) Carbohydrate, protein amino acid status of *Glycine-Glomus-Bradyrhizobium* symbiosis. Physiol Plant 72:733–746

Pearson JN, Schweiger P (1993) *Scutellospora calospora* (Nicol. and Gerd.) Walker and Sanders associated with subterranean clover: dynamics of colonisation, sporulation and soluble carbohydrates. New Phytol 124:215–219

Pearson JN, Schweiger P (1994) *Scuttelospora calospora* (Nicol. and Gerd.) Walker and Sanders associated with subterranean clover produces non-infective hyphae during sporulation. New Phytol 127:697–701

Pearson JN, Abbott LK, Jasper DA (1993) Mediation of competition between two colonizing VA mycorrhizal fungi by the host plant. New Phytol 123:93–98

Pearson JN, Abbott LK, Jasper DA (1994) Phosphate, soluble carbohydrates and the competition between two arbuscular mycorrhizal fungi colonizing subterranean clover. New Phytol 127:101–106

Peterson RL, Guinel FC (2000) The use of plant mutants to study regulation of colonization by AM fungi. In: Kapulnik Y, Douds Jr DD (eds) Arbuscular mycorrhizas: physiology and function. Kluwer, Dordrecht, pp 147–171

Rausch C, Daram P, Brunner S, Jansa J, Laloi M, Leggewie G, Amrhein N, Bucher M (2001) A phosphate transporter expressed in arbuscule-containing cells in potato. Nature 414:462–466

Read DJ (1990) Mycorrhizas in ecosystems – Nature's response to the 'Law of the minimum'. In: Hawksworth DL (ed) Frontiers in mycology. CAB International, Wallingford, pp 101–130

Robson AD, Abbott LK (1989) The effect of soil acidity on microbial activity in soil. In: Robson AD (ed) Soil acidity and plant growth. Academic Press, Sydney, pp 139–165

Rosendahl S, Taylor JW (1997) Development of multiple genetic markers for studies of genetic variation in arbuscular mycorrhizal fungi using AFLP. Mol Ecol 6:821–829

Rosewarne GM, Barker SJ, Smith SE, Smith FA, Schachtman DP (1999) A *Lycopersicon esculentum* phosphate transporter (LePT1) involved in phosphorus uptake from a vesicular-arbuscular mycorrhizal fungus. New Phytol 144:507–516

Sagan M, Morandi D, Tarenghi E, Duc G (1995) Selection of nodulation and mycorrhizal mutants in the model plant *Medicago truncatula* (Gaertn.) after γ-ray mutagenesis. Plant Sci 111:63–71

Saito M (1995) Enzyme activities of the internal hyphae and germinated spores of an arbuscular mycorrhizal fungus, *Gigaspora margarita* Becker and Hall. New Phytol 129:425–431

Saito M (2000) Symbiotic exchange of nutrients in arbuscular mycorrhizas: Transport and transfer of phosphorus. In: Kapulnik Y, Douds Jr DD (eds) Arbuscular mycorrhizas: physiology and function. Kluwer, Dordrecht, pp 85–105

Sanders IR, Fitters AH (1992) Evidence for differential responses between host-fungus combinations of vesicular-arbuscular mycorrhizas from a grassland. Mycol Res 96:415–419

Sano SM, Abbott LK, Solaiman MZ, Robson AD (2002) Influence of liming, inoculum level and inoculum placement on root colonization of subterranean clover. Mycorrhiza 12:285–290

Scheltema MA, Abbott LK, Robson AD (1987) Seasonal variation in infectivity of VA mycorrhizal fungi in annual pastures in a Mediterranean environment. Aust J Agric Res 38:707–715

Schwab SM, Menge JA, Tinker PB (1991) Regulation of nutrient transfer between host and fungus in vesicular mycorrhizas. New Phytol 117:387–398

Senoo K, Solaiman MZ, Kawaguchi M, Imaizumi-Anraku H, Akao S, Tanaka A, Obata H (2000) Isolation of two different phenotypes of mycorrhizal mutants in the model legume plant *Lotus japonicus* after EMS-treatment. Plant Cell Physiol 41:726–732

Shachar-Hill Y, Pfeffer PE, Douds D, Osman SF, Doner LW, Ratcliffe RG (1995) Partitioning of intermediary carbon metabolism in vesicular-arbuscular mycorrhizal leek. Plant Physiol 108:7–15

Sieverding E (1991) Vesicular-arbuscular mycorrhiza management in tropical agroecosystems. Deutsche Gesellschaft für Technische Zusammenarbeit. Eschborn, Germany, pp 371

Smilauer P (2001) Communities of arbuscular mycorrhizal fungi in grassland: seasonal variability and effects of environment and host plants. Folia Geobot 36:243–263

Smith FA, Smith SE (1997) Structural diversity in (vesicular)-arbuscular mycorrhizal symbioses. New Phytol 137:373–388

Smith FA, Jakobsen I, Smith SE (2000) Spatial differences in acquisition of soil phosphate between two arbuscular mycorrhizal fungi in symbiosis with *Medicago truncatula*. New Phytol 147:357–366

Smith SE, Dickson S (1991) Quantification of active vesicular-arbuscular mycorrhizal infection using image analysis and other techniques. Aust J Plant Physiol 18:737–648

Smith SE, Smith FA (1996) Mutualism and parasitism: diversity in function and structure in the "arbuscular" (VA) mycorrhizal symbiosis. Adv Bot Res 22:1–43

Smith SE, Read DJ (1997) Mycorrhizal symbiosis. Academic Press, London, pp 1–605

Smith SE, Dickson S, Smith FA (2001) Nutrient transfer in arbuscular mycorrhizas: how are fungal and plant processes integrated? Aust J Plant Physiol 28:683–694

Solaiman MZ, Hirata H (1997) Effect of arbuscular mycorrhizal fungi inoculation of rice seedlings at the nursery stage upon performance in the paddy field and greenhouse. Plant Soil 191:1–12

Solaiman MZ, Saito M (1997) Use of sugars by intraradical hyphae of arbuscular mycorrhizal fungi revealed by radiorespirometry. New Phytol 136:533–538

Solaiman MZ, Saito M (2001) Phosphate efflux from the intraradical hyphae of an arbuscular mycorrhizal fungus, *Gigaspora margarita*, in vitro and its implication to phosphorus translocation in the hyphae. New Phytol 151:525–533

Solaiman MZ, Abbott LK (2003) Phosphorus uptake by a community of arbuscular mycorrhizal fungi in jarrah forest. Plant Soil 248:313–320

Solaiman MZ, Ezawa T, Kojima T, Saito M (1999) Polyphosphates in intraradical and extraradical hyphae of arbuscular mycorrhizal fungi. Appl Environ Microbiol 65:5604–5606

Solaiman MZ, Senoo K, Kawaguchi M, Imaizumi-Anraku H, Akao S, Tanaka A, Obata H (2000) Characterization of mycorrhizas formed by *Glomus* sp. on roots of hypernodulating mutants of *Lotus japonicus*. J Plant Res 113:443–448

Sylvia DM (1986) Spatial and temporal distribution of vesicular-arbuscular mycorrhizal fungi associated with *Uniola paniculata* in Florida foredunes. Mycologia 78:728–734

Sylvia DM (1988) Activity of external hyphae vesicular arbuscular mycorrhizal fungi. Soil Biol Biochem 20:39–43

Stahl PD, Christensen M (1991) Population variation in the mycorrhizal fungus *Glomus mosseae*: breath of environmental tolerance. Mycol Res 95:300–3007

Sylvia DM, Wilson DO, Graham JH, Maddox JJ, Millner P, Morton JB, Skipper HD, Wright SF, Jarstfer AG (1993) Evaluation of vesicular arbuscular mycorrhizal fungi in diverse plants and soils. Soil Biol Biochem 25:705–713

Tester M, Smith SE, Smith FA (1987) The phenomenon of non-mycorrhizal plants. Can J Bot 65:419–431

Thomson BD, Robson AD, Abbott LK (1986) Effects of phosphorus on the formation of mycorrhizas by *Gigaspora margarita* and *Glomus fasciculatum* in relation to root carbohydrates. New Phytol 103:751–765

Tijssen JPF, Dubbelman TMAR, Van Steveninak J (1983) Isolation and characterization of polyphosphate from the yeast cell surface. Biochem Biophysics Acta 760:143–148

Tilman D (1996) Biodiversity: population versus ecosystem stability. Ecology 77:350–363

Tisdall JM, Oades JM (1982) Organic matter and water-stable aggregates in soils. J Soil Sci 33:141–163

Tommerup IC, Sivasithamparam K (1990) Zygospores and asexual spores of *Gigaspora decipiens*, an arbuscular mycorrhizal fungus. Mycol Res 94:897–900

Van der Heijden MGA, Klironomos JN, Ursic M, Moutoglis P, Streitwolf-Engel, Boller T, Wiemken A, Sanders IR (1998a) Mycorrhizal fungal diversity determines plant biodiversity, ecosystem variability and productivity. Nature 396:69–72

Van der Heijden MGA, Boller T, Wiemken A, Sanders IR (1998b) Different arbuscular mycorrhizal fungal species are potential determinants of plant community structure. Ecology 79:2082–2091

Walker C, Trappe JM (1993) Names and epithets in the glomales and endogonales. Mycol Res 97:339–344

Wardle DA (1999) Is "sampling effect" a problem for experiments investigating biodiversity – ecosystem function relationships? Oikos 87:403–407

Wegel E, Schauser L, Sandal N, Stougaard J, Parniske M (1998) Mycorrhiza mutants of *Lotus japonicus* define genetically independent steps during symbiotic infection. Mol Plant-Microbe Interact 11:933–936

Wright SF, Upadhyaya A (1998) A survey of soils for aggregate stability and glomalin, a glycoprotein produced by hyphae of arbuscular mycorrhizal fungi. Plant Soil 198:97–107

Wright SF, Franke-Snyder M, Morton JB, Upadhyaya A (1996) Time-course study and partial characterization of a protein on hyphae of arbuscular mycorrhizal fungi during active colonization of roots. Plant Soil 181:193–203

Wyss P, Mellor RB, Wiemken A (1990) Vesicular-arbuscular mycorrhizas of wild-type soybean and non-nodulating mutants with *Glomus mossae* contain symbiosis-specific polypeptides (mycorrhizins), immunologically cross-reactive with nodulins. Planta 182:22–26

20 Mycorrhizal Fungi and Plant Growth Promoting Rhizobacteria

José-Miguel Barea, Rosario Azcón
and Concepción Azcón-Aguilar

1 Introduction

Soil microbial communities are crucial in maintaining a biological balance in soil, a key issue for the sustainability of either natural ecosystems or agroecosystems (Kennedy and Smith 1995). When provided with available carbon substrates, soil microorganisms are able to develop a range of activities in the microhabitats where they flourish and some of these activities are of great relevance for plant growth and health and for soil quality (Bowen and Rovira 1999). Soil-borne microbes are found bound to the surface of soil particles or in the soil aggregates, while others interact specifically with plant roots (Glick 1995). Particularly important from the point of view of plant surface microbiology are the interactions at the root – soil interface where microorganisms, plant roots and soil constituents interact (Lynch 1990; Azcón-Aguilar and Barea 1992; Linderman 1992; Kennedy 1998; Bowen and Rovira 1999; Barea 2000; Gryndler 2000) to develop a dynamic environment what is known as the rhizosphere (Hiltner 1904). The rhizosphere, therefore, is the zone of influence of plant roots on the soil microbiota; a microcosm with physical, chemical and biological properties different from those of the root-free bulk soil (Bowen and Rovira 1999; Gryndler 2000; Barea 2000). A characteristic of the rhizosphere is that microbial diversity is altered and that the activity and number of microorganisms is increased (Kennedy 1998).

The supply of photosynthates and decaying plant material to the root-associated microbiota is a key issue for rhizosphere formation and functioning. The release of organic material is known to occur mainly as root exudates, acting as either signals or growth substrates (Werner 1998). However, once the microbial population is established, rhizosphere developments are affected by microbially induced changes on rooting patterns and by the supply of available nutrients to plants, which in turn modify the quality and quantity of root exudates (Bowen and Rovira 1999; Barea 2000; Gryndler 2000). Microbial interactions in the rhizosphere are known to markedly influence plant fitness and soil quality (Lynch 1990; Bethlenfalvay and Schüepp

Plant Surface Microbiology
A. Varma, L. Abbott, D. Werner, R. Hampp (Eds.)

1994). In particular, microorganisms associated with plant roots help the host plant adapt to stress conditions concerning water and mineral nutrition, and soil-borne plant pathogens (Jeffries and Barea 2001).

From the point of view of plant surface microbiology, the compartmentalization of the rhizosphere (broad sense) is important. Kennedy (1998) suggested that there are three separate, but interacting components, namely the rhizosphere (soil), the rhizoplane, and the root itself. The rhizosphere is the zone of soil influenced by roots through the release of substrates that affect microbial activity. The rhizoplane is actually the root surface, but also includes the strongly adhering soil particles. The root itself is a part of the system because certain microorganisms, the endophytes, are able to colonize root tissues. Microbial colonization of the rhizoplane and/or the root tissues is known as root colonization, while the colonization of the adjacent volume of soil under the influence of the root is known as rhizosphere colonization (Kloepper et al. 1991).

2 Main Types of Rhizosphere Microorganisms

In spite of several very different types of microorganisms living in the root – soil interface microhabitats, most studies on rhizosphere microbiology refer to only bacteria and fungi (Bowen and Rovira 1999; Gryndler 2000). Two main groups of microorganisms can be distinguished: saprophytes and symbionts. Both of them comprise detrimental, neutral and beneficial bacteria and fungi. Detrimental microbes include the major plant pathogens, as well as minor parasitic and nonparasitic, deleterious rhizosphere organisms, either bacteria or fungi, (Weller and Thomashow 1994; Nehl et al. 1996). Beneficial microorganisms are known to play fundamental roles in agroecosystem and natural ecosystem sustainability, and some of them can be used as inoculants to benefit plant growth and health (Alabouvette et al. 1997; Barea et al. 1997; Cordier et al. 1999; Barea 2000; Dobbelaere et al. 2001; Probanza et al. 2002).

Saprophytic bacteria and fungi colonize subterranean plant surfaces. Root colonization by rhizosphere bacteria has been extensively studied. It appears to be a strain-specific, active process that is exhibited by a subset of the total rhizosphere bacterial community, termed rhizobacteria, which is known to display a specific ability for root colonization (Kloepper 1994, 1996). The beneficial root colonizing rhizosphere bacteria, the so-called plant growth promoting rhizobacteria (PGPR), carry out important activities in the root/soil interfaces (Probanza et al. 2002).

The endophytic microorganisms colonizing the root tissues develop activities involved in plant growth promotion and plant protection (Kloepper 1994; Chanway 1996; Sturz et al. 2000; Sturz and Novak 2000). Even nonsymbiotic microorganisms may be endophytes and colonize the root tissues (Duijff et al. 1997; Van Loon et al. 1998). *Piriformospora indica* (Basidiomycota) has been described as a plant-growth-promoting root endophyte (Varma et al. 1999). However, because this chapter deals only with rhizobacteria, these, and other fungal endophytes, will not be dealt with further.

Plant symbiotic bacteria and fungi are recognized and can either include pathogens or mutualistic organisms. Mycorrhizal fungi and nitrogen (N_2)-fixing bacteria are the main mutualistic symbionts (Barea 1997). This chapter will focus only on mycorrhizal fungi and PGPRs.

3 Mycorrhizal Fungi

The roots of most plant species associate with certain soil fungi and establish what are known as mycorrhiza (Smith and Read 1997). Mycorrhizal functions include improvement of plant establishment, enhancement of nutrient uptake, protection against cultural and environmental stresses, and the improvement of soil structure (Barea et al. 1997).

Mycorrhizal symbiosis can be found in nearly all types of ecological situations, and most plant species are able to form this symbiosis naturally, the most common type involved in the normal cropping systems is the arbuscular mycorrhizal (AM) type (Smith and Read 1997). The responsible AM fungi belong to the order Glomales in the Zygomycetes (Morton and Redecker 2001), and are a very common group of soil-borne fungi whose origin and divergence date back 100–400 million years ago (Simon et al. 1993; Morton 2000; Redecker et al. 2000). However, a new fungal phylum, the Glomeromycota have recently been proposed (Schüβler et al. 2001). Because this chapter focuses only on arbuscular mycorrhizas, the term "AM fungi" will be used to refer to "arbuscular mycorrhizal fungi".

During the process of AM formation (Giovannetti 2000), in which the plant "accepts" the fungal colonization without any significant rejection reaction (Dumas-Gaudot et al. 2000), a series of root–fungus interactions allows the integration of both organisms. The establishment of the symbiosis is the result of a continuous molecular "dialogue" between plant and fungus, as exerted through the exchange of both recognition and acceptance signals (Vierheilig and Piché 2002). The result of this dialogue will finally depend on the genome expression of both partners (Gianinazzi-Pearson et al. 1996; Franken and Requena 2001).

After the biotrophic colonization of the root cortex, AM fungi develop an external mycelium which is a bridge connecting the root with the surrounding soil microhabitats. Such mycorrhizal (fungal-root) symbiosis is critical in nutrient cycling in soil–plant systems (Smith and Read 1997). In cooperation with other soil organisms, the external AM fungal mycelium forms water-stable aggregates necessary for good soil tilth (Miller and Jastrow 2000; Requena et al. 2001). The AM symbiosis also improves plant health through increased protection against biotic and abiotic stresses (Bethlenfalvay and Linderman

1992; Azcón-Aguilar and Barea 1996; Linderman 2000; Miller and Jastrow 2000; Augé 2001; Requena et al. 2001; Werner et al. 2002).

Recent developments in molecular biology are being applied to the genetic characterization of AM fungi based on PCR-based approaches (Sanders et al. 1996; Helgason et al. 1998; Ferrol et al. 2000). During the last few years, the analysis of ribosomal genes (rRNA) has demonstrated the polymorphism of these genes in AM fungi, particularly those corresponding to the small ribosomal subunit 18S, therefore permitting phylogeny and diversity studies (Clapp et al. 1995; Redecker et al. 1997; van Tuinen et al. 1998; Redecker et al. 2000; Daniell et al. 2001; Schüßler et al. 2001). Novel techniques currently developed for microbial molecular ecology studies, such as PCR-single-strand conformation polymorphism (SSCP) and PCR-temperature gradient gel electrophoresis (TGGE), are being adapted for the characterization of different ecotypes of AM fungi, both in soil and in roots (Kjoller and Rosendahl 2000).

Since AM fungi are obligate symbionts, they must multiply on living roots. This is a limitation for inocula production (Azcón-Aguilar and Barea 1997). However, several substrates and procedures have been described for inoculum production and application in horticulture/fruit culture/forestry (Gianinazzi et al. 1990; Vestberg and Estaun 1994; Lobato et al. 1995; Varma and Schüepp 1995; Calvet et al. 1996; Sylvia 1998; Azcón-Aguilar et al. 2000; Mohammad et al. 2000). The Federation of European Mycorrhizal Inoculum Producers has been established.

4 Plant Growth Promoting Rhizobacteria

The beneficial root colonizing rhizosphere bacteria, the so-called plant growth promoting rhizobacteria (PGPR), are defined by three intrinsic characteristics: (1) they must have the ability to undergo root colonization, (2) they must survive and multiply in microhabitats associated with the root surface, in competition with native microbiota, at least for the time needed to express their plant promotion activities, and (3) they must have the ability to promote plant growth (Kloepper 1994). The PGPR are known to carry out many important ecosystem processes, such as those involved in the biological control of plant pathogens, nutrient cycling and/or seedling establishment (Haas et al. 1991; Kloepper et al. 1991; Lugtenberg et al. 1991; Lemanceau and Alabouvette 1993; O'Gara et al. 1994; Weller and Thomashow 1994; Broek and Vanderleyden 1995; Glick 1995; Bashan and Holguin 1998; Barea 2000; Probanza et al. 2002). Many bacterial taxa include PGPR strains with *Pseudomonas* and *Bacillus* as the most commonly described genera possessing PGPR ability, and some strains from these and other genera are used as seed inoculants (Kloepper 1994; Bertrand et al. 2001; Probanza 2002).

Azospirillum sp. are considered PGPR (Bashan 1999; Bashan and Gonzalez 1999) and are used as seed inoculants under field conditions (Dobbelaere et al. 2001). The main activity of these bacteria is associated with the production of auxin-type phytohormones (Dobbelaere et al. 1999). The production and significance of auxins have been investigated at the molecular level (van de Broek et al. 1999; Lambrecht et al. 2000).

Several systems for studying rhizosphere colonization by PGPR have been proposed, the one described by Simons et al. (1996) appears interesting, not only for testing the PGPR ability, but also to investigate PGPR/plant root interactions.

It is known that several bacterial cell surface properties could be involved in the adhesion of PGPR to roots that may be either nonspecific based on electrostatic forces, or involve specific recognition between the surfaces. In this context, root surface glyco-proteins and several different bacterial exo-polysaccharides could be involved (Weller and Thomashow 1994). In some PGPR, fimbriae (pili) may function in the adherence of cells to roots, but the contribution of flagella to the colonization process apparently depends on the PGPR strain, plant species and type of soil (Kloepper 1994). A sugar-binding protein system has been described in *Azospirillum* related to chemostaxis and root colonization (van Bastelaere et al. 1999).

Novel techniques for microbial community fingerprinting are being developed (Kozdroj and van Elsas, 2000), and these are being adapted for the genetic characterization of PGPR. For example, the approach used by Zinniel et al. (2002), based on the 16S rRNA gene amplification and sequencing, has been proposed for the genetic characterization of endophytic bacteria, while Bertrand et al. (2001) identify PGPR 16S rDNA sequence analysis. The molecular bases of rhizosphere colonization have recently been reviewed (Lugtenberg et al. 2001). Siciliano and Germida (1998) proposed the use of BIOLOG analysis and fatty acid methyl ester profiles to study PGPR behavior after inoculation, particularly the effects on root-associated microbial populations.

Confocal laser scanning microscopy is being used for studying the microorganisms/plant root interactions, for example, to detect a currently used marker such as the green fluorescent protein (Lagopodi et al. 2002) or to localize colonizing bacteria by fluorescence and in situ hybridization, after staining with the fluorescent Live/Dead *Bac*Light dye (Bianciotto et al. 2000, 2001).

The molecular bases of the biocontrol ability of these rhizobacteria have been investigated in the last few years (Keel et al. 1992; O'Gara et al. 1994; Cook et al. 1995; Tomashow and Weller 1995; Chin-A-Woeng et al. 2001; Moenne-Loccos et al. 2001), and systemic induced resistance has been argued as a mechanism of disease suppression by endophytes (Duijff et al. 1998), or other PGPR (Defago and Keel 1995; Chin-A-Woeng et al. 2001).

5 Reasons for Studying Arbuscular Mycorrhizal Fungi and Plant Growth Promoting Rhizobacteria Interactions and Main Scenarios

Since they share common habitats, i.e., the root surface, and common functions, the AM fungi and PGPR have to interact during their processes of root colonization or functioning as root-associated microorganisms. In fact, just like any soil other inhabitant, the AM fungi are immersed in the framework of microbial interactions characteristic of soil microbiota relationships (Barea 1997). Soil microorganisms, particularly PGPR, can influence AM formation and function and, conversely, mycorrhizas can affect the microbial populations, particularly PGPR in the rhizosphere (Azcón-Aguilar and Barea 1992; Linderman 1992, 1994; Fitter and Garbaye 1994; Barea 1997, 2000). The analysis of microbe – microbe interactions is crucial to an understanding of the events which occur at the root – soil interface and, particularly, to those related to the microbial colonization of the root surface, or the processes of root infection/colonization by pathogens or mutualistic symbionts (Lynch 1990).

These interactions must be taken into consideration when trying to manage AM fungi and PGPR for the biological control of plant pathogens or for the biogeochemical cycling of plant nutrients (Barea et al. 1997, 2002).

Information is accumulating with regard to cell-to-cell interactions between AM fungi and PGPR. Bianciotto et al. (1996a, b, 2000, 2001) and Bonfante and Perotto (2000) investigated whether PGPR attach to the structures of the AM fungi by means of a direct cell-to-cell interaction. Attachment of rhizobia and pseudomonads to the spores and fungal mycelium of *Gigaspora margarita* was visualized by a combination of electron and confocal microscopy. The results showed that both rhizobia and pseudomonads adhere to spores and hyphae of AM fungi germinated in vitro, although the degree of attachment depended upon the strain. Bianciotto et al. (1996b) showed that extracellular material of bacterial origin containing cellulose, which was produced around the attached bacteria, may mediate fungal/bacterial interactions. They also support the fact that AM fungi could act as a vehicle for the colonization of plant roots by PGPR, as previously suggested (Boddey et al. 1991).

Bianciotto et al. (1996b) demonstrated that there were no specific receptors for the bacteria on the fungal structures and that physicochemical factors govern attachment to fungal surfaces. Electrostatic interactions may, therefore, play a key role in the early stages of adhesion and cellulose fibrils may be later involved to guarantee a stable attachment. The complex interactions involving the tripartite system composed by AM fungi/bacteria/plant have recently being reviewed (Bonfante and Perotto 2000).

6 Effect of Plant Growth Promoting Rhizobacteria on Mycorrhiza Formation

Microbial populations in the rhizosphere are known to either interfere with or benefit AM establishment (Germida and Walley 1996; Vosátka and Gryndler 1999). Deleterious rhizosphere bacteria (Nehl et al. 1996) and mycoparasitic relationships (Jeffries 1997), have been found to interfere with AM formation, while many microorganisms can improve AM formation and/or functioning (Barea 1997). One example of the beneficial effects is that of the so-called mycorrhiza-helper-bacteria which are known to stimulate mycelial growth of mycorrhizal fungi and/or enhance mycorrhizal formation (Garbaye 1994; Azcón-Aguilar and Barea 1995; Barea 1997; Frey-Klett et al. 1997; Gryndler and Hrselova 1998; Gryndler 2000; Gryndler et al. 2000). Soil microorganisms can produce compounds that increase root cell permeability and are able to increase the rates of root exudation. This, in turn, would stimulate mycorrhizal fungal mycelia in the rhizosphere or facilitate root penetration by the fungus. Plant hormones, as produced by soil microorganisms, are known to affect mycorrhiza establishment (Azcón-Aguilar and Barea 1992, 1995; Barea 1997, 2000).

Rhizosphere microorganisms are also known to affect the pre-symbiotic stages (Giovannetti 2000) of mycorrhizal developments, like spore germination rate and mycelial growth (Azcón-Aguilar and Barea, 1992, 1995).

It is noteworthy that antibiotic-producing *Pseudomonas* sp. (Barea et al. 1998; Vazquez et al. 2000) did not interfere with mycorrhiza formation or functioning.

7 Effect of Mycorrhizas on the Establishment of Plant Growth Promoting Rhizobacteria in the Rhizosphere

The establishment of the AM fungus in the root cortex is known to change many key aspects of plant physiology. These include the mineral nutrient composition in plant tissues, the hormonal balance and the patterns of C allocation (Harley and Smith 1983; Azcón-Aguilar and Bago 1994; Smith et al. 1994). Therefore, the AM symbiotic status changes the chemical composition of root exudates while the development of an AM soil mycelium introduces physical modifications into the environment surrounding the roots. The AM soil mycelium represents a carbon source to microbial communities which is an important contribution through interactions with components of the microbiota to improve plant growth and health, and soil quality (Bethlenfalvay and Schüepp 1994).

Arbuscular mycorrhizal-induced changes in plant physiology affect, both quantitatively and qualitatively, the microbial populations in either the rhizosphere and/or the rhizoplane (Azcón-Aguilar and Barea 1992; Linderman

1992; Barea 1997; Cordier et al. 1999; Barea 2000; Gryndler 2000). This situation creates the "rhizosphere of a mycorrhizal plant". However, there are specific modifications in the environment surrounding the AM mycelium itself, which develop what it known as the mycorrhizosphere (Linderman 1992; Barea 2000; Grynler 2000). In addition to this term, the soil space affected by extraradical hyphae is also called the mycosphere (Linderman 1988) or hyphosphere as an analogy to the term rhizosphere (Gryndler 2000). Large numbers of bacteria (including actinomycetes) and fungi can be associated with AM fungal structures (Filippi et al. 1998; Budi et al. 1999). Since the AM mycelium releases energy-rich organic compounds, an increased growth and activity of microbial saprophytes are expected to occur in the mycorrhizosphere. However, the enrichment of this particular environment with organic compounds is much lower than that of the rhizosphere, which corresponds to lower counts of bacteria in mycorrhizospheric soil, compared to those in the rhizosphere (Andrade et al. 1997). Apparently, there is a preferential establishment of Gram-negative bacteria in the hyphosphere (Vosátka 1996).

It has been demonstrated that mycorrhizal colonization changes some morphological parameters in developing root systems (Atkinson et al. 1994; Berta et al. 1995), with a greater root branching as the most commonly described effect. Undoubtedly these changes must affect the establishment and activity of microorganisms in the mycorrhizosphere environment.

The establishment of PGPR inoculants in the rhizosphere can be affected by AM fungal co-inoculation (Christensen and Jakobsen 1993; Puppi et al. 1994; Barea 1997; Andrade et al. 1998; Ravnskov et al. 1999). In particular, AM inoculation improved the establishment of both inoculated and indigenous phosphate-solubilizing rhizobacteria (Toro et al. 1997; Barea et al. 2002).

Interestingly, mycorrhizal fungi improved rhizosphere colonization by *Pseudomonas* sp. and root colonization by *Azospirillum* sp. (Klyuchnikov and Kozhevin 1990). Moreover, several experiments, reviewed by Nehl et al. (1996), suggest that mycorrhizal colonization may affect whether a given rhizobacterium functions as a PGPR, or as a DRB.

In spite of the fact that most of the reports support the beneficial effects of mycorrhizas on the establishment of PGPR inoculants, detrimental effects have also been found (Waschkies et al. 1994; Marschner and Crowley 1996a, b; Barea et al. 1997).

As indicated previously, an extreme case of close interactions is that of *Burkholderia*-like bacteria as endosymbionts in AM fungi of the Gigasporaceae (Bianciotto et al. 2000; Ruiz-Lozano and Bonfante 2000).

8 Interactions Involved in Nutrient Cycling and Plant Growth Promotion

It is well known that soil microorganisms are able to change the bioavailability of mineral plant nutrients, and this ability has been shown by soil bacteria probed to have PGPR activity and, in some cases, be used as plant inoculants (Barea et al. 1997; Probanza et al. 2002). Several experiments have described the improvement of plant growth and nutrition by means of synergistic interactions between PGPR and AM fungi (Barea 2000). It has also been suggested that a certain level of selectivity ("specificity") is involved in these interactions (Azcón 1989). From the perspective of sustainability, the re-establishment of nutrient cycles after any process of soil degradation is of interest, as is the understanding of the microbial interactions responsible for the subsequent management of such natural resources, either for a low-input agricultural technology (Bethlenfalvay and Linderman 1992; Gianinazzi and Schüepp 1994; Jeffries and Barea 2001), or for the re-establishment of the natural vegetation in a degraded area (Miller and Jastrow 1994, 2000; Requena et al. 2001). Most of the information on this topic concerns N and P cycling (Barea 2000).

In spite of this, *Rhizobium* sp. (general term) are not considered among the PGPR types; due to the relevance of their interaction with AM fungi, it would be interesting to make some comments on this. A great deal of work has been carried out on the tripartite symbiosis legume (general term) – mycorrhiza-*Rhizobium* (Azcón-Aguilar and Barea 1992; Barea et al. 1992; Barea 2000). The inoculation of AM fungi has been shown to improve nodulation and N_2 fixation. Using the isotope ^{15}N has made it possible to ascertain and quantify the amount of N which is actually fixed in a particular situation, and measure the contribution of the AM symbiosis to the process (Barea et al. 1992). The physiological and biochemical mechanisms underlying the AM fungi x *Rhizobium* interactions to improve legume productivity have also been discussed. In spite of the main AM effect in enhancing *Rhizobium* activity mediated by a generalized stimulation of host nutrition, more localized effects may occur at the root or nodule level (Barea et al. 1992). Interactions can also take place at either the pre-colonization stages, when both microorganisms interact as rhizosphere inhabitants, or during the development of the tripartite symbiosis (Azcón-Aguilar and Barea 1992). The influence of host and/or bacterial genotypes in these interactions has also been discussed, suggesting a certain level of specificity (Azcón et al. 1991; Ruiz-Lozano and Azcón 1993; Monzón and Azcón 1996).

Multimicrobial interactions including AM fungi, *Rhizobium* sp. and PGPR have also been tested (Requena et al. 1997). Target microorganisms were isolated from a representative area of a desertification-threatened semi-arid ecosystem in the south-east of Spain. Microbial isolates were characterized and screened for effectiveness in soil microcosms. *Anthyllis cytisoides* L., an

AM-dependent pioneer legume, dominant in the target Mediterranean ecosystem, was the test plant. Several microbial cultures from existing collections were also included in the screening process. In general, the results support the importance of physiological and genetic adaptation of microbes to the environment, thus the use of efficient local isolates is recommended. Several microbial combinations were effective in improving plant development, nutrient uptake, N_2-fixation (^{15}N) and root system quality.

The interactions between AM fungi and Rhizobium have been demonstrated to be beneficial under drought conditions (Goicoechea at al. 1997, 1998; Ruiz-Lozano et al. 2001).

There is also evidence that *Rhizobium* strains are able to colonize the rhizosphere of nonlegume hosts where they establish positive interactions with AM fungi and behave as PGPR (Schloter et al. 1997; Galleguillos et al. 2000).

In spite of the fact that *Azospirillum* are known to benefit plant development, N acquisition and yield under appropriate conditions (Okon 1994; Bashan 1999; Dobbelaere et al. 2001), it has been demonstrated that these bacteria mainly act by influencing the morphology, geometry and physiology of the root system. Interactions between AM fungi and *Azospirillum* have been reviewed by Volpin and Kapulnik (1994) and it has been demonstrated that *Azospirillum* could enhance mycorrhizal formation and response while AM fungi may improve *Azospirillum* establishment in the rhizosphere.

Since some PGPR may improve nodulation by *Rhizobium* sp. (Halverson and Handelsman 1991; Staley et al. 1992; Azcón 1993), certain PGPR-*Rhizobium* interactions could be relevant to mycorrhizosphere interactions.

The interactions related to P-cycling have also received much attention. These are based on the fact that phosphate ions solubilized by free-living microorganisms from sparingly soluble inorganic and organic P compounds (Whitelaw 2000) increase the soil phosphate pools available for the extraradical AM mycelium to benefit plant nutrition (Smith and Read 1997). Several experiments have demonstrated synergistic microbial interactions involving phosphate-solubilizing rhizobacteria (PSB) and mycorrhizal fungi (Barea et al. 1997; Kim et al. 1998). The interactive effect of PSB and mycorrhizal fungi on plant use of soil P sources of low bioavailability was evaluated by using ^{32}P isotopic dilution approaches (Toro et al. 1997, 1998). The PSB behaved as mycorrhiza-helper-bacteria, promoting mycorrhiza establishment by both the indigenous and the inoculated mycorrhiza. Conversely, mycorrhiza formation increased the size of the PSB population. Because the bacteria did not change root weight, length or specific root length, they probably acted by improving the pre-colonization stages of mycorrhiza formation. The dual inoculation treatment significantly increased biomass and N and P accumulation in plant tissues and these dually inoculated plants displayed lower specific activity ($^{32}P/^{31}P$) than their comparable controls, suggesting that the mycorrhizal and bacterized plants were using P sources (endogenous or added as rock phosphate) otherwise unavailable to the plant. It, therefore appears that these rhizosphere/mycorrhizosphere interactions contributed to the biogeochemical P cycling, thereby promoting plant nutrition. The interactive effect of PSB, AM fungi and *Rhizobium* with regard to improving the agronomic efficiency of rock phosphate for legume crops (*Medicago sativa*), was evaluated by using isotopic techniques under controlled conditions, and further validated under field conditions (Barea et al. 2002). It was demonstrated that the microbial interactions tested improved plant growth and N and P acquisition under normal cultivation. Similar results were obtained by using *Medicago arborea*, a woody legume of interest for revegetation and biological reactivation of desertified semi-arid Mediterranean ecosystems (Valdenegro et al. 2001).

9 Interactions for the Biological Control of Root Pathogens

Once the AM status has been established in plant roots, reduced damage caused by soil-borne plant pathogens has been shown. To account for this, several mechanisms have been suggested to explain the enhancement of plant resistance/tolerance in mycorrhizal plants (Linderman 1994, 2000; Azcón-Aguilar and Barea 1992, 1996). One of the proposed mechanisms is based on the microbial changes produced in the mycorrhizosphere. In this context, there is strong evidence that these microbial shifts occur, and that the resulting microbial equilibria could influence the growth and health of the plants. Although this effect has not been assessed specifically as a mechanism for AM-associated biological control, there are indications that such a mechanism could be involved (Azcón-Aguilar and Barea 1992, 1996; Linderman 1994, 2000). In any case, it has been demonstrated that such an effect is dependent on the AM fungus involved, as well as the substrate and host plant (Azcón-Aguilar and Barea 1996; Linderman 2000).

Since specific PGPR antagonistic to root pathogens are being used as biological control agents (Alabouvette et al. 1997), it has been proposed to try to exploit the prophylactic ability of AM fungi in association with these antagonists (Linderman 1994, 2000; Azcón-Aguilar and Barea 1996; Barea et al. 1998; Budi et al. 1999). Experimental evidence is accumulating, but the information is still too scarce to make general conclusions.

Several studies have demonstrated that microbial antagonists of fungal pathogens, either fungi or PGPR, do not exert any anti-microbial effect against AM fungi (Calvet et al. 1993; Barea et al. 1998; Edwards et al. 1998; Vazquez et al. 2000; Werner et al. 2002). This is a key point to exploit the possibilities of dual (AM fungi and PGPR) inoculation in plant defense against root pathogens.

In particular, Barea et al. (1998) carried out a series of experiments to test the effect of *Pseudomonas* spp. producing 2,4-diacetylphloroglucinol (DAPG) on AM formation and functioning. Three *Pseudomonas* strains were tested for their effects on AM fungi: a wild type (F113) producing the antifungal compound DAPG; the genetically modified strain (F113G22), a DAPG-negative mutant of F113; and another genetically modified strain [F113 (pCU203)], a DAPG-overproducer. The results from in vitro and in soil experiments demonstrate no negative effects of these *Pseudomonas* strains on spore germination, and a stimulation of hyphal growth of the AM fungus *Glomus mosseae*. Concentrations of the antifungal compound DAPG which were far in excess of those reached in the rhizosphere of *Pseudomonas*-inoculated plants exhibited negative effects on germination of AM fungal spores, but more realistic concentrations of DAPG did not affect AM fungal development. A soil microcosm system was also used to evaluate the effect of these bacteria on the process of AM formation. No significant difference in AM formation on tomato plants between F113, F113G22 and F113 (pCU203) was observed, with the F113 and F113G22 strains resulting in a significant increase in the percentage of the root system becoming mycorrhizal. Therefore, these strains behaved as MHB. In a field experiment, none of these *Pseudomonas* strains affected: (1) number and diversity of AM fungal native population; (2) the percentage of root length that became mycorrhizal; (3) AM performance. Furthermore, the antifungal *Pseudomonas* improved plant growth and nutrient (N and P) acquisition by the mycorrhizal plants (Barea et al. 1998).

Acknowledgements. This work was supported by CICyT (REN2000-1506 project), Spain, and GENOMYCA (QLK5-2000-01319 project), ECO-SAFE (QLK3-2000-31759 project), and INCO-DEV (ICA4-CT-2001-10057) UE.

References and Selected Reading

Alabouvette C, Schippers B, Lemanceau P, Bakker PAHM (1997) Biological control of fusarium-wilts: towards development of commercial product. In: Boland GJ, Kuykendall LD (eds) Plant microbe interactions and biological control. Marcel Dekker, New York, pp 15–36

Andrade G, Mihara KL, Linderman RG, Bethlenfalvay GJ (1997) Bacteria from rhizosphere and hyphosphere soils of different arbuscular mycorrhizal fungi. Plant Soil 192:71–79

Andrade G, Mihara KL, Linderman RG, Bethlenfalvay GJ (1998) Soil aggregation status and rhizobacteria in the mycorrhizosphere. Plant Soil 202:89–96

Atkinson S, Berta G, Hooker JE (1994) Impact of mycorrhizal colonisation on root architecture, root longevity and the formation of growth regulators. In: Gianinazzi S, Schüepp H (eds) Impact of arbuscular mycorrhizas on sustainable agriculture and natural ecosystems. ALS, Birkhäuser, Basel, Switzerland, pp 47–60

Augé RM (2001) Water relations, drought and vesicular-arbuscular mycorrhizal symbiosis. Mycorrhiza 11:3–42

Azcón R (1989) Selective interaction between free-living rhizosphere bacteria and vesicular-arbuscular mycorrhizal fungi. Soil Biol Biochem 21:639–644

Azcón R (1993) Growth and nutrition of nodulated mycorrhizal and non-mycorrhizal *Hedysarum coronarium* as a result of treatments with fractions from a plant growth-promoting rhizobacteria. Soil Biol Biochem 25:1037–1042

Azcón R, Rubio R, Barea JM (1991) Selective interactions between different species of mycorrhizal fungi and *Rhizobium meliloti* strains, and their effects on growth, N_2-fixation (^{15}N) and nutrition of *Medicago sativa* L. New Phytol 117:399–404

Azcón-Aguilar C, Barea JM (1992) Interactions between mycorrhizal fungi and other rhizosphere microorganisms. In: Allen MJ (eds) Mycorrhizal functioning. An integrative plant-fungal process. Routledge, Chapman and Hall, New York, pp 163–198

Azcón-Aguilar C, Bago B (1994) Physiological characteristics of the host plant promoting an undisturbed functioning of the mycorrhizal symbiosis. In: Gianinazzi S, Schüepp H (eds) Impact of arbuscular mycorrhizas on sustainable agriculture and natural ecosystems. ALS, Birkhäuser, Basel, Switzerland, pp 47–60

Azcón-Aguilar C, Barea JM (1995) Saprophytic growth of arbuscular-mycorrhizal fungi. In: Hock B, Varma A (eds) Mycorrhiza structure function, molecular biology and biotechnology. Springer, Berlin Heidelberg New York, pp 391–407

Azcón-Aguilar C, Barea JM (1996) Arbuscular mycorrhizas and biological control of soil-borne plant pathogens. An overview of the mechanisms involved. Mycorrhiza 6:457–464

Azcón-Aguilar C, Barea JM (1997) Applying mycorrhiza biotechnology to horticulture: significance and potentials. Sci Hortic 68:1–24

Azcón-Aguilar C, Palenzuela EJ, Barea JM (2000) Substrato para la producción de inóculos de hongos formadores de micorrizas. Patente N. 9901814, Spain, CSIC

Barea JM (1997) Mycorrhiza/bacteria interactions on plant growth promotion. In: Ogoshi A, Kobayashi L, Homma Y, Kodama F, Kondon N, Akino S (eds) Plant growth-promoting rhizobacteria, present status and future prospects. OECD, Paris, pp 150–158

Barea JM (2000) Rhizosphere and mycorrhiza of field crops. In: Toutant JP, Balazs E, Galante E, Lynch JM, Schepers JS, Werner D, Werry PA (eds) Biological resource management: connecting science and policy (OECD). INRA, Editions and Springer, Berlin Heidelberg New York, pp 110–125

Barea JM, Azcón R, Azcón-Aguilar C (1992) Vesicular-arbuscular mycorrhizal fungi in nitrogen-fixing systems. In: Norris JR, Read DJ, Varma AK (eds) Methods in microbiology. Academic Press, London, pp 391–416

Barea JM, Azcón-Aguilar C, Azcón R (1997) Interactions between mycorrhizal fungi and rhizosphere microorganisms within the context of sustainable soil-plant systems. In: Gange AC, Brown VK (eds) Multitrophic interactions in terrestrial systems. Blackwell Science, Oxford, pp 65–77

Barea JM, Andrade G, Bianciotto V, Dowling D, Lohrke S, Bonfante P, O'Gara F, Azcón-Aguilar C (1998) Impact on arbuscular mycorrhiza formation of *Pseudomonas* strains used as inoculants for the biocontrol of soil-borne plant fungal pathogens. Appl Environ Microbiol 64:2304–2307

Barea JM, Toro M, Orozco MO, Campos E, Azcón R (2002) The application of isotopic (^{32}P and ^{15}N) dilution techniques to evaluate the interactive effect of phosphate-solubilizing rhizobacteria, mycorrhizal fungi and *Rhizobium* to improve the agronomic efficiency of rock phosphate for legume crops. Nutr Cycl Agroecosyst 63:35–42

Bashan Y (1999) Interactions of *Azospirillum* spp. in soils: a review. Biol Fertil Soils 29:246–256

Bashan Y, Holguin G (1998) Proposal for the division of plant growth-promoting rhizobacteria into two classifications: biocontrol-PGPB (plant growth-promoting bacteria) and PGPB. Soil Biol Biochem 30:1225–1228

Bashan Y, Gonzalez LE (1999) Long-term survival of the plant-growth-promoting bacteria *Azospirillum brasilense* and *Pseudomonas fluorescens* in dry alginate inoculant. Appl Microbiol Biotechnol 51:262–266

Berta G, Trotta A, Fusconi A, Hooker JE, Munro M, Arkinson D, Giovannetti M, Morini S, Fortuna P, Tisserant B, Gianinazzi-Pearson V, Gianinazzi S (1995) Arbuscular mycorrhizal induced changes to plant growth and root system morphology in *Prunus cerasifera*. Tree Physiol 15:281–293

Bertrand H, Nalin R, Bally R, Cleyet-Marel JC (2001) Isolation and identification of the most efficient plant growth-promoting bacteria associated with canola (*Brassica napus*). Biol Fertil Soils 33:152–156

Bethlenfalvay GJ, Linderman RG (1992) Mycorrhizae in sustainable agriculture. ASA Special publication No. 54, Madison, Wisconsin

Bethlenfalvay GJ, Schüepp H (1994) Arbuscular mycorrhizas and agrosystem stability. In: Gianinazzi S, Schüepp H (eds) Impact of arbuscular mycorrhizas on sustainable agriculture and natural ecosystems. Birkhäuser, Basel, pp 117–131

Bianciotto V, Bandi C, Minerdi D, Sironi M, Tichy HV, Bonfante P (1996a) An obligately endosymbiotic mycorrhizal fungus itself harbors obligately intracellular bacteria. Appl Environ Microbiol 62:3005–3010

Bianciotto V, Minerdi D, Perotto S, Bonfante P (1996b). Cellular interactions between arbuscular mycorrhizal fungi and rhizosphere bacteria. Protoplasma 193:123–131

Bianciotto V, Lumini E, Lanfranco L, Minerdi D, Bonfante P, Perotto S (2000) Detection and identification of bacterial endosymbionts in arbuscular mycorrhizal fungi belonging to the family Gigasporaceae. Appl Environ Microbiol 66:4503–4509

Bianciotto V, Andreotti S, Balestrini R, Bonfante P, Perotto S (2001) Mucoid mutants of the biocontrol strain *Pseudomonas fluorescens* CHA0 show increased ability in biofilm formation on mycorrhizal and nonmycorrhizal carrot roots. Mol Plant Microbe Interact 14:255–260

Boddey RM, Urquiaga S, Reis V, Döbereiner J (1991). Biological nitrogen fixation associated with sugar cane. Plant Soil 137:111–117

Bonfante P, Perotto S (2000) Outside and inside the roots: cell-to-cell interactions among arbuscular mycorrhizal fungi, bacteria and host plants. In: Podila GK, Douds DD Jr (eds) Current advances in mycorrhizae research. APS Press, St. Paul, MN, pp 141–155

Bowen GD, Rovira AD (1999) The rhizosphere and its management to improve plant growth. Adv Agron 66:1–102

Broek AV, Vanderleyden J (1995) Genetics of the *Azospirillum*-plant root association. Crit Rev Plant Sci 14:445–466

Budi SW, Van Tuinen D, Martinotti G, Gianinazzi S (1999) Isolation from *Sorghum bicolor* mycorrhizosphere of a bacterium compatible with arbuscular mycorrhiza development and antagonistic towards soilborne fungal pathogens. Appl Environ Microbiol 65:5148–5150

Calvet C, Pera J, Barea JM (1993) Growth response of marigold (*Tagetes erecta* L.) to inoculation with *Glomus mosseae*, *Trichoderma aureoviride* and *Phythium ultimum* in a peat-perlite mixture. Plant Soil 148:1–6

Calvet C, Camprubi A, Rodriguez-Kabana R (1996) Inclusion of arbuscular mycorrhizal fungi in alginate films for experimental studies and plant inoculation. Hortscience 31:285

Chanway CP (1996) Endophytes: they're not just fungi! Can J Bot 74:321–322

Chin-A-Woeng TFC, Thomas-Oates JE, Lugtenberg BJJ, Bloemberg GV (2001) Introduction of the *phzH* gene of *Pseudomonas chlororaphis* PCL 1391 extends the range of biocontrol ability of phenazine-1 carboxylic acid producing *Pseudomonas*. Mol Plant-Microbe Interact 14:1006–1015

Christensen H, Jakobsen I (1993) Reduction of bacterial growth by a vesicular-arbuscular mycorrhizal fungus in the rhizosphere of cucumber (*Cucunis sativus* L). Biol Fertil Soils 15:253–258

Clapp JP, Young JPW, Merryweather JW, Fitter AH (1995) Diversity of fungal symbionts in arbuscular mycorrhizas from a natural community. New Phytol 130:259–265

Cook RJ, Thomashow LS, Weller DM, Fujimoto D, Mazzola M, Bangera G, Kim DS (1995) Molecular mechanisms of defense by rhizobacteria against root disease. Proc Natl Acad Sci USA 92:4197–4201

Cordier C, Lemoine MC, Lemanceau P, Gianinazzi-Pearson V, Gianinazzi S (1999) The beneficial rhizosphere: a necessary strategy for microplant production. Acta Hortic 530:259–265

Daniell TJ, Husband R, Fitter AH, Young JPW (2001) Molecular diversity of arbuscular mycorrhizal fungi colonising arable crops. FEMS Microbiol Ecol 36: 203–209

Defago G, Keel C (1995) Pseudomonads as biocontrol agents of diseases caused by soil-borne pathogens In: Hokkanen HMT, Lynch JM (eds) Benefits and risks of introducing biocontrol agents. University Press, Cambridge

Dobbelaere S, Croonenborghs A, Thys A, Vande Browk A, Vanderleyden J (1999) Phytostimulatory effect *Azospirillum brasilense* strains and auxins on wheat. Plant Soil 212:155–164

Dobbelaere S, Croonenborghs A, Thys A, Ptacek D, Vanderleyden J, Dutto P, Labandera-Gonzalez C, Caballero-Mellado J, Aguirre JF, Kapulnik Y, Brener S, Burdman S, Kadouri D, Sarig S, Okon Y (2001) Response of agronomically important crops to inoculation with *Azospirillum*. Aust J Plant Physiol 28:1–9

Duijff BJ, Gianinazzi-Pearson V, Lemanceau P (1997) Involvement of the outer membrane lipopolysaccharides in the endophytic root colonization of tomato roots by biocontrol *Pseudomonas fluorescens* strain WCS417r. New Phytol 135:325–334

Duijff BJ, Pouhair D, Olivain C, Alabouvette C, Lemanceau P (1998) Implication of systemic induced resistance in the suppression of fusarium wilt of tomato by *Pseudomonas fluorescens* WCS417r and nonpathogenic *Fusarium oxysporum* Fo47. Eur J Plant Pathol. 104:903–910

Dumas-Gaudot E, Gollotte A, Cordier C, Gianinazzi S, Gianinazzi-Pearson V (2000) Modulation of host defence systems In: Kapulnick Y, Douds Jr DD (eds) Arbuscular mycorrhizas: physiology and functions. Kluwer, Dordrecht, pp 121–140

Edwards SG, Young JPW, Fitter AH (1998) Interactions between *Pseudomonas fluorescens* biocontrol agents and *Glomus mosseae*, an arbuscular mycorrhizal fungus, within the rhizosphere. FEMS Microbiol Lett 116:297–303

Ferrol N, Barea JM, Azcón-Aguilar C (2000) The plasma membrane H^+-ATPase genes family in the arbuscular mycorrhizal fungus *Glomus mosseae*. Curr Genet 37:112–118

Filippi C, Bagnoli G, Citernesi AS, Giovannetti M (1998) Ultrastructural spatial distribution of bacteria associated with sporocarps of *Glomus mosseae*. Symbiosis 24:1–12

Fitter AH, Garbaye J (1994). Interactions between mycorrhizal fungi and other soil organisms. Plant Soil 159: 123–132

Franken P, Requena N (2001) Analysis of gene expression in arbuscular mycorrhizas: new approaches and challenges. New Phytol 150:517–523

Frey P, Frey-Klett P, Garbaye J, Berge O, Heulin T (1997) Metabolic and genotypic fingerprinting of fluorescent *Pseudomonads* associated with the Douglas Fir-*Laccaria bicolor* mycorrhizosphere. Appl Environ Microbiol 63:1852–1860

Galleguillos C, Aguirre C, Barea JM, Azcón R (2000) Growth promoting effect of two *Sinorhizobium meliloti* strains (a wild type and its genetically modified derivative) on a non-legume plant species in specific interaction with two arbuscular mycorrhizal fungi. Plant Sci 159:57–63

Garbaye J (1994) Helper bacteria: A new dimension to the mycorrhizal symbiosis. New Phytol 128:197–210

Germida JJ, Walley FL (1996) Plant growth-promoting rhizobacteria alter rooting patterns and arbuscular mycorrhizal fungi colonization of field-grown spring wheat. Biol Fertil Soils 23:113–120

Gianinazzi S, Schüepp H (1994) Impact of arbuscular mycorrhizas on sustainable agriculture and natural ecosystems. ALS, Birkhäuser, Basel

Gianinazzi S, Gianinazzi-Pearson V, Trouvelot A (1990) Potentialities and procedures for the use of endomycorrhizas with special emphasis on high value crops. In: Whipps JM, Lumsden B (eds) Biotechnology of fungi for improving plant growth. Cambridge University Press, Cambridge, pp 41–54

Gianinazzi-Pearson V, Dumas-Gaudot E, Gollotte A, Tahiri-Alaoui A, Gianinazzi S (1996) Cellular and molecular defence-related root responses to invasion by arbuscular mycorrhizal fungi. New Phytol 133:45–57

Giovannetti M (2000) Spore germination and pre-symbiotic mycelial growth In: Kapulnick Y, Douds DD Jr (eds) Arbuscular mycorrhizas: physiology and functions. Kluwer, Dordrecht, pp 47–68

Glick BR (1995) The enhancement of plant growth by free-living bacteria. Can J Microbiol 41:109–117

Goicoechea N, Antolín MC, Sánchez-Díaz M (1997) Influence of arbuscular mycorrhizae and *Rhizobium* on nutrient content and water relations in drought stressed alfalfa. Plant Soil 192:261–268

Goicoechea N, Szalai G, Antolín MC, Sánchez-Díaz M, Paldi E (1998) Influence of arbuscular mycorrhizae and *Rhizobium* on free polyamines and proline levels in water-stressed alfalfa. J Plant Physiol 153:706–711

Gryndler M (2000) Interactions of arbuscular mycorrhizal fungi with other soil organisms In: Kapulnick Y, Douds Jr DD (eds) Arbuscular mycorrhizas: physiology and functions. Kluwer, Dordrecht, pp 239–262

Gryndler M, Hrselova H (1998) Effect of diazotrophic bacteria isolated from a mycelium of arbuscular mycorrhizal fungi on colonization of maize roots by *Glomus fistulosum.* Biol Plant 41:617–621

Gryndler M, Hrselová H, Stríteská D (2000) Effect of soil bacteria on growth of hyphae of the arbuscular mycorrhizal (AM) fungus *Glomus claroideum.* Folia Microbiol 45:545–551

Haas D, Keel C, Laville J, Maurhofer M, Oberliansli T, Schnider U, Voisard C, Wüthrich B, Defago G. (1991) Secondary metabolites of *Pseudomonas fluorescens* strain CHA0 involved in the suppression of root diseases. In: Hennecke H, Verma DPS (eds) Advances in molecular genetics of plant-microbe interactions. Kluwer, Dordrecht, pp 450–456

Halverson LJ, Handelsman J (1991) Enhancement of soybean nodulation by *Bacillus ceresus* UW95 in the field and in a growth chambers. Appl Environ Microbiol 57:2767–2770

Harley JL, Smith SE (1983) Mycorrhizal symbiosis. Academic Press, New York

Helgason T, Daniell TJ, Husband R, Fitter AH, Young JPW (1998) Ploughing up the wood-wide web? Nature 394:431

Hiltner L (1904) Über neuere Erfahrungen und Probleme auf dem Gebiet der Bodenbakteriologie und unter besonderer Berücksichtigung der Gründüngung und Brache. Arb Dtsch Landwirtsch Ges 98:59–78

Jeffries P (1997) Mycoparasitism. In: Wicklow DT, Södertröm BE (eds) The Mycota IV. Environmental and microbial relationships. Springer, Berlin Heidelberg New York, pp 149–164

Jeffries P, Barea JM (2001) Arbuscular mycorrhiza – a key component of sustainable plant-soil ecosystems. In: Hock B (ed) The Mycota. vol. IX. Fungal Associations. Springer, Berlin Heidelberg New York, pp 95–113

Keel C, Schnider U, Maurhofer M, Vorsand C, Laville J, Burger U, Wirthuer P, Hass D, Defago G (1992) Suppression of root diseases by *Pseudomonas fluorescens* CHAO: Importance of the bacterial secondary metabolite 2,4-diacetylphloroglucinol. Mol Plant-Microbe Interact 5:4–13

Kennedy AC (1998) The rhizosphere and spermosphere In: Sylvia DM, Fuhrmann JJ, Hartel PG, Zuberer DA (eds) Principles and applications of soil microbiology. Prentice Hall, Upper Saddle River, New Jersey, pp 389–407

Kennedy AC, Smith KL (1995) Soil microbial diversity and the sustainability of agricultural soils. Plant Soil 170:75–86

Kim KY, Jordan D, McDonald GA (1998) Effect of phosphate-solubilizing bacteria and vesicular-arbuscular mycorrhizae on tomato growth and soil microbial activity. Biol Fertil Soils 26:79–87

Kjoller R, Rosendahl S (2000) Detection of arbuscular mycorrhizal fungi (Glomales) in roots by nested PCR and SSCP (single stranded conformation polymorphism). Plant Soil 226:189–196

Kloepper JW (1994) Plant growth-promoting rhizobacteria (other systems) In: Okon Y (ed) *Azospirillum*/plant associations. CRC Press, Boca Raton, pp 111–118

Kloepper JW (1996) Host specificity in microbe-microbe interactions. BioScience 46:406–409

Kloepper JW, Zablotowick RM, Tipping EM, Lifshitz R (1991) Plant growth promotion mediated by bacterial rhizosphere colonizers. In: Keister DL, Cregan PB (eds) The rhizosphere and plant growth. Kluwer, Dordrecht, pp 315–326

Klyuchnikov AA, Kozhevin PA (1990). Dynamics of *Pseudomonas fluorescens* and *Azospirillum brasiliense* populations in the formation of vesicular arbuscular mycorrhiza. Microbiologia 59:651–655

Kozdrój J, Elsas JD van (2000) Application of polymerase chain reaction-denaturing gradient gel electrophoresis for comparison of direct and indirect extraction methods of soil DNA used for microbial community fingerprinting. Biol Fertil Soils 31:372–378

Lagopodi AL, Ram AFJ, Lamers GEM, Punt PJ, Van den Hondel CAMJJ, Lugtenberg BJJ, Bloemberg GV (2002) Novel aspects of tomato root colonization and infection by *Fusarium oxysporum* f. sp. *radicis-lycopersici* revealed by confocal laser scanning microscopic analysis using the green fluorescent protein as a marker. Mol Plant-Microbe Interact 15:172–179

Lambrecht M, Okon Y, Vande Broek A, Vanderleyden J (2000) Indole-3-acetic acid, a reciprocal signalling molecule in bacteria-plant interactions. Trends Microbiol 8:298–300

Lemanceau P, Alabouvette C (1993). Suppression of fusarium-wilts by fluorescent pseudomonads: mechanisms and applications. Biocontrol Sci Technol 3:219–234

Linderman RG (1988) Mycorrhizal interactions with the rhizosphere microflora. The mycorrhizosphere effects. Phytopathology 78:366–371

Linderman RG (1992) Vesicular-arbuscular mycorrhizae and soil microbial interactions. In: Bethlenfalvay GJ, Linderman RG (eds) Mycorrhizae in sustainable agriculture. ASA Spec, Madison, Wisconsin, pp 45–70

Linderman RG (1994) Role of VAM fungi in biocontrol. In: Pfleger FL, Linderman RG (eds) Mycorrhizae and plant health. APS Press, St Paul, pp 1–26

Linderman RG (2000) Effects of mycorrhizas on plant tolerance to diseases. In: Kapulnik Y, Douds DD Jr (eds) Arbuscular mycorrhizas: physiology and function. Kluwer, Dordrecht, pp 345–365

Lovato PE, Schüepp H, Trouvelot A, Gianinazzi S (1995) Application of arbuscular mycorrhizal fungi (AMF) in orchard and ornamental plants. In: Varma A, Hock B (eds) Mycorrhiza structure, function, molecular biology and biotechnology. Springer, Berlin Heidelberg New York, pp 521–559

Lugtenberg BJJ, Weger de LA, Bennett JW (1991) Microbial stimulation of plant growth and protection from disease. Curr Opin Microbiol 2:457–464

Lugtenberg BJJ, Dekkers L, Bloemberg GV (2001) Molecular determinants of rhizosphere colonization by bacteria. Annu Rev Phytopathol 39:461–490

Lynch JM (1990) The rhizosphere. Wiley, New York, p 462
Marschner P, Crowley DE (1996a) Root colonization of mycorrhizal and nonmycorrhizal pepper (*Capsicum annuum*) by *Pseudomonas fluorescens* 2–79RL. New Phytol 134:115–122
Marschner P, Crowley DE (1996b) Physiological activity of a bioluminescent *Pseudomonas fluorescens* (strain 2–79) in the rhizosphere of mycorrhizal and non-mycorrhizal pepper (*Capsicum annuum* L). Soil Biol Biochem 28:869–876
Miller RM, Jastrow JD (1994) Vesicular-arbuscular mycorrhizae and biogeochemical cycling. In: Pfleger FL, Linderman RG (eds) Mycorrhizae and plant health. APS Press, St. Paul, MN, pp 189–212
Miller RM, Jastrow JD (2000) Mycorrhizal fungi influence soil structure. In: Kapulnik Y, Douds DD Jr (eds) Arbuscular mycorrhizas: physiology and functions. Kluwer, Dordrecht, pp 3–18
Moenne-Loccos Y, Tichy HV, O'Donnell A, Simon R, O'Gara F (2001) Impact of 2,4-diacetylphloroglucinol-producing biocontrol strain *Pseudomonas fluorescens* F113 on intraspecific diver of resident culturable fluorescent pseudomonads associate with the roots of field-grown sugar beet seedlings. Appl Environ Microbiol 67:3418–3425
Mohammad A, Khan AG, Kuek C (2000) Improved aeroponic culture of inocula of arbuscular mycorrhizal fungi. Mycorrhiza 9:337–339
Monzón A, Azcón R (1996) Relevance of mycorrhizal fungal origin and host plant genotype to inducing growth and nutrient uptake in *Medicago* species. Agric Ecosyst Environ 60:9–15
Morton J.B. (2000) Evolution of endophytism in arbuscular mycorrhizal fungi of Glomales. In: Bacon CW, White Jr JE (eds) Microbial endophytes. Marcel Dekker, New York, pp 121–140
Morton JB, Redecker D (2001) Two new families of *Glomales*, *Archaeosporaceae* and *Paraglomaceae*, with two new genera *Archaeospora* and *Paraglomus*, based on concordant molecular and morphological characters. Mycologia 93:181–195
Nehl DB, Allen SJ, Brown JF (1996) Deleterious rhizosphere bacteria: an integrating perspective. Appl Soil Ecol 5:1–20
O'Gara F, Dowling DN, Boesten B (1994) Molecular ecology of rhizosphere microorganisms. VCH, Weinheim, p 173
Okon Y (1994) *Azospirillum*/Plant associations. CRC Press, Boca Raton, p 175
Probanza A, Lucas García JA, Ruiz Palomino M, Ramos B, Gutiérrez Mañero FJ (2002) *Pinus pinea* L. seedling growth and bacterial rhizosphere structure after inoculation with PGPR *Bacillus* (*B. licheniformis* CECT 5106 and *B. pumillus* CECT 5105). Appl Soil Ecol 20:75–84
Puppi G, Azcón R, Höflich G (1994) Management of positive interactions of arbuscular mycorrhizal fungi with essential groups of soil microorganisms. In: Gianinazzi S, Schüepp II (eds) Impact of arbuscular mycorrhizas on sustainable agriculture and natural ecosystems. ALS, Birkhäuser, Basel, pp 201–215
Ravnskov S, Nybroe O, Jakobsen I (1999) Influence of an arbuscular mycorrhizal fungus on *Pseudomonas fluorescens* DF57 in rhizosphere and hyphosphere soil. New Phytol 142:113–122
Redecker D, Thierfelder H, Walker C, Werner D (1997) Restriction analysis of PCR-amplified internal transcribed spacers of ribosomal DNA as a tool for species identification in different genera of the order Glomales. Appl Environ Microbiol 63:1756–1761
Redecker D, Morton JB, Bruns TD (2000) Ancestral lineages of arbuscular mycorrhizal fungi (*Glomales*). Mol Phylogenet Evol 14:276–284
Requena N, Jimenez I, Toro M, Barea JM (1997) Interactions between plant-growth- promoting rhizobacteria (PGPR), arbuscular mycorrhizal fungi and *Rhizobium* spp. in

the rhizosphere of *Anthyllis cytisoides*, a model legume for revegetation in Mediterranean semi-arid ecosystems. New Phytol 136:667–677

Requena N, Perez-Solis E, Azcón-Aguilar C, Jeffries P, Barea JM (2001) Management of indigenous plant-microbe symbioses aids restoration of desertified. Appl Environ Microbiol 67:495–498

Ruiz-Lozano JM, Azcón R (1993) Specificity and functional compatibility of VA mycorrhizal endophytes in association with *Bradyrhizobium* strains in *Cicer arietinum*. Symbiosis 15:217–226

Ruiz-Lozano JM, Bonfante P (2000) A *Burkholderia* strain living inside the arbuscular mycorrhizal fungus *Gigaspora margarita* possesses the *vacB* gene, which is involved in host cell colonization by bacteria. Microbial Ecol 39:137–144

Ruiz-Lozano JM, Collados C, Barea JM, Azcón R (2001) Arbuscular mycorrhizal symbiosis can alleviate drought-induced nodule senescence in soybean plants. New Phytol 151:493–502

Sanders IR, Clapp JP, Wiemken A (1996) The genetic diversity of arbuscular mycorrhizal fungi in natural ecosystems – A key to understanding the ecology and functioning of the mycorrhizal symbiosis. New Phytol 133:123–134

Schloter M, Wiehe W, Assmus B, Steindl H, Becke H, Höflich G, Hartmann A (1997) Root colonization of different plants by plant-growth-promoting *Rhizobium leguminosarum* bv. trifolii R39 studied with monospecific polyclonal antisera. Appl Environ Microbiol 63:2038–2046

Schüβler A, Schwarzott D, Walker C (2001) A new fungal phylum, the Glomeromycota: phylogeny and evolution. Mycol Res 105:1413–1421

Siciliano SD, Germida JJ (1998) BIOLOG analysis and fatty acid methyl ester profiles indicate that pseudomonad inoculants that promote phytoremediation alter the root-associated microbial community of *Bromus biebersteinii*. Soil Biol Biochem 30:1717–1723

Simon L, Bousquet J, Lévesque RC, Lalonde M (1993). Origin and diversification of endomycorrhizal fungi and coincidence with vascular land plants. Nature 363:67–69

Simons M, van der Bij AJ, Brand I, de Weger LA, Wijffelman CA, Lugtenberg BJJ (1996) Gnotobiotic system for studying rhizosphere colonization by plant growth-promoting *Pseudomonas* bacteria. Mol Plant-Microbe Interact 9:600–607

Smith SE, Read DJ (1997) Mycorrhizal symbiosis. Academic Press, London

Smith SE, Gianinazzi-Pearson V, Koide R, Cairney JWG (1994) Nutrient transport in mycorrhizas: structure, physiology and consequences for efficiency of the symbiosis. In: Robson AD, Abbott LK, Malajczuk N (eds) Management of mycorrhizas in agriculture, horticulture and forestry. Kluwer, Dordrecht, pp 103–113

Staley TW, Lawrence EG, Nance EL (1992) Influence of a plant growth-promoting pseudomonad and vesicular-arbuscular mycorrhizal fungus on alfalfa and birdsfoot trefoil growth and nodulation. Biol Fertil Soils 14:175–180

Sturz AV Nowak J (2000) Endophytic communities of rhizobacteria and the strategies required to create yield enhancing associations with crops. Appl Soil Ecol 15:183–190

Sturz AV, Christie BR, Nowak J (2000) Bacterial endophytes: potential role in developing sustainable systems of crop production. Crit Rev Plant Sci 19:1–30

Sylvia DM (1998) Mycorrhizal symbioses. In: Sylvia DM, Fuhrmann JJ, Hartel PG, Zuberer DA (eds) Principles and applications of soil microbiology. Prentice Hall, Upper Saddle River, New Jersey, pp 408–426

Thomashow LS, Weller DM (1995) Current concepts in the use of introduced bacteria for biological control: mechanisms and antifungal metabolites. In: Stacey G, Keen N (eds) Plant-microbe interactions. Chapman and Hall, New York, pp 187–235

Toro M, Azcón R, Barea JM (1997) Improvement of arbuscular mycorrhizal development by inoculation with phosphate-solubilizing rhizobacteria to improve rock phosphate bioavailability (^{32}P) and nutrient cycling. Appl Environ Microbiol 63: 4408–4412

Toro M, Azcón R, Barea JM (1998) The use of isotopic dilution techniques to evaluate the interactive effects of *Rhizobium* genotype, mycorrhizal fungi, phosphate-solubilizing rhizobacteria and rock phosphate on nitrogen and phosphorus acquisition by *Medicago sativa*. New Phytol 138:265–273

Tuinen D van, Jacquot E, Zhao B, Gollotte A, Gianinazzi-Pearson V (1998) Characterization of root colonization profiles by a microcosm community of arbuscular mycorrhizal fungi using 25S rDNA-targeted nested PCR. Mol Ecol 7:879–887

Valdenegro M, Barea JM, Azcón R (2001) Influence of arbuscular-mycorrhizal fungi, *Rhizobium meliloti* strains and PGPR inoculation on the growth of *Medicago arborea* used as model legume for re-vegetation and biological reactivation in a semi-arid Mediterranean area. Plant Growth Regul 34:233–240

Van Bastelaere E, Lambrecht M, Vermeiren H, Keijers V, de Wilde P, Proost P, Van Dommelen A, Varderleyden J (1999) Characterization of a sugar-binding protein from *Azospirillum brasilense* mediating chemotaxis to and uptake of sugars. Mol Microbiol 32:703–714

Van de Broek A, Lambrecht M, Vanderleyden J (1999) Auxins upregulate expression of the indole-3-pyruvate decaboxylase gene from *Azospirillum brasilense*. J Bacteriol 181:1338–1342

Van Loon LC, Bakker PAHM, Pieterse CMJ (1998) Systemic resistance induced by rhizosphere bacteria. Annu Rev Phytopathol 36:453–483

Varma A, Schüepp H (1995) Mycorrhization of the commercially important micropropagated plants. Crit Rev Biotechnol 15:313–328

Varma A, Verma S, Sudha, Sahay N, Bütehorn B, Franken P (1999) *Piriformospora indica*, a cultivable plant-growth-promoting root endophyte. Appl Environ Microbiol 65:2741–2744

Vázquez MM, Cesar S, Azcón R, Barea JM (2000) Interactions between arbuscular mycorrhizal fungi and other microbial inoculants (*Azospirillum*, *Pseudomonas*, *Trichoderma*) and their effects on microbial population and enzyme activities in the rhizosphere of maize plants. Appl Soil Ecol 15:261–272

Vierheilig H, Piché Y (2002) Signalling in arbuscular mycorrhiza: facts and hypotheses In: Buslig B, Manthey J (eds) Flavonoids in cell functions. Kluwer Academic/Plenum Publishers, New York, pp 23–29

Vestberg M, Estaún V (1994) Micropropagated plants, an opportunity to positively manage mycorrhizal activities. In: Gianinazzi S, Schüepp H (eds) Impact of arbuscular mycorrhizas on sustainable agriculture and natural ecosystems. Birkhäuser, Basel, pp 217–226

Volpin H, Kapulnik Y (1994) Interaction of *Azospirillum* with beneficial soil microorganisms. In: Okon Y (ed) *Azospirillum*/plant associations. CRC Press, Boca Raton, pp 111–118

Vosátka M (1996) Soil bacteria – a component of plant, soil and arbuscular mycorrhizal fungal interactions. In: Azcon-Aguilar C, Barea JM (Eds) Mycorrhizas in integrated systems – from genes to plant development. (pp 613–618). European Commission Report EUR 16728, Brussels, Luxembourg

Vosátka M, Gryndler M (1999) Treatment with culture fractions from *Pseudomonas putida* modifies the development of *Glomus fistulosum* mycorrhiza and the response of potato and maize plants to inoculation. Appl Soil Ecol 11:245–251

Waschkies C, Schropp A, Marschner H (1994). Relations between grapevine replant disease and root colonization of grapevine (*Vitis* sp.) by fluorescent pseudomonads and endomycorrhizal fungi. Plant Soil 162:219–227

Weller DM, Thomashow LS (1994) Current challenges in introducing beneficial microorganisms into the rhizosphere. In: O'Gara F, Dowling DN, Boesten B (eds) Molecular ecology of rhizosphere microorganisms biotechnology and the release of GMOs. VCH, Weinheim, pp 1–18

Werner D (1998) Organic signals between plants and microorganisms. In: Pinton R, Varanini Z, Nannipieri P (eds) The rhizosphere: biochemistry and organic substances at the soil-plant interfaces. Marcel Dekker, New York

Werner D, Barea J-M, Brewin NJ, Cooper P, Katinakis P, Lindström K, O'Gara F, Spaink HP, Truchet G, Müller P (2002) Symbiosis and defence in the interaction of plants with microorganisms. Symbiosis 32:83–104

Whitelaw MA (2000) Growth promotion of plants inoculated with phosphate-solubilizing fungi. Adv Agron 69:99–151

Zinniel DK, Lambrecht P, Harris NB, Feng Z, Kuczmarski D, Higley P, Ishimaru CA, Arunakumari A, Barletta RG, Vidaver AK (2002) Isolation and characterization of endophytic colonizing bacteria from agronomic crops and prairie plants. Appl Environ Microbiol 68:2198–2208

21 Carbohydrates and Nitrogen: Nutrients and Signals in Ectomycorrhizas

UWE NEHLS

1 Introduction

Due to plant litter, forest soil is rich in complex carbohydrates (e.g., cellulose and lignin). Nevertheless, these carbohydrates are only slowly degraded by specialized microorganisms and thus forest soils are rather poor in readily cleavable carbohydrates that are necessary for the growth of the majority of microbes including ectomycorrhizal fungi.

Basidiomycetes are able to transfer nutrients and metabolites over long distances. Exploring a rich source of readily utilizable carbohydrates would thus favor the colonization of other soil areas, too. The association with fine roots of woody plants forming ectomycorrhizas is a way that secures exclusive access to such a rich carbohydrate source for ectomycorrhizal fungi.

Organic compounds contained in root exudates are candidates for the carbon transfer from the host to the mycorrhizal fungus. Low-molecular-weight root exudates comprise soluble sugars, carboxylic acids and amino acids (Marschner 1995; Smith and Read 1997; Hampp and Schaeffer 1999). The best growth of ectomycorrhizal fungi (ECM) fungi occurs on the hexoses glucose, fructose, and mannose. Sucrose, which is the preferred transport sugar in most host plants cannot be used by ECM investigated so far (e.g., Salzer and Hager 1991), *Laccaria bicolor* being possibly an exception (Tagu et al. 2000).

Even if plant-derived hexoses are most important for ectomycorrhizal fungi, there is ample evidence that soil carbon sources are also intensively used. Among these are starch, dextrins, glucans, oligosaccharides or sugar alcohols (Palmer and Hacskaylo 1970; Cao and Crawford 1993; Berredjem et al. 1998), proteins (Abuzinadah and Read 1986), or even cellulose or lignin (Norkrans 1950; Trojanowski et al. 1984; Taber and Taber 1987; Haselwandter et al. 1990).

Plant Surface Microbiology
A. Varma, L. Abbott, D. Werner, R. Hampp (Eds.)

2 Trehalose Utilization by Ectomycorrhizal Fungi

Trehalose is a nonreducing disaccharide that is found in a wide variety of organisms including bacteria, fungi, protozoa, nematodes, insects and plants (Elbein 1974; Mellor 1992, Müller et al. 1994). Trehalose has been shown to be a good carbon source for a number of ectomycorrhizal fungi (Lewis and Harley 1965; Palmer and Hacskaylo 1970).

Amanita muscaria hyphae excrete an acid trehalase into the growth medium to break down external trehalose (Wisser et al. 2000). With its apparent molecular mass of 165 kDa, a pI of 3.7, a pH optimum of about 4.0 and an apparent Km value for trehalose of 0.38 mM, it is comparable to that of other fungi.

Excreted acid trehalases, in general, are very stable. Owing to the low Km value of the enzyme, and the acidic pH optimum, trehalose hydrolysis in acidic forest soils should be very efficient. The resulting monosaccharides are then taken up by the highly efficient monosaccharide import system of *A. muscaria* (Chen and Hampp 1993; Nehls et al. 1998, see below).

Carbohydrate-starved mycelia excreted about four times more acid trehalase into the growth medium than mycelia that were well supported with sugar (Wisser 2000), indicating an up-regulated expression of acid trehalase with regard to poor carbon nutrition.

Trehalose is one of the main storage carbohydrates of basidiomycetes, but is also found in bacteria in large quantities. The utilization of trehalose by ectomycorrhizal fungi might thus be important for two reasons:

1. The availability of an additional carbon source would improve ectomycorrhizal fungal growth in soil.
2. By utilization of a soil carbon source that otherwise could be used by other microorganisms, ectomycorrhizal fungi could reduce the growth of their putative competitors for other nutrients (e.g., nitrogen or phosphate).

3 Carbohydrate Uptake

A prerequisite for a rapid uptake of monosaccharides are membrane transport systems. Experiments with suspension-cultured hyphae of ectomycorrhizal fungi (Salzer and Hager 1991) and with protoplasts (Chen and Hampp 1993) indicated that most basidiomycotic ectomycorrhizal fungi have no system for sucrose import or hydrolysis, but for the uptake of glucose and fructose (Palmer and Hacskaylo 1970; Salzer and Hager 1991).

To date, only two hexose transporter genes from *A. muscaria* (Fig. 1), *AmMst1* (Nehls et al. 1998) encoding a protein of 520 amino acids and *AmMst2* encoding a protein of 519 amino acids, and one hexose transporter gene from *Tuber borchii* (Agostini and Stocchi, pers. comm.) have been identified. While *AmMst1* reveals the highest sequence homology with *Hxt1* from

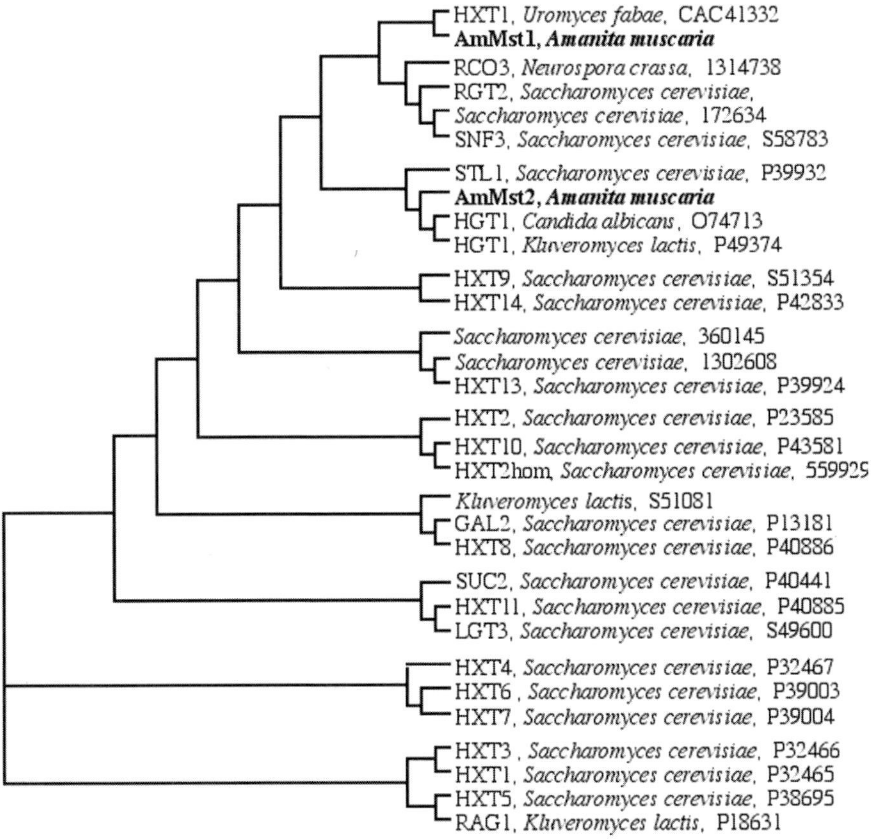

Fig. 1. Dendrogram of the alignment of the deduced protein sequence of *AmMST1* and *AmMst2* with known fungal monosaccharide transporters. The relationship of the deduced protein sequence of *AmMST1* and *AmMst2* to other fungal monosaccharide transporters was determined by multiple alignment using ClustalW (version 1.8)

Uromyces fabae, *AmMst2* has the best homologies to *Stl1* of *S. cerevisiae*. The homology between both deduced *Amanita* proteins is low (33.3 % identity, 56.8 % similarity). Interestingly, the noncoding region of the cDNAs revealed a higher identity (approx. 60 %) than the coding region (approx. 50 %). The expression patterns with regard to carbon and nitrogen nutrition are identical for both genes (Nehls, unpublished). When expressed in yeast, AmMst1 revealed K_M values of 0.46 mM for glucose and 4.2 mM for fructose, indicating a strong preference for glucose (Wiese et al. 2000). Also, *A. muscaria* hyphae strongly favored glucose uptake even in the presence of a large excess of fructose (20 mM vs. 1 mM; Nehls et al. 2001 c). A similar preference for glucose uptake was also observed for the ectomycorrhizal ascomycete *Cenococcum geophilum* (Stülten et al. 1995), indicating that this behavior might be common to ectomycorrhizal fungi. Since *A. muscaria* revealed only one type of hexose uptake kinetic, mostly resembling that of *AmMst1* when expressed in yeast, it is very likely that *AmMst1* and *AmMst2* have similar transport properties. The investigated *A. muscaria* hyphae are heterocaryotic, containing nuclei of two different origins. It could thus be questioned whether *AmMst1* and *AmMst2* are encoding different genes from one nucleus or both genes belong to different nuclei.

4 Carbohydrate Metabolism

As in other organisms phosphofructokinase (ATP-dependent) is the rate-limiting step in fungal glycolysis (Kowallik et al. 1998). In *A. muscaria*, this enzyme is activated by fructose 2,6-bisphosphate (F26BP; k_a about 30 nM; Schaeffer et al. 1996) which is similar to the properties of the corresponding enzyme from yeast or animal cells, but differs from plant phosphofructokinases. It has been shown that *A. muscaria* mycelia grown in the presence of high hexose concentrations as well as mycorrhizal roots have increased amounts of F26BP (Schaeffer et al. 1996; Hoffmann et al. 1997). This could indicate increased rates of glycolysis in hyphae under elevated hexose supply, e.g., hyphae of the Hartig net (Hampp and Schaeffer 1999).

In yeast cells, levels of fructose-2,6 bisphosphate, and thus the predominance of glycolysis over gluconeogenesis, are controlled by the formation of cyclic AMP (cAMP). Increased glucose supply causes an increase of activity of adenylate cyclase (Thevelein 1991) and thus of the cAMP content in hyphae. cAMP activates a cAMP-dependent protein kinase (PKA) which, via phosphorylation, activates F26BP formation while inhibiting F26BP degradation (Thevelein 1991; d'Enfert 1997; RadisBaptista et al. 1998).

At least the initial steps of glucose-dependent regulation of glycolysis also exist in *A. muscaria*. Changes in pool sizes of cAMP were detected in relation to glucose supply (Hoffmann et al. 1997). When suspension cultures of *A. muscaria* were transferred from medium containing low (1 mM) to high (40 mM) glucose concentrations, both cAMP pools as well as rates of activity of protein kinase A increased (Nehls et al. 2001 c).

5 Carbohydrate Storage

In addition to their relevance for carbon storage, storage carbohydrates have additional functions in fungi, e.g., the rapid conversion of carbohydrates that are taken up (to maintain a carbon sink) or membrane and protein protection (e.g., trehalose). Two different pools of storage carbohydrates can be distinguished in ectomycorrhizal fungi: short chain carbohydrates (trehalose) or polyols (mannitol, arabitol, erythritol), and the long chain carbohydrate glycogen.

Hexoses that are taken up by ectomycorrhizal fungi are either introduced into glycolysis (e.g., formation of amino acids) or converted into short chain carbohydrates and polyols (Martin et al. 1985, 1987, 1988, 1998). Growth of *A. muscaria* (suspension culture) on glucose as a carbon source resulted in an increase in the trehalose content until the external glucose concentration was below 4 mM, followed by a depletion of trehalose concentration over time. In contrast, the glycogen content was stable during the investigation (4 weeks; Wallenda 1996). In *Lactarius* sp. the glycogen content was high during winter, declined until summer and was restored during the autumn (Genet et al. 2000). Thus, trehalose and polyols are presumably short term storage compounds, revealing high fluctuation rates, while glycogen is the long term storage carbohydrate of hyphae that is only mobilized when the short term pools are empty.

Ectomycorrhizal fungal colonies could become quite large, and support from plant-derived carbohydrates has been shown to be necessary for fungal growth in soil (Leake et al. 2001). Thus, long-distance transport of carbon is of great interest for fungal physiology. In *P. involutus* ectomycorrhizas, glycogen particles were observed in the Hartig net and the inner and outer hyphal layers of the fungal sheath (Jordy et al. 1998). Since glycogen is stored in the cytoplasm as large, nonmobile granules it is rather likely that glycogen is not the long-distance transport form of carbohydrates. Polyols and trehalose are, in addition to glucose, present in large quantities in fungal hyphae, and are thus good candidates for long-distance transport carbohydrates between different parts of the fungal colony.

6 Carbohydrates as Signal, Regulating Fungal Gene Expression in Ectomycorrhizas

Sugar-regulated gene expression was mainly investigated in saprophytic ascomycetes (Jennings 1995). Here, the external monosaccharide concentration regulates fungal gene expression, e.g., that of monosaccharide transporters at the transcriptional as well as the posttranscriptional (protein degradation rate) level. Two different transcriptional control mechanisms, induction/enhancement or repression of gene expression were described (Felenbok and Kelley 1996; Özcan et al. 1996).

In *A. muscaria*, the expression of the hexose transporter genes is up-regulated by a threshold response mechanism depending on the extracellular concentration of monosaccharides (Nehls et al. 1998). In *A. muscaria* hyphae grown in the presence of glucose concentrations up to 2 mM, the glucose transporter genes *AmMst1* and *AmMst2* are expressed at a basal level, while monosaccharide concentrations above that threshold triggered at least a four-fold increase in the transcript levels. This up-regulation could not be further enhanced by hexose concentrations of up to 100 mM. Since the increase of monosaccharide transporter gene expression is a slow process, it could be interpreted as an adaptation to the elevated hexose concentrations usually only found at the plant/fungus interface, but not in the soil.

In yeast, sugar dependent induction/enhancement of gene expression is controlled by two monosaccharide transporter-like proteins, RGT2 and SNF3, sensing the external sugar concentration (Celenza et al. 1988; Özcan et al. 1996). These transporters have a C-terminal extension containing a conserved amino acid motive thought to be involved in the transduction of the external sugar signal. Due to their different glucose affinities, SNF3 senses low external glucose concentrations while RGT2 senses high concentrations. In contrast to yeast, the putative monosaccharide transporter RCO3 (Madi et al. 1997) that is presumably also acting as monosaccharide sensor in *Neurospora crassa*, does not contain any extension. The signal cascade, transforming the sugar signal into modified gene expression, is not fully understood. To date, two elements have been identified, the transcription factor RGT1 (Özcan et al. 1996) and a signal transduction mediator, the SCF complex (Özcan and Johnston 1999). Without the sugar signal, RGT1 is a repressor for glucose-induced genes while activation via the SCF complex (in response to a sugar signal) modifies RGT1 function to that of a transcriptional activator (Johnston 1999).

The signal regulating the hexose-dependent, enhanced *AmMst1* expression is still unknown, but in contrast to yeast, it seems to be transmitted by an internal and not an external sensor. Glucose analogues that are imported by AmMst1 and phosphorylated, but not further metabolized, did not increase the *AmMst1* transcript level as glucose did (Wiese et al. 2000). Furthermore, the result of these experiments makes it rather likely that the signal must be generated downstream of hexokinase activity, in glycolysis or carbon storage pathways.

While *AmMst1* expression is an example of sugar-dependent enhancement of gene expression in *A. muscaria*, a second gene (*AmPAL*) was identified that revealed sugar-dependent gene repression (Nehls et al. 1999a). PAL is a key enzyme of secondary metabolism and thus of the production of phenolic compounds. ECM-forming fungi have been reported to use phenolic compounds for both their own protection and that of their host against bacterial or fungal attacks (Marx 1969; Chakravarty and Unestam 1987; Garbaye 1991).

In *A. muscaria*, the transcript of *AmPAL* was abundant in hyphae grown at low external glucose concentrations, but exhibited a significant decrease in hyphae cultured at glucose concentrations of above 2 mM (less than 1/30 of the transcript level at low glucose). Unlike *AmMst1*, *AmPAL*-expression is probably regulated by sugar phosphorylation via hexokinase as sugar sensor (Nehls et al. 1999a).

Also in saprophytic ascomycetes the monosaccharide-dependent gene repression is regulated via a hexokinase-dependent signaling pathway (Ronne 1995; Gancedo 1998). The molecular mechanism of signal initiation is still unclear, but a hexokinase (in yeast mainly hex2) initiates the signal in response to the C-flux through the enzyme. Whether hexokinase phosphorylation or its association with other proteins is responsible for signal generation is still a mater of debate, but it is rather likely that the conformational changes of the enzyme during its enzymatic activity are sensed and not the generated hexose phosphates per se. The signaling pathways involve the activation/deactivation of the SNF1 protein complex that is presumably mediated by phosphorylation/dephosphorylation (AMPK kinase and REG1/GLC7 phosphatase complex, respectively; Lesage et al. 1996; Johnston 1999). The sugar-dependent gene repression is mediated by a DNA-binding protein like MIG1 (yeast) or CREA (*Aspergillus*, *Neurospora*; Felenbok and Kelly 1996), acting as repressors.

While in the pure fungal culture *AmMst1* was induced and *AmPAL* repressed by elevated hexose concentrations, both genes were strongly expressed in entire mycorrhizas (Nehls et al. 1998, 1999a). It could thus be concluded that in mycorrhizas the sugar-dependent regulation of both genes is either modified by developmental events, or different in the sheath and Hartig net hyphae. To address this question, ectomycorrhizas were dissected and gene expression was investigated separately for hyphae of the fungal sheath and the Hartig net (Nehls et al. 2001a). Similar to low external hexose concentrations in pure fungal culture, *AmMst1* was expressed only at the basal level in hyphae of the fungal sheath. In contrast, *AmPAL* revealed a high transcript level in this fungal structure. For Hartig net hyphae the opposite expression pattern was observed. As for hyphae in pure culture in the presence of high external hexose concentrations, the transcript level of *AmMst1* was sixfold enhanced while the expression of *AmPAL* was only barely detectable.

Owing to the opposite regulation of both genes in hyphae of fungal sheath and Hartig net that resembles the hexose-dependent expression of these genes in pure culture, different hexose concentrations in the apoplast of the fungus/plant interface (hexose concentration >2 mM) and the apoplast of the fungal sheath (hexose concentration <1 mM) could be supposed (Nehls et al. 2001a).

How could such a hexose gradient (Fig. 2) be generated and maintained? In the symbiotic apoplast at the plant/fungus interface, sucrose is hydrolyzed into equimolar concentrations of glucose and fructose (Salzer and Hager 1991; Schaeffer et al. 1995). Since the hexose concentration is assumed to be mainly determined by fungal activity, and *A. muscaria* takes up preferentially glucose until the concentration drops below 0.5 mM (Wiese et al. 2000), an increased apoplastic fructose concentration (>2 mM) could be assumed in the Hartig net that would trigger the observed hexose-dependent fungal gene expression. Fructose withdrawal from the apoplast presumably takes place mainly within the innermost one or two layers of the fungal sheath since fructose uptake by *A. muscaria* hyphae is rather efficient when the glucose concentration is <0.5 mM. It is thus rather unlikely that hexose concentrations above the threshold of about 2 mM (that would result in an increased expres-

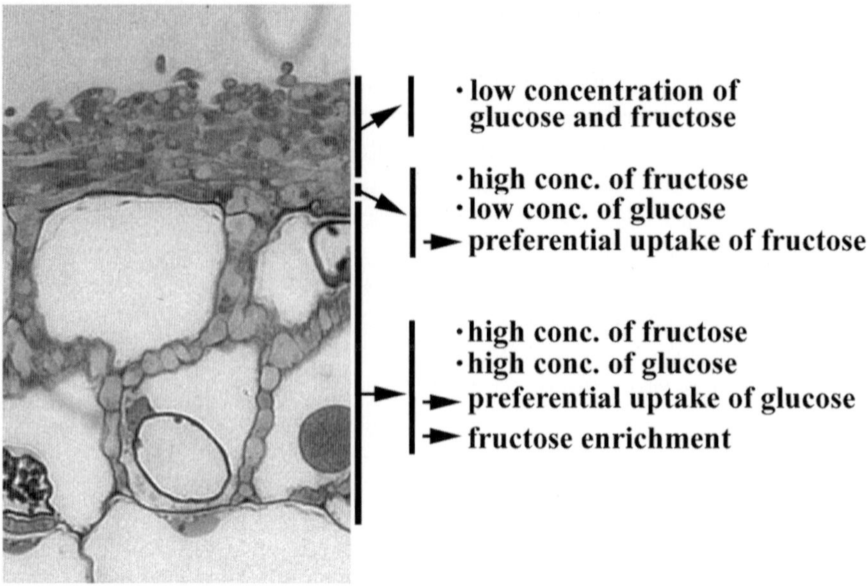

Fig. 2. Spatial distribution of hexose uptake by fungal hyphae in ectomycorrhizas: a model. Sucrose hydrolysis in the apoplast of the Hartig net results in high glucose and fructose concentrations. We assume that here glucose is taken up preferentially, since the uptake of fructose is inhibited (by glucose concentrations above 0.5 mM). In the innermost one or two layers of the fungal sheath glucose concentration is low, due to efficient uptake by fungal hyphae of the Hartig net. Thus, most probably fructose is taken up. In the apoplast of the majority of the fungal sheath, glucose as well as fructose concentrations are low due to the efficient hexose uptake by hyphae of the Hartig net and the inner layers of the sheath

sion of *AmMst1* and a repression of *AmPAL*) are present in the apoplast of the majority of fungal sheath hyphae.

"Metabolic zonation" and "physiological heterogeneity" have already been discussed as important concepts for a functional understanding of ectomycorrhizal symbiosis (Martin et al. 1992; Cairney and Burke 1996; Timonen and Sen 1998). Differences in the apoplastic hexose concentration at the Hartig net vs. fungal sheath could thus be supposed to generate a signal that might regulate fungal physiological heterogeneity in ectomycorrhizas, in addition to the developmental program.

7 Nitrogen

The ability of ectomycorrhizal fungi to take up inorganic nitrogen is well established (Melin and Nilsson 1952; France and Reid 1983; Plassard et al.

1986; Finlay et al. 1988; Chalot and Brun 1998). In accordance with the predominant occurrence of ammonium as inorganic nitrogen source in the soil, most ectomycorrhizal fungi grow better on ammonium than on nitrate in pure culture (France and Reid 1984; Finlay et al. 1992). Nevertheless, even in mature forests nitrate could be present in large amounts as a result of bacterial activity (e.g., open forest areas) or as a result of fertilization in areas with extensive agriculture (Gessler et al. 1998).

In many forest ecosystems, rates of nitrogen mineralization of litter are low and consequently, the supply of inorganic nitrogen is often limited (Read 1991). In addition, nitrification is usually slow and the poorly mobile ammonium ion (Keeney 1980) predominates together with organic nitrogen (e.g., amino acids or protein). Important for the establishment of forest ecosystems is thus, the capability of ectomycorrhizal fungi to exploit (in collaboration with other soil organisms) organic debris (e.g., litter) as a nutrient source (Nasholm and Persson 2001).

8 Utilization of Inorganic Nitrogen

For a number of ectomycorrhizal fungi growth on nitrate as sole nitrogen source (France and Reid 1984; Littke et al. 1984; Plassard et al. 1986) as well as the presence of nitrate reductase activity (Wagner et al. 1989; Sarjala 1990) have been shown. In saprophytic ascomycetes, the expression of nitrate reductase is repressed in the presence of a reduced nitrogen source (e.g., ammonium) and induced only the presence of nitrate. In contrast, nitrate reductase activity of the ectomycorrhizal fungus *Hebeloma cylindrosporum* was similar for nitrate and ammonium-fed hyphae (Scheromm et al. 1990), indicating a different type of regulation. A nitrate transporter gene has been identified so far only from *H. cylindrosporum* (Marmeisse, pers. comm.).

Two ammonium transporter genes of *H. cylindrosporum* (HcAMT2 and HcAMT3) were isolated and functionally characterized in yeast (Javelle et al. 2001). HcAMT2 revealed a K_M value of 58 μM and HcAMT3A of 260 μM. When the fungus was grown under optimal nitrogen conditions (ammonium concentration >2 mM) the expression of both transporter genes was only barely detectable while gene expression strongly increases under nitrogen starvation. Similar results were found in *Paxillus involutus*, where N starvation triggered a fourfold increase in methylamine transport after 2 h incubation in nitrogen-free media (Javelle et al. 1999).

In addition, one ammonium transporter gene (*TbAMT1*) was isolated from the ascomycete *Tuber borchii* (Montanini et al. 2002). Heterologous expression in yeast revealed a K_M value of 2 μM. When exposed to ammonium or nitrate, the gene was expressed at a basal level while nitrogen depletion resulted in a slow and only slight increase in gene expression. This expression profile is quite untypical for fungi where good nitrogen nutrition usually results in a strong repression of ammonium transporter genes.

9 Utilization of Organic Nitrogen

Important for the establishment of forest ecosystems is the capability of ectomycorrhizal fungi to exploit (in collaboration with other soil organisms) organic debris (e.g., litter) as a nutrient source (Nasholm and Persson 2001).

10 Proteolytic Activities of Ectomycorrhizal Fungi

Ericoid fungi (Bajwa et al. 1985; Leake and Read 1990), but also some ectomycorrhizal fungi (Abuzinadah and Read 1986; El-Badaoui and Botton 1989; Zhu 1990; Spägele 1992; Zhu et al. 1994; Bending and Read 1996) are able to utilize protein not only as a nitrogen, but also as a carbon source (for a review, see Smith and Read 1997).

Two proteins with proteolytic activities and molecular masses of about 45 kDa (AmProt1) and 100 kDa (AmProt2) are excreted by *A. muscaria* (Nehls et al. 2001b). AmProt1 was mainly released at pH-values up to pH 5.4 and revealed a narrow pH-optimum around 3.0. It resembles thus, proteases released by *H. crustuliniforme* (Zhu 1990) and the ericoid fungus *Hymenoscyphus ericea* (Leake and Read 1990). AmProt2 was only excreted at pH-values between 5.4 and 6.3 and reveals a broad pH-optimum between 3 and 6. *A. muscaria* is mainly growing in the litter layer of both acidic and less acidic forest soils. Since forest litter layers are, in addition to fungi, intensively colonized by biofilm-forming bacteria (Berg et al. 1998), where the microenvironment is adapted to bacterial growth (e.g., pH 5–6; Fletcher 1996), expression of a protease that is active at a less acidic pH would favor the mobilization of bacteria-derived proteins by ectomycorrhizal fungi.

A cDNA presumably encoding AmProt1 was identified in an EST project (Nehls et al. 2001b). *AmProt1* was not only regulated by the external pH, but also by carbon as well as nitrogen availability. Nitrogen starvation alone increased *AmProt1* expression by a factor of 3 to 4. However, the absence of a carbon source increased the transcript level of the gene by a factor of approximately 12, independent of the presence or absence of nitrogen. The expression of *AmProt1* reflects thus the nutritional status of fungal hyphae with respect to carbon (major regulatory effect) and nitrogen (minor regulatory effect).

11 Uptake of Amino Acids

Amino acids (as a result of protein degradation) are frequently found in forest soils and are thus of great importance for nitrogen nutrition. The ability to take up amino acids with high efficiency has been frequently shown for ectomycorrhizal fungi (Abuzinadah and Read 1988; Chalot et al. 1995, 1996; Wallenda and Read 1999).

Fungal amino acid importer genes have been isolated to date from *A. muscaria* (*AmAAP1*; Nehls et al. 1999b) and *H. cylindrosporum* (Wipf et al. 2002). As determined by heterologous expression in yeast, these genes encode high affinity H^+/amino acid symporter with a broad amino acid spectrum. AmAAP1 has a higher affinity to basic and aromatic amino acids compared to acidic or neutral amino acids. These differences in affinity might reflect the fact that basic amino acids are present in soil in significantly lower concentrations (8–30 µM) than neutral amino acids (70–80 µM; Scheller 1996).

In contrast to *AmProt1* (Nehls et al. 2001b, see above), carbon catabolite repression is not involved in regulation of *AmAAP1* expression (Nehls et al. 1999b). This is in agreement with results obtained for the ectomycorrhizal fungus *P. involutus* (Chalot et al. 1995).

Good nitrogen support of fungal hyphae by amino acids as well as ammonium (not imported by AmAAP1) resulted in a low, constitutive *AmAAP1* expression (Nehls et al. 1999b). In contrast, *AmAAP1* expression increased tenfold at low external nitrogen concentrations. It could thus be concluded that *AmAAP1* expression is regulated by the endogenous nitrogen status of fungal cells, and not by the nitrogen source.

As shown for a yeast mutant lacking arginine uptake activity, the reduced re-import capacity for this amino acid resulted in a net arginine loss of the cells (Grenson 1973). The strongly enhanced expression of *AmAAP1* under nitrogen starvation conditions (even in the absence of amino acids) could also indicate that AmAAP1, in addition to amino acid uptake for nitrogen nutrition, might be important in the reduction of amino acid loss by hyphal leakage.

12 Regulation of Fungal Nitrogen Export in Mycorrhizas by the Nitrogen-Status of Hyphae

The nitrogen-dependent expression profile of nitrogen importer genes of ectomycorrhizal fungi (*A. muscaria*: Nehls et. al. 1999; *H. cylindrosporum*: Javelle et al. 2001; Wipf et al. 2002) resembles that of ascomycetes (yeast: Ter Schure et. al. 1998; *Aspergillus*: Sophianopoulou and Diallinas 1995). Here, nitrogen importer gene expression is regulated at the transcriptional level by two mechanisms: nitrogen repression in the presence of a good nitrogen source (ammonium or glutamine) and the induction of genes necessary for the utilization of alternative nitrogen sources under nitrogen limitation (e.g., Tazebay 1997). Nitrogen-dependent gene repression is presumably regulated by the internal nitrogen status of cells, and not the external nitrogen availability. Either, the intracellular ammonium concentration (Ter Schure et. al. 2000) and/or the activity of the glutamine synthetase (Sophianopoulou and Diallinas 1995) are supposed to sense the endogenous nitrogen status.

In ectomycorrhizal fungi, nitrogen importer gene expression is presumably also regulated by the internal nitrogen status of the hyphae (Nehls et. al. 1999; Javelle et al. 2001; Wipf et al. 2002). This could indicate how nitrogen uptake by soil-growing hyphae and nitrogen export by hyphae of the Hartig net might be managed (Fig. 3). Since the nitrogen content of forest soil is quite low and part of the nitrogen is transported to other parts of the growing fungal colony (e.g., mycorrhizas), soil-growing hyphae are presumably nitrogen-limited, resulting in a low endogenous nitrogen status and a strong expression of nitrogen importer genes. On the other hand, mycorrhizas are well supplied with nitrogen by soil-growing hyphae, thus revealing a high nitrogen status and a strongly reduced nitrogen importer gene expression. This nitrogen-

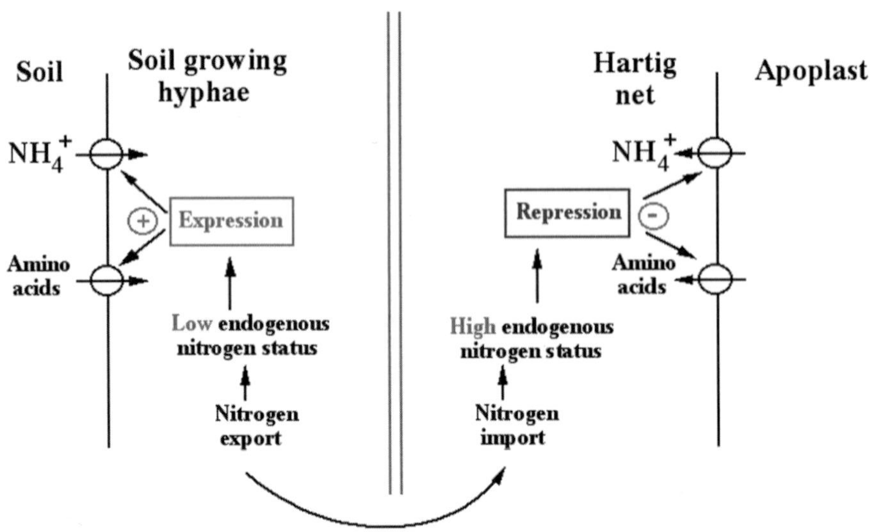

Fig. 3. Regulation of fungal nitrogen uptake from soil and nitrogen excretion at the plant/fungus interface: a model. Nitrogen export to other parts of the fungal colony together with a low nitrogen content in soil results in a low endogenous nitrogen state in soil growing hyphae. In consequence, nitrogen importer genes are highly expressed and nitrogen uptake capacity is high. Nitrogen import from soil growing hyphae causes a high endogenous nitrogen status in hyphae of the Hartig net. This results in a repression of nitrogen importer gene expression and, together with posttranslational inactivation processes, in a low nitrogen uptake capacity. Together with export mechanisms, this leads to a net export of nitrogen at the plant/fungus interface

dependent repression of amino acid transporter gene expression (indicated by *AmAAP1*), together with posttranslational events (e.g., increased degradation of plasma membrane transport proteins) that are described for yeasts (Springael and Andre 1998), could thus result in a highly reduced fungal capacity for re-uptake of amino acids at the plant/fungus interface. In combination with efflux mechanisms (e.g., nitrogen leakage), this would thus result in a net export of nitrogen.

13 Carbohydrate and Nitrogen-Dependent Regulation of Fungal Gene Expression

Carbohydrates as well as nitrogen are essential components of biological molecules (e.g., amino acids or nucleotides), and obviously have a great impact on fungal gene expression (e.g., Gonzales et al. 1997).

With regard to carbon and nitrogen nutrition, four different patterns of regulation have been observed in *A. muscaria*. The amino acid importer gene *AmAAP1* is only regulated by nitrogen nutrition, while the hexose transporters *AmMst1* and *AmMst2* (Nehls et al. 1998) are only regulated by carbohydrate nutrition. On the other hand, *AmProt1* (protease; Nehls et al. 2001b) and *AmTPS1* (trehalose-6-phosphate synthase) are regulated by both nitrogen as well as carbon nutrition. Nevertheless, the impact of carbon and nitrogen nutrition differs significantly for both genes. While *AmProt1* is mainly regulated by nitrogen, *AmTPS1* is mainly regulated by carbon availability. Comparable gene expression patterns have been described for fungi (Gonzales et al. 1997) as well as plants (Coruzzi and Zhou 2001), revealing a universal and phylogenetically old regulation strategy.

14 Conclusions

Since large EST projects of ectomycorrhizal model systems are currently under progress (Tagu and Martin 1995; Johansson et al. 2000; Voiblet et al. 2001; Wipf et al. 2003), macro- and micro-array hybridization will enable an overview of the general impact of carbon and nitrogen nutrition on gene expression for different ectomycorrhizal fungi.

Present data suggest that carbon- and nitrogen-dependent gene repression in ectomycorrhizal fungi is presumably similar to that of saprophytic ascomycetes (yeast, *Neurospora*). Ascomycotic model organisms could thus help to develop working models for ectomycorrhizal function (e.g., nitrogen uptake from soil and release at the plant/fungus interface; see Fig. 3) that could be investigated in turn in an ectomycorrhizal model system. In addition, differences, e.g., in carbon-dependent gene regulation for an ectomycorrhizal fungus (*A. muscaria*) and saprophytic ascomycetes (yeast, *Neurospora*) have been described. They might thus reveal adaptation processes that are necessary for ectomycorrhizal function.

Acknowledgements I am indebted to Magret Ecke and Andrea Bock for excellent technical assistance and to Dr. Mika Tarkka and Dr. Rüdiger Hampp for critical reading of the manuscript. This work was supported by the Deutsche Forschungsgemeinschaft (DFG-Schwerpunkt Mykorrhiza).

References and Selected Reading

Abuzinadah RA, Read DJ (1986) The role of proteins in the nitrogen nutrition of ectomycorrhizal plants. I. Utilization of peptides and proteins by ectomycorrhizal fungi. New Phytol 103:481–493

Abuzinadah RA, Read DJ (1988) Amino acids as nitrogen sources for ectomycorrhizal fungi: utilisation of individual amino acids. Transact Br Mycol Soc 91:473–479

Bajwa R, Abuarghub S, Read DJ (1985) The biology of mycorrhiza in the Ericaceae. X. The utilization of proteins and the production of proteolytic enzymes by the mycorrhizal endophyte and by mycorrhizal plants. New Phytol 101:469–486

Bending GD, Read DJ (1996) Nitrogen mobilization from protein-polyphenol complex by ericoid and ectomycorrhizal fungi. Soil Biol BioChem 28:1603–1612

Berg MP, Kniese JP, Verhoef HA (1998) Dynamics and stratification of bacteria and fungi in the organic layers of a Scots pine forest soil. Biol Fertil Soils 26:313–322

Berredjem A, Garnier A, Putra DP, Botton B (1998) Effect of nitrogen and carbon sources on growth and activities of NAD and NADP dependent isocitrate dehydrogenases of *Laccaria bicolor*. Mycol Res 4:427–434

Cairney JWG, Burke RM (1996) Physiological heterogeneity within fungal mycelia: an important concept for a functional understanding of the ectomycorrhizal symbiosis. New Phytol 134:685–695

Cao W, Crawford DL (1993) Carbon nutrition and hydrolytic and cellulolytic activities in the ectomycorrhizal fungus *Pisolithus tinctorius*. Can J Microbiol 39:529–535

Celenza JL, Marshall-Carlson L, Carlson M (1988) The yeast SNF3 gene encodes a glucose transporter homologous to the mammalian protein. Proc Natl Acad Sci USA 85:2130–2134

Chakravarty P, Unestam T (1987) Differential influence of ectomycorrhizae on plant growth and disease resistance of *Pinus sylvestris* seedlings. J Phytopathol 120:104–120

Chalot M, Brun A (1998) Physiology of organic nitrogen acquisition by ectomycorrhizal fungi and ectomycorrhizas. FEMS Microbiol Rev 22:21–44

Chalot M, Kytöviita MM, Brun A, Finlay RD, Söderström B (1995) Factors affecting amino acid uptake by the ectomycorrhizal fungus *Paxillus involutus*. Mycol Res 99:1131–1138

Chalot M, Brun A, Botton B, Soderstrom B (1996) Kinetics, energetics and specificity of a general amino acid transporter from the ectomycorrhizal fungus *Paxillus involutus*. Microbiol 142:1749–1756

Chen XY, Hampp R (1993) Sugar uptake by protoplasts of the ectomycorrhizal fungus, *Amanita muscaria* (L. ex fr.) Hooker. New Phytol 125:601–608

Coruzzi GM, Zhou L (2001) Carbon and nitrogen sensing and signaling in plants: emerging 'matrix effects'. Curr Opin Plant Biol 4:247–253

d'Enfert C (1997) Fungal spore germination: insights from the molecular genetics of *Aspergillus nidulans* and *Neurospora crassa*. Fungal Genet Biol 21:163–172

El-Badaoui K, Botton B (1989) Production and characterization of exocellular proteases in ectomycorrhizal fungi. Annales des Sciences Foréstières 46:728–730

Elbein A (1974) The metabolism of a,a-trehalose. Adv Carb Chem BioChem 30:227–256

Felenbok B, Kelley JM (1996) Regulation of carbon metabolism in mycelial fungi. In: Brambl R, Marzluf GA (eds) The Mycota III. Biochemistry and molecular biology. Springer, Berlin Heidelberg New York, pp 369–380

Finlay R, Ek H, Odham G, Söderström B (1988) Mycelial uptake, translocation and assimilation of nitrogen from ^{15}N labelled ammonium by *Pinus sylvestris* plants infected with four different ectomycorrhizal fungi. New Phytol 110:59–66

Finlay RD, Frostegärd Ä, Sonnerfeldt A-M (1992) Utilisation of organic and inorganic nitrogen sources by ectomycorrhizal fungi in pure culture and in symbiosis with *Pinus contorta* Dougl. ex Loud. New Phytol 120:105–111

Fletcher M (1996) Bacterial adhesion: Molecular and ecological diversity. Wiley-Liss, New York

France RC, Reid CPP (1983) Interactions of nitrogen and carbon in the physiology of ectomycorrhizas. Can J Bot 61:964–984

France RC, Reid CPP (1984) Pure culture growth of ectomycorrhizal fungi on inorganic nitrogen sources. Microbiol Ecol 10:187–195

Garbaye J (1991) Biological interactions in the mycorrhizosphere. Experientia 47: 370–375

Genet P, Prevost A, Pargney JC (2000) Seasonal variations of symbiotic ultrastructure and relationships of two natural ectomycorrhizae of beech (*Fagus sylvatica*/*Lactarius blennius* var. *viridis* and *Fagus sylvatica*/*Lactarius subdulcis*). Trees 14:465–474

Gessler A, Schneider S, von Sengbusch D, Weber P, Hanemann U, Huber C, Rothe A, Kreutzer K, Rennenberg H (1998) Field and laboratory experiments on net uptake of nitrate and ammonium by the roots of spruce (*Picea abies*) and beech (*Fagus sylvatica*) trees. New Phytol 138:275–285

Gonzalez R, Gavrias V, Gomez D, Scazzocchio C, Cubero B (1997) The integration of nitrogen and carbon catabolite repression in *Aspergillus nidulans* requires the GATA factor AreA and an additional positive-acting element, ADA. EMBO J 16:2937–2944

Grenson M (1973) Specificity and regulation of the uptake and retention of amino acids and pyrimidines in yeast. In: Vanek Z, Hostalek Z, Culdin J (eds) Genetics of industrial microorganisms. Academica, Prague, pp 179–193

Hampp R, Schaeffer C (1999) Mycorrhiza-carbohydrate and energy metabolism. In: Varma A, Hock B (eds) Mycorrhiza: structure, function, molecular biology and biotechnology. Springer, Berlin Heidelberg New York, pp 273–303

Haselwandter K, Bobleter O, Read DJ (1990) Degradation of carbon-14 labelled lignin and dehydropolymer of coniferyl alcohol by ericoid and ectomycorrhizal fungi. Arch Microbiol 153:352–354

Hoffmann E, Wallenda T, Schaeffer C, Hampp R (1997) Cyclic AMP, a possible regulator of glycolysis in the ectomycorrhizal fungus *Amanita muscaria*. New Phytol 137:351–356

Javelle A, Chalot M, Soderstrom B, Botton B (1999) Ammonium and methylamine transport by the ectomycorrhizal fungus *Paxillus involutus* and ectomycorrhizas. FEMS Microbiol Ecol 30:355–366

Javelle A, Rodriguez-Pastrana BR, Jacob C, Botton B, Brun A, Andre B, Marini AM, Chalot M (2001) Molecular characterization of two ammonium transporters from the ectomycorrhizal fungus *Hebeloma cylindrosporum*. FEBS Lett 505:393–398

Jennings DJ (1995) The physiology of fungal nutrition. Cambridge University Press, Cambridge

Johansson A, Le Quéré A, Ahrén D, Lundeberg J, Erlandsson R, Uhlèn M, Söderström B, Tulind A (2000) Transcript profiling during ectomycorrhizal development. In: Transcript profiling during ectomycorrhizal development. Philipps University of Marburg, Marburg, pp 95

Johnston M (1999) Feasting, fasting and fermenting: glucose sensing in yeast and other cells. TIGS 15:29–33

Jordy MN, AzemarLorentz S, Brun A, Botton B, Pargney JC (1998) Cytolocalization of glycogen, starch, and other insoluble polysaccharides during ontogeny of *Paxillus involutus–Betula pendula* ectomycorrhizas. New Phytol 140:331–341

Keeney DR (1980) Prediction of soil nitrogen availability in forest ecosystems: a literature review. For Sci 26:159–171

Kowallik W, Thiemann M, Huang Y, Mutumba G, Beermann L, Broer D, Grotjohann N (1998) Complete sequence of glycolytic enzymes in the mycorrhizal basidiomycete, *Suillus bovinus*. Z Naturforsch 53:818–827

Leake JR, Read DJ (1990) Proteinase activity in mycorrhizal fungi. I. The effect of extracellular pH on the production and activity of proteinase by ericoid endophytes from soil of contrasted pH. New Phytol 115:243–250

Leake JR, Donnelly DP, Saunders EM, Boddy L, Read DJ (2001) Rates and quantities of carbon flux to ectomycorrhizal mycelium following 14C pulse labeling of *Pinus sylvestris* seedlings: effects of litter patches and interaction with a wood-decomposer fungus. Tree Physiol 21:71–82

Lesage P, Yang X, Carlson M (1996) Yeast SNF1 protein kinase interacts with SIP4, a C_6 zink cluster transcriptional activator: a new role for SNF1 in the glucose response. Mol Cell Biol 16:1921–1928

Lewis DH, Harley JL (1965) Carbohydrate physiology of mycorrhizal roots of beech. I. Identity of endogenous sugars and utilization of exogenous sugars. New Phytol 64:224–237

Littke WR, Bledsoe CS, Edmonds RL (1984) Nitrogen uptake and growth in vitro by *Hebeloma crustuliniforme* and other Pacific Northwest mycorrhizal fungi. Can J Bot 62:647–652

Madi L, McBridge SA, Bailey LA, Ebbole DJ (1997) *Rco-3*, a gene involved in glucose transport and conidiation in *Neurospora crassa*. Genet 146:499–508

Marschner H (1995) Mineral nutrition of plants. 2nd edn. Academic Press, London

Martin F, Canet D, Marchal JP (1985) ^{13}C nuclear magnetic resonance study of mannitol cycle and trehalose synthesis during glucose utilization by the ectomycorrhizal ascomycete *Cenococcum geophilum*. Plant Physiol 77:499–502

Martin F, Ramstedt M, Söderhall K (1987) Carbon and nitrogen metabolism in ectomycorrhizal fungi and ectomycorrhizas. Biochimie 69:569–581

Martin F, Ramstedt M, Söderhall K, Canet D (1988) Carbohydrate and amino acid metabolism in the ectomycorrhizal ascomycete *Sphaerosporella brunnea* during glucose utilization a ^{13}C NMR study. Plant Physiol 86:935–940

Martin F, Chalot M, Brun A, Lorillou S, Botton B, Dell B (1992) Spatial distribution of nitrogen assimilation pathways in ectomycorrhizas. In: Read DJ, Lewis DH, Fitter A, Alexander I (eds) Mycorrhizas in ecosystems. CAB International, Wallingford, pp 311–315

Martin F, Boiffin VV, Pfeffer PE (1998) Carbohydrate and amino acid metabolism in the *Eucalyptus globulus–Pisolithus tinctorius* ectomycorrhiza during glucose utilization. Plant Physiol 118:627–635

Marx DH (1969) The influence of ectotrophic ectomycorrhizal fungi on the resistance of pine roots to pathogenic infections. I. Antagonism of mycorrhizal fungi to pathogenic fungi and soil bacteria. Phytopathology 59:153–163

Melin E, Nilsson H (1952) Transport of labelled nitrogen from an ammonium source to pine seedlings through mycorrhizal mycelium. Svensk Bot Tidskr 46:281–285

Mellor R (1992) Is trehalose a symbiotic determinant in symbioses between higher plants and microorganisms? Symbiosis 12:113–129

Montanini B, Moretto N, Soragni E, Percudani R, Ottonello S (2002) A high-affinity ammonium transporter from the mycorrhizal ascomycete *Tuber borchii*. Fungal Genet Biol 36:22–34

Müller J, Xie ZP, Staehelin C, Mellor RB, Boller T, Wiemken A (1994) Trehalose and trehalase in root nodules from various legumes. Physiolog Plant 90:86–92

Nasholm T, Persson J (2001) Plant acquisition of organic nitrogen in boreal forests. Physiol Plant 111:419–426

Nehls U, Wiese J, Guttenberger M, Hampp R (1998) Carbon allocation in ectomycorrhizas: identification and expression analysis of an *Amanita muscaria* monosaccharide transporter. Mol Plant Microbiol Interact 11:167–176

Nehls U, Ecke M, Hampp R (1999a) Sugar- and nitrogen-dependent regulation of an *Amanita muscaria* phenylalanine ammonium lyase gene. J Bacteriol 181:1931–1933

Nehls U, Kleber R, Wiese J, Hampp R (1999b) Isolation and characterization of a general amino acid permease from the ectomycorrhizal fungus *Amanita muscaria*. New Phytol 144:343–349

Nehls U, Bock A, Ecke M, Hampp R (2001a) Differential expression of hexose-regulated fungal genes within *Amanita muscaria*/*Populus tremula* x *tremuloides* ectomycorrhizas. New Phytol 150:583–589

Nehls U, Bock A, Einig W, Hampp R (2001b) Excretion of two proteases by the ectomycorrhizal fungus *Amanita muscaria*. Plant Cell Environ 24:741–747

Nehls U, Mikolajewski S, Magel E, Hampp R (2001 c) The role of carbohydrates in ectomycorrhizal functioning: gene expression and metabolic control. New Phytol 150:533–541

Norkrans B (1950) Studies in growth and cellolytic enzymes of *Tricholoma*. Symb Bot Upsal 11:1–126

Özcan S, Johnston M (1999) Function and regulation of yeast hexose transporters. Microbiol Mol Biol Rev 63:554–569

Özcan S, Dover J, Rosenwald AG, Woelfl S, Johnston M (1996) Two glucose transporters in *Saccharomyces cerevisiae* are glucose sensors that generate a signal for induction of gene expression. Proc Natl Acad Sci USA 93:12428–12432

Palmer JG, Hacskaylo E (1970) Ectomycorrhizal fungi in pure culture. I. Growth on single carbon sources. Physiol Plant 23:1187–1197

Plassard C, Scheromm P, Llamas H (1986) Nitrate assimilation by maritime pine and ectomycorrhizal fungi in pure culture. In: Gianinazzi-Pearson V, Gianinazzi S (eds) Physiological and genetical aspects of mycorrhizae. INRA, Paris, pp 383–388

RadisBaptista G, Valdivia DNU, AbrahaoNeto J (1998) Fructose 2,6-bisphosphate biosynthesis and regulation of carbohydrate metabolism in *Aspergillus oryzae*. Can J Microbiol 44:6–11

Read DJ (1991) Mycorrhizas in ecosystems. Experientia 47:376–391

Ronne H (1995) Glucose repression in fungi. Trends Genet 11:12–17

Salzer P, Hager A (1991) Sucrose utilization of the ectomycorrhizal fungi *Amanita muscaria* and *Hebeloma crustuliniforme* depends on the cell wall-bound invertase activity of their host *Picea abies*. Bot Acta 104:439–445

Sarjala T (1990) Effect of nitrate and ammonium concentration on nitrate reductase activity in five species of mycorrhizal fungi. Physiolog Plant 79:65–70

Schaeffer C, Wallenda T, Guttenberger M, Hampp R (1995) Acid invertase in mycorrhizal and non-mycorrhizal roots of Norway spruce (*Picea abies* [L.] Karst.) seedlings. New Phytol 129:417–424

Schaeffer C, Johann P, Nehls U, Hampp R (1996) Evidence for an up-regulation of the host and a down-regulation of the fungal phosphofructokinase activity in ectomycorrhizas of *Norway spruce* and fly agaric. New Phytol 134:697–702

Scheller E (1996) Aminosäuregehalte von Ap- und Ah-Horizonten verschiedener Böden und deren Huminsäuren- und Fulvosäuren-Fraktion. Mitt Dtsch Bodenkd Ges 81:201–204

Scheromm PS, Plassard C, Salsac L (1990) Regulation of nitrate reductase in the ectomycorrhizal basidiomycete, *Hebeloma cylindrosporum* Romagn., cultured on nitrate or ammonium. New Phytol 114:441–448

Smith SE, Read DJ (1997) Mycorrhizal symbiosis. Academic Press, London

Sophianopoulou V, Diallinas G (1995) Amino acid transporters of lower eukaryotes: regulation, structure and topogenesis. FEMS Microbiol Rev 16:53–75

Spägele S (1992) Charakterisierung der intra- und extrazellulären Proteasenaktivitäten des Fliegenpilzes (*Amanita muscaria* [L. ex Fr.] Hooker). Eberhard-Karls-Universität, Tübingen

Springael JY, Andre B (1998) Nitrogen-regulated ubiquitination of the Gapl permease of *Saccharomyces cerevisiae*. Mol Biol Cell 9:1253–1263

Stülten C, Kong FX, Hampp R (1995) Isolation and regeneration of protoplasts from the ectomycorrhizal ascomycete *Cenococcum geophilum* Fr. Mycorrhiza 5:259–266

Taber WA, Taber RA (1987) Carbon nutrition and respiration of *Pisolithus tinctorius*. Transact Brit Mycol Soc 89:13–26

Tagu D, Martin F (1995) Expressed sequence tags of randomly selected cDNA clones from *Eucalyptus globulus–Pisolithus tinctorius* ectomycorrhiza. Mol Plant Microbiol Interact 8:781–783

Tagu D, Lapeyrie F, Ditengou F, Lagrangem H, Laurent P, Missoum N, Nehls U, Martin F (2000) Molecular aspects of ectomycorrhiza development. In: Poldila G, Douds Jr DD (eds) Current advances in mycorrhizal research. Am Phytopathol Soc, pp 69–90

Tazebay UH, Sophianopoulou V, Scazzocchio C, Diallinas G (1997) The gene encoding the major proline transporter of *Aspergillus nidulans* is upregulated during conidiospore germination and in response to proline induction and amino acid starvation. Mol Microbiol 24:105–117

Ter Schure EG, Sillje HHW, Vermeulen EE, Kalhorn JW, Verkleij AJ, Boonstra J, Verrips CT (1998) Repression of nitrogen catabolic genes by ammonia and glutamine in nitrogen-limited continuous cultures of *Saccharomyces cerevisiae*. Microbiol Reading 144:1451–1462

Ter Schure EG, van Riel NA, Verrips CT (2000) The role of ammonia metabolism in nitrogen catabolite repression in *Saccharomyces cerevisiae*. FEMS Microbiol Rev 24:67–83

Thevelein JM (1991) Fermentable sugars and intracellular acidification as specific activators of the RAS-adenylate cyclase signalling pathway in yeast: the relationship to nutrient-induced cell cycle control. Mol Microbiol 5:1302–1307

Timonen S, Sen R (1998) Heterogeneity of fungal and plant enzyme expression in intact Scots pine-*Suillus bovinus* and -*Paxillus involutus* mycorrhizospheres developed in natural forest humus. New Phytol 138:355–366

Trojanowski J, Haider K, Huttermann A (1984) Decomposition of ^{14}C-labelled lignin, holocellulose and lignocellulose by mycorrhizal fungi. Arch Microbiol 139:202–206

Voiblet C, Duplessis S, Encelot N, Martin F (2001) Identification of symbiosis-regulated genes in *Eucalyptus globulus–Pisolithus tinctorius* ectomycorrhiza by differential hybridization of arrayed cDNAs. Plant J 25:181–191

Wagner F, Gay G, Debaud JC (1989) Genetic variation of nitrate reductase activity in monokaryotic and dikaryotic populations of the ectomycorrhizal fungus, *Hebeloma cylindrosporum* Romagnesi. New Phytol 113:259–264

Wallenda T (1996) Untersuchungen zur Physiologie der Pilzpartner von Ektomykorrhizen der Fichte (*Picea abies* [L.] Karst.). Eberhard-Karls-Universität, Tübingen

Wallenda T, Read DJ (1999) Kinetics of amino acid uptake by ectomycorrhizal roots. Plant Cell Environ 22:179–187

Wiese J, Kleber R, Hampp R, Nehls U (2000) Functional characterization of the *Amanita muscaria* monosaccharide transporter AmMst1. Plant Biol 2:1–5

Wipf D, Benjdia M, Tegeder M, Frommer WB (2002) Characterization of a general amino acid permease from *Hebeloma cylindrosporum*. FEBS Lett 528:119–124

Wipf D, Benjdia M, Rikirsch E, Zimmermann S, Tegeder M, Frommer WB (2003) An expression cDNA library for suppression cloning in yeast mutants, complementation of a yeast his4 mutant, and EST analysis from the symbiotic basidiomycete Hebeloma cylindrosporum. Genome 46(2):177–181

Wisser G (2000) Isolation und Charakterisierung von Trehalasen aus *Amanita muscaria* [L. ex Fr.] Hooker, einem Ektomykorrhizapilz. PhD-thesis, Eberhard-Karls-Universitaet Tuebingen, Germany

Wisser G, Guttenberger M, Hampp R, Nehls U (2000) Identification and characterization of an extracellular acid trehalase from the ectomycorrhizal fungus *Amanita muscaria*. New Phytol 146:169–175

Zhu H (1990) Purification and characterization of an extracellular acid proteinase from the ectomycorrhizal fungus *Hebeloma crustuliniforme*. Appl Environ Microbiol 56:837–843

Zhu H, Dancik BP, Higginbotham KO (1994) Regulation of extracellular proteinase production in an ectomycorrhizal fungus *Hebeloma crustuliniforme*. Mycologia 86:227–234

22 Nitrogen Transport and Metabolism in Mycorrhizal Fungi and Mycorrhizas

Arnaud Javelle, Michel Chalot, Annick Brun
and Bernard Botton

1 Introduction

1.1 Ecological Significance of Ectomycorrhizas

Unlike most other organisms, plants and fungi are restricted to their habitats, creating potential problems when nutritional conditions become limited. To cope with nutrient deficiencies, they have developed a variety of adaptations that enable them to respond to their internal nutritional status as well as to the external availability of nutrients. A strategy for plants is mycorrhizal association, in which expanding mycorrhizal mycelia that grow outward from the mantle into the surrounding soil is a very efficient nitrogen scavenger owing to (1) its capacity to explore a larger soil volume than roots alone (Smith and Read 1997), (2) its ability to provide access to nitrogenous reserves contained in organic horizons (Chalot and Brun 1998) and (3) its greater capacity for uptake of nitrogenous compounds (Javelle et al. 1999; Wallanda and Read 1999).

This interconnected network of hyphae (or specialized aggregates, i.e., rhizomorphs) forms a supracellular compartment for the transport of nutrients from sites of nutrient capture to sites of nutrient utilization and transfer. It has been estimated that the external mycelium makes, by far, the greatest contribution to the overall potential absorbing surface area of pine seedlings inoculated with *Pisolithus tinctorius* or *Cenococcum geophilum* (Rousseau et al. 1994). Fungal hyphae have a number of advantages compared with roots; (1) hyphae have a low ratio of biomass to absorptive surface area and can easily be regenerated (Harley 1989; Rousseau et al. 1994), (2) they have been shown to rapidly colonize nutrient-rich sites (Carleton and Read 1991; Bending and Read 1995) and (3) because of their small diameter, they can exploit small pores inaccessible to roots. The symbiotic association of higher plants with mycorrhizal fungi is considered to have been responsible for the colonization of land by plants (Taylor and Osborn 1995).

Plant Surface Microbiology
A. Varma, L. Abbott, D. Werner, R. Hampp (Eds.)

1.2 Nitrogen Uptake and Translocation by Ectomycorrhizas

Nitrogen plays a critical role in plant and microorganism biochemistry, being needed for the synthesis of many compounds, including amino acids, purines, pyrimidines, some carbohydrates and lipids, enzyme cofactors and proteins, all of which are essential for growth processes. Ammonium and nitrate are believed to be the principal sources of nitrogen in forest soil. When the two compounds are supplied to plants at similar concentrations, ammonium is generally taken up more rapidly than nitrate (Marschner et al. 1991; Kronzucker et al. 1996; Howitt and Udvardi 2000). Attention has also been paid to the utilization of organic nitrogen forms from more complex substrates (Smith and Read 1997; Perez-Moreno and Read 2000), and to the direct mobilization of nutrients from minerals (for a review, see Landeweert et al. 2001). The two processes involved in ammonium assimilation, namely transport and metabolism, have been studied in various ectomycorrhizal models. Increases in nitrogen content of ectomycorrhizal plants, often connected with a growth increase, are well documented (Smith and Read 1997). Studies have demonstrated that the ectomycorrhizal partner plays an integral role in ammonium metabolism in trees (Chalot et al. 1991; Botton and Chalot 1995; Plassard et al. 1997).

Nutrient uptake and transport by extraradical mycelium is suggested to be an important factor for improved nutrient acquisition. The contribution of extraradical mycelium to N nutrition of mycorrhizal Norway spruce was investigated. The addition of N to the hyphal compartment markedly increased dry weight, N concentration and N content in mycorrhizal plants. Calculating the uptake, based on the difference in input and output of nutrients in solution, confirmed a hyphal contribution of 73 % to total N uptake in *Picea abies* seedlings under nitrogen and phosphorus starvation (Brandes et al. 1998). In further studies, Jentschke et al. (2001) have demonstrated in *Picea abies/Paxillus involutus* ectomycorrhizas that hyphal N uptake (NH_4^++NO_3^-) contributed 17 % to total N uptake in mycorrhizal seedlings. Moreover, ammonium is the major source of mineral nitrogen in forest soils (Marschner and Dell 1994), and consequently, ammonium assimilation by extraradical mycelium plays a crucial role for nitrogen transfer in ectomycorrhizal symbiosis. Melin and Nilsson (1952) showed that the mycelia phase of *Suillus variegatus* was capable of absorption and translocation to *Pinus* mycorrhizal seedlings of nitrogen from a labelled ammonium source. Disrupting the external mycelium from ectomycorrhizas greatly decreased [^{15}N]ammonium uptake by birch seedlings (Javelle et al. 1999). Ammonium is incorporated into a range of amino acids and these accumulate in fungal mycelium at considerable distances from plant roots (Finlay et al. 1988). Therefore, external hyphae can be considered as the absorbing structure of ectomycorrhizal roots. These results confirmed the function of extraradical mycelium in translocating N from sources to roots and that it can, therefore, be considered as a nutrient channel (Smith and Read 1997).

2 Nitrate and Nitrite Transport

2.1 Uptake Kinetics

Nitrate uptake rates were estimated in a few ectomycorrhizal fungi and ectomycorrhizas. In the basidiomycete *Rhizopogon roseolus*, NO_3^- uptake measured after incubation of mycelia in 0.05 mM nitrate occurred at the same rate in the absence or presence of NO_3^- in the culture medium, suggesting that no inducible nitrate transporter exists in this species (Gobert and Plassard 2002). These results are in agreement with those of Jargeat (1999) who observed that the mRNA of a high-affinity transport system in the ectomycorrhizal basidiomycete *Hebeloma cylindrosporum*, was found in mycelia grown in N-free medium or in media containing low nitrate concentrations. Km estimates, around 12 µM in *Rhizopogon roseolus* (Gobert and Plassard 2002), and 67 µM in *Hebeloma cylindrosporum* (Plassard et al. 1994) are close to the Michaelis constants found in nonmycorrhizal fungi, with values of 23 µM in *Aspergillus nidulans* (Zhou et al. 2000) and 25 µM in *Neurospora crassa* (Blatt et al. 1997).

In nonmycorrhizal *Pinus pinaster* roots, rates of NO_3^- uptake were enhanced by exposure to external nitrate, as usually found in higher plant species. In the association *Pinus pinaster/Rhizopogon roseolus*, NO_3^- uptake was not modified by external nitrate, but was constantly higher than that measured in nonmycorrhizal roots (Gobert and Plassard 2002). According to these authors, the fungal uptake of nitrate may confer to the mycorrhiza a greater ability to use low and fluctuating concentrations of nitrate in the soil. However, in *Fagus-Laccaria* mycorrhizas, mycorrhization led to reduced rates of NO_3^- net uptake, this effect being caused by reduced influx, plus enhanced efflux of NO_3^- as compared with nonmycorrhizal beech roots (Kreuzwieser et al. 2000).

2.2 Characterization of Nitrate and Nitrite Transporters

Kinetically, two groups of nitrate transporters have been characterized: one with a high affinity, Km in the µM nitrate range, found in filamentous fungi, yeasts, algae and plants (Crawford and Glass 1998; Forde 2000), and one low affinity group, Km in the mM nitrate range, found mainly in plants, although there is indirect evidence of its presence in yeasts and algae (Machin et al. 2000; Navarro et al. 2000).

Aspergillus nidulans possesses two high-affinity nitrate transporters, encoded by the *nrtA* (formerly designated *crnA*) and the *nrtB* genes (Unkles et al. 1991; 2001). Whereas mutants expressing either gene grew normally on nitrate as sole nitrogen source, the double mutant was unable to grow even if the nitrate concentration was increased to 200 mM. This indicates that NRTA and NRTB are the only nitrate transporters in *Aspergillus nidulans*. Both genes were regulated identically under an extensive range of conditions; nevertheless, the transporters revealed different Km and V_{max} values for nitrate. Flux analysis of single gene mutants using $^{13}NO_3^-$ showed that Km values for the NRTA and NRTB proteins were about 100 and 10 µM, respectively, while V_{max} values were approximately 600 and 100 nmol/mg DW/h, respectively (Unkles et al. (2001). This kinetic differentiation may provide the physiological plasticity to acquire sufficient nitrate despite highly variable external concentrations.

In *Hansenula polymorpha*, the genomic DNA containing the nitrate reductase-(*YNR1*) and nitrite reductase-(*YNI1*) encoding genes, revealed an open reading frame of 1524 nucleotides (named *YNT1*, yeast nitrate transporter gene) encoding a putative protein of 508 amino acids with great similarity to the nitrate transporters from *Aspergillus nidulans* and *Chlamydomonas reinhardtii* (Perez et al. 1997). Disruption of the chromosomal *YNT1* copy resulted in an incapacity to grow in nitrate and a significant reduction in the rate of nitrate uptake. The disrupted strain was still sensitive to chlorate and, in the presence of 0.1 mM nitrate, the expression of *YNR1* and *YNI1*, as well as the activity of nitrate reductase and nitrite reductase, were significantly reduced compared to the wild type. Northern-blot analysis showed that *YNT1* was expressed when the yeast was grown in nitrate and nitrite, but not in ammonium solution (Perez et al. 1997).

In *Hansenula polymorpha*, the *YNT1* gene encodes a high affinity nitrate transporter (Km 2–3 µM) which constitutes quantitatively the main nitrate transporter activity in the fungus. The existence of a second nitrate transporter has been inferred from different experimental pieces of evidence, but the gene has not yet been identified (Machin et al. 2000). The protein Ynt1 also transports nitrite with high affinity and belongs to the proposed NNP (nitrate nitrite porter) family involved in nitrate and nitrite transport (Forde 2000). This family, in turn belongs to the major facilitator superfamily (MFS), constituted by transmembrane proteins in which 12 membrane spanning helices connect cytosolic N-terminal and C-terminal domains (Pao et al. 1998). However, in *Hansenula polymorpha*, it is not clear whether nitrite enters through a specific transport system, or if it shares a nitrate transport. Ynt1 presents similarity in sequence with the *Aspergillus nidulans* nitrate transporter NRTA (CRNA) and the high affinity nitrate transporters in plants (Siverio 2002).

In the field of endomycorrhizas, PCR amplifications using tomato DNA and degenerate oligonucleotide primers allowed the identification of a new putative nitrate transporter, named *NRT2* (Hildebrandt et al. 2002). Its sequence showed typical motifs of a high affinity nitrate transporter of the MFS. The formation of its mRNA was positively controlled by nitrate, and negatively by ammonia, but not by glutamine. In situ hybridization experiments showed that this transporter was mainly expressed in rhizodermal cells. In roots colonized by the arbuscular mycorrhizal fungus *Glomus intraradices*, transcript formation of *NRT2* extended to the inner cortical cells where the fungal structures, arbuscules and vesicles, were concentrated. Northern analyses indicated that the expression of the transporter was higher in mycorrhized tomato roots than in noncolonized controls. In addition, mycorrhization caused a significant expression of a nitrate reductase gene of *Glomus intraradices*. According to the authors mentioned above, the results sug-

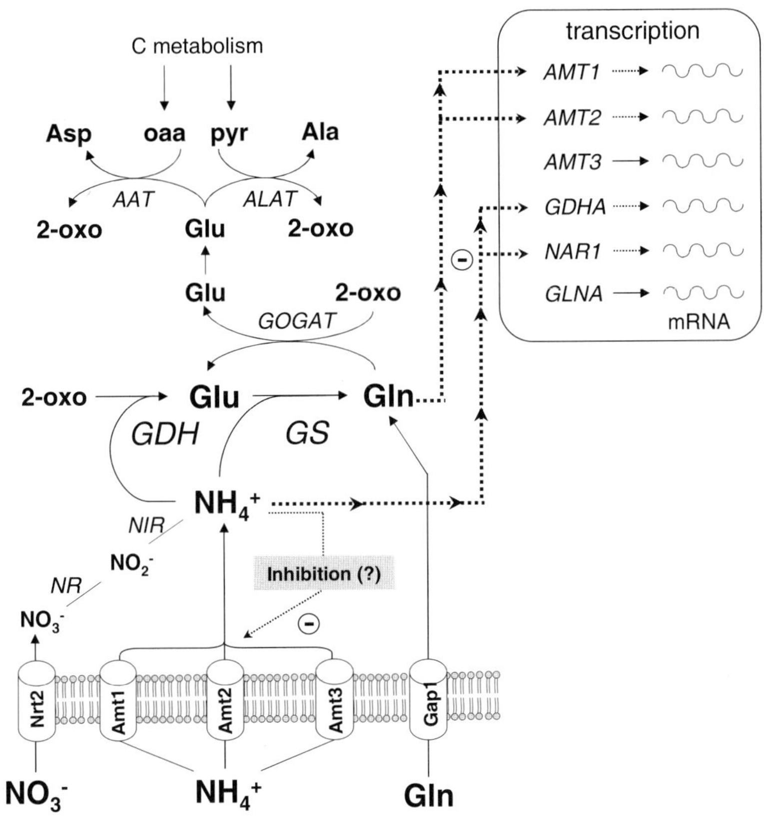

Fig. 1. A model describing the regulation of nitrogen transport and assimilation in *Hebeloma cylindrosporum*. This ectomycorrhizal fungus is able to use nitrate, ammonium and amino acids as nitrogen sources. Under low ammonium status, *AMT1*, *AMT2*, *AMT3*, *GDHA* and *GLNA* are transcribed, which results in elevated ammonium uptake and metabolism capacities. Under ammonium excess, *AMT1*, *AMT2* and *GDHA* are efficiently repressed, which results in reduced ammonium assimilatory capacities. Under these conditions, *AMT3* and *GLNA* would ensure the maintenance of a basal level of ammonium assimilation. *AMT1* and *AMT2* transcript levels are controlled through the effect of intracellular glutamine, whereas the *GDHA* and *NAR1* mRNA level is controlled by ammonium (*bold dotted lines*). Ammonium uptake activity may be controlled by intracellular NH_4^+ through a direct effect (*dotted lines*). *2-oxo* 2-oxoglutarate, *oaa* oxaloacetate, *pyr* pyruvate, *GOGAT* glutamate synthase, *Aat* aspartate aminotransferase, *Alat* alanine aminotransferase, *NR* nitrate reductase, *NIR* nitrite reductase, *Nrt2* nitrate transporter, *GAP1* general amino acid transporter

gest that mycorrhization positively affects nitrate uptake from soil and nitrate allocation to the plant partner, probably mediated preferentially by the transporter. In addition, part of the nitrate taken up is very likely reduced by the fungal partner itself and may then be transferred, when in excess, as glutamine to the plant's symbiotic partner.

Nitrate transporters have not yet been fully characterized in ectomycorrhizal fungi. A gene has been isolated in *Hebeloma cylindrosporum* by Jargeat et al. (2000; Fig. 1), but the molecular mechanism of its regulation is unknown. However, in this fungus, Jargeat et al. (2003) has shown more recently that the nitrate transporter polypeptide is characterized by 12 transmembrane domains and presents both a long putative intracellular loop and a short C-terminal tail, two structural features which distinguish fungal high-affinity transporters from their plant homologues. In addition, in *Hebeloma cylindrosporum*, transcription of the *nrt2* gene (as well as the gene encoding a nitrite reductase) was repressed by ammonium and stimulated, not only in the presence of nitrate, but also in the presence of organic nitrogen sources or under nitrogen deficiency (Jargeat et al. 2003).

3 Ammonium Transport

3.1 Physico-Chemical Properties of Ammonium: Active Uptake Versus Diffusion

Using [^{14}C]methylammonium as an analogue of ammonium, the kinetics and the energetics of NH_4^+ transport were studied in the ectomycorrhizal fungus *Paxillus involutus* (Javelle et al. 1999) and ammonium transporters were first cloned in *Hebeloma cylindrosporum* (*AMT2* and *AMT3*; Javelle et al. 2001) and *Tuber borchii* (*AMT1*; Montanini et al. 2002). Although the process of ammonium uptake is often considered as a rate-limiting step in its acquisition (Jongbloed et al. 1991; Javelle et al. 1999) it has received relatively little attention (Burgstaller 1997).

Ammoniac (NH_3) is a weak base (pK_a of 9.25), with a dipole moment of 1.47D. The neutral molecule, NH_3, dissolves much more rapidly in organic solvents than its ionic counterpart, NH_4^+. Consequently, the permeability of NH_3 across lipid bilayers is three orders of magnitude greater than that of NH_4^+. Whilst diffusion of NH_3 across the lipid portion of membranes is believed to be of biological significance, diffusion of NH_4^+ is not. Reported permeability values for ammoniac, ranging from 2.6 µmol/s (Ritchie and Gibson 1987) to 47 µmol/s (Yip and Kurtz 1995), were found in biomembranes. Therefore, previous investigations have supported the hypothesis that ammonia is transported as a small, uncharged and lipophilic compound across the plasma membrane, a process which does not require specific transporters. However, rates of diffusion do not seem to be sufficient to account for the requirements of plant growth (Burgstaller 1997). At a neutral pH typical of cell cytosol, approximately 99 % of ammonium is present as the cation NH_4^+. By definition, a decrease of one pH unit is accompanied by a tenfold increase of the ratio NH_4^+:NH_3. Therefore, in spite of the general acceptance that NH_3 can readily diffuse across natural membranes, it was postulated that ammonium uptake in cells could also be mediated by other mechanisms.

3.2 Physiology of Ammonium Transport in Ectomycorrhizas

The first evidence that a specific ammonium transport system acts in fungi came from the works of Hackette et al. (1970). They used the ammonium-analogue tracer [^{14}C]methylammonium and suggested that an ammonium transporter acts in the fungus *Penicillium chrysogenum*. The radioactive ammonium analogue [^{14}C]methylammonium has been widely used to assay uptake. Roon et al. (1975) measured an uptake in *Saccharomyces cerevisiae* which resulted in a 1000-fold accumulation. In a further study, Dubois and Grenson (1979) showed that the uptake of ammonium/methylammonium in *S. cerevisiae* is mediated by at least two functionally distinct systems, but this study was hampered by the lack of molecular characterization of the transport systems. The first ammonium transporter genes characterized were *MEP1* cloned in *S. cerevisiae* (Marini et al. 1994), and *AMT1* cloned in *Arabidopsis thaliana* (Ninnemann et al. 1994). They belong to a multigenic family, the so-called Mep/Amt family.

Ammonium mobilization by mycelium from soil sources is directly linked to hyphal uptake capacities. Using [^{14}C]methylamine, kinetics of ammonium/methylammonium transport in ectomycorrhizal fungi have been characterized (Jongbloed et al. 1991; Javelle et al. 1999). A saturable mediated uptake was obtained, which conformed to simple Michaelis-Menten kinetics, and was consistent with a carrier-mediated transport. Both pH dependence and inhibition by protonophores indicate that methylamine transport in *P. involutus* is dependent on the electrochemical H^+-gradient (Javelle et al. 1999). These results suggest that ammonium uptake is an active (energy-requiring) process. Comparing the ammonium uptake capacity of the two partners separately or in symbiosis, it was found that mycelia have much higher capacities for ammonium uptake than nonmycorrhizal roots and ectomycorrhizal fungi increase ammonium uptake capacities of their host roots (Plassard et al. 1997; Javelle et al. 1999).

Nitrogen starvation increased methylamine transport in *P. involutus* (Javelle et al. 1999) and similarly, N-starved plants usually showed a faster NH_4^+ net uptake than N-fed plants (Howitt and Udvardi 2000). However, these studies were hampered by the lack of molecular characterization of the transport systems involved and their regulation at the molecular level remains to be clarified.

3.3 Isolation of Ammonium Transporter Genes

Molecular studies of ammonium transporters in ectomycorrhizal fungi are still scarce and concern only the ectomycorrhizal fungus *Hebeloma cylindrosporum*. Three ammonium transporters, HcAmt1, HcAmt2 and HcAmt3 (Ammonium transporter) were cloned in *H. cylindrosporum*. Both Southern blot experiments and cDNA library screening indicate that *H. cylindrosporum* has only three ammonium transporters, like the yeast *S. cerevisiae* (Marini et al. 1997; Javelle et al. 2001; Javelle et al. 2003b). The hydropathy profiles of HcAmt1, HcAmt2 and HcAmt3 generated with the Kyte and Doolittle algorithm, consist of 11 hydrophobic domains of sufficient length to be considered as potential membrane-spanning domains.

The function of HcAmts in ammonium transport was further characterized by yeast mutant complementation, as previously described for ammonium transporters from plants and animals. *S. cerevisiae* possesses three ammonium transporters, namely Mep1, Mep2 and Mep3 (Methylammonium permease). The yeast strain 31019b, *mep1Δ mep2Δ mep3Δ*, was unable to grow on media containing less than 1 mM ammonium as sole nitrogen source (Marini et al. 1997). Functional expression of HcAmt1, HcAmt2 or HcAmt3 in this triple mutant resulted in complementation of growth defects in the presence of less than 1 mM ammonium as sole nitrogen source. Thus, *HcAMTs* cDNA encode functional NH_4^+ transporters. Kinetic parameters were determined using [^{14}C]methylammonium as a tracer in the transformed yeast strain 31019b. Previous works with mycorrhizal fungi reported Km values in the range 110–180 µM when using methylamine as substrate (Javelle et al. 1999). However, such data could be the result of multiple transporter expressions. In *H. cylindrosporum*, as well as in other organisms (Marini et al. 1997; Gazzarrini et al. 1999; Howitt and Udvardi 2000), multiple Amt transporters with complementary affinities probably allow the fungus to maintain a steady ammonium uptake over a wide range of concentrations. Indeed, in forest soils the quality and quantity of nitrogen sources can vary considerably.

3.4 Regulation of the Ammonium Transporters

Expression levels of the three ammonium transporter (*AMT1, AMT2, AMT3*) genes were studied by Northern blot analysis under different nitrogen conditions. *AMT1* and *AMT2* are high affinity transporters (for example, Km: 58 µM for methylammonium at pH 6.1 for *AMT2*), while *AMT3* is a low affinity transporter (Km: 260 µM for methylammonium at pH 6.1; Javelle et al. 2001). In response to exogenously supplied ammonium or Gln, *AMT1* and *AMT2* were down-regulated, while they were up-regulated upon nitrogen deprivation or in the presence of nitrate. This indicates that these genes are subjected to nitrogen repression in *H. cylindrosporum* (Fig. 2). *AMT3* was poorly regulated at this level.

Expression of *AMT1* only in ammonium-limiting conditions is consistent with a role for the high-affinity ammonium transporter in scavenging low concentrations of ammonium. The low-affinity ammonium transporter Amt3 would be required for growth in ammonium-sufficient conditions.

In order to identify the effector(s) for nitrogen regulation in *H. cylindrosporum*, the correlation coefficient for the relationship between *AMT1*, *AMT2*, *AMT3*, transcript levels and N-compound amounts were calculated. This transcriptional control is driven by intracellular Gln. Indeed, an intracel-

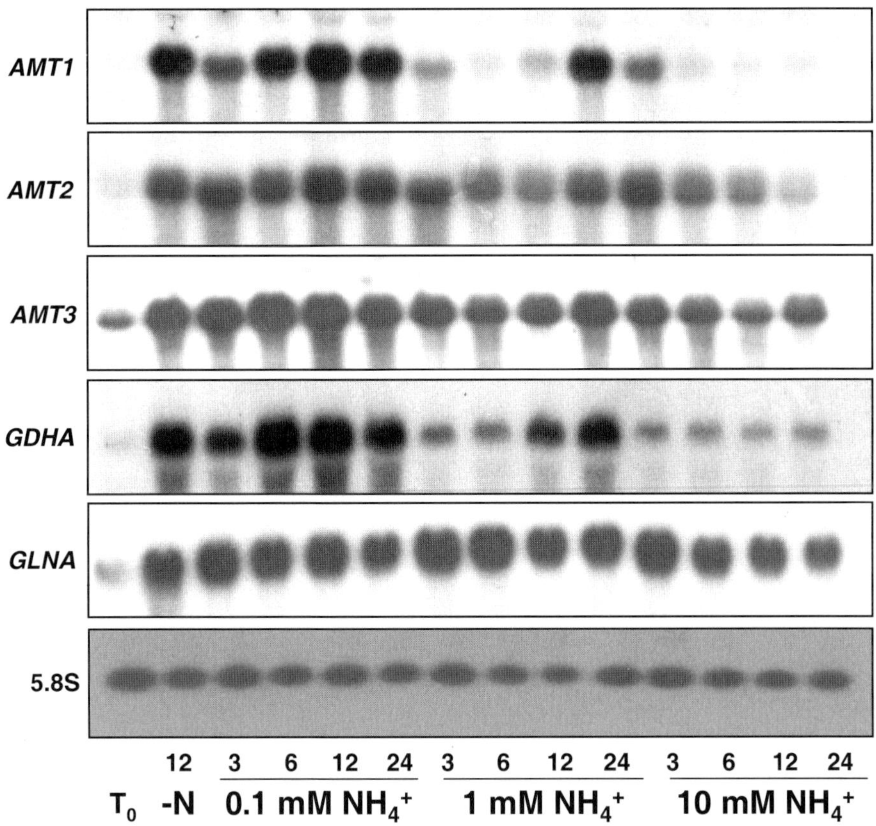

Fig. 2. AMT1, AMT2, AMT3, GDHA and GLNA mRNA levels in *Hebeloma cylindrosporum*. Fungal colonies were grown for 10 days on cellophane-covered agar medium containing 3.78 mM ammonium as sole nitrogen source (T_0) and transferred to a N-free liquid medium for 12 h (*–N*). Some colonies were further transferred to a 0.1, 1 or 10 mM ammonium-containing medium. Total RNA was extracted at 3, 6, 12 and 24 h from 100 mg of mycelium and 20 µg/lane were separated on 1.5% agarose-formaldehyde gel and hybridized to the [α-^{32}P]dCTP labelled cDNA probes or 5.8S rRNA probe as loading control

lular Gln amount higher than 2 nmol/mg DW seems to be sufficient to promote *AMT1* repression in *H. cylindrosporum* (Javelle et al. 2003b). Ammonium influx is inhibited by intracellular ammonium which agrees with other findings from *A. bisporus* (Kersten et al. 1999), and *A. thaliana* (Rawat et al. 1999), but mechanisms responsible for this regulation remain unclear.

3.5 Other Putative Functions of Ammonium Transporters

In addition to their role in ammonium uptake and retrieval, ammonium transporters may have a third putative role. A diploid wild-type strain of the yeast *S. cerevisiae* undergoes a dimorphic transition to filamentous growth in response to nitrogen starvation. Mep2 is one of three related ammonium per-

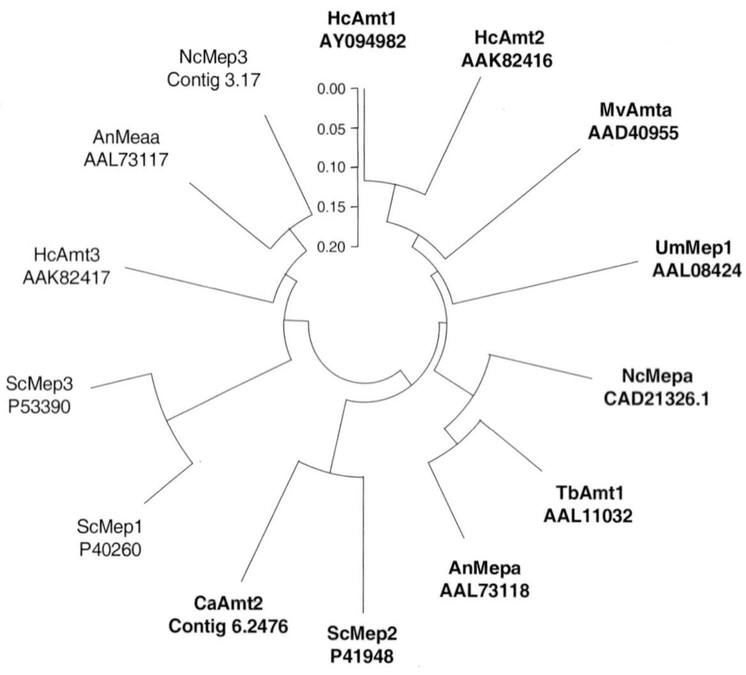

Fig. 3. Phylogenetic relationships among fungal Mep/Amt proteins. Complete amino acid sequences derived from full-length cDNA predicted using TMHMM algorithm were aligned with Clustalw and the tree was constructed by the neighbor-joining method using Mega 2.1. p-distances were estimated between all pairs of sequences using the complete deletion option. Gene names and GenBank accession numbers are indicated. Proteins in *bold* belong to the high affinity ammonium transporter and sensor family (TC 2A 49 3 2), according to the TC classification. Organisms are as follows. *An Aspergillus nidulans, Ca Candida albicans, Hc Hebeloma cylindrosporum, Mv Microbotryum violaceum, Nc Neurospora crassa, Sc Saccharomyces cerevisiae, Tb Tuber borchii, Um Ustilago maydis*

meases which plays a unique role as a nitrogen sensor in the transduction pathway of pseudohyphal differentiation in *S. cerevisiae* not shared with the related Mep1 and Mep3. Interestingly, in the ectomycorrhizal fungus *H. cylindrosporum*, two ammonium transporters (Amt1 and Amt2) are able to complement the pseudohyphal growth defect of a homozygotous *mep2Δ* yeast mutant, whereas the third ammonium transporter (Amt3) is unable to do so (Javelle et al. 2001, 2003b). According to the classification of the transport system available at http://www-biology.ucsd.edu/~msaier/transport/ (TC system), the HcAmts can be divided into two groups. HcAmt1, HcAmt2 and Mep2 belong to the high affinity ammonium transporter and sensor family (TC 2A 49 3 2), whereas HcAmt3 belongs to the low affinity ammonium transporter family (TC 2A 49 3 1; Fig. 3).

We have recently hypothesized that high affinity ammonium transporters from mycorrhizal fungi sense the environment and induce via signal transduction cascades a switch of the fungal growth mode observed during mycorrhiza formation. Upon entering the root depletion zone, mycorrhizal fungi may receive a signal through this sensing mechanism which induces hyphal proliferation around roots, corresponding to the primary events in ectomycorrhiza formation (Javelle et al. 2003a).

4 Amino Acid Transport

4.1 Utilization of Amino Acids by Ectomycorrhizal Partners

It has been well established that ectomycorrhizal fungi can use amino acids as nitrogen and carbon sources (Abuzinadah and Read 1988; Näsholm et al. 1998). Using ^{14}C-labelled compounds, Wallenda and Read (1999) determined the kinetics of uptake of amino acids by excised ectomycorrhizal roots from beech, spruce, and pine. All mycorrhizal types took up amino acids via high-affinity transport systems with Km values ranging from 19 to 233 mM. A comparative analysis for the uptake of amino acids and the ammonium analogue methylammonium showed that ectomycorrhizal roots have similar or even higher affinities for the amino acids, indicating that absorption of these N organic forms can contribute significantly to total N uptake by ectomycorrhizal plants.

Transport of amino acids was investigated in the mycorrhizal fungi *Paxillus involutus* (Chalot et al. 1996), and *Amanita muscaria* (Nehls et al. 1999), which demonstrated their ability to take up a variety of amino acids. In the latter fungus, the uptake characteristics of the encoded transporter protein, as analysed by heterologous expression in yeast, identified the protein as a high-affinity, general amino acid permease (Km: 22 µM for histidine and up to 100 µM for proline). The uptake of amino acids showed characteristic features of active transport.

In *Paxillus involutus*, the apparent Km derived from the Eadie-Hofstee plots ranged from 7 µM for alanine to 27 µM for glutamate. Maximal velocities, expressed as µmol (g dry weight)$^{-1}$ min^{-1}, were between 0.24 for alanine and 0.71 for glutamine. In this fungus, the uptake of amino acids markedly depended on the pH and was optimal at pH 3.9–4.3 for glutamate and glutamine, and at pH 3.9–5.0 for alanine and aspartate.

Both pH dependence and inhibition by protonophores, such as 2,4-dinitrophenol (DNP) and carbonyl cyanide *m*-chlorophenylhydrazone (CCCP), were consistent with a proton symport mechanism for amino acid uptake by *Paxillus involutus*. Competition studies indicated a broad substrate recognition by the uptake system, which resembles the general amino acid permease of yeast (Chalot et al. 1996, 2002).

The impact of birch mycorrhization with *Paxillus involutus* led to a profound alteration of the metabolic fate of exogenously supplied amino acids (Blaudez et al. 2001). Inoculation increased [^{14}C]glutamate and [^{14}C]malate uptake capacities by up to 8 and 17 times, respectively, especially in the early stages of mycorrhiza formation. In addition, it was demonstrated that Gln was the major ^{14}C-sink in mycorrhizal roots and in the free-living fungus. In contrast, citrulline and insoluble compounds were the major ^{14}C compounds in nonmycorrhizal roots (Blaudez et al. 2001).

In order to study how amino acid transport characteristics were affected by mycorrhization, Sokolovsky et al. (2002) used an electrophysiological approach in *Calluna vulgaris* associated or not with the ericoid fungus *Hymenoscyphus ericae*. Both the V_{max} and Km parameters of amino acid uptake were affected by fungal colonization in a manner consistent with an increased availability of amino acid to the plant. Indeed, the transport capacity for asparagine, histidine, ornithine and lysine, in particular, was increased after colonization. Interestingly, α-aminobutyric acid led to a large depolarization only in colonized cells. This implies that mycorrhization triggers a capacity to transport a broader range of substrates, including amino acids that are not metabolized.

4.2 Molecular Regulation of Amino Acid Transport

In *Amanita muscaria*, only a low, constitutive expression of the amino acid transporter was detected in the presence of amino acids and ammonium, which are both sources of N for the fungus (Nehls et al. 1999). By contrast, under N starvation, or in the presence of nitrate or phenylalanine, not utilized by the fungus as N sources, expression of the gene was considerably enhanced. Therefore, in *Amanita muscaria*, as in *S. cerevisiae* or *Aspergillus nidulans* (Sophianopoulou and Diallinas 1995), gene expression of amino acid transporters is regulated at the transcriptional level by N repression. In addition to amino acid uptake for nutrition, the enhanced expression of the gene under conditions of N starvation, suggests that the transporter can also be involved in the prevention of amino acid loss by hyphal leakage in the absence of a suitable N source (Nehls et al. 1999).

A gene named *HcBap1* has recently been isolated from *H. cylindrosporum* by functional complementation of a yeast strain deficient in amino acid transporters (Wipf et al. 2002).

5 Reduction of Nitrate to Nitrite and Ammonium

5.1 Reduction of Nitrate to Nitrite

Nitrate assimilation in fungi follows the same pathway as that described for yeasts and plants. After transport into the cells, nitrate is converted to ammonium by two successive reductions catalysed respectively by nitrate reductase and nitrite reductase. Although nitrate is one of the most abundant nitrogen sources in nature, numerous fungi more readily use ammonium, especially ectomycorrhizal fungi which live predominantly in forest soils where a high organic material content maintains an acidic pH. Under these circumstances, nitrification is inhibited and ammonium is usually the main form of mineral nitrogen (Vitousek and Matson 1985). However, it has been shown that ectomycorrhizal fungi are also able to utilize NO_3^- which, for a few species, is capable of promoting better growth than ammonium (Scheromm et al. 1990; Anderson et al. 1999).

The enzyme complex nitrate reductase which is a molybdoflavoprotein catalyzes the reduction of NO_3^- to NO_2^- by reduced pyridine nucleotides. The enzyme of higher plants has a high molecular weight, varying from 220 to 600 kDa, depending on the organisms in which it occurs (Notton and Hewitt 1978). In fungi, nitrate reductase has been extensively studied in *Neurospora crassa* where it is found as a 228-kDa homodimer (Garrett and Nason 1969) and in *Aspergillus nidulans* where the enzyme has a molecular mass of 180 kDa (Minagawa and Yoshimoto 1982). In plants and fungi, the polypeptide is located in the cytosolic soluble fraction, but is weakly bound to the plasmalemma and tonoplast in *Neurospora crassa* (Roldan et al. 1982).

Nitrate reductase generally appears to be unstable and, due to the difficulties experienced in purifying the enzyme, information on its properties in mycorrhizal fungi is very scarce. However, nitrate reduction by partially purified enzyme preparations has been investigated in *Hebeloma cylindrosporum* by Plassard et al. (1984a). The Michaelis constants for nitrate, NADPH and FAD were found to be 152, 0.185, and 22.7 µM, respectively. In *Pisolithus tinctorius*, nitrate reductase exhibited less affinity for nitrate (Km: 328 µM) and for NADPH (Km: 49.6µM; Aouadj et al. 2000), but the enzyme was similar to those found in nonmycorrhizal fungi. Such values are in the same range as those found in higher plant tissues and suggest that ectomycorrhizal fungi have capabilities of reducing NO_3^- similar to those of most higher plants. However, nitrate reductase activity varies greatly between mycorrhizal species and isolates. For example, in *Rhizopogon vulgaris*, nitrate reductase was 32-fold higher in the S-251 isolate than in the S-219 isolate (Ho and Trappe 1987).

In the ectomycorrhizal basidiomycete *Suillus bovinus*, nitrate reductase proved to be substrate-induced and activity could only be measured after exposure of the mycelia to exogenous nitrate (Grotjohann et al. 2000). Similar results were found in *Scleroderma verrucosum* (Prima Putra et al. 1999), and *Pisolithus tinctorius* (Aouadj et al. 2000), where both nitrate reductases were strongly induced in the presence of nitrate and repressed by ammonium.

5.2 Reduction of Nitrite to Ammonium

Nitrite reductase from the ectomycorrhizal basidiomycete *Hebeloma cylindrosporum* is specific for NADPH and was found to be very unstable (Plassard et al. 1984b). The saturation curve of the enzyme for NO_2^- was biphasic with two apparent Km values at 13 and 350 µM. This suggests that the enzyme of *Hebeloma cylindrosporum* has two types of binding sites for NO_2^- which could make the reaction continuously responsive to concentration changes over a wide range. Nitrite reductase activity measured in *Hebeloma cylindrosporum* was similar to the nitrate reductase activity, ranging from 10 to 30 µmol h^{-1} g^{-1} fresh weight, which is considerably higher than the in vivo NO_3^- uptake capacity of the mycelium (Plassard et al. 1984b). Consequently, nitrite does not accumulate in the fungal cells, and this indicates that nitrite reductase is obviously not a limiting step of NO_3^- assimilation in this ectomycorrhizal fungus.

5.3 Molecular Characterization of Nitrate Reductase and Nitrite Reductase

Genes encoding proteins involved in nitrate assimilation are usually induced by nitrate and subjected to nitrogen catabolite repression. Cloning of two nitrate reductase (NR) genes has been carried out in the ectomycorrhizal fungus *Hebeloma cylindrosporum* (Jargeat et al. 2000). One of these genes (*nar1*) is transcribed and codes for a 908 amino acid polypeptide, while the other gene (*nar2*) for which no mRNA transcripts were detected, is considered to be an ancestral, nonfunctional duplication of *nar1*. It is well known that high nitrate reductase activities are found in mycelia of *Hebeloma cylindrosporum* cultivated in ammonium-containing media, sometimes higher than those exhibited in the presence of nitrate (Plassard et al. 1986). However, Northern analyses showed that *nar1* in *Hebeloma cylindrosporum* was strongly repressed by ammonium, while low nitrogen concentrations or high levels of nitrate, urea, glycine or serine sustained a high level of transcription (Jargeat et al. 2000). The authors have put forward the hypothesis that the nitrate reductase enzyme of the fungus might be extremely stable in vivo and progressively accumulates in the cells growing on ammonium. In addition, such results indicate that in *Hebeloma cylindrosporum*, expression of the nitrate reductase gene is regulated primarily by the availability of ammonium, but not by the presence of nitrate in the medium. This regulation pattern clearly distinguishes this fungus from the other saprophytic and pathogenic species previously studied.

Assimilatory nitrate reductase of higher plants is subjected to a complex regulation of its expression and catalytic properties (Kaiser and Huber 2001). The NR protein is inactivated by phosphorylation combined with a link with a dimeric protein, which may cause a change in NR conformation that interrupts electron transport between the heme and the molybdenum-cofactor domains (Kaiser and Huber 2001). It is known that light as well as CO_2 and oxygen availability are the major external triggers for a rapid and reversible modulation of NR activity, and that sugars and/or sugar phosphates are the internal signals which regulate the protein kinase(s) and phosphatase. In ectomycorrhizal fungi, there is no evidence, so far, for a specific post-translational inactivation of the NR protein. In *Hebeloma cylindrosporum*, the main NR protein named NAR1, like all other fungal NR polypeptides, lacks the short motifs found in the N-terminal and hinge 1 domains of plant NRs, which are both necessary for the post-translational inactivation of these enzymes in response to changes in light or CO_2 status (Su et al. 1996; Jargeat et al. 2000).

Indeed, in *Neurospora crassa* the structural genes that encode nitrogen catabolic enzymes are subject to nitrogen metabolite repression, mediated by the positive-acting NIT2 protein and by the negative-acting NMR protein (for "nitrogen metabolite repression"; Pan et al. 1997). NIT2, a globally acting factor, (or AREA in *Aspergillus nidulans*, or GLN3 in *Saccharomyces cerevisiae*) is a member of the GATA family of regulatory proteins and has a single Cys_2/Cys_2 zinc finger DNA-binding domain. Deletions or certain amino acid substitutions within this zinc finger and the carboxy-terminal tail resulted in a loss of nitrogen metabolite repression (Marzluf 1997). Those mutated forms of NIT2 that were insensitive to nitrogen repression had also lost one of the NIT2-NMR protein–protein interactions. These results provide compelling evidence that the specific NIT2–NMR interactions have a regulatory function and play a central role in establishing nitrogen metabolite repression (Pan et al. 1997).

The different genes involved in nitrate assimilation, as well as putative nitrate transport systems, have been cloned from various saprophytic and pathogenic filamentous ascomycetes; all of these genes are single-copy genes and their transcription is subject to ammonium/glutamine repression and nitrate induction (Kinghorn and Unkles 1994). In the yeast *Hansenula polymorpha*, the genes *YNT1*, *YNR1* and *YNI1*, encoding respectively nitrate transport, nitrate reductase and nitrite reductase (NiR), have been cloned, as well as two other genes encoding transcriptional regulatory factors. Transcriptional regulation is the main regulatory mechanism that controls the levels of the enzymes involved in nitrate metabolism (Siverio 2002). The genetic and molecular bases of repression and induction have been studied in detail in *Aspergillus nidulans* and *Neurospora crassa* (Scazzocchio and Arst 1989; Caddick et al. 1994; Marzluf 1997).

In both species, nitrate induction is mediated by a pathway-specific regulatory gene (*nirA* and *nit-4* in, respectively *Aspergillus nidulans* and *Neurospora crassa*), whose product binds to the promoters of the nitrate pathway genes when NO_3^- is present in the culture medium. Similarly, derepression is mediated by a wide-domain regulatory gene (respectively *areA* and *nit-2*), which encodes a GATA DNA-binding protein. Both *areA* and *nit-2* are responsible, at least in part, for the derepression, when ammonium is absent, of several other genes involved in the use of other nitrogen sources, such as several amino acids or proteins.

In *Neurospora crassa* and *Aspergillus nidulans*, glutamine appears to be the critical metabolite which exerts nitrogen catabolite repression (Chang and Marzluf 1979; Premakumar et al. 1979). Ammonia leads to strong nitrogen repression in these fungi, but is not itself active, since it does not cause repression in mutants lacking glutamine synthetase (Premakumar et al. 1979). Intracellular glutamine, or possibly a metabolite derived from it, leads to repression, but the cellular location of the glutamine pool responsible for this control response, e.g., cytoplasmic or vacuolar, is unknown. An extremely important, but still unknown feature is the identity of the element or signal pathway system that senses the presence of repressing levels of glutamine. It is conceivable that the AREA, NIT2, GLN3, and similar global regulators themselves bind glutamine or that a complex such as a NIT2-NMR heterodimer recognizes the amino acid. However, it is also possible that an as yet unidentified factor detects glutamine and conveys the repression signal to the global activating proteins. Thus, an important goal for future research is the creative use of genetic and biochemical approaches to identify the signalling system that recognizes and processes environmental nitrogen cues.

In the ectomycorrhizal fungus *Hebeloma cylindrosporum*, transcription of *nar1* coding for the NR protein, was repressed in the presence of ammonium, suggesting that the organism might possess a gene homologous to *nit-2* in *Neurospora crassa*. According to Jargeat et al. (2000), inspection of the sequences flanking the NR genes cloned from *Hebeloma cylindrosporum* revealed that they contain several GATA elements to which regulatory GATA proteins could bind.

In *Neurospora crassa*, expression of structural genes which encode the nitrate assimilatory enzymes also has an absolute requirement for nitrate induction mediated by a pathway-specific factor, NIT4 (or NIRA in *Aspergillus nidulans*; Marzluf, 1997). The *Neurospora crassa* NIT4 protein is composed of 1090 amino acids and contains at its amino terminus a Cys_6/Zn_2

binuclear zinc cluster followed by a spacer region and a coiled-coil motif that mediates the formation of a homodimer, the form that is responsible for sequence-specific DNA binding.

In *Hebeloma cylindrosporum*, supply of nitrate is not necessary for the transcription of the NR gene (Jeargeat et al. 2000), suggesting that in this fungus there is no transcription factor such as NIT4 capable of promoting transcription in the presence of nitrate. In agreement with this hypothesis, no motifs resembling the binding sites for NIT4 or NIRA were detected in the promoter regions of the genes cloned in the ectomycorrhizal fungus (Jeargeat et al. 2000).

In the yeast *Hansenula polymorpha* the *YNT1* gene encoding the nitrate transporter is clustered with the structural genes which encode nitrate reductase and nitrite reductase (Perez et al. 1997). Clustering of these three assimilation genes was previously reported in *Aspergillus nidulans* (Johnstone et al. 1990), and more recently in the ectomycorrhizal fungus *Hebeloma cylindrosporum* (Jargeat, Gay, Debaud and Marmeisse, pers. comm.; gene accession number: AJ 238664), which might represent a cell strategy to make the regulation of this important pathway efficient.

The role of arbuscular mycorrhizal fungi in assisting their host plant in nitrate assimilation was studied in the association *Glomus intraradices*/*Zea mays* by Kaldorf et al. (1998). With PCR technology, part of the gene coding for the nitrate reductase apoprotein from either the fungus or from the host-plant was specifically amplified and subsequently cloned and sequenced. Northern blot analysis with these probes indicated that the mRNA level of the maize gene was lower in roots and shoots of mycorrhizal plants than in non-colonized controls, whereas the fungal gene was highly transcribed in roots of mycorrhizal plants.

In agreement with these data, the specific nitrate reductase activity of leaves was significantly lower in endomycorrhizal maize than in the controls. Nitrite formation catalyzed by nitrate reductase was mainly NADPH-dependent in roots of mycorrhizal plants, but not in those of the controls, which is consistent with the fact that these enzymes of fungi preferentially utilize NADPH as reductant. In addition, it has been shown that the fungal nitrate reductase mRNA is detected in arbuscules, but not in vesicles by in situ RNA hybridization experiments (Kaldorf et al. 1998). There is obviously a differential formation of transcripts of a gene coding for the same function in both symbiotic partners.

6 Assimilation of Ammonium

Once inside the cell, NH_4^+ can be incorporated into the key nitrogen donors Glu and Gln for biosynthetic reactions. Glutamate dehydrogenase (NADP-GDH, EC 1.4.1.4) catalyses the reductive amination of 2-oxoglutarate to form

Glu. Glutamine synthetase (GS, EC 6.3.1.2) incorporates ammonium into the carboxyl group of Glu to form Gln. In turn, the Glu and Gln formed serve as donors in transamination and amido nitrogen transfer reactions. Glu is an essential amino N donor for many transaminases and Gln amide nitrogen is used to synthesize many essential metabolites, such as nucleic acids, amino sugars, His, Tyr, Asn, and various cofactors. Both Glu and Gln are essential for protein synthesis. Glutamate synthase (GOGAT) is responsible for the reductive transfer of amide N to α-ketoglutarate for the generation of two molecules of glutamate, one of which is recycled for glutamine biosynthesis. The net result of the combined action of GS and GOGAT is the synthesis of glutamate from ammonium and α-ketoglutarate, frequently referred to as the GS/GOGAT cycle.

6.1 Role and Properties of Glutamate Dehydrogenase

Most of the ascomycete and basidiomycete fungi possess two glutamate dehydrogenases (GDH), each specific for one of the two cofactors. A catabolic role has been assigned to the NAD-specific enzyme (EC 1.4.1.2), whereas the NADP-specific enzyme (EC 1.4.1.4) has been involved in glutamate biosynthesis (Ferguson and Sims 1971). This was confirmed in the ectomycorrhizal fungus *Laccaria laccata* where both enzymes were purified and characterized (Brun et al. 1992; Botton and Chalot 1995; Garnier et al. 1997). Both enzymes revealed biphasic kinetics with two different Km values for glutamate, the NADP-GDH exhibiting a positive cooperativity, and the NAD-GDH a negative cooperativity. At all tested concentrations of glutamate, NAD-GDH showed a higher affinity for this amino acid than the NADP-specific enzyme. This was especially true at low glutamate concentrations where the affinity of NADP-GDH was very low (Km value: 100 mM), while the affinity of NAD-GDH was maximal (Km value: 0.24 mM). In addition, NADP-GDH was found to have a considerably higher affinity for ammonium than the NAD-dependent enzyme and was not calcium-dependent for its activity, contrary to what was found with the latter enzyme. The native NADP-GDHs purified from *Cenococcum geophilum* (Martin et al. 1983), and *Laccaria bicolor* (Ahmad and Hellebust 1991), revealed properties roughly similar to those of the *Laccaria laccata* NADP-GDH.

Activities of glutamate dehydrogenase in conjunction with glutamine synthetase in the free-living *Pezizella ericae*, *Cenococcum geophilum* (Martin et al. 1983), and *Laccaria laccata* (Lorillou et al. 1996), were found to be high and sufficient to sustain high rates of nitrogen assimilation. In cultured *Cenococcum geophilum*, NH_4^+ is assimilated via the glutamate dehydrogenase pathway and the glutamate formed is rapidly used to synthesize glutamine. Ammonium ion assimilation leads to the synthesis of large amounts of glutamine, alanine and arginine (Martin et al. 1987). These amino acids represent the bulk of the free amino acids found in mycelia of ectomycorrhizal fungi. It was suggested that polyphosphate, an impermeant macromolecule, traps the large pool of arginine in the vacuole (Martin 1985), and then reduces the osmotic pressure of the basic amino acid.

The derepression of NADP-GDH specific activity has been observed on nitrate, on low ammonium concentrations, or on nitrogen-free media in *Laccaria bicolor* (Ahmad et al. 1990; Lorillou et al. 1996), and in a wide range of other fungal species such as *Aspergillus nidulans*, *Neurospora crassa*, *Stropharia semiglobata* (Pateman, 1969; Schwartz et al. 1991). The transfer of *Laccaria bicolor* from a NH_4^+-rich medium to either NO_3^- or N-free media caused a rapid, several fold increase in enzyme concentration detected by immunological analysis (Lorillou et al. 1996). These results showed that the changes in NADP-GDH activity were not related to the activation of a constitutive inactive precursor of the enzyme, but to de novo accumulation of newly synthesized GDH. The latter claim was supported by in vivo ^{35}S-labelling experiments which showed that steady-state synthesis of the enzyme increased several fold in mycelia grown in the presence of nitrate or in nitrogen-deficient media (Lorillou et al. 1996).

In the ectomycorrhizal basidiomycete *Suillus bovinus*, cultivated in the presence of ammonium, NADH-dependent glutamate dehydrogenase exhibited high aminating and low deaminating activities, suggesting that this enzyme might also be involved in ammonium assimilation (Grotjohann et al. 2000).

NADP-GDH was found to be located in the cytosol as determined by immunogold labelling carried out in *Cenococcum geophilum* (Chalot et al. 1990) and *Laccaria laccata* (Brun et al. 1993).

GDHA, the gene encoding the NADP-GDH has been cloned and characterized from various fungi (Table 1), including mycorrhizal fungi. In the ectomycorrhizal fungi *Laccaria bicolor* (Lorillou et al. 1996), and *Tuber borchii* (Vallorani et al. 2002), the increased activity of GDH was correlated with its increased synthesis, suggesting that an increased expression of mRNA encoding NADP-GDH occurs under derepressing growth conditions. Quantification of mRNA using a cDNA probe encoding the *Laccaria bicolor* NADP-GDH confirmed that the growth of mycelia on NO_3^- and N-free media, resulted in an increased accumulation of NADP-GDH transcripts (Lorillou et al. 1996). However, the two processes were studied independently in different ectomycorrhizal models and the data obtained until now give only a fragmentary view of ammonium assimilation and its regulation in ectomycorrhizal fungi.

More recently, *GDHA* has been cloned and characterized from *Hebeloma cylindrosporum* and expression of the enzyme was studied in this fungus (Fig. 1; Javelle et al. 2003a). Transfer of the fungus from a 3 mM ammonium to a N-free medium resulted in a 12-fold increase in the GDH transcript level (Fig. 2), corresponding to a similar increase of enzyme activity. On the con-

Table 1. Relationships among fungal NADP-dependent GDH (E.C.1.4.1.4) and GS (E.C.6.3.1.2). Organism, GenBank accession number, sequence length (aa) and molecular weight (MW) are indicated. Sequence identity (ID) using *H. cylindrosporum* GDH or GS sequence as a reference (100 %) is indicated. *A. bisporus, Agaricus bisporus; A. muscaria, Amanita muscaria; A. nidulans, Aspergillus nidulans; B. graminis, Blumeria graminis; F. neoformans, Filobasidiella neoformans; G. fujikuroi, Gibberella fujikuroi, G. cingulata, Glomerella cingulata; L. bicolor, Laccaria bicolor; N. crassa, Neurospora crassa; S. cerevisiae, Saccharomyces cerevisiae; S. pombe, Schizosaccharomyces pombe; S. commune, Schizophyllum commune; S. occidentalis, Schwanniomyces occidentalis; T. borchii, Tuber borchii*

Organism	Accession no.	aa	MW	ID
NADP-dependent glutamate dehydrogenase				
N. crassa	CAD21426	454	48.8	60.9
A. nidulans	S04904	459	49.6	69.2
T. borchii	AAG2878	457	50.1	56.9
S. pombe	T41492	451	48.8	55.6
S. occidentalis	S17907	459	49.8	57.6
S. cerevisiae (*GDH1*)	A25275	454	49.6	56.4
S. cerevisiae (*GDH3*)	AAC04972	457	49.6	56.7
A. bisporus	P54387	457	49.6	83.1
L. bicolor	AAA82936	450	48.5	84.9
H. cylindrosporum	AAL06075	450	48.3	**100**
Glutamine synthetase				
A. bisporus	O00088	354	39.5	90.7
H. cylindrosporum	AAK96111	354	39,2	**100**
A. muscaria	CAD22045	378	41.9	87.0
S. commune	AAF27660	348	38.3	84.2
F. neoformans	CAD10037	358	39.5	72.0
S. cerevisiae	NP015360	370	41.4	68.4
S. pombe	Q09179	359	40.0	63.8
A. nidulans	AAK70354	345	38.5	64.1
G. cingulata	Q12613	360	40.0	64.1
G. fujikuroi	CAC27836	353	39.4	63.4
B. graminis	AAK69535	487	54.1	12.1

trary, feeding the mycelium with ammonium resulted in a rapid decrease of GDH transcripts, which correlated with a decline in GDH-specific activity. Addition of methionine sulfoximine (MSX), an inhibitor of the GS enzyme, to the ammonium-containing medium led to a depletion of glutamine and an accumulation of ammonium in the cells, while a significant decrease of GDH transcript occurred simultaneously (Javelle et al. 2003a). This result strongly suggests that in *Hebeloma cylindrosporum*, GDH repression is controlled by ammonium and not by glutamine, which is obviously different from what was found in *Neurospora crassa* (Chang and Marzluf 1979; Premakumar et al. 1979), and very likely in *Agaricus bisporus* (Kersten et al. 1999), where glutamine or metabolites derived from this amino acid exerted nitrogen catabolite repression.

In *Pleurotus ostreatus*, NADP-dependent glutamate dehydrogenase and glutamate synthase were not detected (Mikes et al. 1994). NAD-GDH was derepressed by ammonia and repressed by high concentrations of L-glutamate. This suggests that this enzyme obviously plays an active role in ammonium assimilation in *Pleurotus ostreatus*. However, a catabolic role of NAD-GDH in the deamination of L-glutamate, due to its very low Km for L-glutamate is not excluded (Mikes et al. 1994).

6.2 Role and Properties of Glutamine Synthetase

Glutamine synthetase (GS; EC 6.3.1.2) is the key enzyme involved in ammonium assimilation in plants (Lea et al. 1990). GS catalyses the ATP-dependent condensation of NH_4^+ with glutamate to produce glutamine. Plant GS is an octameric isozyme with a native molecular mass of approximately 320 or 380 kDa depending on whether it is localized in the cytosol (GS_1) or in plastids/chloroplasts (GS_2; Lea et al.1990).

The in vivo function of GS_2 has been elucidated using genetically modified barley plants (Wallsgrove et al. 1987). The main role is assimilation of NH_4^+ derived from nitrate reduction and photorespiration. The in vivo role of GS_1 depends on the organ in which it is localized. In roots, GS_1 constitutes nearly all GS activity and the main role is assimilation of NH_4^+ for translocation and biosynthesis (Lea et al. 1990).

In gymnosperms, except in the nonconiferous gymnosperm Ginkgo, only cytosolic isoforms of GS (GS_1) have been identified (Suarez et al. 2002). The chloroplastic isoform (GS_2) has not yet been detected by using a number of different molecular approaches including separation of isoforms by ion-exchange chromatography. This implies that in conifers, ammonium is assimilated in the cytosol and therefore, glutamine and glutamate biosynthesis occurs in separate compartments, the GOGAT enzyme being located within chloroplasts. Recent studies indicate the existence of a translocator in the chloroplast membranes of *Pinus pinaster* that may be responsible for the import of glutamine into the organelle, in antiport with glutamate (Suarez et al. 2002).

It is generally assumed that GS activity in plants is regulated at the transcriptional level, and many reports have focused on this aspect (Lam et al. 1996; Oliveira et al. 1997). The dramatic light induction of mRNA for GS_2 is mediated in part by phytochrome and in part by light-induced changes in levels of sucrose (Oliveira and Coruzzi 1999), whereas the transcription of GS_1 in roots depends on the external nitrogen supply level (Finnemann and Schjoerring 2000). Recent work suggests that GS_1 is not only regulated transcriptionally, but also post-translationally by reversible phosphorylation catalysed by protein kinases and microcystin-sensitive serine/threonine protein phosphatase (Finneman and Schjoerring 2000). The more active form is phosphorylated, while the dephosphorylated enzyme is less active and is much more susceptible to degradation. Once phosphorylated, GS reaches its maximal activity through interaction with 14–3–3 proteins, a large group of binding proteins with multiple functions in all eukaryotes (Finneman and Schjoerring 2000). Such a post-translational modulation is similar to that found with nitrate reductase (Kaiser and Huber 2001). However, the activities of NR and GS_1 are oppositely affected by the reversible phosphorylation, as dephosphorylation activates NR, but deactivates GS_1. In addition, phosphorylated NR is an initial step in NR degradation, whereas phosphorylated GS_1 is more protected against degradation than dephosphorylated GS_1. The phosphorylated status of GS_1 changes during light/dark transitions and depends in vitro on the ATP/AMP ratio. However, in leaves of *Brassica napus*, the phosphorylation level increased in darkness and decreased in light, suggesting that the enzyme plays a role in nitrogen remobilization (Finnemann and Schjoerring 2000).

The enzyme was purified and studied from Douglas fir roots (Bedell et al. 1995). The native enzyme had a molecular mass of 460 kDa and was composed of two different subunits of 54 and 64 kDa. The enzyme exhibited a negative cooperativity for ammonium with two Km values which were 11 and 75 µM in the presence of ammonium concentrations lower and higher than 1.3 mM, respectively (Bedell et al. 1995). This possibility for the enzyme to adjust its affinity to the level of ammonium is obviously a very efficient way to assimilate NH_4^+ at different concentrations. However, the enzyme was not investigated after mycorrhization of the Douglas fir roots.

In the fungus *Pleurotus ostreatus*, GS was derepressed by ammonium and L-glutamate, while repression of the enzyme was observed in the presence of L-glutamine (Mikes et al. 1994). This indicates a strong involvement of the enzyme in ammonium assimilation.

GLNA, the gene encoding GS has been cloned and characterized from various fungi (Table 1), including mycorrhizal fungi. Moreover, GS has been purified from the ectomycorrhizal fungus *Laccaria laccata* (Brun et al. 1992). The native enzyme had a molecular weight of approximately 380 kDa and was composed of eight identical subunits of 42 kDa. The enzyme revealed a high affinity for NH_4^+ (24 µM), contrasting with the low affinity of NADP-GDH for this cation (5 mM) in the same fungus. The GS enzyme also represented about 3 % of the total soluble protein pool, which was considerably higher than NADP-GDH, which represented only 0.15 % (Brun et al. 1992). All these results strongly suggest that GS is likely to be the main route of ammonium assimilation in this fungus, especially at low NH_4^+ concentrations.

In ectomycorrhizal fungi, localization studies are more limited than in higher plants. However, immunogold labelling of GS revealed a uniform distribution of the enzyme in the cytosol of *Laccaria laccata* cultivated in pure culture (Brun et al. 1993). In the association Douglas fir-*Laccaria laccata*, the fungal GS was uniformly detected over the entire section of the ectomycorrhizas where the fungal cells were present and no particular accumulation was detected in the mantle, or in the Hartig net fungal cells (Botton and Chalot 1995). The similar patterns of GS distribution observed in the free-living mycelia and in the ectomycorrhizal tissues suggest that the fungal enzyme plays an active role in the primary assimilation of ammonium in ectomycorrhizas.

The expression level of the GS enzyme was studied by Javelle et al. (2003b) in the ectomycorrhizal fungus *Hebeloma cylindrosporum*, where a single mRNA of about 1.2 kb was detected. Transfer of the organism from ammonium-containing media to nitrogen-free media resulted in an increase of GS transcripts, correlating with an increase of GS activity. However, when the culture media were resupplemented with ammonium, up to the concentration of 10 mM, GS transcripts remained almost unchanged or decreased very slowly, indicating that GS in this fungus is not highly regulated, although highly expressed (Javelle et al. 2003b; Fig. 2). Such a regulatory process at the transcriptional level has also been found in *Agaricus bisporus* (Kersten et al. 1997), while in *Saccharomyces cerevisiae*, the enzyme seems to be highly regulated at the post-transcriptional level (ter Schure et al. 1995).

6.3 Role and Properties of Glutamate Synthase

Three classes of glutamate synthases (GOGAT) have been defined, based on their amino acid sequences and the nature of the electron donor (Vanoni and Curti, 1999). (1) Bacterial NADPH-dependent GOGAT consists of two different subunits, the α-subunit of about 150 kDa and the β-subunit of about 50 kDa; (2) Ferredoxin-dependent GOGAT found in photosynthetic cells (higher plants, algae and cyanobacteria) is monomeric and shares considerable homology throughout its sequence with the α-subunit of bacterial GOGAT; (3) plants (especially nonphotosynthetic cells) and fungi including yeasts, as well as lower animals contain a monomeric NAD(P)H-dependent GOGAT of about 200 kDa which results from the fusion of two fragments similar to the α and β-subunits of bacterial GOGAT.

In plants, both enzymes (NADH-GOGAT: EC 1.4.1.14. and ferredoxin (Fd)-GOGAT: EC 1.4.7.1) display different physico-chemical, immunological and regulatory properties and are encoded by separate genes (Ireland and Lea 1999). Fd-GOGAT is an iron-sulphur monomeric flavoprotein, plastid-located and represents the predominant molecular form in photosynthetic tissues although its presence has also been reported in roots and nodules (Temple et al. 1998). In most plants analysed, Fd-GOGAT is encoded by a single gene, however, in *Arabidopsis*, two genes have been characterized (Coschigano et al. 1998). *GLU1* is exclusively expressed in the leaf and is light-regulated, whereas

GLU2 is expressed in leaves and roots and is not regulated by light. The expression pattern of the genes and the physiological characterization of defective mutants support a role of GS_2 and Fd-GOGAT in the assimilation of ammonium derived from the reduction of nitrate and from photorespiration (Coschigano et al. 1998). NADH-GOGAT, also an iron-sulphur monomeric flavoprotein, is present at a low level in leaves, but is more abundant in non-photosynthetic tissues such as roots and nodules, where it is located in nonchlorophyllous plastids (Temple et al. 1998). The structure of the alfalfa gene encoding NADH-GOGAT has been reported by these authors, and its expression is restricted to root nodules where it plays a significant role in the assimilation of ammonium derived from symbiotic N_2 fixation (Trepp et al. 1999). The localization of GS_1 and NADH-GOGAT proteins in the root vascular bundles of rice, and very likely in many other plants, supports the possibility of a co-ordinated function in the assimilation of ammonium in roots (Ishiyama et al. 1998).

In fungi, NADH-GOGAT was purified and studied in *Neurospora crassa* where the enzyme was found as a single polypeptide of 200 kDa (Hummelt and Mora 1980) and in *Saccharomyces cerevisiae* where the enzyme is trimeric, composed of three identical 199-kDa subunits (Cogoni et al. 1995).

In ectomycorrhizal fungi, very little is known about this enzyme. An NADH-dependent GOGAT was, however, detected in *Laccaria bicolor* by Vézina et al. (1989). In *Pisolithus tinctorius*, the kinetics of ^{15}N labelling and the effects of enzyme inhibitors have given evidence that ammonium assimilation occurs through the GS/GOGAT cycle (Kershaw and Stewart 1992). In agreement with these data, Botton and Dell (1994) failed to detect NADP-GDH in this fungus. In *Scleroderma verrucosum*, glutamine synthetase and NAD-glutamate synthase activities were clearly detected, while NADP-GDH was almost undetectable (Prima Putra et al. 1999). This is consistent with the view that ammonium assimilation occurs through the GS/GOGAT cycle in this fungus. In *Cenococcum geophilum*, a number of results based on the use of enzyme-specific inhibitors, enzyme assays and estimation of the amino acid pools are also consistent with the operation of the GS/GOGAT cycle (A. Khalid and B. Botton, unpublished results).

The results obtained by Chalot et al. (1994a) with *Paxillus involutus*, also emphasize a GS/GOGAT cycle in this fungus. Indeed, feeding the fungus with [^{14}C]-glutamine resulted in a significant labelling of glutamate, while addition of azaserine, an inhibitor of the GOGAT enzyme, led to both an accumulation of ^{14}C-glutamine and a reduced pool of labelled glutamate. Interestingly, in these experiments, ^{14}C-aspartate and ^{14}C-alanine did not accumulate under azaserine treatment where ^{14}C-glutamine degradation was inhibited, thus indicating that aspartate and alanine synthesis depends on the carbon skeletons from glutamine (Chalot et al.1994a). In addition, feeding *Paxillus involutus* with ^{14}C-glutamate resulted in a significant accumulation of ^{14}C-glutamine under azaserine treatment, suggesting that the supplied glutamate is used for glutamine synthesis. These results are consistent with the existence of two pools of glutamate in the fungal cells, as previously demonstrated by [^{15}N]amino acid analysis in *Cenococcum geophilum* (Martin et al. 1988). It was thus suggested that newly absorbed glutamate, as well as glutamate synthetized by NADP-GDH are converted to glutamine, whereas glutamate produced by the GOGAT enzyme is utilized by the aminotransferases (Martin et al. 1988; Botton and Chalot 1995).

The glutamine synthetase–glutamate synthase pathway was shown to be the main assimilatory route in beech ectomycorrhizas and glutamate dehydrogenase plays only a minor role, if any, in these tissues (Martin et al. 1986). Glutamine synthetase and glutamate synthase which share immunological similarities with higher plant enzymes were detected in beech ectomycorrhizas by means of Western immunoblotting, whereas a fungal glutamate dehydrogenase could not be detected (Martin, unpubl. results). The absorption of NH_4^+ is associated with glutamine synthesis in beech ectomycorrhizas so that 60–80 % of the nitrogen absorbed is present as this amide after a few hours of absorption (Martin et al. 1986). In addition, there is a rapid and very high ^{15}N-labelling in alanine over the time course of the experiment performed with beech (Martin et al. 1986). These data, together with the measurement of high alanine aminotransferase activity in ectomycorrhizal fungi (Dell et al. 1989), suggest that glutamine and alanine might be the major forms of combined nitrogen exported to the root cells.

7 Amino Acid Metabolism

7.1 Utilization of Proteins by Ectomycorrhizal Fungi and Ectomycorrhizas

As investigated primarily by Lundeberg (1970), it is generally accepted that most ectomycorrhizal fungal strains are unable to metabolize and use humus-bound nitrogen. Several ectomycorrhizal and ericaceous fungi in pure culture are, however, able to grow in nutrient media containing proteins as the sole nitrogen source (Bajwa et al. 1985; Abuzinadah and Read 1986a), and this correlated with the production of exocellular proteinase activities (Botton and Chalot 1991; Leake and Read 1991). In the presence of exogenous proteins (casein, gelatin, albumin, soil proteins), *Cenococcum geophilum* was able to secrete active proteases into the nutrient medium, and ammonia strongly repressed the induction and secretion of these proteases (El-Badaoui and Botton 1989). This capacity of the mycorrhizal fungus to use amino acids as nitrogen sources is retained in the symbiotic state. Melin and Nilsson (1953) have shown that ^{15}N from [^{15}N]glutamate is transferred to *Pinus sylvestris* roots and aerial parts through the mycelia of *Suillus granulatus*. The ability of several ectomycorrhizal fungi to assimilate proteins and to transfer its nitrogen to plants of *Pinus contorta* was also clearly demonstrated (Abuzinadah and Read 1986a, b; Abuzinadah et al.1986). The use of nitrogen sources not available to nonmycorrhizal plants contributes, therefore, to an increased uptake of nitrogen by infected roots.

7.2 Amino Acids Used as Nitrogen and Carbon Sources

Utilization of amino acids by ectomycorrhizal symbionts and ectomycorrhizas may have important implications, not only for their nitrogen metabolism, but also for the overall carbon economy of the plant. Axenic mycelia of the ectomycorrhizal basidiomycete *Suillus bovinus* have been grown in liquid media in the presence of glucose as the only carbohydrate source and under such conditions, they produced similar amounts of dry weight with ammonia, with nitrate or with alanine, 60–80 % more with glutamate or glutamine, but about 35 % less with urea as the only exogenous nitrogen source (Grotjohann et al. 2000).

Recently, the fate of carbon derived from alanine, glutamate and glutamine was investigated in the ectomycorrhizal fungus *Paxillus involutus* (Chalot et al. 1994a, b). The result of the ^{14}C tracer experiments suggested that the carbon skeletons derived from newly absorbed glutamate were mainly used for the synthesis of glutamine. The accumulation of [^{14}C]glutamate and the marked decrease of [^{14}C]glutamine under MSX treatment were consistent with a rapid utilization of glutamate by the glutamine synthetase (GS) enzyme. The newly absorbed, as well as the newly synthesized [^{14}C]glutamine were degraded into [^{14}C]glutamate, suggesting the operation of the glutamate synthase (GOGAT) enzyme. This was confirmed by the striking accumulation of [^{14}C]glutamine when the fungus was cultivated in the presence of azaserine, an inhibitor of GOGAT. In addition, a strong inhibition of glutamine utilization by aminooxyacetate indicated that glutamine catabolism in *Paxillus involutus* might involve a transamination process as an alternative pathway to GOGAT for glutamine degradation (Chalot et al. 1994a).

The use of ^{14}C-labelled amino acids also showed a direct involvement of glutamate and glutamine in the respiration pathways, these amino acids being obviously channelled through the tricarboxylic acid (TCA) cycle and oxidized to CO_2. Feeding the fungus with [^{14}C]alanine resulted in a rapid labelling of pyruvate, citrate, succinate, fumarate and CO_2. Further labelling was detected in glutamate, glutamine and aspartate. The presence of aminooxyacetate completely suppressed $^{14}CO_2$ evolution and decreased the flow of carbon to the Krebs cycle intermediates and amino acids, suggesting that alanine aminotransferase plays a key role in metabolizing alanine in *Paxillus involutus* (Chalot et al. 1994b).

It has been shown by measuring enzyme capacities and metabolite pools that mycorrhization causes a re-arrangement of the main metabolic pathways in the very early stages following contact between the two partners (Blaudez et al. 1998), which correlates with the observed structural changes (Brun et al. 1995). The impact of inoculation with *Paxillus involutus* on the utilization of organic carbon compounds by birch roots was studied by feeding [^{14}C]glutamate or [^{14}C]malate to the partners of the symbiosis, separately or in association, and by monitoring the subsequent distribution of ^{14}C (Blaudez et al. 2001). Inoculation increased [^{14}C]glutamate and [^{14}C]malate absorption capacities by up to 8 and 17 times, respectively. This heterotrophic carbon assimilation by mycorrhizal birch has been estimated using ^{14}C-labelled proteins (Abuzinadah and Read 1989). The authors calculated that 9 % of plant C may be derived from proteins. Moreover, our results demonstrated that inoculation strongly modified the fate of [^{14}C]glutamate and [^{14}C]malate. It was demonstrated that exogenously supplied glutamate and malate might serve as C skeletons for amino acid synthesis in mycorrhizal birch roots and in the free-living fungus. Glutamine was the major ^{14}C-sink in mycorrhizal roots and in the free-living *P. involutus* (Blaudez et al. 2001). In contrast, citrulline and insoluble compounds were the major ^{14}C sinks in nonmycorrhizal roots, whatever the ^{14}C source. Thus, it is obvious that mycorrhiza formation leads to a profound alteration of the metabolic fate of exogenously supplied C compounds.

Translocation through the hyphal network and further transfer of nutrients from fungus to host root has also been discussed in detail (Smith and Read 1997), but the intimate anatomical connections between fungal and root cells presents considerable technical difficulties for unambiguous experimental investigations of nutrient transfer between fungus and host.

8 Conclusion and Future Prospects

After many decades of investigating the anatomical, physiological and biochemical features of ectomycorrhizas, recent years have brought new insights at the molecular level. Considerable knowledge has been gained over the last 10 years on the molecular characteristics and molecular regulation of the transporters and the nitrogen-assimilating enzymes in higher plants and fungi, as well as in ectomycorrhizas. This research has greatly contributed to our understanding of how organic and inorganic nitrogen is taken up by the cells and assimilated in the organisms. However, the available information is still limited and efforts should be made to increase basic research on nitrogen metabolism and to integrate new advances in biotechnology.

A current focus in plant improvement is the modification of the expression of genes involved in metabolism. Recent studies have shown that important characteristics can be introduced in transgenic herbaceous plants by the expression of heterologous GS isoenzymes. Thus, a higher capacity for photorespiration (Migge et al. 2000), and increase in tolerance to salt stress

(Hoshida et al. 2000), have been reported using engineered plants which overexpress chloroplastic GS_2. Furthermore, an increase in growth has been observed in leguminous plants, which overexpress cytosolic GS_1 (Limami et al. 1999). The modification of N assimilation efficiency has recently been approached in trees by the overexpression of pine GS_1 in a hybrid poplar (Gallardo et al. 1999). Poplar is considered as a model in molecular investigations because of its small genome size, easy vegetative propagation and the possibility of in vitro culture, and its amenability to transformation via *Agrobacterium tumefaciens* (Gallardo et al. 1999).

Considerable knowledge has been gained over the last decade on the molecular characteristics and molecular regulation of N-assimilating enzymes in woody plants, including angiosperm and gymnosperm species. This research has greatly contributed to our understanding of how inorganic N is assimilated and utilized in trees. However, the available information is still limited and efforts should be made to increase basic research on N metabolism and to integrate new advances in biotechnology to improve growth and development of economically important woody species. Although all new studies will contribute to this goal, the concentration of efforts in model trees, such as poplar for angiosperms and pine for gymnosperms, is advisable. In future years, the availability of new molecular tools for biological studies of trees will permit characterization of new genes involved in N metabolism and determination of their specific physiological roles. Functional studies are now possible in woody plants because routine transformation protocols via *Agrobacterium* are available for poplar and rapid progress has been reported in the last few years for conifers. The use of somatic embryogenic cell lines is critical for the generation of transgenic trees. For example, genomic technologies have recently been used to study the effect of a variety of N regimes on plant metabolism (Wang et al. 2000). Results from this study indicate that changes in N supply influence not only expression of genes involved in N assimilation, but also those involved in other metabolic pathways. Similar studies of gene expression at the organ or tissue levels are now feasible in tree models with the existence of EST databases from poplar.

Another promising line of research will be to study at the molecular level, the genetic basis of important traits, such as N use efficiency and growth efficiency in the presence of the mycorrhizal fungus. Genetic maps for poplar and pine have been established and now genes involved in N metabolism can be localized in the genome. The possible association of specific genes with quantitative trait loci (QTL) are currently being investigated in a number of laboratories. This will allow molecular characterization of gene clusters involved in traits of interest in forestry and tree management.

Transformation of ectomycorrhizal fungi is more limited. Indeed, the assignment of functions to genes and their products has been limited to deduction based on sequence homologies, subcellular localization studies and expression in heterologous hosts, since transformation techniques for the vast majority of ectomycorrhizal basidiomycetes have not been readily available. Exceptions are *Laccaria laccata* (Barret et al. 1990), and *Hebeloma cylindrosporum* (Marmeisse et al. 1992), which have been transformed by the protoplast method, and *Paxillus involutus* (Bills et al. 1995), and *Laccaria bicolor* (Bills et al. 1999), which have been transformed by particle bombardment. Since the first report on successful genetic transfer from *Agrobacterium tumefaciens* to the yeast *Saccharomyces cerevisiae* (Bundock et al. 1995), a number of ascomycetous filamentous fungi were also shown to be amenable to this transformation system (Abuodeh et al. 2000; Chen et al. 2000).

Our understanding of metabolite regulation of gene expression supports the notion that ammonium and N-assimilation products such as amino acids might act as signals whose levels are sensed as an indicator for a high internal N status. Along these lines, putative sensors of glutamate in plants, glutamate receptor genes, have been identified in *Arabidopsis* (Lam et al. 1998). The emerging tools of genomics and bioinformatics should allow us, in the near future, to identify the interacting pathways that control gene expression in response to mycorrhization.

References and Selected Reading

Abuodeh RO, Orbach MJ, Mandel MA, Das A, Galgiani JN (2000) Genetic transformation of *Coccidioides immitis* facilitated by *Agrobacterium tumefaciens*. J Infect Dis 181:2106–2110

Abuzinadah RA, Read DJ (1986a) The role of proteins in the nitrogen nutrition of ectomycorrhizal plants. I. Utilization of peptides and proteins by ectomycorrhizal fungi. New Phytol 103:481–493

Abuzinadah RA, Read DJ (1986b) The role of proteins in the nitrogen nutrition of ectomycorrhizal plants. III. Protein utilization by *Betula pendula* and *Pinus* mycorrhizal association with *Hebeloma cylindrosporum*. New Phytol 103:506–514

Abuzinadah RA, Read DJ (1988) Amino acids as nitrogen sources for ectomycorrhizal fungi: utilization of individual amino acids. Trans Br Mycol Soc 91:473–479

Abuzinadah RA, Read DJ (1989) Carbon transfer associated with assimilation of organic nitrogen sources by silver birch (*Betula pendula* Roth.). Trees 3:17–23

Abuzinadah RA, Finlay RD, Read DJ (1986) The role of proteins in the nitrogen nutrition of ectomycorrhizal plants. II. Utilization of proteins by mycorrhizal plants of *Pinus contorta*. New Phytol 103:495–506

Ahmad I, Hellebust JA (1991) Enzymology of nitrogen assimilation in mycorrhiza. In: Norris JR, Read DJ, Varma AK (eds) Methods in microbiology no 23, Academic Press, New York, pp 181–202

Ahmad I, Carleton TJ, Malloch DW, Hellebust JA (1990) Nitrogen metabolism in the ectomycorrhizal fungus *Laccaria bicolor* (R. Mre) Orton. New Phytol 116:431–441

Anderson IC, Chambers SM, Cairney JWG (1999) Intra- and interspecific variation in patterns of organic and inorganic nitrogen utilisation by three Australian *Pisolithus* species. Mycol Res 103:1579–1587

Aouadj R, Es-Sgaouri A, Botton B (2000) Etude de la stabilité et de quelques propriétés de la nitrate réductase du champignon ectomycorrhizien *Pisolithus tinctorius*. Cryptog Mycol 21:187–202

Bajwa R, Abuarghub S, Read DJ (1985) The biology of mycorrhiza in the Ericaceae. X. The utilization of proteins and the production of proteolytic enzymes by the mycorrhizal endophyte and by mycorrhizal plants. New Phytol 101:469–486

Barret V, Dixon RK, Lemke PA (1990) Genetic transformation of a mycorrhizal fungus. Appl Microbiol Biotechnol 33:313–316

Bedell J-P, Chalot M, Brun A, Botton B (1995) Purification and properties of glutamine synthetase from Douglas-fir roots. Physiol Plant 94:597–604

Bending GD, Read DJ (1995) The structure and function of the vegetative mycelium of ectomycorrhizal plants. V. Foraging behaviour and translocation of nutrients from exploited litter. New Phytol 130:401–409

Bills SN, Richter DL, Podila GK (1995) Genetic transformation of the ectomycorrhizal fungus *Paxillus involutus* by particle bombardment. Mycol Res 99:557–561

Bills SN, Podila GK, Hiremath ST (1999) Genetic engineering of the ectomycorrhizal fungus *Laccaria bicolor* for use as a biological control agent. Mycologia 91:237–242

Blatt MR, Maurousset L, Meharg A (1997) High-affinity NO_3^- H^+ co-transport in the fungus *Neurospora*: induction and control by pH and membrane voltage. J Membr Biol 160:59–76

Blaudez D, Chalot M, Dizengremel P, Botton B (1998) Structure and function of the ectomycorrhizal association between *Paxillus involutus* and *Betula pendula*. II. Metabolic changes during mycorrhiza formation. New Phytol 138:543–552

Blaudez D, Botton B, Dizengremel P, Chalot M (2001) The fate of [^{14}C]glutamate and [^{14}C]malate in birch roots is strongly modified under inoculation with *Paxillus involutus*. Plant Cell Environ 24:449–457

Botton B, Chalot M (1991) Techniques for the studies of nitrogen metabolism in ectomycorrhiza. In: Norris JR, Read DJ, Varma AK (eds) Methods in microbiology no 23, Academic Press, New York, pp 204–244

Botton B, Dell B (1994) Expression of glutamate dehydrogenase and aspartate aminotransferase in Eucalypt ectomycorrhizas. New Phytol 126:249–257

Botton B, Chalot M (1995) Nitrogen assimilation: enzymology in ectomycorrhizas. In: Varma AK, Hock B (eds) Mycorrhiza, structure, function, molecular biology and biotechnology. Springer, Berlin Heidelberg New York, pp 325–363

Brandes B, Godbold DL, Kuhn AJ, Jentschke G (1998) Nitrogen and phosphorus acquisition by the mycelium of the ectomycorrhizal fungus *Paxillus involutus* and its effect on host nutrition. New Phytol 140:735–743

Brun A, Chalot M, Botton B, Martin F (1992) Purification and characterization of glutamine synthetase and NADP-glutamate dehydrogenase from the ectomycorrhizal fungus *Laccaria laccata*. Plant Physiol 99:938–944

Brun A, Chalot M, Botton B (1993) Glutamate dehydrogenase and glutamine synthetase of the ectomycorrhizal fungus *Laccaria laccata*: occurrence and immunogold localization in the free living mycelium. Life Sci Adv Plant Physiol 12:53–60

Brun A, Chalot M, Finlay RD, Söderström B (1995) Structure and function of the ectomycorrhizal association between *Paxillus involutus* (Batsch) Fr. and *Betula pendula* (Roth.). I. Dynamics of mycorrhiza formation. New Phytol 129:487–493

Bundock P, den Dulk-Ras A, Beijersbergen A, Hoykaas PJJ (1995) Trans-kingdom T-DNA transfer from *Agrobacterium tumefasciens* to *Saccharomyces cerevisiae*. EMBO J 14:3206–3214

Burgstaller W (1997) Transport of small ions and molecules through the plasma membrane of filamentous fungi. Crit Rev Microbiol 23:1–46

Caddick MX, Peters D, Platt A (1994) Nitrogen regulation in fungi. Antonie van Leeuwenhoek 65:169–177

Carleton TJ, Read DJ (1991) Ectomycorrhizas and nutrient transfer in conifer-feather moss ecosystems. Can J Bot 69:778–785

Chalot M, Brun A (1998) Physiology of organic nitrogen acquisition by ectomycorrhizal fungi and ectomycorrhizas. FEMS Microbiol Rev 22:21–44

Chalot M, Brun A, Botton B (1990) Occurrence and distribution of the fungal NADP-dependent glutamate dehydrogenase in spruce and beech ectomycorrhizas. In: Werner D, Müller P (eds) Fast growing trees and nitrogen fixing trees. Gustav Fischer Verlag, Jena, Germany, pp 324–327

Chalot M, Stewart GR, Brun A, Martin F, Botton B (1991) Ammonium assimilation by spruce-*Hebeloma* sp. ectomycorrhizas. New Phytol 119:541–550

Chalot M, Brun A, Finlay RD, Söderström B (1994a) Metabolism of [^{14}C]glutamate and [^{14}C]glutamine by the ectomycorrhizal fungus *Paxillus involutus*. Microbiology 140:1641–1649

Chalot M, Brun A, Finlay RD, Söderström B (1994b) Respiration of [^{14}C]alanine by the ectomycorrhizal fungus *Paxillus involutus*. FEMS Microbiol Lett 121:87–92

Chalot M, Brun A, Botton B, Söderström B (1996) Kinetics, energetics and specificity of the general amino acid transporter from the ectomycorrhizal fungus *Paxillus involutus*. Microbiology 142:1749–1756

Chalot M, Javelle A, Blaudez D, Lambilliote R, Cooke R, Sentenac H, Wipf D, Botton B (2002) An update on nutrient transport processes in ectomycorrhizas. Plant Soil 244:165–175

Chang L, Marzluf GA (1979) Nitrogen regulation of uricase synthesis in *Neurospora crassa*. Mol Gen Genet 176:385–392

Chen X, Stone M, Schlagnhaufer C, Romaine CP (2000) A fruiting body tissue method for efficient *Agrobacterium*-mediated transformation of *Agaricus bisporus*. Appl Environ Microbiol 66:4510–4513

Cogoni C, Valenzuela L, Gonzales-Halphen D, Oliveira H, Macino G, Ballario P, Gonzalez A (1995) *Saccharomyces cerevisiae* has a single glutamate synthase gene coding for a plant-like high-molecular-weight polypeptide. J Bacteriol 177:792–798

Coschigano KT, Melo-Oliveira R, Lim J (1998) *Arabidopsis* glts mutants and distinct Fdx-GOGAT genes: implications for photorespiration and primary nitrogen assimilation. Plant Cell 10:741–752

Crawford NM, Glass AD (1998) Molecular and physiological aspects of nitrate uptake in plants. Trends Plant Sci 3:389–395

Dell B, Botton B, Martin F, Le Tacon F (1989) Glutamate dehydrogenases in ectomycorrhizas of spruce (*Picea excelsa* L.) and beech (*Fagus sylvatica* L.). New Phytol 111: 683–692

Dubois E, Grenson M (1979) Methylamine ammonia uptake systems in *Saccharomyces cerevisiae*. Multiplicity and regulation. Mol Gen Genet 175:67–76

El-Badaoui K, Botton B. 1989. Production and characterization of exocellular proteases in ectomycorrhizal fungi. Ann Sci For 46:728–730

Fergusson AR, Sims AP (1971) Inactivation in vivo of glutamine synthetase and NAD-specific glutamate dehydrogenase: its role in the regulation of glutamine synthesis in yeasts. J Gen Microbiol 69:423–427

Finlay RD, Ek H, Odham G, Söderström B (1988) Mycelial uptake, translocation and assimilation of nitrogen from ^{15}N-labelled ammonium by *Pinus sylvestris* plants infected with four different ectomycorrhizal fungi. New Phytol 110:59–66

Finnemann J, Schjoerring JK (2000) Post-translational regulation of cytosolic glutamine synthetase by reversible phosphorylation and 14-3-3 protein interaction. Plant J 24:171–177

Forde B (2000) Nitrate transporters in plants: structure, function and regulation. Biochim Biophys Acta 1465:219–235

Gallardo F, Fu J, Cantón FR, García-Gutiérrez A, Cánovas FM, Kirby EG (1999) Expression of a conifer glutamine synthetase gene in transgenic poplar. Planta 210:19–26

Garnier A, Berredjem A, Botton B (1997) Purification and characterization of the NAD-dependent glutamate dehydrogenase in the ectomycorrhizal fungus *Laccaria bicolor* (Maire) Orton. Fungal Genet Biol 22:168–176
Garrett RH, Nason A (1969) Further purification and properties of *Neurospora* nitrate reductase. J Biol Chem 244:2870–2882
Gazzarrini S, Lejay L, Gojon A, Ninnemann O, Frommer WB, von Wiren N (1999) Three functional transporters for constitutive, diurnally regulated, and starvation-induced uptake of ammonium into *Arabidopsis* roots. Plant Cell 11:937–947
Gobert A, Plassard C (2002) Differential NO_3^- dependent patterns of NO_3^- uptake in *Pinus pinaster*, *Rhizopogon roseolus* and their ectomycorrhizal association. New Phytol 154:509–516
Grotjohann N, Kowallik W, Huang Y, Schulte in den Baumen A (2000) Investigations into enzymes of nitrogen metabolism of the ectomycorrhizal basidiomycete, *Suillus bovinus*. Z Naturforsch C, 55:203–212
Guan C, Ribeiro A, Akkermans ADL, Jing Y, van Kammen A, Bisseling T, Pawlowski K (1996) Nitrogen metabolism in actinorhizal nodules of *Alnus glutinosa*: expression of glutamine synthetase and acetylornithine transaminase. Plant Mol Biol 32:1177–1184
Hackette SL, Skye GE, Burton C, Segel IH (1970) Characterization of an ammonium transport system in filamentous fungi with methylammonium-^{14}C as the substrate. J Biol Chem 245:4241–4250
Harley JL (1989) The significance of mycorrhizas. Mycol Res 92:129–139
Hildebrandt U, Schmelzer E, Bothe H (2002) Expression of nitrate transporter genes in tomato colonized by an arbuscular mycorrhizal fungus. Physiol Plant 115:125–136
Ho I, Trappe JM (1987) Enzymes and growth substances of *Rhizopogon* species in relation to mycorrhizal host and infrageneric taxonomy. Mycologia 79:553–558
Hoshida H, Tanaka Y, Hibino T, Hayashi Y, Tanaka A, Takabe T, Takabe T (2000) Enhanced tolerance to salt stress in transgenic rice that overexpresses chloroplast glutamine synthetase. Plant Mol Biol 43:103–111
Howitt SM, Udvardi MK (2000) Structure, function and regulation of ammonium transporters in plants. Biochim Biophys Acta 1465:152–170
Hummelt G, Mora J (1980) Regulation and function of glutamate synthase. Biochem Biophys Res Commun 96:1688–1694
Ireland RJ, Lea PJ (1999) The enzymes of glutamine, glutamate, asparagine, and aspartate metabolism. In: Singh BK (ed) Plant amino acids, biochemistry and biotechnology. Marcel Dekker, New York, pp 49–109
Ishiyama K, Hayakawa T, Yamaya T (1998) Expression of NADH-dependent glutamate synthase protein in the epidermis and exodermis of rice root in response to the supply of ammonium ions. Planta 204:288–294
Jargeat P (1999) Caractérisation et manipulation génétique de la voie d'assimilation du nitrate du champignon symbiotique *Hebeloma cylindrosporum*. PhD Thesis, University Claude-Bernard, Lyon I, Lyon
Jargeat P, Gay G, Debaud JC, Marmeisse R (2000) Transcription of a nitrate reductase gene isolated from the symbiotic basidiomycete fungus *Hebeloma cylindrosporum* does not require induction by nitrate. Mol Gen Genet 263:948–956
Jargeat P, Rekangalt D, Verner MC, Gay G, Debaud JC, Marmeisse R, Fraissinet-Tacher L (2003) Characterisation and expression analysis of a nitrate transporter and nitrite reductase genes, two members of a gene cluster for nitrate assimilation from the symbiotic basidiomycete *Hebeloma cylindrosporum*. Curr Genet 43:199–205
Javelle A, Chalot M, Söderström B, Botton B (1999) Ammonium and methylamine transport by the ectomycorrhizal fungus *Paxillus involutus* and ectomycorrhizas. FEMS Microbiol Ecol 30:355–366

Javelle A, Rodriguez-Pastrana BR, Jacob C, Botton B, Brun A, André B, Marini AM, Chalot M (2001) Molecular characterization of two ammonium transporters from the ectomycorrhizal fungus *Hebeloma cylindrosporum*. FEBS Lett 505:393–398

Javelle A, André B, Marini A, Chalot M (2003a) High affinity ammonium transporters and nitrogen sensing in mycorrhizas. Trends Microbiol 11:53–55

Javelle A, Morel M, Rodriguez-Pastrana BR, Botton B, André B, Marini AM, Brun A, Chalot M (2003b) Molecular characterization, function and regulation of ammonium transporters (Amt) and ammonium-metabolizing enzymes (GS, NADP-GDH) in the ectomycorrhizal fungus *Hebeloma cylindrosporum*. Mol Microbiol 47:411–430

Jentschke G, Brandes B, Kuhn AJ, Schröder WH, Godbold DL (2001) Interdependence of phosphorus, nitrogen, potassium and magnesium translocation by the ectomycorrhizal fungus *Paxillus involutus*. New Phytol 149:327–338

Johnstone IL, McCabe PC, Greaves P, Gurr SJ, Cole GE, Brow MAD, Unkles SE, Clutterbuck AJ, Kinghorn JR, Innis MA (1990) Isolation and characterisation of the *crnA-niiA-niaD* gene cluster for nitrate assimilation in *Aspergillus nidulans*. Gene 90:181–192

Jongbloed RH, Clement JMAM, Borst-Pauwels GWFH (1991) Kinetics of NH_4^+ and K^+ uptake by ectomycorrhizal fungi. Effect of NH_4^+ on K^+ uptake. Physiol Plant 83:427–432

Kaiser WM, Huber SC (2001) Post-translational regulation of nitrate reductase: mechanism, physiological relevance and environmental triggers. J Exp Bot 363:1981–1989

Kaldorf M, Schmelzer E, Bothe H (1998) Expression of maize and fungal nitrate reductase genes in arbuscular mycorrhiza. Mol Plant Microbe Interact 11:439–448

Kershaw JC, Stewart GR (1992) Metabolism of ^{15}N-labelled ammonium by the ectomycorrhizal fungus *Pisolithus tinctorius* (Pers), Coker & Couch. Mycorrhiza 1:71–77

Kersten MA, Muller Y, Op den Camp HJ, Vogels GD, Van Griensven LJ, Visser J, Schaap PJ (1997) Molecular characterization of the glnA gene encoding glutamine synthetase from the edible mushroom *Agaricus bisporus*. Biochim Biophys Acta 1428:260–272

Kersten MA, Arninkhof MJ, Op den Camp HJ, Van Griensven LJ, van der Drift C (1999) Transport of amino acids and ammonium in mycelium of *Agaricus bisporus*. Biochim Biophys Acta 1428:260–272

Kinghorn JR, Unkles SE (1994) Inorganic nitrogen assimilation: molecular aspects. In: Martinelli SD, Kinghorn JR (eds) Aspergillus: 50 years on. Elsevier, Amsterdam, pp 181–194

Kreuzwiezer J, Stulen I, Wiersema P, Vaalburg W, Rennenberg H (2000) Nitrate transport in *Fagus-Laccaria* mycorrhiza. Plant Soil 220:107–117

Kronzucker HJ, Siddiqi MY, Glass ADM (1996) Kinetics of $NH4^+$ uptake in spruce. Plant Physiol 110:773–779

Lam HM, Coschigano KT, Oliveira IC, Melo-Oliveira R, Coruzzi GM (1996) The molecular-genetics of nitrogen assimilation into amino acids in higher plants. Ann Rev Plant Physiol Plant Mol Biol 47:569–593

Lam HM, Chiu J, Hsieh M, Meisel L, Oliveira I, Shin M, Coruzzi GM (1998) Glutamate-receptor genes in plants. Nature 396:125–126

Landeweert R, Hoffland E, Finlay RD, Kuyper TW, van Breemen N (2001) Linking plants to rocks: ectomycorrhizal fungi mobilize nutrients from minerals. Trends Ecol Evol 16:248–254

Lea PJ, Blackwell RD, Chen FL, Hecht U (1990) Enzymes of nitrogen assimilation. In: Dey PM, Harborne JB (eds) Methods in plant biochemistry(). Academic Press, London, pp 257–276

Leake JR, Read D (1991) Proteinase activity in mycorrhizal fungi. III. Effects of protein, protein hydrolysate, glucose and ammonium on production of extracellular proteinase by *Hymenoscyphus ericae* Read Korf and Kernan. New Phytol 117:309–318

Limami A, Phillipson B, Ameziane R, Pernollet N, Jiang Q, Roy R, Deleens E, Chaumont-Bonnet M, Gresshoff PM, Hirel B (1999) Does root glutamine synthetase control plant biomass production in *Lotus japonicus* L.? Planta 209:495–502
Lorillou S, Botton B, Martin F (1996) Nitrogen source regulates the biosynthesis of NADP-glutamate dehydrogenase in the ectomycorrhizal basidiomycete *Laccaria bicolor*. New Phytol 132:289–296
Lundeberg G (1970) Utilization of various nitrogen sources, in particular bound soil nitrogen, by mycorrhizal fungi. Studia Forestalia Suecica 79:1–95
Machin F, Perdomo G, Pérez MD, Brito N, Siverio JM (2000) Evidence for multiple nitrate uptake systems in *Hansenula polymorpha*. FEMS Microbiol Lett 194:171–174
Marini AM, Vissers S, Urrestarazu A, André B (1994) Cloning and expression of the MEP1 gene encoding a transporter of ammonium in *Saccharomyces cerevisiae*. EMBO J 13:3456–3463
Marini AM, Soussi-Boudekou S, Vissers S, André B (1997) A family of ammonium transporters in *Saccharomyces cerevisiae*. Mol Cell Biol 17:4282–4293
Marschner H, Dell B (1994) Nutrient uptake in mycorrhizal symbiosis. Plant Soil 159:89–102
Marschner H, Haussling M, George E (1991) Ammonium and nitrate rates and rhizosphere pH in non-mycorrhizal roots of Norway spruce [*Picea abies* (L.) Karst]. Plant Soil 178:239–245
Martin F (1985) ^{15}N-NMR studies of nitrogen assimilation and amino acid biosynthesis in the ectomycorrhizal fungus *Cenococcum geophilum*. FEBS Lett 182:350–354
Martin F, Msatef Y, Botton B (1983) Nitrogen assimilation in mycorrhizas. I. Purification and properties of the nicotinamide adenine dinucleotide phosphate-specific glutamate dehydrogenase of the ectomycorrhizal fungus *Cenococcum graniforme*. New Phytol 93:415–422
Martin F, Stewart GR, Genetet I, Le Tacon F (1986) Assimilation of $^{15}NH_4$ by beech (*Fagus sylvatica* L.) ectomycorrhizas. New Phytol 102:85–94
Martin F, Ramstedt M, Söderhäll K (1987) Carbon and nitrogen metabolism in ectomycorrhizal fungi and ectomycorrhizas. Biochimie 69:569–581
Martin F, Stewart GR, Genetet I, Mourot B (1988) The involvement of glutamate dehydrogenase and glutamine synthetase in ammonia assimilation by the rapidly growing ectomycorrhizal ascomycete *Cenococcum geophilum* Fr. New Phytol 110:541–550
Marmeisse R, Gay G, Debaud JC, Casselton LA (1992) Genetic transformation of the symbiotic basidiomycete fungus *Hebeloma cylindrosporum*. Curr Genet 22:41–45
Marzluf GA (1997) Genetic regulation of nitrogen metabolism in the fungi. Microbiol Mol Biol Rev 61:17–32
Melin E, Nilsson H (1952) Transport of labelled nitrogen from an ammonium source to pine seedling through mycorrhizal mycelium. Sven Bot Tidkr 46:281–285
Melin E, Nilsson H (1953) Transfer of labelled nitrogen from glutamic acid to pine seedlings through the mycelium of *Boletus variegatus* (Sw.) Fr. Nature 171:134
Migge A, Carrayol E, Hirel B, Becker TW (2000) Leaf-specific overexpression of plastidic glutamine synthetase stimulates the growth of transgenic tobacco seedlings. Planta 210:252–260
Mikes V, Zofall M, Chytil M, Fulnecek J, Schanel L (1994) Ammonia-assimilating enzymes in the basidiomycete fungus *Pleurotus ostreatus*. Microbiology 140:977–982
Minagawa N, Yoshimoto A (1982) Purification and characterization of the assimilatory NADH-nitrate reductase of *Aspergillus nidulans*. J Biochem 91:761–774
Montanini B, Moretto N, Soragni E, Percudani R, Ottonello S (2002) A high-affinity ammonium transporter from the mycorrhizal ascomycete *Tuber borchii*. Fungal Geneti Biol 36:22–34
Näsholm T, Ekblad A, Nordin A, Giesler R, Högberg M, Högberg P (1998) Boreal forest plants take up organic nitrogen. Nature 392:914–916

Navarro MT, Guerra E, Fernandez E, Galvan A (2000) Nitrite reductase mutants as an approach to understanding nitrate assimilation in *Chlamydomonas reinhardtii*. Plant Physiol 122:283–290

Nehls U, Kleber R, Wiese J, Hampp R (1999) Isolation and characterization of a general amino acid permease from the ectomycorrhizal fungus *Amanita muscaria*. New Phytol 144:343–349

Ninnemann O, Jauniaux JC, Frommer W (1994) Identification of a high affinity ammonium transporter from plants. EMBO J 13:3464–3471

Notton BA, Hewitt EJ (1978) Structure and properties of higher plant nitrate reductase, especially *Spinacea oleracea*. In: Hewitt EJ, Cuttings CV (eds) Nitrogen assimilation of plants. Academic Press, New York, pp 227–244

Oliveira IC, Coruzzi GM (1999) Carbon and amino acids reciprocally modulate the expression of glutamine synthetase in *Arabidopsis*. Plant Physiol 121:301–309

Oliveira IC, Lam HM, Coschigano K, Melo-Oliveira R, Corruzi GM (1997) Molecular-genetic dissection of ammonium assimilation in *Arabidopsis thaliana*. Plant Physiol Biochem 35:185–198

Pan H, Feng B, Marzluf GA (1997) Two distinct protein-protein interactions between the NIT2 and NMR regulatory proteins are required to establish nitrogen metabolite repression in *Neurospora crassa*. Mol Microbiol 26:721–729

Pao SS, Paulsen IT, Saier MH (1998) Major facilitator superfamily. Microbiol Mol Biol Rev 62:1–34

Pateman JA (1969) Regulation of synthesis of glutamate dehydrogenase and glutamine synthetase in micro-organisms. Biochem J 115:769–775

Perez MD, Gonzales C, Avila J, Brito N, Siverio JM (1997) The YNT1 gene encoding the nitrate transporter in the yeast *Hansenula polymorpha* is clustered with genes YNI1 and YNR1 encoding nitrite reductase and nitrate reductase, and its disruption causes inability to grow in nitrate. Biochem J 321:397–403

Perez-Moreno J, Read DJ (2000) Mobilization and transfer of nutrients from litter to tree seedlings via the vegetative mycelium of ectomycorrhizal plants. New Phytol 145: 301–309

Plassard C, Mousain D, Salsac L (1984a) Mesure in vitro de l'activité nitrate réductase dans les thalles de *Hebeloma cylindrosporum*, champignon basidiomycète. Physiol Vég 22:67–74

Plassard C, Mousain D, Salsac L (1984b) Mesure in vivo and in vitro de l'activité nitrite réductase dans les thalles de *Hebeloma cylindrosporum*, champignon basidiomycète. Physiol Vég 22:147–154

Plassard C, Martin F, Mousain D, Salsac L (1986) Physiology of nitrogen assimilation by mycorrhiza. In: Gianinazzi S, Gianinazzi-Pearson V (eds) Les mycorhizes: physiologie et génétique. INRA, Paris, pp 111–120

Plassard C, Barry D, Eltrop L, Mousain D (1994) Nitrate uptake in maritime pine (*Pinus pinaster*) and the ectomycorrhizal fungus *Hebeloma cylindrosporum*: effect of ectomycorrhizal symbiosis. Can J Bot 72:189–197

Plassard C, Chalot M, Botton B, Martin F (1997) Le rôle des ectomycorhizes dans la nutrition azotée des arbres forestiers. Rev For Fr 49:82–98

Premakumar RG, Sorger J, Gooden D (1979) Nitrogen metabolite repression of nitrate reductase in *Neurospora crassa*. J Bacteriol 137:1119–1126

Prima Putra D, Berredjem A, Chalot M, Dell B, Botton B (1999) Growth characteristics, nitrogen uptake and enzyme activities of the nitrate-utilizing ectomycorrhizal fungus *Scleroderma verrucosum*. Mycol Res 103:997–1002

Rawat SR, Silim SN, Kronzucler HJ, Siddiqi MY, Glass AD (1999) AtAMT1 gene expression and NH_4^+ uptake in roots of *Arabidopsis thaliana*: evidence for regulation by root glutamine levels. Plant J 19:143–52

Ritchie RJ, Gibson J (1987) Permeability of ammonia and amines in *Rhodobacter sphaeroides* and *Bacillus firmus*. Arch Biochem Biophys 258:332–341

Roldan JM, Verbelen J, Waren LB, Tokuiyasu K (1982) Intracellular localization of nitrate reductase in *Neurospora crassa*. Plant Physiol 70:872–874

Roon RJ, Even HL, Dunlop P, Larimore FL (1975) Methylamine and ammonia transport in *Saccharomyces cerevisiae*. J Bacteriol 122:502–509

Rousseau JVD, Sylvia DM, Fox AJ (1994) Contribution of ectomycorrhiza to the potential nutrient-absorbing surface of pine. New Phytol 128:639–644

Scazzocchio C, Arst HN Jr (1989) Regulation of nitrate assimilation in *Aspergillus nidulans*. In: Wray JL, Kinghorn JR (eds) Molecular and genetic aspects of nitrate assimilation. Oxford University Press, Oxford, pp 314–363

Scheromm P., Plassard C., Salsac L (1990) Effect of nitrate and ammonium nutrition on the metabolism of the ectomycorrhizal fungus *Hebeloma cylindrosporum*. New Phytol 114:227–234

Schwartz T, Kusnan MB, Fock HP (1991) The involvement of glutamate dehydrogenase and glutamine synthetase/glutamate synthase in ammonia assimilation by the basidiomycete fungus *Stropharia semiglobata*. J Gen Microbiol 137:2253–2258

Siverio JM (2002) Assimilation of nitrate by yeasts. FEMS Microbiol Rev 26:277–284

Smith SE, Read DJ (1997) Mycorrhizal symbiosis. Academic Press, London, 521 pp

Sokolovski SG, Meharg AA, Maathuis FJM (2002) *Calluna vulgaris* root cells show increased capacity for amino acid uptake when colonized with the mycorrhizal fungus *Hymenoscyphus ericae*. New Phytol 155:525–530

Sophianopoulou V, Diallinas G (1995) Amino acid transporters of lower eucaryotes: regulation, structure and topogenesis. FEMS Microbiol Rev 16:53–75

Su W, Huber SC, Crawford NM. 1996. Identification in vitro of a post-translational regulatory site in the hinge 1 region of *Arabidopsis* nitrate reductase. Plant Cell 8, 519–527

Suarez MF, Avila C, Gallardo F, Canton FR, Garcia-Gutiérrez A, Claros MG, Canovas FM (2002) Molecular and enzymatic analysis of ammonium assimilation in woody plants. J Exp Bot 370:891–904

Taylor TM, Osborn JM (1995) The importance of fungi in shaping the paleoecosystem. Rev Palaeobot Palynol 90:249–262

Temple SJ, Vance CP, Gantt JS (1998) Glutamate synthase and nitrogen assimilation. Trends Plant Sci 3:51–56

ter Schure EG, Sillje HH, Raeven LJ, Boonstra J, Verkleij AJ, Verrips CT (1995) Nitrogen-regulated transcription and enzyme activities in continuous cultures of *Saccharomyces cerevisiae*. Microbiology 141:1101–1108

Trepp GB, Plank DW, Gantt S, Vance CP (1999) NADH-glutamate synthase in alfalfa root nodules. Immuno-cytochemical localization. Plant Physiol 119:829–837

Unkles SE, Hawker C, Grieve EI, Campbell EI, Montague P, Kinghorn JR (1991) *crnA* encodes a nitrate transporter in *Aspergillus nidulans*. Proc Natl Acad Sci USA 88:204–208

Unkles SE, Zhou D, Siddiqi MY, Kinghorn JR, Glass ADM (2001) Apparent genetic redundancy facilitates ecological plasticity for nitrate transport. EMBO J 20:6246–6255

Vallorani L, Polidori E, Sacconi C, Agostini D, Pierleoni R, Piccoli G, Zepa S, Stocchi V (2002) Biochemical and molecular characterization of NADP-glutamate dehydrogenase from the ectomycorrhizal fungus *Tuber borchii*. New Phytol 154:779–790

Vanoni MA, Curti B (1999) Glutamate synthase: a complex iron-sulfur flavoprotein. Cell Mol Life Sci 55:617–638

Vézina LP, Margolis HA, McAfee BJ, Delanay S(1989) Changes in the activity of enzymes involved with primary nitrogen metabolism due to ectomycorrhizal symbiosis on jack pine seedlings. Physiol Plant 75:55–62

Vitousek PM ., Matson PA (1985) Causes of delayed nitrate production in two Indiana forests. Forest Sci 31:122–131

Wallenda T, Read DJ (1999) Kinetics of amino acid uptake by ectomycorrhizal roots. Plant, Cell Environ 22:179–187

Wallsgrove RM, Turner JC, Hall NP, Kendall AC, Bright SWJ (1987) Barley mutants lacking chloroplast glutamine synthetase. Biochemical and genetic analysis. Plant Physiol 83:155–158

Wang R, Guegler K, LaBrie ST, Crawford N (2000) Genomic analysis of a nutrient response in *Arabidopsis* reveals diverse expression patterns and novel metabolic and potential regulatory genes induced. Plant Cell 12:1491–1509

Wipf D, Benjdia M, Tegeder M, Frommer WB (2002) Characterization of a general amino acid permease from *Hebeloma cylindrosporum*. FEBS Lett 528:119–24

Yip KP, Kurtz I (1995) NH_3 permeability of principal cells and intercalated cells measured by confocal fluorescence imaging. Am J Physiol 269:545–50

Zhou JJ, Trueman LJ, Boorer KJ, Theodoulou FL, Fordes BG, Miller AJ (2000) A high affinity fungal nitrate carrier with two transport mechanisms. J Biol Chem 275:39894–39899

23 Visualisation of Rhizosphere Interactions of *Pseudomonas* and *Bacillus* Biocontrol Strains

Thomas F.C. Chin-A-Woeng, Anastasia L. Lagopodi,
Ine H.M. Mulders, Guido V. Bloemberg and Ben J.J. Lugtenberg

1 Introduction

This chapter provides hands-on protocols as well as theoretical background information for the selection of *Pseudomonas* and *Bacillus* strains from the rhizosphere antagonistic to phytopathogens. These strains can be evaluated in a bioassay for their beneficial properties. The strains can be marked with a reporter gene after selection and used to study cellular and molecular interactions between one or more beneficial strains and a soil-borne phytopathogen in the rhizosphere of a host plant. Autofluorescent proteins can be used for the non-invasive study of rhizosphere interactions using epifluorescence and confocal laser scanning microscopy (CLSM). Autofluorescent proteins have become an outstanding and convenient tool for studying rhizosphere and other in situ environmental interactions and have allowed microbiologists to visualise the spatial distribution of various microorganisms. Intricate molecular mechanisms and relationships in the rhizosphere can now be studied. Methods to mark rhizosphere bacteria as well as fungi are provided.

2 Tomato Foot and Root Rot and the Need for Biological Control

Tomato (*Esculentum lycopersicum*) foot and root rot caused by the fungus *Fusarium oxysporum* Schlechtend.:Fr. f. sp. *radicis-lycopersici* W.R. Jarvis and Shoemaker (*F.o.r.l.*) is a disease which causes considerable losses to tomato crops. The disease differs from fusarium wilt caused by *Fusarium oxysporum* Schlechtend.:Fr. f. sp. *lycopersici* (Sacc.) W.C. Snyder & H.N. Hans. Plants with *Fusarium* foot and root rot show yellowing along the margin of the oldest leaves, followed by necrosis. Dry brown lesions develop in the cortex of the tap or main lateral roots. Necrotic lesions may also develop on the surface of the stem from the soil line to 10–30 cm above it. Infected plants may be stunted

Plant Surface Microbiology
A. Varma, L. Abbott, D. Werner, R. Hampp (Eds.)

and wilted. Cool soil temperatures favour the disease. The fungus lives over winter and survives for many years in the soil as chlamydospores. Long distance spread is caused by transplants and by soil on farm machinery. Spores are air-borne in greenhouses. The disease causes losses in tomato cropping in agricultural fields, glasshouses, and hydroponic growth. The fungus forms a problem for hydroponic tomato growth in glasshouses in the Netherlands. In the southwest of Florida it is one of the most important tomato diseases and it is emerging at new locations in the United States. Until now only partially resistant varieties have been identified and preplant fumigation with, e.g. methylbromide, which is a management practice often used for many soil-borne diseases, does not completely control the fungus. This practice is also deprecated in view of sustainable agricultural practices. Hence, an efficient way to control the disease is important.

An alternative to chemical control of plant diseases is the use of bacteria (biocontrol). They have the potential to displace or antagonise phytopathogenic or deleterious microorganisms in the rhizosphere. Biocontrol bacteria also produce chemicals, but these are degradable and only produced in low amounts at targeted locations. The latter approach fits well in the worldwide strategy to grow healthier plants in a sustainable way and, therefore produce high quality food.

To use biocontrol strains efficiently, the molecular interactions between plant, biocontrol agent, pathogen and their environment need to be understood. Genetic engineering is an important tool in helping us to define the molecular basis of pathogenicity and is also useful in helping us to identify the mechanisms in the action of biocontrol strains. Molecular genetic modification of microorganisms requires the development of plasmid-mediated transformation systems that include: (1) introduction of exogenous DNA into recipient cells, (2) expression of transformed genes, and (3) stable maintenance and replication of the inserted DNA leading to expression of the desired phenotypic trait. In this chapter, a practical approach to the analysis of biocontrol strains including the isolation, testing, and tagging of these strains, and transformation systems for pathogenic fungi to express reporter genes to track and visualise them in the rhizosphere, are discussed in relation to the pathogenic fungus *Fusarium oxysporum* f. sp. *radicis-lycopersici*.

3 Selection of Antagonistic Strains

3.1 Selection of Antagonistic *Pseudomonas* and *Bacillus* sp.

Pseudomonas and *Bacillus* species constitute, together with *Streptomyces* species, a substantial fraction of the bacterial community isolated from the rhizosphere. Their presence is sometimes correlated with disease suppression. These beneficial bacteria can be exploited as biological pesticides to be used either as an alternative to, or in combination with chemicals to reduce the dose of these chemicals. *Pseudomonas* and *Bacillus* spp. are often abundantly present in the rhizosphere and surrounding soil of many crop plants. Many of these species produce secondary metabolites that inhibit growth of, or kill, soil-borne phytopathogens. These antagonistic bacteria can either be isolated from the rhizosphere or from the soil in which plants have been grown.

In the following isolation procedure, tomato plants harvested at the end of the growing season were picked randomly. Plant roots (0.3–0.4 g fresh weight) were vigorously shaken in phosphate buffer saline (PBS) for 1 h to detach the rhizosphere bacteria from the roots. The resulting bacterial suspensions from individual root systems were diluted and plated on one tenth strength tryptic soy agar (TSA) supplemented with the fungicide cycloheximide (50 µg/ml). The use of a nutrient-poor medium was reported to yield the highest numbers of isolates. After an incubation period of 2–7 days at 28 °C, a large variety of colonies with different morphologies were observed.

The number of fluorescent pseudomonads found in the rhizosphere is very often variable. In some studies they were reported to be a dominant group, whereas other studies report that their numbers did not exceed 1 % of the total rhizosphere population isolated. The variations may be due to differences in plant species or cultivars, soil type, age of the plant roots, or the isolation method. Recently, it was also found that the percentage of antagonistic pseudomonads from a maize rhizosphere grown without chemical pesticides in Totontepec, Oaxaca State, Mexico, was 20 times higher than that from a rhizosphere grown in a commercial tomato field treated with chemicals in Andalusia, Spain (van den Broek et al., unpubl. data). No single medium is definitely suited for an unbiased selection of all culturable rhizosphere bacteria. *Pseudomonas* isolation (PI) agar can be used to specifically favour the growth of pseudomonads. One should also keep in mind that only the culturable part of the rhizosphere population will be obtained.

Putative *Bacillus* strains are isolated by heating root samples at 80 °C for 10 min prior to washing the bacteria from the roots. The bacterial solution is plated on Luria-Bertani (LB) agar plates supplemented with cycloheximide (50 µg/ml) and incubated for 2–5 days at 28 °C. Colonies with a *Bacillus*-like morphology are then compared to *Bacillus*-type strains. To determine whether one is dealing with Gram-positive or Gram-negative organisms, a first identification of colonies can be performed by determining the ability to form mucoid threads after pulling a toothpick out of a bacterial suspension in 3 % KOH, which is indicative for Gram-negative organisms. A definite determination requires a standard Gram stain. Further characterisation methods include the use of Biolog, which is based on the ability of a strain to oxidise particular carbon sources, amplified ribosomal DNA restriction analysis (ARDRA), or PCR amplification of 16S ribosomal DNA fragments with specific primers followed by nucleotide sequencing and homology studies. In the

Biolog method, data sets derived from the carbon source utilisation patterns can be analysed with an appropriate software program (depending on the Gram character of the strain) and compared to known patterns of species present in commercially available databases. The latter two methods are based on specific sequences conserved between closely related species in the ribosomal rRNA gene fragments encompassing the 16S rDNA, the 16S–23S spacer region, and part of the 23S rDNA.

3.2 In Vitro Antifungal Activity Test

A simple in vitro assay to determine the activity against fungi can be performed by growing single bacterial colonies on agar medium in the presence of the fungus. The fungus is stab-inoculated in the centre of a Petri dish and bacterial strains are spot-inoculated at 2–2.5 cm distance from the fungus. The bacteria and fungus are allowed to grow concentrically and the formation of an inhibitory zone around the bacterial colony is an indication that the strain secretes a diffusible compound which inhibits growth of the fungus.

A large scale identification of antifungal activity in growth supernatants of bacterial cultures can be performed in 96-well microtiter plates in the presence of *F.o.r.l.* The assay allows the convenient screening of a large number of strains in a reproducible and quantitative way. Strains to be tested are grown in a 96-well microtiter plate in a volume of 200 µl. After growth, the cells are sedimented by centrifugation at 5000 rpm for 10 min and the culture supernatants are passed through a 0.45 or 0.22-µm pore size filter. A volume of 75 µl supernatant is mixed with an equal volume of an agarose-spore suspension (2x malt extract broth, $1.3x10^4$ spores/ml, 1.5 % (w/v) agarose). The final concentration of the spores in the wells is 1000 spores/well. The wells are sealed either with 75 µl paraffin oil (filter-sterile), with an oxygen-permeable plate seal, or with a piece of Saran wrap. Germination and mycelium growth is followed by measuring optical density (OD_{620}) of the wells using a microtiter plate reader (every hour) for approximately 72 h during growth at 28 °C. When an automated stack reader is used, many plates can be screened simultaneously in this way.

4 In Vivo Biocontrol Assays

4.1 *Fusarium oxysporum*–Tomato Biocontrol Assay in a Potting Soil System

Biocontrol of *Pseudomonas* and *Bacillus* rhizosphere isolates can be tested in a bioassay in which tomato seedlings grown from seeds coated with biocontrol bacteria are grown in potting soil infected with *F.o.r.l.* spores. Spores are obtained from liquid cultures and mixed with the soil prior to planting the seeds.

To isolate spores, *F.o.r.l.* is stab-inoculated onto potato dextrose agar medium and grown at 24 °C until the fungal mycelium covers the entire plate after a few days. One third of a PDA agar plate with *F.o.r.l.* is minced and used for inoculation of 200 ml Czapek-Dox medium in a 1-l Erlenmeyer flask. The fungus is grown for 2–3 days at room temperature under shaking at 110 rpm. Fungal mycelium and spore growth should be clearly visible at this stage. The *F.o.r.l.* inoculum is passed through Miracloth (Calbiochem-Novabiochem Corporation, La Jolla, CA, USA) or glass wool to remove the mycelium. The spore concentration is determined with a haemocytometer (with a depth of 100 µm).

The spore suspension is diluted in water to $1x10^6$ spores/ml and added to potting soil to a final concentration of $6x10^6$ spores/kg of soil. Spores are thoroughly mixed through the potting soil and the pots are filled with the infected soil. Seeds are sown in 8 plots of 12 pots, one seed per pot at a depth of 1–2 cm. Plants are watered from below to prevent disturbance of the root colonisation process.

Bacterial strains are coated onto the tomato seeds in a simple procedure using methylcellulose. *Pseudomonas* strains are grown overnight in 3 ml King's medium B at 28 °C. Bacilli are grown in 3 ml tryptic soy broth (TSB) for 3 days at 28 °C. The overnight culture of bacteria is washed with PBS to remove the growth medium and diluted to a concentration of $2x10^9$ CFU/ml. For bacilli the concentration is adjusted to $2x10^7$ CFU/ml. Then equal volumes of the bacterial suspension and a 2.0% (w/v) methylcellulose solution are mixed (methylcellulose is dissolved in water by vigorous stirring or by using a blender). Seeds are dipped into the mixture and dried in a sterile air stream on a filter paper. The coated seeds can be sown directly or kept at 4 °C for 1 or 2 days. The number of bacteria recovered from tomato seeds after coating is approximately 10^4 CFU/seed. Seedling germination is determined 1 week after sowing. Between 2–3 weeks after sowing, depending upon the disease pressure, the percentage of diseased plants is determined. A percentage of diseased plants of approximately 60% is preferred to perform statistical analyses.

4.2 Gnotobiotic *Fusarium oxysporum–Pythium ultimum* and *Rhizoctonia solani*–Tomato Bioassays

The gnotobiotic system used for this bioassay has been extensively used to study root colonisation. Briefly, tomato seeds are surface-sterilised in a 5% household sodium hypochlorite solution for 3 min, followed by four thorough rinses with 20 ml sterile water for 2 h. Incubation of sterilised tomato seeds on KB medium, in our hands, shows that this method consistently yields seeds free of contamination. After incubation for 24 h on agar-solidified plant nutrient solution (PNS) medium at 4 °C, seeds are allowed to germinate at 28 °C. Seedlings are inoculated 2 days later. A *F.o.r.l.* spore suspension, prepared as described previously, is added to the plant nutrient solution to a final concentration of $5x10^2$ spores/ml, which is than mixed through the sterile sand to 10 % (v/w) PNS.

Rhizoctonia solani was grown on 2 % water agar for 5 days. Discs of approximately 4 mm in diameter were cut from the edge of a growing colony and blended in PNS. *P. ultimum* was grown for 3–4 weeks in clarified V8 medium or hemp seed extract in water for 1–2 weeks. Oospores were collected free of the mycelium by washing them three times with sterile water and blending in 0.1 M sucrose. The blended culture was incubated for 2 h on a shaker (130 rpm) at 28 °C, sedimented, and resuspended in 1 M sucrose. To kill the mycelium fragments, the suspension was incubated at –20 °C for 12 h. The culture was washed, layered over 1 M sucrose and centrifuged at 2300 rpm. Oospores were added in a final concentration of 5–25 oospores/g of sand.

Germinated tomato seeds were incubated in a bacterial suspension with a concentration of 10^7 CFU/ml (*Pseudomonas*) or 10^9 CFU/ml (*Bacillus*) for 10 min, after which the germinated seeds were planted in the sand at a depth of approximately 5 mm. Seed inoculation is preferred above inoculation from soil since commercial biocontrol of tomato pathogens is also based on seed coating while the pathogen is already present in the soil. This form of inoculation also results in more reproducible experimental data. Plants were grown in a growth chamber or a greenhouse for 7 days and the disease index was determined by scoring the plants according a fixed disease index (Table 1). The data can be analysed statistically using a standard χ^2 analysis.

To confirm the presence of the fungus on plants, suspected diseased root parts can be placed in 0.005 % household bleach for 30 s, thoroughly rinsed with sterile water, and placed on a rich (LC or PDA) medium. Plates are inspected for fungal growth after incubation at 28 °C for 2 days.

Table 1. Example of *Pythium ultimum* disease indices

Disease symptoms	Disease index
No visible symptoms	0
Small brown spots on the main root and/or the crown	1
Brown spots on the central root and extensive discoloration of crown	2
Damping-off	3
Dead	4

5 Microscope Analysis of Infection and Biocontrol

5.1 Marking Fungi with Autofluorescent Proteins

5.1.1 Transformation of Pathogenic Fungi

5.1.1.1 Growth of Fungal Mycelium

Protoplasts are usually used for transformations of fungi. The removal of the cell wall is achieved by treating mycelia or germlings in the presence of lytic enzymes. The osmotic balance of protoplasts in a suspension is usually maintained using sugars such as sucrose and sorbitol and salts such as magnesium chloride, potassium chloride, and ammonium sulphate.

The following polyethylene/$CaCl_2$-mediated transformation procedure has been successfully applied to mycelium of *F.o.r.l.* Growing mycelium is prepared by inoculation of 100 ml potato dextrose broth in a 300-ml Erlenmeyer flask with a 5x4 mm size inoculum of mycelium. Depending on the particular *F.o.r.l.* strain, the fungus is grown between 2 and 5 days at 28 °C and 160 rpm. For example, *F. oxysporum* Fo47 is grown for 5 days, *F. oxysporum* f. sp. *radicis-lycopersici* ZUM2407 (IPO-DLO, Wageningen, The Netherlands) is grown for 2 days. Subsequently, the culture is passed through two layers of Miracloth and the filtrate is collected and sedimented by centrifugation at 5000 rpm for 10 min. The supernatant is immediately discarded and washed three times with 50 ml of sterile water and sedimented. The characteristic purple upper layer is discarded and the pellet is resuspended in 2–5 ml of sterile water. The spore concentration is determined with a haemocytometer. From this spore suspension, a number of $5x10^8$ conidia is inoculated into 40 ml potato dextrose broth and grown at 25 °C and 300 rpm for approximately 18 h or until the length of the germ tubes is at least ten times the size of a spore. The overall percentage of germinated spores should be higher than 95 %.

5.1.1.2 Preparation of Protoplasts

Germlings to be converted into protoplasts are sedimented by centrifugation at 2000 rpm for 10 min, after which the supernatant is carefully removed and the pellet resuspended in 25 ml magnesium sulphate solution (1.2 M $MgSO_4$, 50 mM sodium citrate, pH 5.8). The suspension is then passed through three layers of Miracloth. The mycelium trapped in the Miracloth is washed twice with magnesium sulphate solution and then transferred to a new tube with a cotton swab. The mycelium is then incubated in a protoplasting mix (10 mg/l Lysing Enzyme (Sigma L-2265, Sigma Chemicals Co., St. Louis, MO, USA), 15 mg/ml Driselase (Sigma Chemicals Co., St. Louis, MO, USA) in magnesium sulphate solution). The enzyme solution should be centrifuged to remove any solid particles prior to use. The mixture is incubated for 24 h at 30 °C on a shaker (65 rpm). The conversion of cells into protoplasts can be followed by phase contrast microscopy and, when the protoplastation nears completion, the protoplasts are collected on three layers of Miracloth, transferred to a new tube and washed with a sterile cold sorbitol solution (1 M sorbitol, 50 mM $CaCl_2$, 10 mM Tris-HCl, pH 7.4). The protoplasts are sedimented by centrifugation at 850xg (2100 rpm) at 4 °C and the number of protoplasts is determined with a haemocytometer.

5.1.1.3 Transformation of Protoplasts

Protoplast are transformed by addition of up to 15 µg of DNA to 200 µl of protoplast suspension and incubated on ice for 15 min, or stored at 4 °C. A volume of 1.0 ml PEG solution (60% (w/v) polyethylene glycol 6000, 50 mM $CaCl_2$, 10 mM Tris-HCl, pH 7.4) is slowly added while shaking gently. The mixture is incubated on ice for 30 min after which the protoplasts are washed with a magnesium sulphate/potato dextrose broth solution at 4 °C. The protoplasts are sedimented by centrifugation at 2500 rpm at 4 °C for 10 min and resuspended in the remaining fluid after discarding the supernatant. The protoplasts are incubated 30 min at room temperature and portions (50–1200 µl) are plated onto selective media containing 0.8 M sucrose, 10 mM Tris-HCl pH 7.4, 100 µg/ml hygromycin, and 1.5% (w/v) agar. Plates are incubated for 2 or 3 days at the appropriate growth temperature.

5.2 Marking Rhizosphere Bacteria with Autofluorescent Proteins

The green fluorescent protein (GFP) of the jellyfish *Aequorea victoria* has been rapidly and successfully adopted as an important marker for investigating processes in the rhizosphere. GFP is a 27-kDa polypeptide which converts the blue chemiluminescence of the Ca^{2+}-sensitive photoprotein aequorin into green light. The active chromophore is a tripeptide, the formation of which is oxygen-dependent and occurs gradually after translation by undergoing an autocatalytic reaction. GFP emits bright green light (λ_{max}=510 nm) when excited with ultraviolet (UV) or blue light (λ_{max}=395 nm) in vivo and in vitro.

GFP allows the non-destructive localisation and monitoring of individual cells on the root surface and does not require, unlike other biomarkers, exogenously added substrates, energy sources, or cofactors other than molecular oxygen. GFP fluorescence is stable, species-independent, requires no processing by the cells and fixing or staining is not necessary so artefacts cannot be introduced. However, if required, GFP allows fixation since it is unaffected by paraformaldehyde treatment. It is also stable under many other denaturing conditions such as the presence of denaturants or proteolytic enzymes, high temperatures (65 °C), and pH levels (6–12). Expression can be easily detected using epifluorescence or confocal laser scanning microscopy. Other optical methods that can be used to detect GFP-marked bacteria include the use of charge couple device (CCD) microscopy and cell sorting by fluorescent-activated cell sorters (FACS), which allows the sampling and identification of subpopulations of bacteria in a non-destructive way at the single cell level. Autofluorescently labelled colonies on agar plates can be detected under a hand-held UV-lamp or a low-resolution binocular microscope equipped with a UV lamp.

Since *gfp* is eukaryotic in origin, optimised constructs for the expression of *gfp* in bacteria have been constructed and successfully applied. This was achieved by expression of *gfp* under the control of strong constitutive promoters or using red-shifted and UV-optimised mutant derivates. These GFP variants provide an increased fluorescent signal intensity in bacteria, faster rates of oxidative chromophore formation, resistance to photobleaching and excitation maximums better suited to conventional detection instruments. GFPuv emits bright green light (maximum at 509 nm) when exposed to UV or blue light (395 or 470 nm). Mutant proteins GFPmut2 and GFPmut3 have emission maximums of 507 and 511 nm when excited by blue light (481 and 501 nm, respectively).

Stable plasmid vectors (multicopy) and transposon vectors (single copy) for marking with fluorescent proteins are available for use in Gram-negative as well as Gram-positive bacteria. They can be used for tagging bacteria with a biomarker, construction of fusion proteins, assaying gene activity, or promoter probing. Plasmids pGB5, carrying *gfp* driven by a *tac* promoter, was shown to be 100 % stably maintained in *Pseudomonas* in the tomato rhizosphere and resulted in constitutive expression in *Pseudomonas* without addition of an inducer. Dandie et al. (2001) constructed transposon-based tagging vectors using a *gfp* marker gene under control of either constitutive or inducible promoters.

Plasmids pFPV1 and pFPV2 direct high levels of *gfp* expression in *E. coli*, *Salmonella typhimurium*, and *Yersinia pseudotuberculosis* and in different mycobacterial species. The high levels of *gfp* expression were achieved by expression under control of the *lacZpo* and *hsp60* heat-shock promoters, respectively. They have been used to visualise the infection process of mammalian cells by the three species. Transposon plasmid Tn5GFP1 was successfully used to follow *Pseudomonas putida* cells during water transport through a sand matrix. To study the colonisation pattern of *P. chlororaphis* MA342 on barley seeds, the strain was tagged using a plasmid pUTgfp2X harbouring *gfp*.

For many applications, such as the analysis of chromosomal genes under physiological (monocopy) conditions using transcriptional fusions, stable integration of the reporter, or reduction of the risk of transfer of the genetic marker to other microorganisms, it is necessary to integrate the *gfp* transcriptional fusion into the chromosome of target bacteria by site-specific recombination or by random insertion, e.g. by means of transposons. A *gfp* cloning cassette vector, pGreenTIR, was designed specifically for use in the construction of prokaryotic transcriptional fusions. The cassette confers sufficient fluorescence to recipient cells to be used in low copy-number plasmids with promoters conferring low levels of transcription in *E. coli* and *Pseudomonas*. The bacterial transposon Tn*7* inserts at a high frequency into a specific intergenic site *att*Tn*7* on the chromosome in a number of Gram-negative bacteria. Tn*7*-based systems allow stable single-copy insertion of marker genes and insertion of transcriptional fusions in a single copy on the chromosome for gene expression studies at a neutral, intergenic site. Koch et al. developed a panel of flexible mini-Tn*7* delivery vectors, including cloning vectors with an increased number of unique cloning sites, the lack of which has limited the use of Tn*7* systems so far. A Tn*10*-based transposon was successfully used for fluorescence tagging of marine bacteria.

Based on mini-Tn*5* transposon derivatives, a *gfp* containing promoter-probe mini transposon was constructed for use in *Pseudomonas* species. Another set of vectors containing a mutated *gfp* gene was constructed for use with Gram-negative bacteria other than *E. coli*. pTn*3gfp* can be used for making random promoter probe *gfp* insertions into cloned DNA in *E. coli* for subsequent introduction into host strains. pUTmini-Tn*5gfp* can be used for making random promoter probe insertions directly into host strains. Plasmids p519*gfp* and p519*nfp* are broad host range mobilisable plasmids with *gfp* expressed from a *lac* and an *npt2* promoter, respectively.

Fluorescent markers can also be used to study viability and metabolic activity of bacteria in the rhizosphere. Normander et al. used *gfp*mut3b (Ser-64 Gly) to visualise the effect of indigenous populations on the distribution and activity of inoculated *P. fluorescens* DR54-BN14 in the barley rhizosphere. Using *gfp*-marked strains, they demonstrated that microcolonies of the inoculant strain were closely associated with cells of indigenous populations and that the majority of the cells have properties similar to those of starved cells.

Mutagenesis and protein engineering of the original GFP from the jellyfish *Aequorea* has yielded variants with different excitation and emission spectra that can be used for dual colour imaging. Many engineered variants also appear to be improved in other aspects such as photostability, codon usage, and thermosensitivity. The first dual colour imaging of bacteria in a mixed population of *E. coli* cells was achieved by selective excitation of wild-type GFP and mutant derivatives with a red-shift in the excitation spectrum.

Fluorescent proteins can also be successfully combined with the use of other biomarkers such as luciferase. To monitor cell numbers and metabolic activity of specific bacterial populations in liquid cultures and soil samples, a dual *gfp-luxAB* under control of the *psbA* promoter was integrated into the chromosomes of *E. coli* DH5a and *P. fluorescens* SBW25. Since luciferase output from *luxAB*-tagged bacteria decreases during starvation, *lux* expression was used as a marker for metabolic activity, while the much more stable *gfp* expression was used as an indicator for biomass. Alternatively, unstable variants of autofluorescent proteins with shorter half-lives can be used.

Variants, fluorescent in colours ranging from blue to yellow, namely blue fluorescent protein (BFP), yellow fluorescent protein (YFP), and cyan fluorescent protein (CFP), and optimised counterparts of EGFP and EBFP were created by mutagenesis. By labelling microorganisms differently, these variants can be used to track multiple microorganisms simultaneously. The major problem with using GFP variants to label strains for simultaneous detection is the complicated separation of the spectral overlap of the different GFP-isoforms. Recently, red fluorescent protein (drFP583 or DsRed), isolated from the tropical Indo Pacific reef coral *Discosoma* sp., has been cloned. With an emission maximum at 583 nm, DsRed is suitable for almost crossover-free dual colour labelling in combination with EGFP (emission 509 nm) upon simultaneous excitation. Similarly, combination of cells tagged with ECFP and EGFP or a mixture of cells labelled with ECFP and EYFP allows them to be clearly distinguished from each other in the tomato rhizosphere. In addition, DsRed can be combined with any other autofluorescent protein since the emission spectrum of DsRed does not overlap that of the others. Using different colours of fluorescent proteins, up to three labels (e.g. EGFP, ECFP and DsRed) can be simultaneously traced in the rhizosphere. These variants have also been used to visualise interactions of a DsRed-labelled biocontrol bacterium *P. chlororaphis* PCL1391 with *gfp*-labelled *F.o.r.l.* strain in the tomato rhizosphere (Lagopodi et al., unpubl. data). Bacteria were dually labelled merely to localise them in the rhizosphere.

The *gfp* genes can also be used as reporters for gene expression in the rhizosphere or for genes involved in quorum sensing. The estimated half-life of wild-type GFP is estimated to be at least 1 day. Since fluorescent proteins are extremely stable, they cannot be used for transient (real time) gene expression studies. Less stable variants have been constructed that can be used for analysis of transient gene expression in bacteria and, hence, promoter activity in the rhizosphere. Unstable variants of fluorescent proteins can be produced by addition of C-terminal degradation domains to the protein that are targets of natural protein degradation systems in cells. One such system exploits the action of intracellular tail-specific protein via the *ssrA*-mediated peptide degradation of prematurely terminated polypeptides at the C-terminal end. Homologues of *ssrA* have been identified in both Gram-negative and Gram-positive bacteria. Gfpmut3 derivatives carrying these degradation domains have half-lives between 40 min and 2 h, while the estimated half-life of wild *gfp*mut3 is estimated to be at least 1 day.

GFP can also be used and expressed in Gram-positive species such as *Bacillus* spp. pAD213 was constructed as a promoter-trap plasmid for *Bacillus cereus*. It allows screening of large libraries for identifying regulatory sequences and screening using flow cytometry and cell sorting. Plasmid vectors have been described that enable routine production of GFP, YFP and CFP fusions in Gram-positive bacteria.

One disadvantage of the use of fluorescent proteins is the maturation time of the protein, particularly that of DsRed. Although EGFP requires ~ 4 h for efficient microscopic visualisation, visualisation of DsRed requires longer periods. This delay is not due to inefficient expression of the DsRed protein since the protein can be detected in high quantities very soon, but it is rather due to an extended maturation time of the protein (20–48 h). DsRed is in fact brighter than first reported, but the fluorescence matures very slowly and the protein naturally forms a tetramer. More rapidly maturing and soluble variants of DsRed have been generated by mutagenesis (Brooke and Glick 2002). Furthermore, *E. coli* cells expressing DsRed protein are in general smaller than cells expressing EGFP or untransformed bacteria, indicating that DsRed might have a toxic effect. Another problem with the use of fluorescent proteins is the variability of expression in different bacterial species. GFP expressed from the same constructs is two to ten times higher expressed in *E. coli* than in pseudomonads. Interference by other fluorescent particles, bacteria, or root autofluorescence may also introduce artefacts or complicate the observations.

5.3 Confocal Laser Scanning Microscopy of Rhizosphere Interactions

The advent of fluorescent proteins offers a broad range of applications to track bacteria and study gene expression in the rhizosphere. By labelling different strains with different flavours of fluorescent proteins such as green, red, blue, or yellow fluorescent protein, multiple bacterial strains and their interactions with pathogens can be tracked simultaneously in the rhizosphere.

To express *gfp* in *F.o.r.l.*, pGFDGFP on which the *sgfp* gene is cloned between the *A. nidulans gpdA* promoter and the *trpC* terminator sequences was transformed to *F.o.r.l.*. The fungus was transformed by the previously described polyethyleneglycol/$CaCl_2$-mediated transformation of protoplasts in the presence of pAN7–1, which allows selection for hygromycin B resistance (100 µg/ml). The level of *gfp* expression was high in the mycelium, micro- and macroconidia, and chlamydospores. The labelled isolates were equally pathogenic to tomato as the wild type. The marked fungus was introduced into the gnotobiotic sand system by mixing spores with sand. First, the interactions between fungal pathogens and the tomato root were studied. CLSM observations show that after 2 days the main root is surrounded by hyphae, which are interwoven with the root hairs. The contact between hyphae and the root was initiated at or via the root hairs. After 3 days, spot attachments of hyphae to the root surface are observed, predominantly at the crown and hyphae grow along the junctions of the epidermal cells after attachment. The first infection events take place 4 days after inoculation, as observed by penetration of epidermal cells by hyphae. No penetration structures are observed except for swollen hyphae at the penetration site. Five days after planting, at which the first disease symptoms can be observed, a tight network of hyphae has grown around the root surface and epidermal cells are intercellularly colonised by hyphae. After complete destruction of the root system, the fungus forms macroconidia and starts colonising the cotyledons.

After introduction of biocontrol bacteria to the test system, observations show that in the *F.o.r.l.* -tomato biocontrol system *Pseudomonas* bacteria not only colonise the tomato root surface, but also fungal hyphae (Bolwerk and Lagopodi, unpublished). These are indications that biocontrol bacteria not only protect the roots against fungi by niche exclusion and production of antibiotics, but that they actively attack the pathogen. Still, there is much to be discovered from these rhizosphere studies. The use of autofluorescent proteins has shown to be a promising way of visualising and understanding the interactions taking place in the rhizosphere between *Pseudomonas* and *Bacillus* biocontrol strains and fungal pathogens.

6 Conclusions

The whole procedure of isolation, screening for antifungal activity, and determining disease suppression in bioassays allows fast isolation of potential biocontrol strains. The gnotobiotic test system has proven to be a valuable test system to study interactions between biocontrol bacteria, phytopathogen, and host plant. Combined with the use of autofluorescent proteins, it provides us with an extraordinary opportunity to study the intricate cellular and molecular interactions that the key players use to mediate their actions in the rhizosphere.

References and Selected Reading

Alabouvette C (1986) Fusarium wilt suppressive soils from the Chateaurenard region: reviews of a 10 year study. Agronomie 6:273–284

Alexeyev MF, Shokolenko IN, Croughan TP (1995) New mini-Tn5 derivatives for insertion mutagenesis and genetic engineering in gram-negative bacteria. Can J Microbiol 41:1053–1055

Andersen JB, Sternberg C, Poulsen LK, Bjorn SP, Givskov M, Molin S (1998) New unstable variants of green fluorescent protein for studies of transient gene expression in bacteria. Appl Environ Microbiol 64:2240–2246

Anderson M, Pollitt CE, Roberts IS, Eastgate JA (1998) Identification and characterization of the *Erwinia amylovora rpoS* gene: RpoS is not involved in induction of fireblight disease symptoms. J Bacteriol 180:6789–6792

Baird GS, Zacharias DA, Tsien RY (2000) Biochemistry, mutagenesis, and oligomerization of DsRed, a red fluorescent protein from coral. Proc Natl Acad Sci USA 97:11984–11989

Ballance DJ, Buxton FP, Turner G (1983) Transformation of *Aspergillus nidulans* by the orotidine-5′-phosphate decarboxylase gene of *Neurospora crassa*. Biochem Biophys Res Commun 112:284–289

Beckman CH (1987) The nature of wilt diseases of plants. The American Phytopathological Society Press, Saint Paul, MN

Bloemberg GV, O'Toole GA, Lugtenberg BJJ, Kolter R (1997) Green fluorescent protein as a marker for *Pseudomonas* spp. Appl Environ Microbiol 63:4543–4551

Bloemberg GV, Wijfjes AH, Lamers GE, Stuurman N, Lugtenberg BJ (2000) Simultaneous imaging of *Pseudomonas fluorescens* WCS365 populations expressing three different autofluorescent proteins in the rhizosphere: new perspectives for studying microbial communities. Mol Plant-Microbe Interact 13:1170–1176

Bochner BR (1989) Sleuthing out bacterial identities. Nature 339:157–158

Brooke JB, Glick BS (2002) Rapidly maturing variants of the *Discosoma* red fluorescent protein (DsRed). Nat Biotechnol 20:83–87

Brown JW, Hunt DA, Pace NR (1990) Nucleotide sequence of the 10Sa RNA gene of the beta-purple eubacterium *Alcaligenes eutrophus*. Nucleic Acids Res 18:2820

Burlage RS, Yang ZK, Mehlhorn T (1996) A transposon for green fluorescent protein transcriptional fusions: application for bacterial transport experiments. Gene 173:53–58

Chalfie M, Tu Y, Euskirchen G, Ward WW, Prasher DC (1994) Green fluorescent protein as a marker for gene expression. Science 263:802–805

Chin-A-Woeng TFC, Bloemberg GV, van der Bij AJ, van der Drift KMGM, Schripsema J, Kroon B, Scheffer RJ, Keel C, Bakker PAHM, Tichy HV, de Bruijn FJ, Thomas-Oates JE, Lugtenberg BJJ (1998) Biocontrol by phenazine-1-carboxamide-producing *Pseudomonas chlororaphis* PCL1391 of tomato root rot caused by *Fusarium oxysporum* f. sp. *radicis-lycopersici*. Mol Plant-Microbe Interact 11:1069–1077

Chiu WL, Niwa Y, Zeng W, Hirano T, Kobayashi H, Sheen J (1996) Engineered GFP as a vital reporter in plants. Curr Biol 6:325–330

Cody CW, Prasher DC, Westler WM, Prendergast FG, Ward WW (1993) Chemical structure of the hexapeptide chromophore of the *Aequorea* green-fluorescent protein. Biochemistry 32:1212–1218

Cormack BP, Valdivia RH, Falkow S (1996) FACS-optimized mutants of the green fluorescent protein (GFP). Gene 173:33–38

Cowan SE, Gilbert E, Khlebnikov A, Keasling JD (2000) Dual labeling with green fluorescent proteins for confocal microscopy. Appl Environ Microbiol 66:413–418

Craig NL (1989) Transposon Tn*7*. In: Berg DE, Howe MM (eds) Mobile DNA. American Society for Microbiology, Washington, pp 211–225

Crameri A, Whitehorn EA, Tate E, Stemmer WP (1996) Improved green fluorescent protein by molecular evolution using DNA shuffling. Nat Biotechnol 14:315–319

Cuppers HGAM, Oomes S, Brul S (1997) A model for the combined effects of temperature and salt concentration on growth rate of food spoilage molds. Appl Environ Microbiol 63:3764–3769

Dandie CE, Thomas SM, McClure NC (2001) Comparison of a range of green fluorescent protein-tagging vectors for monitoring a microbial inoculant in soil. Lett Appl Microbiol 32:26–30

Daubner SC, Astorga AM, Leisman GB, Baldwin TO (1987) Yellow light emission of *Vibrio fischeri* strain Y-1: purification and characterization of the energy-accepting yellow fluorescent protein. Proc Natl Acad Sci USA 84:8912–8916

de Lorenzo V, Timmis KN (1994) Analysis and construction of stable phenotypes in gram-negative bacteria with Tn*5*- and Tn*10*-derived minitransposons. Methods Enzymol 235:386–405 de Lorenzo V, Herrero M, Jakubzik U, Timmis KN (1990) Mini-Tn*5* transposon derivatives for insertion mutagenesis, promotor probing, and chromosomal insertion of cloned DNA in Gram-negative Eubacteria. J Bacteriol 172:6568–6572

Dekkers LC, Mulders IH, Phoelich CC, Chin AWT, Wijfjes AH, Lugtenberg BJ (2000) The *sss* colonization gene of the tomato-*Fusarium oxysporum* f. sp. *radicis-lycopersici* biocontrol strain *Pseudomonas fluorescens* WCS365 can improve root colonization of other wild-type *Pseudomonas* spp. bacteria. Mol Plant Microbe Interact 13:1177–1183

Dunn AK, Handelsman J (1999) A vector for promoter trapping in *Bacillus cereus.* Gene 226:297–305

Emmert EAB, Handelsman J (1999) Biocontrol of plant disease: a (Gram-) positive perspective. FEMS Microbiol Lett 171:1–9

Feucht A, Lewis PJ (2001) Improved plasmid vectors for the production of multiple fluorescent protein fusions in *Bacillus subtilis.* Gene 264:289–297

Geels FP, Schippers G (1983) Selection of antagonistic fluorescent *Pseudomonas* spp. and their root colonization and persistence following treatment of seed potatoes. Phytopath Z 108:193–206

Glandorf DCM, Sluis I, Anderson AJ, Bakker PAHM, Schippers B (1994) Agglutination, adherence, and root colonization by fluorescent pseudomonads. Appl Environ Microbiol 60:1726–1733

Gutterson N (1990) Microbial fungicides: recent approaches to elucidating mechanisms. Crit Rev Biotechnol 10:69–91

Haidinger W, Szostak MP, Beisker W, Lubitz W (2001) Green fluorescent protein (GFP)-dependent separation of bacterial ghosts from intact cells by FACS. Cytometry 44:106–112

Halebian S, Harris B, Finegold SM, Rolfe RD (1981) Rapid method that aids in distinguishing gram-positive from gram-negative anaerobic bacteria. J Clin Microbiol 13:444–448

Handelsman J, Raffel SJ, Mester EH, Wunderlich L, Grau CR (1999) Biological control of damping-off of alfalfa seedlings with *Bacillus cereus* UW85. Appl Environ Microbiol 56:713–718

Heim R, Prasher DC, Tsien RY (1994) Wavelength mutations and posttranslational autoxidation of green fluorescent protein. Proc Natl Acad Sci USA 91:12501–12504

Hoffland E, Findenegg GR, Nelemans JA (1989) Solubilization of rock phosphate by rape. Plant Soil 113:161–165

Inouye S, Tsuji FI (1994) Evidence for redox forms of the *Aequorea* green fluorescent protein. FEBS Lett 351:211–214

Johnstone IL, Hughes SG, Clutterbuck AJ (1985) Cloning an *Aspergillus nidulans* developmental gene by transformation. EMBO J 4:1307–1311

Jakobs S, Subramaniam V, Schonle A, Jovin TM, Hell SW (2000) EFGP and DsRed expressing cultures of *Escherichia coli* imaged by confocal, two-photon and fluorescence lifetime microscopy. FEBS Lett 479:131–135

Karatani H, Wilson T, Hastings JW (1992) A blue fluorescent protein from a yellow-emitting luminous bacterium. Photochem Photobiol 55:293–299

Keiler KC, Waller PR, Sauer RT (1996) Role of a peptide tagging system in degradation of proteins synthesized from damaged messenger RNA. Science 271:990–993

King EO, Ward MK, Raney DE (1954) Two simple media for the demonstration of pyocyanin and fluorescein. J Lab Clin Med 44:301–307

Kistler HC, Benny UK (1988) Genetic transformation of the fungal wilt pathogen, *Fusarium oxysporum.* Curr Genet 13:145–149

Koch B, Jensen LE, Nybroe O (2001) A panel of Tn*7*-based vectors for insertion of the *gfp* marker gene or for delivery of cloned DNA into Gram-negative bacteria at a neutral chromosomal site. J Microbiol Methods 45:187–195

Kremer RJ, Begonia MFT, Stanley L, Lanham ET (1990) Characterization of rhizobacteria associated with weed seedlings. Appl Environ Microbiol 56:1649–1655

Lagopodi AL, Ram AFJ, Lamers GEM, Punt PJ, van den Hondel CAMJJ, Lugtenberg BJJ, Bloemberg GV (2002) Novel aspects of tomato root colonization and infection by *Fusarium oxysporum* f. sp. *radicis-lycopersici* revealed by confocal laser scanning microscopic analysis using the green fluorescent protein as a marker. Mol Plant-Microbe Interact 15:172–179

Lambert B, Leyns F, Rooyen L, Gossele F, Papon Y, Swings J (1987) Rhizobacteria of maize and their antifungal activities. Appl Environ Microbiol 53:1866–1871

Lambert B, Meire P, Joos H, Lens P, Swings J (1990) Fast-growing, aerobic, heterotrophic bacteria from the rhizosphere of young sugar beet plants. Appl Environ Microbiol 56:3375–3381

Lewis PJ, Errington J (1996) Use of green fluorescent protein for detection of cell-specific gene expression and subcellular protein localization during sporulation in *Bacillus subtilis*. Microbiology 142(Pt 4):733–740

Lewis PJ, Marston AL (1999) GFP vectors for controlled expression and dual labelling of protein fusions in *Bacillus subtilis*. Gene 227:101–110

Macheroux P, Schmidt KU, Steinerstauch P, Ghisla S, Colepicolo P, Buntic R, Hastings JW (1987) Purification of the yellow fluorescent protein from *Vibrio fischeri* and identity of the flavin chromophore. Biochem Biophys Res Commun 146:101–106

Matthysse AG, Stretton S, Dandie C, McClure NC, Goodman AE (1996) Construction of GFP vectors for use in gram-negative bacteria other than *Escherichia coli*. FEMS Microbiol Lett 145:87–94

Matz MV, Fradkov AF, Labas YA, Savitsky AP, Zaraisky AG, Markelov ML, Lukyanov SA (1999) Fluorescent proteins from nonbioluminescent *Anthozoa* species. Nat Biotechnol 17:969–973

Mess JJ, Wit R, Testerink CS, de Groot F, Haring MA, Cornelissen BJC (1999) Loss of avirulence and reduced pathogenicity of a gamma-irradiated mutant of *Fusarium oxysporum* f. sp. *lycopersici*. Phytopathology 89:1131–1137

Miller DM III, Desai NS, Hardin DC, Piston DW, Patterson GH, Fleenor J, Xu S, Fire A (1999) Two-color GFP expression system for *C. elegans*. BioTechniques 26:914–916

Miller HJ, Henken G, van Veen JA (1989) Variation and composition of bacterial populations in the rhizospheres of maize, wheat, and grass cultivars. Can J Microbiol 35:656–660

Miller HJ, Liljeroth E, Henken G, van Veen JA (1990) Fluctuations in the fluorescent pseudomonad and actinomycete populations of the rhizosphere and rhizoplane during growth of spring wheat. Can J Microbiol 36:254–258

Miller WG, Lindow SE (1997) An improved GFP cloning cassette designed for prokaryotic transcriptional fusions. Gene 191:149–153

Normander B, Hendriksen NB, Nybroe O (1999) Green fluorescent protein-marked *Pseudomonas fluorescens*: localization, viability, and activity in the natural barley rhizosphere. Appl Environ Microbiol 65:4646–4651

Oakley BR, Rinehart JE, Mitchell BL, Oakley CE, Carmona C, Gray GL, May GS (1987) Cloning, mapping and molecular analysis of the *pyrG* (orotidine-5′- phosphate decarboxylase) gene of *Aspergillus nidulans*. Gene 61:385–399

Osmani SA, May GS, Morris NR (1987) Regulation of the mRNA levels of *nimA*, a gene required for the G2-M transition in *Aspergillus nidulans*. J Cell Biol 104:1495–1504

Pusey PL (1999) Use of *Bacillus subtilis* and related organisms as biofungicides. Pestic Sci 27:133–140

Pusey PL, Wilson CL (1984) Postharvest biological control of stone fruit brown rot by *Bacillus subtilis*. Plant Dis 68:753–756

Scher FM, Baker R (1980) Mechanism of biological control in a fusarium-suppressive soil. Phytopathology 72:1567–1573

Silo-suh LA, Lethbridge BJ, Raffel SJ, He H, Clardy J, Handelsman J (1994) Biological activities of two fungistatic antibiotics produced by *Bacillus cereus* UW85. Appl Environ Microbiol 60:2023–2030

Simons M, van der Bij AJ, Brand J, de Weger LA, Wijffelman CA, Lugtenberg BJJ (1996) Gnotobiotic system for studying rhizosphere colonization by plant growth-promoting *Pseudomonas* bacteria. Mol Plant-Microbe Interact 9:600–607

Steidle A, Sigl K, Schuhegger R, Ihring A, Schmid M, Gantner S, Stoffels M, Riedel K, Givskov M, Hartmann A, Langebartels C, Eberl L (2001) Visualization of *N*-acylhomoserine lactone-mediated cell-cell communication between bacteria colonizing the tomato rhizosphere. Appl Environ Microbiol 67:5761–5770

Stretton S, Techkarnjanaruk S, McLennan AM, Goodman AE (1998) Use of green fluorescent protein to tag and investigate gene expression in marine bacteria. Appl Environ Microbiol 64:2554–2559

Stutz EW, Defago G, Kern H (1986) Naturally occurring fluorescent pseudomonads involved in the suppression of black root rot of tobacco. Phytopathology 76:181–185

Stuurman N, Bras CP, Schlaman HR, Wijfjes AH, Bloemberg G, Spaink HP (2000) Use of green fluorescent protein color variants expressed on stable broad-host-range vectors to visualize rhizobia interacting with plants. Mol Plant Microbe Interact 13:1163–1169

Suarez A, Guttler A, Stratz M, Staendner LH, Timmis KN, Guzman CA (1997) Green fluorescent protein-based reporter systems for genetic analysis of bacteria including monocopy applications. Gene 196:69–74

Tilburn J, Scazzocchio C, Taylor GG, Zabicky-Zissman JH, Lockington RA, Davies RW (1983) Transformation by integration in *Aspergillus nidulans*. Gene 26:205–221

Tombolini R, Unge A, Davey ME, deBruijn FJ, Jansson JK (1997) Flow cytometric and microscopic analysis of GFP-tagged *Pseudomonas fluorescens* bacteria. FEMS Microbiol Ecol 22:17–28

Tombolini R, van der Gaag DJ, Gerhardson B, Jansson JK (1999) Colonization pattern of the biocontrol strain *Pseudomonas chlororaphis* MA 342 on barley seeds visualized by using green fluorescent protein. Appl Environ Microbiol 65:3674–3680

Tyagi JS, Kinger AK (1992) Identification of the 10Sa RNA structural gene of *Mycobacterium tuberculosis*. Nucleic Acids Res 20:138

Unge A, Tombolini R, Davey ME, de Bruijn FJ, Jansson JK (1998) GFP as a marker gene. In: Akkermans AD, van Elsas JD, de Bruijn FJ (eds) Molecular microbial ecology manual. Kluwer, Dordrecht, pp 1–16

Unge A, Tombolini R, Molbak L, Jansson JK (1999) Simultaneous monitoring of cell number and metabolic activity of specific bacterial populations with a dual *gfp-luxAB* marker system. Appl Environ Microbiol 65:813–821

Ushida C, Himeno H, Watanabe T, Muto A (1994) tRNA-like structures in 10Sa RNAs of *Mycoplasma capricolum* and *Bacillus subtilis*. Nucleic Acids Res 22:3392–3396

Valdivia RH, Falkow S (1996) Bacterial genetics by flow cytometry: rapid isolation of *Salmonella typhimurium* acid-inducible promoters by differential fluorescence induction. Mol Microbiol 22:367–378

Valdivia RH, Hromockyj AE, Monack D, Ramakrishnan L, Falkow S (1996) Applications for green fluorescent protein (GFP) in the study of host–pathogen interactions. Gene 173:47–52

Vaneechoutte M, Boerlin P, Tichy HV, Bannerman E, Jager B, Bille J (1998) Comparison of PCR-based DNA fingerprinting techniques for the identification of *Listeria* species and their use for atypical *Listeria* isolates. Int J Syst Bacteriol 48:127–139

Ward DM (1989) Molecular probes for analysis of microbial communities. In: Characklis WG, Wilderer PA (eds) Structure and function of biofilms. Wiley, New York, pp 145–155

Waterhouse RN, Buhariwalla H, Bourn D, Rattray EAS, Glover LA (1996) CCD detection of *lux*-marked *Pseudomonas syringae* pv. *phaseolicola* forms associated with Chinese-cabbage and the resulting disease protection against *Xanthomonas campestris*. Lett Appl Microbiol 22:262–266

Weller DM, Cook RJ (1983) Suppression of take-all of wheat by seed treatments with fluorescent pseudomonads. Phytopathology 73:463–469

Weller DM, Zhang BX, Cook RJ (1985) Application of a rapid screening test for selection of bacteria suppressive to take-all of wheat. Plant Dis 69:710–713

Williams JG, Kubelik AR, Livak KJ, Rafalski JA, Tingey SV (1990) DNA polymorphisms amplified by arbitrary primers are useful as genetic markers. Nucleic Acids Res 18:6531–6535

Yang TT, Kain SR, Kitts P, Kondepudi A, Yang MM, Youvan DC (1996a) Dual color microscopic imagery of cells expressing the green fluorescent protein and a red-shifted variant. Gene 173:19–23

Yang TT, Cheng L, Kain SR (1996b) Optimized codon usage and chromophore mutations provide enhanced sensitivity with the green fluorescent protein. Nucleic Acids Res 24:4592–4593

Yang TT, Sinai P, Green G, Kitts PA, Chen YT, Lybarger L, Chervenak R, Patterson GH, Piston DW, Kain SR (1998) Improved fluorescence and dual color detection with enhanced blue and green variants of the green fluorescent protein. J Biol Chem 273:8212–8216

Yelton MM, Hamer JE, Timberlake WE (1984) Transformation of *Aspergillus nidulans* by using a *trpC* plasmid. Proc Natl Acad Sci USA 81:1470–1474

24 Microbial Community Analysis in the Rhizosphere by in Situ and ex Situ Application of Molecular Probing, Biomarker and Cultivation Techniques

Anton Hartmann, Rüdiger Pukall, Michael Rothballer, Stephan Gantner, Sigrun Metz, Michael Schloter and Bernhard Mogge

1 Introduction

It is well known that the bacterial diversity in soil habitats is much greater compared to the artificial cultivation techniques (Torsvik et al. 1996; Chatzinotas et al. 1998). It is generally accepted that only a combination of methods including cultivation and several cultivation-independent techniques is able to provide a more representative picture of the microbial diversity in environmental habitats (Wagner et al. 1993; Liesack et al. 1997). This is also true for the plant/soil compartment, although the degree of culturability is thought to be higher on the root surface. Supposedly, rhizosphere microbes respond to the presence of easily consumable substrates on the root surface with fast growth rates, which is indicative for r-strategy; successful colonization of the rhizosphere is the final result of this behavior.

In-depth characterization of bacterial communities residing in environmental habitats has been greatly stimulated by the application of molecular phylogenetic tools, such as 16S ribosomal RNA-directed oligonucleotide probes derived from extensive 16S rDNA sequence analysis. These phylogenetic probes can be successfully applied in diverse microbial habitats using the fluorescence in situ hybridization (FISH) technique (Giovannoni et al. 1988; Amann et al. 1995; Tas and Lindström 2001). In addition, the application of the immunofluorescence techniques to detect specific subpopulations of enzymes and of fluorescence marker-tagged bacteria or reporter constructs enables a highly resolving population and functional analysis (Hartmann et al. 1997; Unge et al. 1999). Phylogenetic in situ studies of the population structure can thus be supplemented with functional or phenotypic in situ investigation approaches.

Plant Surface Microbiology
A. Varma, L. Abbott, D. Werner, R. Hampp (Eds.)

The rhizosphere is defined as the soil compartment which is greatly influenced by plant roots (Campbell and Greaves 1990a). The rhizosphere microbial community is shaped by the effect of root exudates (Brimecomb et al. 2001). Several methodological approaches are available to study the rhizosphere carbon flow and the microbial population dynamics induced by root-born carbon sources (Morgan and Whipps 2001). In addition, multiple communicative links exist between the rhizosphere microflora and the roots on the basis of highly specific organic signals (Werner 2001). It is appropriate to distinguish the root itself (with the endorhizosphere and the root surface, the rhizoplane) from the soil compartment surrounding the root (bulk soil and ectorhizosphere). In the following sections, two experimental approaches to investigate root-associated bacterial communities are presented. Figure 1 provides a flow diagram of the separation of the rhizosphere compartments and the various in situ and ex situ methods applied. On one hand, population and functional studies can be conducted directly in the rhizoplane (in situ) by combining specific fluorescence probing with confocal laser scanning microscopy yielding detailed information about the localization and small scale distribution of bacterial cells and their activities on the root surface (Sect. 2). On the other hand, the separated rhizosphere compartments and the bacteria extracted from these different compartments allow a variety of subsequent ex situ-studies (Sect. 3). Studies, such as cultivation of bacteria on plates and microscopic counting of bacteria on filters after FISH analysis provide quan-

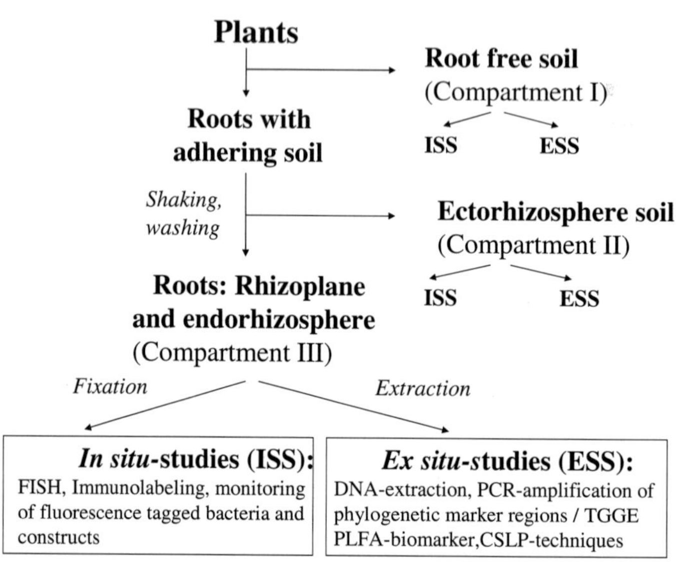

Fig. 1. Flow diagram of separation of rhizosphere compartments and overview of in situ and ex situ analyses using molecular probing, biomarker and cultivation techniques

titative data about the community composition. In addition, the bacterial diversity can be investigated using PCR-amplification of phylogenetic marker genes combined with subsequent electrophoretic fingerprint analysis or cloning and sequencing studies. These approaches can be supplemented by a general microbial structural and functional diversity analysis using community phospholipid fatty acid and substrate utilization pattern analysis, respectively.

2 In Situ Studies of Microbial Communities Using Specific Fluorescence Labeling and Confocal Laser Scanning Microscopy

A detailed understanding of the ecology of bacterial populations requires in situ information about the localization of the colonization sites at specific areas on root surfaces and also about neighboring populations. Therefore, true in situ studies need to be performed and these must include an identification of the bacteria on a phylogenetic level and also information about their in situ activity. Since soil and plant surfaces are very complex in microstructure and optical appearance, special microscopic techniques have to be applied. Confocal laser scanning microscopy enables us to circumvent to a great degree disturbing autofluorescence from out-of-focus-planes by performing optical sections (xy and xz scans) through the sample (Hartmann et al. 1998). It has been demonstrated that CSLM studies combined with the application of specific fluorescent probes considerably improve microbial ecology studies in the rhizosphere (Schloter et al. 1993; Aßmus et al. 1995). The confocal pinhole cuts off all out-of-focus fluorescence to reach the amplifiers. The application of several lasers with different excitation wavelengths in combination with differently fluoro-labeled probes allow the simultaneous analysis of different populations and/or activities (Amann et al. 1995; Stoffels et al. 2001). If possible, nested approaches with overlapping probe specificities should be used to improve the fidelity of the in situ identification, e.g., by fluorescence in situ hybridization. In addition, the use of the green fluorescent protein (GFP) as a structural and functional autofluorescence marker has successfully lightened up the biology and ecology of diverse biota, including bacteria, fungi, protozoa and plants (Lorang et al. 2001).

2.1 Fluorescence in Situ Hybridization

Root samples are fixed overnight at 4 °C in 3 % paraformaldehyde containing PBS (phosphate-buffered saline, composed of 0.13 M NaCl, 7 mM Na_2HPO_4 and 3 mM NaH_2PO_4 [pH 7.2]). Root pieces are washed in PBS, mixed with 0.3 % agarose, dropped onto glass slides and dried at room temperature.

Table 1. Phylogenetic oligonucleotide probes for fluorescence in situ hybridization (FISH) and dot blot hybridization

Probe	Probe sequence (5′–3′)	Target site, rRNA position[a]	Specificity	Reference
EUB338	GCTGCCTCCCGTAGGAGT	16S rRNA, 338–355	*Bacteria*	Amann et al. (1990)
EUB788[b]	CTACCAGGGTATCTAATCC	16S rRNA, 785–803	*Bacteria*	Lee et al. (1993)
EUB927[b]	ACCGCTTGTGCGGGCCC	16S rRNA, 927–942	*Bacteria*	Giovannoni et al. (1988)
EUB1055[b]	CACGAGCTGACGACAGCCAT	16S rRNA, 1055–1074	*Bacteria*	Lee et al. (1993)
EUB1088[b]	GCTCGTTGCGGGACTTAACC	16S rRNA, 1088–1107	*Bacteria*	Lee et al. (1993)
ALF1b	CGTTCG(C/T)TCTGAGCCAG	16S rRNA, 19–35	Alpha subclass of Proteobacteria	Manz et al. (1992)
BET42a[c]	GCCTTCCCACTTCGTTT	23S rRNA, 1027–1043	Beta subclass of Proteobacteria	Manz et al. (1992)
CF319a	TGGTCCGTGTCTCAGTAC	16 SrRNA, 319–336	*Cytophaga-Flavobacterium* cluster	Manz et al. (1996)
GAM42a[c]	GCCTTCCCACATCGTTT	23S rRNA, 1027–1043	Gamma subclass of Proteobacteria	Manz et al. (1992)
Rhi1247	TCGCTGCCCACTGTC	16S rRNA, 1247–1261	*Rhizobium, Ochrobactrum*	Ludwig et al. (1998)
GP[d]	TCATCATGCCCCTTATG	16S rRNA, 1199–1215	Gram-positive bacteria	Rheims et al. (1996)
HM[d]	CCCTGAGTTATTCCGAAC	16S rRNA, 142–159	*Hyphomicrobium*, methylotrophs	Tsien et al. (1990)
PLA[d]	GGC(GA)TGGATTAGGCATGC	16S rRNA, 41–58	*Planctomycetaceae*	Liesack and Stackebrandt (1992)
PS-MG[d]	CCTTCCTCCCAACTT	16S rRNA, 440–454	*Pseudomonas aeruginosa*	Braun-Howland et al. (1993)

[a] *E. coli* numbering, Brosius et al. (1981)
[b] Used in combination with probe EUB338 and three other domain-specific probes for quantification of bacterial cells on filters (EUB-MIX)
[c] Used with an equimolar amount of unlabeled competitor oligonucleotide GAM42a or BET42a, respectively
[d] Used for dot blot hybridization only

These glass slides are immersed in 50, 80 and 96 % ethanol for 3 min each and stored at room temperature. Oligonucleotide probes (Table 1) labeled with Cy3, Cy5 or 5(6)-carboxyfluorescein-*N*-hydroxysuccinimide ester (FLUOS) at the 5″-end are used. The oligonucleotides are stored in distilled water at a concentration of 50 ng/µl (Amann et al. 1990).

FISH was performed as described in detail, e.g., by Wagner et al. (1993) at 46 °C for 90 min in hybridization buffer (20 mM Tris-HCl, pH 7.2, 0.01 % SDS and 5 mM EDTA) containing 0.9 M NaCl and formamide at the percentages shown in Table 1. Hybridization was followed by a stringent washing step at 48 °C for 15 min. The washing buffer was removed by rinsing the slides with distilled water. Counterstaining with DAPI and mounting in Citifluor AF1 (Citifluor Ltd., London, UK) was performed as described previously (Aßmus et al. 1995).

The microscopic in situ analysis can be performed using an LSM 410 or LSM 510 inverted confocal laser scanning microscope (Zeiss, Jena, Germany), equipped with two lasers (Ar-ion UV; Ar-ion visible; HeNe) supplying excitation wavelengths at 365, 488, 543 and 633 nm, respectively. Sequentially recorded images are assigned to the respective fluorescence color and then merged into a true color display. All image combining and processing is performed with the standard software provided by Zeiss.

Using the general cell/DNA staining with DAPI and FISH with probes specific for the domain bacteria and group-specific probes (Table 1), bacteria can simultaneously be localized and identified at the rhizoplane. In addition to the group-specific probes, in situ binding genus- and species-specific oligonucleotide probes are available for a number of root-associated and symbiotic bacteria (Ludwig et al. 1998; Hartmann et al. 2000). Figure 2A shows the localization of *Azospirillum brasilense* in the wheat rhizosphere by FISH (combination of two differently labeled oligonucleotide probes Eub338-Cy3 and Abras1420-Cy5) and CSLM. The application software "orthogonal view" of the LSM 510 (Zeiss, Germany) allows the display of optical cuts through the sample in xz and yz sections (Fig. 2B). The localization within the tissue is clearly visible.

2.2 Immunofluorescence Labeling Combined with Fluorescence in Situ Hybridization

The combination of FISH, which allows a phylogenetic identification of bacteria from the phylum down to the species level, with immunological approaches extends the in situ identification to the individual strain level, if strain-specific antisera or special monoclonal antibodies are applied. Antibodies directed against bacterial surface antigens can be created by using, e.g., UV-inactivated bacteria as antigens (Schloter et al. 1995; Hartmann et al. 1997). In addition, antibodies can also be created to identify specific enzymes, e.g., denitrifying enzymes (Bothe et al. 2000) and thus add a phenotypic or

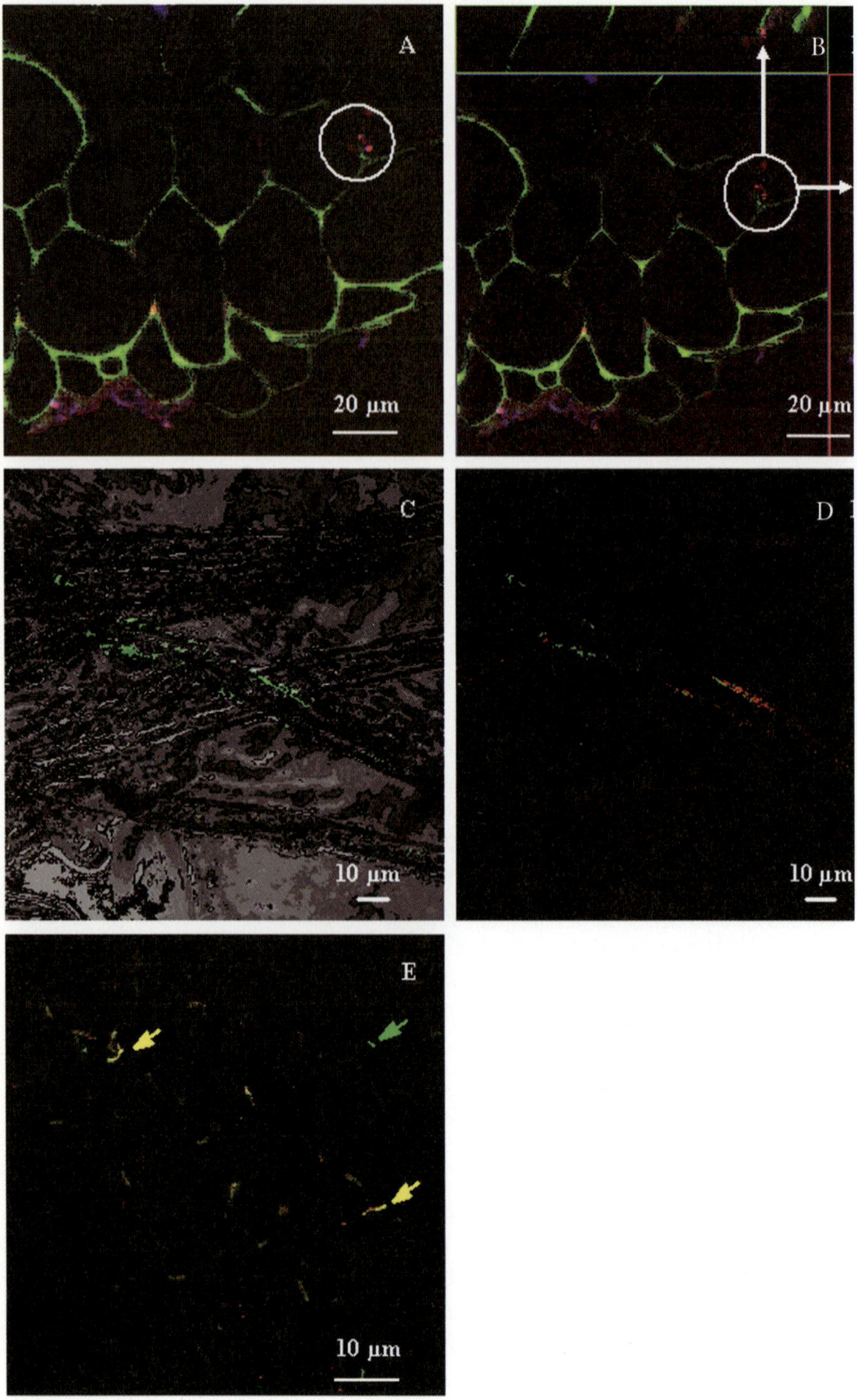
A
20 µm
B
20 µm
C
10 µm
D
10 µm
E
10 µm

expression level approach to organismic identification. As a basic protocol for combining FISH with immunofluorescence labeling, the procedure in Aßmus et al. (1997) can be used with some modifications in specific cases.

After fixation of the sample and FISH analysis (see Sect. 2.1), the immunolabeling is performed with solutions containing 0.9 M NaCl. The presence of NaCl is necessary for the stability of the rRNA-oligonucleotide probe complex (Metz 2002). In addition, all incubation steps are performed at 4 °C. As usual, the immunolabeling procedure starts with a 1-h incubation of the slides carrying the samples with 3 % BSA (Frac. V) in 1/10 PBS+0.9 M NaCl to block unspecific binding of the antibody. After rinsing in washing solution (0.5 % BSA, 0.5 % Tween 80, 1/10 PBS, 0.9 M NaCl), the slides are incubated for 2.5 h at 4 °C with the specific antibody to be applied. After two washing steps, the second antibody (e.g., antimouse-FLUOS-Fab-Fragment) is applied at 4 °C for 1.5 h. After washing, the slides are mounted in Citifluor AF1 (Citifluor Ltd., London, UK). It has to be noted that not all monoclonal antibodies or polyclonal antisera are applicable to this protocol, because the antigen–antibody complex may not be stable at 0.9 M NaCl. Alternatively, the original protocol of Aßmus et al. (1997) can be applied, using the antibody treatment first and the fixation and FISH analysis second. Using this approach, strain-specific monoclonal antibodies against a specific *Azospirillum brasilense* strain were applied in situ together with the FISH analysis (Aßmus et al. 1997). Thus, the

Fig. 2. In situ identification of bacteria in the rhizosphere using fluorescence-labeling techniques and CLSM. **A** Rhizosphere of wheat (Brazilian cultivar PF839197) inoculated with *Azospirillum brasilense* strain Sp245 (rgb-laser scanning image). Roots of inoculated, soil-grown wheat plants were harvested 4 weeks after inoculation. After thorough washing in PBS, the root was cut manually, fixation by heat was performed for 30 min at 70 °C and fixation in 3 % paraformaldehyde was done for 2 h at room temperature. Fluorescence in situ hybridization (FISH) was performed using 45 % formamide and the probes Eub338Mix-Cy3 and Abras1420-Cy5. *A. brasilense* Sp245 cells appear *violet*, because they bind two probes (red and blue emission color code) simultaneously. Plant cell walls have a different emission light, giving a green color code. **B** Same picture as **A**, but in the "orthogonal view", providing insight into optical sections of the sample; z-scan density: 21 µm. **C** In situ localization of GFP-labeled *Serratia liquefaciens* MG44 on root hairs of tomato plants. Using 488-nm excitation wavelength, the GFP-labeled bacteria are clearly visible in the bright field picture. **D** Laser scanning microscopic picture of the same sample as C, but here two excitation wavelengths (488 and 560 nm) were used simultaneously, making the RFP-labeled *Pseudomonas putida* IsoF also visible. **E** Laser scanning microscopic picture of bacteria extracted from roots of *Medicago sativa*, inoculated with *Sinorhizobium meliloti* L33. The bacteria were treated as described and finally concentrated on polycarbonate filters. The fluorescence-labeled probes EuB338Mix-FLUOS and Rhi1247-TRITC were used in FISH analysis. Active bacteria with high ribosome content were labeled green (*green arrow*), while Rhizobia – obviously bacteroids released from nodules – appear yellow (*yellow arrow*), binding both probes simultaneously root surface colonization by a particular bacterial strain could be investigated in a background of other members of this species, identified by using rRNA-targeting probes and FISH.

2.3 Application of Fluorescence Tagging and Reporter Constructs

The fate of particular bacterial inocula in the rhizosphere can also be monitored using molecular-tagged bacteria. In addition to the use of the visually detectable *lux*- and *gus*-markers (Lux: luciferase, Gus: β-glucuronidase), the exploitation of the green fluorescent protein (GFP) from the jellyfish *Aequorea victoria* has brought further progress into the field. GFP is a protein that contains a fluorescent cyclic tripeptide sequence. It requires only molecular oxygen for fluorescence, which means that GFP will fluoresce in virtually any aerobic organism (Lorang et al. 2001). Therefore, GFP-labeled bacteria can be observed by CLSM or by regular fluorescence microscopy. Figure 2C, D shows a localization of GFP-labeled *Serratia liquefaciens* MG44 in the rhizoplane of tomato. Furthermore, the application of DsRed from *Discosoma* sp. provides a red fluorescing molecular marker (Christensen et al. 1999; Tolker-Nielsen et al. 2000). In addition, a mutated form of GFP (ASV) with a short half-life enables real-time in situ expression studies (Andersen et al. 1998; Ramos et al. 2000).

The application of GFP-labeling in expression studies using promotor-*gfp* fusions and GFP fusion proteins has revolutionized the in situ activity studies, because of the relative ease of recording the fluorescence microscopically. The bacteria carrying the gene constructs either on a plasmid or integrated into the chromosome are applied to sense or report conditions in the microhabitat they have been introduced. As in the case of simple tagging of organisms, not only *lux*- and *gus*-reporter (Kragelund et al. 1997) were used, but also constructs using the ice-nucleation gene (Loper and Henkels 1997), or the ferrichrom iron receptor (FhuA; Stubner et al. 1994). These constructs allowed the in situ sensing of N-, P- and C-starvation response (Kragelund et al. 1997; Koch et al. 2001), expression of nitrogen fixation genes (Egener et al. 1999), presence of oxygen (Hojberg et al. 1999), availability of iron (Loper and Henkels 1997) general activity and cell number (Unge et al. 1999), genotoxic effects (Stubner et al. 1994) or the presence of quorum-sensing signal molecules of the N-acylhomoserine lactone type (Steidle et al. 2001). Figure 2C provides an example of in situ localization of GFP-labeled *Serratia liquefaciens* MG44 on root hairs in the rhizosphere of tomato as a bright field picture with 488-nm excitation wavelength, while Fig. 2D shows the same sample as CLSM-picture with two excitation wavelengths (560 and 488 nm) making the RFP-labeled *Pseudomonas putida* isoF also visible.

In some of these studies, bacterial cells with reporter constructs need to be extracted from the habitat for analysis (Koch et al. 2001). Although these reporter cells monitor in situ conditions, the tests are performed ex situ. For this purpose, a separation of the bacteria from the soil was accomplished by applying formaldehyde (1 %)-fixed extracts to density gradient centrifugation with Nycodenz (Nycomed Pharma, Oslo, Norway) with a density of 1.3 g/ml. After a centrifugation step (10,000xg, 30 min, 4 °C) the bacteria on the top of the Nycodenz layer were used for further analysis (Unge et al. 1999).

Monitoring of in situ bacterial growth activity in the plant rhizosphere is suggested by Ramos et al. (2001) using ribosome content and synthesis rate measurements.

3 Ex Situ Studies of Microbial Communities After Separation of Rhizosphere Compartments

For the desorption of bacteria from surfaces, Campbell and Greaves (1990b) recommended the use of a stomacher. Sodium cholate and the ion exchange resin beads Dowex A1 or Chelex 100 were recommended for the treatment of soil particles or root pieces by Macdonald (1986) or Hopkins et al. (1991), respectively, to obtain the bacteria adsorbed. Herron and Wellington (1990) developed a method to extract streptomycete spores from soil particles and used polyethylene glycol (PEG) 6000 for reducing hydrophobic interactions. Each extraction protocol for root-associated bacteria has to be optimized for the system under investigation with the appropriate controls to prove its success. Mogge et al. (2000) described a standardized protocol for the differentiation of the rhizosphere compartments ectorhizosphere and rhizoplane/endorhizosphere and the extraction of the adsorbed bacteria from the rhizoplane of *Medicago sativa europae*. This procedure used the recommendations by Macdonald (1986) and Herron and Wellington (1990) in a modified form. FISH in combination with CLSM was applied for the proof of desorption efficiency in root surface studies.

3.1 Recovery of Bacteria from Bulk Soil, Ecto- and Endorhizosphere

Roots are carefully separated from the soil using sterile tweezers. The soil should be rather dry at the time of harvest to facilitate the separation of roots from the adhering soil. All steps are conducted with sterile solutions on ice. Bulk soil (compartment I) and root-attached soil particles which have been collected by shaking the roots (ectorhizosphere: compartment II) are suspended 1:9 (w/v) in 0.01 M phosphate buffer (Na_2HPO_4/KH_2PO_4, pH 7.4) and dispersed for 1 min at the highest speed in a Stomacher 80 (Seward Medical, UK). To extract rhizoplane and endorhizosphere bacteria (compartment III), 1 g (fresh weight) of roots that have been cleaned from adhering soil particles (see above) and washed in phosphate buffer is suspended in 20 ml 0.1% sodium-cholate buffer (Macdonald 1986). The suspension is treated in a Stomacher 80 at the highest speed for 4 min to disrupt polymers. After transfer into Erlenmeyer flasks, 0.5 g of polyethylene glycol 6000 (Sigma, Deisenhofen) and 0.4 g of cation change polystyrene beads (chelex 100: Sigma, Deisenhofen) are added and the suspension is stirred at 50 rpm/min for 1 h at 4 °C. The stomacher/stirring procedure is repeated three times, whereby the roots are transferred to "fresh" 0.1% sodium cholate buffer with PEG 6000 and chelex 100 after each extraction step (compartment IIIa-c). Finally, aliquots of the obtained suspensions are combined. Root and soil particles are removed by filtration through gauze (40-µm mesh width) and subsequent filtration through 5-µm syringe filters (Sartorius No. 17549, Göttingen, Germany).

In the case of *Medicago sativa* grown in sandy loam, this approach yielded total counts of $3.3 x 10^9$ to $6.5 x 10^8$/g root dry weight from the first to the third treatment, while hybridizing bacteria remained constant at $1.5 x 10^8$/g root dry weight (Mogge et al. 2000). It was calculated that about 88% of the bacteria had been desorbed from the rhizoplane by this technique. This result was confirmed by in situ studies of roots applying confocal laser scanning microscopy. The roots usually harbor large numbers of phylogenetically different bacteria, belonging, e.g., to the α-, β- and γ-subclasses of proteobacteria. However, after the third extraction step, no bacteria could be detected any more on the root surface (20 root pieces of 2–3 cm length were scanned).

The suspensions obtained from bulk soil (I), ectorhizosphere (II), and rhizoplane/endorhizosphere (IIIa-c: merged suspension) can be used for cultivation and dot blot-hybridizations (see Sect. 3.2). DAPI-staining and FISH can be applied for counting total and hybridizing bacteria in the three compartments collected on polycarbonate filters (see Sect. 3.3). PCR-amplification of 16S rDNA and subsequent electrophoretic fingerprinting of the amplification products as well as clone bank studies can be performed with these fractions too (see Sect. 3.4). In addition, these compartments can be investigated for structural and functional microbial diversity by community fatty acid analysis and community level physiological profiling (see Sect. 3.5).

3.2 Community Analysis by Cultivation and Dot Blot Studies

Serial dilutions (0.85% NaCl) from bulk soil (compartment I), ectorhizosphere (compartment II), and rhizoplane/endorhizosphere (compartment III) suspensions (Fig. 1) were plated onto agar media containing different nutritional levels (Table 2). The selection of media used for the isolation of soil and ectorhizosphere-associated bacteria was made to allow the growth of oligotrophic, slow growing strains as well as fast growers. Minimal media were suggested because of the sensitivity of soil bacteria to salts (NaCl) or organic compounds (yeast extracts, casamino acids) as described by Hattori and Hattori (1980). On the other hand, depending on the lower growth rate and a longer incubation period, exuberant growth of the fast growers was reduced, giving the slow growing strains a chance to develop (Gorlach et al. 1994; Mitsui et al. 1997). In addition, minimal media like M9, were supplemented with compounds described as root exudates, and with soil or root extracts (Table 2). Plates were incubated at 20 °C for up to 4 weeks. Cell and colony morphology was recorded and Gram-test, oxidase and catalase tests performed according to Gerhardt et al. (1994). Genomic DNA of these isolates was extracted and purified as described previously (Pukall et al. 1998). The primer pair 27f and 1500r can be used for the amplification of the almost complete 16S rRNA gene of the bacterial isolates (Lane 1991). PCR-amplification of a part of the 23S rDNA was performed using the primer pair 2053r and 990 f.

Using this approach, about 70 % of the bacterial isolates from bulk soil and ectorhizosphere were identified as Gram-positive bacteria using the oligonucleotide GP (Rheims et al. 1996), whereas their numbers were reduced to 17 % in the rhizoplane/endorhizosphere compartment of *Medicago sativa* (Mogge et al. 2000). A similar result was obtained by Lilley et al. (1996) and Mahaffee and Kloepper (1997). On the other hand, the numbers of isolates belonging to the α-, β-, and γ-subclasses of proteobacteria were increased in the rhizoplane

Table 2. Composition of media used to retrieve bacteria from bulk soil, ectorhizosphere and rhizoplane/endorhizosphere samples

Medium	Company or reference
King's B agar; R2A agar; Actinomycete isolation agar; nutrient agar	Difco
CASO agar	Merck
Yeast extract mannitol agar	Dunger and Fiedler (1997)
Starch agar with and without root extract	Dunger and Fiedler (1997)
Cellulose agar supplemented with soil extract	Stotzky et al. (1993)
Planctomyces isolation agar(+N-acetylglucosamin)	Schlesner (1994)
Hyphomicrobium isolation agar	Moore and Marshall (1981)
Caulobacter isolation agar	Poindexter (1964)
Glucose-yeast extract malt agar (GYM)	Shirling and Gottlieb (1966)
M9 minimal medium[a] (+ carbon source[b]/ + trace elements[c])	Sambrook et al. (1989, modified)

[a] Composed of Na_2HPO_4 10.2, KH_2PO_4 3.0, NaCl 0.6, and NH_4Cl 1.2 g/l

[b] 5 g/l carbohydrates (glucose, glucose and vitamin solution No.6 (Staley 1968), fructose, sucrose, arabinose) or 2 g/l organic acids (fumaric acid, oxal acetic acid)

[c] 1 ml of sterile filtered trace element stock solution composed of $CaCl_2x6\ H_2O$ 2.7 g, $MgSO_4x7\ H_2O$ 15 g, $FeCl_3$ 0.02 g/l

to 13, 26 and 35 % as compared to 4.2, 8.5 and 0.8 % respectively in the ectorhizosphere as was shown by using the probes ALF1b, BET42a and GAM42 respectively to group the isolates obtained. No differences were found for isolates of the *Cytophaga-Flavobacteria* group, which were only a minor portion in both compartments (3.5 %).

Quantitative population analyses in soil and rhizosphere environments were also conducted by using strains carrying unique selectable markers. This was aimed to enumerate one particular introduced strain in the presence of a large excess of other microbes. Since the usually suitable selectable markers are missing in wild-type strains, spontaneous or transposon-induced mutants, which are, e.g., resistant to an antibiotic, are frequently used for selective plating assays. However, these mutants may be less fit than the wild type and, therefore, the results of the surveys are biased. De Leij et al. (1998) demonstrated such effects on environmental fitness in several mutants of *Pseudomonas fluorescens* SBW25, constructed by site-directed genomic insertions of marker genes. Recently, Hirano et al. (2001) selected a site in the *gacS-cysM* intergenic region in *Pseudomonas syringae* pv. syringae B728, in which the insertion of an antibiotic resistance marker cassette did not affect the fitness of the bacterium in the field. They concluded that carefully selected intergenic regions, which are suitable for the integration of specific marker cassettes, exist in any bacterium.

3.3 Community Analysis by Fluorescence in Situ Hybridization on Polycarbonate Filters

Bacterial suspensions (extract of the rhizosphere compartments, Fig. 1) are fixed overnight at 4 °C with 3 % formaldehyde and concentrated in three parallels onto 0.2-µm polycarbonate filters (100-µl aliquots). Dehydration of cells is performed with 50, 80 and 96 % ethanol for 3 min each. For details on the FISH protocol see Sect. 2.1. The slides are finally mounted with Citifluor AF1 to reduce photobleaching. A Zeiss Axiophot 2 epifluorescence microscope (Zeiss, Jena, Germany) equipped with filter sets F31–000, F41–001 and F41–007 (Chroma Tech. Corp., Battleboro, VT, USA) can be used for the enumeration of bacteria on filters. Total cell counts (DAPI) and hybridizing bacteria using a set of domain-specific probes (Table 1) are determined by evaluating at least 10 microscopic fields with 20–100 cells per field.

In the case of the *M. sativa* roots, the extraction method was also applied to the rhizoplane/endorhizosphere of roots inoculated with *Sinorhizobium meliloti* as well as to inoculated roots after the nodules had been removed with a sterile scalpel. During the three repeated stomacher/stirring-treatments, nodules cracked and *S. meliloti*-bacteroids were released (Mogge et al. 2000). Figure 2E shows a representative photomicrograph of bacteria concentrated on polycarbonate filters after extraction of roots with nodules. Large (up to 10-µm long) pleomorphic cells hybridized with a set of FLUOS-labeled oligonucleotide probes directed against the domain Bacteria and the TRITC-labeled oligonucleotide probe Rhi1247 directed against *Rhizobium* (Table 1). Obviously, these large cells were bacteroids originating from crushed nodules and were missing when the nodules had been removed before the application of the extraction procedure.

3.4 Community Analysis by (RT) PCR-Amplification of Phylogenetic Marker Genes, D/TGGE-Fingerprinting and Clone Bank Studies

The differentiated rhizosphere compartments can also be used to isolate rRNA and genomic DNA following previously described protocols (Felske et al. 1996; Miethling et al. 2000). A further purification of the DNA extracts, e.g., with the Wizard DNA clean-up (Promega, Madison, WI), may be necessary, before PCR can be applied. For amplification, the highly conserved bacterial 16S rRNA primers U968-GC and L1346 are used. Amplification of 16S rDNA is performed as described by Felske et al. (1996) using the following PCR-program: 1 cycle at 94 °C for 5 min, 35 cycles at 94 °C for 90 s (denaturation), 61 °C for 40 s (annealing), 70 °C for 40 s (extension), and a single final extension at 70 °C for 5 min. Amplification of 16S rRNA as well as denaturing temperature gradient gel electrophoretic (D/TGGE) separation of the PCR-products of DNA and RNA is performed as described by Miethling et al. (2000).

D/TGGE profiles represent the frequency distribution of PCR-amplified segments of rDNA or rRNA separated due to their melting behavior in the electric field of a temperature gradient gel. The resulting profiles represent the frequency distribution of the most prominent community members in a first approximation (Muyzer and Smalla 1998). Since the ratio of 16S rDNA and 16S rRNA is dependent on cellular activity (Wagner 1994), comparisons of TGGE patterns derived from 16S rRNA and 16S rDNA amplicons can provide interesting information about the active members of the community. Variations in the relation of band intensities (rRNA/rDNA) indicated shifts in the relative activity of the respective dominant DNA sequences. In particular, the composition of the communities are changing along the gradient from bulk soil to the rhizoplane/endorhizosphere (Mogge et al. 2000, Wieland et al. 2001). Additional sequences show higher evenness visible by larger band formation in the rhizoplane/endorhizosphere compartment, which is clearly different from all the other examined habitats. In this compartment, a larger fraction of the community seems to be active, as deduced from the fraction of bands common to the rDNA and rRNA patterns of the communities. Using the same methodological approach, Wieland et al. (2001) have demonstrated recently that the TGGE-patterns of 16S rRNA did not change during the plant development in the bulk soil, whereas some pattern variation could be correlated to plant development in the rhizosphere and rhizoplane habitats. On the root surface of different plants and plants growing in different soils, more apparent differences in the complete TGGE-pattern was obvious. The frequency distribution of target sequences from the total and active community members appeared to be mostly identical at the rhizoplane/endorhizosphere where the most prominent bands of the rRNA-derived pattern are also dominant in the DNA pattern. However, it has to be taken into account that a high ribosome content does not always indicate a high physiological activity of bacterial cells, because different bacteria inherently contain different ribosome numbers (Fegatella et al. 1998). It is likely that both phenomena play a role, and this may be different for different bacterial groups (Duarte et al. 1998). Duineveld et al. (2001) applied a similar 16S rDNA/rRNA PCR-amplification approach followed by DGGE analysis in the *Chrysantemum* rhizosphere, but found very little difference between the bacterial community of root-adhering soil and bulk soil. Heuer et al. (2002) used not only general PCR-primers for the amplification of bacterial 16S rDNA (between positions 968 and 1401, *E. coli* numbering according to Brosius et al. 1981), but also the taxon-specific primers F203alpha for alpha-proteobacteria and F964β for β-proteobacteria. Using this approach, these authors revealed a more differentiated fingerprint for rhizosphere bacterial communities in DGGE-electrophoresis. A PCR approach targeting the ribosomal 16S–23S rDNA intergenic spacer region, called ribosomal intergenic spacer analysis (RISA), can also reveal insight into the bacterial diversity, because this spacer region varies considerably in different species. Baudoin et al. (2001) applied this approach for the assessment of the bacterial community structure along maize roots and in different growth stages. Weidner et al. (1996) applied restriction fragment length polymorphism (RFLP) analysis of cloned 16S rDNA from the roots of the seagrass *Halophila stipulacea* to investigate unculturable bacterial rhizosphere communities. Finally, a strain-specific detection of certain bacterial strains in the rhizosphere based on a highly specific PCR-amplification of the 16S–23S intergenic spacer (IGS) region was recently developed by Tan et al. (2001). The sequence variability in this region was used to differentially identify *Bradyrhizobium* and *Rhizobium* strains colonizing rice roots by a nested PCR approach and analysis of the amplification products on simple agarose gels.

The genomic DNA extracted from the rhizosphere compartments I–III (Fig. 2) can also be used to create 16S rDNA clone banks or dot blot experiments with 16S rDNA fragments or probing with specific oligonucleotides. When the oligonucleotide GP (Rheims et al. 1996) was used, a reduced number of 16S rDNA clones related to Gram-positive bacteria was detected in the library generated from the rhizoplane/endorhizosphere of *Medicago sativa* (12 %) as compared to the library generated from the bulk soil fraction (26 %; Mogge et al. 2002). Thus, the results of community analysis using cultivation techniques and FISH analysis (see Sects. 3.2 and 3.3) were, in general, confirmed by this PCR-based cultivation independent technique.

3.5 Community Analysis by Fatty Acid Pattern and Community Level Physiological Profile Studies

The overall microbial diversity in environmental habits can be assessed by cultivation independent biomarker analysis, different from the phylogenetic ribosomal genes or other genetic markers. As is the case in chemotaxonomic studies, the fatty acid patterns are used for this purpose. In one type of analysis, the fatty acid methyl esters (FAME) are obtained from the fatty acids after saponification of 5 g of soil or root with adhering soil in methanoic NaOH (at 100 °C, 30 min; Dunfield and Germida 2001). Alternatively, the lipids are extracted from 5 g of soil with methanol:chloroform (2:1), the phospholipids are separated by chromatography, and finally hydrolyzed to liberate the phospholipid fatty acids (PLFA; White and Ringelberg 1998). The PLFA analysis has the advantage of giving insight into the living community, because PFLA are efficiently hydrolyzed in dead biomass, while the direct FAME analysis may contain fatty acids from dead organisms too. The GC-MSanalysis finally provides much information on the diversity of this biomarker (Zelles 1997; White and Ringelberg 1998). Using the FAME analysis, Germida et al. (1998) investigated the diversity of root-associated bacterial communities in canola and wheat, and Dunfiled and Germida (2001) compared the bacterial communities in the rhizosphere and endorhizosphere of field-grown genetically modified varieties of canola (*Brassica napus*). An example of a recent application of the PFLA approach in rhizosphere studies is the investigation of the microbial community response in the rhizosphere of *Spartina alterniflora* to changing environmental conditions by Lovell et al. (2001).

An investigation targeting the analysis of the functional abilities of a complex community is the substrate utilization profile assays using the Biolog^R-plates. Baudoin et al. (2001) applied this approach recently to characterize the functional microbial diversity in different rhizosphere compartments of maize plants. The differences between the rhizosphere and nonrhizosphere soil samples were more pronounced in 4-week-old compared to 2-week-old plants. In addition, adhering soil from different root zones (ramification, root hair-elongation, root tip) revealed dissimilar community level physiological profiles (CLPP). However, this approach needs to be regarded as reflecting the potential rather than the in situ-activity of most culturable microbes, because these are known to respond and contribute most to the activity at the incubation conditions of the CLPP-assay (Garland et al. 1997).

4 Conclusions

Using a polyphasic approach including cultivation-dependent and different cultivation-independent methods, it could be shown that a high proportion of culturable bacteria is present in the rhizoplane when a variety of appropriate media are applied. This corroborates the findings of Hengstmann et al. (1999), who reported similar results in their studies on the microbial community of the rice rhizosphere. The separation into the three compartments, bulk soil, ectorhizosphere and rhizoplane/endorhizosphere has to be performed with great care and actually needs an optimization for each plant and soil type under study. The degree to which adhering soil particles (ectorhizosphere) are included in the rhizosphere studies considerably influences the outcome of the study, since these soil particles are carrying a microbial community resembling, to a varying extent, the soil situation compared to the root surface or rhizoplane situation. The microbial population colonizing the root surface should be approached only after washing the roots free of adhering soil particles. In conclusion, the way "rhizosphere" is defined by the experimental protocol is of crucial importance for the results of root colonization studies. Certainly, in situ and ex situ studies (with the separated rhizosphere compartments) both complement each other to give a more comprehensive picture. Although the microscopic in situ approach has the great advantage of providing detailed spatial information about root surface colonization, quantitative and qualitative data about the structural and functional diversity of root colonization can be obtained by a variety of complementary ex situ approaches.

References and Selected Reading

Amann RI, Binder BJ, Olson RJ, Chisholm SW, Devereux R, Stahl DA (1990) Combination of 16S rRNA-targeted oligonucleotide probes with flow cytometry for analyzing mixed microbial populations. Appl Environ Microbiol 56:1919–1925

Amann RI, Ludwig W, Schleifer KH (1995) Phylogenetic identification and in situ detection of individual microbial cells without cultivation. Microbiol Rev 59:143–169

Andersen JB, Sternberg C, Poulsen LK, Bjorn SP, Givskov M, Molin S (1998) New unstable variants of green fluorescent protein for studies of transient gene expression in bacteria. Appl Environ Microbiol 64:2240–2246

Aßmus B, Hutzler P, Kirchhof G, Amann RI, Lawrence JR, Hartmann A (1995) In situ localization of *Azospirillum brasilense* in the rhizosphere of wheat with fluorescently labeled, rRNA-targeted oligonucleotide probes and scanning confocal laser microscopy. Appl Environ Microbiol 61:1013–1019

Aßmus B, Schloter M, Kirchhof G, Hutzler P, Hartmann A (1997) Improved in situ tracking of rhizosphere bacteria using dual staining with fluorescence-labeled antibodies and rRNA-targeted oligonucleotides. Microbial Ecol 33:32–40

Baudoin E, Benizri E, Guckert A (2001) Impact of growth stage on the bacterial community structure along maize roots, as determined by metabolic and genetic fingerprinting. Appl Soil Ecol 52:1–11

Braun-Howland EB, Vescio PA, Nierzwicki-Bauer SA (1993) Use of a simplified cell blot technique and 16S rRNA-directed probes for identification of common environmental isolates. Appl Environ Microbiol 59:3219–3224

Beringer JE (1974) R factor transfer in *Rhizobium leguminosarum*. J Gen Microbiol 84:188–198

Bothe H, Jost G, Schloter M, Ward BB, Witzel KP (2000) Molecular analysis of ammonia oxidation and denitrification in natural environments. FEMS Microbiol Rev 24: 673–690

Brimecombe MJ, De Leij FA, Lynch JM (2001) The effect of root exudates on rhizosphere microbial populations. In: Pinton R, Varanini Z, Nannipieri P (eds) The rhizosphere. Marcel Dekker, New York, pp 95–140

Brosius J, Dull TJ, Sleeter DD, Noller HF (1981) Gene organization and primary structure of a ribosomal RNA operon from *Escherichia coli.* J Mol Biol 148:107–127

Campbell R, Greaves MP (1990a) Anatomy and community structure of the rhizosphere. In: Lynch JM (ed) The rhizosphere. Wiley, Chichester, pp 11–34

Campbell R, Greaves MP (1990b) Methods for studying the microbial ecology of the rhizosphere. Meth Microbiol 22:447–477

Chatzinotas A, Sandaa RA, Schönhuber W, Amann R, Daae FL, Torsvik V, Zeyer J, Hahn D (1998) Analysis of broad-scale differences in microbial community composition of two pristine forest soils. Syst Appl Microbiol 21:579–587

Christensen BB, Sternberg C, Andersen JB, Palmer Jr RJ, Nielsen JJ, Givskov M, Molin S (1999) Molecular tools for study of biofilm physiology. Meth Enzymol 310:20–42

De Leij FAAM, Thomas CE, Bailey MJ, Whipps JM, Lynch JM (1998) Effect of insertion site and metabolic load on the environmental fitness of a genetically modified *Pseudomonas fluorescens* isolate. Appl Environ Microbiol 64:2634–2638

Duarte GF, Rosado AS, Seldin L, Keijzer-Wolter AC, Van Elsas JD (1998) Extraction of ribosomal RNA and genomic DNA from soil for studying the diversity of the indigenous bacterial community. J Microbiol Meth 32:21–29

Duineveld BM, Kowalchuk GA, Keijzer A, van Elsas JD, van Veen J (2001) Analysis of bacterial communities in the rhizosphere of *Chrysanthemum* via denaturing gradient gel electrophoresis of PCR-amplified 16S rRNA as well as DNA fragments coding for 16S rRNA. Appl Environ Microbiol 67:172–178

Dunfield KE, Germida JJ (2001) Diversity of bacterial communities in the rhizosphere and root interior of field-grown genetically modified *Brassica napus.* FEMS Microbiol Rev 38:1–9

Dunger W, Fiedler HJ (1997) Methoden der Bodenbiologie. Gustav Fischer-Verlag, Jena, pp 89–107

Egener T, Hurek T, Reinhold-Hurek B (1999) Endophytic expression of *nif* genes of *Azoarcus* sp. strain BH72 in rice roots. Mol Plant-Microbe Interact 12:813–819

Fegatella F, Lim J, Kjelleberg S, Cavicchiolli R (1998) Implications of rRNA operon copy number and ribosome content in the marine oligotrophic ultramicrobacterium *Sphingomonas* sp. strain RB2256. Appl Environ Microbiol 64:4433–4438

Felske A, Engelen B, Nübel U, Backhaus H (1996) Direct ribosome isolation from soil to extract bacterial rRNA for community analysis. Appl Environ Microbiol 62:4162–4167

Garland JL, Cook KL, Loader CA, Hungate BA (1997) The influence of microbial community structure and function on community-level physiological profiles. In: Insam H, Rangger A (eds) Microbial communities: functional versus structural approaches. Springer, Berlin Heidelberg New York, pp 171–183

Gerhardt P, Murray RGE, Wood WA, Krieg NR (1994) Methods for general molecular bacteriology. American Society for Microbiology, Washington, DC

Germida JJ, Siciliano SD, de Freitas JR, Seib AM (1998) Diversity of root-associated bacteria associated with field-grown canola (*Brassica napus* L.) and wheat (*Triticum aestivum* L.) FEMS Microbiol Ecol 26:43–50

Giovannoni SJ, DeLong EF, Olsen GJ, Pace NR (1988) Phylogenetic group-specific oligodeoxynucleotide probes for identification of single microbial cells. J Bacteriol 170:720–726

Gorlach K, Shingaki R, Morisaki H, Hattori T (1994) Construction of eco-collection of paddy field soil bacteria for population analysis. J Gen Microbiol 40:509–517

Hartmann A, Aßmus B, Kirchhof G, Schloter M (1997) Direct approaches to study soil microflora. In: van Elsas JD, Trevors JT, Wellington EMH (eds) Modern soil microbiology. Marcel Dekker, New York, pp 279–309

Hartmann A, Lawrence JR, Aßmus B, Schloter M (1998) Detection of microbes by laser confocal microscopy. In: Akkermans ADL, van Elsas JD, de Bruijn FJ (eds) Molecular microbial ecology manual, Supplement 3. Kluwer, Dordrecht, Chap. 4.1.10

Hartmann A, Stoffels M, Eckert B, Kirchhof G, Schloter M (2000) Analysis of the presence and diversity of diazotrophic endophytes. In: Triplett EW (ed) Prokaryotic nitrogen fixation: A model system for analysis of a biological process. Horizon Scientific Press, Wymondham, USA, pp 727–736

Hattori R, Hattori T (1980) Sensitivity to salts and organic compounds of soil bacteria isolated on diluted media. J Gen Appl Microbiol 26:1–14

Hengstmann U, Chin KJ, Janssen PH, Liesack W (1999) Comparative phylogenetic assignment of environmental sequences of genes encoding 16S rRNA and numerically abundant culturable bacteria from an anoxic rice paddy soil. Appl Environ Microbiol 65:5050–5058

Herron PR, Wellington EMH (1990) New method for extraction of streptomycete spores from soil and application to the study of lysogene in sterile amended and nonsterile soil. Appl Environ Microbiol 56:1406–1412

Heuer H, Kroppenstedt RM, Lottmann J, Berg G, Smalla K (2002) Effects of T4 lysozyme release from transgenic potato roots on bacterial rhizosphere communities are negligible relative to natural factors. Appl Environ Microbiol 68:1325–1335

Hirano SS, Willis DK, Clayton MK, Upper CD (2001) Use of an intergenic region in *Pseudomonas syringae* pv. syringae B728a for site-directed genomic marking of bacterial strains for field experiments. Appl Environ Microbiol 67:3735–3738

Hojberg O, Schnider U, Winteler HV, Sorensen J, Haas D (1999) Oxygen-sensing reporter strain of *Pseudomonas fluorescens* for monitoring the distribution of low-oxygen habitats in soil. Appl Environ Microbiol 65:4085–4093

Hopkins DW, MacNaughton SJ, O'Donnell AG (1991) A dispersion and differential centrifugation technique for representatively sampling microorganisms from soil. Soil Biol Biochem 23:217–225

Koch B, Worm J, Jensen LE, Hojberg O, Nybroe O (2001) Carbon limitation induces sigma^{s}-dependent gene expression in *Pseudomonas fluorescens* in soil. Appl Environ Microbiol 67:3363–3370

Kragelund L, Hosbond C, Nybroe O (1997) Distribution of metabolic activity and phosphate starvation response of *lux*-tagged *Pseudomonas fluorescens* reporter bacteria in the barley rhizosphere. Appl Environ Microbiol 63:4920–4928

Lane DJ (1991) 16S/23S rRNA sequencing. In: Stackebrandt E, Goodfellow M (eds) Nucleic acid techniques in bacterial systematics. Wiley, Chichester, pp 125–175

Lee S, Malone C, Kemp PF (1993) Use of multiple 16S rRNA-targeted fluorescent probes to increase signal strength and measure cellular RNA from natural planktonic bacteria. Mar Ecol Prog Ser 101:193–201

Liesack W, Stackebrandt E (1992) Occurrence of novel groups of the domain bacteria as revealed by analysis of genetic material isolated from an Australian terrestrial environment. J Bacteriol 174:5072–5078

Liesack W, Janssen PH, Rainey FA, Ward-Rainey N, Stackebrandt E (1997) Microbial diversity in soil: the need for a combined approach using molecular and cultivation techniques. In: van Elsas JD, Trevors JT, Wellington EMH (eds) Modern soil microbiology. Marcel Dekker, New York, pp 375–439

Lilley AK, Fry JC, Bailey MJ, Day MJ (1996) Comparison of aerobic heterotrophic taxa isolated from four root domains of mature sugar beet (*Beta vulgaris*). FEMS Microbiol Ecol 21:231–242

Loper JE, Henkels MD (1997) Availability of iron to *Pseudomonas fluorescens* in rhizosphere and bulk soil evaluated with an ice nucleation reporter gene. Appl Environ Microbiol 60:2944–2948

Lorang JM, Tuori RP, Martinez JP, Sawyer TL, Redman RS, Rollins JA, Wolpert TJ, Johnson KB, Rodriguez RJ, Dickman MB, Ciuffetti LM (2001) Green fluorescent protein is lighting up fungal biology. Appl Environ Microbiol 67:1987–1994

Lovell CR, Bagwell CE, Czákó M, Márton L, Piceno YM, Ringelberg DB (2001) Stability of a rhizosphere microbial community exposed to natural and manipulated environmental variability. FEMS Microbiol Ecol 38:69–76

Ludwig W, Amann R, Martinez-Romero E, Schönhuber W, Bauer S, Neef A, Schleifer KH (1998) rRNA based identification and detection systems for rhizobia and other bacteria. Plant Soil 204:1–19

Macdonald RM (1986) Sampling soil microfloras: dispersion of soil by ion exchange and extraction of specific microorganisms from suspension by elutriation. Soil Biol Biochem 18:399–406

Mahaffee WF, Kloepper JW (1997) Temporal changes in the bacterial communities of soil, rhizosphere, and endorhiza associated with field-grown cucumber (*Cucumis sativus* L.). Microb Ecol 34:210–223

Manz W, Amann R, Ludwig W, Wagner M, Schleifer KH (1992) Phylogenetic oligodeoxynucleotide probes for the major subclasses of proteobacteria: problems and solutions. System Appl Microbiol 15:593–600

Manz W, Amann R, Ludwig W, Vancanneyt M, Schleifer KH (1996) Application of a suite of 16S rRNA-specific oligonucleotide probes designed to investigate bacteria of the phylum cytophaga-flavobacter-bacteroides in the natural environment. Microbiology 142:1097–1106

Metz S (2001) Herstellung von monoklonalen Antikörpern gegen die Cu-abhängige dissimilatorische Nitritreduktase und deren Anwendung zum in situ-Nachweis der Denitrifikationsaktivität von Bakterien. Doctoral Thesis, Ludwig-Maximilians-Universität München, Fakultät für Biologie

Miethling R, Wieland G, Backhaus H, Tebbe CC (2000) Variation of microbial rhizosphere communities in response to crop species, soil origin and inoculation with the marker gene-tagged *Sinorhizobium meliloti* L33. Microb Ecol 40:43–56

Mitsui H, Gorlach K, Lee HJ, Hattori R, Hattori T (1997) Incubation time and media requirements of culturable bacteria from different phylogenetic groups. J Microbiol Methods 30:103–110

Mogge B, Lebhuhn M, Schloter M, Stoffels M, Pukall R, Stackebrandt E, Wieland G, Backhaus H, Hartmann A (2000) Erfassung des mikrobiellen Populationsgradienten vom Boden zur Rhizoplane von Luzerne (*Medicago sativa*). In: Hartmann A (ed) Biologische Sicherheit: Biomonitor und Molekulare Mikrobenökologie. Projektträger BEO, Jülich, pp 217–224

Moore RL, Marshall KC (1981) Attachment and rosette formation by hyphomicrobia. Appl Environ Microbiol 42:751–757

Morgan JAW, Whipps JM (2001) Methodological approaches to the study of rhizosphere carbon flow and microbial population dynamics. In: Pinton R, Varanini Z, Nannipieri P (eds) The rhizosphere. Marcel Dekker, New York, pp 373–409

Muyzer G, Smalla K (1998) Application of denaturing gradient gel electrophoresis (DGGE) and temperature gradient gel electrophoresis (TGGE) in microbial ecology. Antonie van Leuwenhook 73:127–141

Poindexter JS (1964) Biological properties and classification of the Caulobacter group. Bacteriol Rev 28:231–295

Pukall R, Brambilla E, Stackebrandt E (1998) Automated fragment length analysis of fluorescently-labeled 16S rDNA after digestion with 4-base cutting restriction enzymes. J Micobiol Meth 32:55–63

Ramos C, Molbak L, Molin S (2000) Bacterial activity in the rhizosphere analyzed at the single-cell level by monitoring ribosome contents and synthesis rates. Appl Environ Microbiol 66:801–809

Ramos C, Licht TR, Sternberg C, Krogfelt KA, Molin S (2001) Monitoring bacterial growth activity in biofilms from laboratory flow-chambers, plant rhizosphere and animal intestine. Methods Enzymol 337:21–42

Rheims H, Sproer C, Rainey FA, Stackebrandt E (1996) Molecular biological evidence for the occurrence of uncultured members of the actinomycete line of descent in different environments and geographical locations. Microbiology 142:2863–2870

Sambrook J, Fritsch EF, Maniatis T (1989) Molecular cloning. A laboratory manual, 2nd edn. Cold Spring Harbor Laboratory Press, New York

Schlesner H (1994) Development of media suitable for the microorganisms morphologically resembling *Planctomycetes* spp., *Pirellula* spp., other *Planctomycetales* from various aquatic habitats using dilute media. System Appl Microbiol 17:135–145

Schloter M, Borlinghaus R, Bode W, Hartmann A (1993) Direct identification, and localization of *Azospirillum* in the rhizosphere of wheat using fluorescence-labelled monoclonal antibodies and confocal scanning laser microscopy. J Microsc 171:173–176

Schloter M, Assmus B, Hartmann A (1995) The use of immunological methods to detect and identify bacteria in the environment. Biotechnol Adv 13:75–90

Shirling EB, Gottlieb D (1966) Methods for characterization of *Streptomyces* species. Int J Syst Bacteriol 16:313–340

Staley JT (1968) *Prosthecomicrobium* and *Ancalomicrobium*: new prosthecate freshwater bacteria. J Bacteriol 95:1921–1942

Steidle A, Sigl K, Schuhegger R, Ihring A, Schmid M, Gantner S, Stoffels M, Riedel K, Givskov M, Hartmann A, Langebartels C, Eberl L (2001) Visualization of N-acylhomoserine lactone-mediated cell-cell communication between bacteria colonizing the tomato rhizosphere. Appl Environ Microbiol 67:5761–5770

Stoffels M, Castellanos T, Hartmann A (2001) Design and application of new 16S rRNA-targeted oligonucleotide probes for the *Azospirillum-Skermanella-Rhodocista*-cluster. Syst Appl Microbiol 24:83–97

Stotzky G, Broder MW, Doyle JD, Jones RA (1993) Selected methods for the detection and assessment of ecological effects resulting from the release of genetically engineered microorganisms to the terrestrial environment. Adv Appl Microbiol 38:1–98

Stubner S, Schloter M, Moeck GS, Coulton JW, Ahne F, Hartmann A (1994) Construction of *umu-fhuA* operon fusions to detect genotoxic potential by an antibody-cell surface reaction. Environ Tox Water Qual 9:285–291

Tan Z, Hurek T, Vinuesa P, Müller P, Ladha JK, Reinhold-Hurek B (2001) Specific detection of *Bradyrhizobium* and *Rhizobium* strains colonizing rice (*Oryza sativa*) roots by 16S-23S ribosomal DNA intergenic spacer-targeted PCR. Appl Environ Microbiol 67:3655–3664

Tas É, Lindström K (2001) Identification of bacteria by their intrinsic sequences: Probe design and testing of their specificity. In: Akkermans ADL, Van Elsas JD, De Bruijn FJ (eds) Molecular microbial ecology manual, Suppl. 5, Kluwer Academic Press, Dordrecht

Tolker-Nielsen T, Brinch UC, Ragas PC, Andersen JB, Jacobsen CS, Molin S (2000) Development and dynamics of *Pseudomonas* sp. biofilms. J Bacteriol 182:6482–6489

Torsvik V, Sorheim R, Goksoyr J (1996) Total bacterial diversity in soil and sediment communities: a review. J Industr Microbiol 17:170–178

Tsien HC, Bratina BJ, Tsuji K, Hanson RS (1990) Use of oligodeoxynucleotide signature probes for identification of physiological groups of methylotrophic bacteria. Appl Environ Microbiol 56:2858–2865

Unge A, Tombolini R, Molbak L, Jansson JK (1999) Simultaneous monitoring of cell number and metabolic activity of specific bacterial populations with a dual *gfp-luxAB* marker system. Appl Environ Microbiol 65:813–821

Wagner M, Amann R, Lemmer H, Schleifer KH (1993) Probing activated sludge with oligonucleotides specific for proteobacteria: inadequacy of culture-dependent methods for describing microbial community structure. Appl Environ Microbiol 59:1520–1525

Wagner R (1994) The regulation of ribosomal rRNA synthesis and bacterial cell growth. Arch Microbiol 161:100–106

Weidner S, Arnold W, Pühler A (1996) Diversity of uncultured microorganisms associated with the seagrass *Halophila stipulacea* estimated from restriction fragment length polymorphism analysis of PCR-amplified 16S rRNA genes. Appl Environ Microbiol 62:766–771

Werner D (2001) Organic signals between plants and microorganisms. In: Pinton R, Varanini Z, Nannipieri P (eds) The rhizosphere. Marcel Dekker, New York, pp 197–222

White DC, Ringelberg DB (1998) Signature lipid biomarker analysis. In: Burlage RS, Atlas R, Stahl D, Geesey G, Sayler G (eds) Techniques in microbial ecology. Oxford University Press, New York, pp 255–272

Wieland G, Neumann R, Backhaus H (2001) Variation of microbial communities in soil, rhizosphere, and rhizoplane in response to crop species, soil type, and crop development. Appl Environ Microbiol 67:5849–5854

Zelles L (1997) Phospholipid fatty acid profiles in selected members of soil microbial communities. Chemosphere 35:275–294

25 Methods for Analysing the Interactions Between Epiphyllic Microorganisms and Leaf Cuticles

Daniel Knoll and Lukas Schreiber

1 Introduction

The plant cuticle forms the solid surface environment for epiphyllic microorganisms. This chapter presents newly developed techniques for analysing the interactions between epiphyllic microorganisms and leaf cuticles. The methods take into account the unique physical, chemical and functional characteristics of the cuticular interface of leaves. Furthermore, a new experimental approach simulating leaf surface microbe interactions on the basis of isolated cuticular membranes (CM) will be presented. Changes in cuticular properties in relation to microbial growth can be assessed in vitro under controlled conditions.

2 Physical Characterisation of Cuticle Surfaces by Contact Angle Measurements

Surface wetting can be determined quantitatively by measuring the contact angle σ of an aqueous droplet applied to a surface. The contact angle σ is defined by the angle (°) between the flat leaf surface and the line tangent to a water droplet through the point of contact as demonstrated in Fig. 1. The size of the contact angle σ is directly related to the hydrophobic properties of a surface. Low contact angles indicate well wettable surfaces (left-hand side of Fig. 1), whereas high contact angles indicate little wettable surfaces (right-hand side of Fig. 1). Generally, advancing contact angles are measured with the aid of a goniometer within the first minute after application of a droplet onto the surface. The droplet volume may vary from 1 to 10 µl, since it has been previously shown that contact angles were independent of the droplet size (Schreiber 1996). However, contact angles can be significantly dependent on the pH values of the buffered aqueous solutions. So-called contact angle titration measuring contact angles at different pH values ranging between pH

Plant Surface Microbiology
A. Varma, L. Abbott, D. Werner, R. Hampp (Eds.)

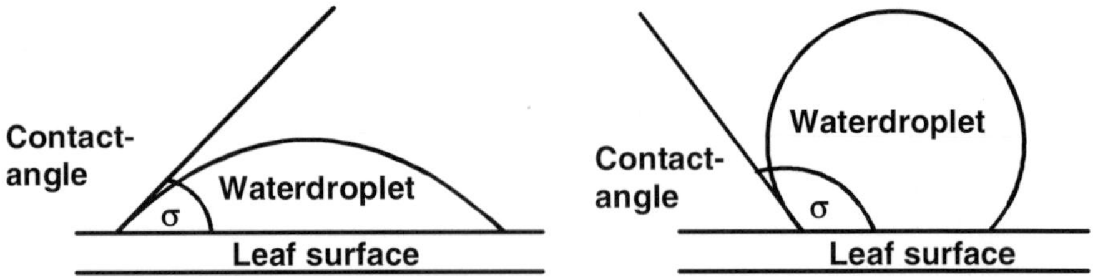

Fig. 1. A scheme of contact angles from aqueous droplets on surfaces of different hydrophobicity. The contact angle σ is related to wetting properties of surfaces. Low contact angles indicate well wettable surfaces (*left*), whereas high contact angles indicate rarely wettable surfaces (*right*)

3.0 and 11.0 can reveal important additional information about the chemical nature of interfacial molecules.

Contact angles can be measured on leaf surfaces and a variety of different model surfaces (Knoll and Schreiber 1998, 2000). Prior to the contact angle measurement on leaf surfaces, leaves have to be immersed in deionised water for 10 s and carefully blotted with filter paper. This washing step removes any deposits and dust particles weakly adsorbed to the leaf surface, which might dissolve in the aqueous drops used for the contact angle measurements. Leaf strips are cut out from the leaf avoiding central veins and necrotic lesions. Then leaf strips are attached to microscope slides that are placed in the goniometer to measure the contact angle of the applied droplet. Contact angles can be measured on leaf surfaces that are naturally or artificially colonised in different degrees with microorganisms.

In order to analyse the impact of cuticular waxes and of epiphytic microorganisms on wetting properties of leaf surfaces, both components can be isolated and applied separately to microscope slides as artificial supports. Isolated wax is recrystallised from the melt on chloroform-washed microscope slides. For details about wax extraction, refer to the second part of this chapter. Wetting properties of different species of epiphytic microorganisms can be determined after cell adherence to artificial glass supports (Fig. 2). Washed cell suspensions are incubated with hydrophilic chloroform-washed glass slides and with highly hydrophobic slides that were obtained by chemical silanisation of the slides (Leibnitz and Struppe 1984). Washed cell suspensions (25 ml) are transferred onto sterilised microscope slides in sterile tissue culture dishes. After incubation for 24 h at 25 °C, microscope slides are carefully washed using a gentle stream of sterile deionised water and remaining amounts of water are allowed to evaporate. Contact angles are measured immediately after drying of the surfaces. In order to measure contact angles as a function of cell density glass slides are incubated at 25 °C with different cell concentrations for 6 and 48 h, respectively.

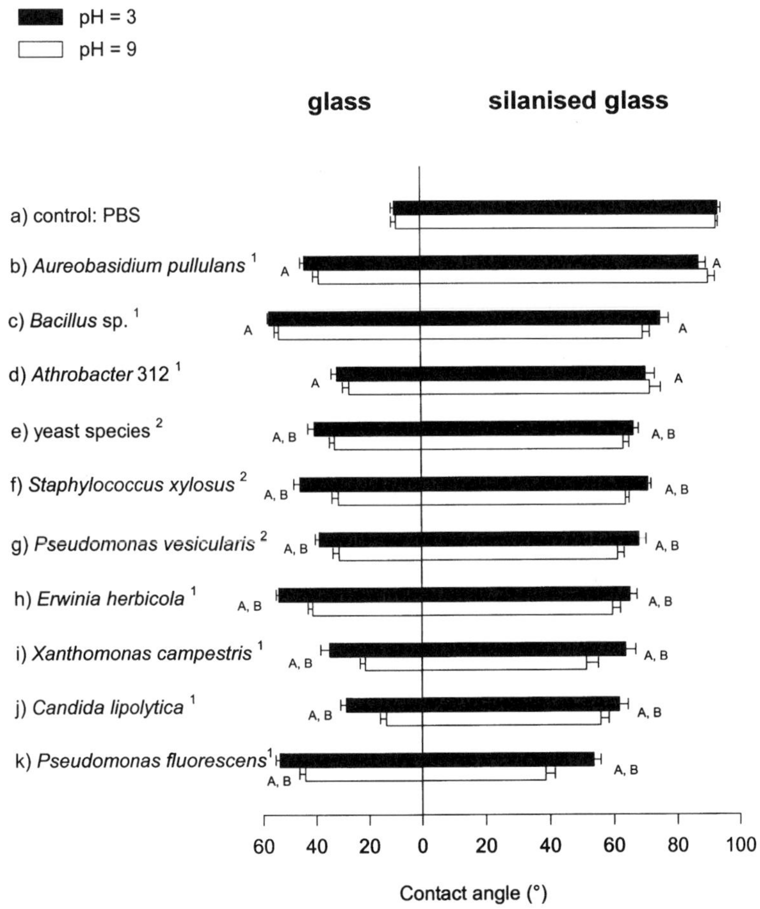

Fig. 2. Contact angles of aqueous solutions of different pH values measured on colonised glass surfaces of different hydrophobicity. Untreated, polar and silanised, unpolar glass surfaces were inoculated with various microbial cell suspensions for 24 h at 25 °C. As a control, glass and silanised glass surfaces were incubated with PBS buffer. Values are means with 95 % confidence intervals (ci) from at least 20 contact angle measurements with 10 mM citric buffer (pH 3.0) and 10 mM borate buffer (pH 9.0)

3 Chemical Characterisation of Cuticle Surfaces

The chemical composition of cutin and cuticular waxes is determined via gas chromatography coupled with flame ionisation, infrared or mass spectrometric detectors. Further information on chemical wax and cutin chemistry can be obtained from a series of reviews (Kolattukudy 1996; Holloway 1982; Walton 1990; Riederer and Markstädter 1996). In the following, a brief outline of the principal steps necessary for wax analysis is given. Sample preparation for chemical analysis generally includes extraction with organic solvents, concentration of the samples by solvent evaporation, derivatisation of alcoholic and carboxylic groups and analysis by gas chromatography.

Cuticular waxes can be easily extracted from plant surfaces using organic solvents like chloroform. Brief extractions of fresh foliage of around 10 s have been shown to be sufficient to remove all of the surface wax and most of the embedded wax (Schreiber and Schönherr 1993). After evaporation of the chloroform, the wax concentration is adjusted to 1 mg/ml and 100 µl of the extract is transferred into 1-ml reactivials for chemical analysis. In order to quantify wax components, known amounts of highly pure alkane standards (e.g., 5 µg Dotriacontane) are added to the sample. Derivatisation is necessary in order to convert free hydroxyl and carboxyl groups into their corresponding trimethylsilyl ethers and esters. This is done by treating the dried extracts with 10–30 µl of pyridine and of N,N-trimethylsilyl-trifluoroacetamide (BSTFA) at 70 °C for 30 min. Of the silylated samples, 1 µl is then injected into a gas chromatograph equipped with a flame ionisation detector. Optimised temperature and pressure programs as well as special fused silica capillary columns gain the best separation of the larger-molecular-weight aliphatic components based on their different C-carbon chain lengths. An example of a

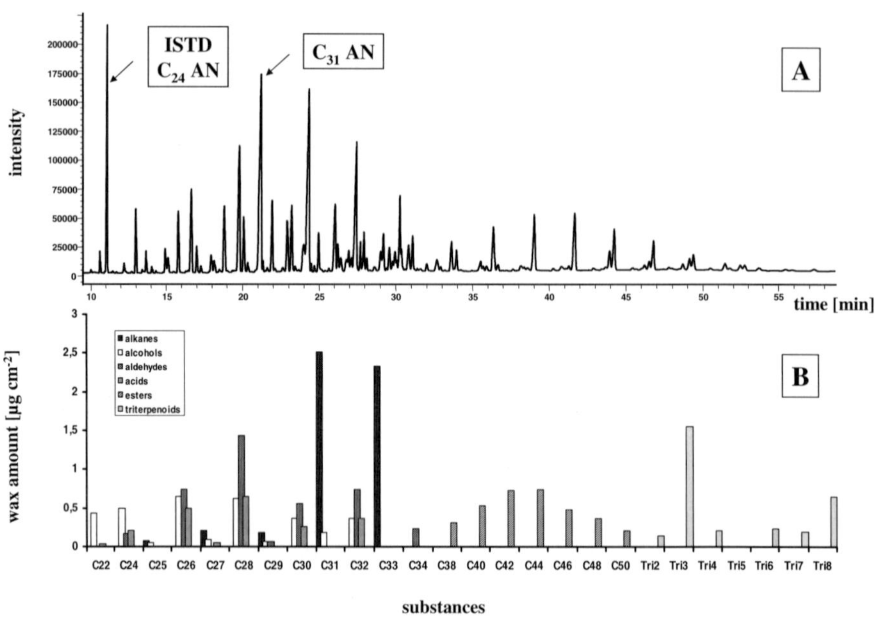

Fig. 3. Gas chromatographic analysis of the leaf surface wax of strawberry (*Fragaria* x *ananassa* cv. Elsanta). **A** Example of an original gas chromatogram of the strawberry wax analysed on a gas chromatograph equipped with a flame ionisation detector: *ISTD* internal standard, $C_{24}AN$ tetracosane, $C_{31}AN$ untriacontane). **B** Chain lengths distribution and quantitative wax coverage of the leaf surface of strawberry

gas chromatogram is shown in Fig. 3. The qualitative wax analysis is performed by gas chromatography combined with a mass spectrometric detector. Identification of wax components relies on the specific mass spectra of the molecules. The wax coverage and the wax composition is usually given per unit area of plant surface. Therefore, the total area of extracted leaves or cuticles needs to be determined after wax extraction.

4 A New in Vitro System for the Study of Interactions Between Microbes and Cuticles

4.1 Isolated Cuticles as Model Surfaces for Phyllosphere Studies

This new experimental system for in vitro studies of leaf surface–microbe interactions is based on isolated cuticles as colonisation surfaces. Isolated cuticles are ideal model surface for simulation of the phylloplane habitat as the special interfacial character of the phyllosphere is retained. Surfaces of cuticular membranes reflect the topography of epidermal cells with anticlinal cell wall depressions and the course of leaf veins like an reverse imprint of the

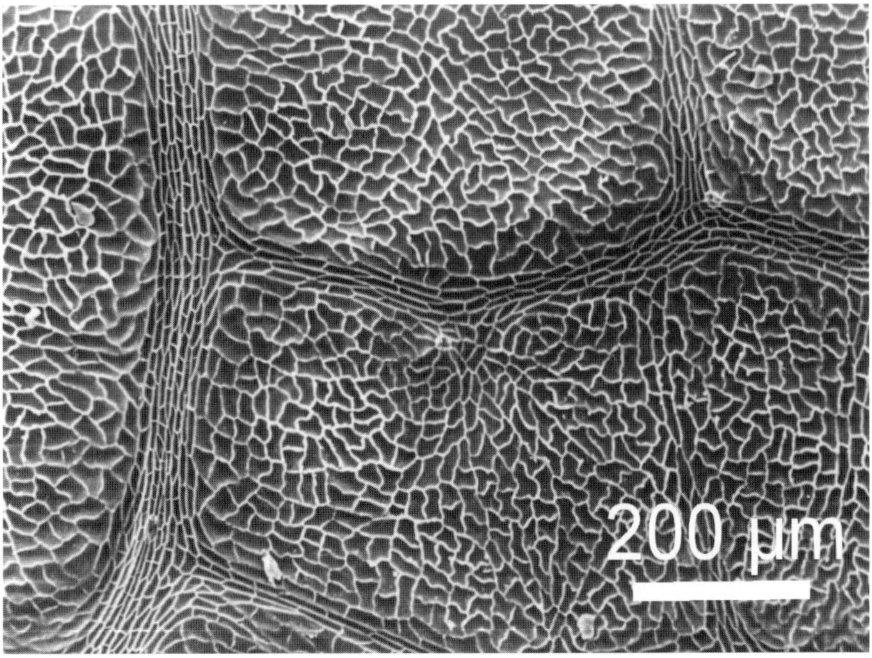

Fig. 4. Scanning electron microscope picture of an isolated cuticular membrane of ivy (*Hedera helix* L.). View of the physiological inner side of the cuticle. The pattern of epidermal cell walls and leaf veins is clearly visible

leaf surface (Fig. 4). Furthermore, isolated cuticles have a functionally intact wax layer that leads to an extreme high surface hydrophobicity and a reduction of solute transport across the cuticle. However, cuticular membranes are still permeable to a lesser extent for water and for anorganic as well as polar organic molecules. Thus, a boundary layer with higher humidity is formed above the cuticle surface and the naturally occurring leaching process of minerals or sugars through the cuticle is simulated. Important properties of the plant cuticle, like the cuticular permeability or the barrier function against microbial penetration, can be measured directly in relation to colonisation of the cuticle by microorganisms under strictly controlled conditions during incubation. This allows deeper insight into the mechanisms of possible interactions. In parallel, microbial population densities can be monitored by determining the colony forming units (cfu) and by microscope visualisation of the colonised cuticle surface. All methods were established with a strain of the commonly found epiphyllic leaf bacteria *Pseudomonas fluorescens* and can be adapted easily for the studies of other species.

4.2 Enzymatic Isolation of Plant Cuticles

Cuticular membranes are enzymatically isolated from astomatous leaf sides according to the method of Schönherr and Riederer (1986). Punched leaf disks with a diameter of 20 mm are vacuum-infiltrated with an enzyme solution containing 2 % (v/v) cellulase (Celluclast, Novo Nordisk, Bagsvaerd, Denmark) and 2 % (v/v) pectinase (Trenolin Super DF, Erbslöh, Geisenheim) dissolved in 10^{-2} M citric buffer. 10^{-3} M NaN_3 (Sigma, Deisenhofen, Germany) is added in order to inhibit microbial aerobic growth. After an incubation period of several days at room temperature, cuticles can be completely separated from adhering leaf tissue by washing carefully with deionised water. Subsequently, isolated cuticles are air-dried and stored at room temperature. For still unknown reasons the enzymatic isolation of cuticular membranes is limited to a certain number of plant species to which *Prunus laurocerasus* L., *Hedera helix* L. and *Juglans regia* L. belong.

4.3 The Experimental Set-Up of the System

The experimental set-up of the system consists of stainless steel chambers (Fig. 5A) that were originally designed to measure cuticular permeability of volatile chemicals (Bauer 1991). Isolated cuticles are placed on the top of the chamber and fixed with a metal ring sealing the cuticle/steel interfaces with high-vacuum silicone grease (Wacker Chemie, Burghausen, Germany). Prior to assembly, chambers and rings coated with silicone grease at the chamber/ring interfaces are sterilised by dry hot air at 180 °C for 3 h. Cuticles are sterilised by UV radiation for 30 min on each side. Sterilised cuticles are then mounted in the chambers under sterile conditions. Care is taken that the physiological outer side of the cuticles is orientated to the outside. The physiological inner side of the cuticle faces 800 µl of a highly concentrated nutrition solution consisting of 20 % (w/v) glucose and 5 % (w/v) yeast extract or simply water. The inner volume of the chamber is accessible by sampling ports that can be closed by metal stoppers. Using a sterile plastic syringe, the solution inside the chamber can be replaced several times during the course of the experiment. Chambers were incubated upside down on a metal grid in a climate-controlled incubation box for some hours at 25 °C before the inoculation with microbial cells. Incubation boxes are 10x20 cm in size and can be closed with an air-tight lid. Boxes are sterilised with 70 % (v/v) ethanol and with UV radiation. Sterile pressurised air is conducted through the incubation box. Air humidity is set by simply changing the temperature of the water reservoir. At a temperature of 25 °C, the air has a humidity of 100 %. Lower moisture levels can be set in the incubation box by reducing the temperature of the water reservoir under 25 °C as the saturation vapour pressure of water in air is dependent on temperature (Nobel 1991). One incubation box is equipped with a hygrometer and a temperature sensor in order to verify the actual climate conditions inside the box.

4.4 Inoculation of Cuticular Membranes with Epiphytic Microorganisms

A cell culture of *P. fluorescens* is cultivated in glucose-yeast-medium overnight at 25 °C. Cells are harvested by centrifugation (2120xg, 20 min), resuspended and washed twice in 10^{-2} M phosphate buffered saline (PBS, pH 7.4; Sigma Chemicals). Prior to inoculation the cell suspension is adjusted to an optical density of 1.0 that corresponds to $2.5 \cdot 10^8$ cfu/ml. The outer cuticle surface is inoculated with bacteria by spreading 200 µl of a washed cell suspension of *P. fluorescens* evenly over the entire exposed cuticle surface (Fig. 5B). Chambers are incubated for 6 h at 25 °C in a sterile glass Petri dish containing PBS-buffer-moistened filter papers at the bottom in order to avoid evaporation of water from the inoculation solution. During the inoculation period, bacterial cells adhere to the cuticle. After 6 h the suspension is withdrawn and the surface is carefully washed five times with 200 µl sterile deionised water to remove unbound bacteria. Chambers are left in a laminar flow hood until dry. Immediately after the drying of the washed cuticle surface, the chambers are transferred upside down in the incubation box (Fig. 5C). Furthermore, two control experiments are performed. One control is necessary for checking sterile conditions during the course of experiment. Therefore, cuticles are incubated with 200 µl of sterile PBS and treated in the same way as described above. Another control is to verify that during the inoculation period bacteria are not able to pass through the silicone grease from the outer cuticle surface

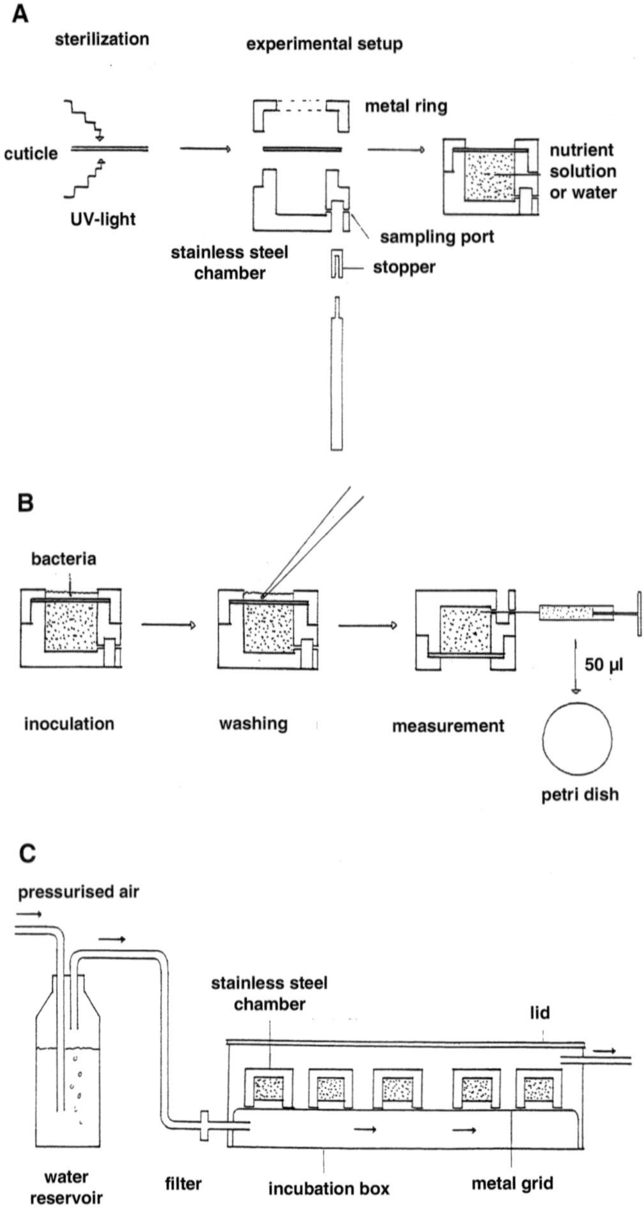

Fig. 5. Scheme of the experimental set-up for the in vitro study of microorganisms–leaf cuticle interactions. **A** Enzymatic isolated cuticular membranes are sterilised by UV radiation and mounted in a stainless steel chamber. The chamber is filled with nutrient solution or water. **B** The physiological outer side of the cuticle is inoculated with a microbial cell suspension for 6 h at 25 °C. Microbial cells not bound to the cuticle surface are removed by washing the cuticle with deionised water. Samples of the solution inside the chamber can be taken with a sterile syringe via closable sampling ports. **C** Inoculated cuticles are incubated up-side down on a metal grid in sterile incubation boxes at 25 °C. Pressurised air of the desired moisture level is conducted through the incubation box

into the nutrition solution inside the chamber volume. Therefore, round glass cover slips that definitely cannot be breached by bacteria are mounted in place of cuticles in the chambers and inoculated with 200 µl of the cell solution.

4.5 Measurement of Changes in Cuticular Transport Properties

4.5.1 Determination of Cuticular Water Permeability

Cuticular water permeability is measured according to a gravimetric method of Schönherr und Lendzian (1981). The permeability coefficients P (m/s) for water are calculated using the equation:

$$P = \frac{F}{A \times \Delta C}$$

where F is the water flow across the cuticular membrane (g/s), A is the area of the exposed cuticle surface (m^2) and ΔC represents the difference in the water concentration between the aqueous phase inside the chamber and the outer atmosphere of the incubation box. The water flow across the cuticular membrane can be measured by weighing the chambers at periodic intervals on an electronic balance with an accuracy of ±0.1 mg. The weight loss from the chambers is plotted against the incubation time and the water flow is calculated by linear regression analysis (Fig. 6). The sampling ports of the cham-

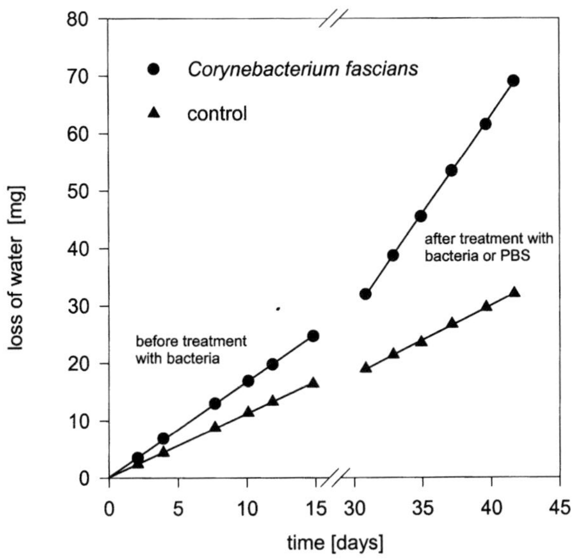

Fig. 6. Effect of *Corynebacterium fascians* on the cuticular water permeability of *Prunus laurocerasus*. The flow of water through the cuticular membrane was increased by a factor of 2 after treatment with bacteria, whereas treatment with PBS did not significantly change the cuticular water flow

bers are additionally sealed with adhesive tape to avoid diffusion of water through the sampling ports. Chambers are incubated upside down on dried silica gel in an air-tight polyethylene box at 25 °C. The silica gel adsorbs all free water of the air resulting in a water concentration inside the polyethylene box constantly held at 0 %. Thus, the driving force ΔC for the water flow across the cuticle corresponds to the density of water (10^3 kg m^{-3}). The salt and sugar concentration of the nutrition solution can be neglected as it does not affect significantly the water activity a_w. Control experiments showed that there was no significant change in cuticular water permeability when using deionised water or nutrition solution as the aqueous solution inside the chamber volume. Sterilisation of cuticles by UV radiation also did not significantly change water permeability.

4.5.2 *Effect of Bacteria on Cuticular Water Permeability*

Isolated cuticles are mounted in stainless steel chambers and permeability coefficients P1 for water are determined for each sample as described above. Cuticular permeability coefficients P1 determined after UV radiation ranged between $1.44 \cdot 10^{-10}$ m/s for *Vinca major* leaf cuticles and 10.8×10^{-9} m/s for *Lycopersicon esculentum* fruit cuticles (Table 1). Then cuticles are inoculated with bacteria and incubated for 12 days in the incubation box at 25 °C at air humidity close to 100 %. Control experiments are conducted by inoculating the cuticles with 200 µl PBS in place of the bacterial cell solution. After incubation with bacteria chambers are again transferred onto dried silica gel and cuticular water permeability coefficients P2 are determined after an equilibrium period of 1 day. The effects of bacteria on water permeability of the respective cuticular membrane are calculated from the permeance of the cuticle after treatment with bacteria (P2) divided by the initial permeance (P1).

$$\text{Effect} = \frac{P2}{P1}$$

Table 1. Cuticular permeability coefficients for water P_{water} (m/s) from different plant species. Values are arithmetic means with 95 % confidence intervals (ci) of 14 measured permeability coefficients for each plant species

Species	$P_{water} \times 10^{-10}$ (m/s)	ci 95 %$\times 10^{-10}$ (m/s)
Vinca major	1.44	1.26–1.64
Hedera helix can.	2.16	1.76–2.65
Prunus laurocerasus	2.93	2.34–3.68
Citrus aurantium	4.53	2.97–6.92
Lycopersicon esculentum	10.80	8.85–13.17

An example for the change in water permeability of one cuticular membrane before and after treatment with bacteria is shown in Fig. 6. The effects on water permeability for an entire sample unit consisting of at least 12 cuticles are given as mean values of the effects measured for individual membranes. Some results are presented in the chapter "Interactions between Epiphyllic Microorganisms and Leaf Cuticles" by Schreiber et al. (Chap. 9, this Vol.).

The effects on water permeability need not necessarily be measured before and after treatment with bacteria, but can also be measured during the incubation with bacteria by lowering the air humidity inside the incubation box to, e.g. 90 %. As the driving force for the water flow across the cuticle is reduced to 1/10, periodical intervals in between weighing the chamber are increased to 4 days in order to measure a significant weight loss.

4.6 Measuring Penetration of Microorganisms Through Cuticular Membranes

Penetration of microorganisms through cuticular membranes can be measured as well using the described in vitro system. The outer side of the cuticle is inoculated with bacterial cells, whereas the inner side faces a sterile nutrition solution. If the cuticle, located between microbial cells and nutrition solution inside the chamber, is penetrated by bacterial cells, microbial growth will be detectable in the nutrition solution. Thus, penetration of isolated cuticles by bacteria can be easily monitored by transferring 50 µl of the nutrition solution inside the chamber onto glucose-supplemented yeast extract agar plates using a sterile syringe. Subsequent microbial growth on the agar plates indicates that a penetration event through the cuticular membrane has occurred. In that way, the amount of cuticles penetrated is determined in daily intervals. The amount of cuticles penetrated after different periods of incubation is given in percent of the total amount of inoculated membranes.

$$\%CM_{penetrated} = \frac{\text{Number of CMpenetrated}}{\text{Number of CMtotal}} \times 100$$

An example for a penetration kinetic is shown in Fig. 7. The amount of penetrated cuticular membranes increased over the incubation period of 12 days. Some typical characteristics of a penetration kinetic can be used to describe the barrier function of cuticles quantitatively: (1) at the end of the incubation period there was a steady increase of penetrated cuticular membranes versus incubation time. Rates of penetration ($\% \, CM_{penetrated}$/day) can be calculated from the slopes of the linear regression. (2) Another meaningful parameter is the time needed by the microorganisms to penetrate 50 % of inoculated membranes ($T_{50\,\%}$). High rates of penetration and small $T_{50\,\%}$ values indicate low

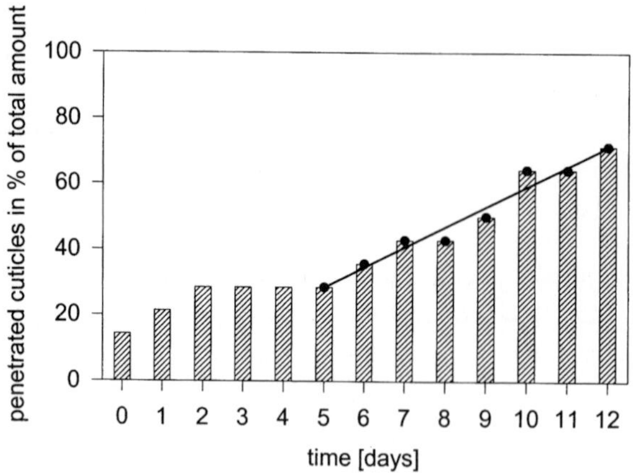

Fig. 7. Penetration of *Pseudomonas fluorescens* through cuticular membranes of *Vinca major*. The amount of penetrated cuticles increases with incubation time. After 9.5 days 50 % of the inoculated cuticles are penetrated by *P. fluorescens*. At the end of the kinetic there is a linear increase of penetrated cuticles with a rate of 6.1 % penetrated CM per day in relation to the total amount of inoculated cuticles

barrier functions of the cuticle for microbial penetration. Once these values are known, barrier properties of cuticles of different plant species that differ in their morphology like cuticle thickness or chemistry like wax composition can be compared. Another attractive application is to measure penetration of different microbial strains or mutants that differ in their array of extracellular enzymes like cutinase activity.

Several control experiments need to be conducted to ensure bacterial penetration through isolated cuticles. (1) When glass slides are mounted into the chambers in place of cuticles there was never any bacterial growth detectable in the nutrition solution. This gives evidence that bacteria are not able to bypass the glass surface via the silicone grease seal. (2) In addition, no bacterial growth was detected in the nutrition solution when cuticles were inoculated with sterile PBS indicating that the system itself is sterile and no other origins for bacterial growth are possible except from the inoculus on the outer cuticle surface. (3) Finally, a third control consists of applying 200 µl of dead bacteria. Cells are cultivated as described above and subsequently killed with paraformaldehyde and stained with the fluorescent dye DAPI. It was checked that all bacterial cells were killed. After the inoculation period of 6 h the nutrition solution is checked for the presence of DAPI-stained cells with fluorescence microscopy. A fraction of about 10 % of the inoculated cuticles was apparently leaky for dead cells. This might be due to mechanical injuries to the cuticular membranes during the process of isolation or during the mounting of cuticular membranes in the chambers. Those membranes were sorted out and not considered any further. Furthermore, cuticular water permeability measured prior to inoculation with bacterial cells was very low (Table 1), indicating that the membranes form high effective barriers for the transport of water on the molecular level. This also suggests that they build intact barriers for microbial cells as well. Basically, all control experiments confirmed that after the inoculation period of 6 h bacterial cells are solely present on the inoculated outer cuticle surface.

4.7 Determination of the Viable Cell Number on the Cuticle Surface

In order to document the microbial development on isolated cuticular membranes, the cfu is determined. The initial cfu on isolated cuticles is determined directly after inoculation of membranes with microorganisms. As an example, the initial cfu of *P. fluorescens* attached to cuticles of *V. major* was $2.85x10^5 \pm 0.98x10^5$ cfu/CM. Then cfu measurements are done in daily intervals during the incubation period. First, the nutrition solution inside the chamber is totally removed with a sterile syringe and kept in sterile glass tubes to check for microbial growth (see below). After having removed the nutrition solution, the membrane is cut out of the chamber with a sterile scalpel blade and transferred in a 1.5-ml tube containing 0.05 g of sterile sand. The cuticle is ground in 100 μl PBS with a micropestle for 2 min. After homogenisation of the cuticle, 900 μl PBS is added and the tube contents mixed. Serial dilutions of 100 μl are incubated on glucose yeast extract agar plates at 25 °C for 2 days before colonies have been counted.

In order to determine the microbial cfu exclusively on the outer cuticle surface, it is very important to check the nutrition solution inside the chamber for microbial growth. Therefore, the nutrition solution removed from the inner chamber volume is simply incubated at 25 °C for 2 days. Only if there is no microbial growth detectable, is the cfu determined for that cuticle considered to describe the microbial development on the outer cuticle surface.

4.8 Microscope Visualisation of Microorganisms on the Cuticle

Microscopic detection of microbial cells on isolated cuticles gives information about the colonisation pattern and development. The fluorescent dyes acridine orange and DAPI are used to stain bacteria. Both dyes are polar substances with a very high affinity to bind nucleic acids. Thus, microbial cells adhering to the cuticle surface are specifically stained, whereas the hydrophobic cuticle surface itself is not stained. 0.02 % (w/v) acridine orange and 0.001 % (w/v) DAPI are dissolved in deionised water and filtered through 0.2 μm membrane filters to remove dye crystals and dust particles. Care is taken that staining solutions are protected from daylight. For better handling cuticles are left mounted in the chambers for staining of bacterial cells. Staining solution (200 μl) is evenly distributed over the outer cuticle surface. Chambers are incubated in the dark at room temperature on a horizontal shaker (30 rpm). After different staining times of 5, 20, 40 and 60 min, respectively, the cuticle surfaces are washed twice with 200 μl of sterile-filtered deionised water to remove unbound dye molecules. Cuticles are left over silica gel until dry. The dried cuticle surfaces are excised from the chambers with a scalpel blade and cut into four parts. Cuticle pieces are transferred onto a thin hydrophobic layer of silicon grease on a microscopic slide. A cover slip together with one drop of immersion oil is put on the top of the cuticle prior to microscopic examination. Due to the hydrophobic layer of silicon grease and the immersion oil, the entire surfaces of the cuticle pieces are spread totally flat minimising problems with depth focus. Furthermore, the refraction of light is markedly reduced allowing fluorescence microscopy with isolated cuticles. Samples can be viewed with a Zeiss Axioplan microscope (Zeiss, Oberkochen, Germany) equipped with a 50 W mercury high pressure bulb, a 40x objective (Zeiss, Plan-Neofluar) and a Zeiss filter set No. 09 (excitation: 450–490 nm; dichroic beamsplitter ≥510 nm; emission ≥520 nm). One examples of a fluorescence microscopy micrograph of a colonised cuticle surface is shown in Fig. 8. The surface coverage of the cuticle colonised by bacte-

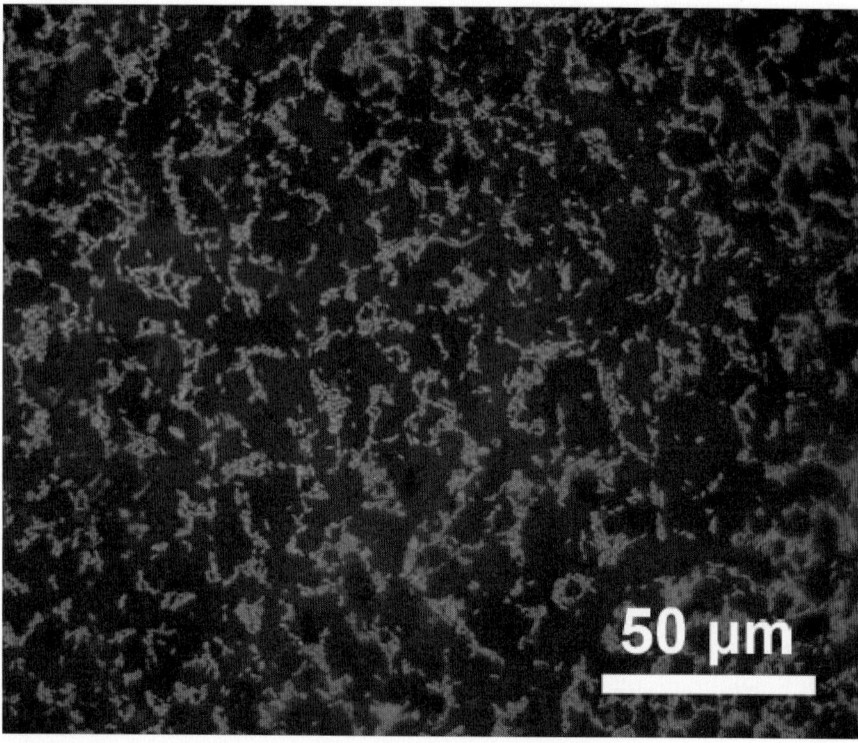

Fig. 8. Epifluorescent microscope image of an isolated cuticular membrane of *Prunus laurocerasus* artificially colonised with *Pseudomonas fluorescens* (magnification ×400). Bacterial cells are stained with acridine orange and viewed at an excitation of 450–490 nm. Approximately 27.4 % of the cuticle surface is covered by bacteria. Bacterial cells are accumulated in small clusters over the entire cuticle surface

rial cells can be quantified by digital image analysis. Digitised video images are analysed for the pixel size of stained bacterial cells using Adobe Photoshop software. Percentage coverage of bacterial cells is calculated as follows:

$$\%\ \text{coverage} = \frac{\text{No. of pixel of bacterial cells of digitized image}}{\text{No. of total pixel size of digitized image}} \times 100$$

Percentage coverage by bacteria is given as the mean value of 12 analysed digitised images at 400x magnification from randomly chosen sites of at least three cuticles per sampling point. The influence of staining time with acridine orange on the area coverage can be seen in Fig. 9A. The optimal staining time is 20 m. An adhesion kinetic of cells of P. *fluorescens* to cuticle surfaces of *P.*

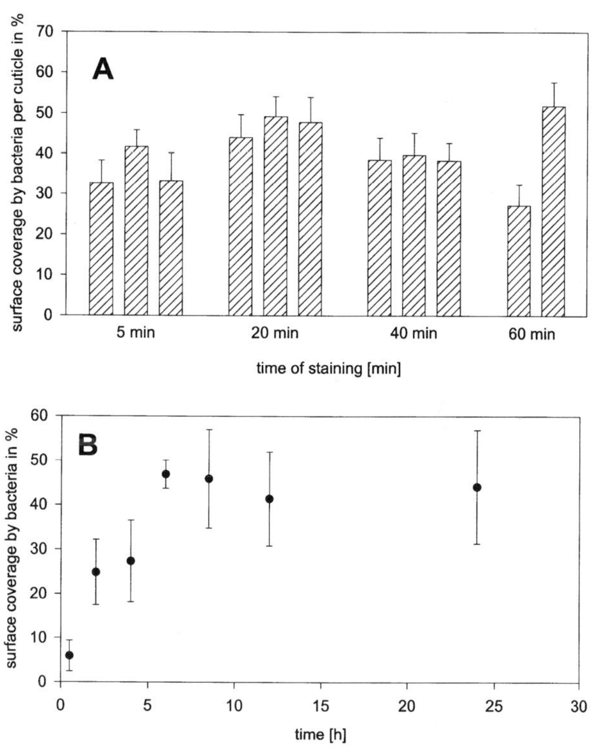

Fig. 9. Surface coverage of cuticles from *Prunus laurocerasus* with *Pseudomonas fluorescens*. **A** Dependence of the surface coverage by bacterial cells on the staining time with acridine orange. The optimal staining time was 20 min. **B** Adhesion of bacterial cells to the cuticle surface over time. Maximal adhesion of 46.9 % occurred after 6 h of inoculation. Percentage coverage by bacterial cells is given as the mean value with 95 % confidence intervals of 12 analysed digitised images at x400 magnification from randomly chosen sites of each of three examined cuticles. Only two membranes could be analysed for the 60-min time sample

laurocerasus is shown in Fig. 9B. Maximal surface coverage of 46.9 % was reached after 6 h of inoculation with bacterial cell solution.

5 Conclusions

The presented methods allow a detailed analysis of a variety of microbe–cuticle interactions combining physicochemical, ecophysiological and microbial ecological aspects. Isolated cuticles are excellent model surfaces to study the mechanisms of such interactions. Using the presented in vitro system, even minor changes in cuticular wax composition or permeability can be examined in relation to microbial growth. When working with entire leaves, such changes would probably be masked by the physiological influence of the leaf. Therefore, this new approach might be very helpful to reveal possible mechanisms of interactions that occur in reality only in the scale of microhabitats. The impact of cuticular features will help us to understand the observed heterogeneous colonisation of the leaf habitat and the formation of microcolonies. Vice versa, the capacity of microbial cells to change cuticular properties might be of crucial importance for a successful colonisation of the leaf surfaces and could contribute substantially to the microbial fitness of individual epiphyllic species.

Acknowledgements. The authors gratefully acknowledge financial support of this work by the Deutsche Forschungsgemeinschaft and the FCI.

References and Selected Reading

Bauer H (1991) Mobilität organischer Moleküle in der pflanzlichen Kutikula. PhD Thesis, Technical University of Munich, Germany

Holloway PJ (1982) The chemical constitution of plant cutins. In: Cutler DF, Alvin KL, Price CE (eds) The plant cuticle. Academic Press, London

Knoll D (1998) Die Bedeutung der Kutikula bei der Interaktion zwischen epiphyllen Mikroorganismen und Blattoberflächen. PhD Thesis, University of Würzburg, Germany

Knoll D, Schreiber L (1998) Influence of epiphytic micro-organisms on leaf wettability: wetting of the upper leaf surface of *Juglans regia* and of model surfaces in relation to colonization by microorganisms. New Phytol 140:271–282

Knoll D, Schreiber L (2000) Plant-microbe interactions: wetting of ivy (*Hedera helix* L.) leaf surfaces in relation to colonization by epiphytic microorganisms. Microb Ecol 41:33–42

Kolattukudy PE (1996) Biosynthetic pathways of cutin and waxes. In: Kerstiens G (ed) Plant cuticles: an integrated functional approach. BIOS Scientific Publishers, Oxford, pp 83–108

Leibnitz E, Struppe HG (1984) Handbuch der Gaschromatographie. Akademische Verlagsgesellschaft, Leipzig

Nobel PS (1991) Physicochemical and environmental plant physiology. Academic Press, San Diego

Riederer M, Markstädter C (1996) Cuticular waxes: a critical assessment of current knowledge. In: Kerstiens G (ed) Plant cuticles: an integrated functional approach. BIOS Scientific Publishers, Oxford, pp 189–200

Schönherr J, Lendzian K (1981) A simple and inexpensive method of measuring water permeability of isolated plant cuticular membranes. Z Pflanzenphysiol 102:321–327

Schönherr J, Riederer M (1986) Plant cuticles sorb lipophilic compounds during enzymatic isolation. Plant Cell Environ 4:459–466

Schreiber L (1996) Wetting of the upper needle surface of *Abies grandis*: influence of pH, wax chemistry and epiphyllic microflora on contact angles. Plant Cell Environ 19:455–463

Schreiber L, Schönherr J (1993) Mobilities of organic compounds in reconstituted cuticular wax of barley leaves: Determination of diffusion coefficients. Pestic Sci 38:353–361

Walton TJ (1990) Waxes, cutin and suberin. Meth Plant Biochem 4:105–158

26 Quantifying the Impact of ACC Deaminase-Containing Bacteria on Plants

DONNA M. PENROSE and BERNARD R. GLICK

1 Introduction

In 1994, we reported that the bacterium, *Pseudomonas putida* GR12-2 (Lifshitz et al. 1986), a well-known plant growth promoting strain, contained the enzyme, 1-aminocyclopropane-1-carboxylic acid (ACC) deaminase (Jacobson et al. 1994). This enzyme hydrolyzes ACC, the immediate precursor of ethylene, in plant tissues (Yang and Hoffman 1984). Ethylene is required for seed germination by many plant species and the rate of ethylene production increases during germination and seedling growth (Abeles et al. 1992). Although low levels of ethylene appear to enhance root initiation and growth, and promote root extension, high levels of ethylene produced by fast growing roots can lead to inhibition of root elongation (Mattoo and Suttle 1991; Ma et al. 1998). We have proposed a model that suggests that ACC deaminase-containing plant growth promoting bacteria can lower ethylene levels and thus stimulate plant growth (Glick et al. 1998). It is quite likely that much of the ACC produced during ethylene biosynthesis is taken up by the bacterium and subsequently hydrolyzed to α-ketobutyrate and ammonia by ACC deaminase. The uptake and cleavage of ACC by ACC deaminase would decrease the amount of ACC, as well as ethylene.

2 Selection of Bacterial Strains that Contain ACC Deaminase

We developed a rapid and novel procedure for the isolation of ACC deaminase-containing bacteria and used this technique to identify and isolate seven plant growth promoting strains based on their ability to utilize ACC as the sole source of nitrogen (Glick et al. 1995). These bacterial strains were isolated from soil samples collected during late summer in Waterloo, Ontario, Canada and various locations in California, USA from the rhizosphere of seven different plants (Table 1). Originally, these strains were designated as *Pseudomonas* sp., but were re-classified following fatty acid analysis (Shah et al. 1997).

Plant Surface Microbiology
A. Varma, L. Abbott, D. Werner, R. Hampp (Eds.)

Table 1. ACC-utilizing bacterial strains isolated from Waterloo, Ontario, Canada and California, USA

Genus and species	Strain	Soil location	Plant source
Pseudomonas putida	UW1	Waterloo, Ontario, Canada	Bean
Enterobacter cloacae	UW2	Waterloo, Ontario, Canada	Clover
Pseudomonas putida	UW3	Waterloo, Ontario, Canada	Maize
Enterobacter cloacae	UW4	Waterloo, Ontario, Canada	Reeds
Pseudomonas fluorescens	CAL1	San Benito, California, USA	Oats
Enterobacter cloacae	CAL2	King City, California, USA	Tomato
Enterobacter cloacae	CAL3	Fresco, California, USA	Cotton

Our method of isolating bacteria entails screening soil bacteria for the ability to use ACC as a sole nitrogen source, a trait that is a consequence of the presence of the activity of the enzyme, ACC deaminase. One gram of soil is added to 50 ml of sterile medium containing 10 g proteose peptone, 10 g casein hydrolysate, 1.5 g anhydrous $MgSO_4$, 1.5 g K_2HPO_4 and 10 ml glycerol (PAF medium) in a 250-ml flask. The flask and its contents are incubated in a shaking water bath (200 rpm) at either 25 or 30 °C depending on the geographic location of the soil samples, i.e., the samples collected in the cooler Canadian climate of Waterloo, Ontario were grown at 25 °C and those from the warmer weather of California, USA were grown at 30 °C. After 24-h, a 1-ml aliquot is removed from the growing culture, transferred to 50 ml of sterile PAF medium in a 250-ml flask and incubated at 200 rpm in a shaking water bath for 24 h, at either 25 or 30 °C, the same temperature as the first incubation. Following these two incubations, the population of pseudomonads is enriched and the number of fungi in the culture is reduced.

A 1-ml aliquot is removed from the second culture and transferred to a 250-ml flask containing 50 ml of sterile minimal medium, DF salts (Dworkin and Foster 1958; per litre): 4.0 g KH_2PO_4, 6.0 g Na_2HPO_4, 0.2 g $MgSO_4{\cdot}7H_2O$, 2.0 g glucose, 2.0 g gluconic acid and 2.0 g citric acid with trace elements: 1 mg $FeSO_4{\cdot}7H_2O$, 10 µg H_3BO_3, 11.19 µg $MnSO_4{\cdot}H_2O$, 124.6 µg $ZnSO_4{\cdot}7H_2O$, 78.22 µg $CuSO_4{\cdot}5H_2O$, 10 µg MoO_3, pH 7.2 and 2.0 g $(NH_4)_2SO_4$ as a nitrogen source. In our lab the DF minimal medium is prepared as follows: (1) the trace elements (10 mg H_3BO_3, 11.19 mg $MnSO_4{\cdot}H_2O$, 124.6 mg $ZnSO_4{\cdot}7H_2O$, 78.22 mg $CuSO_4{\cdot}5H_2O$, and 10 mg MoO_3) are dissolved in 100 ml of sterile distilled water and then stored in the refrigerator for up to several months; (2) $FeSO_4{\cdot}7H_2O$ (1 mg) is dissolved in 10 ml of sterile distilled water and is stored in the refrigerator for up to several months; (3) all of the other ingredients including 4.0 g KH_2PO_4, 6.0 g Na_2HPO_4, 0.2 g $MgSO_4{\cdot}7H_2O$, 2.0 g glucose, 2.0 g gluconic acid, 2.0 g citric acid, 2.0 g $(NH_4)_2SO_4$ and 0.1 ml of each of the solutions of trace elements and $FeSO_4{\cdot}7H_2O$ are dissolved in 1 l of distilled water and autoclaved for no more than 20 min. If this medium is prepared by dissolving one ingredient at a time, i.e., by not adding another ingredient until the previous one is completely dissolved, this medium should not contain a precipitate. Following an incubation of 24 h in a shaking water bath at 200 rpm at either 25 or 30 °C, the same temperature as the first incubation, a 1-ml aliquot is removed from this culture and transferred to 50 ml of sterile DF salts minimal medium in a 250-ml flask containing 3.0 mM ACC (instead of $(NH_4)_2SO_4$) as the source of nitrogen. A 0.5 M-solution of ACC (Calbiochem-Novobiochem Corp., La Jolla, CA, USA), which is very labile in solution, is filter-sterilized through a 0.2 µm membrane and the filtrate collected, aliquoted and frozen at –20 °C. Just prior to inoculation, the ACC solution is thawed and a 300-µl aliquot added to 50 ml of sterile DF salts minimal medium; following inoculation, the culture is placed in a shaking water bath at 200 rpm and grown for 24 h at either 25 or 30 °C, the same temperature as the previous incubation.

Dilutions of this final culture are plated onto solid DF salts minimal medium and incubated for 48 h at either 25 or 30 °C, the same temperature as the previous incubations. These plates are prepared with 1.8 % Bacto-Agar (Difco Laboratories, Detroit, MI, USA), which has a very low nitrogen content, and are spread with ACC (30 µmol/plate) just prior to use. Before streaking with either a loopful of bacterium or an individual colony, the ACC is allowed to dry fully. The inoculated plates are incubated at the appropriate temperature – no higher than 35 °C because all of the known ACC deaminases are inhibited above this temperature – for 3 days and the growth on the plates is checked daily. Even when apparently nitrogen-free agar is used, and no additional source of nitrogen is included in the medium, it is almost impossible to obtain plates with absolutely no growth, but it is possible to obtain plates with very, very light growth.

The colonies isolated from each of the seven soil samples displayed a similar colony morphology and rate of growth. In order to avoid isolating multiple copies of the same bacterium, only a single colony from each soil sample is selected for further testing. Each selected colony is tested for the synthesis of siderophores, antibiotics and indole acetic acid, as well as for plant growth stimulation and ACC deaminase activity. It is interesting to note that Belimov et al. (submitted for publication) used a variant of the procedure described above to isolate ACC deaminase-containing strains of *Bacillus*.

3 Culture Conditions for the Induction of Bacterial ACC Deaminase Activity

The assessment of bacterial ACC deaminase activity and root growth enhancement both require growth conditions that favor the induction of ACC deaminase. The bacteria are cultured first in rich medium and then transferred to minimal medium with ACC as the sole source of nitrogen. Bacterial cells are grown to mid- up to late-log phase in 15 ml of rich medium, e.g., tryptic soybean broth (TSB; Difco Laboratories, Detroit, MI, USA) divided between two culture tubes: each tube is inoculated with 5 µl of the appropriate strain. Cultures are incubated overnight in a shaking water bath at 200 rpm at either 25 or 30 °C – the temperature most suitable for the bacterial strain. The accumulated biomass is harvested by centrifugation of the contents of the combined tubes at 8000x*g* for 10 min at 4 °C in a Sorvall RC5B/C centrifuge using an SS34 rotor. The supernatant is removed and the cells are washed with 5 ml of DF salts minimal medium. Following an additional centrifugation for 10 min at 8000x*g* in the same rotor at 4 °C, the cells are suspended in 7.5 ml of DF salts minimal medium, in a fresh culture tube. Just prior to incubation, the frozen 0.5 M ACC solution (prepared as described in Sect. 2) is thawed, and an aliquot of 45 µl is added to the cell suspension; the final ACC concentration is 3.0 mM. The bacterial cells are returned to the shaking water bath to induce the activity of ACC deaminase – at 200 rpm for 24 h at the same temperature as the overnight incubation, either 25 or 30 °C. The bacteria are harvested by centrifugation at 8000x*g* for 10 min at 4 °C in an SS34 rotor in a Sorvall RC5B/C centrifuge. The supernatant is removed, and the cells are washed by suspending the cell pellet in 5 ml of either 0.1 M Tris-HCl, pH 7.6 if the cells are to be assayed for ACC deaminase activity, or 0.03 M $MgSO_4$ if they are to be used as a bacterial treatment in the gnotobiotic root elongation assay or the high performance liquid chromatography (HPLC) protocol for measuring ACC. Following centrifugation at 8000x*g* at 4 °C for 10 min in the same rotor and centrifuge, the supernatant is discarded. The washing procedure is repeated twice to ensure that the pellet is free of medium. The pelleted cells are stored at either –20 °C for measurement of ACC deaminase activity, or at 4 °C for seed treatment in the gnotobiotic root elongation assay or HPLC measurement of ACC.

4 Gnotobiotic Root Elongation Assay

The gnotobiotic root elongation assay is used as a method of assessing the effect of various bacterial strains on the growth of canola seedlings. Each of the seven strains of ACC deaminase-containing soil bacteria isolated in our lab was assayed by the root elongation assay and was shown to promote canola seedling growth under gnotobiotic conditions. The protocol described below is a modification of the procedure developed by Lifshitz et al. (1987) and is used to measure the elongation of canola roots from seeds treated with different strains of bacteria or chemical ethylene inhibitors. The bacterial cell pellet, prepared as described in Section 3, is suspended in 0.5 ml of sterile 0.03 M $MgSO_4$ and then placed on ice. A 0.5-ml sample is removed from the cell suspension and diluted eight to ten times in 0.03 M $MgSO_4$; the absorbance of the sample is measured at 600 nm. This measurement is used to adjust the absorbance at 600 nm, of the bacterial suspension, to 0.15 with sterile 0.03 M $MgSO_4$.

Seed-pack growth pouches (Northrup King Co., Minneapolis, MN, USA) are prepared for the gnotobiotic assay of canola root elongation. Following the addition of 12 ml of distilled water to each one, the growth pouches are wrapped in aluminum foil in groups of ten, placed in an upright position to prevent water loss, and autoclaved at 121 °C for 15 min.

Canola seeds (*Brassica campestris*) are disinfected immediately before use. (Tomato seeds may also be used in this assay.) The seeds (approximately 0.2 g/treatment) are soaked in 70 % ethanol for 1 min in glass Petri dishes (60×15 mm); the ethanol is removed and replaced with 1 % sodium hypochlorite (household bleach). After 10 min the bleach solution is suctioned off and the seeds are thoroughly rinsed with sterile distilled water at least five times, sterile distilled water is added to the dish of seeds, swirled and removed by suction. Each dish is incubated at room temperature for 1 h with the appropriate treatment: sterile 0.03 M $MgSO_4$ (used as a negative control) or bacterial suspensions in sterile 0.03 M $MgSO_4$. Following incubation with each treatment, the seeds are placed in growth pouches with sterilized forceps: six seeds are set in each growth pouch and ten pouches are used for each treatment. The pouches are grouped together according to treatment and placed upright in a rack (Northrup King Co., Minneapolis, MN, USA) ensuring that the pouches are not touching. Two empty pouches are placed at the ends of each rack. Racks are placed in a clean plastic bin containing sterile distilled water, to a depth of approximately 3 cm, and covered loosely with clear plastic wrap to prevent dehydration. Pouches are incubated in a growth chamber (Conviron CMP 3244, Controlled Environments Ltd., Winnipeg, MB, Canada) which is maintained at 20±1 °C with a cycle beginning with 12 h of dark followed by 12 h of light (18 μmol m^{-1} s^{-1}). Each rack is positioned such that the center of the row of pouches is 8in. below and 5 in. lateral to the light source. The primary root lengths are measured on the fifth day of growth and the data are analyzed. Seeds that fail to germinate 2 days after they were sown are marked and the roots that subsequently develop from these seeds are not measured.

5 Measurement of ACC Deaminase Activity

ACC deaminase activity is assayed according to the method of Honma and Shimomura (1978) which measures the amount of α-ketobutyrate when the enzyme, ACC deaminase, cleaves ACC. The number of μmoles of α-ketobutyrate produced by this reaction is determined by comparing the absorbance at 540 nm of a sample to a standard curve of α-ketobutyrate ranging between 0.1 and 1.0 μmol (Fig. 1). A stock solution of 100 mM α-ketobutyrate (Sigma-Aldrich Co.) is prepared in 0.1 M Tris-HCl pH 8.5 and stored at 4 °C. Just prior

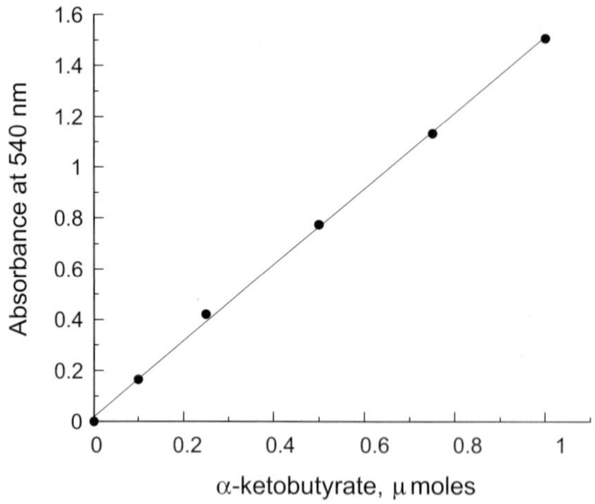

Fig. 1. Standard curve of α-ketobutyrate versus absorbance at 540 nm

to use, the stock solution is diluted with the same buffer to make a 10-mM solution from which a standard concentration curve is generated. Each in a series of known α-ketobutyrate concentrations is prepared in a volume of 200 μl and transferred to a glass test tube (100x13 mm); each point in the series is assayed in duplicate. Three hundred μl of the 2,4-dinitrophenylhydrazine reagent (0.2% 2,4-dinitrophenyl-hydrazine in 2 N HCl; Sigma-Aldrich Co.) is added to each glass tube and the contents are vortexed and incubated at 30 °C for 30 min during which time the α-ketobutyrate is derivatized as a phenylhydrazone. The color of the phenylhydrazone is developed by the addition of 2.0 ml of 2 N NaOH; after mixing, the absorbance of the mixture is measured at 540 nm.

5.1 Assay of ACC Deaminase Activity in Bacterial Extracts

5.1.1 Preparation of Bacterial Extracts

ACC deaminase activity is measured in bacterial extracts prepared in the following manner. Bacterial cell pellets, prepared as described in Section 3, are each suspended in 1 ml of 0.1 M Tris-HCl, pH 7.6 and transferred to a 1.5-ml microcentrifuge tube. The contents of the 1.5-ml microcentrifuge tube are centrifuged at 16,000xg for 5 min in a Brinkmann microcentrifuge and the supernatant is removed with a fine-tip transfer pipette. The pellet is suspended in 600 μl of 0.1 M Tris-HCl, pH 8.5. Thirty μl of toluene is added to the cell suspension and vortexed at the highest setting for 30 s. At this point, a 100-μl aliquot of the "toluenized cells" is set aside and stored at 4 °C for protein assay at a later time. The remaining toluenized cell suspension is immediately assayed for ACC deaminase activity.

5.1.2 *Measurement of ACC Deaminase Activity*

All sample measurements are carried out in duplicate. Two hundred µl of the toluenized cells is placed in a fresh 1.5-ml microcentrifuge tube; 20 µl of 0.5 M ACC is added to the suspension, briefly vortexed, and then incubated at 30 °C for 15 min. Following the addition of 1 ml of 0.56 N HCl, the mixture is vortexed and centrifuged for 5 min at 16,000x*g* in a Brinkmann microcentrifuge at room temperature. One ml of the supernatant is vortexed together with 800 µl of 0.56 N HCl in a clean glass tube (100x13 mm). Thereupon, 300 µl of the 2,4-dinitrophenylhydrazine reagent (0.2 % 2,4-dinitrophenylhydrazine in 2 N HCl) is added to the glass tube, the contents vortexed and then incubated at 30 °C for 30 min. Following the addition and mixing of 2 ml of 2 N NaOH, the absorbance of the mixture is measured at 540 nm.

The absorbance of the assay reagents including the substrate, ACC, and the bacterial extract are taken into account. After the indicated incubations, the absorbance at 540 nm of the assay reagents in the presence of ACC is used as a reference for the spectrophotometric readings; it is subtracted from the absorbance of the bacterial extract plus the assay reagents in the presence of ACC. The contribution of the extract, i.e., the absorbance at 540 nm of extract and the assay reagents without ACC, is determined and subtracted from the absorbance value calculated above. This value is used to calculate the amount of α-ketobutyrate generated by the activity of ACC deaminase.

6 Measurement of ACC in Plant Roots, Seed Tissues and Seed Exudates

In order to be able to test the model described earlier, we required a method of measuring ACC in plant tissues. Since all of the available methods for ACC quantification had problems and limitations associated with their use, we adapted the Waters AccQ•Tag Method, designed to measure amino acids, for ACC analysis. This procedure is simple and relatively sensitive. ACC, which is an amino acid, is derivatized with the Waters AccQ•Fluor reagent; the ACC derivatives are separated by reversed phase HPLC and quantified by fluorescence. We have used this procedure to quantify the amount of ACC in extracts of germinating canola seeds, seedling roots, and seed exudate (Penrose et al. 2001; Penrose and Glick 2001).

6.1 Collection of Canola Seed Tissue and Exudate During Germination

Canola seed tissue and exudate is collected from 200-seed samples exposed to various treatments and then incubated in the dark for up to 50 h. The seeds are disinfected immediately before use. Two hundred seeds (0.400±0.008 g) are measured into an aluminum weigh boat and soaked in 5 ml of 10 % hydrogen peroxide at room temperature (Bayliss et al. 1997). After 2 min, the hydrogen peroxide solution is removed by suction and the seeds are rinsed with sterile distilled water at least four times. Each dish is then incubated at room temperature for 1 h with 5 ml of the appropriate treatment: 0.03 M $MgSO_4$, (used as a negative control) or bacterial cells (grown as described in Sect. 3) suspended in 0.03 M $MgSO_4$ and diluted to an absorbance of 0.15 at 600 nm. Following incubation, the solution used for seed treatment (0.03 M $MgSO_4$ or bacterial suspension) is removed from the seeds and they are rinsed twice with sterile distilled water. After the water is removed by suction, the seeds are transferred to a 100-μm nylon sterile cell strainer (Becton Dickinson Labware, Franklin Lakes, NJ, USA) set into a sterile disposable polypropylene Petri plate (60x15 mm). One ml of autoclaved distilled water is added to each Petri dish and the Petri plates are placed in loosely covered plastic containers. The containers are incubated in the dark at 20±1 °C. After 20 h of incubation, 1 ml of sterile water is added to the remaining Petri dishes and following 44 h of incubation, another1 ml of water is added to the samples.

At specified times after seed treatment, duplicate Petri plates are removed from the growth chamber. The cell sieve is removed from each Petri plate and the seeds transferred by sterile forceps to autoclaved screw-capped 1.5-ml microcentrifuge tubes (VWR Canlab, Canada). The tubes are immediately placed in liquid nitrogen and the frozen seeds stored at –80 °C.

After the germinating seedlings have been gathered from the strainers at each time point, the seedling exudate is removed from the Petri plate (and any clinging to the cell strainer) with a 1-ml sterile disposable syringe fitted with a #20 gauge needle (Becton Dickinson Labware, Franklin Lakes, NJ, USA). The exudate is filtered through a 0.2-μm sterile syringe filter (Gelman Sciences, Ann Arbour, MI, USA), pre-wetted with sterile distilled water. The filtrate is collected into 1.5-ml glass vials (12x32 mm) capped with silicon septa (75/10) and polypropylene open top lids (Chromatographic Specialities Inc., Brockville, ON, Canada) and immediately frozen at –80 °C.

6.2 Preparation of Plant Extracts

We used a modification of the protocol described by Siefert et al. (1994) to make extracts of the canola seed-samples and the roots of the 4.5-day-old seedlings grown for the root elongation assay. Roots excised from the approximately 60 seedlings grown for the root elongation assay, are set in aluminum weigh boats, immediately frozen in liquid nitrogen and stored at –80 °C. All of the glassware used in the preparation of crude plant extracts, i.e., mortars and pestles, solution bottles, centrifuge tubes, pipettes, Pasteur pipets, and glass vials and silicon septa, is heated overnight at 275 °C and cooled to room temperature just prior to use. Each of the frozen tissue samples is ground in a prechilled mortar and pestle, suspended in 2.5 ml of 0.1 M sodium acetate pH 5.5 and kept on ice for 15 min. The contents of the mortar are scraped into a 15-ml glass centrifuge tube and the mortar and pestle are rinsed with 0.5 ml of the same buffer. The ground tissue suspension, together with the rinses, is centrifuged in an SS34 rotor at 17,500x*g* in a Sorvall R5C/B centrifuge for 15 min at 4 °C to remove cell debris. The supernatant is collected and clarified by centrifugation in a Beckman L8–70 ultracentrifuge at 100,000×g in a 70.1 Ti rotor for 1 h at 4 °C and then, if necessary, by an additional centrifugation at 100,000x*g* for 15 min. The clarified supernatant is collected and distributed into 1-ml aliquots, some of which are stored at –80 °C in glass vials for ACC determination by HPLC, and the remainder stored in 1.5-ml microcentrifuge tubes at 4 °C for protein determination.

6.3 Protein Concentration Assay

The protein concentrations are measured according to a protocol based on the method of Bradford (1976) and BSA (bovine serum albumin) is used as the standard protein. Each point on the standard curve and all of the samples are assayed in triplicate.

6.3.1 Protein Concentration Assay of Bacterial Extracts

The 100-µl aliquots of toluenized cell suspensions, which have been set aside and stored at 4 °C during the preparation of crude bacterial cell extracts, are each mixed with 100 µl of 0.1 N NaOH and incubated for 10 min at 100 °C. After the mixtures have cooled, between 20 and 50 µl of each sample is transferred to a clean glass test tube (100x13 mm); the volume is adjusted to 100 µl with 0.1 M Tris-HCl pH 8.5, and 5 ml of the diluted dye reagent is added to the tube. The contents of the tube are vortexed and incubated for 5–20 min at room temperature. The absorbance of the samples is measured at 595 nm.

6.3.2 Protein Concentration Assay of Plant Extracts

Aliquots of the plant extracts, set aside and stored at 4 °C, are each transferred to clean glass test tubes (100x13 mm) and the volume is adjusted to 100 µl with 0.1 M sodium acetate pH 5.5. Varying amounts of the different extracts are transferred to the tubes, depending on the concentration of the extract: routinely, 30 µl of seed extract and 100 µl of root extract are used. Sufficient buffer is added to each tube to bring the volume up to 100 µL. After 5 ml of the diluted dye reagent are added to each test tube, it is vortexed and incubated at room temperature between 5 and 20 min. The absorbance of each sample is measured at 595 nm.

6.4 Measurement of ACC by HPLC

6.4.1 *Chemicals*

The Waters AccQ•Fluor Reagent Kit, AccQ•Tag eluent A concentrate (a premixed concentrated acetate-phosphate buffer) and the amino acid standard, a mixture of 17 amino acids (tryptophan, glutamine, and asparagine not included) each at a concentration of 2.5 mM with the exception of cysteine which is 1.25 mM, are supplied by Waters Limited. The Waters AccQ•Fluor Reagent Kit contains the chemicals for derivatization: AccQ•Fluor reagent powder (6-aminoquinolyl-*N*-hydroxysuccinimidyl carbamate; AQC), AccQ•Fluor reagent borate buffer and AccQ•Fluor reagent diluent (acetonitrile). ACC, β- and γ-aminobutyric acid are purchased from Calbiochem-Novabiochem Corp. (La Jolla, CA, USA), HPLC grade acetonitrile from Caledon Laboratories (Georgetown, ON, Canada), α-aminobutyric acid from Fisher Scientific, and L-α-(2-amino-ethoxyvinyl) glycine hydrochloride (AVG) from Sigma-Aldrich Co. All water used is purified by a Milli-Q Water System (Millipore Co. Bedford, MA, USA), autoclaved and then filtered through a 0.45-μm HA filter (Millipore Co. Bedford, MA, USA).

6.4.2 *Treatment of Glassware*

All glassware used in this procedure is washed and then flushed at least six times with tap water, twice with deionized water and twice more with distilled water. Just prior to use, the cleansed glassware is wrapped in aluminum foil, heated overnight at 275 °C and cooled to room temperature. Solutions and samples are stored in heat-treated bottles and vials (including septa and lids).

6.4.3 *Preparation of Standard Solutions*

Stock 2.5-mM solutions of ACC, α-aminobutyric acid, β-aminobutyric acid, γ-aminobutyric acid, and a mixture of 17 amino acids are prepared in 25 ml of 0.1 N HCl in a 25-ml volumetric flask. These solutions are diluted with sterile distilled water to yield a concentration of 0.1 mM. The 2.5-mM and 0.1-mM stock solutions are divided into 0.5-ml aliquots, frozen at –20 °C, thawed once when needed and then discarded. With the exception of ACC, the 0.1-mM solutions are further diluted with sterile distilled water to generate concentrations between 5 and 25 pmol/20 μl injection. Dilutions of the 0.1-mM solutions of ACC yield between 1 and 25 pmol ACC/20 μl injection. Standard mixtures of ACC, α-, β-, and γ - aminobutyric acids are prepared in sterile distilled water to yield 12.5 pmol/20 μl injection. The amino acid standard is diluted such that each 20-μl injection included 25 pmol of each of the 17 amino acids with the exception of the amount of cysteine which was 12.5 pmol. Aliquots of the standard solutions are frozen at –20 °C, and when required, thawed once and used.

6.4.4 Derivatization Procedure

Standard solutions of ACC; ACC, α-, β- and γ -aminobutyric acids; the amino acids and plant extracts are coupled with ACQ according to the directions in the Waters AccQ•Fluor Reagent Kit Instruction Manual. The AccQ•Fluor derivatization reagent, once reconstituted, is stable for 1 week. The derivatization reagent is reconstituted by adding 1 ml of acetonitrile (vial 2B) to the AccQ•Fluor reagent powder, vortexing for 10 s, and heating on top of a 55 °C heating block for no more than 10 min to dissolve the powder. The concentration of the reconstituted AccQ•Fluor reagent is approximately 10 mM in acetonitrile; amino acid derivatization is optimal when the reconstituted AccQ•Fluor reagent is in excess and the pH is between pH 8.2 and 10. The derivatization reactions are carried out in duplicate in 6x55 mm glass sample tubes (Waters Limited). Ten µl of standard or sample solution is placed in each tube; 70 µl of AccQ•Fluor borate buffer is added to it and the mixture is immediately vortexed for several seconds. Following the addition of 20 µl of reconstituted AccQ•Fluor, the mixture is briefly vortexed again, allowed to stand at room temperature for 1 min and then heated at 55 °C for 2 min in a heating block. Once cooled to room temperature (5–10 min) the solution may be injected immediately or sometime during the next week. Amino acids derivatized by this procedure are quite stable and can be stored at room temperature for at least 1 week.

6.4.5 HPLC Determination of ACC Content

The AccQ•Tag Column, a high-efficiency 4 µm Nova-Pak C_{18} column specifically certified for use with the AccQ•Tag Method (Waters Limited) is used to separate the amino acid derivatives produced by the AccQ•Fluor derivatization reaction, and a Hewlett Packard column heater is used to maintain the column temperature at 37 °C. Amino acid derivatives are detected and measured by using a Hewlett Packard HPLC system which consists of a 1050 Series Quaternary Pump and a 104a Programmable Fluorescence Detector. A PC computer system (DTK 3300 386/33) is used to run the supporting computer software, i.e., Hewlett Packard's ChemStation (DOS Series).

The solvent system includes eluent A, a diluted solution of Waters AccQ•Tag acetate-phosphate buffer concentrate prepared daily, (50 ml concentrate diluted with 500 ml 18 Megohm Milli–Q water), eluent B, HPLC-grade acetonitrile, and eluent C, 18 Megohm Milli-Q water. The solvents are continuously sparged with helium and the solvent lines are purged for at least 60 s prior to use to remove any air bubbles present. The AccQ•Tag column is conditioned with 60 % eluent B/40 % eluent C at a flow rate of 1 ml/min for 30 min and then equilibrated with 100 % eluent A for 10 min at a flow rate of 1 ml/min before injection of the first sample. The gradient recommended by Waters Limited for separation of the AccQ•Tag-labelled amino acids was modified to enhance resolution of the ACC peak (Table 2).

Table 2. Gradient table for Waters AccQ•Tag system modified for ACC elution

Time (min)	Flow rate (ml/min)	A (%)	B (%)	C (%)
0	1.0	100.0	0	0
0.5	1.0	99.0	1.0	0
3.0	1.0	91.0	9.0	0
13.0	1.0	88.0	12.0	0
14.0	1.0	83.0	17.0	0
16.0[a]	1.0	0	60	40
18.0	1.0	100.0	0	0
23.0	1.0	100.0	0	0

Abbreviations: A, Waters AccQ•Tag acetate-phosphate buffer concentrate (50 ml diluted with 500 ml 18 Megohm Milli-Q water); B, HPLC-grade acetonitrile; C, 18 Megohm Milli-Q water

[a] From this point in the gradient, the column is washed and conditioned for the next sample

The Hewlett Packard 104a Programmable Fluorescence Detector is set up according to the Waters AccQ•Tag Amino Acid Analysis Method and is turned on at least 40 min prior to sample injection. The settings are as follows: excitation wavelength, 250 nm; emission wavelength, 395 nm; response time, 4; pmt gain, 15, and lamp setting, 3–5 W/220 Hz.

Once the column is conditioned and equilibrated, and the detector is warmed up, a standard solution, containing 12.5 pmol of α-, β-, and γ -aminobutyric acid, is injected. Following the injection of standard solutions, samples are injected and analyzed; the run time for each sample is 23 min and includes washing and re-equilibrating the column following the separation of the derivatized amino acids. Duplicates of each standard and sample are derivatized and injected. The needle port is rinsed with eluent A prior to each injection in order to reduce contamination from previously injected samples.

The injection volume of all samples including blanks, standards and plant extracts is 20 µl. Plant tissue extracts are diluted just prior to derivatization. The quantity of sample hydrolyzed and derivatized in 20 µl is estimated to be 0.1–1.0 µg (4–40 pmol) of protein, based on a protein average molecular weight of 25,000 Daltons.

6.4.6 *Quantification of ACC*

The amount of ACC in samples is quantified by using an ACC standard curve that is linear between 1 and 25 pmol of ACC per sample (Fig. 2). The ACC standard curve is prepared from a fresh stock solution of ACC (0.1 mM) diluted with sterile distilled water to yield between 1 and 25 pmol of ACC/20-µl injection. The ACC dilutions are derivatized, and following injection, are eluted

Fig. 2. Standard curve of ACC measured in fluorescence units

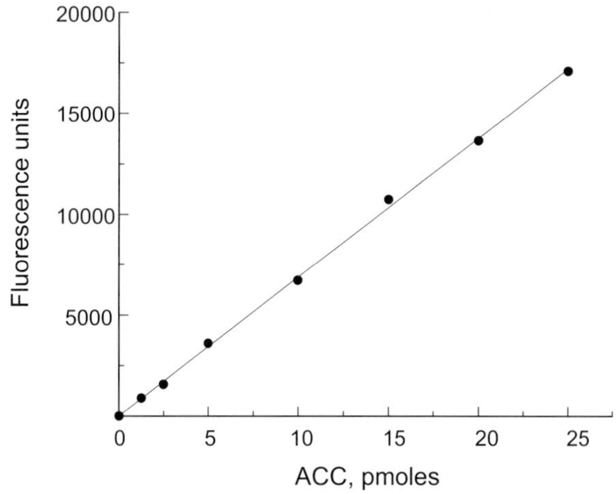

from the AccQ•Tag column at approximately 7.6 min. Similar standard curves may be prepared for α-, β- and γ- aminobutyric acid, metabolites of ACC, which are eluted from the AccQ•Tag column at 8.2, 8.7 and 9.2 min, respectively.

References and Selected Reading

Abeles FB, Morgan PW, Saltveit ME Jr (1992) Ethylene in plant biology, 2nd edn. Academic Press, New York

Bayliss C, Bent E, Culham DE, MacLellan S, Clarke AJ, Brown GL, Wood JM (1997) Bacterial genetic loci implicated in the *Pseudomonas putida* GR12-2R3 – canola mutualism: identification of an exudate-inducible sugar transporter. Can J Microbiol 43:809–818

Bradford M (1976) A rapid and sensitive method for the quantitation of microgram quantities of protein utilizing the principle of protein-dye binding. Anal Biochem 73:248–258

Dworkin M, Foster J (1958) Experiments with some microorganisms which utilize ethane and hydrogen. J Bacteriol 75:592–601

Glick BR, Karaturovíc DM, Newell PC (1995) A novel procedure for rapid isolation of plant growth promoting pseudomonads. Can J Microbiol 41:533–536

Glick BR, Penrose DM, Li J (1998) A model for the lowering of plant ethylene concentrations by plant growth-promoting bacteria. J Theor Biol 190:63–68

Honma M, Shimomura T (1978) Metabolism of 1-aminocyclopropane-1-carboxylic acid. Agric Biol Chem 42:1825–1831

Jacobson CB, Pasternak JJ, Glick BR (1994) Partial purification and characterization of 1-aminocyclopropane-1-carboxylate deaminase from the plant growth promoting rhizobacterium *Pseudomonas putida* GR12-2. Can J Microbiol 40:1019–1025

Lifshitz R, Kloepper JW, Scher FM, Tipping EM, Laliberté M (1986) Nitrogen-fixing Pseudomonads isolated from roots of plants grown in the Canadian High Arctic. Appl Environ Microbiol 51:251–255

Lifshitz R, Kloepper JW, Kozlowski M, Simonson C, Carlson J, Tipping EM, Zaleska I (1987) Growth promotion of canola (rapeseed) seedlings by a strain of *Pseudomonas putida* under gnotobiotic conditions. Can J Microbiol 33:390–395

Ma J-H, Yao J-L, Cohen D, Morris B (1998) Ethylene inhibitors enhance in vitro formation from apple shoot cultures. Plant Cell Rep. 17:211–214

Mattoo AK, Suttle CS (1991) The plant hormone ethylene. CRC Press, Boca Raton, FL, p 337

Penrose DM, Glick BR (2001) Levels of 1-aminocyclopropane-1-carboxylic acid (ACC) in exudates and extracts of canola seeds treated with plant growth-promoting bacteria. Can J Microbiol 47:368–372

Penrose DM, Moffatt BA, Glick BR (2001) Determination of 1-aminocyclopropane-1-carboxylic acid (ACC) to assess the effects of ACC deaminase-containing bacteria on roots of canola seedlings. Can J Microbiol 47:77–80

Shah S, Li J, Moffatt BA, Glick BR (1997) ACC deaminase genes from plant growth promoting bacteria. In: Ogoshi A, Kobayashi K, Homma Y, Kodama F, Kondo N, Akino S (eds) Plant growth-promoting rhizobacteria: present status and future prospects. OECD, Paris, pp 320–324

Siefert F, Langebartels C, Boller T, Grossmann K (1994) Are ethylene and 1-aminocyclopropane-1-carboxylic acid involved in the induction of chitinase and β-1,-3-glucanase activity in sunflower cell-suspension cultures? Planta 192:431–440

Yang SF, Hoffman NE (1984) Ethylene biosynthesis and its regulation in higher plants. Annu Rev Plant Physiol 35:155–189

27 Applications of Quantitative Microscopy in Studies of Plant Surface Microbiology

Frank B. Dazzo

"Sometimes what counts can't be counted, and what can be counted doesn't count."
(Albert Einstein)

1 Introduction

Whereas the animal carries its major community of indigenous microflora (generally of a beneficial kind) on the moist warm walls of its peristaltic gut, the plant does likewise, but on its entire exposed surfaces, from apical tip to root cap. These plant surfaces represent an oozing, flaking layer of integument which discharges a wide range of substances that support a vast number of spatially discrete and specialized microbial communities, including parasites and symbionts that can have a major impact on plant growth and development. A modern view of the plant surface is now seen as a dynamic adaptable envelope, flexible in both its import and export of materials, forming a plant–microbe ecosystem in its own right and the first barrier between the moist, concentrated, balanced plant cell and a hostile ever-changing external environment.

Manipulation of the plant surface microflora to improve its health is a long-standing goal in plant microbiology. However, efforts to exploit this type of biological control have frequently been impeded because of major technical difficulties that must be overcome to fully understand the microbial ecology of this ecosystem, especially the lack of ability to extract in situ data that are both informative and quantifiable at spatial scales relevant to the ecological niches of the microorganisms involved.

Most of this chapter describes the author's development and utilization of quantitative microscopy in studies of plant surface microbiology. The majority of this work has been done to gain a better understanding of the *Rhizobium*-legume root-nodule symbiosis. Various types of microscopy have been employed, including brightfield, phase-contrast, Nomarski-interference contrast, polarized light, real-time and time-lapse video, darkfield, conventional and laser scanning confocal epifluorescence, scanning electron, transmission

Plant Surface Microbiology
A. Varma, L. Abbott, D. Werner, R. Hampp (Eds.)

electron, and field-emission scanning/transmission electron microscopies combined with visual counting techniques and manual interactive applications of image analysis. More recently, the author has led a team of scientists to develop a new generation of innovative, customized image analysis software designed specifically to analyze digital images of microbial populations and communities and extract all the informative, quantitative data of in situ microbial ecology from them at spatial scales relevant to the microbes themselves. We have begun to apply this new computer-assisted imaging technology to the fascinating field of plant surface microbiology. The chapter includes many figures that exemplify how the awesome resolving power of the microscope has significantly enhanced our understanding of plant surface microbiology, and richly illustrates how this topic area is even more enhanced with the added dimension of quantitation using computer-assisted digital image analysis.

2 Quantitation of Symbiotic Interactions Between Rhizobium and Legumes by Visual Counting Techniques

2.1 The Modified Fåhraeus Slide Culture Technique for Studies of the Root–Nodule Symbiosis

The slide culture technique of Fåhraeus (1957) was the single, most important method developed to facilitate the microscopical examination of the infection process in the *Rhizobium*-legume symbiosis, especially with small-seeded legumes like white clover in symbiosis with its root-nodule endosymbiont, *R. leguminosarum* bv. trifolii. This simple method of culturing the symbionts under microbiologically controlled conditions made it possible to examine the interactions between the plant and microbial symbionts by various types of microscopy, including a classic time-lapse cinema depicting the developmental morphology of clover root hair infection (Nutman et al. 1973). Phase contrast microscopy using this slide culture technique also revealed the paramount importance of host specificity in the infection process at the stage of infection thread formation within host root hairs (Li and Hubbell 1969).

The original Fåhraeus slide method involved vertical cultivation of a seedling on a microscope slide within a large enclosed tube containing an isotonic nitrogen-free plant culture medium, and with its root inoculated with rhizobia embedded in an agar medium beneath a large cover slip (Fåhraeus 1957). Various modifications of this slide culture technique have been made to further facilitate detailed microscopical examinations of the infection process. For instance, the embedding agar was found unnecessary even for cultivation of two seedlings per slide. Elimination of the embedding agar permitted the symbionts to interact unimpeded by this fibrous matrix, the roots to be processed more consistently and efficiently after an appropriate period of incubation, and more detailed microscopy to be performed with a cleaner, phase-transparent background. These added features have significantly improved the signal-to-noise ratio of image quality, making it possible to accurately quantify many of the pre-infection and post-infection events occurring on root hairs in vivo at single bacterial cell resolution, including rhizobial attachment to root hairs (phenotype Roa) of constant length by phase-contrast microscopy (Dazzo et al. 1976; Dazzo 1982). The results of studies using this quantitative microscopical counting technique revealed important spatio-temporal aspects of the Roa phenotype, including its distinct cellular orientations/patterns/phases of adhesion, the positive relationship of certain patterns of attachment to host specificity, the importance of cell-surface glycoconjugates and saccharide-binding host lectins to symbiont recognition, the inhibition of symbiont recognition and infection by combined nitrogen, and the manipulation of rhizobial genes affecting cell surface components and rhizobial attachment to host root hairs (Dazzo and Hubbell 1975; Dazzo et al. 1976, 1978, 1984; Dazzo and Brill 1978, Sherwood et al. 1984; Rolfe et al. 1996). This same modification of the slide culture technique also made it possible to quantitate clover root hair infection. Thus, quantitative microscopy of the infection process resulted in the discovery of potent, stimulating infection-related biological activities of various purified rhizobial components required for primary host infection by *R. leguminosarum* bv. trifolii, including its clover lectin-binding acidic heteropolysaccharides and corresponding oligosaccharide repeat unit fragments which retained their affinity for the clover lectin, its clover lectin-binding lipopolysaccharide glycoform, and its diverse family of membrane chitolipooligosaccharides that modulate cell wall architecture and growth physiology of these target differentiated host cells (Abe et al. 1984; Dazzo et al. 1991, 1996). Further applications of this modified Fåhraeus slide technique to study the *R. leguminosarum* bv. trifolii-white clover symbiosis have utilized real-time video microscopy and digital image analysis of track-reconstructions to define the quantitative influence of root secretions on rhizobial motility in situ in the aqueous, external clover root environment (Dazzo and Petersen 1989), and of cells and purified lectin-binding lipopolysaccharide of *Rhizobium* on cytoplasmic streaming in root hairs indicating activation of their cytoskeleton activity (Dazzo and Petersen 1989, Dazzo et al. 1991). Another modification of the Fåhraeus slide technique was to culture seedlings vertically and flat on small agarose-solidified plates with a portion of their roots covered with the same nitrogen-free medium and small coverslips. This modification plus the customized construction of a "horizontal growth station" created the opportunity to perform real-time and time-lapse video microscopy of seedling roots grown axenically and geotropically with as little as 10 µl volumes of bacterial test solutions. Applications of this technique resulted in the detection and quantitation of symbiosis-related growth responses of clover root hairs to minute quantities of several different types of bioactive metabolites made by the microsymbiont, *R. leguminosarum* bv. trifolii under strict microbiologically controlled conditions (Dazzo et al. 1987, 1996; Dazzo and Petersen 1989, Hollingsworth et al. 1989, Philip-Hollingsworth et al. 1991; Orgambide et al. 1994, 1996). This technique was also used in conjunction with engineered rhizobia containing reporter gene fusions to locate attached rhizobial cells expressing pSym *nod* genes in situ on root hair tips (Dazzo et al. 1988).

2.2 Attachment of Rhizobia to Legume Root Hairs

Although attachment of rhizobia to legume root hairs (Roa [Root attachment] phenotype) has often been described as a simple, one-step event lacking any form of specificity, this is a gross oversimplification of the real case. Instead, quantitative time-resolved microscopy at single bacterial-cell resolution reveals that Roa is a dynamic, multiphase process including distinct nonspecific and host-specific events. Figure 1 summarizes a unified view of this dynamic sequence of events involved in attachment of encapsulated rhizobia to host legume root hairs (Dazzo et al. 1984). This model culminates in the development of the specific Roa-3 pattern of *R. leguminosarum* bv. trifolii attachment to white clover root hairs in modified Fåhraeus slide cultures prepared with a relatively small, defined size inoculum of fully encapsulated cells (Dazzo et al. 1984). This pattern of rhizobial attachment to root hairs (an immobilized aggregate of cells at the root hair tip and individual polarly attached cells along the shaft of the same root hair) requires the intervention of bacterial proteins and polysaccharides, host lectin, and enzymes that

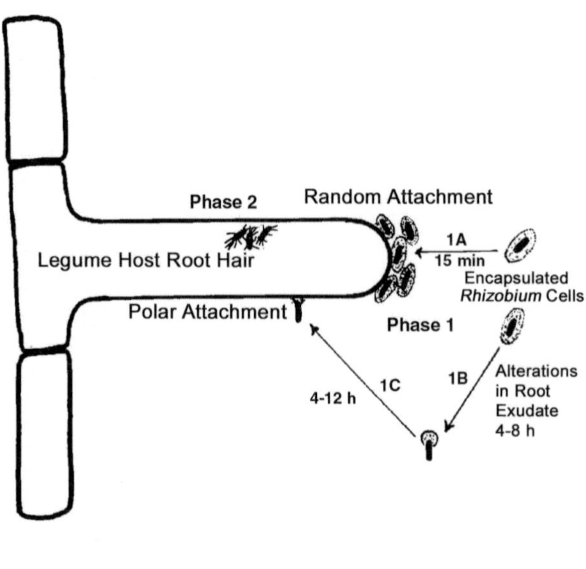

Fig. 1. Diagram of the dynamic phases of rhizobial attachment to host root hairs (Roa), based on studies using phase contrast light microscopy, scanning electron microscopy, and transmission electron microscopy. Cell sizes are approximately proportional. Reprinted with permission from the American Society for Microbiology

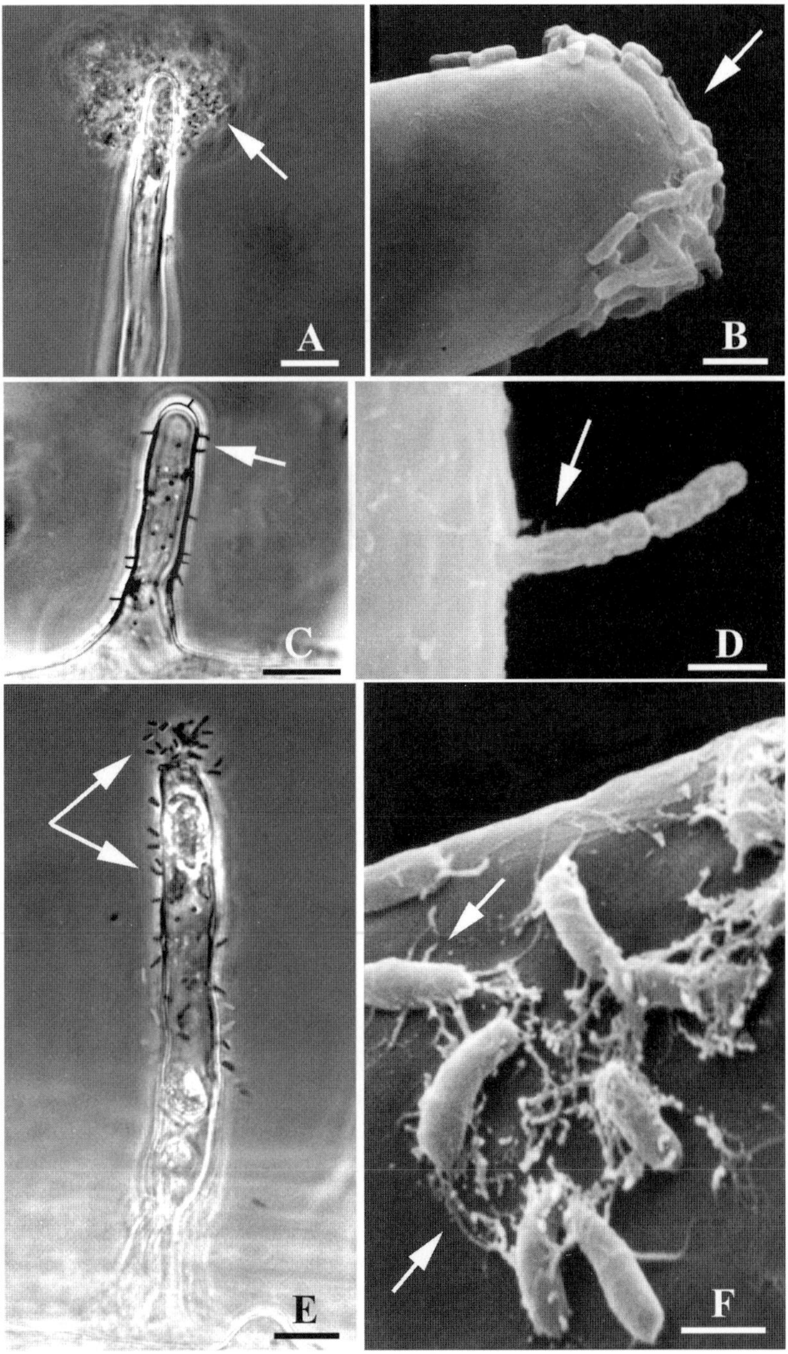

Fig. 2. Phase contrast microscopy (**A**, **C**, **E**) and scanning electron microscopy (**B**, **D**, **F**) of distinct patterns of attachment of *R. leguminosarum* bv. trifolii to white clover root hairs. **A, B** Phase 1A=Roa-1, **C, D** phase 1C=Roa-2, **E** phase 1A+1C=Roa-3, **F** phase 2 with associated microfibrils. *Scale bar* **A** and **C** 20 μm, **B** 2 μm, **D, F** 1 μm, **E** 15 μm

degrade the bacterial polysaccharides; it exhibits host-selectivity and is found on approximately 95 % of successfully infected root hairs in the *Rhizobium*-white clover symbiosis (Dazzo and Hubbell 1975; Dazzo et al. 1976, 1982, 1984; Dazzo and Brill 1979; Sherwood et al. 1984; Rolfe et al. 1996; Smit et al. 1992). The Phase 1A pattern of randomly oriented attachment occurs within 15 min of inoculation, and involves an initial nonhost-specific interaction of a rhizobial surface protein "rhicadhesin" on individual bacteria with the root hair tip (Smit et al. 1992), followed within the first hour by a more host-specific aggregation of bacterial cells immobilized at the root hair tip and mediated by an excreted, multivalent host lectin. Cells that have not yet attached to the host root become polarly encapsulated in the external root environment during the next 4–8 h (Phase 1B), due to the combined action of "polarase" enzymes in root exudate and de novo synthesis of a new capsule at one cell pole (Dazzo et al. 1982; Sherwood et al. 1984). Beginning approximately 4 h after inoculation, these polarly encapsulated cells attach "end-on", i.e., perpendicular to the surface along the sides of the same root hair (phase 1C). Phase 1 attachment is distinguished from phase 2 adhesion by the significantly increased strength of adhesion of attached cells detected approximately 12 h after inoculation, concurrent with the elaboration of extracellular microfibrils that increase the degree of contact of the attached bacteria to the root hair surface (Dazzo et al. 1984). Indeed, this strength of Phase 2 rhizobial adhesion to legume host root hairs is immense, exceeding that which anchors some root hairs onto the root itself! Figure 2A–F is a series of phase contrast light micrographs and scanning electron micrographs that illustrate each of these distinct patterns of rhizobial attachment to white clover root hairs (Dazzo and Brill 1979; Dazzo et al. 1984).

2.3 Rhizobium-Induced Root Hair Deformations

Root hairs on axenic seedlings are straight, but become deformed (Had [Hair deformation] phenotype) during growth in response to various bioactive metabolites made by rhizobia. Four different morphotypes of white clover Had are induced under axenic conditions by minute quantities of purified bioactive Nod metabolites made by *R. leguminosarum* bv. trifolii. These are root hair distortions, tip swellings, branches, and corkscrews induced by rhizobial membrane chitolipooligosaccharides, N-acetylglutamic acid, and diglycosyl diacylglycerol glycolipids (Philip-Hollingsworth et al. 1991; Orgambide et al. 1994, 1996; Dazzo et al. 1996a, b;). Collectively called moderate Had, these various types of root hair deformations are less symbiont-specific than marked curling of the root hair tip (commonly referred to as the "Shepherd's crook" Hac [Hair curling] phenotype). This Hac morphotype is illustrated in Fig. 3 and requires close proximity of viable cells of the homologous symbiont (Li and Hubbell 1969; Yao and Vincent 1976). This figure is a

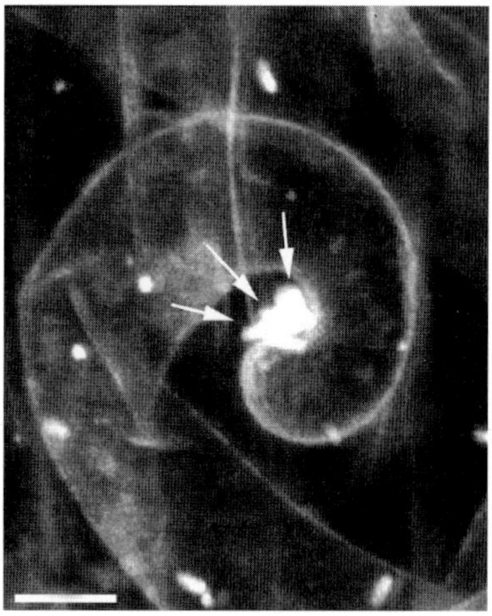

Fig. 3. Portion of a white clover root hair that has undergone a markedly curled deformation induced by *Rhizobium leguminosarum* bv. *trifolii*. This optisection obtained using laser scanning confocal microscopy and immunofluorescence staining with a strain-specific monoclonal antibody to the bacterial LPS provides direct evidence that the center of the shepherd's crook overlap contains a clump of rhizobia. *Scale bar* 7.5 µm

laser scanning confocal epifluorescence micrograph that elegantly provides direct evidence that the overlap of the shepherd's crook entraps a clump of rhizobial cells, as has long been predicted, but not convincingly shown before. In this case, the confocal image is an optisection located at the optical median plane of the curled root hair cell, and it definitively shows the immunofluorescent rhizobia detected by using a fluorescent monoclonal antibody to their lipopolysaccaride (LPS) somatic O-antigen. It has been predicted that the confining morphological structure of the shepherd's crook serves to concentrate in a localized region the metabolic events of microsymbiont penetration while preventing lysis of the root hair during primary host infection (Napoli et al. 1975a; Napoli and Hubbell 1976; Dazzo and Hubbell 1982).

2.4 Primary Entry of Rhizobia into Legume Roots

Figure 4 illustrates a rhizobial-induced infection thread in white clover root hairs. Successful infections of this type typically exhibit a bright refractile spot in the center overlap of markedly curled root hair tips, and infection threads that have elongated through the root hair to its base. A central event of this infection process in the *Rhizobium*-legume symbiosis is the modification of the host cell wall barrier to form a portal of entry large enough for bacterial penetration. Transmission electron microscopy indicates that rhizobia enter the legume root hair through a completely eroded hole that is slightly larger than the bacterial cell (generally 2–3 µm in diameter) and is presumably created by localized enzymatic hydrolysis of the host cell wall (Napoli and Hubbell 1976; Callaham and Torrey 1981). Time-lapse cinema microscopy (Nutman et al. 1973) has elegantly shown that the root hair ceases to

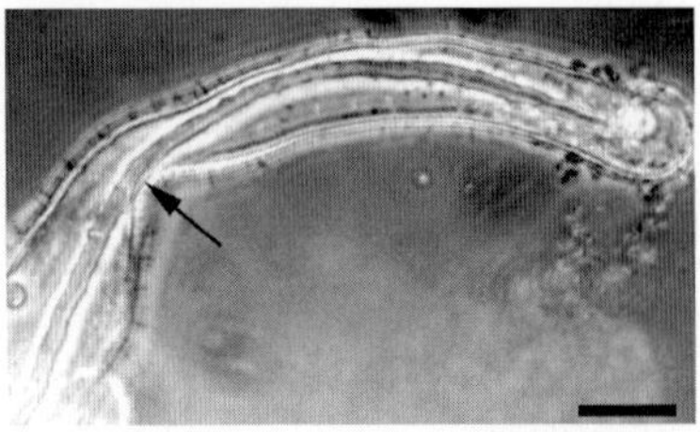

Fig. 4. Phase contrast micrograph of primary host infection in the *Rhizobium*-legume symbiosis. Note the prominent infection thread (*arrow*) within the deformed root hair cell. *Scale bar* 10 µm

elongate during the inward growth of the infection thread, which proceeds at approximately the same elongation rate. This inward growth of the infection thread is led by a mobile nucleus and a flurry of cytoplasmic streaming within the root hair (Nutman et al. 1973). Successful infections are best quantitated by visual counting while viewed by phase contrast microscopy; light staining of the infection thread with methylene blue can enhance contrast to detect them. Distinctions of successful vs. unsuccessful infections can be made by detailed microscopical examination to assess whether the infection thread has grown to the root hair base and penetrated into the underlying subepidermal cortical cell. Infective rhizobia engineered with Gus or green fluorescent protein reporter genes can facilitate the detection of infected root hairs, but this is overkill for skilled microscopists.

An alternate primitive route of primary host infection of legumes leading to effective nodule formation is the crack entry of rhizobia into natural wounds of the host plant epidermis. This commonly occurs in many tropical legumes (Napoli et al. 1975b) and some temperate legumes, but can also occur infrequently in anomalous ineffective nodulations by rhizobia outside their normal cross-inoculation group (Hrabak et al. 1985). In the aquatic legume *Neptunia natans* where root hairs do not normally develop, the natural splitting of the epidermis during development of the spongy aerenchyma and emergence of adventitious and lateral roots create openings that allow "crack entry" of the rhizobial symbiont, *Allorhizobium undicola, Rhizobium undicola,* or *Devosia neptuniae,* as the normal mode of primary host infection (Subba-Rao et al. 1995).

Recently, an interesting novel combination of infection events has been found to occur in development of the root-nodule symbiosis of rhizobia with tagasaste (*Chamaecytisus proliferus* L.), a legume indigenous to the Canary Islands near the west coast of Africa. In this symbiosis, primary host infection initially involves rhizobial deformation and penetration of host root hairs, but all these primary host infections abort and the rhizobia then revert to a crack entry mode of invasion directly into the emerging root nodules without development of infection threads (Vega-Hernandez et al. 2001). Quite a remarkable, unique mode of plant infection by surface rhizobia!

2.5 In Situ Molecular Interactions Between Legumes Roots and Surface-Colonizing *Rhizobia*

Microscopy has played a central role in elucidating molecular events important to the development of the *Rhizobium*-legume root-nodule symbiosis. The use of various molecular probes combined with the awesome resolving power of the microscope has made it possible to dissect and locate key molecules that participate in primary host infection, including the cell surface interfaces during symbiotic recognition, attachment, deformation, and root hair penetration, and also in root nodule development. Various types of microscopy that can view intact living cells noninvasively have added new dimensions to unraveling the symbiotic interactions of potent rhizobial signal molecules with host cells, including the precise localization of specific binding receptor sites on the host root surface, the rapid internalization of certain rhizobial signal communication molecules within root hairs and their transfer to underlying cortical cells, and various other infection-related host cell responses. The significance of all of these studies is improved when the various microscopical techniques are accompanied by quantitative methods of data acquisition. Some examples of in situ "molecular microscopy" in studies of plant surface microbiology are illustrated here.

2.6 Cross-Reactive Surface Antigens and Trifoliin A Host Lectin

Rhizobium leguminosarum bv. trifolii and white clover roots share related surface components that are antigenically cross-reactive (Dazzo and Hubbell 1975; Dazzo and Brill 1979). Quantitative immunofluorescence microscopy indicates that these cell-surface antigens are transient, symbiont-specific, infection-related, and participate in the host lectin-mediated stage of symbiont recognition on the clover root hair surface (Dazzo and Hubbell 1975; Dazzo and Brill 1979; Dazzo et al. 1979). Transformation of *Azotobacter vinelandii* with DNA from *R. leguminosarum* bv. trifolii resulted in hybrid recombinants that expressed these symbiotic cross-reactive antigens (Bishop et al. 1977), and these recombinants gained the ability to carry out the phase 1A pattern of bacterial cell attachment to white clover root hair tips (Dazzo and Brill 1979). The cell surface location of these epitopes plus their infection-related symbiont-specificity, interaction with the multivalent white clover root lectin, and role in cell attachment formed the basis for proposing their involvement as cell-surface receptors in a lectin cross-bridging model of symbiont recognition during early stages of primary host infection (Dazzo and Hubbell 1975; Dazzo and Brill 1979). Recent studies using plant molecular biology techniques have provided substantial evidence supporting the validity of this cross-bridging model (van Rhijn et al. 1998; Hirsch 1999).

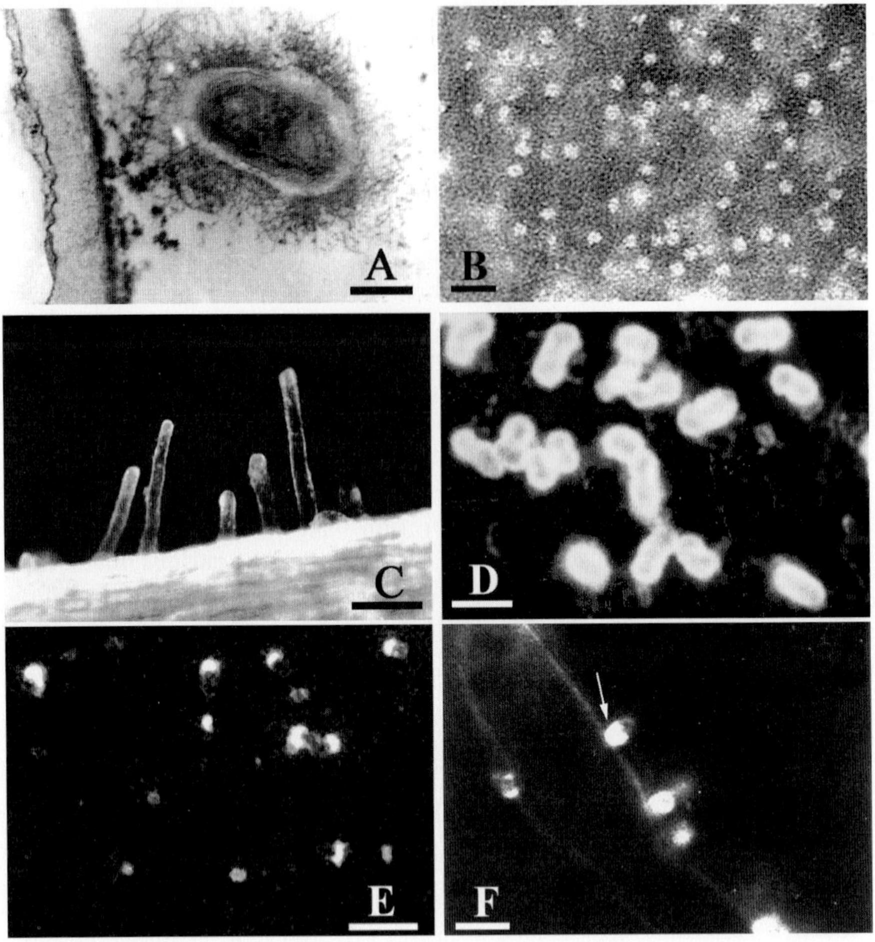

Fig. 5. Symbiont-specific interaction of trifoliin A white clover lectin and *R. leguminosarum* bv. trifolii. **A, B** Transmission electron microscopy, **C–F** conventional immunofluorescence microscopy using antibody to purified trifoliin A. **A** The historical micrograph which suggested the involvement of a particulate cross-bridging clover lectin in the attachment of encapsulated *R. leguminosarum* bv. trifolii cells to host root hairs. **B** Negatively stained particles of purified trifoliin A white clover lectin. **C** Distribution of trifoliin A on root hair tips of white clover seedlings. **D** Intense binding of root-derived trifoliin A to *R. leguminosarum* bv. trifolii. **E** In situ binding of trifoliin A to the polar capsule of *R. leguminosarum* bv. trifolii cultured in the external clover root environment. **F** Direct detection of trifoliin A at the contact interface (*arrow*) of rhizobial cells polarly attached to a white clover root hair. *Scale bar* **A** 1 µm, **B** 25 nm, **C** 50 µm, **D, E F** 2 µm. Reprinted with permission from the American Society for Microbiology

The ultrastructure of the docking stage of rhizobial attachment to the clover root hair surface is illustrated in Fig. 5A. This transmission electron micrograph revealed the electron-dense granules accumulated on the outer face of the hair wall that interact with the fibrillar capsule of *R. leguminosarum* bv. trifolii (Dazzo and Hubbell 1975). Since this granular material also occurred on the surface of axenic root hairs, it was presumably of host origin and predictably a carbohydrate-binding lectin (Dazzo and Hubbell 1975). In follow-up studies, a lectin was purified from white clover seed, shown to exist as an aggregated particle of glycoprotein and to accumulate on white clover root hairs, especially at their tips, as shown by transmission electron microscopy and immunofluorescence microscopy (Dazzo et al. 1978; Gerhold et al. 1985; Fig. 5B, C). This white clover lectin displayed symbiont-specificity in agglutination of *R. leguminosarum* bv. trifolii and was named *trifoliin A* (Dazzo et al. 1978). The intense, saccharide-inhibitable binding of root trifoliin A to encapsulated cells of *R. leguminosarum* bv. trifolii cells is illustrated in the immunofluorescence micrograph of Fig. 5D. Subsequently, it was shown that most of the trifoliin A glycoprotein synthesized de novo in roots of white clover seedlings was excreted into the external root environment where it interacted in situ with encapsulated cells of *R. leguminosarum* bv. trifolii (Dazzo and Hrabak 1981, Dazzo et al. 1982; Sherwood et al. 1984; Truchet et al. 1986; Fig. 5E). Direct evidence indicating that trifoliin A accumulated at the contact interface between polarly attached *R. leguminosarum* bv. trifolii cells and the surface of the white clover root hair wall was shown by conventional immunofluorescence microscopy viewed with the pre-confocal optics of a high magnification objective having a narrow depth of focus (Dazzo et al. 1984; Fig. 5F). Quantitative immunofluorescence microscopy indicated that hybrid recombinants of *R. leguminosarum* bv. viciae carrying multicopy plasmids of cloned pSym *nod* genes of *R. leguminosarum* bv. trifolii controlling clover host specificity acquired the ability to bind trifoliin A in situ in the external white clover root environment (Philip-Hollingsworth et al. 1989b). All of these findings contributed to the proposal that host lectin mediates symbiont recognition during host-specific events that precede primary host infection in the *Rhizobium*-legume symbiosis. Subsequent elegant plant molecular biology studies by Kijne and colleagues (Diaz et al. 1989), and more recently Hirsch and colleagues (van Rhijn et al. 1998; Hirsch 1999), have confirmed that the host-encoded lectins play a crucial role in microsymbiont recognition and host specificity in the *Rhizobium*-legume symbiosis, as originally predicted.

2.7 *Rhizobium* Acidic Heteropolysaccharides

Rhizobium leguminosarum bv. trifolii normally produces a profound true capsule that is revealed by ruthenium red staining and transmission electron microscopy. The bulk of this capsule consists of a large acidic heteropolysaccharide (Dazzo and Hubbell 1975). Bioassays scored by quantitative phase contrast microscopy indicate that oligosaccharide fragments produced by enzymatic depolymerization of this polysaccharide are biologically active in promoting root hair infectibility in white clover seedlings inoculated with *R. leguminosarum* bv. trifolii (Abe et al. 1984; Hollingsworth et al. 1984). The complete structures of the acidic heteropolysaccharides of several strains of *R. leguminosarum* bv. trifolii have been elucidated and shown to consist of repeated octasaccharide units of 5Glc:2GlcA:1Gal containing a tetrasaccharide backbone of 2Glc:2GlcA substituted with O-acetate and a tetrasaccharide sidechain of 3Glc:1Gal bearing pyruvyl substitutions on the terminal Gal and penultimate Glc, and a O-hydroxybutyrate substitution on the terminal Gal (Hollingsworth et al. 1988; Philip-Hollingsworth et al. 1989a). Trifoliin A binds selectively to this acidic heteropolysaccharide, and the symbiont-specificity in this protein–carbohydrate interaction involves recognition of the sites of linkage and stoichiometry of noncarbohydrate substitutions in the octasaccharide repeat unit (Abe et al. 1984; Hollingsworth et al. 1984, 1988; Philip-Hollingsworth et al. 1989b). Subsequent biochemical studies revealed host-range related structural features of *R. leguminosarum* bv. trifolii acidic heteropolysaccharides that distinguish these cell surface polymers and those of the closely related pea symbiont, *R. leguminosarum* bv. viciae, based on subtle differences in molar stoichiometry and positions of attachment of these noncarbohydrate substitutions (Philip-Hollingsworth et al. 1989a, b). Other studies have shown a link between rhizobial genes involved in determining the acidic heteropolysaccharide structures and the legume host-range in *R. leguminosarum* and *Rhizobium* sp. (*Acacia*; Philip-Hollingsworth et al. 1989b; Lopez-Lara et al. 1993, 1995). This relationship is expressed in some, but not all genetic backgrounds of *R. leguminosarum* (Orgambide et al. 1992). Recently, we have presented a micrograph of a portion of an isolated molecule of the *R. leguminosarum* bv. trifolii acidic polysaccharide acquired using a field-emission scanning/transmission electron microscope at extremely high magnification (Dazzo and Wopereis 2000). Image analysis of the branches projecting perpendicular to the main polymer backbone in that micrograph indicate that they are within the same size range as the predicted 20±2 angstrom length of the substituted tetrasaccharide side-chain. Molecular microscopy!

A role of the capsular polysaccharide from *R. leguminosarum* bv. trifolii in symbiotic recognition was clearly shown by labeling this polymer with the fluorochrome FITC and documenting its direct interaction with white clover roots using epifluorescence microscopy (Dazzo and Brill 1977). Figure 6A illustrates the result, providing direct evidence for the existence and distribution of receptor sites on clover root hairs that specifically recognized the capsular polysaccharide of this rhizobial microsymbiont. Further studies using fluorescence microscopy indicated that these receptor sites are saturable,

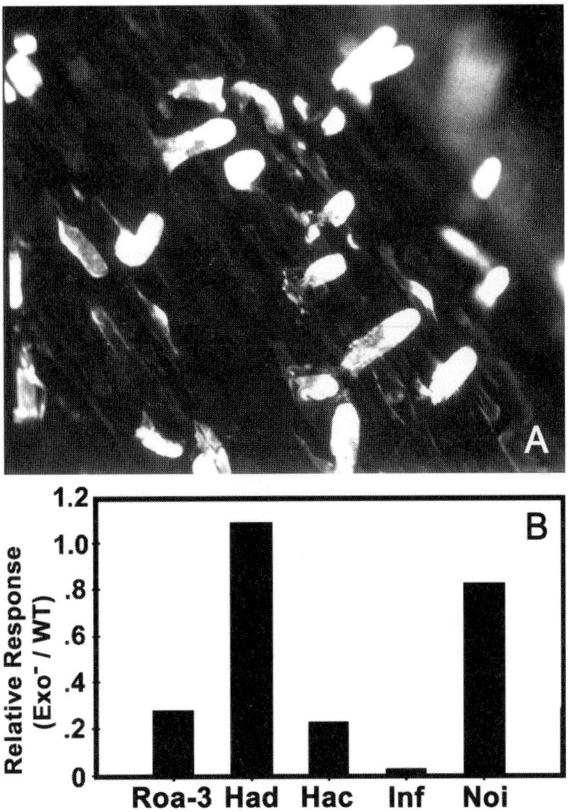

Fig. 6. Role of *Rhizobium* acidic heteropolysaccharide in symbiotic development with legumes. **A** Direct detection of symbiont-specific receptor sites for *R. leguminosarum* bv. trifolii acidic heteropolysaccharide on root hairs of white clover seedlings. **B** Quantitative microscopy of symbiotic phenotypes of an *R. leguminosarum* bv. trifolii ANU437 Exo^- mutant relative to its Exo^+ wild-type ANU794 parent scored on the white clover host. The significant requirement of the bacterial acidic heteropolysaccharide in expression of its important Roa-3, Hac, and Inf symbiotic phenotypes is clearly indicated. Reprinted with permission from the American Society for Microbiology

match the cellular distribution of trifoliin A on the root surface, and are specifically hapten-inhibitable, thus implicating an involvement of this root hair lectin in recognition of the rhizobial acidic heteropolysaccharide (Dazzo and Brill 1977; Dazzo et al. 1978).

Further symbiotic roles of the acidic heteropolysaccharide from *R. leguminosarum* bv. trifolii in clover root nodulation were shown by detailed microscopy of the phenotypes exhibited by mutants blocked in its synthesis. A common symbiotic phenotype of "exo-minus" mutants of many fast-growing rhizobia is their defective ability to invade nodules on their respective host plant (Leigh et al. 1987; Lopez-Lara et al. 1993, 1995; Rolfe et al. 1996; Sanchez et al. 1997). Figure 6B summarizes the results of detailed, quantitative microscopical analysis of symbiotic phenotypes in exo-minus mutants of *R. leguminosarum* bv. trifolii scored on white clover seedling roots *prior to* nodule invasion (Rolfe et al. 1996). These quantitative microscopy results clearly indicate that the acidic heteropolysaccharide of *R. leguminosarum* bv. trifolii plays a crucial role in several early events of the infection process, including the rhizobial expression of the symbiont-specific (Roa-3) pattern of attachment to root hairs, the induction of markedly curled shepherd's crooks at root hair tips (Hac), and the formation of successful infection threads in root hairs (Inf), but not the induction of moderate root hair deformations (Had) or root nodule primordia (Noi). Thus, the acidic heteropolysaccharide of *R. leguminosarum* bv. trifolii is a very important cell surface component needed to accomplish symbiont recognition, Roa-3, Hac, and Inf events crucial to primary host infection in the *Rhizobium*-clover symbiosis, as predicted (Dazzo and Hubbell 1975; Dazzo and Brill 1977, 1979; Dazzo et al. 1984; Sherwood et al. 1984; Philip-Hollingsworth et al. 1989a, b; Orgambide et al. 1992; Rolfe et al. 1996). In concurrence with our earlier findings using *R. leguminosarum* bv. trifolii and white clover, detailed microscopy has more recently revealed the importance and essential requirement of extracellular acidic heteropolysaccharide from wild-type *Rhizobium leguminosarum* bv. viciae and *Sinorhizobium meliloti* in successful root hair infection of their corresponding hosts, vetch and alfalfa (van Workum et al. 1998; Cheng and Walker 1998; Pellock et al. 2000).

2.8 *Rhizobium* Lipopolysaccharides

The lipopolysaccharide (LPS) is another cell surface component of *R. leguminosarum* bv. trifolii that was predicted to play a role in symbiotic infection when it was found to bind trifoliin A and contain the glycosyl component quinovosamine (2-amino-2,6-dideoxyglucose) in its structure, which turned out to be a potent saccharide hapten inhibitor of trifoliin A-*Rhizobium* polysaccharide interactions (Dazzo and Brill 1979; Hrabak et al. 1981; Sherwood et al. 1984; Dazzo et al. 1991). Quantitative bioassays of root hair infections on white clover scored directly by phase contrast microscopy indicated a role of a transient, trifoliin A-binding glycoform (K90) of *R. leguminosarum* bv. trifolii LPS in activating the infection process (Dazzo et al. 1991). This infection-related biological activity significantly increased the frequency of successful infection threads that grew the entire length of the root hair and penetrated into the underlying cortical cells (Dazzo et al. 1991). Further studies using immunofluorescence and immunoelectron microscopy revealed the direct interaction between this bioactive LPS glycoform and white clover root hairs (Dazzo et al. 1991), including its localized binding to root hair tips where trifoliin A accumulates (Fig. 7A, B), and its uptake and internalization within the root hair cell (Fig. 7C). Real-time video microscopy and quantitative image analysis revealed that this specific interaction of the trifoliin A-binding glycoform of *R. leguminosarum* bv. trifolii LPS and white clover root hairs induced rapid changes in cytoplasmic streaming indicative of altered cytoskeleton activity, and 2-D gel electrophoresis revealed changes in levels of several specific root hair proteins made in response to LPS exposure (Dazzo et al. 1991).

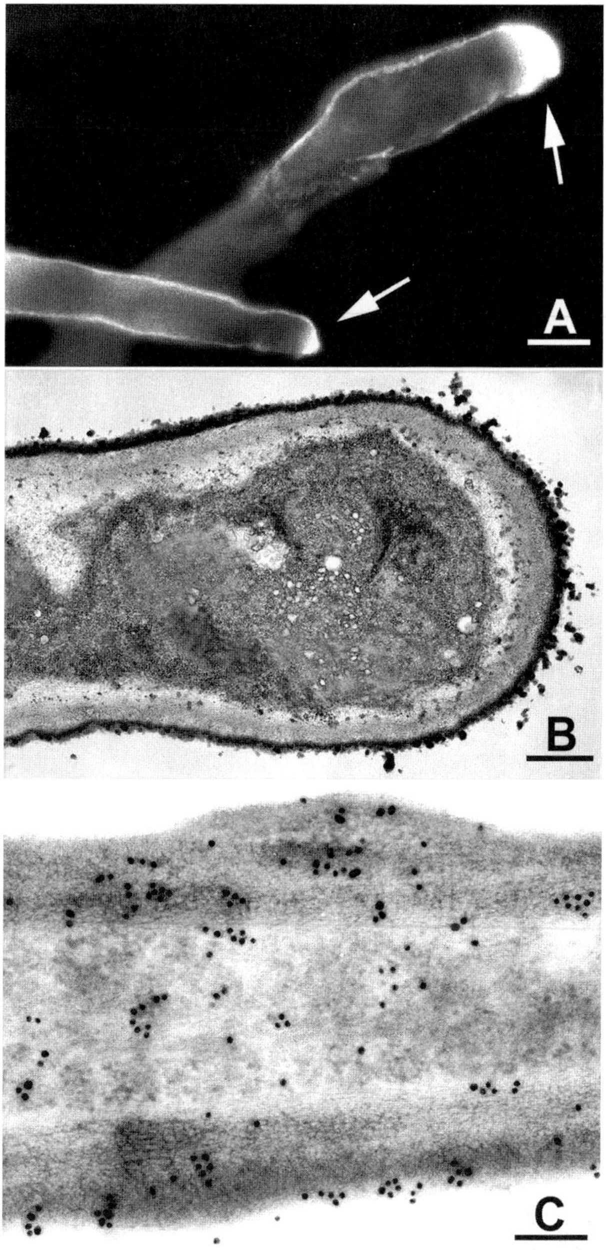

Fig. 7. Direct interaction of *Rhizobium* lipopolysaccharide with host root hairs. Adsorption of the trifoliin A-binding glycoform of LPS from *R. leguminosarum* bv. trifolii to the tips of white clover root hairs, and their internalization of this bioactive *Rhizobium* signal molecule are shown by immunofluorescence microscopy (**A**), conventional transmission electron microscopy (**B**), and immunoelectron microscopy (**C**). *Scale bar* **A** 10 μm, **B**, **C** 3 μm. Reprinted with permission from the American Society for Microbiology

In contrast, quantitative microscopy revealed that similar treatment of white clover roots with LPS from heterologous wild-type rhizobia (e.g., *R. leguminosarum* bv. viciae or *S. meliloti*) resulted in very incompatible root hair responses (Dazzo et al. 1991). These included a reduction in frequency of successful infections made by wild-type *R. leguminosarum* bv. trifolii, a corresponding increase in proportion of aborted infections accompanied by accumulation of intensely autofluorescent material at the arrested infection thread within the root hairs, and the suppression in levels of some of the newly synthesized root hair proteins plus elevation in levels of other specific root hair proteins (Dazzo et al. 1991). These results indicate that *Rhizobium* LPS is a potent signal molecule that rapidly communicates with host root hairs before bacterial penetration, triggering signal transduction of various molecular and physiological changes in these host cells that modulate infection thread development and compatibility/incompatibility events during primary host infection (Dazzo et al. 1991).

2.9 Chitolipooligosaccharide Nod Factors

Microscopy has played a major role in showing that chitolipooligosaccharides (CLOS), first described by Lerouge et al. in *S. meliloti* (Lerouge et al. 1990), are one group of several different types of Nod factor molecules made by *R. leguminosarum* bv. trifolii capable of inducing Had and Ccd/Noi on white clover roots (Hollingsworth et al. 1989; Philip-Hollingsworth et al. 1991, 1997; Orgambide et al. 1994, 1995, 1996; Dazzo et al. 1996a; Dazzo et al. 1996b;). Consistent with their amphiphilic physicochemistry, CLOSs of *true wild type* (i.e., not genetically manipulated) *R. leguminosarum* bv. trifolii accumulate three log cycles higher in their cellular membranes rather than in the extracellular milieu, and comprise a diverse family of at least 23 different types of CLOS that vary in O-acetyl and N-fattyacyl substitution, and in degree of oligomerization (Orgambide et al. 1995; Philip-Hollingsworth et al. 1995).

Because these wild-type Nod factors are primarily associated with membranes rather than secreted extracellularly (contrary to dogma), it was important to establish if they represent the symbiotically relevant forms. Quantitative microscopy bioassays on axenic seedlings showed that this was definitely the case. The family of wild-type membrane CLOSs from *R. leguminosarum* bv. trifolii was fully active in its ability to induce Had, Ccd and Noi in white clover roots at subnanomolar concentrations (Orgambide et al. 1996). Furthermore, these symbiotic activities of *R. leguminosarum* bv. trifolii membrane CLOSs were host-specific in that they elicited no mitogenic Ccd or Noi activity in hairy vetch or alfalfa roots (heterologous legumes of different cross-inoculation groups), no Had in alfalfa at any concentration tested, and only elicited a weak Had response in hairy vetch requiring a 10^4-fold higher threshold concentration than in the homologous host white clover (Orgambide et al. 1996).

We combined organic chemical synthesis and quantitative microscopy approaches to dissect the molecular structural features of wild-type CLOS molecules required for their Had and Ccd/Noi symbiotic activities. A variety of small analog molecules bearing various motifs of CLOS glycolipids were chemically synthesized and bioassayed on axenic legume plants (Philip-Hollingsworth et al. 1997). The results of this study were straightforward and very informative. Quantitative brightfield microscopy indicated that nanomolar concentrations of a single glucosamine residue bearing a long-chain fatty N-acyl substitution were required and sufficient to induce Had and Ccd/Noi activity on both white clover and alfalfa, without any structural requirements for sulfation, O-acetylation, oligomerization of the glucosamine backbone, or unsaturation of the N-acyl fatty acid moiety of CLOSs (Philip-Hollingsworth et al. 1997). Further molecular dissection of the polar head group (e.g., removal of the C5 and C6 groups from the pyranose ring) rendered the amphiphilic CLOS analog inactive in these Had and Ccd/Noi bioassays (Philip-Hollingsworth et al. 1997). Contrary to "dogma", these studies on the molecular determinants of CLOS action showed that the minimal portion of the native CLOS molecule that is both essential and sufficient for these symbiotic activities resides simply at the nonreducing glucosamine terminus substituted with an N-acylated long-chain fatty acid, and the remaining variations in components of the CLOS molecule leading to their native diverse family restrict which host (white clover or alfalfa) will respond to them rather than serve as required, positive effectors of their Had and Noi bioactivities per se (Philip-Hollingsworth et al. 1997). These key results which show that N-fatty acyl polyunsaturation and sulfation (for alfalfa) are not essential components of the minimal active structural component of CLOS for Ccd/Noi in legumes have been independently confirmed (Vernoud et al. 1999; Diaz et al. 2000). Consistent with these findings, other related studies show that perception of NodRm CLOS factors by membrane fractions of alfalfa have no significant structural requirement for N-fatty acyl polyunsaturation nor sulfation (Bono et al. 1995). Collectively, these significant findings have profound impact on the validity of models that assign the physiological location of CLOS in rhizobia, as well as their structural requirements for perception and symbiotic bioactivities in legume hosts like white clover, alfalfa, and vetch.

In this same study (Philip-Hollingsworth et al. 1997), we developed various fluorescent molecular probes to investigate the in vivo fate and uptake of bioactive CLOS molecules into living root cells of intact white clover seedlings. By chemically labeling the reducing N-acetylglucosamine terminus of wild-type *R. leguminosarum* bv. trifolii CLOSs with NBD fluorochrome, we were able to produce a family of fluorescent NBD-CLOS derivatives with minimal molecular perturbation that retained their Had and Ccd/Noi inducing biological activities on white clover roots (Philip-Hollingsworth et al. 1997). This approach is far superior to conjugation of CLOSs with certain alternative fluorochromes, e.g., biodipi, whose relatively large and hydrophobic molecular structure could significantly perturb the physiological bioactivity of the CLOS molecules. This NBD-CLOS molecular probe was applied to axenic seedling roots under microbiologically controlled conditions. At various time points thereafter, the specimens were rinsed free of unbound conjugate and examined in vivo by laser scanning confocal microscopy, with results acquired in real time at subcellular resolution (Philip-Hollingsworth et al. 1997). Figure 8A–H illustrates the key in vivo results of these studies, providing direct microscopical evidence that the NBD-CLOSs made by wild-type *R. leguminosarum* bv. trifolii interact rapidly with clover root hairs, traverse their cell walls, absorb to their cell membrane, and within minutes are then internalized within these living cells, where they migrate to the base of the root hairs and translocate to underlying cortical cells in a discrete region of the root. Quantitative fluorescence microscopy indicated that NBD-CLOSs from wild-type *R. leguminosarum* bv. trifolii were internalized by a significantly higher proportion of root hairs from the host legume white clover than from the nonhost legume alfalfa (Philip-Hollingsworth et al. 1997). As predicted, the structural requirements for internalization of NBD-CLOS analogs in living root hairs matched those required for Had and Noi bioactivities of CLOSs in white clover and alfalfa as described above. In contrast, the fluorescent analog NBD-chitotriose (without a linked lipid) was not taken up by living clover root hairs or cortical cells, indicating that in vivo internalization of

Fig. 8. Laser scanning confocal microscopy of the direct, dynamic interaction of chitolipooligosaccharides (CLOSs) from wild-type *R. leguminosarum* bv. trifolii ANU843 with living cells of white clover roots. Purified CLOSs were conjugated with the fluorochrome NBD to produce a fluorescent molecular probe with minimal molecular perturbation that retained Had and Noi bioactivities on white clover roots. **A** When applied to roots, these labeled Nod factors rapidly adsorbed to the root hairs. Closer examination of a time-series sequence of images showed that the NBD-tagged CLOSs adsorbed to the root hair cell membrane, and then within minutes were internalized within these epidermal cells (**B–F**), some migrating to the base of the root hair cell (**B–D**) and others remaining on the cell membrane or inside the root hair nucleus (**E, F**). Within 30 min, some NBD-CLOSs were translocated to a discrete region of the underlying root cortex and internalized within selected cortical cells (**G, H**). *Arrowheads* in the paired micrographs of (**E, F**) point to the root hair nucleus that internalized some labeled CLOSs. The NBD-CLOSs of ANU843 were internalized by a significantly higher proportion of the root hairs on white clover than alfalfa roots. Further studies using synthetic CLOS analogs and axenic seedling bioassays evaluated by these microscopy techniques established the minimal structural features of these Nod factor molecules that are required and sufficient for uptake and Had/Noi-inducing activities on both white clover and alfalfa roots. *Scale bar* A 50 μm, **B–D** 15 μm, **E, F** 10 μm, **H** 100 μm. Reprinted with permission from Lipid Research, Inc.

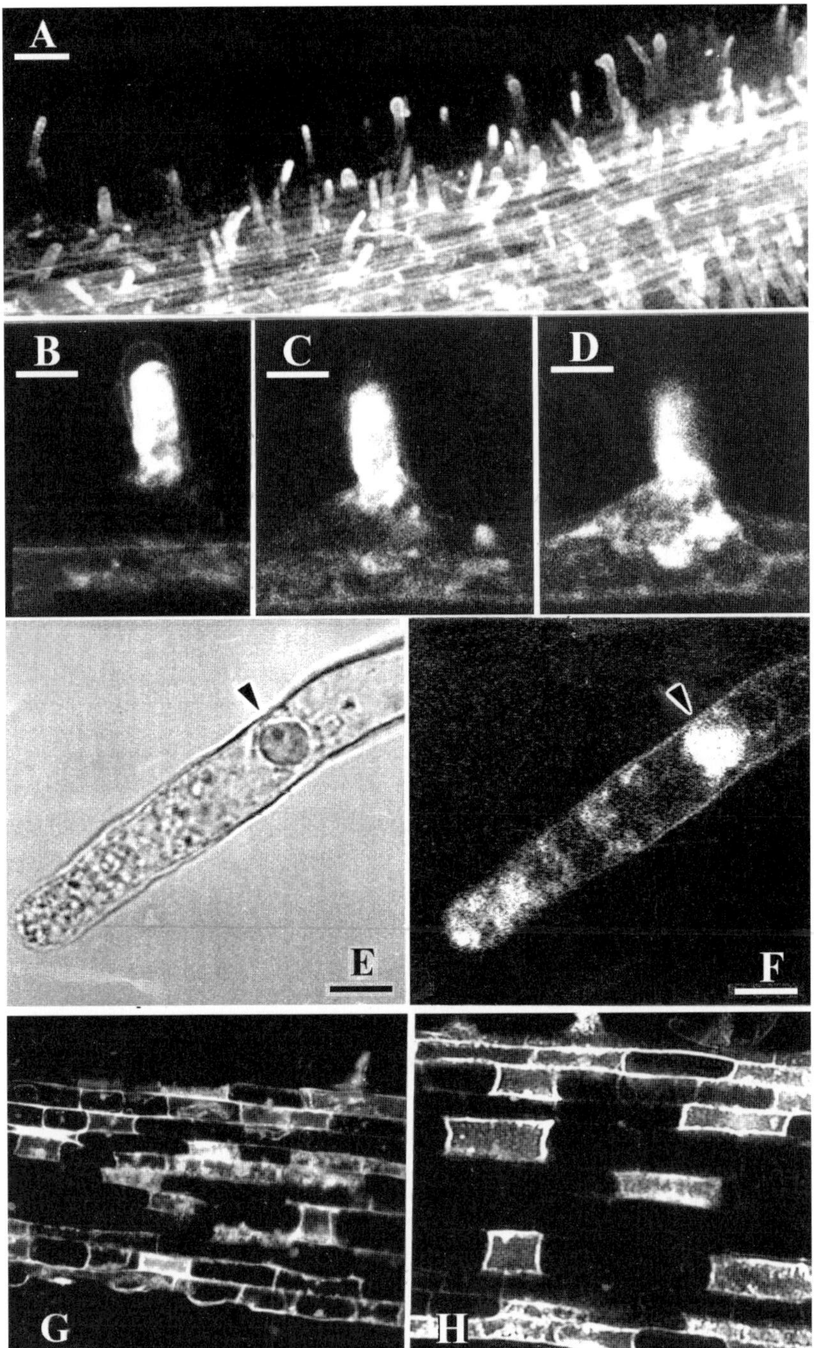
A
B
C
D
E
F
G
H

NBD-labeled CLOSs and CLOS analogs by these host cells require the long chain N-acyl fatty acid moiety, but it does not have to be polyunsaturated. These results also indicated that the observed fluorescence was not due to autofluorescence of root cells, nor to uptake of a cleavage product of NBD-CLOSs degraded by plant chitinases, and that the root epidermis of seedlings used in these experiments had no open cracks through which NBD-CLOSs could passively diffuse into the root. Another interesting finding was that the interior of the clover root hair nucleus was a specific target reached rapidly by some of the internalized fluorescent NBD-CLOSs applied to white clover roots, as illustrated in the paired images of a root hair using phase contrast light microscopy (Fig. 8E) and the corresponding, longitudinal epifluorescence optisection obtained by laser scanning confocal microscopy that samples through the fluorescent nucleoplasm of its nucleus (Fig. 8F). These findings (Philip-Hollingsworth et al. 1997) impact profoundly on our understanding of the very early fate of rhizobial CLOS molecules before primary infection and nodule induction, and on the nature, location, and molecular specificity of putative host receptor sites for these Nod factors in the host legume root.

2.10 Epidermal Pit Erosions

Recently, we used various types of microscopy and enzymology to further clarify how rhizobia modify root epidermal cell walls in order to shed new light on the mechanism of primary host infection in the *Rhizobium*-legume symbiosis (Mateos et al. 2001). A thorough scanning electron microscope (SEM) examination of the epidermal surface of white clover roots inoculated with *R. leguminosarum* bv. trifolii revealed a nonuniform distribution of eroded pits that follow the contour of the *Rhizobium* cell (Fig. 9A). Their localized structure suggested that rhizobia have cell-bound wall-degrading enzymes, and indeed, follow-up biochemical studies confirmed that rhizobia produce multiple cell-bound isozymes of cellulase and polygalacturonase (Mateos et al. 1992, 1996; Jiminez-Zurdo et al. 1996). Quantitative SEM indi-

Fig. 9. Epidermal eroded pits induced by *Rhizobium leguminosarum* bv. trifolii on white clover roots. **A** Scanning electron micrograph of the root epidermis pitted by attached cells of rhizobia (*arrows*). **B** Transmission electron micrograph showing ultrastructural details of the pitted interface between an attached cell of rhizobia and the clover epidermal root cell wall. Note that the localized erosion is restricted to amorphous regions and not the ordered microfibrillar wall layer (*arrows*). **C** (control), **E**, **G** Phase contrast microscopy and **D**, **F** Nomarski interference contrast microscopy of the **Hot** (Hole on the tip) reaction representing the complete erosion of a *transmuro* hole made by purified cellulase from *R. leguminosarum* bv. trifolii through the noncrystalline wall at root hair tips (*arrows*). Reprinted with permission from the Canadian National Research Council

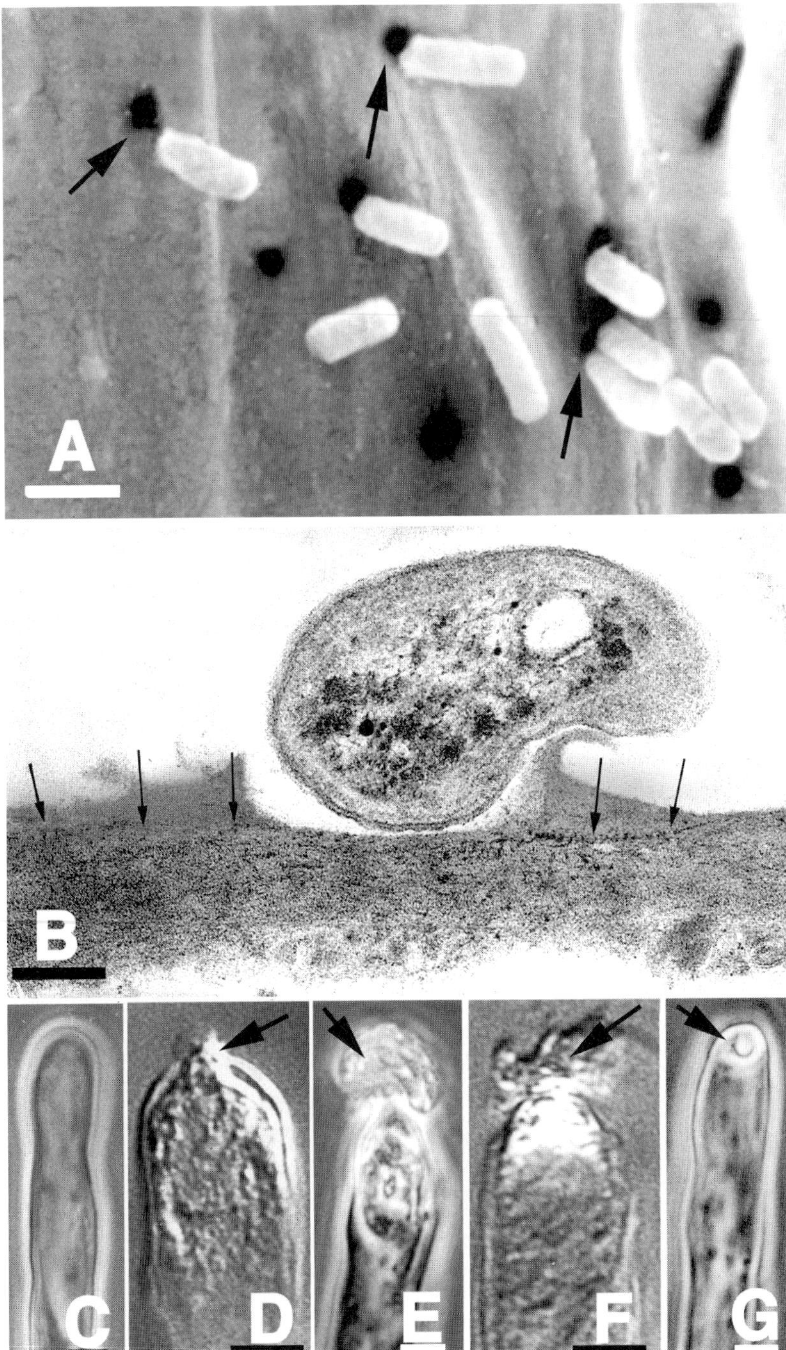
A
B
C
D
E
F
G

cated that the spatial density of these rhizobia-associated eroded pits was significantly higher on the root epidermis of host rather than nonhost legume combinations, was inhibited by high nitrate supply, and was not induced by immobilized wild-type *R. leguminosarum* bv. trifolii chitolipooligosaccharide Nod factors reversibly adsorbed to latex beads. Transmission electron microscope (TEM) examination of these highly localized epidermal pits indicated that they were only partially eroded, i.e., only the outer amorphous region of the plant wall in direct contact with the bacterial cell was disrupted, whereas the underlying highly ordered portion(s) of the wall remained ultrastructurally intact (Fig. 9B). Further studies using phase contrast and polarized light microscopy indicated that (1) the structural integrity of clover root hair walls is dependent on wall polymers that are valid substrates for the purified cell-bound polysaccharide-degrading enzymes (e.g., C2 cellulase isozyme) from rhizobia (Fig. 9C–G); (2) the major site where these rhizobial cell-bound enzymes can completely erode through the root hair wall is highly localized at the isotropic, noncrystalline apex of the root hair tip (Fig. 9C–G), and (3) the degradability of clover root hair walls by these rhizobial polysaccharide-degrading enzymes is enhanced by modifications induced during growth in the presence of CLOS Nod factors from wild-type clover rhizobia. These results suggest that these eroded plant structures represent incomplete attempts of bacterial penetration that had only progressed through isotropic, noncrystalline layers of the plant cell wall, and that the rhizobial cell-bound glycanases and chitolipooligosaccharides participate in complementary roles that ultimately create the localized *transmuro* portal of entry for successful primary host infection (Munoz et al. 1998; Mateos et al. 2001).

2.11 Elicitation of Root Hair Wall Peroxidase by Rhizobia

Many investigators have proposed that successful infection of legumes by rhizobia may depend on the microsymbiont's ability to escape, suppress, or avoid host defense responses that normally protect plants against invasive microorganisms (Vance 1983; Djordjevic et al. 1987; Parniske et al. 1990, 1991). To test this hypothesis, we performed in situ enzyme cytochemistry at subcellular resolution using brightfield microscopy followed by in vitro enzyme assays to detect changes in activity of plant wall-bound peroxidase as an indication of a localized host defense response following inoculation of white clover and pea roots with compatible and incompatible combinations of rhizobial symbionts (*R. leguminosarum* biovars trifolii and viciae; Salzwedel and Dazzo 1993). For compatible combinations, elevated peroxidase activity was initially delayed, but subsequently located precisely at infection-related sites: the center of markedly deformed shepherd's crooks and at penetration sites of incipient infection thread formation, but not elsewhere on the infected root hairs including the intracellular infection thread itself. In contrast, the incompatible combinations rapidly elicited elevated plant peroxidase activity over larger areas of the uninfected root hairs corresponding to their entire irregularly deformed root hair tips. Studies using various pSym *nod* mutant strains (provided by B. Rolfe, Australian National University) indicated a role of extracellular factors and the host-specific nodulation genes *nodEL* in this *Rhizobium*-controlled modulation of root hair peroxidase activity (Salzwedel and Dazzo 1993). Thus, active suppression of host defense responses by compatible rhizobia prior to primary host infection are implicated by these studies. Induction of white clover root peroxidase by compatible and incompatible rhizobial symbionts has been independently confirmed by differential display plant molecular biology techniques (Crockard et al. 1999).

2.12 In Situ Gene Expression

Reporter strains of *Rhizobium* with gene fusions encoding β-galactosidase, β-glucuronidase, and green fluorescent protein are expanding the contribution of microscopy in unraveling many mysteries of the fascinating infection process in the *Rhizobium*-legume symbiosis. A common application of this technology is the use of reporter strains to locate primary host infections since they occur infrequently. Another informative application is the use of merodiploid reporter strains to locate at single cell resolution where, and at what stage of infection do rhizobia express symbiotic genes in situ with minimal risk of disturbing their symbiotic phenotypes. This application was used in quantitative microscopy studies that documented the in situ expression of pSym *nodA* by *R. leguminosarum* bv. trifolii cells during their early interaction with the root surface of the white clover host, especially those bacterial cells that have been clumped together on white clover root hair tips by trifoliin A during the first few hours of phase 1 attachment (Dazzo et al. 1988).

Several methods have been used to detect expression of host symbiotic genes during early interactions of rhizobia with their legume host. One approach has been to use darkfield microscopy with in situ hybridization of DNA probes to specific mRNAs in plant tissue to locate which legume root cells express early nodulins ["*Enods*"] in response to inoculation with rhizobia (McKhann and Hirsch 1993). Such in situ localization studies can be enhanced even further if accompanied by immunofluorescence microscopy at single cell resolution (Dazzo and Wright 1996; McDermott and Dazzo 2002), to determine if the antigenic gene product of interest remains with the same cell(s) expressing the gene and/or is redistributed to other cells in the tissue. A second method to examine the cellular location and timing of expression of symbiotically important host genes induced by rhizobia makes use of chimeric fusions of the Gus-reporter gene in transgenic plants. For instance, recent microscopical examination of transgenic alfalfa plants stained for GUS activity has shown that *nod* mutants of *S. meliloti* although blocked in ability to introduce polyunsaturation of the N-acyl fatty acid moiety, in O-acetylation and in sulfation of CLOS Nod factors, are still capable of inducing *ENOD20* (a marker of cortical cell activation) and (most importantly) eliciting cortical cell divisions in this legume host (Vernoud et al. 1999). This result is fully consistent with our studies described earlier that defined the minimal structural requirements for uptake and bioactivity of rhizobial CLOS analogs in legume roots, including induction of alfalfa and white clover cortical cell divisions (Philip-Hollingsworth et al. 1997), contrary to the dogma indicating that those structural features of CLOS dictate host specificity in the *S. meliloti*-alfalfa symbiosis. Finally, a third powerful approach to detect target mRNA is based on staining tissue sections for in situ PCR-amplified antisense riboprobes. This approach has recently been used to detect a novel Enod [dd23b] in white clover roots induced within 6 h after inoculation with wild-type *R. leguminosarum* bv. trifolii or the corresponding purified wild-type CLOS (Crockard et al. 2002).

3 Quantitation of Symbiotic Interactions Between Rhizobium and Legumes by Image Analysis

The value of quantitative microscopy for plant surface microbiology can be enhanced even further when coupled with computer-assisted digital image analysis (Hollingsworth et al. 1989; Orgambide et al. 1996). This fast-growing technology utilizes the digital computer to derive numerical information regarding selected image features. Although image analysis technology cannot add anything that is not already present, its ability to extract the maximum amount of data from the image, as well as to quickly store, retrieve, and electronically transmit that data makes it an invaluable research tool for the microscopist. Computer-assisted microscopy has been used to enhance developmental morphology studies of the *Rhizobium*-legume symbiosis since 1989 (Dazzo and Petersen 1989). Here, I highlight a few examples of new information on the *Rhizobium*-legume symbiosis derived from microscopical studies utilizing digital image analysis, and later illustrate how we have opened new ground in plant surface microbiology by development and implementation of innovative image analysis software tailored to studies of in situ microbial ecology.

3.1 Definitive Elucidation of the Nature of *Rhizobium* Extracellular Microfibrils

The extracellular microfibrils made by *R. leguminosarum* bv. trifolii in pure culture were isolated and shown by chemical analysis to consist of microcrystalline cellulose (Napoli et al. 1975a). However, the nature of the microfibrils associated with rhizobia that highlight the beginning of their Phase 2 firm adhesion to the legume root epidermis (Fig. 2F) was more difficult to define. The combined use of scanning electron microscopy, enzyme cytochemistry, and computer-assisted digital image analysis provided direct in situ evidence of the cellulosic nature of the extracellular microfibrils extending from *R. leguminosarum* bv. trifolii cells colonized on the white clover root epidermis (Mateos et al. 1995).

3.2 Rhizobial Modulation of Root Hair Cytoplasmic Streaming

Many studies have shown that rhizobia influence the cytoplasmic streaming of host root hairs (beginning with the classic microscopical studies of root hair infection by Fåhraeus (1957) and Nutman et al. (1973), but very few have gone the extra mile to quantitate the changes in velocity of this early host cytoskeletal event in vivo. We utilized high resolution, phase-contrast video microscopy and play-back digital image analysis in real time to establish that the velocity of cytoplasmic streaming within living root hairs of white clover is increased by 35 and 63 % soon after exposure to cells or isolated clover lectin-binding lipopolysaccharide *of R. leguminosarum* bv. trifolii, respectively (Dazzo and Petersen 1989; Dazzo et al. 1991).

3.3 Motility of Rhizobia in the External Root Environment

Rhizobium leguminosarum bv. trifolii is peritrichously flagellated. How fast does it swim in the external root environment of its host, white clover? By focusing the phase objective lens just below the coverslip in modified Fåhraeus slide cultures without the agar matrix, it is possible to record enough examples of long swimming runs of individual cells within the depth of focus to perform image analysis on track reconstructions of real-time video recorded images played back in slow motion. Quantitation of this activity by digital image analysis showed that *R. leguminosarum* bv. trifolii swims in this external clover root environment at an average velocity of 52 μm/s (around 40 times its cell length, compared to around 60 body lengths for *E. coli* under ideal testing conditions), and cells tethered by their lateral flagella to the underside of the coverslip rotate at a frequency of 5–6 Hz/s in this slide culture environment (Dazzo and Petersen 1989).

When Fåhraeus slide cultures of white clover seedlings and *R. leguminosarum* bv. trifolii are prepared using 0.4 % agarose, two zones of bacterial chemotropic swarming can be visualized by darkfield illumination. Digital image analysis indicates that one of these bacterial chemotropic responses forms a hollow sphere whose center is at the root tip and an intercept of radius approximately 4 mm above the root tip. The second chemotropic response is less structured, but accumulates in a cylindrical zone surrounding the root 2–4 mm from the root tip. These microscopical observations suggest that *Rhizobium* responds chemotactically in situ to different, multiple chemical gradients in the external environment surrounding the clover root.

3.4 Root Hair Alterations Affecting Their Dynamic Growth Extension and Primary Host Infection

Quantitative microscopy has played a major role in analyzing the developmental morphology of white clover root hairs to elucidate the mechanisms of rhizobial CLOS action in modulating the growth dynamics and symbiont infectibility of these target host cells (Dazzo et al. 1996a). We performed time-lapse video microscopy of axenic seedling roots treated with nanomolar concentrations of wild-type *R. leguminosarum* bv. trifolii CLOSs and grown geotropically under microbiologically controlled conditions, followed by a quantitative time-series image analysis of individual root hair growth in the acquired video-recorded images at 4-s resolution (Dazzo et al. 1996a). This analysis indicated that the earliest discernible root hair deformations occur within 2.12±0.65 h after application of the wild-type CLOS, and that the morphological basis of the dominant type of CLOS-induced Had is a short-range alteration in direction of polar extension growth of the root hair tip rather than distortion of an already elongated root hair wall, resulting in a redirection of tip growth that deviates from the medial axis of the root hair cylinder. Further studies of quantitative microscopy indicated that CLOS action extends the growing period of active root hair elongation for ~ 5.2 h beyond its normal duration without affecting the elongation rate per se (~19 µm/h), resulting in mature root hairs that are on average about 100 µm longer. This extended growth period predictably increases the duration in which the root hair's "window of infectibility" remains open before cessation of growth. Consistent with this hypothesis, CLOS action was shown by polarized light microscopy to induce localized isotropic alterations in the otherwise anisotropic, ordered crystalline architecture of root hair walls and shown by phase contrast light microscopy to significantly increase the number of potential infection sites and promote their infectibility by wild-type *R. leguminosarum* bv. trifolii (Dazzo et al. 1996a). These studies gave new information on the mechanisms of CLOS action that participate in activating root hair infectibility in the *Rhizobium*-legume symbiosis.

4 A Working Model for Very Early Stages of Root Hair Infection by Rhizobia

These various studies that capitalize on the added dimension of quantitative microscopy at cellular and subcellular resolution (Abe et al. 1984; Dazzo et al. 1982, 1991, 1996; Mateos et al. 1992, 1996, 2001; Salzwedel and Dazzo 1993; Rolfe et al. 1996; Sanchez et al. 1997) have led to a working model for primary infection of white clover root hairs by the N_2-fixing symbiont, *R. leguminosarum* bv. trifolii. This model includes a transient, rapid *nodEL*-dependent suppression of host peroxidase activity during the initial period in which root hair infectibility is activated by trifoliin A-binding CPS oligosaccharides and K90 LPS, and CLOS-induced growth extension and disruptions in crystalline architecture of the growing root hair wall. The infection-related pattern of rhizobial attachment allows for the short-range combined action of these bioactive molecules to result in an increased localized susceptibility of this host wall barrier to a highly controlled degradation by cell-bound rhizobial enzymes that eventually form a small, but complete *transmuro* erosion site that ultimately becomes the primary portal of bacterial entry while still enclosed within the center overlap of the root hair shepherd's crook. An increased flurry of cytoplasmic streaming within the root hair stimulated by the rhizobial symbiont is proposed to facilitate the delivery of new host cell components involved in initiation and continued inward growth of the walled tubular infection thread, while simultaneously directing the traffic of internalized membrane-associated CLOS signal molecules to the root hair nucleus. Later, a localized host wound response at the site of incipient bacterial penetration elevates peroxidase activity that cross-links structural polymers of the eroded wall in order to avoid lysis of the root hair protoplast after bacterial entry and infection thread formation. In contrast, rapid elicitation of clover peroxidase activity in the incompatible combinations (rhizobia with heterologous *nodEL*) may represent a localized discriminating host defense response that rapidly increases cross-linking of wall polymers, thus making the primary host barrier of the root hair wall more resistant to bacterial penetration. This unifying model assigns the ability of *Rhizobium* to modulate the plasticity (i.e., the summation of softening and hardening processes) of the root hair wall as a major symbiotic event controlling successful host infection.

5 Improvements in Specimen Preparation and Imaging Optics for Plant Rhizoplane Microbiology

Residual rhizosphere soil remaining on plant roots after gentle washing significantly obscures the underlying rhizoplane microflora. We have addressed this major limitation in plant surface microbiology using very young white clover seedlings (≤2 days old) grown in a sandy loam soil. By empirically optimizing the gyrorotary angular velocity and duration of gentle washing of excavated roots of white clover in isotonic Fåhraeus medium, we have largely solved this technical problem to expose the underlying pioneer microflora that develops rapidly on root surfaces of seedlings germinated in soil. The optimal conditions to solve this problem vary with the density and length of root hairs, hence the plant species used. Quantitative image analysis of the white clover seedling roots indicate that this optimized washing procedure uncovers the vast majority (≥80 %) of the rhizoplane surface for viewing microbes without fragmentation of the root hairs.

A major limitation of conventional epifluorescence microscopy used to examine the natural rhizoplane microflora that develops on soil-grown roots stained with fluorochromes is the background of blurred fluorescence outside the plane of focus used to produce the image. In this case, useful morphological information can only be extracted from images of cells lacking a background of out-of-focus fluorescence. A significant development in microscopy is the use of laser scanning confocal microscopy (LSCM) combined with digital image processing techniques. The unique feature of LSCM is that it utilizes pinholes at the laser light source and at the detection of the object's image. This optical design eliminates the stray and out-of-focus light that interferes with the formation of the object's image (a major limitation of the conventional fluorescence microscope), thereby only allowing signals from the focused plane to be detected (McDermott and Dazzo 2002). This optical design also improves the resolution and contrast of microbial cells in natural environments by greatly diminishing objectionable background fluorescence arising from plant tissue, soil particles, or organic debris. Because light from outside the plane of focus is not included in image formation, the 2-D (x–y) image becomes an accurate optidigital thin section with a thickness approaching the theoretical 0.2-µm resolution of the light microscope. Also, by digitizing a sequential series of 2-D images while focusing through the specimen in the third (z) dimension, a 3-D computer-reconstructed digital image can be produced, rotated, 'resectioned' in another plane, displayed, and quantitatively analyzed.

Because LSCM imaging technology solves so many problems inherent in conventional fluorescence microscopy, it is receiving wide application for in situ studies of microbial ecology. The first LSCM examination of the general rhizoplane microflora in situ was done with acridine orange-stained roots of white clover seedlings grown in soil (Dazzo et al. 1993). This approach eliminated the major background fluorescence due to dye absorption into the root interior, which makes conventional epifluorescence microscopy impossible for this type of specimen. Subsequently, Schloter et al. (1993) demonstrated the usefulness of LSCM for immunofluorescence examination of *Azospirillum* on wheat roots. They used a dual laser system to produce the green autofluorescence of the root background upon which the distinctive red immunofluorescence of *Azospirillum* (probed with tetramethylrhodamine isothiocyanate-labeled monoclonal antibodies) could be easily seen. They also utilized the noninvasive optical sectioning ability of the confocal microscope to locate the *Azospirillum* cells within the root mucigel layer. More recently, we have used optical sectioning by LSCM to document the entry of neptunia-nodulating rhizobia into crevices at lateral root emergence of the aquatic legume *Neptunia natans* (Subba-Rao et al. 1995), and azorhizobia colonized on the root surface and within cortical cells of intact rice roots (Reddy et al. 1997).

6 CMEIAS: A New Generation of Image Analysis Software for in Situ Studies of Microbial Ecology

6.1 CMEIAS v. 1.27: Major Advancements in Bacterial Morphotype Classification

A major challenge in microbial ecology is to develop reliable methods of computer-assisted microscopy that can analyze digital images of microbial populations and complex microbial communities at single cell resolution, and compute useful ecological characteristics of their organization and structure in situ without cultivation. To address this challenge, we are developing customized semi-automated image analysis software capable of extracting the full information content in digital images of actively growing microbial populations and communities. This analytical tool, called CMEIAS (Center for Microbial Ecology Image Analysis System) consists of plug-in files for the free downloadable program UTHSCSA ImageTool (Wilcox et al. 1997) operating in a personal computer running Windows NT 4.0/2000. The first release version of CMEIAS was developed primarily to perform morphotype classification of bacteria in segmented digital images of complex microbial communities (Liu et al. 2001). This CMEIAS version 1.27 uses pattern recognition algorithms optimized by us to recognize bacterial morphotypes with an overall classification accuracy of 97 %, and a sensitivity that can classify morphotypes present in the community at a frequency as low as ~0.1 % (Liu et al. 2001). CMEIAS v. 1.27 can recognize 11 major morphotypes, including cocci, spirals, curved rods, U-shaped rods, regular straight rods, clubs, ellipsoids, prosthecates, unbranched filaments, rudimentary branched rods, and branched filaments, representing a complexity level of morphological diversity equivalent to 98 % of the genera described in the 9th edition of Bergey's *Manual of Determinative Bacteriology* (Holt et al. 1994). An interactive edit feature is included in CMEIAS v. 1.27 to revise the output of automatic classification data if necessary (occurring at a 3 % error rate), and add up to five additional morphotypes not included in the automatic classification routine (Liu et al. 2001). Our first major application of CMEIAS v. 1.27 was to contribute data on dynamic changes in community structure, including its resistance, resilience, and ecological succession in a polyphasic taxonomy study of microbial community responses to nutrient perturbation, using complex anaerobic bioreactors as the model system (Fernandez et al. 2000; Hashsham et al. 2000). CMEIAS v. 1.27 will soon be released for free Internet download at a website linked to the Michigan State University Center for Microbial Ecology (http://cme.msu.edu/cmeias).

6.2 CMEIAS v. 3.0: Comprehensive Image Analysis of Microbial Communities

A significantly upgraded version of CMEIAS is being developed with several new analytical modules designed to extract four ecologically relevant, in situ features of microbial communities in digital images: (1) morphotype classification and diversity, (2) microbial abundance for both filamentous and nonfilamentous morphotypes, (3) in situ studies of microbial phylogeny/autecology/metabolism, and (4) in situ spatial distribution analysis of microbial colonization on various surfaces. Significant new features will include an advanced morphotype classifier that incorporates default size and shape dimensional borders that are taxonomically relevant and has user-defined flexibility to discriminate any customized level of morphological diversity; various computations of cell density, biovolume, biomass carbon, biosurface area, and filamentous length; color recognition of foreground objects stained with fluorescent molecular probes; various measurement features of plot-less, plot-based, and georeferenced patterns of spatial distribution analysis; spreadsheet macros for automatic data preparation, sampling statistics and spatial statistics analyses; and automated image editing routines (Reddy et al. 2002a, b; see http://lter.kbs.msu.edu/Meetings/2003_All_inv_Meeting/Abstracts.dazzo.htm). Data extracted from images by CMEIAS can be used in other advanced ecological statistics programs, e.g., EcoStat (Towner 1999), and GS+Geostatistics (Robertson 2002), to compute numerous other statistical indices that further characterize microbial community structure. Our vision is for CMEIAS to become an accurate, robust and user-friendly software tool that can analyze microbial communities without cultivation, thereby creating many new approaches to study microbial ecology in situ at spatial scales physiologically relevant to the individual microbes.

To illustrate some of the awesome computational power of CMEIAS that can be applied to in situ studies of plant surface microbiology, examples of analyses have been performed on (1) the abundance and spatial distribution of *Rhizobium leguminosarum* bv. trifolii cells colonized on a white clover seedling root in gnotobiotic culture; (2) a comparison of the morphological diversity and distribution of abundance in natural microbial communities that colonize the phylloplane leaf surface of two different varieties of field-grown corn, and (3) the in situ spatial patterns of root colonization by the pioneer microflora that develop on the white clover seedling rhizoplane during their first 2 days of growth in soil, and (4) the spatial scale of quorum sensing of signal molecules by rhizobacteria colonized on the root surface.

6.3 CMEIAS v. 3.0: Plotless and Plot-Based Spatial Distribution Analysis of Root Colonization

For this first example, CMEIAS image analysis was performed on a scanning electron micrograph of a region of the white clover seedling root surface colonized by cells of *Rhizobium leguminosarum* bv. trifolii wild-type strain ANU843 to extract many different types of quantitative data relevant to plant surface microbiology (Fig. 10A). Figure 10B illustrates the frequency distribution of their first and second nearest neighbor distances (distance between object centroids), as two examples of their spatial distribution in a plot-less analysis. Table 1 lists 15 quantitative features relevant to plant surface micro-

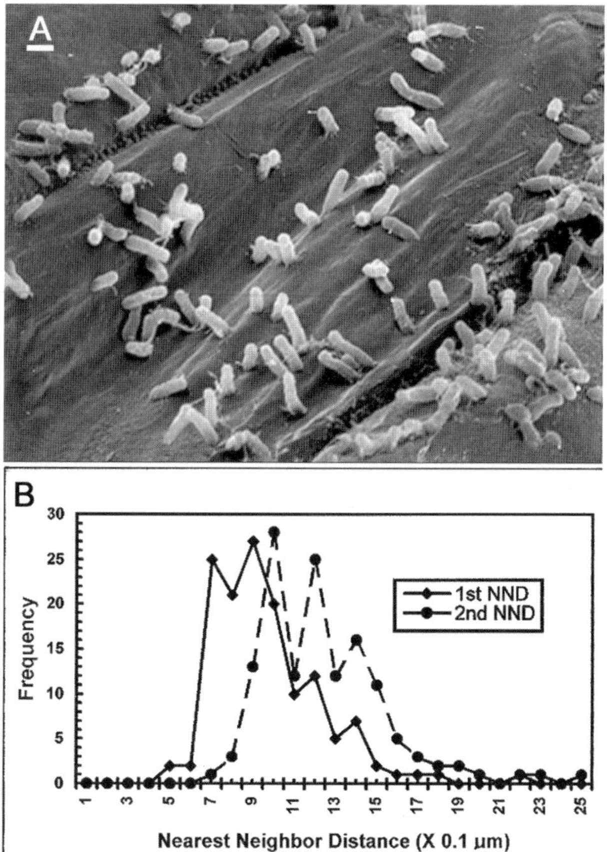

Fig. 10. Colonization of the white clover root surface by wild-type *R. leguminosarum* bv. trifolii ANU843 in gnotobiotic culture. **A** Scanning electron micrograph of a region of the root surface colonized by the bacteria. *Bar scale* 1 µm. **B** CMEIAS plot-less spatial distribution analysis of bacterial cells in (**A**) measured as the frequency distribution of each cell's first and second nearest neighbor distance

Table 1. Quantitative data relevant to plant surface microbiology extracted from Fig. 10A by CMEIAS image analysis

Measurement feature	Type of analysis	Value
Number of cells	Microbial abundance	138
Avg. cell biovolume (μm³)	Microbial abundance	0.17
Avg. cell biomass C (fg)	Microbial abundance	34.66
Avg. cell biosurface area (μm²)	Microbial abundance	1.72
Cumulative bacterial biovolume (μm³)	Microbial abundance	23.91
Cumulative bacterial biomass C (fg)	Microbial abundance	4783.18
Cumulative bacterial biosurface area (μm²)	Microbial abundance	237.86
Cumulative area covered by bacteria (μm²)	Microbial abundance	60.25
Avg. first nearest neighbor distance (μm)	Plotless spatial distribution	0.84±0.30
Avg. second nearest neighbor distance (μm)	Plotless spatial	1.10±0.32
Average aggregation (cluster) index (μm⁻¹)	Plotless spatial distribution	1.31±0.38
Holgate's A value of spatial randomness	Plotless spatial distribution	0.622 =clumped
Significance value of Holgate's A (*p*)	Plotless spatial distribution	0.001
Spatial density of bacteria (cells/mm²)	Plot-based spatial distribution	427,245
Microbial cover (%)	Plot-based spatial distribution	18.7
Uncovered root surface area (%)	Plot-based spatial distribution	81.3

biology that were extracted from this same image, eight features that measure microbial abundance, and seven (four plot-less plus three plot-based) features that measure their spatial distribution. The Aggregation (Cluster) Index is a plot-less spatial distribution measurement feature that we have introduced, equal to the inverse of the first nearest neighbor distance (Dazzo et al. 2003). The Holgate's method for plot-less spatial analysis is a statistical test for spatial randomness requiring that *n* random points be selected and that the distance to the two nearest individuals be measured. This method computes Holgate's A, a measure of aggregation. Values of A are 0.5 for randomly spaced populations, >0.5 for clumped populations, and <0.5 for uniformly spaced populations. Since the computed A value for this population of bacteria in the image is >0.5, their spatial distribution is clumped, and the Z-test for spatial randomness is rejected at the statistically significant level of 99.9 %. Definitive quantitative spatial distribution data acquired by computer-assisted microscopy!

6.4 CMEIAS v. 3.0: In Situ Analysis of Microbial Communities on Plant Phylloplanes

In the second example, CMEIAS was used to perform an in situ image analysis of the microbial communities that developed on phylloplane surfaces of two corn varieties grown under field conditions: one was a genetically modified corn genotype engineered to express the insecticide protein made by the bacterium *Bacillus thuriengensis* (BT-corn variety Pioneer 3573), and the second was a control corn (non-BT variety Pioneer 36N05) receiving no insecticide. Corn leaf disks (4 mm in diameter) were sampled in July, 1999 from mature field-grown plants cultivated in an Long-Term Ecological Research [LTER] experimental site at the Michigan State University Kellogg Biological Station (KBS). Adjacent quadrats (n=26) of digital images were acquired by scanning electron microscopy at ×1000 and at ×100 to resolve the prokaryotic

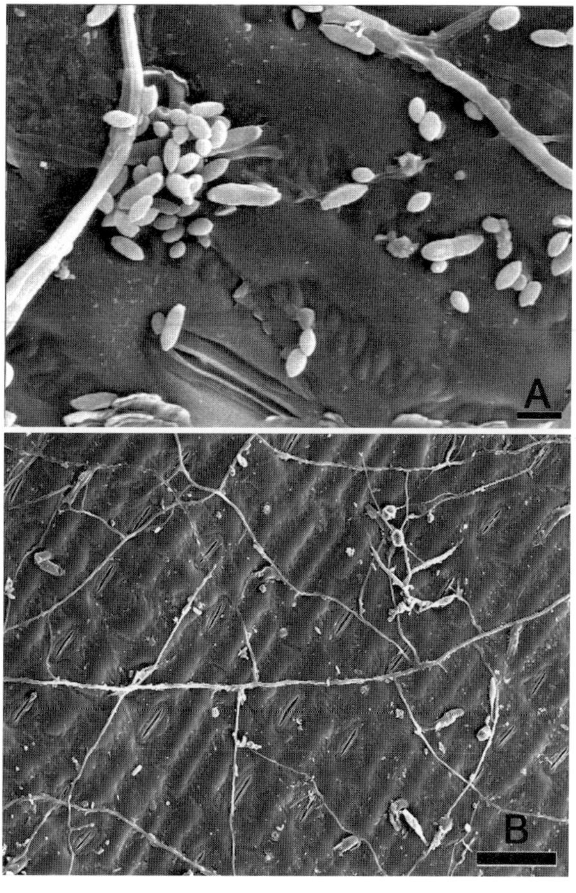

Fig. 11. Scanning electron micrographs of the phylloplane microflora developing on the leaf surface of field-grown corn. Images were acquired at 1000x (**A**) and 100x (**B**) to locate and analyze the prokaryotic (bacteria) and eukaryotic (fungi) microorganisms in the phylloplane community, respectively. *Scale bar* 1 µm in **A** and 100 µm in **B**

and eukaryotic components of the microbial communities, respectively, on the upper corn leaf surfaces (Fig. 11A, B), and then analyzed by CMEIAS to characterize their community structures in situ. Figure 12 compares the morphological diversity of the prokaryotic microorganisms in these two communities, with data presented in a rank-order pareto plot of their richness and percent numerical abundance of operational morphological units (OMU utilizing both shape and size classification schemes), plus a table insert of their morphological diversity index (based on Shannon's Diversity Index H' using nearly equivalent community sample sizes and OMUs rather than species), J evenness distribution of OMUs, and a proportional similarity index that is weighted according to OMU dominance. The latter three indices are derived from computations of the CMEIAS data in EcoStat. These results indicate that the prokaryotic component of the phylloplane communities developed on these two corn varieties had quite similar values of OMU richness, morphological diversity indices, and J evenness in distribution of OMUs, but deeper CMEIAS data mining indicate that they have a proportional similarity index in prokaryotic morphological diversity of only 64.2 % due to major differ-

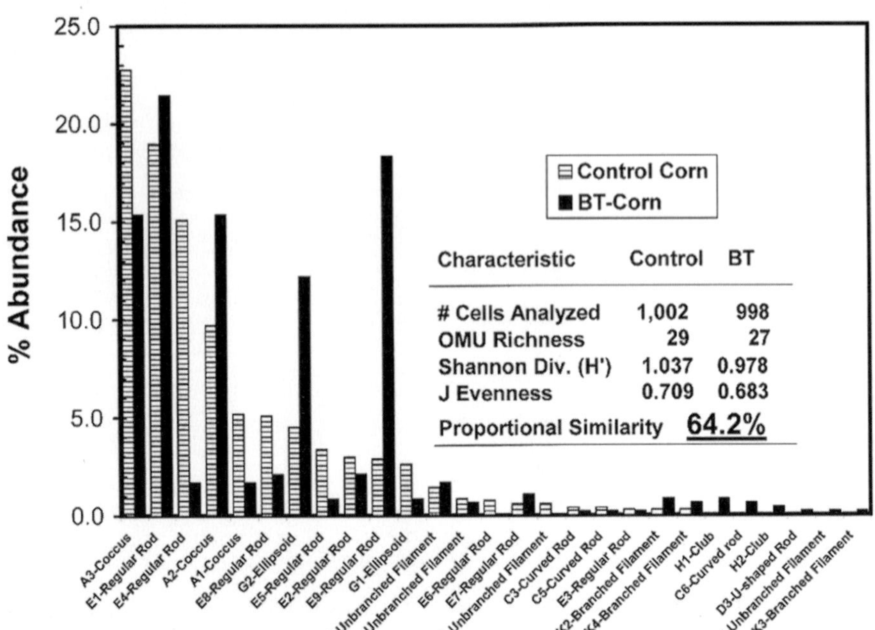

Fig. 12. Rank-abundance diversity plots of CMEIAS morphotype classification data among prokaryotic microorganisms that colonize the phylloplane surface of control (non-BT) corn and BT-corn expressing the bacterial insecticide. The *insert table* reports the similarities and differences in indices of community structure based on morphological analyses using CMEIAS

ences in relative abundance of various sized regular rods, cocci, and ellipsoid OMUs (Fig. 12). One should be able to readily appreciate from this example how CMEIAS can augment other methods of polyphasic taxonomy (e.g., 16S rDNA sequence analysis) to analyze and quantitatively compare complex microbial communities in situ without cultivation.

CMEIAS offers several different algorithms to compute microbial biovolume, with the most accurate overall being adaptive to shape, i.e., CMEIAS first classifies the cell shape and then applies the most appropriate formula to compute its volume based on that particular shape. Figure 13 summarizes the CMEIAS analysis of total abundance and relative distribution of biovolume among the various prokaryotic morphotypes in these two different corn phylloplane communities. The results clearly indicate a significantly greater abundance of prokaryotic biovolume per unit of phylloplane surface area for the Pioneer 36NO5 (control) variety than the Pioneer 3573 (BT-corn) variety (Fig. 13A), and substantial differences in relative distribution of prokaryotic biovolume for certain dominant morphotypes in these communities (Fig. 13B).

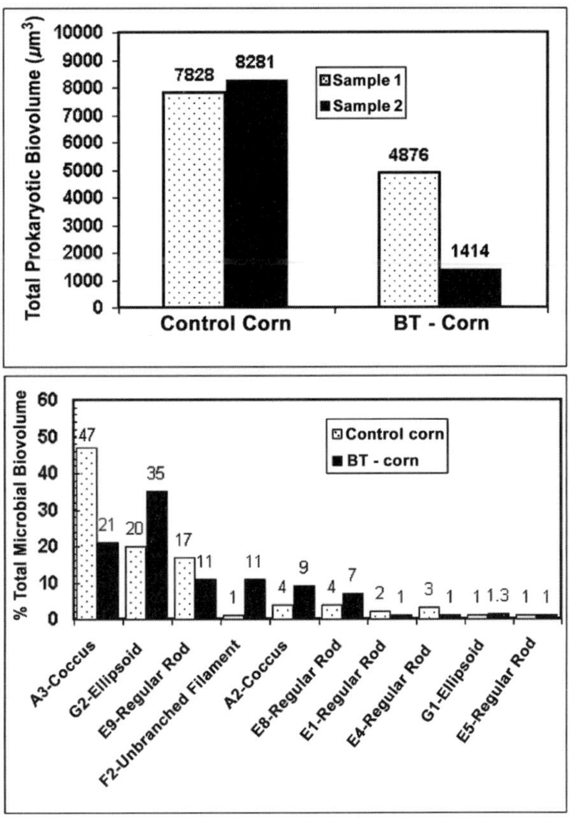

Fig. 13. CMEIAS analysis of biovolume abundance in microbial communities developed on the phylloplane of field-grown control corn and BT-corn. *Above* Total standing crop of prokaryotic biovolume. *Below* Distribution of community biovolume among different prokaryotic morphotypes

Table 2. In situ plot-based spatial distribution analysis of corn phyllosphere prokaryotic microbial communities

Parameter	Control corn (Pioneer 36N05)	BT corn (Pioneer 3573)	Interpretation
Spatial density (cells/mm^2)	214,615	172,692	Higher on control corn
Morista dispersion value	1.3086	1.7805	Clumped distribution
Variance/mean ratio	7.7346	13.6624	Clumped distribution
Negative binomial K distribution	1.9238	0.6779	Clumped distribution
Lloyd's patchiness value	1.3144	1.7931	Clumped distribution
Nonfilamentous microbial cover (%)	2.2	0.4	Higher on control corn

The spatial distribution of prokaryotic microorganisms on the phylloplane surface of these two corn varieties was compared by analyzing several CMEIAS in situ plot-based measurement features on a sample set of 104 quadrat images (52 quadrats each). The mean values of their spatial density (cells/unit of surface area) and percent nonfilamentous microbial cover indicated a significantly higher level of bacterial colonization on the phylloplane of the Pioneer 36N05 (control corn) variety (Table 2). An ascending sort plot of the entire range of spatial density for each image quadrat provided further insight into the basis for this difference in spatial distribution, with clear indication that the overall density of bacteria on the BT-corn phylloplane was lower because that habitat contained more image quadrats with no bacterial cover (Fig. 14).

Table 2 lists several other computations that define the patterns of spatial distribution for microbes that colonize these plant leaf surfaces, all derived from in situ plot-based data extracted from image quadrats by CMEIAS and computed in EcoStat. The Morista Index measures the degree of dispersion, with values <1 for an overdispersed distribution, ≅1 for a random pattern of distribution, and >1 for a clumped pattern. The variance/mean ratio from the observed pattern of frequencies (proportion of quadrats that contain organisms) to those predicted by a Poisson distribution is approximately 1 for randomly spaced populations, significantly >1 for clumped spacing, and <1 for overdispersed or uniformly spaced populations. The negative binomial K distribution is computed from an iterated maximum likelihood method, with small (<8) K values indicating clumping and large K values (>8) indicating random spacing. Lloyd's patchiness value is a statistic similar to Morista's index and computes a "mean crowding" value of spatial patchiness. All these computations indicate that the pattern of spatial distribution for the prokaryotic microbes in these phylloplane communities is clumped rather than random or uniform, and this distribution is more pronounced for the Pioneer 3573 (BT-corn) variety.

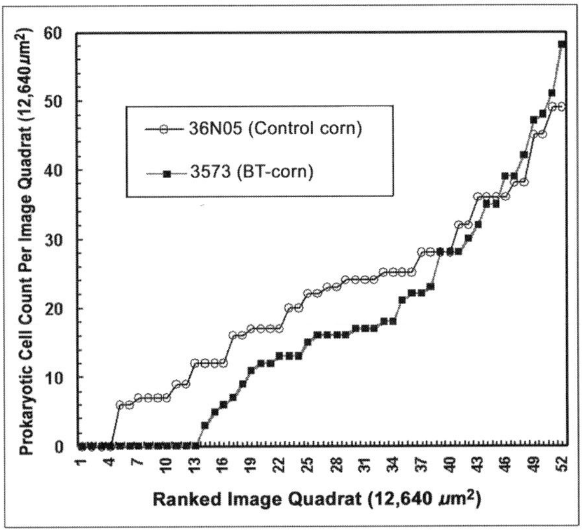

Fig. 14. CMEIAS spatial density analysis of prokaryotic microorganisms colonized on the phylloplane of field-grown control corn (36N05) and BT-corn (3573)

Table 3. Spatial abundance of filamentous fungi on the phylloplane of field-grown corn. Values reported are per mm² phylloplane surface area

Spatial abundance parameter	Control (36N05) corn	BT (3573) corn
Filamentous hyphae cover (%)	12.7	6.4
Cumulative hyphal length (µm)	1737	803
Hyphal biosurface area (µm²)	4244	2035
Hyphal biovolume (µm³)	816	439
Hyphal biomass carbon (fg)	163	88

Table 3 summarizes the CMEIAS-based analyses of spatial abundance for the filamentous fungi on these corn phylloplanes. CMEIAS recognizes filamentous microorganisms in digital images based on their shape characteristic of a length/width ratio >16 without wave form periodicity, and classifies unbranched and branched filaments separately based on whether they have more than two cell poles (Liu et al. 2001). All five CMEIAS measurement parameters indicated an approximate 99 % higher spatial abundance of filamentous fungi biomass on the Pioneer 36N05 control corn variety.

These results illustrated in this second example indicate that CMEIAS performs admirably in the in situ analysis of phylloplane microbial communities. One could definitively conclude from the results that different microbial communities developed on the phylloplane sampled from field-grown Pioneer 36N05 and Pioneer 3573 varieties of corn, but more thorough studies would be necessary before reaching any firm conclusion regarding the possible involvement of the BT insecticide itself in influencing how these microbial communities developed to these different states.

6.5 CMEIAS v. 3.0: In Situ Geostatistical Analysis of Root Colonization by Pioneer Rhizobacteria

In the third example, CMEIAS was used to analyze the pattern of spatial distribution for the pioneer rhizobacterial community that first colonizes seedling roots grown in soil. For this study, Dutch white clover seeds were planted in a wetted sandy loam soil sampled at the KBS-LTER field site. Seedling roots were harvested after 2 days of germination, then gently washed free of rhizosphere soil by optimized gyrorotary rotation in Fåhraeus medium, stained briefly with a 1:10,000 aqueous solution of acridine orange, rinsed in 1 % Na pyrophosphate, and mounted in Vectashield photobleaching retardant reagent. Fluorescent micrographs of the natural pioneer rhizobacterial communities that developed on the clover rhizoplane were acquired as a series of optisection, grayscale images georeferenced to the root tip landmark using laser scanning confocal microscopy with the 63x oil immersion objective and direct through-the-ocular confocal viewing. These digital images were segmented and used to produce a large continuous montage in Adobe Photoshop. The montage images were analyzed by CMEIAS to locate the x, y Cartesian coordinates of each individual microbial cell on the rhizoplane and compute its Cluster index (inverse of first nearest neighbor distance) in situ as the z variate. These CMEIAS data were then analyzed by the spatial geostatistics modeling techniques of semivariogram autocorrelation and kriging analyses (Murray 2002) using GS+ software (Robertson 2002).

The variogram of Fig. 15 is the first of its kind in showing that an isotropic exponential model best fits the semivariance autocorrelation data of spatial dependence for pioneer root colonization by microorganisms in soil. It further clearly indicates that there is a spatial dependence in the nearest neighbor distribution of rhizoplane colonization for microorganisms, with a spatial scale of influence up to a separation distance of ~52 µm. Thus, microbes separated from each other by distances up to this spatial limit do influence each other's root colonization pattern. Such information is fundamentally new in that it provides a real world perspective of the in situ spatial scales that are truly relevant to microbial colonization of plant root surfaces in soil. A first for in situ microbial ecology!

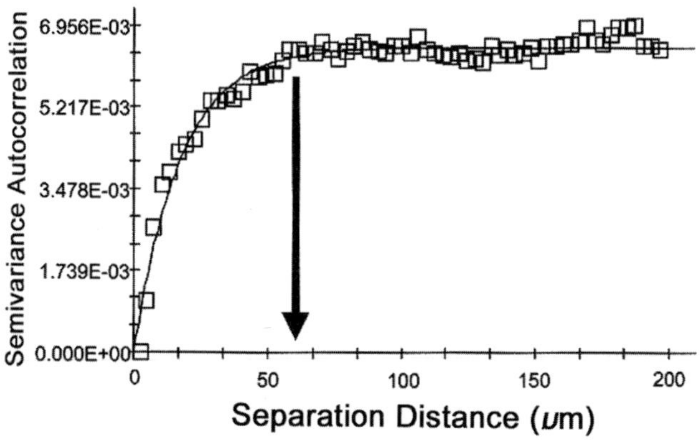

Fig. 15. CMEIAS/GS+ analysis of spatial geostatistics (autocorrelation semivariogram) for rhizobacteria during pioneer colonization of white clover seedlings grown in soil. This graph indicates the highly significant, fundamentally new finding that pioneer colonization of seedling roots by bacteria in soil has an in situ spatial dependence over a spatial scale up to 52 μm

6.6 CMEIAS v. 3.0: Quantitative Autecological Biogeography of the Rhizobium–Rice Association

In the fourth example, CMEIAS is being used to study the biogeography of *R. leguminosarum* bv. trifolii strain E11, a plant growth-promoting endocolonizer of rice roots isolated in the Nile delta where rice and berseem clover have been rotated since antiquity (Yanni et al. 1997). We are using this strain in a model study designed to define the autecological biogeography of rhizobial PGPR endophytes of rice at two spatial scales, one relevant to the organisms (its colonization of rice roots), and second relevant to the rice farmer who would be using such strains as rice biofertilizer inoculants to enhance rice production with less dependence on chemical fertilizer N (Yanni et al. 2001). Figure 16A is an SEM image quadrat of the rice root surface after gnotobiotic cultivation with strain E11. Note that the root hair cells above the plane of focus have obscured some of the root surface, and therefore the full distribution of bacteria in this sampled area cannot be examined directly. This problem in microbial biogeography is solved by a geostatistical analysis of the spatial distribution data acquired by CMEIAS using a kriging analysis to interpolate spatial dependence information on a continuous scale even in areas not sampled. Figure 16B shows the 2-D krig map that provides a statistically defendable interpolation of the spatial density of bacteria in a continuous mode, even in these areas obscured by the overlying root hairs (Fig. 16B). The power of CMEIAS geostatistical analysis!

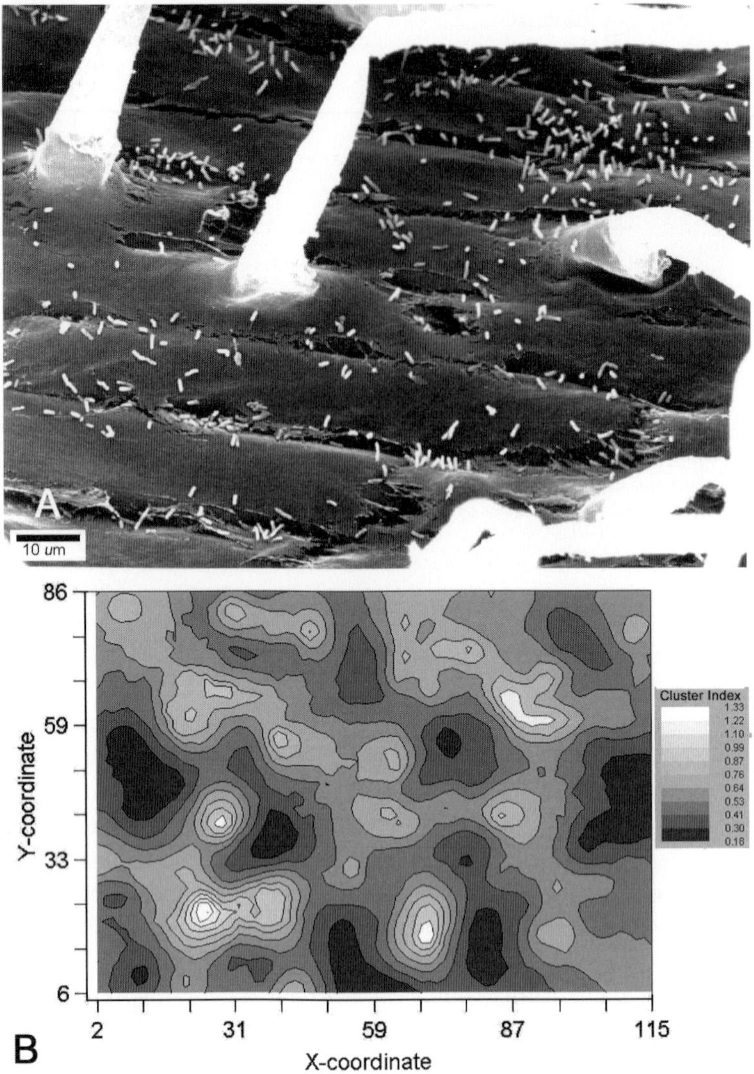

Fig. 16. Geostatistical analysis of the spatial distribution of a plant growth-promotive strain of *Rhizobium leguminosarum* bv. trifolii colonized on the rice root surface. **A** Typical colonization pattern as shown by scanning electron microscopy. *Scale bar* 10 μm. **B** 2-D interpolation kriging map of the spatial density of bacterial cells in **A**

6.7 CMEIAS v. 3.0: Spatial Scale Analysis of in Situ Quorum Sensing by Root-Colonizing Bacteria

In the fifth example, CMEIAS is being used to extract information that sheds new light on the spatial scale at which cell-cell communication of quorum sensing occurs in situ during bacterial colonization of roots. This work is being done in a collaboration of the author with Prof. Anton Hartmann, Stephan Gantner and Christine Duerr in Germany. Confocal fluorescence images of roots are acquired to locate the positions of the red fluorescent protein reporter strain of bacteria that produces and secretes the acyl homoserine lactone quorum signal (**source** cells) and the green GFP-reporter strain of bacteria that cannot produce these signal molecules, but is nevertheless activated by them (**sensor** cells). The range of distances between each green sensor cell and its nearest red source cell neighbor then becomes a measure of the spatial scale at which the cell-to-cell communication of quorum-sensing signal molecules occurs in situ during root colonization. Early indications are that this spatial scale is close to the same range found for spatial dependence in root colonization as described in Fig. 15 above. Figure 17 further illustrates

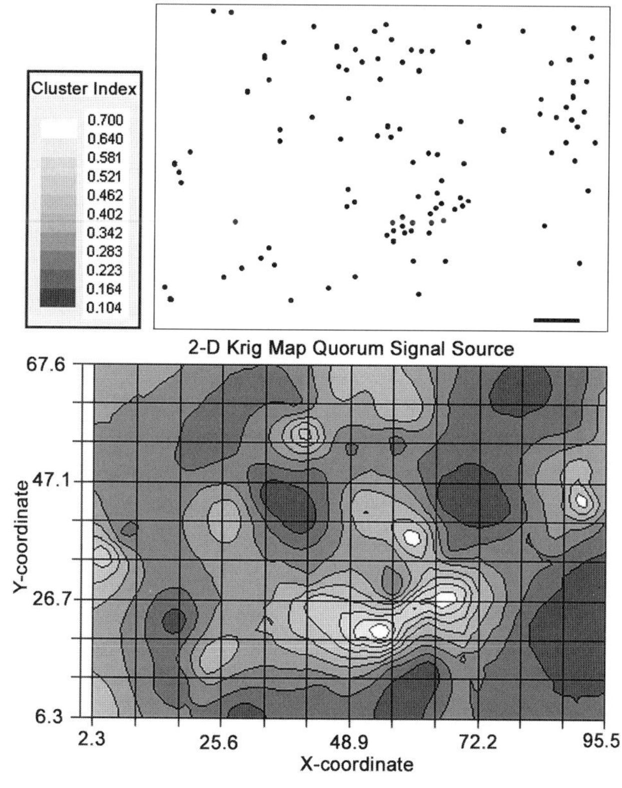

Fig. 17. CMEIAS/GS+ spatial geostatistical analysis of in situ quorum sensing among neighboring bacteria colonizing the root surface. **A** Dot map indicating the location of bacteria that provide the source of the extracellular quorum signal molecule (N-acyl homoserine lactone). *Scale bar* 10 µm. **B** 2-D interpolation kriging map of the predicted gradients of the quorum sensing molecule in situ on the root surface based on the localized cluster indices of colonized bacteria

the power of geostatistical kriging as a spatial modeling technique that can provide a statistically defendable graphical display of the predicted gradients of quorum sensing signals that would diffuse from aggregates of "source cell" bacteria colonized on the root surface. Figure 17A shows the sample point location of signal source bacteria in an image quadrat and Fig. 17B is a 2-D kriging map of their cluster index on a continuous scale. This new technique in computer-assisted quantitative microscopy made possible by CMEIAS image analysis will undoubtedly impact on our understanding of plant surface microbiology and rhizoplane microbial ecology.

7 Conclusions

This chapter has illustrated with many examples from the author's work on the *Rhizobium*-legume symbiosis how quantitative microscopy can make important contributions to the field of plant surface microbiology. In addition, numerous examples illustrate how our CMEIAS software can "count what really counts" to enhance the quantitative analysis of microbial communities and populations in situ without cultivation. This opportunity created by development of CMEIAS will undoubtedly yield fundamentally new information on plant-microbe interactions, and by so doing, expand our understanding of this fascinating subject of plant surface microbiology.

Acknowledgments. Funds to support portions of the research reported in this chapter were provided by the Michigan State University Center for Microbial Ecology (National Science Foundation Grant NO. DEB-91-20006 and the MSU Research Excellence Fund), the MSU Kellogg Biological Station Long-Term Ecological Research project, the US-Egypt Science and Technology Joint Fund (projects BI02-001-017-98 and BI05-001-015), and the Michigan Agricultural Experiment Station. The author thanks Jim Tiedje, Phil Robertson, Rawle Hollingsworth, Youssef Yanni, Howard Towner, Dominic Trione, and Edward Marshall for advice and assistance, and the MSU Center for Advanced Microscopy for use of their facilities.

References and Selected Reading

Abe M, Sherwood JE, Hollingsworth RI, Dazzo FB (1984) Stimulation of clover root hair infection by lectin-binding oligosaccharides from the capsular and extracellular polysaccharides of *Rhizobium trifolii.* J Bacteriol 160:517–520

Bishop P, Dazzo FB, Applebaum E, Maier R, Brill W (1977) Intergeneric transfer of symbiotic genes from *Rhizobium trifolii* to *Azotobacter vinelandii.* Science 198:938–940

Bono JJ, Riond J, Nicolaou KC, Bockovich NJ, Estevez VA, Cullimore JV, Ranjeva R (1995) Characterization of a binding site for chemically synthesized lipo-oligosaccharidic NodRm factors in particulate fractions prepared from roots. Plant J 7:253–260

Callaham D, Torrey J (1981) The structural basis for infection of root hairs of *Trifolium repens* by *Rhizobium.* Can J Bot 59:1647–1664

Cheng HP, Walker GC (1998) Succinoglycan is required for initiation and elongation of infection threads during nodulation of alfalfa by *Rhizobium meliloti.* J Bacteriol 180:5183–5191

Crockard MA, Bjourson AJ, Cooper JE (1999) A new peroxidase cDNA from white clover: Its characterization and expression in root tissue challenged with homologous rhizobia, heterologous rhizobia or *Pseudomonas syringae.* Mol Plant-Microbe Interact 12:825–828

Crockard MA, Bjourson AJ, Dazzo FB, Cooper JE (2002) A white clover nodulin gene, *dd23b*, encoding a cysteine cluster protein (CCP), is expressed in roots during the very early stages of interaction with *Rhizobium leguminosarum* biovar trifolii and after treatment with chitolipooligosaccharide Nod factors. J Plant Res 115:439–447

Dazzo FB (1982) Leguminous root nodules. In: Burns R, Slater J (eds) Experimental microbial ecology. Blackwell, Cambridge, pp 431–446

Dazzo FB, Hubbell DH (1975) Cross-reactive antigens and lectin as determinants of symbiotic specificity in the *Rhizobium*-clover association. Appl Microbiol 30:1017–1033

Dazzo FB, Brill WJ (1977) Receptor sites on clover and alfalfa roots for *Rhizobium.* Appl Environ Microbiol 33:132–136

Dazzo FB, Brill W (1978) Regulation by fixed nitrogen of host-symbiont recognition in the *Rhizobium*-clover symbiosis. Plant Physiol 62:18–21

Dazzo FB, Brill W (1979) Bacterial polysaccharide which binds *Rhizobium trifolii* to white clover root hairs. J Bacteriol 137:1362–1373

Dazzo FB, Hrabak EM (1981) Presence of trifoliin A, a *Rhizobium*-binding lectin, in clover root exudate. J Supramol Struct Cell Biochem 16:133–138

Dazzo F, Hubbell DH (1982) Control of root hair infection. In: Broughton W (ed) Ecology of nitrogen fixation: vol II. *Rhizobium.* Oxford University Press, Oxford, pp 274–310

Dazzo FB, Petersen MA (1989) Applications of computer-assisted image analysis for microscopic studies of the *Rhizobium*-legume symbiosis. Symbiosis 7:193–210

Dazzo FB, Wright SF (1996) Production of anti-microbial antibodies and their use in immunofluorescence microscopy. In: Akkermans A, van Elsas J, de Bruijn F (eds) Molecular microbial ecology manual. vol 4.12. Kluwer, Dordrecht, pp 1–27

Dazzo FB, Wopereis J (2000) Unraveling the infection process in the *Rhizobium*-legume symbiosis by microscopy. In: Triplett E (ed) Prokaryotic nitrogen fixation: a model system for the analysis of a biological process. Horizon Scientific Press, Wymondham, UK, pp 295–347

Dazzo FB, Napoli C, Hubbell DH (1976) Adsorption of bacteria to roots as related to host specificity in the *Rhizobium*-clover symbiosis. Appl Environ Microbiol 32:166–177

Dazzo FB, Yanke W, Brill W (1978) Trifoliin: a *Rhizobium* recognition protein from white clover. Biochim Biophys Acta 536:276–286

Dazzo FB, Urbano MR, Brill WJ (1979) Transient appearance of lectin receptors on *Rhizobium trifolii.* Curr Microbiol 2:15–20

Dazzo FB, Truchet GL, Sherwood JE, Hrabak EM, Gardiol AE (1982) Alteration of the trifoliin A-binding capsule of *Rhizobium trifolii* 0403 by enzymes released from clover roots. Appl Environ Microbiol 44:478–490

Dazzo FB, Truchet G, Sherwood J, Hrabak E, Abe M, Pankratz HS (1984) Specific phases of root hair attachment in the *Rhizobium trifolii*-clover symbiosis. Appl Environ Microbiol 48:1140–1150

Dazzo FB, Hollingsworth RI, Abe M, Smith KB, Welsch M, Morris PJ, Philip-Hollingsworth S, Salzwedel JL, Castillo RM (1987) *Rhizobium trifolii* polysaccharides, oligosaccharides, and other metabolites affecting development and symbiotic infection of clover root hairs. In: Steffens G, Rumsey T (eds) Biomechanisms regulating growth and development: keys to progress. Beltsville Symposium XII on Agricultural Research. Kluwer, Dordrecht, pp 343–355

Dazzo FB, Hollingsworth R, Philip-Hollingsworth S, Robeles M, Olen T, Salzwedel J, Djordjevic M, Rolfe B (1988) Recognition processes in the *Rhizobium trifolii*-white clover symbiosis. In: Bothe H, de Bruijn F, Newton W (eds) Nitrogen fixation: hundred years after. Gustav Fischer, Stuttgart, Germany, pp 431–436

Dazzo F, Truchet G, Hollingsworth R, Hrabak E, Pankratz H, Philip-Hollingsworth S, Salzwedel J, Chapman K, Appenzeller L, Squartini A, Gerhold D, Orgambide G (1991) *Rhizobium* lipopolysaccharide modulates infection thread development in white clover root hairs. J Bacteriol 173:5371–5384

Dazzo FB, Mateos P, Orgambide G, Philip-Hollingsworth S, Squartini A, Subba-Rao NS, Pankratz HS, Baker D, Hollingsworth R, Whallon J (1993) The infection process in the *Rhizobium*-legume symbiosis and visualization of rhizoplane microorganisms by laser scanning confocal microscopy. In: Guerrero R, Pedros-Alio C (eds) Trends in microbial ecology. Spanish Society for Microbiology, Barcelona, pp 259–262

Dazzo FB, Orgambide G, Philip-Hollingsworth S, Hollingsworth RI, Ninke K, Salzwedel JL (1996a) Modulation of development, growth dynamics, wall crystallinity, and infection thread formation in white clover root hairs by membrane chitolipooligosaccharides from *Rhizobium leguminosarum* bv. trifolii. J Bacteriol 178:3621–3627

Dazzo FB, Orgambide G, Philip-Hollingsworth S, Hollingsworth RI, Ninke K, Smith D, Mateos PF, Squartini A, Bjourson AJ, Cooper JE, Wopereis J (1996b) Involvement of membrane chitolipo-oligosaccharides in the *Rhizobium*-white clover symbiosis. In: Chordi-Corbo A, Martinez-Molina E, Mateos P, Carpio-Santos M (eds) Advances in the investigation on biological nitrogen fixation. Excma Diputacion Provincal De Salamanca, Salamanca, Spain, pp 29–33

Dazzo FB, Joseph AR, Gomma AB, Yanni YG, Robertson GP (2003) Quantitative indices for the autecological biogeography of a *Rhizobium* endophyte of rice at macro and micro spatial scales. Symbiosis 35:147–158

Diaz CL, Melchers LS, Hooykaas PJ, Lugtenberg BJ, Kijne JW (1989) Root lectin as a determinant of host-plant specificity in the *Rhizobium*-legume symbiosis. Nature 338:579–581

Diaz CL, Spaink HP, Kijne JW (2000) Heterologous rhizobial lipochitin and chitin oligomers induced cortical cell divisions in red clover roots transformed with the pea lectin gene. Molec Plant-Microbe Interact 13:268–276

Djordjevic MA, Gabriel DW, Rolfe BG (1987) *Rhizobium*: the refined parasite of legumes. Annu Rev Phytopathol 25:145–168

Fåhraeus G (1957) The infection of clover roots by nodule bacteria studied by a simple glass slide technique. J Gen Microbiol 16:374–381

Fernandez A, Hashsham S, Dollhopf D, Raskin L, Glagoleva O, Dazzo FB, Hickey R, Criddle C, Tiedje JM (2000) Flexible community structure correlates with stable community function in methanogenic bioreactor communities perturbed by glucose. Appl Environ Microbiol 66:4058–4067

Gerhold DL, Dazzo FB, Gresshoff PM (1985) Selective removal of seedling root hairs for studies of the *Rhizobium*-legume symbiosis. J Microbiol Meth 4:95–102

Hashsham S, Fernandez A, Dollhopf S, Dazzo FB, Hickey R, Tiedje JM, Criddle CS (2000) Parallel processing of substrate correlates with greater functional stability in methanogenic bioreactor communities perturbed by glucose. Appl Environ Microbiol 66:4050–4057

Hirsch A (1999) Role of lectins (and rhizobial exopolysaccharides) in legume nodulation. Curr Opin Plant Biol 2:320–326

Hollingsworth RI, Abe M, Sherwood JE, Dazzo FB (1984) Bacteriophage-induced acidic heteropolysaccharide lyases that convert acidic heteropolysaccharides of *Rhizobium trifolii* into oligosaccharide units. J Bacteriol 160:510–516

Hollingsworth RI, Dazzo FB, Hallenga K, Musselman B (1988) The complete structure of the trifoliin A lectin-binding capsular polysaccharide of *Rhizobium trifolii* 843. Carbohydr Res 172:97–112

Hollingsworth RI, Squartini A, Philip-Hollingsworth S, Dazzo FB (1989) Root hair deforming and nodule initiating factors from *Rhizobium trifolii*. In: Lugtenberg B (ed) Signal molecules in plants and plant-microbe interactions. Springer, Berlin Heidelberg New York, pp 387–393

Holt JJ, Krieg NR, Sneath PH, Staley JT, Williams ST (1994) Bergey's manual of determinative bacteriology 9th edn. Williams and Wilkins, Baltimore, 787 pp

Hrabak EM, Urbano MR, Dazzo FB (1981) Growth-phase dependent immunodeterminants of *Rhizobium trifolii* lipopolysaccharide which bind trifoliin A, a white clover lectin. J Bacteriol 148:697–711

Hrabak EM, Truchet GL, Dazzo FB, Govers F (1985) Characterization of the anomalous infection and nodulation of subterranean clover roots by *Rhizobium leguminosarum* 1020. J Gen Microbiol 131:3287–3302

Jiminez-Zurdo J, Mateos P, Dazzo FB, Martinez-Molina E (1996) Cell-bound cellulase and polygalacturonase production by *Rhizobium* and *Bradyrhizobium* species. Soil Biol Biochem 28:917–921

Leigh J, Reed J, Hanks J, Hirsch A, Walker G (1987) *Rhizobium meliloti* mutants that fail to succinylate their calcofluor-binding exopolysaccharide are defective in nodule invasion. Cell 51:579–587

Lerouge P, Roche P, Faucher C, Maillet F, Truchet G, Prome J, Denarie J (1990) Symbiotic host-specificity of *Rhizobium meliloti* is determined by a sulfated and acylated glucosamine oligosaccharide signal. Nature 344:781–784

Li D, Hubbell DH (1969) Infection thread formation as a basis for nodulation specificity in *Rhizobium*-strawberry clover associations. Can J Microbiol 15:1133–1136

Liu J, Dazzo FB, Glagoleva O, Yu B, Jain A (2001) CMEIAS©: a computer-aided system for image analysis of microbial communities. Microbial Ecology 41:173–194, 42:215

Lopez-Lara I, Orgambide G, Dazzo FB, Olivares J, Toro N (1993) Characterization and symbiotic importance of acidic extracellular polysaccharides of *Rhizobium* sp. strain GRH2 isolated from *Acacia* nodules. J Bacteriol 175:2826–2832

Lopez-Lara I, Orgambide G, Dazzo FB, Olivares J, Toro N (1995) Surface polysaccharide mutants of *Rhizobium* sp. (*Acacia)* strain GRH2: major requirement of lipopolysaccharide and acidic exopolysaccharide for successful invasion of *Acacia* nodules and host range determination. Microbiology (UK) 141:573–581

Mateos P, Jiminez J, Chen J, Squartini A, Martinez-Molina E, Hubbell DH, Dazzo FB (1992) Cell-associated pectinolytic and cellulolytic enzymes in *Rhizobium trifolii*. Appl Environ Microbiol 58:1816–1822

Mateos P, Baker D, Philip-Hollingsworth S, Squartini A, Peruffo A, Nuti M, Dazzo FB (1995) Direct in situ identification of cellulose microfibrils associated with *Rhizobium leguminosarum* biovar trifolii attached to the root epidermis of white clover. Can J Microbiol 41:202–207

Mateos PF, Zurdo J, Molina-Blanco J, Velazquez A, Dazzo FB, Martinez-Molina E (1996) Implication of cellulase production by *Rhizobium* in the establishment of the symbiosis with legumes. In: Chordi-Corbo A, Martinez-Molina E, Mateos P, Capri-Santos M (eds) Advances in the investigation on biological nitrogen fixation, Excma Diputacion Provincal De Salamanca, Salamanca, Spain, pp 45–48

Mateos P, Baker DL, Petersen M, Velázquez E, Jiménez-Zurdo JI, Martínez-Molina E, Squartini A, Orgambide G, Hubbell DH, Dazzo FB (2001) Erosion of root epidermal cell walls by *Rhizobium* polysaccharide-degrading enzymes as related to primary host infection in the *Rhizobium*-legume symbiosis. Can J Microbiol 47:475–487

McDermott TR, Dazzo FB (2002) Use of fluorescent antibodies for studying the ecology of soil- and plant-associated microbes. In: Hurst C, Crawford RC, Knudsen GR, McInerney MJ, Stetzenbach LD (eds), Manual of environmental microbiology, Chap. 28, American Society for Microbiology Press, Washington, DC, pp 615–626

McKhann HI, Hirsch AM (1993) In situ localization of specific mRNAs in plant tissues. In: Thompson J, Glick B (eds) Methods in plant molecular biology and biotechnology. CRC Press, Boca Raton, pp 173–205

Munoz J, Coronado C, Perez-Hormeache J, Kondorosi A, Ratet P, Palomares AJ (1998) *MsPG3,* a *Medicago sativa* polygalacturonase gene expressed during the alfalfa-*Rhizobium meliloti* interaction. Proc Natl Acad Sci USA 95:9686–9692

Murray CJ (2002) Sampling and data analysis for environmental microbiology. In: Manual of environmental microbiology, American Society for Microbiology Press, Washington, DC, pp 166–177

Napoli C, Hubbell DH (1976) Ultrastructure of *Rhizobium*-induced infection threads in clover root hairs. Appl Microbiol 30:1003–1009

Napoli C, Dazzo FB, Hubbell DH (1975a) Production of cellulose microfibrils by *Rhizobium.* Appl Microbiol 30:123–131

Napoli CA, Dazzo FB, Hubbell DH (1975b) Ultrastructure of infection and common antigen relationships in the *Rhizobium-Aeschynomene* symbiosis. In: Vincent J (ed) Proceedings of the 5th Australian Legume Nodulation Conference. Brisbane, Australia, pp 35–37

Nutman P, Doncaster C, Dart P (1973) Infection of Clover by Root-Nodule Bacteria. British Film Institute, London

Orgambide G, Philip-Hollingsworth S, Cargill L, Dazzo FB (1992) Evaluation of acidic heteropolysaccharide structures in *Rhizobium leguminosarum* biovars altered in nodulation genes and host range. Mol Plant-Microbe Interact 5:482–488

Orgambide GG, Philip-Hollingsworth S, Hollingsworth RI, Dazzo FB (1994) Flavone-enhanced accumulation and symbiosis-related biological activity of a diglycosyl diacylglycerol membrane glycolipid from *Rhizobium leguminosarum* biovar trifolii. J Bacteriol 176:4338–4347

Orgambide G, Lee J, Hollingsworth R, Dazzo FB (1995) Structurally diverse chitolipooligosaccharide Nod factors accumulate primarily in membranes of wild type *Rhizobium leguminosarum* bv. trifolii. Biochemistry 34:3832–3840

Orgambide G, Philip-Hollingsworth S, Mateos P, Hollingsworth RI, Dazzo FB (1996) Subnanomolar concentrations of membrane chitolipooligosaccharides from *Rhizobium leguminosarum* biovar trifolii are fully capable of eliciting symbiosis-related responses on white clover. Plant Soil 186:93–98

Parniske M, Zimmermann C, Cregan PB, Werner D (1990) Hypersensitive reaction of nodule cells in the *Glycine max* sp./*Bradyrhizobium japonicum* symbiosis occurs at the genotype-specific level. Botanica Acta 103:143–148

Parniske M, Ahlborn B, Werner D (1991) Isoflavonoid inducible resistance to the phytoalexine glyceollin in soybean rhizobia. J Bacteriol 173:3432–3439

Pellock BJ, Cheng HP, Walker GC (2000) Alfalfa root nodule invasion efficiency is dependent on *Sinorhizobium meliloti* polysaccharides. J Bacteriol 182:4310–4318

Philip-Hollingsworth S, Hollingsworth RI, Dazzo FB (1989a) Host-range related structural features of the acidic extracellular polysaccharides of *Rhizobium trifolii* and *Rhizobium leguminosarum.* J Biol Chem 264:1461–1466

Philip-Hollingsworth S, Hollingsworth RI, Dazzo FB, Djordjevic M, Rolfe BG (1989b) The effect of interspecies transfer of *Rhizobium* host-specific nodulation genes on acidic polysaccharide structure and in situ binding by host lectin. J Biol Chem 264:5710–5714

Philip-Hollingsworth S, Hollingsworth RI, Dazzo F (1991) N-acetylglutamic acid: an extracellular Nod signal of *Rhizobium trifolii* ANU843 which induces root hair branching and nodule-like primordia in white clover roots. J Biol Chem 266:16854–16858

Philip-Hollingsworth S, Orgambide G, Bradford J, Smith D, Hollingsworth R, Dazzo FB (1995) Mutation or increased copy number of *nodE* has no effect on the spectrum of chitolipooligosaccharide Nod factors made by *Rhizobium leguminosarum* bv. trifolii. J Biol Chem 270:20968–20977

Philip-Hollingsworth S, Hollingsworth RI, Dazzo FB (1997) Structural requirements of chitolipooligosaccharides from *Rhizobium leguminosarum* bv. trifolii for uptake and mitogenic activity in legume roots as revealed by synthetic analogs and bioreactive fluorescent probes. J Lipid Res 38:1229–1241

Reddy PM, Ladha JK, So R, Hernandez R, Dazzo FB, Angeles O, Ramos M, de Bruijn F (1997) Rhizobial communication with rice: induction of phenotypic changes, mode of invasion and extent of colonization. Plant Soil 194:81–98

Reddy C, Liu J, Wadekar M, Prabhu A, Trione D, Marshall E, Zurdo J, Liu F-I, Urbance J, Dazzo FB (2002a) New features of CMEIAS innovative software for computer-assisted microscopy of microorganisms and their ecology. 2002 Ann. Mtg., Long-Term Ecological Research in Row-Crop Agriculture, KBS-LTER Site. Abstract at <http://lter.kbs.msu.edu/Meetings/2002_All_Inv_Meeting/Abstracts/Reddy_etal2.htm>

Reddy C, Liu F-I, Zurdo J, Dazzo FB (2002b) A new CMEIAS color recognition program for digital microbial ecology. 2002 Ann. Mtg., Long-Term Ecological Research in Row-Crop Agriculture, KBS-LTER Site. Abstract at <http://lter.kbs.msu.edu/Meetings/2002_All_Inv_Meeting/Abstracts/Reddy_etal.htm>

Robertson P (2002) GS+ Geostatistics for the environmental sciences. Gamma Design Software, http://www.gammadesign.com

Rolfe BG, Carlson RW, Ridge RW, Dazzo FB, Mateos PF, Pankhurst CE (1996) Defective infection and nodulation of clovers by exopolysaccharide mutants of *Rhizobium leguminosarum* bv. trifolii. Aust J Plant Physiol 23:285–303

Salzwedel J, Dazzo FB (1993) pSym *nod* gene influence on elicitation of peroxidase activity from white clover and pea roots by Rhizobia and their cell-free supernatants. Mol Plant-Microbe Interact 6:127–134

Sanchez B, Coronado C, Philip-Hollingsworth S, Dazzo FB, Palomares A (1997) Structure and role in symbiosis of the *exoB* gene of *Rhizobium leguminosarum* bv. trifolii. Mol Gen Genet 255:131–140

Schloter M, Borlinghaus R, Bode W, Hartmann A (1993) Direct identification and localization of *Azospirillum* in the rhizosphere of wheat using fluorescence-labeled monoclonal antibodies and confocal scanning laser microscopy. J Microsc 171:173–177

Sherwood JE, Truchet GL, Dazzo FB (1984a) Effect of nitrate supply on in vivo synthesis and distribution of trifoliin A, a *Rhizobium-trifolii* binding lectin, in *Trifolium repens* seedlings. Planta 162:540–547

Sherwood JE, Vasse JM, Dazzo FB, Truchet GL (1984b) Development and trifoliin A-binding ability of the capsule of *Rhizobium trifolii*. J Bacteriol 159:145–152

Smit G, Swart S, Lugtenberg B, Kijne JW (1992) Molecular mechanisms of attachment of bacteria to plant roots. Mol Microbiol 6:2897–2903

Subba-Rao NS, Mateos PF, Baker D, Pankratz HS, Palma J, Dazzo FB, Sprent JI (1995) The unique root-nodule symbiosis between *Rhizobium* and the aquatic legume, *Neptunia natans* (L. f.) Druce. Planta 196:311–320

Towner H (1999) EcoStat ecological analysis program for windows, Ver. 1.03, Trinity Software, Campton, NH

Truchet GL, Sherwood JE, Pankratz HS, Dazzo FB (1986) Clover root exudate contains a particulate form of the lectin, trifoliin A, which binds *Rhizobium trifolii*. Physiol Plant 66:575–582

Vance CP (1983) *Rhizobium* infection and nodulation: A beneficial plant disease? Annu Rev Microbiol 37:399–424

van Rhijn P, Goldberg R, Hirsch A1 (1998) *Lotus corniculatus* nodulation specificity is changed by the presence of a soybean lectin gene. Plant Cell 10:1233–1249 van Workum WAT, van Slogeren S, van Brussel AA, Kijne JW (1998) Role of exopolysaccharides of *Rhizobium leguminosarum* bv. viciae as host-specific molecules required for infection thread formation during nodulation of *Vicia sativa.* Molec Plant-Microbe Interact 11:1233–1241

Vega-Hernández MC, Pérez-Galdona R, Dazzo FB, Jarabo-Lorenzo A, Alfayate MC, León-Barrios M (2001) Novel infection process in the indeterminate root nodule symbiosis between tagasaste (*Chamaecytisus proliferus*) and *Bradyrhizobium* sp. (*Chamaecytisus*). New Phytol 150:707–721

Vernoud V, Journet EP, Barker DG (1999) *MtENOD20,* a Nod factor-inducible molecular marker for root cortical cell activation. Mol Plant-Microbe Interact 12:604–614

Wilcox CD, Dove SB, Doss-McDavid W, Greer DB (1997) UTHSCSA ImageTool© Ver. 1.27, http://www.uthscsa.edu/dig/itdesc.html, Univ. Texas Health Science Center, San Antonio, TX

Yanni Y, Rizk R, Corich V, Squartini A, Ninke K, Philip-Hollingsworth S, Orgambide G, deBruijn F, Stoltzfus J, Buckley D, Schmidt T, Mateos P, Ladha JK, Dazzo FB (1997) Natural endophytic association between *Rhizobium leguminosarum* bv. trifolii and rice roots and assessment of its potential to promote rice growth. Plant Soil 194:99–114

Yanni YG, Rizk RY, Abd El-Fattah FK, Squartini A, Corich V, Giacomini A, de Bruijn F, Rademaker J, Maya-Flores J, Ostrom P, Vega-Hernandez M, Hollingsworth RI, Martinez-Molina E, Mateos P, Velazquez E, Wopereis J, Triplett E, Umali-Garcia M, Anarna JA, Rolfe BG, Ladha JK, Hill J, Mujoo R, Ng PK, Dazzo FB (2001) The beneficial plant growth-promoting association of *Rhizobium leguminosarum* bv. trifolii with rice roots. Aust J Plant Physiol 28:845–870

Yao Y, Vincent JM (1976) Factors responsible for the curling and branching of clover root hair by *Rhizobium.* Plant Soil 45:1–16

28 Analysis of Microbial Population Genetics

Emanuele G. Biondi, Alessio Mengoni and Marco Bazzicalupo

1 Introduction

The knowledge of genetic diversity in bacterial population has increased considerably over the last 15 years, due to the application of molecular techniques to microbial ecological studies. Quantitative resolution has improved as a large number of haplotypic markers are found within each sample and as a large number of samples can be simultaneously investigated.

Among the molecular methods, the PCR-based techniques provide a powerful and high throughput approach for the study of genetic diversity in bacterial populations. PCR fingerprinting methods for the analysis of biodiversity are numerous and usually very effective. Some of the most commons are the PCR-RFLP of specific sequences (16S rDNA, intergenic transcribed spacer, ITS) (Laguerre et al. 1996), the Repetitive Extragenic Palindromic-PCR (Woods et al. 1992) and the BOX-PCR (Louws et al. 1994) based on the presence of repetitive elements within the bacterial genome, the DNA amplification fingerprintings (DAF; Caetano-Anollés and Bassam 1993), RAPDs (random amplified polymorphic DNA; Williams et al. 1990, Welsh and McClelland 1990) and AFLPs (amplified fragment length polymorphism; Vos et al. 1995). Each method has advantages and disadvantages and the choice of the appropriate one depends on the expected degree of polymorphism within the population, the selection of the specific genomic region and the possibility of automation for screening of large samples. ITS, RAPD and AFLP have been shown to be particularly relevant for the study of genetic diversity within populations of bacteria belonging to the same or closely related species.

Plant Surface Microbiology
A. Varma, L. Abbott, D. Werner, R. Hampp (Eds.)

2 Materials for RAPD, AFLP and ITS

Equipment:

- Thermal cycler
- Gel electrophoresis apparatus with power supply, agarose and polyacrylamide (sequencing)
- Automated sequencer for capillary electrophoresis equipped with discrete band analysis software
- UV transilluminator and gel documentation system

Caution: UV rays are dangerous. Protect eyes with a plastic shield

Reagents and solutions:

- double distilled water (ddH_2O) sterilised by autoclaving. Prepare 100 µl aliquots before sterilisation and keep at –20 °C. Discard the aliquot after each use
- 50 mM $MgCl_2$ stock solution usually supplied with the *Taq* enzyme
- dNTPs stock solution (2 mM of each dNTP in ddH_2O)
- *Taq* DNA polymerase
- Restriction enzymes: *Eco*RI and *Mse*I, a single restriction buffer compatible with both enzymes
- T4 DNA ligase and specific ligation buffer, 5x stock solution supplied with the ligase enzyme
- Double-stranded adapters for AFLP (use 50 pmol for each adapter in the ligation mixture). The sequence of two single stranded oligonucleotides (5′–3′) corresponding to double-stranded adapters are reported. To prepare the double-strand molecule, incubate 100 pmol/µl of each oligonucleotide at 94 °C for 10 min and then slowly decrease the temperature down to 4 °C. Keep at –20 °C
  *Eco*RI adapter oligonucleotides:
  5′–CTC GTA GAC TGC GTA CC–3′
  5′–AAT TGG TAC GCA GTC TAC–3′
  *Eco*RI double stranded adapter:
  5′–CTC GTA GAC TGC GTA CC–3′
  5′–AAT TGG TAC GCA GTC TAC–3′
  *Mse*I adapter oligonucleotides:
  5′–GAC GAT GAG TCC TGA G–3′
  5′–TAC TCA GGA CTC AT–3′
  *Mse*I double stranded adapter:
  5′–GAC GAT GAG TCC TGA G–3′
  5′–TC TCA GGA CTC TA–3′
- Primers for AFLP (without selective bases): p*Eco*RI-T (5′–GAC TGC GTA CCA ATT C-T–3′), 5′ labelled with 6-carboxifluorescein (6-FAM); p*Mse*I-A (5′–GAT GAG TCC TGA GTA-A–3′), 5′ labelled with 4,7,2′,4′,5′,7′-hexachloro-6-carboxyfluorescein (HEX). Prepare 10 µM stock solution

- Primers for ITS amplification: FGPS1490 (5′–TGCGGCTGGATCACCTC-CTT–3′) and FGPS132′ (5′–CCGGGTTTCCCCATTCGG–3′), 10 µM stock solution in ddH_2O. These primers are selected for Rhizobia and may apply to other bacterial groups, however specific primers for particular genera can be designed on known 16S and 23S sequences retrieved from GenBank or Ribosomal Database (RDP)
- 10-base random primers for RAPD (series OP from Operon Technologies), 80 µM stock solution in ddH_2O. The choice of the primers is highly relevant for the usefulness of the results obtained (see the RAPD principles section for details)
- DNA size marker: good examples are a 100-bp ladder for agarose gel electrophoresis and TAMRA 500 (Applied Biosysem, PE) for capillary electrophoresis
- Genomic DNA: for RAPD and ITS concentrations at 10 ng/µl in ddH_2O. For the AFLP, use concentrations at 50 ng/µl. For general extraction protocols, see Bazzicalupo and Fancelli (1997). Alternatively, use BIO 101 DNA extraction kit

Note: All the above reagents should be kept at –20 °C

- TAE buffer: 40 mM Tris/Acetate, 1 mM EDTA, pH 8. Prepare a 50x stock solution
- ethidium bromide stock solution: 10 mg/µl; store in a dark bottle
- agarose
- 10x loading buffer: 70 % (w/v) glycerol, 0.5 % (w/v) bromophenol blue; store at 4 °C

Caution: Ethidium bromide is a powerful mutagen: wear gloves when handling this compound; wear mask when weighing it

3 RAPD

Principle

The RAPD assay (Welch and McClelland 1990; Williams et al. 1990) is a PCR amplification performed on genomic DNA template using a single short, arbitrary oligonucleotide primer and low annealing temperature, conditions that ensure the generation of several discrete DNA products. Each of these fragments is derived from a region of the genome that contains two primer binding sites on opposite strands and at an amplifiable distance. Polymorphism between strains results from sequence differences which inhibit or enhance primer binding or otherwise affect amplification. Single base mutations, insertions and deletions are molecular events that produce RAPD polymorphism. The large number of bands amplified with a single arbitrary primer generates a complex fingerprinting that can be utilised to detect relative differences in the random amplified DNA sequences from two different genomes. RAPDs have been applied to bacterial population genetics for several species which live in association with plants, such as *Sinorhizobium meliloti* (Paffetti et al. 1996; Carelli et al. 2000), *Burkholderia cepacia* (Di Cello et al. 1997) and *Pseudomonas* (Picard et al. 2000).

Although the sequence of RAPD primers is arbitrarily chosen, two basic criteria must be met: a minimum of 40 % G+C content (50–80 % G+C content is generally used) and the absence of palindromic sequences. Primers can also be purchased as a specific set for RAPD reactions from Operon Technologies (http://www.operon.com/).

Experimental procedure

In order to minimise the risk of contamination, the reaction should be prepared with a set of pipettes and tips used exclusively for this purpose in a clean environment (laminar flow hood being optimal) and wearing gloves.

1. Prepare a master mix in ice of those reagents common to all the programmed reactions, i.e. dNTPs, $MgCl_2$, primer, buffer and *Taq* DNA polymerase. Mix all reagents well. Prepare a quantity sufficient for the samples and for control reactions in which template DNA is omitted. Usually RAPD reactions are carried out in 25 µl total volume in the 0.2-ml PCR tube. The following concentrations are required:
   - Template DNA: 1 ng/µl
   - dNTPs: 200 µM
   - Primer: 6.4 µM
   - $MgCl_2$: 3 mM
   - *Taq* buffer: 1x strength
   - *Taq* DNA polymerase: 0.032 U/µl

   For 10 samples, 11 reactions should be prepared using the following volumes:
   - ddH_2O: 149.6 µl
   - 10x *Taq* buffer: 27.5 µl
   - 2 mM dNTPs: 27.5 µl
   - 80 µM primer: 22 µl
   - 50 mM $MgCl_2$: 16.5 µl
   - 2 U/µl *Taq* DNA polymerase: 4.4 µl
2. Aliquot the DNA (25 ng=2.5 µl) in the PCR tubes and then add the required volume of master mix (22.5 µl per tube)
3. Place the tubes in the thermal cycler and perform an initial denaturation step at 94 °C for 5 min
4. Cycle the reactions 45 times with the following temperature profile: denaturation at 94 °C for 1 min, annealing at 36 °C for 1 min and extension at 72 °C for 2 min. After the last cycle perform an extension step of 10 min
5. Store samples at 4 °C for a few hours (or –20 °C if longer)
6. Prepare a 2 % agarose gel in TAE buffer with 1 µg/ml of ethidium bromide. It is advisable to use a comb with teeth as thin as possible: the thinner the

teeth, the sharper the bands will appear. Caution: ethidium bromide is a mutagen! Wear gloves when handling

7. Add 1 µl of 10x loading buffer to 9 µl of each sample
8. Load the samples and the required amount of size marker
9. Run the gel at 10 V/cm for 1 h 30 min
10. Document the gel on UV transilluminator

Results and Comments

RAPD is a fast, cheap and powerful technique, which generates a high amount of polymorphism, being able to distinguish among isolates of the same populations in a very effective manner. The RAPD assay, according to the described protocol, generates reproducible fingerprints. Usually the size of RAPD products ranges from a few hundreds to about 2000 bp (Fig. 1). As a rule, the highest and lowest bands should be avoided as they are less reproducible. Before starting the analysis, a collection of primers, usually 20–30, should be tested on a selected subsample of strains in order to choose those that appear more suitable for the purpose and exclude those that did not show polymorphism. In general, four to six primers with different degrees of polymorphism are used for population analysis.

Troubleshooting

- Low polymorphism: use different primers.
- Reproducibility: use the same enzyme brand, the same thermal cycler for all the experiments. Poor quality or insufficient amounts of template DNA are most likely involved for low reproducibility. Perform RAPD reactions twice on at least some of the samples to check the reproducibility of all the recorded bands.
- Smearing: an excessive amount of template or primer or *Taq* DNA polymerase has most likely been used. Perform test reaction with reduced amount of each of these components at a time.

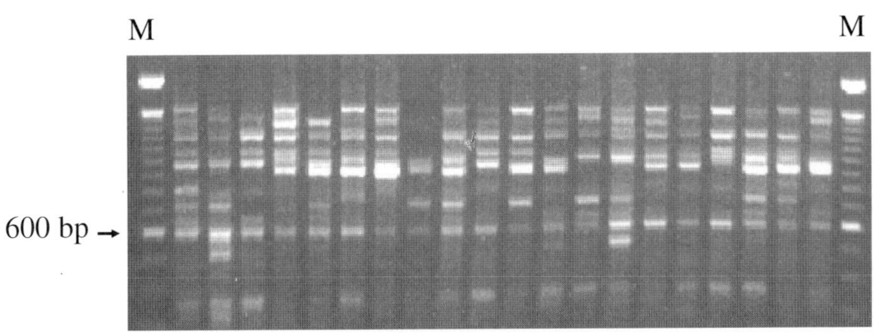

Fig. 1. RAPD pattern of different isolates of *Sinorhizobium meliloti*. *M* Ladder 100 bp (Life Technologies)

- Low intensity of the bands: insufficient amount of primer or dNTPs. Try increasing the amount of each one of these components at a time.

3 AFLP

Principle

Amplified Fragment Length Polymorphism (AFLP) is a recently developed technique based on restriction and amplification (Zabeau et al, 1993; Vos et al. 1995; Fig. 2). Using this method it is possible to generate up to 100 genomic markers with a single combination of restriction enzyme and primers. In particular, the application to microbial population analysis has been used to differentiate bacteria at strain and species levels from the taxonomic, phylogenetic or the population genetics point of view (Biondi et al. 2003). Moreover,

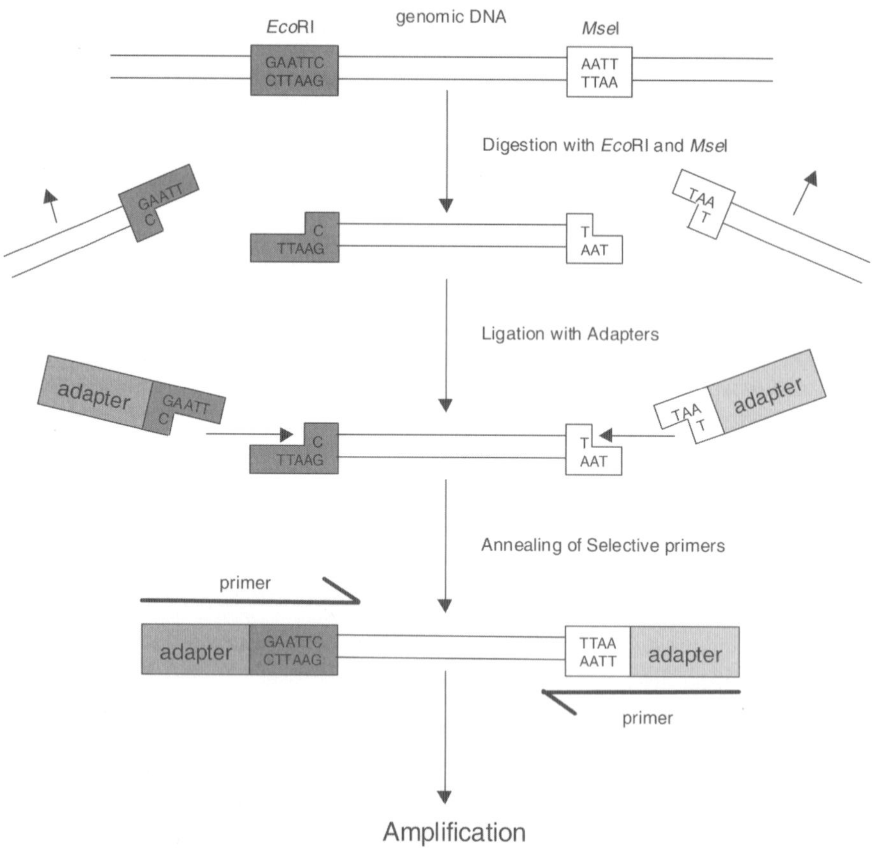

Fig. 2. Outline of AFLP technique. In *dark grey* the *Eco*RI restriction site and in *light grey* the *Mse*I restriction site. See text for details

the AFLP analysis can be applied to map phenotypic traits in eukaryotic organisms.

The genomic DNA is digested with restriction enzymes chosen to obtain fragments whose size is less than 1 Kb. After the digestion, all the fragments are ligated with adapters which recognise the digested ends. During this passage all DNA molecules acquire the same sequence at the ends. The ligated DNA is used as a template for PCR amplification; the primers used in this amplification are complementary to the adapter's sequence. Moreover, by adding one or two bases to the 3′ end of the primers sequence, it is possible to obtain different numbers of genetic markers: more selective bases result in the reduction of the number of amplified fragments. Finally, the detection of the amplified fragments and the estimation of their size can be obtained by two methods: polyacrylamide gel electrophoresis and capillary electrophoresis. This second method is more powerful and easier to handle and, therefore we will discuss only this method to analyse AFLP results.

Experimental procedures

1. Prepare the DNA using an extraction method that preserves the integrity of high molecular weight molecules (see material for reference)
2. Digest 200 ng aliquots of extracted genomic DNA in 25 µl final volume with 5 U of *Eco*RI and 5 U of *Mse*I using as enzyme buffer the *Mse*I buffer supplied by the manufacturer, incubate 2 h at 37 °C. Heat-inactivate the enzymes at 70 °C for 15 min
3. Ligate the adapters to the restriction products by adding 25 µl of the 2x ligation solution (1 unit of T4 DNA ligase, 50 pmoles of each adapter) to the digestion mixture (50 µl final volume) using double-stranded adapters with single-stranded overhang complementary to 5′ and 3′ ends generated during digestion. The ligation solution is incubated for 2 h at 20 °C
4. Perform the amplification reactions in a 50 µl total volume containing, 10x reaction buffer, 2.5 mM $MgCl_2$, 0.2 mM of each dNTP, 1.6 U of *Taq* DNA polymerase, 10 pmoles of each primer, 1 µl of template DNA (corresponding to approximately 4 ng of digested and ligated genomic DNA). For example: for 10 samples, consider a master mix solution for 11 single PCR reactions; add 1µl of template derived from the AFLP ligation to a 0.2-ml PCR tube; prepare a master mix (MM) with 408.1 µl of ddH_2O, 11µl of each primer solution, 55µl PCR buffer 10x, 27.5 µl of a 50 mM $MgCl_2$ solution and finally 9.9 µl of *Taq* DNA polymerase solution (3.5 U/µl); mix gently and aliquot 50 µl of the MM solution in each tube. The PCR conditions have been optimised in a Perkin-Elmer 9600 thermocycler (Perkin-Elmer, Norfolk, CT, USA), using the following amplification program: (94 °C for 30 s + 65 °C for 30 s' + 72 °C for 60 s) repeated for 13 cycles, decreasing the annealing temperature by 0.7 °°C each cycle and 23 cycles as follows: 94 °C for 30 s + 56 °C for 30 s + 72 °C for 60 s. Several combinations of primers can be selected, but good results were obtained with: p*Eco*RI-T (5′–GAC TGC GTA

CCA ATT C**T**–3′), labelled with 5′-6-carboxifluorescein (6-FAM) and p*Mse*I-A (5′–GAT GAG TCC TGA GTA A**A**–3′), labelled with 5′-4,7,2′,4′,5′,7′-hexachloro-6-carboxyfluorescein (HEX) (in bold the selective bases)

5. Check the amplification by running a 5 μl aliquot in a 1.5 % agarose gel
6. Size the product on an automatic capillary electrophoresis sequencer (Perkin-Elmer ABI 310 analyser). Load the capillary with 1.5 μl volume of AFLP PCR product and 0.5 μl of GenScan internal size standard TAMRA-500 (PE Biosystems) with 12.5 μl of deionised formamide and perform the electrophoresis as recommended by the manufacturer for fragment sizing.

Results and Comments

The AFLP technique usually produces a large amount of data (up to 100 different molecular markers) and for this reason it is recommended to use a computer-based system to manage the results. In this section we will discuss only the DNA sequencer output data analysis which gives data corresponding to fragments between 50 and 500 bp (range of TAMRA 500 molecular marker). The first step is the selection of useful data from the raw results. First, reject all peaks derived from single fluorochromes and analyse only the signals derived from both fluorochromes. After that, a threshold has to be introduced in order to continue only with real amplification signals. Usually only peaks having an intensity higher than 50 Fluorescence Units will be selected. After the selection, signals will be used to compute a distance matrix from which the genetic structure of the population can be analysed.

Troubleshooting

- Low intensity of the amplified AFLP bands: check the purity and the amount of DNA (100–300-ng range), try a different extraction method and different amount of DNA, test the reagents and the procedure with control DNA and control primers. Check that the primers used are correctly labelled. For the PCR reaction try a different amount of ligated DNA (1–4 μl) as template. Magnesium chloride concentration and annealing temperature are most likely involved in poor amplification, perform test reactions modifying the amount/value of these variables. Load different amounts of the PCR product on the automatic sequencer to optimise the fluorescent signal.
- Too many or too few bands: test different combinations of primers. If the bands are fewer than expected, remove the extra bases from the adapter complementary primers. On the contrary, if the bands are too many, add selective bases of up to two for each primer.

6 ITS-RFLP analysis

Principle

The 16S–23S rRNA intergenic transcribed spacer (Fig. 3; ITS, the spacer sequence between 16S and 23S rRNA bacterial genes synonymous with IGS, inter genic spacer) is a sequence that exhibits large variability useful in identifying genomic groups at the intraspecific level (Barry et al. 1991; Jensen et al. 1993; Laguerre et al. 1996; Doignon-Bourcier et al. 2000). The genetic variability of this particular region derives from: (1) the presence of t-RNA genes inside the ITS and (2) the mutation rate of ITS higher than that of ribosomal genes. Restriction fragment length polymorphism of PCR-amplified ITS (ITS-RFLP) is a fingerprinting method for the characterisation of bacterial strains with a higher discriminating power than the 16S rDNA RFLP (ARDRA method). For the amplification of the ITS, different primer pairs, designed on the coding regions of 16SrRNA and 23SrRNA genes, can be chosen depending on the bacterial group to be analysed. In general, the forward primer corresponds to the internal region of the 16S gene while the reverse primer corresponds to the beginning of the 23S gene. Information on primers for specific bacterial groups can be retrieved from the specific literature or from GenBank or RibosomalDataBase (http://www.ncbi.nlm.nih.gov/ or http://rdp.cme.msu.edu/html/). For the amplification of the ITS region of rhizobia we used primers FGPS1490 (Navarro et al., 1992) and FGPS132′ (Ponsonnet and Nesme 1994). FGPS1490 is designed on conserved sequences of the 3′ end of the 16S rRNA gene (corresponding to *Eschericha coli* numbering positions 1525–1541), and reverse primer FGP132′ is designed on the 5′ end of the 23S rRNA gene (corresponding to the *E. coli* numbering positions 115–132). For ITS-RFLP the amplified intergenic region is digested with four-base recognition site restriction enzymes in order to generate specific patterns of bands. Depending on the type of the samples and on the aim of the study, from two to five or more different restriction enzymes are used. The more enzymes used, the higher the number of bands, i.e. molecular markers produced. The restriction of the amplification product should be performed using enzymes which cut several times in the intergenic spacer, thus, before starting to analyse the IGS of a particular species, a number of enzymes should be tested to select the best combination. Some restriction enzymes frequently used are: *Alu*I, *Mse*I, *Hha*I, *Taq*I, *Sau*3A.

Fig. 3. Structure of the bacterial ribosomal operon showing the position of ITS region. Primers are indicated by *arrows*

Experimental procedures:

1. Perform a PCR amplification reaction in a 50 µl total volume containing, 10x reaction buffer, 2.5 mM $MgCl_2$, 0.2 mM of each dNTP, 1.6 U of *Taq* DNA polymerase, 10 pmoles of each primer (FGPS1490 and FGPS132′), 25 ng of template DNA concentrated to 25 ng/µl. For nine samples consider a master mix solution for ten single PCR reactions with the following volumes:
   - ddH_2O: 337µl
   - 10x *Taq* buffer: 50 µl
   - 2 mM dNTPs: 50 µl
   - 10 µM primer FGPS1490 : 10 µl
   - 10 µM primer FGPS132′: 10 µl
   - 50 mM $MgCl_2$: 25 µl
   - 2 U/µl *Taq* DNA polymerase: 8 µl
   - 1 µl of template in each tube before aliquoting 49 µl of the master mix
2. Cycle the reactions through the following temperature profiles: initial melting at 94 °C for 5 min followed by 35 cycles at 94 °C for 1 min, 55 °C for 55 s, 72 °C for 2 min. Perform a final extension step at 72 °C for 10 min
3. Analyse 5 µl of each amplification mixture by agarose gel (1.2 % w/v) electrophoresis in TAE buffer containing 1 µg/ml (w/v) of ethidium bromide. **Caution:** ethidium bromide is mutagenic: wear gloves when handling. The result of the electrophoresis will ensure that amplification has been successful and will also help to quantify the amount of amplified DNA
4. Digest approximately 500–600 ng (5–6 µl) of the amplified IGS, with 2 units of the restriction enzyme in a total volume of 15 µl for 2 h. Use the buffer and incubation conditions recommended by the manufacturer of the restriction enzyme. Inactivate the enzyme. Make a separate digestion for each restriction enzyme to be used
5. Resolve the reaction products (15 µl) by agarose gel (2.5 % w/v) electrophoresis in TAE buffer run at 10 V/cm and stained with 1 µg/ml (w/v) of ethidium bromide.
   Caution: ethidium bromide is mutagenic: wear gloves when handling

Troubleshooting

- Low intensity of the amplified ITS: check PCR reaction conditions. Magnesium chloride concentration and annealing temperature are most likely involved, perform test reactions modifying these variables.
- Partially digested products: excessive amount of amplified ITS, low restriction enzyme concentration, incubation time too short. Perform test reactions with a reduced amount of DNA, or add more restriction enzyme or incubate for a longer time.

7 Statistical analysis

Introduction

The studies of microbial population genetics with molecular methods are often characterised by an extremely high number of samples and by a high number of molecular markers. As a consequence, an immediate interpretation of the results can be difficult unless powerful statistical techniques are used in order to describe the structure of the populations and to highlight the contributions of its components (Mengoni and Bazzicalupo 2002).

Methods and Procedure

Statistical treatment of data in microbial population genetics include at least four different levels of analysis:

1. Quantification of the genetic diversity within the population
2. Measurement of genetic distances between strains
3. Analysis of the genetic structure
4. Analysis of the genetic relationships among populations.

Several methods can be used to address each of these points. Here, a brief summary of the principal parameters and software used is provided.

The molecular data obtained from RAPD, AFLP, and ITS-RFLP analyses are usually bands in a gel or peaks in a chromatogram. These data are transformed into a matrix of state binary vectors (molecular haplotype) for each individual isolate using a compiler such as Microsoft Excel or similar. Bands and peaks of equal sizes are interpreted as identical and intensity is not considered as a difference. The molecular haplotype of each isolate is expressed as a vector of zeroes (for the absence of the band) or ones (for its presence), assuming that bands represent independent loci.

1. The quantification of genetic diversity within the population can be done using several parameters. The most commonly used are the gene diversity, the average gene diversity over loci and the mean number of pairwise differences between haplotypes.
   The gene diversity is equivalent to the expected heterozygosity for diploid data. It is defined as the probability that two randomly chosen molecular haplotypes are different in the sample.
   The average gene diversity over loci is defined as the probability that two individuals are different for a randomly chosen locus. These two parameters vary from 0 (all isolates identical) to 1 (maximum diversity).
   The mean number of pair-wise differences simply calculates the mean number of differences between all pairs of molecular haplotypes in the population. The computation of these three parameters is performed with Arlequin software (Schneider et al. 1997).
2. For the measurement of the genetic distances between strains, several methods can be applied. The basic principle is the ratio of bands shared by

two strains with respect to the total ones. One commonly used parameter is the Euclidean distance whose formalisation is $E=n(1-2n_{xy}/2n)$, where n is the total number of bands of strain x and y and n_{xy} the number of bands shared by strains x and y. Another widely used parameter is the Nei's distance which, using the same notation, can be formalised as $D=1-[2n_{xy}/(n_x+n_y)]$, where n_x and n_y are the number of bands present in the strains x and y, respectively.

3. Examples of techniques for ordering the genetic diversity to analyse the genetic structure of a population are the analysis of molecular variance (AMOVA) and the principal component analysis (PCA). The AMOVA is a methodology for the analysis of variance which makes use of molecular data. AMOVA allows us to uncover the structure of the population and to test the validity of the hypotheses on the subdivision of the analysed population. AMOVA was designed by Excoffier, Smouse and Quattro in 1992 (Excoffier et al. 1992) as "an alternative methodology that makes use of available molecular information gathered in population surveys, while remaining flexible enough to accommodate different types of assumptions about the evolutionary genetic system" (Excoffier et al. 1992). Assuming that a set of samples belongs to different populations and that these populations could be arranged in genetically distinguishable groups, the aim of AMOVA is to perform statistical tests on the hypothesised genetic structure. A hierarchical analysis of variance splits the total genetic variance into components due to intra-population differences among individual samples, inter-population differences and inter-group differences.
   The PCA is an analysis in which a data set is searched for some significant independent variables, with respect to all possible variables. These variables are termed 'components' and interest attaches especially to the principal, or most important, components, hence the name 'principal component analysis'. The output of the analysis is a plot in which the samples are dispersed in a two- or three-dimensional space allowing the recognition of the clusterisation pattern with respect to one of the dimensions (components).
4. The genetic relationships among populations can be estimated as the results of AMOVA with respect to the variance between populations. The parameter of the genetic separation between populations is F_{ST} (Wright 1965) which derives directly from the analysis of variance. The F_{ST} values can be used to construct a matrix of distances whose representation takes the form of a dendrogram or tree. Two tree-building methods are applicable to the distance matrix: UPGMA and Neighbor-Joining (Saitou and Nei 1987). The UPGMA is based on a simple mathematical algorithm in which a step-wise clusterisation is made. The Neighbor-Joining method is a simplified version of a minimal evolution method; a star-like tree is made and then the topology is reconstructed on the basis of the minimisation of the overall length of the tree.

Software requirements

- Scoring of the bands: MICROSOFT EXCEL or similar.
- Quantification of genetic diversity within population: ARLEQUIN (Schneider et al. 1997).
- Measurement of the genetic distances between strains: ARLEQUIN (Schneider et al. 1997);
  NTSYS-pc (Rohlf 1990)
  RAPDistance (freely downloaded from http://life.anu.edu.au/molecular/software/rapd.html)
- Analysis of the genetic structure:
  (1) AMOVA: ARLEQUIN (Schneider et al. 1997);
  (2) PCA: NTSYS-pc (Rohlf 1990)
- Estimation of genetic relationships among populations: ARLEQUIN (Schneider et al. 1997)
  MEGA (Kumar et al. 1993)
  NTSYS-pc (Rohlf 1990)
  PHYLIP (Freely downloaded from
  http://evolution.genetics.washington.edu/phylip.html)

Many of these softwares exist either as DOS/Windows and as Mac versions. For ARLEQUIN a Linux version also has been developed.

8 Concluding Remarks

Several techniques for the analysis of genetic diversity of bacterial populations have been proposed. RAPD, ITS and AFLP are effective technologies able to show intra-population polymorphism and to detect phylogenetic relationships among strains belonging to the same or closely related bacterial species. RAPD is a suitable technique in that it is fast, cheap and the amount of polymorphism displayed is high. RAPD has the disadvantage of requiring accurate setting up of the conditions to obtain high reproducibility. ITS-RFLP analysis on the contrary, shows less polymorphism, which is linked to a defined DNA region, being more suitable to define phylogenetic relationships among strains.

AFLP shows some advantages over the other methods: (1) the high stringency of the PCR conditions gives robust reproducibility; (2) easy application to plant, animal and bacterial genomic DNA. AFLP requires more DNA than RAPD and ITS-RFLP and a more laborious procedure. Nevertheless, AFLP has a high informational content per single reaction, in fact, up to 100 different bands can be displayed in a single lane and the scoring can be done with an automatic sequencer.

References and Selected Reading

Barry T, Colleran G, Glennon M, Dunican LK, Gannon F (1991) The 16 S/23 S ribosomal spacer region as a target for DNA probes to identify eubacteria. PCR Methods Appl 1:51–56

Bazzicalupo M, Fancelli S (1997) DNA extraction from bacterial colonies. In: Micheli MR, Bova R (eds) Fingerprinting methods based on arbitrary primed PCR. Springer, Berlin Heidelberg New York, pp 41–46

Biondi EG, Pilli E, Giuntini E, Roumiantseva ML, Andronov EE, Onichtchouk OP, Kurchak ON, Simarov BV, Dzyubenko NI, Mengoni A, Bazzicalupo M (2003) Evolutionary relationship of *Sinorhizobium meliloti* and *Sinorhizobium medicae* strains isolated from Caucasian region. FEMS Lett 220:207–213

Caetano-Anollés G, Bassam BJ (1993) DNA amplification fingerprinting using arbitrary oligonucleotide primers. Appl Biochem Biotech 42:189–200

Carelli M, Gnocchi S, Fancelli S, Mengoni A, Paffetti D, Scotti C, Bazzicalupo M (2000) Genetic diversity and dynamics of *Sinorhizobium meliloti* populations nodulating different alfalfa varieties in Italian soils. Appl Environ Microbiol 66:4785–4789

Di Cello F, Bevivino A, Chiarini L, Fani R, Paffetti D, Tabacchioni S, Dalmastri C (1997) Biodiversity of a *Burkholderia cepacia* population isolated from the maize rhizosphere at different plant growth stages. Appl Environ Microbiol 63:4485–4493

Doignon-Bourcier F, Willems A, Coopman R, Laguerre G, Gillis M, De Lajudie P (2000) Genotypic characterization of *Bradyrhizobium* strains nodulating small Senegalese legumes by 16S-23S rRNA intergenic gene spacers and amplified fragment length polymorphism fingerprint analyses. Appl Environ Microbiol 66:3987–3997

Ellsworth DL, Rittenhouse KD, Honeycutt EL (1993) Artifactual variation in randomly amplified polymorphic DNA banding patterns. BioTechniques 14:214–217

Excoffier L, Smouse PE, Quattro JM (1992) Analysis of molecular variance inferred from metric distances among DNA haplotypes: application to human mitochondrial DNA restriction data. Genetics 131:479–491

Jensen MA, Webster JA, Strauss N (1993) Rapid identification of bacteria on the basis of polymerase chain reaction-amplified ribosomal DNA spacer polymorphisms. Appl Environ Microbiol 59:945–952

Kumar S, Tamura K, Nei M (1993) MEGA: Molecular Evolutionary Genetics Analysis, version 2.0. The Pennsylvania State University, University Park, PA 16802. Freely downloadable from: http://www.megasoftware.net/

Laguerre G, Mavingui P, Allard MR, Charnay MP, Louvrier P, Mazurier SI, Rigottier-Gois L, Amarger N (1996) Typing of rhizobia by PCR DNA fingerprinting and PCR-restriction fragment length polymorphism analysis of chromosomal. Appl Environ Microbiol 62:2029–2036

Louws FJ, Fulbright DW, Stephens CT, de Bruijn FJ (1994) Specific genomic fingerprints of phytopathogenic *Xanthomonas* and *Pseudomonas* pathovars and strains generated with repetitive sequences and PCR. Appl Environ Microbiol 60:2286–2295

Mengoni A, Bazzicalupo M (2002) The statistical treatment of data and the analysis of molecular variance (AMOVA) in molecular microbial ecology. Ann Microbiol 52:95–101

Navarro E, Simonet P, Normand P, Bardi R (1992) Characterization on natural populations of *Nitrobacter* spp. Using PCR/RFLP analysis of the ribosomal intergenic spacer. Arch Microbiol 157:107–115

Paffetti D, Scotti C, Gnocchi S, Fancelli S, Bazzicalupo M (1996) Genetic diversity of an Italian *Rhizobium meliloti* population from different *Medicago sativa* varieties. Appl Environ Microbiol 62:2279–85

Paffetti D, Daguin F, Fancelli S, Gnocchi S, Lippi F, Scotti C, Bazzicalupo M (1998) Influence of plant genotype on the selection of nodulating *Sinorhizobium meliloti* strains by *Medicago sativa*. Antonie Van Leeuwenhoek 73:3–8

Picard C, Di Cello F, Ventura M, Fani R, Guckert A (2000) Frequency and biodiversity of 2,4-diacetylphloroglucinol-producing bacteria isolated from the maize rhizosphere at different stages of plant growth. Appl Environ Microbiol 66:948–955

Ponsonnet C, Nesme X (1994) Identification of Agrobacterium strains by PCR-RFLP analysis of pTi and chromosomal regions. Arch Microbiol 161:300–309

Rohlf FJ (1990) NTSYS-pc. Numerical Taxonomy and Multivariate Analysis System. Version 2.0. Exeter Software, New York

Saitou N, Nei M (1987) The neighbour-joining method: A new method for reconstructing phylogenetic trees. Molec Biol Evol 4:406–425

Schneider S, Kueffer JM, Roessli D, Excoffier L (1997) ARLEQUIN: a software for population genetics data analysis. Version 1.1. University of Geneva. Freely downloadable from http://lgb.unige.ch/arlequin/

Vos P, Hogers R, Bleeker M, Reijans M, van de Lee T, Hornes M, Frijters A, Pot J, Peleman J, Kuiper M, Zabeau M (1995) AFLP: a new technique for DNA fingerprinting. Nucleic Acids Res 23:4407–4414

Welsh J, McClelland M (1990) Fingerprinting genomes using PCR with arbitrary primers. Nucleic Acids Res 18:7213–7218

Williams JGK, Kubelik AR, Livak KJ, Rafalski JA, Tingey SV (1990) DNA polymorphism amplified by arbitrary primers are useful as genetic markers. Nucleic Acids Res 18:6531–6535

Woods CR Jr, Versalovic J, Koeuth T, Lupski JR (1992) Analysis of relationships among isolates of *Citrobacter diversus* by using DNA fingerprints generated by repetitive sequence-based primers in the polymerase chain reaction. J Clin Microbiol 30:2921–2929

Wright S (1965) The interpretation of population structure by *F*-statistics with special regards to systems of mating. Evolution 19:395–420

Zabeau M, Vos P (1993) Selective restriction fragment amplification: a general method for DNA fingerprinting. Publication no. 0 534 858 A1. European Patent Office, Munich, Germany

29 Functional Genomic Approaches for Studies of Mycorrhizal Symbiosis

Gopi K. Podila and Luisa Lanfranco

1 Introduction

Mycorrhizal fungi, one of the principal biological components of the rhizosphere, interact with the roots of about 90 % of land plants to form different types of symbiotic associations (Smith and Read 1997). On the basis of the colonization pattern of host cells, two main types of mycorrhizas can be identified: ectomycorrhizas and arbuscular mycorrhizas. In the ectomycorrhizas, the fungus does not penetrate the host cells, whereas in endomycorrhizas the fungal hyphae form intracellular structures like coils or arbuscules (Smith and Read 1997). Mycorrhizal fungi are commonly beneficial due to a wide network of external hyphae that extend beyond the depletion zone, allowing host plants to have improved access to limited soil resources. On the other hand, mycorrhizal fungi receive carbon compounds from host plants to sustain their metabolism and complete the life cycle and this may lead to reductions in plant growth under some circumstances (Graham and Eissenstat 1998; Graham 2000).

While there is a considerable amount of knowledge based on the ecology and physiology of mycorrhizal fungi and their uses, the knowledge about cellular and molecular aspects leading to the growth and the development of a mycorrhizal fungus as well as the establishment of a functioning symbiosis is still limited (Harrison 1999; Martin et al. 2001; Podila et al. 2002). The development of molecular techniques has offered new opportunities: automatic high-throughput sequencing methods has made it possible to determine the complete sequence of even eucaryotic genomes. While many ectomycorrhizal fungal genomes are supposedly of reasonable size (Doudrick 1995), some mycorrhizal fungi including arbuscular mycorrhizal fungi (AMF), have a large genome size (Bianciotto and Bonfante 1992; Hosny et al. 1998; for a review Gianinazzi-Pearson 2001). The presence of repetitive DNA, regulative regions and introns makes the analysis of genomic sequences relatively complex. The sequencing of complete genomes for mycorrhizal fungi is still years away until better methods for application towards mycorrhizal fungi are available.

Plant Surface Microbiology
A. Varma, L. Abbott, D. Werner, R. Hampp (Eds.)

An appropriate approach to the study of mycorrhizal fungi is to understand the molecular process leading to the host recognition, development and functioning of mycorrhiza through the analysis of expressed sequences. With the advent of many high throughput techniques that have been successfully applied to the functional analysis of genes from many organisms, it is now possible to apply similar strategies to study the various aspects of the mycorrhizal symbiosis. In this chapter, we describe protocols leading to (1) expressed sequence tags (EST) and (2) macroarray techniques.

The EST methods allow for rapid identification of sequences through single-run sequencing of 250–700 bases on randomly picked cDNA clones. Comparisons with sequence databases frequently allow the assignation of potential functions to the corresponding gene products. Since its introduction (Adams et al. 1991), this technique has been successfully applied to several organisms to provide an overview of the gene repertoire expressed in a particular stage of development or in a particular tissue (Hofte et al. 1993; Nelson et al. 1997; Kamoun et al. 1999; Lee et al. 2002). The macroarray or membrane array methods allows the study of genome-wide expression patterns. Macroarrays require considerably less RNA for target preparation compared to microarrays and do not involve costly set-ups. With more refined protocols macroarrays can be as sensitive as microarrays and also are more easily accessible for academic laboratories (see Bertucci et al. 1999; Jordan 1998). Macroarrays are also more suitable for gene expression studies, where only small subsets of genes (unigene sets) need to be tested for their expression, for example, genes involved in carbon or nitrogen metabolism, signal transduction, ion transport, etc.

In this chapter, we describe the experimental procedures for the establishment of EST collections from mycorrhizal fungi and also macroarray-based techniques for gene expression profiling of symbiosis process. These procedures can be applied even in cases of limited amount of biological starting material.

2 Material and Methods

2.1 Equipment

Micro-centrifuge and high-speed centrifuge with proper rotors
Sterile hood
Chemical hood
–20 and –80 °C freezers
VP scientific 384 pin multiblot replicator
Thermal Cycler with a heated lid (Hybaid or Eppendorf Master Cycler or similar)
37 °C shaking incubator
37 °C gravity convection incubator

Boekel cooler or similar to maintain 16 °C temperature
Gel electrophoresis equipment and power supplies
Hybridization oven (Amersham or Fisher biotech or similar)
Variable volume pipettes
High resolution scanner with transparency adapter
Phosphorimager (Bio-Rad Personal Imager FX)

2.2 Biological Material

Biological material used for the RNA extraction is collected as quickly as possible, immediately frozen in liquid nitrogen and stored at –80 °C.

2.3 RNA Extraction

The protocol is modified from the one described by Chomoczynski and Sacchi (1987).

Reagents
Extraction buffer:
4 M Guanidinium thiocyanate
25 mM Na-citrate pH 7
0.5 % Na-laurylsarcosine
0.1 M β-mercaptoethanol (added just prior to use)
2 M Na-acetate pH 4
Phenol water-saturated
Chloroform:isoamylalcohol (49:1, v/v)
Isopropanol
8 M LiCl
80 % ethanol

Procedure
Grind the material (100 mg) in liquid nitrogen (with a pestle in a microfuge tube or in a mortar)

1. Add 200 µl of extraction buffer and place on ice.
2. Add 40 µl of 2 M Na-acetate pH 4, mix thoroughly by inversion.
3. Add 400 µl of phenol and mix by inversion.
4. Add 150 µl of chloroform/isoamylalcohol, mix by inversion and place on ice 10 min. Centrifuge at 10,000 g for 20 min at 4 °C.
5. Transfer the aqueous phase into a new tube and extract with an equal volume of chloroform/isoamylalcohol.
6. Transfer the aqueous phase into a new tube and add 1 volume of isopropanol.

7. Incubate 2 h at –20 °C.
8. Centrifuge at 10,000 *g* for 20 min at 4 °C. Remove the supernatant.
9. Wash the pellet with 80 % ethanol.
10. Resuspend in 50–100 μl of sterile water and store at –80 °C.

Note: *As an alternative to the RNA extraction protocol explained above, commercial kits from several companies are available. These usually have no need of phenol:chloroform manipulations and are relatively rapid. RNA obtained with these kits often needs to be treated with RNase-free DNase to remove DNA.*

DNase Treatment

Incubate the RNA sample in DNase 1× buffer (100 mM Tris-HCl pH 7.5, 10 mM $MgCl_2$, 1 % BSA) with units of DNase (RNase-free; Promega, Madison, WI, USA) for 30 min at 37 °C. Add EDTA for 2 mM final concentration. Extract with an equal volume of phenol/chloroform/isoamylalcohol (25/24/1; v/v/v). Precipitate the RNA with Na-acetate (0.3 M final concentration) and ethanol (2.5 volumes). As an alternative, to remove contaminant DNA, a precipitation with LiCl (final concentration 2 M, overnight at 4 °C) can be performed.

3 RNA Quantification

RNA quantification can be determined with a spectrophotometer (A 260/280) or fluorometer (Amersham Pharmacia Biotech). The quality of RNA should be checked on a denaturing agarose gel (Sambrook and Russel 2001) to make sure that the integrity of RNA is good.

3.1 Construction of a cDNA Library

There are many kits available for the construction of a cDNA library. If there is plenty of total RNA available to purify poly-A RNA, standard cDNA synthesis kits can be used such as lambda zap kits (Stratagene, CA, USA). However, if the availability of the amounts of RNA is limited, it is advisable to use a kit that can use either a small amount of total RNA or poly-A RNA to synthesize the cDNA library. Because the amount of tissue and RNA available from mycorrhizal tissues or mycorrhizal fungi is often limited, we describe here the method of synthesizing a cDNA library using the SMART cDNA library construction kit (Clontech, CA, USA). This kit can work on as little as 50 ng of total RNA since it uses an amplification step after the first strand cDNA synthesis that compensates for small amounts of starting RNA material.

3.1.1 cDNA Synthesis (Total volume: 10 ml)

1. Combine the following reagents:
   1–3 μl of RNA (50 ng–1 μg)
   1 μl SMART III Oligonucleotide (10 μM)
   5′AAGCAGTGGTATCAACGCAGAGTGGCCATTATGGCCGGG 3′
   1 μl CDS III/3′ PCR Primer (10 μM)
   5″ATTCTAGAGGCCGAGGCGGCCGACATG $-d(T)_{30}N*N$ 3′
   (N*:A, G or C; N: A, G, C or T)
2. Incubate at 70 °C for 2 min, snap cool the tube on ice for 2 min
3. Add
   2 μl x5 First strand buffer
   (250 mM Tris-HCl pH 8.3, 30 mM $MgCl_2$, 375 mM KCl)
   1 μl DTT (20 mM)
   1 μl SuperScript II 200 U/μl (Invitrogen, CA, USA)
4. Incubate at 42 °C for 1 h
5. First strand cDNA can be stored at –20 °C for up to 3 months

3.1.2 Long-Distance PCR and Synthesis of Double-Stranded cDNA

1. Combine the following reagents:
   2 μl first strand cDNA
   80 μl sterile H_2O
   10 μl cDNA PCR buffer
   2 μl dNTPs (10 mM)
   2 μl 5′PCR Primer (10 μM) 5′ AAGCAGTGGTATCAACGCAGAGT 3′
   2 μl CDS/3′ PCR primer
   2 μl 50x Advantage cDNA Polymerase Mix (Clontech, CA, USA)
   100 μl total volume
2. Run a PCR program on a thermal cycler (Perkin Elmer 2400/9600 with a heated lid) following these parameters:
   1 cycle: 5 °C 20 s
   18–26 cycles:
   95 °C 5 s
   68 °C 6 min

Note: *The number of cycles depends on the amount of RNA starting material. If 1 μg of RNA is used, usually 10–15 cycles should be enough. If you start with 0.05–0.25 μg total RNA, 25 cycles are recommended. It is critical not to over-cycle in order to retain the proportion of rare cDNAs. Over-cycling will result in a disproportionate amplification of abundant cDNAs.*

3. Check an aliquot (5 μl) of the PCR product (double-stranded cDNA) on a 1 % agarose gel: a smear of DNA fragments of molecular weight between 0.1 and 4 kbp should appear (Fig. 1). At this stage, the ds cDNA can be stored at –20 °C up to 3 months.

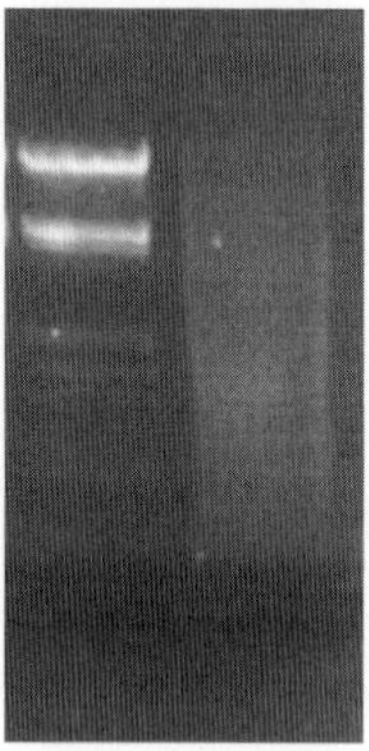

Fig. 1. Analysis of double stranded cDNA synthesis products. *Lane MW* is molecular weight markers in kilobase pairs. The bright smear ranging from 4–1 kb in *lane dscDNA* shows a good spread of cDNA fragment sizes

3.1.3 Reparation of cDNAs for Ligation: Proteinase K Treatment and SfiI Digestion

1. Transfer 50 µl of the ds cDNA into a new tube, add
   2 µl of proteinase K (20 µg/µl)
   Incubate at 45 °C for 20 min.
2. Add 50 µl of H_2O.
3. Mix contents and spin the tube briefly.
4. Incubate at 45 °C for 20 min. Spin the tube briefly.
5. Add 50 µl of deionized H_2O to the tube.
6. Add 100 µl of phenol:chloroform:isoamyl alcohol (25:24:1;v/v/v) and mix by continuous gentle inversion for 1–2 min.
7. Centrifuge at 10,000 *g* for 5 min to separate the phases.
8. Remove the top (aqueous) layer to a clean 0.5-ml tube.
9. Add 100 µl of chloroform:isoamylalcohol (24:1, v/v) to the aqueous layer. Mix by continuous gentle inversion for 1–2 min.
10. Centrifuge at 10,000 *g* for 5 min to separate the phases.
11. Remove the top (aqueous) layer to a clean 0.5-ml tube.
12. Add 10 µl of 3 M sodium acetate, 1.3 µl of glycogen (20 µg/ml) and 260 µl of room-temperature 95 % ethanol. Immediately centrifuge at 10,000 *g* for 20 min at room temperature.
13. Carefully remove the supernatant with a pipette. Do not disturb the pellet.
14. Wash pellet with 100 µl of 80 % ethanol.
15. Air-dry the pellet (~10 min) to evaporate residual ethanol.
16. Add 79 µl of deionized H_2O to resuspend the pellet.

***Note:** Proteinase K treatment is necessary to inactivate the DNA polymerase activity.*

17. *SfiI I digestion*
    Combine the following components in a fresh 0.5-ml tube:
    79 µl cDNA (Step 15, above)
    10 µl 10x *SfiI* I buffer
    10 µl *SfiI* I enzyme
    1 µl 100x BSA
    100 µl total volume
18. Mix well. Incubate the tube at 50 °C for 2 h.

Note: *SfiI I-digested cDNA should be fractionated to remove small fragments which would otherwise compromise the quality of the cDNA library.*

3.1.4 cDNA Size Fractionation by CHROMA SPIN-400

1. Label 16 1.5-ml tubes and arrange them in a rack in order.
2. Prepare the CHROMA SPIN-400 column (Clontech, CA, USA) for drip procedure:
   CHROMA SPIN column should be warmed to room temperature before use. Invert the column several times to completely resuspend the gel matrix. Remove air bubbles from the column. Use a 1000-µl pipette to resuspend the matrix gently; avoid generating air bubbles. Remove the bottom cap and let the column drip.
3. Attach the column to a ring stand. Let the storage buffer drain through the column by gravity flow until you can see the surface of the gel beads in the column matrix. The top of the column matrix should be at the 1.0-ml mark on the wall of the column. The flow rate should be approximately 1 drop/40–60 s. The volume of 1 drop should be approximately 40 µl.
4. When the storage buffer stops dripping out, carefully and gently (along the column inner wall) add 700 µl of column buffer to the top of the column and allow it to drain out.
5. When this buffer stops dripping (~15–20 min), carefully and evenly apply ~100 µl mixture of *SfiI* I-digested cDNA mixed with 2 µl xylene cyanol dye (1 %) to the top-center surface of the matrix.
6. Allow the sample to be fully absorbed into the surface of the matrix (i.e., there should be no liquid remaining above the surface).
7. With 100 µl of column buffer, wash the tube that contained the cDNA and gently apply this material to the surface of the matrix.
8. Allow the buffer to drain out of the column until there is no liquid left above the resin.
9. Place the rack containing the collection tubes under the column, so that the first tube is directly under the column outlet.
10. Add 600 µl of column buffer and immediately begin collecting single-drop fractions in tubes #1–16 (approximately 35 µl per tube). Cap each tube after each fraction is collected. Recap the column after fraction #16 has been collected.

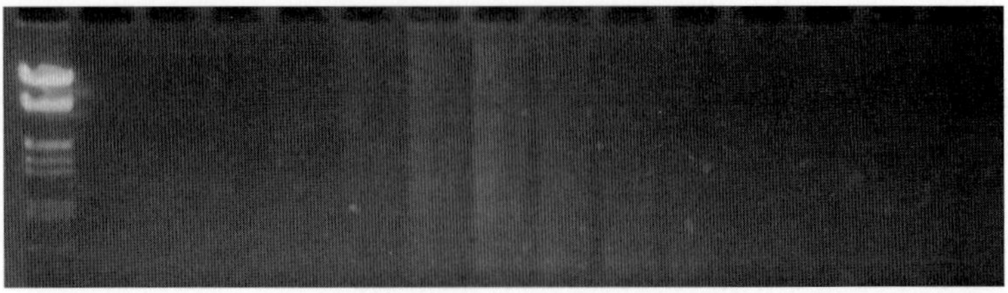

Fig. 2. Analysis of cDNA fractions on an agarose gel. In this particular case, fractions 6, 7, and 8 are collected as they seem to represent a good spread of cDNA sizes. *Lane MW* is the molecular weight markers in kilobase pairs.

11. Check the profile of the fractions before proceeding with the experiment on a 1.1% agarose/EtBr gel; run 3 μl of each fraction in adjacent wells, alongside a 1-kb DNA size marker (0.1 μg). Run the gel at 150 V for 10 min (running the gel longer will make it difficult to see the cDNA bands). Determine the peak fractions by visualizing the intensity of the bands under UV (see Fig. 2).
12. Collect the fractions containing cDNA fraction that matches your desired size distribution. Pool the above fractions in a clean 1.5-ml tube.
13. Add the following reagents to the tube with 3–4 pooled fractions containing the cDNA: (105–140 μl, respectively):
    1/10 vol sodium acetate (3 M; pH 4.8)
    1.3 μl glycogen (20 mg/ml)
    2.5 vol 95% ethanol (–20 °C)
14. Mix by gently rocking the tube back and forth.
15. Store the tube at –20 °C overnight.
16. Centrifuge the tube at 10,000 *g* for 20 min at room temperature.
17. Carefully remove the supernatant with a pipette. Do not disturb the pellet.
18. Briefly centrifuge the tube to bring all remaining liquid to the bottom.
19. Carefully remove all liquid and allow the pellet to air-dry for ~10 min.
20. Resuspend the pellet in 7 μl of deionized H_2O and mix gently. The *SfiI* I-digested cDNA is now ready to be ligated to the *SfiI* I-digested, dephosphorylated λTriplEx2 vector provided with the kit or the cDNA can be stored at –20 °C until the ligation step.

3.1.5 Ligation of cDNA to λTriplEx2 vector

Note: *The ratio of cDNA to vector in the ligation reaction is a critical factor in determining transformation efficiency, and ultimately the number of independent clones in the library. The optimal ratio of cDNA to vector in ligation reactions must be determined empirically for each vector/cDNA combination. To* *ensure that you obtain the best possible library from your cDNA, set up three parallel ligations using three different ratios of cDNA to vector, as shown below.*

1. Label three 0.5-ml tubes and add the indicated reagents. Mix the reagents gently; avoid producing air bubbles. Spin tubes briefly to bring contents to the bottom of the tube.

   Ligations using three different ratios of cDNA to phage vector

Component	1st ligation	2nd ligation	3rd ligation
cDNA	0.5	1.0	1.5
Vector (500 ng/µl)	1.0	1.0	1.0
10× Ligation buffer*	0.5	0.5	0.5
ATP (10 mM)	0.5	0.5	0.5
T4 DNA Ligase	0.5	0.5	0.5
Deionized H_2O	2.0	1.5	1.0
Total volume (µl)	5.0	5.0	5.0

   *x10 ligation buffer: 300 mM Tris-HCl, pH 7.8, 100 mM $MgCl_2$, 100 mM DTT
2. Incubate tubes at 16 °C overnight.

3.1.6 Packaging of Ligated cDNA and Preparation of cDNA Library

Perform a separate, λ-phage packaging reaction for each of the ligations as per manufacturer's instructions. We used Gigapack packaging extracts (Stratagene, CA, USA) and also MaxPlaq packaging extracts (Epicenter, WI, USA) with very good success.

1. Thaw three packaging extracts (50 µl per extract) on ice.
2. Immediately after the extracts have thawed, add 5 µl of each ligation mixture to one tube, mix gently.
3. Incubate at 22 °C for 4 h and add phage buffer (20 mM Tris-HCl, pH 7.4; 100 mM NaCl; 10 mM $MgSO_4$) to 250 µl and 10 µl of chloroform. Gently mix well and allow the chloroform to settle down. This packaged mix can be stored at 4 °C up to 4 weeks.
4. Titer each of the resulting libraries. From the three ligations combined, you should obtain 1–2x10^6 independent clones.

Note: *If you obtained <1–2x10^6 clones, you may want to perform another ligation with the remaining cDNA. Repeat the ligation using the ratio of cDNA to vector (of the initial three ligations) that gave the best results. Scale up the volumes of all reagents according to the amount of cDNA used. Then package and titer this scaled-up ligation.*

3.1.7 Titration of Packaged Phage Library

Usually, the cDNA synthesis kit is provided with the bacterial strains such as E. coli XL1-Blue to titer the library.

1. To recover the frozen cells, streak a small portion (~5 µl) of the frozen stock onto an LB agar plate containing the appropriate antibiotic. This is the primary streak plate. Use LB/tet for XL1-Blue stock plates.
2. Incubate at 37 °C overnight. Wrap plate in Parafilm and store at 4 °C for up to 2 weeks.
   To prepare a working stock plate, pick a single isolated colony from the primary streak plate and streak it onto another LB/$MgSO_4$ agar plate (with antibiotics).
3. Inoculate one colony into 5 ml of LB medium supplemented with 50 µl of 20 % maltose (filter-sterilized) and 50 µl of 1 M $MgSO_4$ solution. Shake at 160 rpm at 37 °C for 6–9 h or until OD_{600}=0.6.
4. Dilute each packaged library sample 1000×, 5000×, or 10,000× with phage buffer.
5. Add 20 µl of $MgSO_4$ solution and 2.8 ml of melted top agar to sterile glass tubes in a sterile laminar flow hood. Cap the tubes and keep them in a water bath at 50 °C for at least 30 min.
6. Mix 0.1 ml of the diluted phage with 0.1 ml of fresh bacterial cells in a microfuge tube and allow the phage to adsorb to the cells in an incubator at 37 °C for 30 min.
7. After incubation add the phage/bacterial mixture to the specified tubes of top agar in the water bath. Vortex gently to mix the contents and pour immediately onto LB agar plates. Rotate the plates and gently spread the top agar uniformly on the surface of LB agar.
8. Cool the plates at room temperature for 10 min to allow the top agar to harden. Invert the plates and incubate them at 37 °C for 6–18 h. Periodically check the plates to ensure that plaques are developing.
9. Count the plaque forming units (pfu) and calculate the titer of the phage

$$pfu/ml = number\ of\ plaques\ per\ plate \times dilution\ factor \times 10$$

Determining the percentage of recombinant clones

10. In λTriplEx2, as in many other λ expression vectors, the cloning site is embedded in the coding sequence for the a-polypeptide of β-galactosidase (lacZ). This makes it possible to use lacZ a-complementation (Sambrook and Russel 2001) to easily identify insert-containing phage by transducing an appropriate host strain (such as *E. coli* XL1-Blue) and screening for blue on medium containing IPTG and X-gal.
11. To perform blue/white screening in *E. coli* XL1-Blue, follow the procedure for titering on LB/$MgSO_4$ plates, but add IPTG and X-gal to the melted top agar before plating the phage + bacteria mixtures. For every 2 ml of melted top agar, use 50 µl each of the IPTG (10 mM stock) and X-gal (2 % stock). Aim for 500–1000 plaques/90-mm plate. Incubate plates at 37 °C for 6–18 h, or until plaques and blue color develop.

12. The ratio of white (recombinant) to blue (nonrecombinant) plaques will give a quick estimate of recombination efficiency. A successful ligation will result in at least 80 % recombinants.
13. Single plaques are isolated in 500 µl of SM buffer+20 µl of chloroform and stored at 4 °C.

***Note:** λTriplEx2 phages from isolated single plaques can be converted into pTriplEx2 plasmids by in vivo excision using the E. coli strain BM25.8 and following the manufacturer's instructions (Clontech, CA, USA). A brief protocol is given below.*

4 Conversion Protocol

1. Pick a single, isolated colony from the working stock plate of BM25.8 host cells (prepared similarly to XL-blue cells as described in phage titration protocol above) and use it to inoculate 10 ml of LB broth in a 50-ml test tube or Erlenmeyer flask. Incubate at 31 °C overnight with shaking (at 150 rpm) until the OD 600 of the culture reaches 1.1–1.4.
2. Add 100 µl of 1 M $MgCl_2$ to the 10-ml overnight culture of BM25.8 (10 mM final concentration of $MgCl_2$).
3. Pick each well-isolated plaque from the titration plates and place it in 350 µl of phage buffer in 96-well plates. Mix the contents thoroughly using an 8 or 12 channel pipette and allow phage to elute at 4 °C overnight.
4. In a deep 96-well plate combine 100 µl of overnight cell culture with 100 µl of the eluted plaque from each well. (Save the remainder of the eluted plaques in case you need to repeat the conversion.)
5. Incubate at 31 °C for 30 min without shaking.
6. Add 200 µl of LB broth and cover the plate with the lid provided.
7. Incubate at 31 °C for an additional 1 h with shaking (225 rpm).
8. Using a multichannel pipette transfer 1–5 µl of infected cell suspension into a 96-well LB/carbenicillin plate to obtain colonies and grow at 31 °C.
9. Pick bacterial growth from each clone and prepare plasmid DNA. For high throughput processing use Qiagen (Qiagen, CA, USA) or Eppendorf (Eppendorf, MA, USA) 96 format plasmid Miniprep kits. The isolated plasmid DNA should be pure enough for direct sequencing. The pTriplEx2 sequencing primers provided may be used with standard ds-DNA-sequencing protocols.

4.1 Evaluation of the Quality of the cDNA Library

To check the quality of the cDNA library two factors must be considered: (1) the number of primary recombinants (at least 10^5–10^6) and (2) the insert length. Insert size can be estimated by PCR with primers flanking the insertion site.

1. Insert DNAs can be PCR-amplified directly from bacterial colonies with oligonucleotides designed on sequences flanking the cloning site 5T (5′ CTCGGGAAGCGCGCCATTGTGTTGG 3′) and 3T (5′ ATACGACTCACTATAGGGCGAATTGGCC 3′). PCR reactions carried out in a final volume of 50 μl containing 10 mM Tris-HCl pH 8.3, 50 mM KCl, 1.1 mM $MgCl_2$, 0.01 % gelatin, 200 μM dNTPs, 50 pmol of each primer and 2 U of Red*Taq* DNA polymerase (Sigma, St. Louis, MO, USA). The following amplification program is run in a Hybaid thermal cycler: 3 min at 95 °C (1 cycle); 45 s at 92 °C, 45 s at 55 °C, 2 min at 72 °C (30 cycles).
2. PCR products should be separated by gel electrophoresis (Sambrook and Russel 2001).

5 Troubleshooting

Possible contaminations by ribosomal sequences (18S and 28S rRNAs might contain stretches of A that can complement oligo d-T).
Remedial actions: when a sufficient amount of total RNA is available, purify the poly-A RNA by using oligo d-T affinity columns (Sambrook and Russel 2001).
Inserts with small size
Remedial actions: enrichment of high molecular weight cDNAs through fractionation into a column.

6 Sequencing Strategies

Single run sequences of 250–700 bases are determined in most cases using an automated sequencer such as ABI Prism or Beckman CEQ 8 or other sequencers, whose accuracy has been estimated to be greater than 95 %. Because of the large number of sequences that needs to be processed in a relatively short time, it is advisable to outsource the sequencing process (private companies can do the service for a high number of sequences for a relatively low price). For people who have access to their own automated sequencers, we found Big Dye cycle sequencing kit from ABI (ABI, Foster City, CA, USA) or the DynamicET cycle sequencing kit from Amersham (Amersham Pharmacia Biotech, Piscataway, NJ, USA) give very good results. Both kits can be used to scale down the reactions to quarter reactions and still produce very good sequence reads, and makes it very economical. Since cDNAs are cloned into the pTriplex vector in a defined orientation, it is advisable to carry out the sequencing from the 5′-end first so as to obtain coding region information from each EST.

Note: *If you use quarter reactions, purity of plasmid DNA and quantity are critical. Sequencing of the 3′-end could help to identify cDNAs derived from the same mRNA, but which are truncated at different positions at the 5′-end.*

6.1 Data Analysis

Sequence similarity of each EST can be detected by BLASTX alignment (Altschul et al. 1997) of amino acid translations of the six possible open reading frames against the NCBI (National Center for Biotechnology Information) databank. Sequences of less than 100 bp should be removed from data analysis as these are usually not very useful in finding matches. Also, vector sequences or linker sequences should be filtered from the EST sequence before performing a similarity analysis. There are many programs such as DNA sequencher (Gene Codes, Ann Arbor, MI, US) that can automatically remove vector or linker sequences and clean up the EST sequences before thorough analysis.

6.2 Sequence Homology Comparisons

Organize all EST sequences into batches in Microsoft Word (Microsoft Corp., CA) in FASTA format. Batch nucleotide-protein searches can be done using BLASTX against all protein databases at GenomeNet http://www.blast.genome.ad.jp (Japan). cDNA sequences yielding an E value=10^{-5} can be considered to classify known genes or have partial similarity to known genes. BLAST scores should be further analyzed visually to confirm significant similarity and not solely determined by numerical value. In order to remove redundancies from EST sequences analysed, one can use Multalin, a multiple sequence alignment with hierarchical clustering, http://www.prodes.toulose.inra.fr/multalin/multalin.html (Corpet 1988). DNA sequences with approximately 90% identity to another clone may be eliminated as redundant sequences. The sequence with the most accurate information should be kept for further analysis. All nonredundants (sequences or ESTs) may be then submitted to GenBank at the National Center for Biotechnology Information (http://www.ncbi.nih.gov/Genbank).

6.3 Examples of Expressed Sequence Tag Data Analysis

6.3.1 Expressed Sequence Tags from the Asymbiotic Phase of an Arbuscular Mycorrhizal Fungus

A cDNA library was constructed from about 100 germinated spores of the endomycorrhizal fungus *Gigaspora margarita* (BEG 34). Insert lengths ranged from 100 to 800 bp with an average of 500 bp. Randomly selected cDNAs were characterized by sequencing at the 5′ end and comparison with databases. BLASTX searches were performed through the NCBI and clones were grouped on the basis of the E value (Table 1; Fig. 3).

Table 1. Selected list of EST clones from *G. margarita* and their similarity to known genes and E values to indicate the level of similarity.

Clone	BLASTX similarity	Species	E value
1	14-3-3 Protein	*Lentinula edodes*	e^{-101}
2	Endopeptidase	*Arabidopsis thaliana*	$1e^{-88}$
3	Heat shock protein HSS1	*Puccinia graminis*	$1e^{-71}$
4	Cell cycle switch protein	*Medicago sativa*	$7e^{-68}$
5	Protein involved in phosphate metabolism	*Saccharomyces cerevisiae*	$2e^{-52}$
6	Cu-Zn Superoxide dismutase	*Ovis aries*	$7e^{-51}$
7	Pre-mRNA cleavage factor	*Homo sapiens*	$4e^{-49}$
8	Spliceosome-associated protein	*Schizosaccharomyces pombe*	$6e^{-47}$
9	Aldehyde dehydrogenase	*Caenorhabditis elegans*	$3e^{-46}$
10	Polyubiquitine	*Neurospora crassa*	$5e^{-44}$
11	Histone H4	*Styela plicata*	$3e^{-39}$
12	Maleylacetate isomerase 2	*Drosophila melanogaster*	$1e^{-37}$
13	Cytidindeaminase	*Homo sapiens*	$1e^{-30}$
14	Glutathione S transferase	*Naegleria fowleri*	$7e^{-24}$
15	Ornithine carbamoyltransferase	*Aspergillus terreus*	$7e^{-21}$
16	Transcriptional factor StuA	*Aspergillus nidulans*	$4e^{-21}$
17	Subunit G of vacuolar ATP synthase	*Neurospora crassa*	$5e^{-20}$
18	Isocitrate lyase	*Dendrobium crumenatum*	$1e^{-20}$

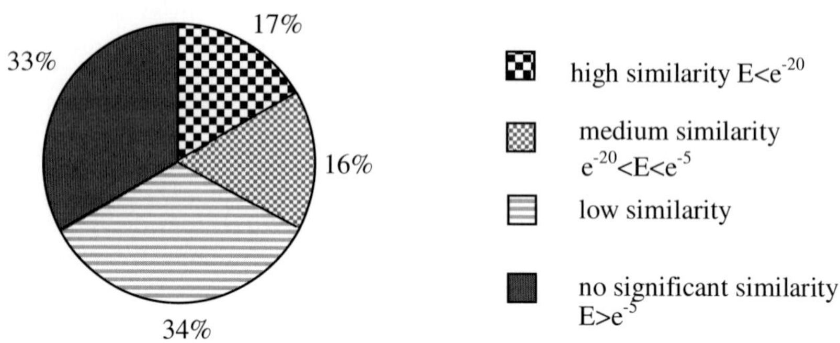

Fig. 3. Distribution of *G. margarita* EST clones based on the level of BLASTX similarity

ESTs presenting an E value $<e^{-5}$ were analyzed to estimate the percentage of clones showing similarity to proteins from different group of organisms: bacteria, fungi, plants or animals (Fig. 4). Most of the sequences showed similarity to proteins from the fungal kingdom (44%), although plants and animals were also well represented (21 and 30%, respectively). Clones were also grouped on the basis of their putative functions within the cell (Fig. 5) as deduced from the BLASTX searches. Most of the sequences were related to basal metabolism and protein synthesis. Interestingly, 9% of the clones

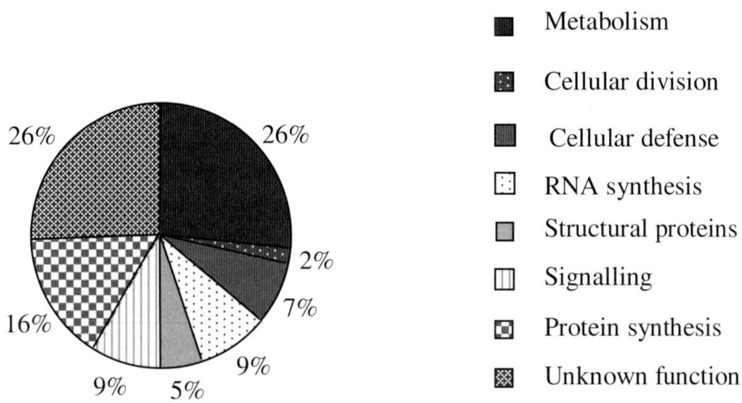

Fig. 4. Distribution of *G. margarita* EST clones based on matches to various organisms

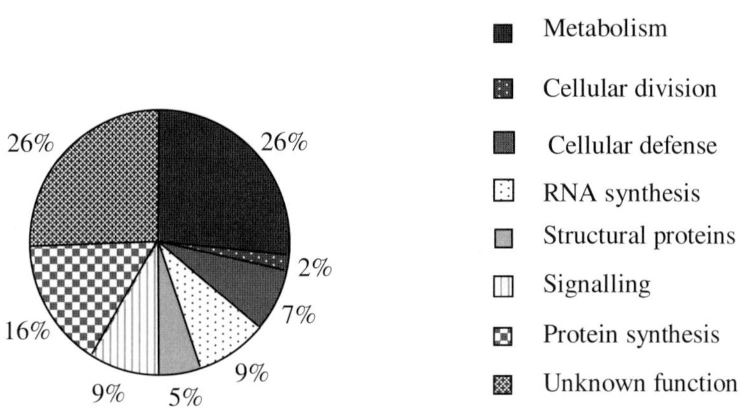

Fig. 5. Distribution of *G. margarita* EST clones based on similarity to functional groups using BLASTX analysis

showed similarity to proteins involved in defence responses to stresses. This result could suggest that the in vitro growth conditions are not favorable and could mimic a stress situation.

6.3.2 Expressed Sequence Tags from Early Symbiotic Interactions Between the Ectomycorrhizal Fungus Laccaria bicolor and Red Pine

Over 500 random EST clones from a cDNA library made from pooled RNA samples from various stages of interaction were sequenced. Out of these, over 400 nonredundant clones were obtained based on sequence analysis. Based on the BLAST analysis (Altschul et al. 1997), 33 % of the clones showed no significant similarity to any sequences in the NCBI database. The sequences of the remaining 67 %, however, suggested that they were homologues of genes previously identified in other systems. These were classified into groups based on their probable function. About 13 % were related to signal transduction, 15 % to metabolism, 10 % to cellular protein synthesis/processing and turnover, 9 % to transport and movement of ions/peptides and amino acids, 7 % to structural proteins, 5 % to RNA/DNA processing, 6 % to transcriptional regulation, and about 10 % to hypothetical proteins with no known function. In addition, one clone sequence suggested it was related to apoptosis. The majority of the matches for *L. bicolor* ESTs came from animal systems rather than fungi.

7 Macroarrays

7.1 PCR Amplification of cDNA Inserts

cDNA inserts from plasmid templates of EST clones need to be amplified, purified and quantified before used for printing macroarrays. The following protocol describes the general methods to obtain PCR products for printing macroarrays. Due to the large numbers of clones to be amplified, it is best to use 96-well formatted PCR plates, which will also facilitate printing macroarrays using either 96 or 384 pin manual or robotic arrayer. Conversely, 8 or 12 PCR strip tubes can also be used for rapid manipulation in setting up the PCR reactions.

1. For each 96-well plate to be amplified, prepare a PCR reaction mixture containing the following ingredients:

1000 µl	10× PCR buffer
20 µl	dATP (100 mM)
20 µl	dGTP (100 mM)
20 µl	dCTP (100 mM)
20 µl	dTTP (100 mM)
5 µl	forward primer* (1 mM)
5 µl	reverse primer* (1 mM)
100 µl	Red-Taq polymerase (1 U/µl)
8800 µl	H_2O

Note: ** primers used for PCR amplification depend on the vector in which the cDNA inserts are. Keep all reagents on ice and return the enzyme tube promptly to the freezer.*

2. Label 96-well PCR plates and aliquot 100 µl of PCR reaction mix to each well. Gently tap plates to insure that no air bubbles are trapped at the bottom of the wells.
3. Add 1 µl (10 ng) of purified EST plasmid template to each well. Mix well with pipette.

Note: *Mark the donor and recipient plates at the corner near the A1 well to facilitate correct orientation during transfer of the template. It is important to watch* *that the pipette tips are all submerged in the PCR reaction mix when delivering the template. Mixing the liquid is easier when multi-channel pipettes are used. Always use sterile filtered tips to avoid contamination.*

4. Replace PCR plate covers and centrifuge the plates at 2700 rpm for 1 min.
5. Place the PCR plates in a thermal cycler (such as Eppendorf Master Cycler) and run the following cycling program.

Initial denaturation	96 °Cx×2 min
Denaturation	94 °Cx30 s×30 cycles
Primer annealing	55 °Cx30 s×30 cycles
Primer extension	72 °Cx2 min×30 cycles
Final extension	72 °Cx5 min

***Note:** After PCR, plates can be held at 4 °C while quality controls are performed on PCR products. To check the quality of the amplified products, analyze 2 µl each of PCR products on 2% TAE agarose gel as described (Sambrook and Russel 2001). Take a digital photo of the gel on a UV table and store the image for future reference. The gels should show bands of fairly uniform brightness distributed in size between 600 and 2000 base-pairs depending on the sizes of cDNAs amplified. Further computer analysis of such images can be carried out with image analysis packages to provide a list of the number and size of bands. Ideally this information can be made available during analysis of the data from hybridizations involving these PCR products*

7.2 Purification and Quantification of PCR Products

1. Spin down PCR reaction plates and then transfer the PCR products (100 µl) to a Multiscreen filter plate and place the filter on a vacuum manifold filtration system (e.g., Millipore; Cat # MAVM0960R).
2. Apply a vacuum pressure of approx. 10–15 in. Hg (250–380 mm Hg) for 10 min or until plate is dry.
3. Remove plate from manifold filtration system and add 100 µl of MilliQ water to each well. Place filter plate on a shaker and shake vigorously for 20 min to resuspend the DNA.
4. Pipette the purified PCR product to a new U-bottom 96 well plate. Seal PCR storage plates with a plastic cap mat or adhesive foil lid and store at –20 °C until needed for printing macroarrays.

7.3 Printing of Macroarrays

1. Transfer PCR products to 384-well source plate at a concentration of 100 ng/µl.

***Note:** The concentration of the source plate is critical, we have shown spot intensity is directly related to source plate concentrations.*

2. Include a group of negative and positive control clones with the other clones to be printed onto the membrane.
3. Denature the PCR products in 5× denaturation solution (2 N NaOH, 50 mM EDTA) diluted to 1x final concentration at 37 °C for 30 min (Jordan 1998).
4. Presoak Hybond N+ nylon membrane filters (Amersham Pharmacia Biotech, Piscataway, NJ, USA) in 0.1 M NaOH for 1 min, then place on 3-mm filter paper.

Note: *When using a manual arrayer, one layer of filter paper on top of a mouse pad seems to be an optimal surface for printing. Spotting can be done with a 384-pin dot-blot tool (V&P Scientific, San Diego, CA, USA) with 0.9–0.5 mm diameter flat tip pins. If you are going to print many copies, use a multi-print replication device (V&P Scientific, San Diego, CA, USA) to give consistent alignment between membranes and to allow spacing for printing up to 4x384 spots on the same membrane (Schummer et al. 1997).*

5. Dip the pins into the 384-well plate containing the denatured DNA. The pins will deliver ~50 nl of sample yielding spots consisting of approximately 5 ng (the linearity of delivery can be tested by using different concentrations of PCR products in the same volume).
6. Cross-link membranes for 30 s in UV-crosslinker at optimal setting (Fisher Scientific, Pittsburgh, PA).
7. Neutralize the array for 5 min in a solution of 0.5 M Tris pH 7.8, 1.5 M NaCl followed by rinsing in ddH_2O for 5 min.
8. Air dry the arrays on filter paper and wrap in Saran wrap until use.

7.4 Generation of Exponential cDNA Probes from RNA for Macroarrays and Hybridization Analysis

We found the protocols described by Gonzalez et al. (1998) work very well. We use components from SMART cDNA synthesis kit (Clontech, CA, USA) for this purpose.

1. Assemble the following in a 0.2-ml PCR tube

0.5–1 µg RNA	1 µl
10 µM oligo dT primer (CDS from SMART cDNA kit)	1 µl
10 µM of SMART IV oligonucleotide	1 µl
ddH_2O	2 µl

Heat the mixture to 70 °C for 2 min, spin briefly and cool at room temperature to anneal the primers.

2. Add

5x Reverse transcription buffer	2 µl
20 mM DTT	1 µl
10 mM dNTPs	1 µl
Powerscript Reverse Transcriptase 200 U (Clontech, USA)	1 µl
Total volume	10 µl

Incubate at 42 °C for 1 h.
Add 40 µl of TE buffer (10 mM Tris-HCl, pH 7.2, 1 mM EDTA) to stop the reaction. All reactions can be done in a PCR machine.

3. To determine the number of cycles required to obtain a population of representative dscDNAs, 1 µl from each sscDNA reaction should be amplified following the protocol given below.

sscDNA	1 µl
ddH_2O	41 µl
10× PCR buffer	5 µl
10 µM PCR anchor primer	1 µl
10 mM dNTPs	1 µl
Advantage Taq 2 U (Clontech, USA)	1 µl
Total volume	50 µl

4. Set up three reactions and amplify for 17, 20, and 23 cycles (95 °C for 15 s, 65 °C for 30 s, and 68 °C for 6 min).
5. Run aliquots from each reaction on an agarose gel and stain with ethidium bromide. Select the cycle number before the reaction's plateau.

7.5 Exponential Amplification of the sscDNAs

1. Amplify 2 µl of sscDNA from the RT reactions using the number of cycles selected from above. Use same conditions for amplification.
2. Clean the PCR products using QIAquick columns (Qiagen, Chatsworth, CA, USA) into a final volume of 50 µl.

8 Generation of Radiolabeled Probes

1. Denature the dscDNAs prepared above by heating the tube in boiling water for 5 min and snap-cool the tube on ice.
2. Add to Prime-A-Gene (Promega, Madison, WI, USA) random primer labeling mixture containing 50 µCi each of ^{32}P-dATP and ^{32}Pd-CTP in a 50 µl reaction volume with twice as much Klenow DNA polymerase than the kit recommends.
3. Incubate at 37 °C for 2 h and purify the probes using Qiagen nucleotide removal kit (Qiagen, Chatsworth, CA, USA) as per manufacturer's instructions.

Note: *The double labelling with ^{32}P-dATP and ^{32}P-dCTP not only helps in getting high specific activity targets, but also eliminates problems associated with labeling GC-rich sequences.*

9 Hybridization of Macroarrays to Radiolabeled Probes

1. Prehybridize membranes in 10 ml of prehybridization solution (5×SSC, 10×Denhardt's solution, 0.5 % SDS, 100 μg/ml sheared salmon sperm DNA), at 65 °C for 4 h.
2. Add denatured probe and continue incubation for 22 h at 65 °C.
3. Wash the hybridized membranes successively for 3x5 min in 2xSSC at room temperature, 2x20 min in 2×SSC containing 0.5 % SDS, 2x20 min in 1xSSC containing 0.1 % SDS, and 2x20 min in 0.1xSSC containing 0.1 % SDS, all at 65 °C.

Note: *All washes are recommended even if signal intensity seems to drop.*

4. Wrap the membranes in Saran wrap and expose to X-ray film (Kodak Biomax MR) at –80 °C for varying periods (7 h to 3 days).

Note: *Exposing membranes to film without an intensifying screen yields clearer spots.*

5. Alternatively, capture the image on a Kodak storage phosphor screen (Eastman Kodak Company, Rochester, NY, USA) and scan the screens using a Bio-Rad FX Phosphorimager (Bio-Rad, Hercules, CA) at 100 μM resolution.

Note: *It is preferable to use a Phosphorimager as it produces better resolution and automates the acquisition of data from macroarrays for downstream processing.*

10 Data Analysis

Using Phosphorimager:

1. Transfer the raw image data obtained with the phosphorimager imaging system into a computer.
2. Define each spot on the image by making a grid using QUANTITY ONE software (Bio-Rad, Hercules, CA, USA).
3. For each image, determine the average pixel intensity (representing the hybridized DNA) within each spot in each grid square.
4. Generate a data table using QUANTITY ONE and export the data to Excel worksheet (Microsoft Corporation, Redmond, WA, USA).
5. Calculate background for each membrane by averaging over ten positions on the image where there are no DNA spots.
6. Calculate net signal for each spot by subtracting the average background value from the spot intensity.

Note: *If any spot values fall below the set threshold value (twofold less than the background) assign a arbitrary value of 0.1.*

7. Probe to probe variance can be filtered out using signal intensities of positive or negative controls used in the macroarray. In addition, to take into account experimental variations in specific activity of the cDNA probe

preparations or exposure time that might alter the signal intensity, normalize the data obtained from different hybridizations by dividing the intensity for each spot by the average of the intensities of all the spots present on the filter, to obtain a centered, normalized value (Eisen et al. 1998).

8. It is important to take average values from multiple experiments to reduce the variation from experiment to experiment. The final data can be analyzed using Cluster and Treeview software (http://rana.lbl.gov) to obtain more normalized data.

Note: *You can also use k-means analysis and hierarchical clustering on the net via the website: http://ep.ebi.ac.uk/EP/EPCLUST/ dedicated to statistical analysis of gene expression data from macroarrays.*

10.1 Data Analysis Autoradiography Images on X-ray Films

Note: *If X-ray film is used to capture the image, it is important to do multiple time exposures to obtain more reliable spot intensities for further analysis. It is also important not to overexpose the X-ray film where the signal intensities are not saturated. This will prevent the calculation of any subtle differences in the expression levels.*

1. Scan X-ray films at high resolution (1200x1200 dpi) using a scanner with transparency adapter and save the images as TIFF files.

Note: *These images take up a substantial amount of hard disk space (on average 25–30 MB).*

2. Open the image in a quantification program such as One D-scan (Scanalytics Inc., Fairfax, VA, USA).
3. Scale down the image to fit the screen using the scale and rotate option.
4. Draw a grid over the image using a preset size of 16 rows x 24 columns and a numbering scheme to match the EST database.
5. Place the grid such that all spots are in the center of each cell. It is possible to remove segments if artifacts or defects or over-intensity occur on the image where the neighboring spots may overlap.
6. Calculate the spot intensity values using the volumes option in the analyzing tool bar.
7. Calculate background using the boundary of each segment option, this takes a value from each pixel bordering the cell and averages them to yield the background value.
8. Using the volumes tool create a spread sheet of the data containing the segment number with it's background value and volume, along with high and low values found within each cell.
9. Calculate the volume, or intensity, by adding the intensity of each pixel within a given segment.
10. Transfer these data to Microsoft Excel (Microsoft Corp., CA) for further manipulation.

11. Calculate a normalized value for each segment by dividing each spot by the average intensity of all spots on the film to account for probe–probe variance (Eisen et al. 1998).
12. Correct all values by subtracting the average intensity of the four negative controls on the membrane.
13. Calculate fold increase or decrease in expression and create another column for these data.
14. Now the data can be incorporated into a graphical form between the control and treatments by giving the control a value of 1. Data for a treatment can then be combined into one spreadsheet along with control data and imported into GeneCluster at:
http://www-genome.wi.mit.edu/cancer/software/software.html.

This output shows genes that are expressed at approximately the same ratios, thus creating a cluster of genes that are most likely to be linked in some biochemical pathway or genes that are relevant to interaction response.

11 Example of *Laccaria bicolor* Macroarrays

We examined quantitative changes in the expression of ESTs from *L. bicolor* using the membrane array technique. We prepared a membrane array consisting of 384 EST clones selected from *L. bicolor* interaction cDNA library (Kim et al. 1999) and probed it with control free-living mycelium mRNA probes and mRNA probes prepared from various time points of preinfection stage interaction with red pine. A typical membrane array image obtained after hybridization with control mRNA and 72-h interaction mRNA probes is

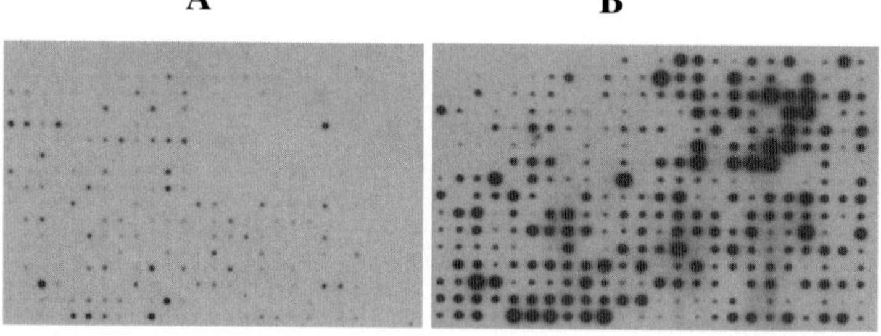

Fig. 6. Example of *L. bicolor* EST macroarray prepared using the 384 pin manual unit and expression profiling of interaction-related gene expression in *L. bicolor*. Macroarrays were printed using 0.9-mm diameter pins containing 384 ESTs and hybridized to probes prepared from RNA from free living *L. bicolor* (**A**) or from 72 h interaction (**B**). Image obtained from X-ray films exposed to the hybridized membrane. Image is captured using Saphir high-resolution scanner (Linotype-Hell, Heidelberg Inc. NY. USA) at 1200x1200 dpi. Quantification of signal intensities from spots and gene expression levels is determined using Mac 1-D software (Scanalytics, VA, USA)

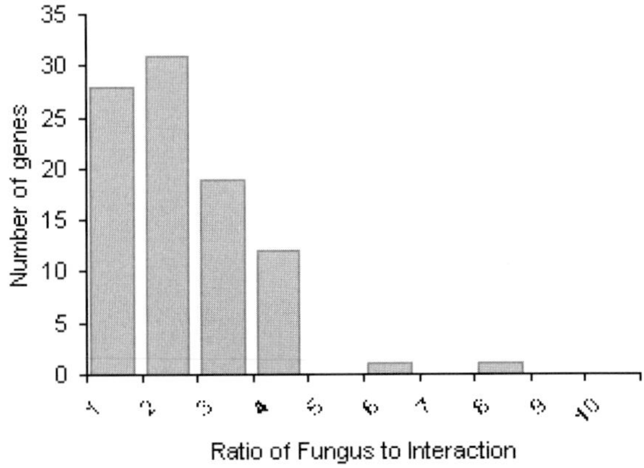

Fig. 7. Histogram showing clustering of genes from macroarray analysis based on levels of expression. The interaction (72 h)-related expression is shown on the *x-axis* and the number of genes that are upregulated at a given expression ratio is shown on the *y-axis*. Only genes that are upregulated are shown from the macroarray in Fig. 6A. For identification of some of the genes upregulated by interaction see Table 2

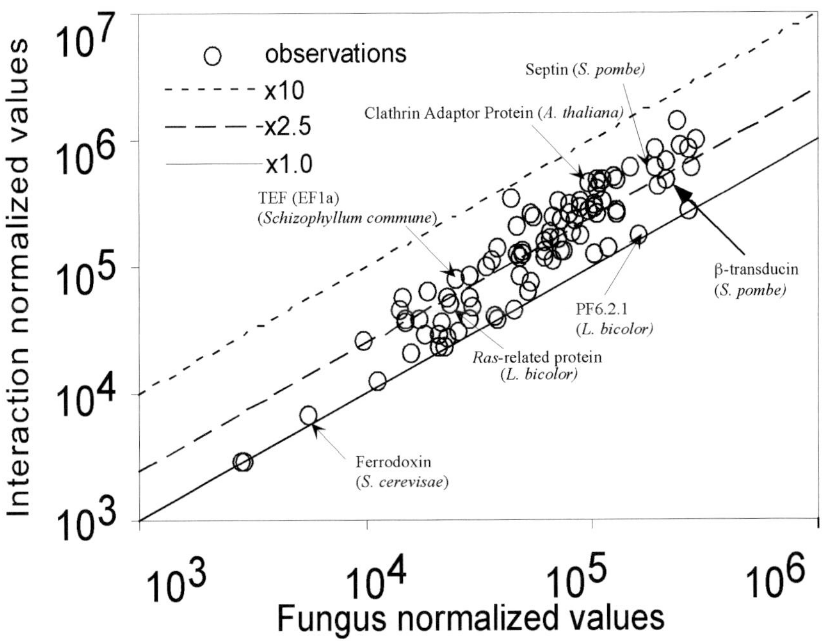

Fig. 8. Scatter plot analysis of interaction-specific genes between free-living *Laccaria bicolor* and *L. bicolor* interacted 72 h with *Pinus resinosa* seedling roots. For each gene, transcript levels were calculated for the free-living mycelium and the mycelium that interacted with red pine seedling root signals. *Solid lines* indicate an expression level of onefold or above the free-living mycelium, *dashed lines* 2.5–6-fold increase and *dotted line* eightfold or higher levels of expression. Only nonredundant clones are represented in the plot. Selected clones that showed significant differential expression are *highlighted*

Table 2. Differential expression of selected interaction clones from *L. bicolor*. Clones are selected from macroarray analysis. The E ratio was calculated by comparing the expression levels in the interacted fungal tissue with those in the free-living fungus using macroarrays

GenBank no.	E value	Database match	Expression ratio
BI094576	3e^{-69}	BiP protein (*Aspergillus nidulans*)	4.10
BI094582	2e^{-69}	PF6.2.1 (*Laccaria bicolor*)	8.00
BI094583	2e^{-48}	α-tubulin (*Ustilago maydis*)	4.51
BI094587	1e^{-10}	Homeobox genes Hox-2.6 (*Mus musculus*)	3.30
BI094592	2e^{-10}	PEP carboxykinase (*Mus musculus*)	4.33
BI094601	2e^{-63}	LbAut7 (*L. bicolor*)	3.73
BI094606	3e^{-09}	β-importin (*Schizosaccharomyces pombe*)	4.43
BI094612	2e^{-35}	Malate synthase (*L. bicolor*)	3.82
BI094615	9e^{-36}	TEF (EF1a) (*Schizophyllum commune*)	3.61
BI094619	1e^{-06}	IRS 1-like protein (*Xenopus laevis*)	3.13
BI094621	1e^{-31}	Ras related protein (*L. bicolor*)	3.81
BI094622	3e^{-86}	AAD (*Phanerochate. chrysosporium*)	2.61
BI094623	1e^{-06}	LZK protein kinase (*Homo sapiens*)	2.90
BI094629	7e^{-13}	Lactonohydrolase (*Fusarium oxysporum*)	3.91
BI094632	9e^{-27}	E-MAP-115 (*H. sapiens*)	4.21
BI094635	1e^{-26}	SUG1 subunit 8 (*S. cerevisiae*)	4.61
BI094639	1e^{-16}	Septin Spn3 (*S. pombe*)	3.92
BI094653	5e^{-62}	Rho GTPase (*S. cerevisiae*)	3.21
BI094657	1e^{-10}	Clathrin adapter protein (*A. thaliana*)	2.08
BI094660	6e^{-21}	AcetylCoA acetyltransferase (*L. bicolor*)	4.25
BI094667	1e^{-20}	β-transducin (*S. pombe*)	4.11
BI094676	2e^{-37}	Chitin synthase I (*U. maydis*)	3.71

shown in Fig. 6. An E ratio that indicates the relative increase in the expression of each gene in the interaction over the free-living state is used to quantitate differential expression. There is an overall increase in levels of expression of several clones tested (Fig. 7). The scatter plot of the normalized data from signal analysis of the membranes is presented in Fig. 8, which shows global changes in the expression of interaction related genes. Levels of expression of selected genes from 72-h interaction are listed in Table 2.

12 Conclusions

The EST and macroarray approaches provide efficient tools for mycorrhizal symbiosis research. These approaches have the resolution and ability to obtain a more comprehensive view of various stages of mycorrhiza development or treatment effects due to nutritional changes or differences due to host responses. In addition, since they require a relatively modest budget, compared to genome sequencing or microarray-based methods, they are easily accessible for many academic research groups. With increased use of these techniques using a variety of mycorrhizal symbiosis models, data can be exchanged and compared between different laboratories and eventually will provide a platform to understand the key players (genes) that are markers for ectomycorrhizal or AM fungal symbioses. In the last couple of years, several laboratories have begun using these approaches to unravel the mycorrhizal symbiosis (Martin et al. 2001; Voiblet et al. 2001; Podila et al. 2002; Polidori et al. 2002).

References and Selected Reading

Adams MD, Kelley JM, Gocayne JD, Dubnick M, Polymeropoulos M, Xiao H, Merril C, Wu A, Olde B, Moreno R (1991) Complementary DNA sequencing, expressed sequence tags and humane genome project. Science 252:1651–1656

Altschul SF, Madden TL, Schaffer AA, Zhang J, Zhang Z, Miller W, Lipman DJ (1997) Gapped Blast and PSI-BLAST: a new generation of protein database search programs. Nucl Acids Res 25:3389–3402

Bertucci F, Bernard K, Loriod B, Chang YC, Granjeaud S, Birnbaum D, Nguyen C, Peck K, Jordan BR. (1999) Sensitivity issues in DNA array-based expression measurements and performance of Nylon microarrays for small samples. Hum Mol Genet 8(9):1715–22

Bianciotto V, Bonfante P (1992) Quantification of the nuclear content of two arbuscular mycorrhizal fungi. Mycol Res 96:1071–1076

Chomoczynski P, Sacchi N (1987) Single-step method of RNA isolation by acid guanidinium thiocyanate-phenol-chloroform extraction. Anal Biochem 162:156–159

Corpet F (1988) Multiple sequence alignment with hierarchical clustering. Nucleic Acids Res 16:10881–10890

Doudrick RL, Raffle VL, NelsonCD, Fournier GR (1995) Genetic analysis of homokaryons from a basidiome of *Laccaria bicolor* using random amplified polymorphic DNA (RAPD) markers. Mycol Res 99:1361–1365

Eisen MB, Spellman PT, Brown PO, Botstein D (1998) Cluster analysis and display of genome-wide expression patterns. Proc Natl Acad Sci USA 95:14863–14868

Gianinazzi-Pearson V, van Tuinen D, Dumas-Gaudot E, Dulieu H (2001) Exploring the genome of Glomalean fungi. In: Hock B (ed) The Mycota, vol IX. Fungal Associations. Springer, Berlin Heidelberg New York, pp 3–17

Gonzalez P, Zigler S, Epstein DL, Borras T (1998) Identification and isolation of differentially expressed genes from very small tissue samples. Biotechniques 26:884–892

Graham JH (2000) Assessing costs of arbuscular mycorrhizal symbiosis in agroecosystems. In: Podila GK, Douds DD (eds) Current advances in mycorrhizae research. APS Press, St. Paul, MN, pp 127–140

Graham JH, Eissenstat DM (1998) Field evidence for carbon costs of citrus mycorrhizas. New Phytol 140:103–110

Harrison MJ (1999) Molecular and cellular aspects of the arbuscular mycorrhizal symbiosis. Ann Rev Plant Physiol Plant Mol Biol 50:361–389

Hofte H, Desprez T, Amselem J, et al. (1993) An inventory of 1152 expressed sequence tags obtained by partial sequencing of cDNA from *Arabidopsis thaliana*. Plant J 4:1051–1061

Hosny M, Gianinazzi-Pearson, V, Dulieu H (1998) Nuclear DNA contents of 11 fungal species in Glomales. Genome 41:422–429

Jordan BR (1998) Large-scale expression measurement by hybridization methods: from high-density membranes to "DNA chips". J Biochem (Tokyo) 124(2):251–258

Kamoun S, Hraber P, Sobral B, Nuss D, Govers F (1999) Initial assessment of gene diversity for the oomycete pathogen *Phytophthora infestans* based on expressed sequence tags. Fungal Genet Biol 28:94–106

Lee SH, Kim BG, Kim KJ, Lee JS, Yun DW, Hahn JH, Kim GH, Lee KH, Suh DS, Kwon ST, Lee CS, Yoo YB (2002) Comparative analysis of sequences expressed during the liquid-cultured mycelia and fruit body stages of *Pleurotus ostreatus*. Fungal Genet Biol 35(2):115–134

Martin F, Duplessis S, Ditengou F, Lagrange H, Voiblet C, Lapeyrie F (2001) Development of cross talking in the ectomycorrhizal symbiosis: Signals and communication genes. New Phytol 151:145–154

Nelson MA, Kang S, Braun EL, Crawford ME, Dolan PL, Leonard PM, Mitchell J, Armijo AM et al. (1997) Expressed sequences from conidial, mycelial and sexual stages of *Neurospora crassa*. Fungal Genet Biol 21:348–363

Podila GK, Zheng, J, Balasubramanian S, Sundaram S, Hiremath S, Brand J, Hymes M (2002) Molecular interactions in ectomycorrhizas: identification of fungal genes involved in early symbiotic interactions between *Laccaria bicolor* and red pine. Plant Soil 244:117–128

Polidori E, Agostini D, Zeppa S, Potenza L, Palms F, Sisti D, Stocchi V (2002) Identification of differentially expressed cDNA clones in *Tilia platyphyllos – Tuber borchii* ectomycorrhizae using a differential screening approach. Mol Gen Genomics 266:858–864

Sambrook J, Russel DW (2001) Molecular Cloning. A Laboratory Manual. 3rd Edition. Cold Spring Harbor Laboratory Press, Cold Spring Harbor, New York

Schummer M, Ng, WL, Nelson PS, Bumgarner RB, Hood L (1997) A simple high-performance DNA arraying device for comparative expression analysis of a large number of genes. BioTechniques 23:1087–1092

Smith SE, Read DJ (1997) Mycorrhizal symbiosis, 2nd edn. Academic Press, London

Voiblet C, Duplessis D, Encelot N, Martin F (2001) Identification of symbiosis-regulated genes in *Eucalyptus globulus-Pisolithus* ectomycorrhiza by differential hybridization of arrayed cDNAs. Plant J 25:181–191

30 Axenic Culture of Symbiotic Fungus *Piriformospora indica*

Giang Huong Pham, Rina Kumari, Anjana Singh, Rajani Malla, Ram Prasad, Minu Sachdev, Michael Kaldorf, François Buscot, Ralf Oelmüller, Rüdiger Hampp, Anil Kumar Saxena, Karl-Heinz Rexer, Gerhard Kost and Ajit Varma

1 Introduction

A large number of media compositions are available in the literature for the cultivation of various groups of fungi, but almost no literature is available for axenic cultivation of symbiotic fungi. In this chapter, we have made efforts to provide the documentary evidence for growth and multiplication of *Piriformospora indica* (see Chap. 15, this Vol. for characteristic features of the fungus). This new fungus named *P. indica*, due to its characteristic spore morphology, improves the growth and overall biomass production of different plants, herbs and trees, etc., and can easily be cultivated on a number of complex and synthetic media (Varma et al. 1999, 2001; Singh An et al. 2003a, b). Significant quantitative and morphological changes were detected when the fungus was grown on different nutrient compositions with no apparent negative effect on plants. It is relevant to mention here that different media can be used to understand the morphological and functional properties, or to test possible biotechnological applications.

2 Morphology

Young mycelia were white and almost hyaline, but inconspicuous zonations were recorded in other cultures. The mycelium was mostly flat and submerged into the substratum. Hyphae were thin-walled and of different diameters ranging from 0.7 to 3.5 µm. The hyphae were highly interwoven, often adhered together and gave the appearance of simple intertwined cords. The hyphae often showed anastomoses and were irregularly septated. They often intertwined and overlapped each other. In older cultures and on the root surface, hyphae were often irregularly inflated, showing a nodose to coralloid

Plant Surface Microbiology
A. Varma, L. Abbott, D. Werner, R. Hampp (Eds.)

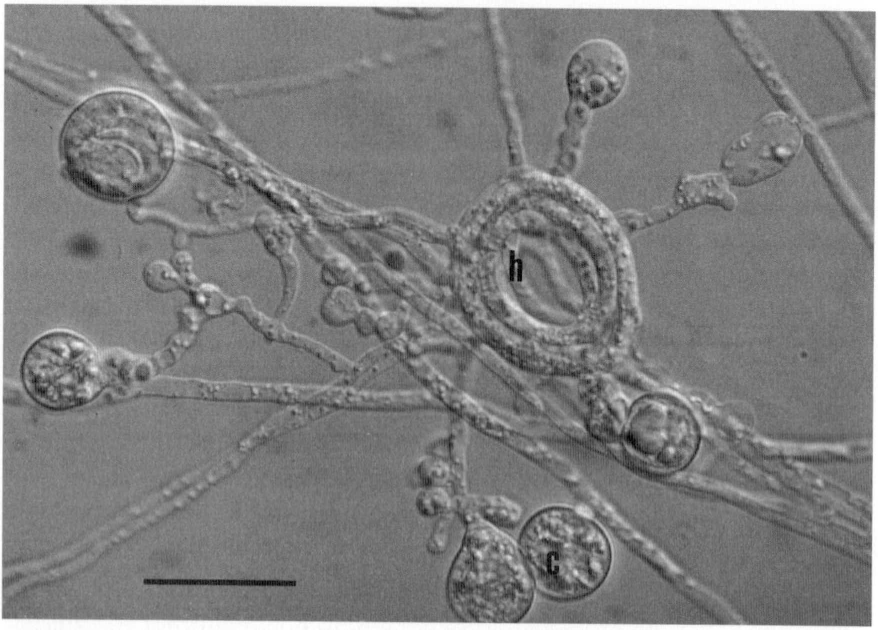

Fig. 1. An overall view of *P. indica*, grown on solidified MYP medium for 7 days. Note the distinct hyphal coils (h) and pear-shaped chlamydospores (c) Bar = 20 μm. By courtesy of Oliver Blechert

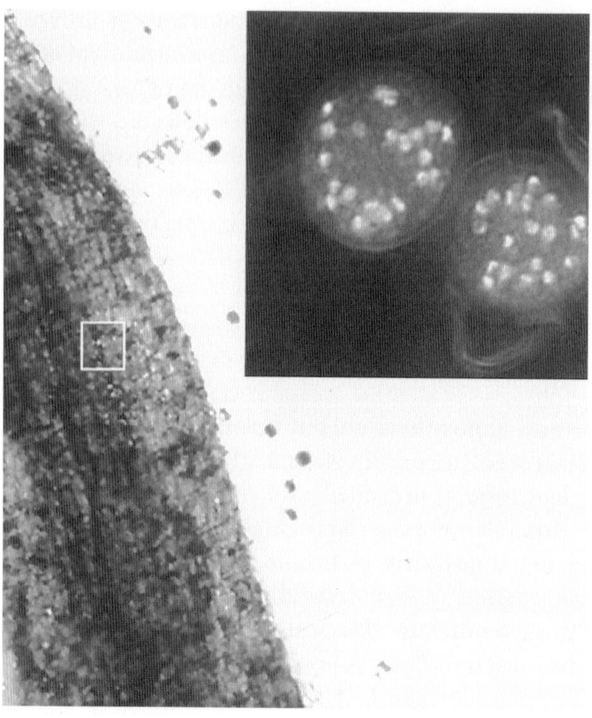

Fig. 2. *P. indica* colonized maize root segment covered by numerous chlamydospores on the surface and scattering away from the root. Enlarged view of chlamydospores showing nuclei. Chlamydospores were stained with DAPI and observed in epifluorescence. Different optical planes were assembled in one picture using the IMPROVISION software package (IMPROVISION, Govenny, UK)

shape and granulated dense bodies were observed. Many cells contained more than one nucleus. Chlamydospores were formed from thin-walled vesicles at the tips of the hyphae. The chlamydospores appeared singly or in clusters and were distinctive due to their pear-shaped appearance (Fig. 1). The chlamydospores were (14-) 16–25 (-33) μm in length and (9-) 10–17 (-20) μm in width. Figure 2 shows the maize root colonization. The cytoplasm of the chlamydospores was densely packed with granular material and usually contained 8–25 nuclei (Fig. 2, inset). Very young spores had thin, hyaline walls. At maturity, these spores had walls up to 1.5 μm thick, which appeared two-layered, smooth and pale yellow. Neither clamp connections nor sexual structures could be observed.

3 Taxonomy of the Fungus

Different kinds of substrates were tested to induce sexual development, such as young and mature leaves of *Cynodon dactylon* and pollen grains, oat meal, potato, carrot or tomato dextrose agar and soil-on-agar culture methods. No apparent adverse affect was seen on cultivation in light. It is not necessary to grow the fungus in the dark. Growth under light and dark conditions did not promote sexuality. May be the fungus is heterothallic in nature and one has to work for compatible strains. Since all these efforts did not lead to the desired results, there were only a few features to characterize the fungus morphologically and group it according to the classical species concept. In order to obtain more information about the systematic position of the new fungus, the ultrastructures of the septal pore and the cell wall were examined. The cell walls were very thin and multilayered structures. The septal pores consisted of dolipores with continuous parenthosomes. The dolipores were very prominent, with a multilayered cross wall and a median swelling mainly consisting of electron-transparent material. The electron-transparent layer of the cross walls extended deep into the median swellings, but did not fan out. In median sections of the septal pores, the parenthosomes were always straight and had the same diameter as the corresponding dolipore. Parenthosomes were flat discs without any detectable perforation. The parenthosomes consisted of an electron-dense outer layer, which showed an inconspicuous dark line in the median region. The parenthosomes were in contact with the ER membranes, which were mostly found near the dolipore (Verma et al. 1998).

The ultrastructural data proof that *P. indica* is a menber of the Hymenomycetes (Basidiomycota). Studies on the moleclar phylogeny will help to reveal the closest relatives of this species (Fig. 3).

Interestingly, immunological characterization showed a strong cross-reactivity with the members of Zygomycota (Glomerales) instead of species of Basidiomycota (Table 1). This aspect needs further critical appraisal.

Table 1. Cross-reactivities of polyclonal antisera raised against *P. indica* (total hyphal homogenate) as determined by ELISA. Optical density ($OD_{405\,nm}$) values are given as the mean of three replicates after correction of control ($OD_{405\,nm}$)±SD. Statistical analysis was done by ANOVA. (n.d., not detectable)

Antigens	$OD_{405\,nm}$ 1:1600	Source
Piriformospora indica	**0.49±0.005**	Ajit Varma, JNU, New Delhi
Nonmycorrhizal fungi		
Agaricus bisporus	0.08±0.002	AK Sarbhoy, IARI, New Delhi
Beauvaria sp.	0.003±n.d.	AK Sarbhoy, IARI, New Delhi
Candida albicans	**0.11±0.004**	R Prasad, JNU, New Delhi
Cladosporium sp.	0.004±n.d.	AK Sarbhoy, IARI, New Delhi
Cunninghamella echinulata	0.03±0.001	G Kost, Marburg, Germany
Fusarium solani	0.03±0.002	AK Sarbhoy, IARI, New Delhi
Rhizoctonia bataticola	0.04±0.002	G Kost, Marburg, Germany
Rhizoctonia solani	0.013±0.001	A K Sarbhoy, IARI, New Delhi
Rhizopus sp.	0.06±0.001	AK Sarbhoy, IARI, New Delhi
Saccharomyces cerevisiae	**0.17±0.020**	R Prasad, JNU, New Delhi
Schizophyllum commune	0.005±0.004	G Kost, Marburg, Germany
Sclerotinia sclerotiorum	**0.16±0.006**	AK Sarbhoy, IARI, New Delhi
Sclerotium solani	0.05±0.001	G Kost, Marburg, Germany
Ustilago maydis	0.08±0.007	AK Sarbhoy, IARI, New Delhi
Ectomycorrhizal fungi		
Amanita muscaria	**0.18±0.007**	T Satyanarayana, South Campus, Delhi University
Lactarius torminosus	0.03±0.004	Erika Kothe, Jena, Germany
Lentinus edodes	0.02±0.001	T Satyanarayana, South Campus, Delhi University
Paxillus involutus	0.02±0.001	T Satyanarayana, South Campus, Delhi University
Pisolithus tinctorius	**0.15±0.007**	Erika Kothe, Jena, Germany
Rhizopogon roseolus	**0.12±0.019**	T Satyanarayana, South Campus, Delhi University
Rhizopogon vulgaris	0.01±0.001	T Satyanarayana, South Campus, Delhi University
Suillus variegatus	0.003±0.004	Erika Kothe, Jena, Germany
Endomycorrhizal fungi		
Gigaspora margarita	**0.41±0.005**	Alok Adholeya, TERI, New Delhi
Gi. gigantia	**0.46±0.002**	KVBR Tilak, IARI, New Delhi
Glomus caledonium	**0.20±0.039**	François Buscot, Jena, Germany
G. coronatium	0.07±0.011	François Buscot, Jena, Germany
G. geosporura	0.16±0.019	François Buscot, Jena, Germany
G. intraradices	0.003±0.003	François Buscot, Jena, Germany
G. lamellosum	0.02±0.004	François Buscot, Jena, Germany
G. mosseae	0.15±0.010	François Buscot, Jena, Germany
G. mosseae 376	0.10±0.027	François Buscot, Jena, Germany
G. proliferum	0.24±0.023	François Buscot, Jena, Germany
Scutellospora gilmorei	0.40±0.002	Ajay Shanker, JNU, New Delhi
AMF-like		
Sebacina vermifera var *sensu*	0.39±0.049	Karl-Hein Rexer, Marburg, Germany
Sebacina sp.	0.23±0.013	Karl-Hein Rexer, Marburg, Germany

Statistical analysis of the data shows the *P* values, which are significant ($P<0.001$; cf. Varma et al. 2001, 2002; Singh et al. 2003a)

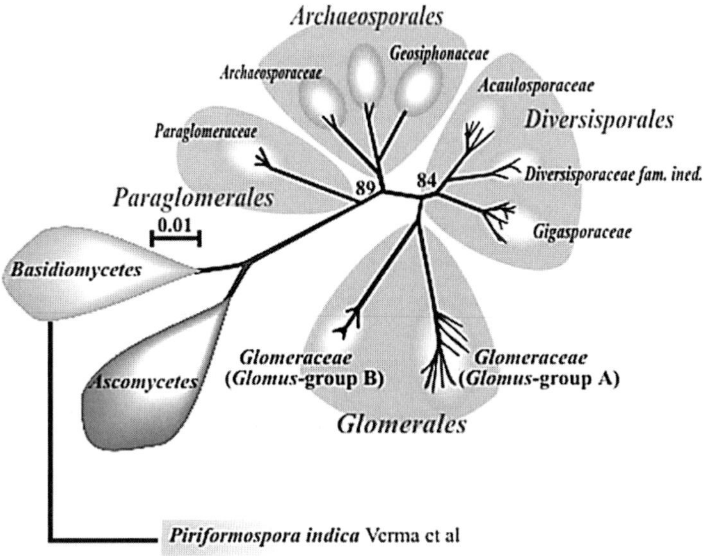

Fig. 3. An overall view of the molecular taxonomic position of *P. indica* (modified after Schüßler et al. 2001)

4 Chlamydospore Formation and Germination

The fungus produces chlamydospores at the apex of hyphae, which were mostly irregular undulated in shape. These chlamydospores can be easily germinated on various synthetic media (Verma et al. 1998). On solidified agar (2 %) medium, a tendency for cluster formation of chlamydospores was observed. Temperature (low to high and/or vice versa), pH (alkaline to acid or vice versa), and shock treatment also induced excessive sporulation. These spores were viable for over a year when preserved at room temperature. Loss in viability of the dormant spores was the least when germinated after 1 year. Dormant spores germinated within 1 day of their placement on nutrient agar medium and incubated at 40 °C under high humidity (>90 %). The first step of germination was the formation of germ tubes at the protruded zone of the spore, followed by hyphal emergence. Most of the nuclei followed the hyphae and seldom were one or two nuclei left behind in the spore. Soon branching appeared with a short and long branch (Fig. 4).

5 Cultivation

Fungi are heterotrophic for carbon compounds and these serve two essential functions in fungal metabolism. The first function is to supply the carbon needed for the synthesis of compounds which comprise living cells. Proteins, nucleic acids, reserve and food materials, etc., would be included here. Second, the oxidation of carbon compounds produces appreciable amounts of energy. Fungi can utilize a wide range of carbon sources such as monosaccharides, disaccharides, oligosaccharides, polysaccharides, organic acids and lipids. Carbon dioxide can be fixed by some fungi, but cannot be used as an exclusive source of carbon for metabolism. *P. indica* can be successfully cultivated on a wide range of synthetic solidified and broth media, e.g., MMN1/10, modified aspergillus, M4N, MMNC, MS, WPM, MMN, Malt-Yeast Extract, MYP, PDA and aspergillus (Fig. 5). Among the tested media, aspergillus (Kae-

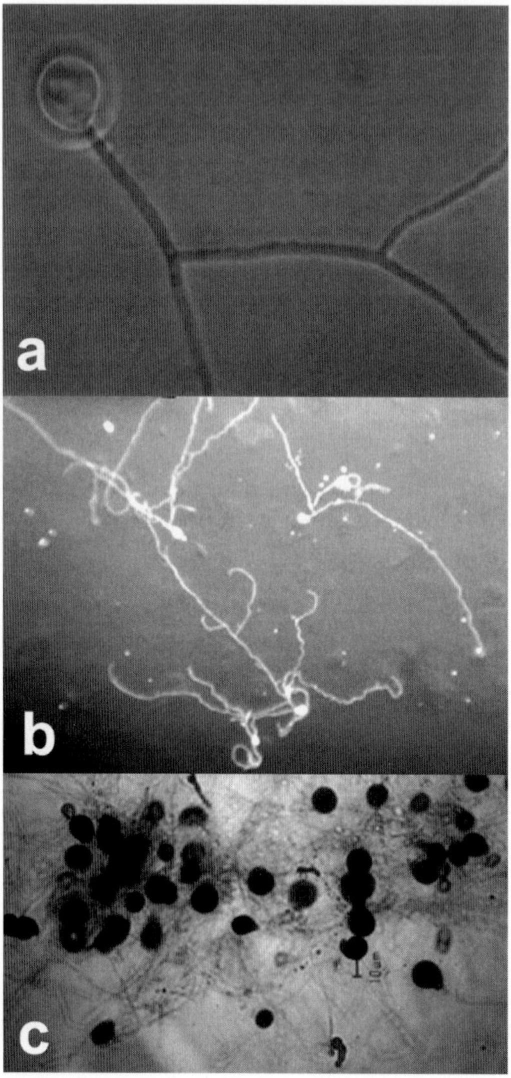

Fig. 4. Chlamydospores of *P. indica*. **a** Germinating chlamydospore showing initial branching after 12 h, **b** mature chlamydospores were germinated on a glass slide coated with thin nutrient agar, photographed after 24 h, **c** scattered spores and thin, irregular, undulating hyphae

fer 1977) was the best. However, other media were helpful in carrying out several physiological and molecular experiments (see Chap. 15, this Vol.). Figure 6 shows typical growth on solidified aspergillus medium after 28 days. Rhythmic growth was often recorded. The mycelium stopped its growth for some time and produced a large number of chlamydospores of different dimensions. After 24–48 h, the mycelium started its growth again, producing normal amount of chlamydospores. This resulted in the formation of rythmic rings. The physiological reason for this phenomenon is not yet known, although this tendency has been recorded for several other members of Basidiomycetes. The fungus grows profusely upon shaking broth aspergillus medium. The temperature range of the fungal growth is 25–35 °C; the optimum temperature and pH being 30 °C and 5.8 (4.8–6.8), respectively. Figure 7 gives a view of the cultivation on broth media. Colonies were large and small depicting sea urchin-like radial growth. The maximum surface growth was recorded after 10 days. The colony diameter is indicated in Fig. 8. The fungal biomass is indicated in Table 2. The optimum growth was recorded after 5

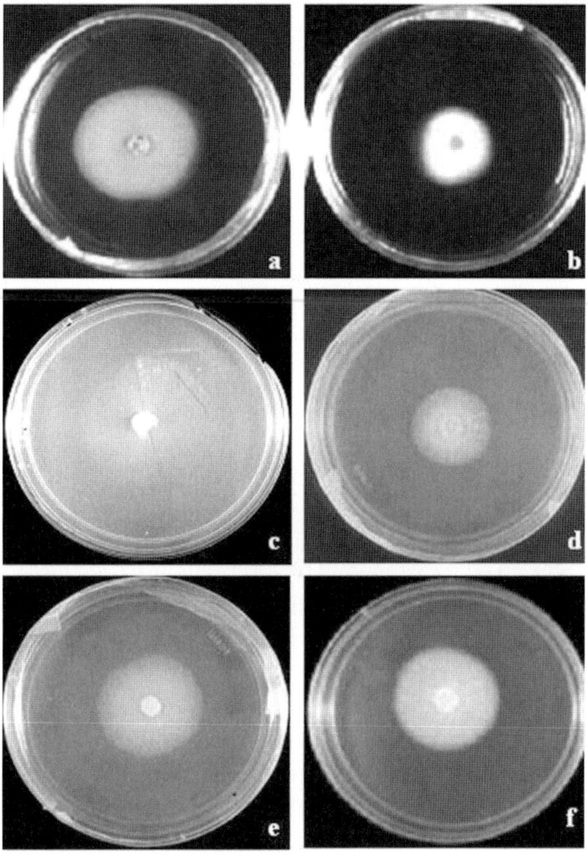

Fig. 5. *P. indica* was grown on the following solidified media. **a** MS, **b** WPM, **c** MMN, **d** M4N, **e** PDA, **f** aspergillus

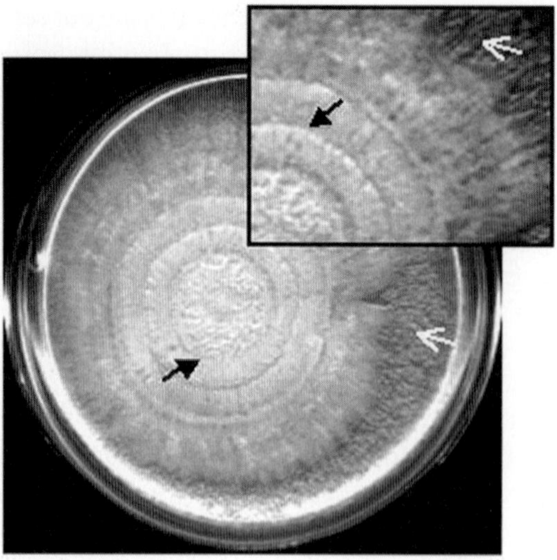

Fig. 6. An overall view of *P. indica* grown on solidified aspergillus medium. Inset shows enlarged view of a small portion. On an agar concentration of 2 % w/v concentric rings often appeared (*arrows*) indicating the rhythmic growth of the fungus. The black arrows point on the regions with slow growth and high amount of chlamydospores, the white arrows point on thin mycelial mats resulting from fast growth of the hyphae

days with a gradual decrease in fresh and dry biomass after prolong incubation. Linear growth of the fungus on different solidified agar media is represented in Table 3. On modified Melin-Norkrans (MMN) medium sparsely running hyaline hyphae on the agar surface were seen, while on Potato Dextrose Agar deep furrows with strong adhesion to the agar surface were apparent. This sharp mode of growth was not observed when fortified with malt extract and normal aspergillus medium. In contrast to aspergillus medium, shaking conditions on MMN broth medium invariably inhibited the growth. The explanation for this observation is not known. Fungal growth acidifies the medium within 10 days to pH 4.4. Buffered medium prevented the reduction of pH (Table 4). 2-(N-morpholine) ethane sulfonic acid (MES) in the range of 25–100 mM was used.

6 Carbon and Energy Sources

Individual sugars were uniformly added to the minimal broth at a rate of 1.0 % (w/v) in all treatments. They were included in the medium separately after sterilization. In all the sugar-supplied media, growth was better than the control (Table 5). There were not many changes in the growth except for rafinose and fructose. There were no changes in the color of the mycelium. Good growth was recorded in media containing maltose followed by xylose, sucrose, rhamnose, arabinose, glucose, lactose and mannose, respectively. The final pH was not altered significantly, but was lower than that of the control (Table 5).

In a further study, fungal growth was best when glucose (1 % w/v) was used as a carbon source as compared to sucrose, and followed by fructose. A

Table 2. The data represent an average of *P. indica* biomass of 5 replicates grown in 100 ml aspergillus broth medium in 250 ml capacity Erlenmeyer flasks. Incubation was done on a rotary shaker (GFL 3.19, Germany) at 144 rpm at 30 °C

Days	Biomass (g)	
	Fresh	Dry
5	3.67±0.84	0.06±0.02
7	2.99±0.38	0.06±0.01
10	2.39±0.01	0.07±0.01

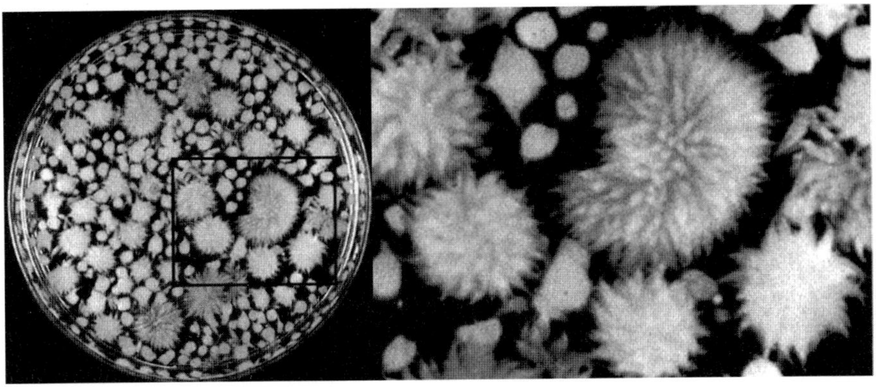

Fig. 7. Growth of *P. indica* on aspergillus broth medium under constant shaking condition at 25 °C for 7 days. Colonies of different developmental stages were shown. Mature colonies have the appearance of sea urchins

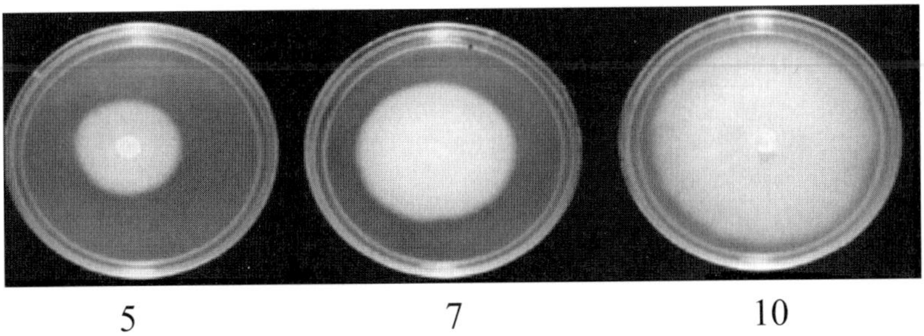

5		7		10	
LM	A	LM	A	LM	A
3.6 ± 0.15	10.2	5.4 ± 0.36	22.9	7.5 ± 0.16	43.9

Fig. 8. A comparative linear growth of *P. indica* on aspergillus solidified medium. Measurements were made after 5, 7 and 10 days, respectively. Incubation was conducted in dark at 25 °C. Parameter selected was the diameter of 5 replicates of the linear measurement (*LM*). Readings are given in cm standard deviation and area (*A*) on agar medium. Statistical analysis of the data showed $P<0.05$ (Jandel Scientific, Statistical software, version 2.0, and copyright 1992–1995)

Table 3. Comparative linear growth of *P. indica* on different solidified agar media. The data represent an average colony diameter of five replicates, measured after 5 days of incubation

Media	Linear measurement of the growth (cm)
MMN	4.2±0.05
M4N	2.8±0.09
PDA	3.5±0.11
aspergillus	3.6±0.15

Table 4. Change of pH of the aspergillus broth medium incubated with *P. indica*

Medium conditions (pH)	Incubation days				
	0	3	5	7	10
Unbuffered	6.5	6.0	5.7	5.1	4.4
Buffered	6.5	6.5	6.5	6.5	6.3

Initial pH was adjusted to 6.5. 25–100 mM MES was used as buffering agent

Table 5. End pH and biomass of *P. indica* grown on minimal aspergillus broth medium containing different sugars (each 1 % w/v)

Sugars	End pH	Biomass (mg/10 ml)
Control (no addition)	5.18	7.5
Glucose	4.39	10.0
Fructose	5.25	8.0
Maltose	4.46	12.0
Rhamnose	4.60	10.3
Mannose	4.30	9.5
Lactose	4.44	10.0
Sucrose	4.39	10.0
Xylose	4.31	11.0
Arabinose	4.28	10.0
Raffinose	4.43	9.0

One agar disc (1 cm in diameter loaded with hyphae and chlamydospores) was transferred to individual test tubes containing 10 ml minimal broth. Sterile sugar solution (microsyringe-filtered, 0.22 μm Schleicher & Schuell) was included. Incubation was done under constant shaking conditions (GFL, 3026, Germany) for 7 days at 25 °C. Fungal biomass was removed and end pH was measured

Table 6. Growth of *P. indica* on unbuffered aspergillus broth medium supplemented with different carbon sources

Sugars (w/v)	Fresh weight (g/l)	Remarks
Sucrose(0.5 %)	148.8	Compact and numerous chlamydospores
Glucose (1 %)	184.8	Loose, peg-like bodies, few chlamydospores
Fructose (1 %)	78.0	Compact, numerous tiny chlamydospores
Glucose + fructose (0.5 % each)	109.6	Loose, a few chlamydospores, turned slimy

Data represent an average of three replicates; biomass measured after 7 days.

combination of glucose and fructose (each 0.5 % w/v) led to a medium increase in the biomass of *P. indica* (Table 6). On supplementation of glucose by a mixture of glucose, fructose and sucrose (each 0.5 % w/v), the former was consumed completely and then the sucrose was metabolized by production of invertase. This led to an increase of the fructose concentration of the medium. After the complete consumption of free glucose there was a slow utilization of fructose.

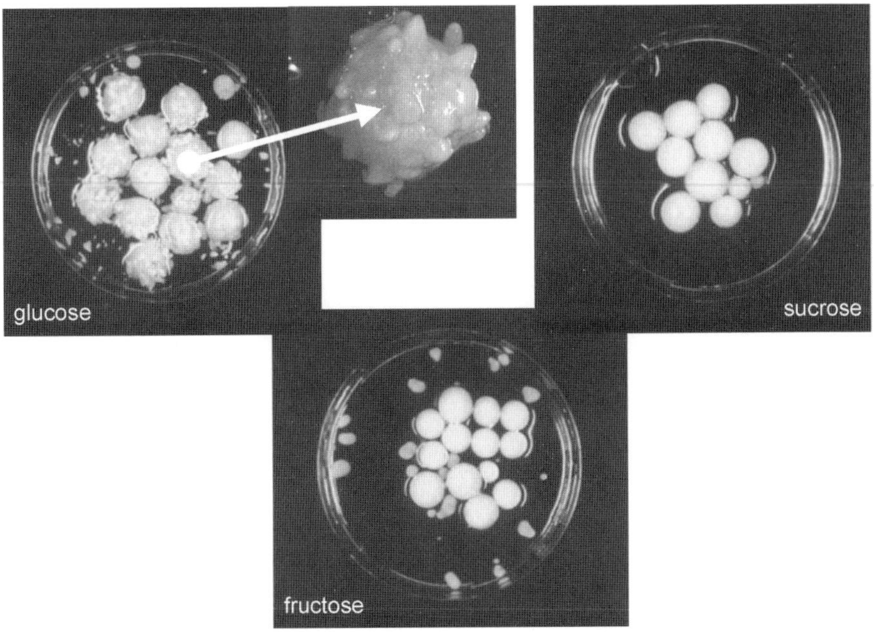

Fig. 9. *P. indica* colonies produced in aspergillus broth medium fortified with glucose, sucrose or fructose. An enlarged view of a colony showing protuberances and peg-like structures on glucose medium

The morphology of the colonies differed according to the sugar supply. In fructose and sucrose, the colonies were roundish and compact, in glucose they were large and irregular with short and long protrusions (Fig. 9).

7 Biomass on Individual Amino Acids

The addition of glycine, methionine, serine, alanine promoted fungal growth to different extents (Table 7). Not much difference in mycelial growth was observed in media containing glutamine, asparagine and histidine, although a substantial difference in the end pH of these amino acid-fortified culture broths was recorded (Table 7).

8 Growth on Complex Media

P. indica grown in minimal broth was transferred onto one set of fresh minimal media containing agar. In the minimal broth, the complex mixtures such as soil-extract, malt-extract, peptone, beef-extract, yeast-extract and casein-hydrolysate were added individually to an amount of 1 % (w/v). Before autoclaving, the pH of the media was adjusted to 6.5. Compared to all other media used, excellent growth of mycelium was recorded in the incubation broth fortified with casamino hydrolysate-HCl. Growth in beef, yeast, malt extracts and peptone was moderate. Soil extracts did not support fungal growth (Table 8).

Table 7. End pH and fungal biomass grown on minimal broth medium supplemented with amino acids (each 0.5 % w/v)

Amino acids	End pH	Biomass (mg/10 ml)
Control (no addition)	5.14	3.8
Alanine	6.94	6.2
Phenyl alanine	4.31	5.8
Methionine	4.71	7.0
Serine	5.82	6.9
Asparagine	5.86	4.2
Glutamine	4.76	4.3
Cysteine	1.90	3.8
Glycine	5.01	7.8
Aspartic acid	2.91	3.8
Arginine	8.93	3.8
Histidine	7.05	4.0

pH was re-adjusted after the addition of microsyringe-filtered amino acids to 6.5. Incubation was done under constant shaking condition (GFL, 3026, Germany) for 7 days at 25 °C. Fungal biomass was removed and end pH was measured

9 Phosphatic Nutrients

Phosphorus is an essential mineral for the growth of *P. indica*. Optimum growth was obtained on supplementing the modified aspergillus medium with Di-potassium hydrogen phosphate in equimolar concentrations (Table 9). Interestingly, the fungus utilized tri-poly phosphate and solubilized insoluble calcium-hydrogen phosphate. Acid phosphatases were observed to be active in *P. indica* mycelium (Varma et al. 2001). The fungus was able to utilize a variety of inorganic and organic phosphate sources which is in accordance with the broad range of the substrates utilized by the acid phosphatases of many fungi. Moreover, phosphate starvation of *P. indica* led to an overall (27 %) increase in the intracellular acid phosphatase activity. This increase was probably due to the appearance of a P-repressible isoform of acid phos-

Table 8. Mycelial biomass of *P. indica* grown on complex modified aspergillus medium

Complexes	Mycelial biomass (mg/10 ml)
Control (no addition)	10
Soil-extract	6
Malt-extract	8
Peptone	9
Beef-extract	12
Yeast-extract	11.5
Casein hydrolysate	5

Complex chemicals (obtained from Difco or Hi media) were included at the rate of 1 % (w/v); and soil-extract 15 % v/v. pH was readjusted to 6.5. Incubation conditions were the same as described earlier

Table 9. Biomass of *P. indica* after 24 days of growth on modified aspergillus medium supplemented with equimolar (10 mM) concentrations of phosphatic nutrient sources

Phosphate source	Dry biomass g/1000 ml	Final pH of the medium
Control (P-)	2.9±0.001	4.10±0.013
Di-hydrogen potassium phosphate	6.9±0.009	4.62±0.05
Di-potassium hydrogen phosphate	7.9±0.002	4.59±0.024
Calcium-hydrogen phosphate	4.8±0.003	4.17±0.008
Di-hydrogen sodium phosphate	6.3±0.005	4.47±0.068
Di-potassium hydrogen phosphate	6.8±0.003	4.34±0.053
Di-hydrogen ammonium phosphate	5.5±0.004	4.36±0.056
Tetra-hydrogen ammonium phosphate	5.2±0.003	4.26±0.04
Tri-polyphosphate	7.9±0.002	5.49±0.074
Tri-metaphosphate	5.8±0.001	4.40±0.008

Aspergillus medium was modified by reducing the concentration of peptone, yeast extract and casein hydrolysate to ten times the normals phatase in addition to the constitutive one observed in the enzyme staining of the native polyacrylamide gels. The significance of these enzymes in the phosphate transport needs to be further substantiated by the studies on the plant roots colonized with *P. indica*.

10 Composition of Media

a Aspergillus (Kaefer 1977)

Composition	(g/l)
Glucose	20.0
Peptone	2.0
Yeast extract	1.0
Casein hydrolysate	1.0
Vitamin stock solution	1.0 ml
Macro-elements from stock	50.0 ml
Micro-elements from stock	2.5 ml
Agar	10.0
$CaCl_2$ 0.1 M	1.0 ml
$FeCl_3$ 0.1 M	1.0 ml
pH	6.5
Macro-elements	
(Major elements)	Stock (g/l)
$NaNO_3$	120.0
KCl	10.4
$MgSO_4{\cdot}7H_2O$	10.4
KH_2PO_4	30.4
Micro-elements	
Trace elements	Stock (g/l)
$ZnSO_4{\cdot}7H_2O$	22.0
H_3BO_3	11.0
$MnCl_2{\cdot}4H_2O$	5.0
$FeSO_4{\cdot}7H_2O$	5.0
$CoCl_2{\cdot}6H_2O$	1.6
$CuSO_4{\cdot}5H_2O$	1.6
$(NH_4)_6\ Mo_7O_{27}\ 4H_2O$	1.1
Na_2EDTA	50.0
Vitamins	% (w/v)
Biotin	0.05
Nicotinamide	0.5
Pyridoxal phosphate	0.1
Amino benzoic acid	0.1
Riboflavin	0.25

The pH was adjusted to 6.5 with 1 N HCl. All the stocks were stored at 4 °C except the vitamins which were stored at –20 °C

b Modified aspergillus medium (Varma et al. 2001)
The media composition was the same, except that yeast extract, peptone and casein hydrolysate were reduced to 1/10 in quantity

c M4N (Mukerji et al. 1998)

Composition	(g/l)
D-Glucose	10.0
$(NH_4)_2HPO_4$	0.25
KH_2PO_4	0.50
$MgSO_4 \cdot 7H_2O$	0.15
$CaCl_2 \cdot 2H_2O$	0.05
Ferric citrate	
(2 % Ferric citrate,	
2 % Citric acid w/v)	7.0 ml
NaCl	0.025
Thiamine HCl	100.0 µg
MES	2.5
Malt extract	1.5
Yeast extract	1.5
Agar	15.0
pH	5.6

d Malt Extract (Galloway and Burgess 1952)

Composition	(g/l)
Malt extract	30.0
Mycological peptone	5.0
Agar	15.0
pH	5.4

e MMN (Modified Melin-Norkrans) (Johnson et al. 1957)

Composition	(g/l)
NaCl	0.025
KH_2PO_4	0.5
$(NH_4)_2HPO_4$	0.25
$CaCl_2$	0.05
$MgSO_4$	0.15
$FeCl_3$	0.001
Thiamine HCl	83.0 µl
Tryticase peptone	0.1 % (w/v)
Glucose monohydrate	1.0 % (w/v)
Malt extract	5.0 % (w/v)
Trace elements from stock 10.0 ml/l	
Trace elements (stock)	(g/l)
KCl	3.73

H_3BO_3	1.55
$MnSO_4{\cdot}H_2O$	0.85
$ZnSO_4$	0.56
$CuSO_4$	0.13

pH adjusted to 5.8 with 1 N HCl/NaOH. All stocks were stored at 4 °C except thiamine hydrochloride which was stored at –20 °C

f MMN 1/10 (Herrmann et al. 1998)

Composition	(g/l)
$CaCl_2{\cdot}2H_2O$	0.07
$MgSO_4{\cdot}7H_2O$	0.15
NaCl	0.03
$(NH_4)_2HPO_4$	0.03
KH_2PO_4	0.05
Trace elements (stock)	(mg/l)
$(NH_4)_6Mo_7O_{24}{\cdot}4H_2O$	0.09
H_3BO_4	1.55
$CuSO_4{\cdot}5H_2O$	0.13
KCl	3.73
$MnSO_4{\cdot}H_2O$	0.84
$ZnSO_4{\cdot}7H_2O$	0.58
Fe-EDTA	(mg/l)
$FeSO_4$	8.50
EDTA	1.50
Agar	20.0 g/l

g MMNC (Marx 1969; Kottke et al. 1987)

Composition	(g/l)
Glucose	10.0
$CaCl_2{\cdot}2H_2O$	0.07
$MgSO_4{\cdot}7H_2O$	0.15
NaCl	0.03
$(NH_4)_2HPO_4$	0.25
KH_2PO_4	0.5
Casein hydrolysate	1.0
Malt extract	5.0
Trace elements	(mg/l)
$(NH_4)_6Mo_7O_{24}{\cdot}4H_2O$	0.02
H_3BO_4	1.55
$CuSO_4{\cdot}5H_2O$	0.13
KCl	3.73
$MnSO_4{\cdot}H_2O$	0.85
$ZnSO_4{\cdot}7H_2O$	0.58
Fe-EDTA	(mg/l)

$FeSO_4$	8.5
EDTA	1.5
Vitamins	(mg/l)
Thiamine HCl	0.1
Riboflavin	0.1
pH	5.6
Agar	20.0 g/l

h Moser b (modified after Moser 1960)

Macro-elements	(g/l)
Glucose	10
Sucrose	10
Maltose	10
Malt extract	10
Peptone	2
K_2HPO_4	0.15
KH_2PO_4	0.35
NH_4NO_3	1
$NaNO_3$	0.3
$MgSO_4 \cdot 7H_2O$	0.5
$CaCl_2$	0.1
Micro-elements	mg/l
Thiamine	50
Biotine	1
Inositol	50
$ZnSO_4$	1
$FeCl_3$	10
$MnSO_4$	5
Agar	20 g/l

i MS (Murashige and Skoog 1962)

Composition	(mg/l)
Macro-nutrients	
NH_4NO_3	0.5
KNO_3	1650.0
$CaCl_2 \cdot 2H_2O$	900.0
$MgSO_4 \cdot 7H_2O$	440.0
KH_2PO_4	370.0
Micro-nutrients	
KI	170.0
H_3BO_3	0.83
$MnSO_4 \cdot H_2O$	6.20
$ZnSO_4 \cdot 7H_2O$	15.60
$NaMoO_4 \cdot 2H_2O$	8.60

$CuSO_4 \cdot 5H_2O$	0.25
$CoCl_2 \cdot H_2O$	0.025
Iron source	
Na_2EDTA	0.025
$FeSO_4 \cdot 7H_2O$	37.30
Vitamins	
Nicotinic acid	27.8
Pyridoxine HCl	0.5
Thiamine HCl	0.1
Glycine	2.0
Myo-inositol	100.0
Agar	0.7% (w/v)
Sucrose	3.0% (w/v)
pH, 5.6–5.7	

Each chemical was dissolved in bidistilled water. The pH of the medium was adjusted using 1 N NaOH/HCl before autoclaving at 121 °C, for 20 min. Stock solutions were stored at 4 °C except organic supplements, which were stored at –20 °C

j MYP (Bandoni 1972)

Composition	(g/l)
Malt extract	7
Soytone (Difco)	1
Yeast extract	0.5
Agar	15

k Potato Dextrose Agar (PDA) (Martin 1950)

Composition	(g/l)
Potato peel	200.0
Dextrose	20.0
Agar	15.0

The periderm (skin) of potatoes (200 g) was peeled-off, cut into small pieces and boiled in 500 ml of water, until they were easily penetrated by a glass rod. After filtration through cheese cloth, dextrose was added. Agar was dissolved and the required volume (1 l) was made up by the addition of water. The medium was autoclaved at 121 °C for 20 min.

l WPM ("Woody Plant Medium" for *Populus*) (Ahuja et al. 1986)

Composition	(g/l)
Sucrose	20.0
K_2SO_4	1.00
$Ca(NO_3)_2 \cdot 4H_2O$	0.73
NH_4NO_3	0.40

$MgSO_4 \cdot 7H_2O$	0.37
Myo-inositol	0.10
Agar	7.00

Add 700 ml H_2O, adjust pH to 5.8 using 3.7 % HCl (ca. 9.5 µl),
Add after autoclaving sterile phosphate solution (0.17 g KH_2PO_4 dissolved in 270 ml H_2O+15 µl NaOH (saturated).
10 ml of trace element stock solution (see below)
10 ml Fe-EDTA (see below)
10 ml glycine stock solution (100x: dissolve 20 mg in 100 ml)
1 ml thiamine – stock solution (1000x: dissolve 10 mg in 100 ml)
1 ml nicotinic acid – stock solution (1000x: dissolve 50 mg in 100 ml)
1 ml $CaCl_2$ – stock solution (1000×: dissolve 3.6 g in 50 ml)
250 µl Pyridoxine – stock solution (4000x: dissolve 40 mg in 100 ml)
100 µl $CuSO_4$ – stock solution (10,000x: dissolve 25 mg in 100 ml)
sterilize by filtration before adding 100× trace element stock solution (autoclave, store at 4 °C)

$MnSO_4 \cdot H_2O$	2.23
$ZnSO_4 \cdot 7H_2O$	0.86
H_3BO_4	0.62
Ammonium molybdate	0.10
KI	0.09

100x Fe-EDTA stock solution: dissolve 0.128 g $FeSO_4$ and 0.172 g EDTA at 60 °C in 100 ml H_2O, store at 4 °C

$CaCl_2 \cdot 2H_2O$	0.07
$MgSO_4 \cdot 7H_2O$	0.15
NaCl	0.03
$(NH_4)_2HPO_4$	0.03
KH_2PO_4	0.05
Trace elements	(mg/l)
$(NH_4)_6Mo_7O_{24} \cdot 4H_2O$	0.018
H_3BO_4	0.62

The fungus grew on a wide range of synthetic and complex media. Significant quantitative and morphological changes were detected when the fungus was challenged to grow on different media. Shaking during incubation retarded growth in MMN broth cultures (7–12 g fresh wt./l, after 2 weeks at 30 °C), whereas no such negative effect was ever observed during cultivation on any other substrates. There was practically no growth when mycelia were incubated under shaking conditions, whereas in stationary conditions, normal growth was obtained. Hyphae did not adjust to even a slow rate of shaking. In fact, the fungal biomass was considerably enhanced on shaking cultures with aspergillus medium, sometimes up to 50 g fresh wt./l after 2 weeks at 30 °C. On aspergillus and Moser b media, the colonies appeared compact, wrinkled with furrows and constricted. The mycelium produced fine zonation and a great amount of white aerial hyphae. Hyphae were highly interwoven, often adhered together and gave the appearance of simple cords. New branches emerged irregularly and the hyphal walls showed some external deposits at regular intervals, which stained deeply with toluidine blue. Since septation was irregular, the single compartment could contain more than one nucleus. The chlamydospores appeared singly or in clusters at the apex of hyphae. They were distinctive due to their pear-shaped habit.

11 Conclusions

Mycorrhiza does not always promote the growth of agricultural crops. In phosphorus-rich soils, they can parasitize plants such as citrus, wheat and maize by tapping sugars from these plants without giving anything back. Researchers ignore this darker side of the mycorrhiza. Theoretically, mycorrhiza can also harm biodiversity. In the long run, specific mycorrhizas can promote the growth of one plant at the expense of another. "What exactly happens probably depends on the system itself," states Van der Heijden (2002). In any case, the interaction between plants and mycorrhiza forming fungi clearly has at least as great an effect on the ecosystem's species composition as the interaction/competition between plants themselves.

P. indica, the fungus treated in this chapter, acts as biofertilizer, bioregulator and bioprotector, and can be easily mass-multiplied on defined synthetic media. It is thus, an interesting model fungus with respect to studies on endomycorrhiza. In addition, commercial production of this fungus under aseptic conditions could support biological hardening of tissue-culture-raised plants as well as plant survival in general on poor soils.

Acknowledgements. The Indian authors are thankful to DBT, DST, CSIR, UGC, and the Government of India for partial financial assistance.

References and Selected Reading

Ahuja MR (1986) In: Evans DA, Sharp WR and Ammirato PJ (eds) Handbook of plant cell culture 4, techniques and applications. Macmillan, New York, pp 626–651

Badoni RJ (1972) Terrestrial occurrence of some aquatic Hyphomycetes. Can J Bot 50:2283–2288

Galloway LD, Burgess R (1952) Applied mycology and bacteriology, 3rd edn. Leonard Hill, London, pp 54–57

Herrmann S, Munch JC, Buscot F (1998) A gnotobiotic culture system with oak micro-cuttings to study specific effects of mycobionts on plant morphology before, and in the early phase of, ectomycorrhiza formation by *Paxillus involutus* and *Piloderma croceum*. New Phytol 138:203–212

Johnson CN, Stout PR, Broyer RC, Carlton AB (1957) Comparative chlorine requirements of different plant species. Plant Soil 8:337–353

Kaefer E (1977) Meiotic and mitotic recombination in *Aspergillus* and its chromosomal aberrations. Adv Genet 19:33–131

Kottke I, Guttenberger M, Hampp R, Oberwinkler F (1987) An in vitro method for establishing mycorrhizae on coniferous tree seedlings. Trees 1:191–194

Martin JP (1950) Use of acid, rose bengal and streptomycin in the plate method for estimating soil fungi. Soil Sci 69:215–232

Marx, DH (1969) The influence of ectotrophic mycorrhizal fungi on the resistance of pine roots to pathogenic infections. I. Antagonism of mycorrhizal fungi to root pathogenic fungi and soil bacteria. Phytopathology 59:153–163

Moser M (1960) Die Gattung *Phlegmacium*. J Klinkhardt, Bad Heilbrunn, Austria

Mukerji KG, Mandeep, Varma A (1998) Mycorrhizosphere microorganisms: screening and evaluation. In: Varma A (ed) Mycorrhiza manual. Springer, Berlin Heidelberg New York, pp 85–98

Murashige T, Skoog F (1962) A revised medium for rapid growth and bioassay with tobacco tissue cultures. Physiol Plant 15:431–487

Schüßler A, Schwarzott D, Walker C (2001) A new fungal phylum, the Glomeromycota: phylogeny and evolution. Mycol Res 105:1413–1421

Singh An, Singh Ar, Kumari M, Rai MK, Varma A (2003a) Biotechnological importance of *Piriformospora indica* Verma et al. a novel symbiotic mycorrhiza-like fungus: an overview. Indian J Biotechnol 2:65–75

Singh An, Singh Ar, Kumari M, Kumar S, Rai MK, Sharma AP and Varma A (2003b) Unmassing the accessible treasures of the hidden unexplored microbial world. In: Prasad BN (ed) Biotechnology in sustainable biodiversity and food security. Science Publishers, Inc. Enfield, NH, USA, pp 101–124

Van der Heijden MAG (2002) Arbuscular mycorrhizal fungi as a determinant of plant diversity: in search for underlying mechanisms and general principles. In: Van der Heijden MGA and Sanders IR (eds) Mycorrhizal ecology. Ecological Studies 157. Springer, Berlin Heidelberg New York, pp 243–266

Varma A, Verma S, Sudha, Sahay NS, Franken P (1999) *Piriformospora indica*, a cultivable plant growth promoting root endophyte with similarities to arbuscular mycorrhizal fungi. Appl Environ Microbiol 65:2741–2744

Varma A, Singh A, Sudha, Sahay NS, Sharma J, Roy A, Kumari M, Rana D, Thakran S, Deka D, Bharati K, Hurek T, Blechert O, Rexer KH, Kost G, Hahn A, Hock B, Maier W, Walter M, Strack D, Kranner I (2001) *Piriformospora indica*: An axenically culturable mycorrhiza-like endosymbiotic fungus. In: Hock B (ed) Mycota IX. Springer, Berlin Heidelberg New York, pp 123–150

Varma A, Singh A, Sudha, Sahay NS, Kumari M, Bharti K, Sarbhoy AK, Maier W, Walter MH, Strack D, Franken P, Singh An, Malla R, Hurek T (2002) *Piriformospora indica*: A plant stimulator and pathogen inhibitor arbuscular mycorrhizal-like fungus. In: Markandey DK, Markandey NR (eds) Microorganisms in bioremediation. Capital Publishing Company Ltd., New Delhi, pp 71–89

Verma S, Varma A, Rexer KH, Hassel A, Kost G, Sarbhoy A, Bisen P, Bütehorn B, Franken P (1998) *Piriformospora indica* gen. et sp. nov., a new root-colonizing fungus. Mycologia 90:895–909

Subject Index

T

Printing: Saladruck, Berlin
Binding: Stein+Lehmann, Berlin